CONFERENCE ON HARMONIC ANALYSIS

THE WADSWORTH MATHEMATICS SERIES

W. Beckner, A. Calderón, R. Fefferman, P. Jones, *Conference on Harmonic Analysis in Honor of Antoni Zygmund*

M. Behzad, G. Chartrand, L. Lesniak-Foster, *Graphs and Digraphs*

J. Cochran, *Applied Mathematics: Principles, Techniques, and Applications*

A. Garsia, *Topics in Almost Everywhere Convergence*

K. Stromberg, *An Introduction to Classical Real Analysis*

CONFERENCE ON HARMONIC ANALYSIS

IN HONOR OF ANTONI ZYGMUND

VOLUME I

Edited by

William Beckner
Alberto P. Calderón
Robert Fefferman
Peter W. Jones

University of Chicago

Wadsworth International Group
Belmont, California
A Division of Wadsworth, Inc.

Mathematics Editor: John Kimmel
Production Editor: Diane Sipes
Copy Editor: Janet Greenblatt
Typists: Elizabeth J. Ritter, Jo Ann Vine
Photographs: Patricia Evans

Printed in the United States of America

1 2 3 4 5 6 7 8 9 10---87 86 85 84 83

Library of Congress Cataloging in Publication Data

Conference on Harmonic Analysis (1981: Chicago, Ill.)
Conference on Harmonic Analysis in honor of Antoni Zygmund.

(Wadsworth international mathematics series)
Papers presented at the Chicago Conference on Harmonic Analysis, March 23-28, 1981; sponsored by the National Science Foundation and the University of Chicago.
1. Harmonic analysis—Addresses, essays, lectures. 2. Zygmund, Antoni, 1900- . I. Zygmund, Antoni, 1900- . II. Beckner, William. III. National Science Foundation (U.S.) IV. University of Chicago. V. Title. VI Series.
QA403.C64 1981 515'.2433 82-11172
ISBN 0-534-98040-6 (v. 1)
ISBN 0-534-98041-4 (v. 2)
ISBN 0-534-98043-0 (set)

These proceedings are dedicated to
Professor Antoni Zygmund
in celebration of his eightieth birthday.

Antoni Zygmund

PREFACE

Between March 23 and 28, 1981, some two hundred mathematicians gathered at the University of Chicago to honor Professor Antoni Zygmund on the occasion of his eightieth birthday. It was a public tribute to the undisputed highest authority in Fourier series and one of the foremost analysts in the world.

Ever since the publication of its first edition in 1935, Professor Zygmund's famous *Trigonometric Series* has been one of the most influential books in the modern mathematical literature, making a permanent imprint on the mathematical work and outlook of many of the most distinguished present-day analysts.

His extensive research has covered many topics in Fourier series, real variables, and related fields. Many of his results—for example, those in the fine theory of differentiation, interpolation of linear operations, and singular integrals—now occupy a permanent and central place in analysis and have deeply influenced the modern development of other important areas of analysis, such as the theory of partial differential equations.

Professor Zygmund has also been an extraordinary teacher. He has had a large number of students—many of whom became top-ranking mathematicians themselves—and has had a profound influence in the development of analysis in several countries, including the United States, France, Argentina, Spain, and, of course, his native Poland. It is largely because of this influence that the United States has become the leader in the mathematical areas he cultivates.

These proceedings reflect the impact of Professor Zygmund's work and teaching. Most of the contributions to the conference are on topics he has either initiated or permanently shaped.

The Editors

CONTENTS

PART THREE: FOURIER ANALYSIS ON R^n AND REAL ANALYSIS 123

PART FOUR: SINGULAR INTEGRALS AND PSEUDODIFFERENTIAL OPERATORS 257

PART FIVE: HARDY SPACES 347

PART SIX: DIFFERENTIATION THEORY 507

PART SEVEN: PARTIAL DIFFERENTIAL EQUATIONS 555

PART EIGHT: OTHER TOPICS RELATED TO HARMONIC ANALYSIS 651

ANTONI ZYGMUND

Antoni Zygmund was born in Warsaw, Poland, on December 26, 1900, the son of Wincenty Zygmund and Antonina Perkowska Zygmund. After elementary school, he entered high school (gymnasium) in Warsaw in 1912. His early interests were in science, especially in astronomy, and he planned to become an astronomer.

At the outbreak of the First World War, his family was evacuated to Poltava, in the Ukraine, where it remained until autumn of 1918 before returning to (by then) independent Poland. In 1919 Zygmund entered the University of Warsaw, which had become Polish in 1915. The university was in its initial stages of development and the teaching of astronomy was still being organized. This situation, together with the interest in mathematical problems he had developed through independent study of astronomy, pushed Zygmund toward pure mathematics.

At that time there had emerged in Poland a group of talented young mathematicians who exerted a powerful attraction on university students. They were to develop what was later known as the Polish Mathematical School. Prominent among these young mathematicians were Waclaw Sierpinski (1882-1961), Stefan Mazurkiewicz (1888-1945), and Zygmunt Janiszewski (1888-1920). They founded a new mathematical periodical, *Fundamenta Mathematicae*, devoted primarily to the theory of sets and related topics, which had a strong impact on the subsequent development of mathematics in Poland. Zygmund was caught up in all this mathematical fervor. In addition, he was influenced by two of his older schoolmates, Stanislaw Saks and Zygmunt Zalcwasser. Saks was to play a very important role in the development of analysis in Poland. His primary interest was real variables, mainly differentiability and integrability of functions. His *Theory of the Integral*, first published in French in 1933, is a classic on the subject, and his *Analytic Functions*, written together with Zygmund, is still considered one of the best books on this topic. Alexander Rajchman, then an instructor and docent at the University of Warsaw, had made very important contributions to the theory of formal multiplication and uniqueness of trigonometric series, and Zygmund became particularly interested in the problems of this area. One may thus say that Zygmund was primarily a disciple of Saks and Rajchman.

Zygmund obtained his doctorate in 1923. A year earlier, he had been appointed instructor at the Polytechnical School in Warsaw, a position he kept until his departure to Wilno in 1930. In 1926 he was appointed docent at the University of Warsaw and also was awarded a Rockefeller Fellowship to spend a year in England with Hardy in Oxford and Littlewood in Cambridge. Unexpectedly, he encountered there one more mathematician who influenced him very deeply: R. E. A. C. Paley, a student of Littlewood, with whom Zygmund maintained a fruitful collaboration until Paley's early accidental death in 1933, a severe blow to Zygmund both personally and professionally.

In 1930, after his return from England, Zygmund was appointed professor at the University of Wilno. There he encountered Józef Marcinkiewicz, who became first his student and then his collaborator. Marcinkiewicz, whose name is familiar to everyone interested in functional analysis and Fourier series, was an extraordinary mathematician. His collaboration with Zygmund lasted almost ten years and produced a number of important results.

At the outbreak of the Second World War, in 1939, both Marcinkiewicz and Zygmund were mobilized as reserve officers in the Polish Army. Marcinkiewicz was taken prisoner in the East and disappeared without trace, most likely executed in captivity. Zygmund succeeded in returning to Wilno to his wife and son. Through the help of friends, mainly Professors Jacob Tamarkin and Jerzy Neyman, he managed to leave Poland and find his way to the United States, where he arrived early in 1940. The specter of impending mobilization and war weakened the need for university teachers and made jobs difficult to obtain. Nevertheless, Zygmund received an appointment as professor at Mount Holyoke College in Massachusetts.

In 1945 Zygmund moved to the University of Pennsylvania in Philadelphia. Two years later he was invited by Professor Marshall Stone to join the Department of Mathematics of the University of Chicago. A most impressive group of outstanding mathematicians was to gather there, A. A. Albert, S. S. Chern, S. MacLane, L. M. Graves, A. Weil, I. Kaplansky, P. Halmos, I. Segal, and E. Spanier among them. Zygmund went to the University of Chicago and remained there until he retired in 1980 as the Gustave and Ann Swift Distinguished Service Professor of Mathematics. At Chicago he had a large following and developed what has come to be known as the Chicago School of Analysis. The mathematical influence Zygmund exerted through his work, his teaching, and his students has spread across the United States and beyond, to South America and Europe.

Professor Antoni Zygmund has been awarded many honors and has held important positions in the mathematical community. He has honorary degrees from Washington University, the University of Toruń in Poland, the University of Paris, and the University of Uppsala in Sweden. The learned societies of which he is a member include the National Academy of Sciences of the United States; the Polish Academy of Sciences; the National Academy of Sciences of Buenos Aires, Argentina; the Royal Academy of Sciences of Madrid; and the National Academy of Sciences of Palermo, Italy.

Alberto P. Calderón

ACKNOWLEDGMENT

The Chicago Conference on Harmonic Analysis
March 23-28, 1981

was sponsored by

The National Science Foundation
grant number
MCS-8019742

and

The University of Chicago.

PART ONE

Introductory Paper

THE DEVELOPMENT OF SQUARE FUNCTIONS IN THE WORK OF A. ZYGMUND

E. M. Stein
Princeton University

I've decided to write this essay about "square functions" for two reasons. First, their development has been so intertwined with the scientific work of A. Zygmund that it seems highly appropriate to do so now on the occasion of his eightieth birthday. Also these functions are of fundamental importance in analysis, standing as they do at the crossing of three important roads many of us have travelled by: complex function theory, the Fourier transform (or orthogonality in its various guises), and real-variable methods. In fact, the more recent applications of these ideas, described at the end of this essay, can be seen as confirmation of the significance Zygmund always attached to square functions.

This is going to be a partly historical survey, and so I hope you will allow me to take the usual liberties associated with this kind of enterprise: I will break up the exposition into certain "historical periods," five to be precise; and by doing this I will be able to suggest my own views as to what might have been the key influences and ideas that brought about these developments.

One word of explanation about "square functions" is called for. A deep concept in mathematics is usually not an idea in its

---

Reprinted from *Bulletin (New Series) of the American*

pure form, but rather takes various shapes depending on the uses it is put to. The same is true of square functions. These appear in a variety of forms, and while in spirit they are all the same, in actual practice they can be quite different. Thus the metamorphosis of square functions is all important.

FIRST PERIOD (1922–1926): THE PRIMORDIAL SQUARE FUNCTIONS

It appears that square functions arose first in an explicit form in a beautiful theorem of Kaczmarz and Zygmund dealing with the almost everywhere summability of orthogonal expansions. The theorem was proved in 1926 as the culmination of several papers each had written at about that time. The theorem itself was an outgrowth of what certainly was one of the main preoccupations of analysts at that time, namely the question of convergence of Fourier series. The problem was the following. Suppose $f = f(\theta)$ is a continuous function on the circle, $0 \leq \theta \leq 2\pi$, or more generally assume that f is in $L^2(0,2\pi)$ or even that f is merely integrable; then does its Fourier series

$$\sum a_n e^{in\theta}, \quad \text{with } a_n = \frac{1}{2\pi} \int_0^{2\pi} f(\theta) e^{-in\theta}\, d\theta, \tag{1}$$

converge almost everywhere?

A related parallel issue was the corresponding question for a general orthonormal expansion, but now limited to $f \in L^2$. Thus if $\{\phi_n\}$ is an orthonormal system, and if $f \sim \Sigma\, a_n\phi_n$ with $a_n = \int f\overline{\phi}_n$, where $\Sigma\, |a_n|^2 < \infty$, then what could be said about the convergence almost everywhere of

$$\sum_{n=1}^{\infty} a_n\phi_n(x)? \tag{2}$$

The period we are dealing with (1922-1926) was marked by several striking achievements in this area, whose essential interest

---

is not diminished even when viewed from the distant perspective of more than a half century. The first result to mention was the construction by Kolmogorov in 1923 of an L^1 function whose Fourier series (1) diverged almost everywhere.[†]

This construction made even more pressing the question of whether the Fourier series (1) converges almost everywhere when (say) f belongs to L^2, a problem that was not solved till more than forty years later. We shall turn to that in a moment, but now we point out that Kolmogorov's example put into sharper relief the L^2 results for general orthonormal developments that had been obtained (in 1922 and 1923) by Rademacher and Menshov. They showed that if

$$\sum |a_n|^2 \ (\log n)^2 < \infty \tag{3}$$

then the series (2) converges a.e.

Moreover the condition (3) is best possible in the sense that if $\{\lambda_n\}$ is monotonic and $\lambda_n/\log n \to 0$, then there exists an orthonormal system $\{\phi_n\}$ and expansion (2) which diverged a.e., while $\Sigma\ |a_n|^2\lambda_n^2 < \infty$.

For ordinary Fourier series it was proved [22, 23] that the condition (3) could be relaxed and be replaced by

$$\sum_{-\infty}^{\infty} |a_n|^2 \log(|n| + 2) < \infty. \tag{4}$$

This last result stood unsurpassed for forty years until Carleson in 1966 showed that indeed the Fourier Series of an L^2 function converged almost everywhere. It may be interesting to note here that the basic tools required for Carleson's theorem—the properties of the Hilbert transform and their relation with partial sums of Fourier series—were first brought to light in this early period: Kolmogorov's proof of the weak-type (1,1)

[†]Kolmogorov later even outdid himself when in 1926 he refined his construction to show the existence of an L^1 function whose Fourier series diverged *everywhere*.

property in 1925; M. Riesz's paper of 1927 containing the L^p inequalities for conjugate functions and partial sums; and Besicovitch's work (in 1923 and 1926) which began the development of "real-variable" methods for Hilbert transforms.

Against this background we can now state the idea of Kaczmarz and Zygmund. It asserts as a general principle that for an L^2 orthonormal expansion (i.e., one where $\Sigma\,|a_n|^2 < \infty$), at almost all points the summability of the series $\Sigma\, a_n\phi_n(x)$ by one method one has as a consequence the summability by any method which is essentially stronger than convergence. A special (but typical) case is as follows:

Theorem 1. Suppose $\Sigma|a_n|^2 < \infty$*. Then* $\Sigma\, a_n\phi_n(x)$ *is Cesàro summable at almost each point* x *where it is Abel summable.*

Recall that the series is Abel summable at x if $\lim_{r\to 1} \Sigma\, a_n r^n \phi_n(x)$ exists. In addition, setting $s_n = \sum_{k=0}^n a_k\phi_k$, and $\sigma_n = (s_0 + s_1 + \cdots + s_{n-1})/n$, the Cesàro summability at x means the existence of the limit $\lim_{n\to\infty} \sigma_n(x)$.

If a series is Cesàro summable it is automatically Abel summable (an exercise!), but the converse is in general not true. To gain a better idea of the scope of Theorem 1 let us point out that

$$\sigma_n(x) = \sum_{k=0}^{n} (1 - k/n)a_k\phi_k(x)$$

and a result similar to Theorem 1 holds when $\sigma_n(x)$ is replaced by $\sigma_n^{\varepsilon}(x) = \sum_{k=0}^n (1 - k/n)^{\varepsilon} a_k\phi_k(x)$, with $\varepsilon > 0$ (which corresponds essentially to (C,ε) summability), but not for $\varepsilon = 0$ which of course would give the usual convergence.

For the proof of Theorem 1 Kaczmarz and Zygmund used a square function which they introduced for this purpose, namely

$$K(f) = \left(\sum_{n=2}^{\infty} n|\sigma_n - \sigma_{n-1}|^2 \right)^{1/2} \tag{5}$$

with $f \sim \Sigma\, a_n\phi_n$. The basic fact was the L^2 inequality.

Lemma.

$$\| K(f) \|_{L^2} \leq c \| f \|_{L^2}.$$

Clearly

$$\sigma_n = \sigma_{n-1} = \frac{1}{n(n-1)} \sum_{k=0}^{n-1} k a_k \phi_k ,$$

$$\text{so } \| \sigma_n - \sigma_{n-1} \|_2^2 \leq \frac{c}{n^4} \sum_{k<n} k^2 |a_k|^2 , \ n \geq 2,$$

and thus $\sum_2^\infty n \| \sigma_n - \sigma_{n-1} \|_2^2 \leq c' \sum |a_k|^2 = c' \| f \|_2^2$, which proves the lemma.

To prove the theorem one invokes a variant of the classical Tauberian argument, namely, if ΣA_n is Abel summable and $\Sigma n A_n^2 < \infty$, then ΣA_n converges. Now set $A_n = \sigma_n - \sigma_{n-1}$; then the Abel summability of ΣA_n follows from the corresponding Abel summability of $\Sigma a_n \phi_n$. The Tauberian condition holds at almost all points because of the lemma, and hence one obtains a.e. the convergence of $\Sigma(\sigma_n - \sigma_{n-1})$, proving the theorem.

We have seen the first example of a square function, namely (5). While here it plays a minor role, its basic character is already revealed: Because of the agility of its quadratic nature it can exploit easily any situation in which orthogonality might be important.

SECOND PERIOD (1931–1937): LITTLEWOOD AND PALEY

Our scene shifts now from the Continent to England, and to the work of Littlewood and Paley. Our attention will be focused on two important series of connected papers: three jointly by Littlewood and Paley [1931 - 1937], and two by Paley [1932]. The investigations described in these papers were initiated simultaneously (the first paper in each series was submitted in April 1931), but because of Paley's death in 1933 the final versions of several of the papers were probably Littlewood's work alone.

It is also interesting to note that no reference is made in these papers to the results described above, and so it is a reasonable guess that they were not aware of the possible relevance of the ideas of Kaczmarz and Zygmund.

The main theme of the Littlewood-Paley work was to consider the "dyadic decomposition" of Fourier series, namely

$$f(\theta) = \sum_{k=0}^{\infty} \Delta_k(\theta),$$

with

$$\Delta_k(\theta) = \sum_{2^{k-1} \leq |n| < 2^k} a_n e^{in\theta}, \quad k \geq 1; \ \Delta_0 = a_0.$$

Their basic result was that the L^p norm of a function was equivalent with the L^p norm of the square function associated with its dyadic decomposition.

Theorem 2. *For* $1 < p < \infty$,

$$\left\| \left(\sum_{k=0}^{\infty} |\Delta_k(\theta)|^2 \right)^{1/2} \right\|_p \simeq \|f\|_p.$$

To prove this theorem they needed and thus formulated an "Abelian" analogue, where partial sums are replaced by Abel means, i.e., the Poisson integral of $f = u(r,\theta)$. Thus given f, let Φ be the holomorphic function in the unit disc with $\mathrm{Re}(\Phi) = u$, and $\mathrm{Im}(\Phi(0)) = 0$. They defined another square function the "g-function" of f by

$$g(f)(\theta) = \left(\int_0^1 (1 - r) |\Phi'(re^{i\theta})|^2 \, dr \right)^{1/2}$$

and proved the following

Theorem 3. *With* $1 < p < \infty$

$$\|g(f)\|_p \simeq \|f\|_p, \quad \text{if } a_0 = 0. \tag{6}$$

Paley sought a better understanding of the nature of these problems by considering variants of Theorem 2 where the Fourier series expansion is replaced by the Walsh-Paley expansion. The Walsh-Paley functions (called Walsh-Kaczmarz functions at that time) are now usually described as follows. We identify the interval [0,1] with the compact group consisting of an infinite product of copies of the two-element group (via the usual binary expansion). The characters of that group are the Walsh-Paley functions. Writing each integer as a sum of powers of 2 gives a natural enumeration of the characters $\{\phi_n\}_{n=0}^{\infty}$. If we set $f \sim \sum a_n\phi_n$, and $\Delta_k = s_{2^k} - s_{2^{k-1}} = \sum_{2^{k-1}<n\leq 2^k} a_n\phi_n$, with $\Delta_0 = a_0$, then Paley's theorem reads as

Theorem 4. For the Walsh-Paley series, with $1 < p < \infty$

$$\left\| \left(\sum |\Delta_k|^2\right)^{1/2} \right\|_p \simeq \|f\|_p.$$

What makes the proof of Theorem 4 easier than that of Theorem 2 are the various simplifications inherent in the fact that $\{s_{2^k}(f)\}$ is a martingale sequence. The name "martingale" had not yet been coined. Moreover, a systematic extension of Theorem 4 from the point of view of martingales, and its further exploration in the magical world of Brownian motion—all these came much later, as we shall see. However in Paley's time some of the arguments typical of martingale theory were already understood. Thus it had been observed that $s_{2^k}(f)$ was constant on each 2^k intervals (of length 2^{-k}) of the form $((\ell - 1)/2^k, 1/2^k)$, $\ell = 1, \ldots, 2^k$, and that the values of $s_{2^k}(f)$ on each of these intervals was the mean-value of f there. From this it is obvious when $f \in L^p$, $1 \leq p \leq \infty$, then $\{s_{2^k}(f)\}$ are bounded in L^p norm; the analogue for Fourier series is definitely nonobvious when $1 < p < \infty$, and in fact false when $p = 1$ or $p = \infty$.

We shall now describe the main device Paley used in his proof of Theorem 4. Paley was, from what one can learn about his life, a man of courage and almost reckless daring. A hint of that spirit can be found in his approach to difficult mathematical problems. When faced by the proof of an inequality like

$$\int \left(\sum |\Delta_k|^2\right)^{p/2} dx \leq A_p^p \int |f|^p \, dx \tag{7}$$

where p is e.g. an even integer $2r$, he instinctively sought to face the problem head-on by multiplying out the r infinite sums, and then coming to grips directly with the resulting multitude of terms. This kind of audacious attack is not so common in our time when it is easier to rely on a variety of sophisticated gadgets which are household items for the working analyst. But given Paley's resourcefulness this approach worked marvelously well. His key observation was that

$$\sum_{i_r} \int \Delta_{i_1}^2 \Delta_{i_2}^2 \cdots \Delta_{i_r}^2 \, dx \leq \int \Delta_{i_1}^2 \cdots \Delta_{i_{r-1}}^2 f^2 \, dx \tag{8}$$

where the summation is taken over those i_r for which $i_r > \max(i_1, \ldots, i_{r-1})$, which in turn follows from the martingale property that

$$\int g(x)\Delta_k(x) \, dx = 0 \tag{9}$$

whenever g is "measurable with respect to the past." From (8) Paley was able to achieve the proof of (7) in a few strokes.

The same idea inspired Littlewood and Paley's proof of Theorem 3, although the execution is more complicated; a more recondite form of (8) must be proved, and here nothing as simple as (9) holds. The appropriate substitute must be fashioned with care out of Green's theorem in conjugation with the identity $\Delta(|\Phi|^2) = 4|\Phi'|^2$. With Theorem 3 proved, Littlewood and Paley were able to deduce Theorem 4, but here also the steps required were not easy. It was only after their theory was reexamined by Zygmund and his student Marcinkiewicz that a clearer and broader view of the whole subject began to emerge. To this we shall now turn.

THIRD PERIOD (1938–1945): MARCINKIEWICZ AND ZYGMUND

There are two significant events that marked the period we are now concerned with. The first, which even predated the Littlewood-Paley collaboration, was the introduction by Lusin in 1930 of his "area integral." The idea of Lusin seems to have sparked no further interest until Marcinkiewicz and Zygmund took up the subject again about 8 years later. There began a brief but very creative period of work by them—a flowering of the theory where connections with a variety of other ideas were brought to light. The second event, a tragic one, followed soon thereafter with the death of Marcinkiewicz in 1940, and it was left to Zygmund alone to resolve some of the issues that their work had led them to.

It may help to clarify the description of the principal ideas that Marcinkiewicz and Zygmund contributed to the study of square functions if we organize our presentation in terms of the four main lines along which their work proceeded.

The first subject we shall treat (and the only one that was, strictly speaking, joint work) deals with the area integral of Lusin. The definition of this is as follows. Suppose $\Phi(z)$ is holomorphic in the unit disc and define $A(\Phi)(\theta)$ by

$$(A(\Phi)(\theta))^2 = \int_{\Gamma(\theta)} |\Phi'(z)|^2 \, dx \, dy \tag{10}$$

with $\Gamma(\theta)$ a standard "triangle" (nontangential approach region) in the unit disc with vertex at $e^{i\theta}$. Observe that the expression represents the area of the image of $\Gamma(\theta)$ under the mapping $z \to \Phi(z)$, with points counted according to their multiplicity. Lusin's discovery was that if Φ is bounded, then $A(\Phi)(\theta)$ is finite for almost any θ; more generally that

$$\| A(\Phi)(\theta)\|_2 \simeq \|\Phi\|_2, \quad \text{if } \Phi(0) = 0. \tag{11}$$

Marcinkiewicz and Zygmund realized that on the one hand there was a close analogy between the Littlewood-Paley g-function and $A(\Phi)$ (in fact A is a pointwise majorant of g, and the same kind of L^p inequalities held for A as for g); but on the other hand they surmised that the parallel between these two square functions should not be pushed too far. The main result they obtained for A was a localized version of Lusin's result. This can be stated as follows. Let $\Phi^*(\theta) = \sup_{z \in \Gamma(\theta)} |\Phi(z)|$.

Theorem 5a. *If Φ is holomorphic in the unit disc, then for almost every θ, $\Phi^*(\theta) < \infty$ implies $A(\Phi)(\theta) < \infty$.*

The converse was proved five years later by Spencer[†], namely

Theorem 5b. *If Φ is holomorphic in the unit disc, then for almost every θ, $A(\Phi)(\theta) < \infty$ implies $\Phi^*(\theta) < \infty$.*

A corresponding converse for g-functions is false, and so the area integral A has some special affinities with the boundary behavior of Φ, going beyond what it shares with g.

The second line of investigation was Zygmund's reexamination of the Littlewood-Paley theorem for the dyadic decomposition of Fourier series. His analysis led him to recast and simplify the ideas of the proof. These simplifications had important consequences for later work, as we shall see; but their immediate interest was that it allowed him to connect the square function $(\sum |\Delta_k|^2)^{1/2}$ with the one he and Kaczmarz had considered a dozen years earlier in their study of summability of orthogonal series (see (5)). We suppose that we take the Fourier expansion and set $f(\theta) \sim \sum_{n \geq 0} a_n e^{in\theta}$, $f \in L^p$, so that $f \in H^p$. If we write as before

---

[†]Incidentally, the work of Spencer [36] seems to be the only major exception to the monopoly Marcinkiewicz and Zygmund held during that period on subjects relating to square functions.

$$K(f)(\theta) = \left(\sum_{n\geq 1} n|\sigma_n(\theta) - \sigma_{n-1}(\theta)|^2\right)^{1/2}, \text{ where}$$

$$\sigma_n(\theta) = \sum_{0\leq k<n} (1 - k/n)a_k e^{ik\theta},$$

then we can state the following theorem:

Theorem 6. $\|K(f)\|_p \leq A_p\|f\|_p$, $1 < p < \infty$.†

The proof of this theorem required two steps. First, like that of Theorem 2, one needed the L^p inequalities for the g-function (see (6)). Here the major simplification was made by Zygmund some years later [51] and it came in the proof of the fact that $\|g(f)\|_p \leq A_p\|f\|_p$, when $p > 2$. (The case $p = 2$ was easy, and the range $p < 2$ was reducible to $p = 2$ by the artifice standard in those days of using Blaschke product decompositions for H^p functions.) For the difficult case $p > 2$ a "square duality" was used. An ingenious argument shows that whenever $\phi \geq 0$,

$$\int g(f)^2\phi \, d\theta \leq c\left\{\int g(f)g(\phi)M(f) \, d\theta + \int |f|^2\phi \, d\theta\right\} \qquad (12)$$

where M is the Hardy-Littlewood maximal function. For $p \geq 4$, (12) then gives the desired result as a consequence of the case $p \leq 2$ applied to $g(\phi)$. Incidentally, the notion of square duality which seems to have originated in this context continues to find other applications of interest.

The second simplification Zygmund made was in the manner one could reduce the L^p control of $(\Sigma|\Delta_k|^2)^{1/2}$ to that of the g-function; and in fact a whole list of other square functions (in particular $(\Sigma\, n|\sigma_n - \sigma_{n-1}|^2)^{1/2}$) could be handled in the same way.‡ This streamlining of the proof he found can be said to have led directly to the "Marcinkiewicz multiplier theorem."

---

†There was also a result for $p = 1$, i.e., $\|K(f)\|_p \leq A_p\|f\|_1$ if $p < 1$.

‡This is the presentation given in Zygmund's book [52], Chapter 15.

In its one-dimensional form the celebrated theorem that bears Marcinkiewicz's name can be stated as follows. Suppose we consider a transformation T given by a multiplier sequence $\{\lambda_n\}_{-\infty}^{\infty}$, defined by

$$Tf \sim \sum \lambda_n a_n e^{in\theta}, \quad \text{whenever } f \sim \sum a_n e^{in\theta}.$$

Then T is bounded on L^p, $1 < p < \infty$, if (i) the sequence $\{\lambda_n\}$ is bounded, and (ii) if it varies boundedly over each dyadic block; more precisely, $\sum_{2^k \leq |j| < 2^{k+1}} |\lambda_j - \lambda_{j-1}| \leq M$. (Note that the special case when the sequence is constant on each dyadic block is an immediate consequence of Theorem 2.) In one dimension the theorem's greatest merit is, I believe, in its formulation rather than its proof; the latter is much the same as that of Theorem 6.

It is in the passage to higher dimensions, however, that one finds the great significance of Marcinkiewicz's work on multipliers. Its importance was not only the fact that one could use hitherto one-dimensional methods to prove n-dimensional results; even more profound were the applications to other questions, such as estimates for partial differential equations, already envisaged at that time. We can now see in retrospect that Marcinkiewicz thus anticipated some of the basic inequalities later proved by the theory of singular integrals.†

For simplicity of notation we shall state the Marcinkiewicz multiplier theorem in the case of two dimensions. Consider the multiplier operator T given by $Tf \sim \sum \lambda_{nm} a_{nm} e^{i(n\theta+m\phi)}$ for $f \sim \sum a_{nm} e^{i(n\theta+m\phi)}$. Let I_k denote the dyadic interval = $\{n \mid 2^{k-1} \leq |n| < 2^k\}$, and $J_\ell = \{m \mid 2^{\ell-1} \leq |m| < 2^\ell\}$. Write $\Delta_1\lambda_{nm} = \lambda_{n+1,m} - \lambda_{n,m}$, $\Delta_2\lambda_{n,m} = \lambda_{n,m+1} - \lambda_{n,m}$, and $\Delta_{1,2} = \Delta_1 \cdot \Delta_2$. Now assume the finiteness of the following four quantities:

---

†This becomes clearer upon reading Marcinkiewicz's paper [26] and taking into account the continuous analogue Mihlin deduced from it in 1956 [30].

(i) $\sup_{n,m}|\lambda_{n,m}|$;

(ii) $\sup_{k,m}\sum_{n\in I_k}|\Delta_1\lambda_{n,m}|$, and $\sup_{m,\ell}\sum_{m\in J_\ell}|\Delta_2\lambda_{n,m}|$; and

(iii) $\sup_{k,\ell}\sum_{n\in J_\ell}|\Delta_1\Delta_2\lambda_{n,m}|$.

Theorem 7. *Under the assumption made above, T is bounded on L^p, $1 < p < \infty$.*

The last of the four major lines of investigation concerning square functions that Marcinkiewicz and Zygmund undertook dealt with the attempt to find a completely "real-variable" analogue of the functions of Lusin and Littlewood-Paley. Starting with a function f on the circle, the area integral and g-functions are defined in terms of holomorphic (or harmonic) functions whose boundary values are related to f. Also the dyadic square function of Theorem 2 requires the Fourier expansion of f. What was desired was a variant that could be defined more directly in terms of the basic real-variable operations such as integration, differentiation, etc.

After some experimentation Marcinkiewicz hit upon the idea of considering

$$\mu(F)(x) = \left(\int_0^\pi |F(x+t) + F(x-t) - 2F(x)|^2 \frac{dt}{t^3}\right)^{1/2} \tag{13}$$

with

$$F(x) = \int^x f(t)\, dt.$$

It was not difficult to see that

$$\|\mu(F)\|_{L^2} \simeq \|f\|_{L^2}, \quad \text{if } \int_0^{2\pi} f(x)\, dx = 0.$$

With this, and using the real-variable tools he had already developed, he was able to prove the analogue of the theorem he and Zygmund had found for the area integral (Theorem 5a). The result was as follows.

Theorem 8a. *Suppose $F \in L^2$. If $F'(x)$ exists in a set E, then $\mu(F)(x) < \infty$ for almost every $x \in E$.*

The questions that arose were first, whether some of the other properties of the area integral or g-function held as well for μ; and, more interestingly, what was the real significance of the Marcinkiewicz function. Zygmund found an answer to the first question in 1944 when he proved

Theorem 8b. For $1 < p < \infty$,

$$\|\mu(F)\|_{L^p} \simeq \|f\|_{L^p}, \quad \text{if } \int_0^{2\pi} f(x)\,dx = 0.$$

The argument he developed to show this was not an easy one. He was required to invoke the most arcane of the square functions, the function g^*, which Littlewood and Paley had also studied. He established the L^p inequalities for it and showed that it actually was a majorant of the Marcinkiewicz function. Incidentally g^* is defined by

$$(g^*(\Phi)(\theta))^2 = \int_0^1\int_0^{2\pi} |\Phi'(re^{i(\theta+\phi)})|^2 \left|\frac{1-r}{1-re^{i\phi}}\right|^2 d\phi\, dr,$$

and so majorizes also the area integral (10), but it takes into account "the tangential" approach to the boundary.†

The problem that remained was to discover whether there was a converse to the local result given by Theorem 8a, or to put the question more broadly, to find the meaning of the Marcinkiewicz function. It was to be almost twenty more years before an answer to that question would be found.

FOURTH PERIOD (1950–1964): ZYGMUND AND HIS STUDENTS

Starting about 1950 a new direction of considerable importance began to emerge in force. Hinted at in earlier work (of Besicovitch

---

†It is an ironic fact that a dozen years later (in 1956), Zygmund [50] and Sunouchi [46] independently showed that g^* is essentially identical with the square function K of Theorem 6.

and Marcinkiewicz, among others), its thrust was the development of "real-variable" methods to replace complex function theory—that favored ally of one-dimensional Fourier analysis. What made this new emphasis particularly timely, in fact indispensable, was that only with techniques coming from real-variable theory could one hope to come to grips with many interesting n-dimensional analogues of the one-dimensional theory.

The mathematician animating this development was Antoni Zygmund. In many ways he set the broad outlines of the effort, he mastered by his work some of the crucial difficulties, and was throughout the source of inspiration for his students and collaborators.

The Area Integral

A pioneering result in this new direction was Calderón's extension to $\mathbf{R}^n$ of the theorem of Marcinkiewicz and Zygmund concerning the area integral, a subject he had taken up at the suggestion of Zygmund. The setting for this is as follows. We let $\mathbf{R}_+^{n+1} = \{(x,y),\ x = (x_1, \ldots, x_n) \in \mathbf{R}^n,\ y \in \mathbf{R}^+\}$ be the upper half-space, and suppose that $u(x,y)$ is harmonic (with respect to the $n + 1$ variables $x_1, \ldots, x_n, y$). Sometimes we shall assume that u is in fact the Poisson integral of an appropriate function f defined on $\mathbf{R}^n$, and then we shall write $u = \mathrm{PI}(f)$. We let $\Gamma = \{(x,y),\ |x| < y\}$ be a standard cone with vertex at the origin, Γ' its truncated version, $\Gamma' = \Gamma \cap \{y < 1\}$. For any $\bar{x} \in \mathbf{R}^n$, $\Gamma(\bar{x})$ and $\Gamma'(\bar{x})$ will be the corresponding cones with vertices at $\bar{x}$. The area integral of u is defined by

$$(A(u)(\bar{x}))^2 = \int_{\Gamma(\bar{x})} |\nabla u|^2 y^{1-n}\, dx\, dy \tag{14}$$

where $|\nabla u|^2 = |\partial u/\partial y|^2 + \sum_{j=1}^{n} |\partial u/\partial x_\partial|^2$.

Similarly for the local theory one needs the analogue of (14) where $\Gamma(\bar{x})$ is replaced by $\Gamma'(\bar{x})$; this defines $A_{\mathrm{loc}}(u)(\bar{x})$. The maximal function u^* is defined by $*(\bar{x}) = \sup_{(x,y)\in\Gamma(\bar{x})} |u(x,y)|$, and its local analogue u^*_{loc} is given by replacing $\Gamma(\bar{x})$ by $\Gamma'(\bar{x})$ in the definition.

Theorem 9a. Suppose u is harmonic in $\mathbf{R}^{n+1}_+$. *Then* $A_{\mathrm{loc}}u(\bar{x}) < \infty$ *at almost every point* $\bar{x} \in \mathbf{R}^n$ *where* $u^*_{\mathrm{loc}}(\bar{x}) < \infty$.

Calderón's proof of this theorem was published at the same time (1950) as another important result he found, namely the extension of Privalov's theorem: u has a nontangential limit at almost every $\bar{x} \in \mathbf{R}^n$, where $u^*_{\mathrm{loc}}(\bar{x}) < \infty$. We shall discuss the ideas behind the proof of Theorem 9a later when we take up its converse. Now we turn to the "global" version, i.e., the higher-dimensional analogue of the Littlewood-Paley theorem (Theorem 3).

Theorem 9b. Suppose $u = \mathrm{PI}(f)$, *then*

$$\|A(u)\|_{L^p} \simeq \|f\|_{L^p}, \quad 1 < p < \infty.$$

It would be difficult after 25 years to recall the precise thoughts that motivated the proof of Theorem 9b, nor would it be easy now for one to appreciate the difficulties that seemed then to stand in the way. But I do remember that those of us who were graduate students of Zygmund in the middle 1950's were shaped by the event, akin to the Creation, which appeared to some of us to be the beginning of everything important: the 1952 Acta paper which developed via the Calderón-Zygmund lemma, the real variable methods giving the extension of the Hilbert transform to n-dimensions. What was more natural, therefore, than to attempt to prove the L^p boundedness of $f \to A(u)$ by adapting these methods? This idea indeed worked, although the initial complicated proofs were later much simplified. The analysis succeeded as well for the Marcinkiewicz function (13), and proved also that the mappings $f \to A(u)$ and $f \to u(F)$ were of weak-type (1,1).

We turn now to the proof of Theorem 9a. Its one-dimensional version (Theorem 5a) had been done by using complex function theory, in particular conformal mappings. So a completely different approach was needed. The idea behind it can be understood by examining the case $p = 2$ of Theorem 9b, which has an easy proof. A direct calculation shows that

$$\int_{\mathbf{R}^n} A^2(u)\, dx = c \int_{\mathbf{R}^{n+1}_+} y|\nabla u|^2\, dx\, dy, \tag{15}$$

where c is the volume of the unit ball. Next we can use the fact that $|\nabla u|^2 = \frac{1}{2}\Delta(|u|^2)$, and so by Green's theorem

$$\int_{\mathbf{R}^n} A^2(u)\, dx = \frac{c}{2}\int\int_{\mathbf{R}^{n+1}_+} y\Delta(|u|^2)\, dx\, dy = \frac{c}{2}\int |u(x,0)|^2\, dx,$$

which proves Theorem 9b for $p = 2$, since $u(x,0) = f(x)$. Thus in order to control $A_{\text{loc}}(u)(x)$ on a set E, it is natural to consider $\int_E A^2_{\text{loc}}\, u(x)\, dx$ which in turn is dominated by $c \int_{R(E)} y|\nabla u|^2\, dx\, dy$, where $R(E)$ is a standard "sawtooth" region in $\mathbf{R}^{n+1}_+$ based on E. At this stage (which is the turning point of the proof) Calderón invoked Green's theorem for another region containing $R(E)$, whose Green's function he could essentially bound from below by $c'y$.

To prove the converse of Theorem 9a along these lines appeared to require, among other things, appropriate bounds from above for Green's function for such regions, and that seemed much beyond what could be done then.† What turned out to be the right course of action was to finesse the problem of Green's function and to proceed directly with estimates that followed from the finiteness of $\int_{R(E)} y|\nabla u|^2\, dx\, dy$. These arguments also proved to be useful in other situations, as we shall see later. The result obtained was

Theorem 9c. Suppose u is harmonic in $\mathbf{R}^{n+1}_+$. *Then* $u^*_{\text{loc}}(\bar{x}) < \infty$ *for almost all points* $\bar{x} \in \mathbf{R}^n$ *where* $A_{\text{loc}}(u)(\bar{x}) < \infty$.

I remember quite vividly the excitement surrounding the events at the time of this work. It was March 1959, and I had returned to the University of Chicago the fall before. Frequently

---

†In fact, good estimates for Green's function and the Poisson kernel for such regions, and more generally "Lipschitz domains," originate in the work of Dahlberg in 1977 [15].

I met with my friends Guido Weiss and Mary Weiss, and together we often found ourselves in Zygmund's office (Eckhart 309, two doors from mine). With our teacher our conversations ranged over a wide variety of topics (not all mathematical) and more than once the subject of square functions arose. When this happened the mood would change, if only slightly, as if in deference to their special status, and the enigma that surrounded them. I had an idea which seemed promising. But before we could see where it might lead came the spring break. Further work would have to be held in abeyance since we were each going our own ways: Zygmund travelled to Boston to visit Calderón; Guido and Mary Weiss, having borrowed my Chevrolet, drove to Virginia for a vacation trip; and I went to New York to be married.

The Marcinkiewicz Function

Influenced by the renewed interest in area integrals, and encouraged by some recent work he had done with Mary Weiss [47], Zygmund returned to the study of the Marcinkiewicz integral (13) and the problem of finding a converse to Theorem 8a. He was convinced that now (more than 20 years after Marcinkiewicz's original work) the time was ripe to see matters to a conclusion. He suggested to me that we work on the problem together, and of course I was very happy to accept his offer. For me this was a unique and rewarding collaboration—not just because of the special satisfaction one derives when accepted as an equal by one's teacher—but also because as it turned out he did most of the work that really counted!

We realized first that Theorem 8a itself could be somewhat strengthened; what was required was the notion of the derivative $F'(x)$ existing (at x) "in the L^2 sense." Thus $F'(x)$ existed in this generalized sense if

$$\frac{1}{h}\int_0^h \left| \frac{F(x+t)-F(x)}{t} - F'(x) \right|^2 dt \to 0, \quad \text{as } h \to 0.^{\dagger} \tag{16}$$

The finer version of Theorem 8a was then: If $F \in L^2$ had a derivative in the sense of (16) at each $x \in E$, then $\mu(F)(x) < \infty$ for almost every $x \in E$. It was in this form that one might seek a converse. The basic plan was to try to make matters turn on the analogous situation which held for the area integral, where one can pass from the finiteness of a quadratic expression to the existence of a limit. After a series of reductions we were able to show that at each point x where $\mu(F)(x) < \infty$ one had

$$\int_{|t|\le y} \left| \frac{\partial^2 u}{\partial y^2}(x+t, y) + \frac{\partial^2 u}{\partial y^2}(x-t, y) \right|^2 dt\, dy < \infty \tag{17}$$

with $u = \mathrm{PI}(F)$. On the other hand we could show (using Theorem 5b) that at almost every x where

$$\int_{|t|\le y} \left| \frac{\partial^2 u}{\partial y^2}(x+t, y) \right|^2 dt\, dy < \infty \tag{18}$$

the conclusion (16) actually held.

The basic difficulty, the passage from (17) to (18), was overcome by Zygmund using a clever "desymmetrization" argument; several weeks later he presented me with an essentially final draft of the paper which he had typed himself!

There were several variants of the final result—involving extensions to n-dimensions, or higher derivatives, or even fractional derivatives. The simplest version, however, was the following:

---

†This notion had at that time (1961) been introduced by Calderón and Zygmund [10].

Theorem 10. Let $F \in L^2(0, 2\pi)$. *Then the set of points* x *where* $\int_0^\pi |F(x+t) + F(x-t) - 2F(x)|^2 \, dt/t^3 < \infty$ *and the set of points where* $F'(x)$ *exists in the* L^2 *sense (i.e. (16)) differ by a set of measure zero.*

FIFTH PERIOD (1966–PRESENT): FURTHER APPLICATIONS OF SQUARE FUNCTIONS

We have traced the development of square functions from their beginnings to a stage where their nature was much better understood, in terms of a series of deep theorems that had been obtained. Yet it is only more recently that their central role in several fields of analysis has become more apparent. I shall try to describe this very briefly in terms of three specific areas: H^p spaces, symmetric diffusion semigroups, and differentiation theory in $\mathbf{R}^n$.

H^p Theory

Beginning in about 1966 two separate directions of research involving square functions were undertaken, and when brought together these ultimately led to a rich harvest in the theory of H^p spaces. The first started with Burkholder's extension of Paley's theorem (Theorem 4 for Walsh-Paley series) to general martingales. He observed that Paley's argument extended to this general setting, but also found his own approach which was very different. He showed that if $E_k = E(\cdot|\mathscr{F}_k)$ are the conditional expectations for an increasing sequence of σ-fields $\{\mathscr{F}_k\}_{k=0}^\infty$, then with $E_{-1}(f) \equiv 0$,

$$\left\| \left(\sum_{k=0}^{\infty} |(E_k - E_{k-1})(f)|^2 \right)^{1/2} \right\|_p \simeq \lim_{k\to\infty} \| E_k(f) \|_p, \quad 1 < p < \infty. \quad (19)$$

Next, in work with Gundy [6], and later also with Silverstein [7], the following advances were made: It was shown that (19) extended to $p \leq 1$ if $\lim_{k\to\infty} \| E_k(f) \|_p$ was replaced with $\| \sup_k E_k(f) \|_p$, for a large class of martingales. This class incidentally includes those occurring for the Walsh-Paley series, but more importantly these results went over to the (continuous parameter) martingales arising from Brownian motion applied to harmonic functions. To be more precise, let $z_t(\omega)$ denote the standard Brownian motion in the complex z-plane, starting at the origin and stopped when reaching the unit circle. Here $0 \leq t < \infty$ is the time parameter, and ω labels the Brownian path, with $\omega \in \Omega$, Ω being the probability space. If u is harmonic in the unit disc, $t \to u(z_t(\omega))$ is a continuous-time martingale. Let $M_B(u)(\omega) = \sup_{0 \leq t < \infty} |u(z_t(\omega))|$ be the Brownian maximal function, and $S(u)(\omega)$ the martingale square function, $S(u)(\omega) = (\int_0^\infty |\nabla u(z_t(\omega))|^2 \, dt)^{1/2}$. Their result then was that

$$\| Su \|_{L^p(\Omega)} \simeq \| M_B(u) \|_{L^p(\Omega)}, \qquad 0 < p < \infty, \tag{20}$$

whenever $u(0) = 0$.

The most striking application of this circle of ideas was a conclusion drawn from (20), to wit, whenever $F = u + iv$ is holomorphic in the unit disc, then $F \in H^p$ if and only if $u^* \in L^p$, $0 < p < \infty$.

The second line of research began when a more direct connection between standard multiplier operators and square function was discovered. The result was easy to state. Whenever T is a multiplier operator of the Marcinkiewicz type on $\mathbf{R}^n$ (more precisely one that satisfies the kind of conditions put in Hörmander's version of that multiplier theorem), then the area integral corresponding to $T(f)$ is *pointwise* dominated by a g^* function of f, i.e.,

$$A(Tf)(x) \leq c g_\lambda^*(f)(x), \tag{21}$$

where

$$g_\lambda^*(f)(x) = \left(\int |\nabla u(x - t, y)|^2 \left(\frac{y}{y + |t|} \right)^{n\lambda} y^{1-n} \, dy \, dt \right)^{1/2},$$

and λ is a parameter which depends on the nature of the multiplier. An H^p theory in $\mathbf{R}^n$ had already been initiated several years before (by the efforts of G. Weiss and others), and using it and (21) it followed that these multipliers also extended to bounded operators on H^p.

From these considerations it might be guessed that a basic tool for H^p theory is the relation between square functions and maximal properties of (harmonic) functions. Here important contributions were made by C. Fefferman. One of the results obtained in this direction was the following theorem [16]:

Theorem 11. Suppose that u is harmonic in $\mathbf{R}_+^{n+1}$, *and* $u(x,y) \to 0$, *as* $y \to \infty$. *Then* $\|A(u)\|_p \simeq \|u^*\|_p$, $0 < p < \infty$.

Incidentally it should be remarked that the proof used the same approach as its "local" analogue, Theorem 9c, but additional arguments of a quantitative nature were of course needed. More recently some of these results for square functions have been extended to product domains, and in this context generalizations of Theorems 9 and 11 have been found [25, 4, 17].

Symmetric Diffusion Semigroups

The semigroups which are the subject of the title are a family of operators $\{T^t\}_{t\geq 0}$, each bounded and selfadjoint on L^2, with T^t having norm ≤ 1 on every L^p, $1 \leq p \leq \infty$, and $T^{t_1+t_2} = T^{t_1}T^{t_2}$, with $\lim_{t\to 0} T^t_f = f$, for $f \in L^2$. Sometimes the additional hypotheses are made that $T^t(1) = 1$, and T^t is positivity-preserving.

The significance of this notion derives from the many important examples of such semigroups in analysis, and the many rich properties that they share. In fact some of the basic results discussed above have sessions valid in this context. Here we mention two, a maximal theorem, and a multiplier theorem in the spirit of Marcinkiewicz's theorem (Theorem 7).

Theorem 11a. $\|\sup_{t>0}|T^t f|\|_p \leq A_p\|f\|_p$, $1 < p \leq \infty$.

To formulate the multiplier theorem we write T^t in terms of its spectral decomposition, $T^t = \int_0^\infty e^{-\lambda t}\, dE(\lambda)$, where $E(\lambda)$ is a spectral resolution on L^2. For each bounded Borel measurable function m on $(0,\infty)$, consider the "multiplier" operator T_m given by $T_m = \int_0^\infty m(\lambda)\, dE(\lambda)$. Here we assume that m is of the form $m(\lambda) = \lambda \int_0^\infty M(s) e^{-\lambda s}\, ds$, with M a bounded function.

Theorem 11b. $\| T_m(f) \|_p \leq A_p \| f \|_p$, $1 < p < \infty$.

A key tool used for the proof of both these theorems is the Littlewood-Paley type functions

$$g_k(f)(x) = \left(\int_0^\infty t^{2k-1} \left| \frac{\partial^k}{\partial t^k} T^t(f) \right|^2 dt \right)^{1/2}, \text{ with } k = 1,2, \ldots .$$

Also for T_m a relation of the same kind as (21) holds [41].

Differentiation Theorems in $\mathbf{R}^n$

Probably the most dramatic applications of square functions occur in differentiation theory. The general problem here is to prove that

$$\lim_{\operatorname{diam} R \to 0} \frac{1}{\mu(R)} \int_R f(x - y)\, d\mu(y) = f(x) \quad \text{a.e.} \tag{22}$$

where R ranges over a suitable collection $\mathscr{R}$ of sets "centered" at the origin. The classical examples of these are (i) where $\mathscr{R}$ is the collection of all balls (or cubes) containing the origin, and (ii) where $\mathscr{R}$ is the collection of all rectangles containing the origin, with sides parallel to the axes. For each of these results a Vitali-type covering theorem has played a decisive result. Thus it may seem surprising that the alien notion of square functions would turn out to be the appropriate idea in related situations, where covering arguments were unavailing. In formulating the results obtained this way we shall, as is usual, deal with the corresponding maximal function

$$M_{\mathscr{R}}(f)(x) = \sup_{R \in \mathscr{R}} \frac{1}{\mu(R)} \left| \int_R f(x-y)\, d\mu(y) \right| ;$$

and the possibility of asserting inequalities of the type

$$\| M_{\mathscr{R}}(f) \|_p \leq A_p \| f \|_p . \tag{23}$$

Theorem 12. The inequality (23) holds in the following cases:

(a) $\mathscr{R}$ is the collection of spheres centered at the origin; $d\mu$ is the uniform surface measure; and $n \geq 3$, with $p > n/n - 1$.

(b) $\mathscr{R}$ is the collection of initial segments $\{\gamma(t),\ 0 \leq t \leq h\}$ of a smooth curve $t \to \gamma(t)$, with $\gamma(0) = 0$, and γ having nonzero "curvature" at the origin; here $d\mu$ is arc-length, $n \geq 1$ and $p > 1$.

(c) $\mathscr{R}$ is the collection of rectangles (in $\mathbf{R}^2$) containing the origin, which make an angle θ_k with a fixed direction, where $\{\theta_k\}$ is a sequence of numbers tending rapidly to zero; here $p > 1$.

The proof of each part of this theorem requires its own square function. We shall not describe these rather complicated quadratic functions here, but refer the reader to the literature for further details.†

EPILOGUE

Since the original draft of this essay was written two new results were found which use square functions in a decisive way.

The first is the solution of the problem of Cauchy's integral for Lipschitz curves by Coifman, McIntosh, and Meyer [13]. It is to be noted that in Calderón's initial work on this problem [8],

---

†For (a), see Stein [38]; for (b), see Stein and Wainger [42]; (c) was first proved for $p \geq 2$ by covering arguments; see Stromberg [45] and Cordoba and Fefferman [14]. The general case is in Nagel, Stein, and Wainger [31].

square functions were already used in a crucial way. In particular the inequality $c\|F\|_{H^p} \le \|A(F)\|_p$, $p \le 1$, was proved there for this purpose.

The second result deals with the standard maximal function in $\mathbf{R}^n$

$$M_n(f)(x) = \sup_{r>0} \frac{1}{c_n r^n} \left| \int_{|y|\le r} f(x-y)\,dy \right|,$$

where c_n is the volume of the unit ball in $\mathbf{R}^n$.

The question that arises is, how does the L^p norm of M_n behave for large n? The best that can be proved by the usual Vitali covering arguments gives $\|M_n(f)\|_p \le A(p,n)\|f\|_p$ $(1 < p)$, with $A(p,n) \le A(p)2^{n/p}$, which is a large growth as $n \to \infty$. However much more can be said.

Theorem 13. $\|M_n(f)\|_p \le A_p\|f\|_p$, $1 < p \le \infty$, *with* A_p *independent of* n.

The idea of the proof is to consider in $\mathbf{R}^m$ the maximal functions $M_{m,k}$ defined by

$$M_{m,k}(f)(x) = \sup_{r>0} \frac{\left| \int_{|y|\le r} f(x-y)|y|^k\,dy \right|}{\int_{|y|\le r} |y|^k\,dy}, \qquad k \ge 0.$$

Then if m is so large that $p > m/(m-1)$,

$$\|M_{m,k}(f)\|_p \le A_{p,m}\|f\|_p \tag{24}$$

with $A_{p,m}$ independent of k, $k \ge 0$. This follows from Theorem 12, Part (a). From this Theorem 13 is obtained by lifting the m-dimensional result (24) into $\mathbf{R}^n$, where $n \ge m$ (and $k = n - m$), by integrating over the Grassmannian of m-planes in $\mathbf{R}^n$ through the origin.

REFERENCES

1. Benedck, A.; Calderón, A. P.; and Panzone, R. Convolution operators on Banach space-valued functions. *Proc. Nat. Acad. Sci.* 48 (1962): 356-365.

2. Besicovitch, A. S. On a general metric property of summable functions. *J. London Math. Soc.* 1 (1926): 120-128.

3. Besicovitch, A. S. Sur la nature des fonctions à carré sommable et des ensembles mesurables. *Fund. Math.* 4 (1923): 172-195.

4. Brossard, J. Comportement des fonctions biharmoniques là ou l'intégrale d'aire est finie. *Bull. Sci. Math.* 103 (1979): 77-95.

5. Burkholder, D. L. Martingale transforms. *Ann. Math. Statistics* 37 (1966): 1494-1504.

6. Burkholder, D. L., and Gundy, R. F. Extrapolation and interpolation of quasi-linear operators on martingales. *Acta Math.* 124 (1970): 249-304.

7. Burkholder, D. L.; Gundy, R. F.; and Silverstein, M. L. A maximal function characterization of the class H^p. *Trans. Amer. Math. Soc.* 157 (1971): 137-153.

8. Calderón, A. P. Commutators of singular integral operators. *Proc. Nat. Acad. Sci.* 53 (1965): 1092-1099.

9. Calderón, A. P. On the behavior of harmonic functions on the boundary. *Trans. Amer. Math. Soc.* 68 (1950): 47-54. Also, On a theorem of Marcinkiewicz and Zygmund. Ibid., 55-61.

10. Calderón, A. P., and Zygmund, A. Local properties of solutions of elliptic partial differential equations. *Studia Math.* 20 (1961): 171-225.

11. Calderón, A. P., and Zygmund, A. On the existence of certain singular integrals. *Acta Math.* 88 (1952): 85-139.

12. Carleson, L. On the convergence and growth of partial sums of Fourier series. *Acta Math.* 116 (1966): 135-157.

13. Coifman, R. R.; McIntosh, A.; and Meyer, Y. L'Integrale de Cauchy definit un operateur borné sur L^2 pour les courbes Lipschitziennes. *Ann. Math.*, to appear.

14. Córdoba, A., and Fefferman, R. On differentiation of integrals. *Proc. Nat. Acad. Sci. U.S.A.* 74 (1977): 2211-2213.

15. Dahlberg, B. E. Estimates of harmonic measure. *Arch. Rational Mech. Anal.* 65 (1977): 275-288.

16. Fefferman, C., and Stein, E. M. H^p spaces of several variables. *Acta Math.* 129 (1972): 137-193.

17. Gundy, R. F., and Stein, E. M. H^p theory for the poly-disc. *Proc. Nat. Acad. Sci.* 76 (1979): 1026-1029.

18. Kaczmarz, S. Uber die Konvergenz der Reihen von Orthogonal-Funktionen. *Math. Zeit.* 23 (1925): 263-270.

19. Kolmogorov, A. N. Sur les fonctions harmoniques conjuguées et les Séries de Fourier. *Fund. Math.* 7 (1925): 23-28.

20. Kolmogorov, A. N. Une série de Fourier-Lebesgue divergente partout. *Comptes Rendus Acad. Sci.* 178 (1926): 303-305.

21. Kolmogorov, A. N. Une série de Fourier-Lebesgue divergente presque-partout. *Fund. Math.* 4 (1923): 324-328.

22. Kolmogorov, A. N., and Seliverstov, G. Sur la convergence des séries de Fourier. *Rend. Acc. Linca.* 3 (1926): 307-310.

23. Littlewood, J. E., and Paley, R.E.A.C. Theorems on Fourier series and power series. *J. London Soc.* 6 (1931): 230-233; *Proc. London Math. Soc.* 42 (1936): 52-89; 43 (1937): 105-126.

24. Lusin, N. Sur une propriete des fonctions a carré sommable. *Bull. Calcutta Math. Soc.* 20 (1930): 139-154.

25. Malliavin, M. P., and Malliavin, P. Intégrales de Lusin-Calderón pour les fonctions biharmoniques. *Bull. Sci. Math.* 701 (1977): 357-384.

26. Marcinkiewicz, J. Sur les multiplicateurs des séries de Fourier. *Studia Math.* 8 (1939): 78-91.

27. Marcinkiewicz, J. Sur quelques intégrales du type de Dini. *Ann. de la Soc. Polonaise de Math.* 17 (1938): 42-50.

28. Marcinkiewicz, J., and Zygmund, A. A theorem of Lusin. *Duke Math. J.* 4 (1938): 473-485.

29. Menshov, D. Sur les séries de fonctions orthogonales. *Fund. Math.* 4 (1923): 82-105; 8 (1926): 56-108; 10 (1927): 375-420.

30. Mihlin, S. G. On multipliers of Fourier integrals. *Dokl. Akad. Nauk.* 109 (1956): 701-703.

31. Nagel, A.; Stein, E. M.; and Wainger, S. Differentiation in lacunary directions. *Proc. Nat. Acad. Sci.* 75 (1978): 1060-1062.

32. Paley, R.E.A.C. A remarkable system of orthogonal functions. *Proc. London Math. Soc.* 34 (1932): 241-264, 265-279.

33. Plessner, A. Über die Konvergenz von trigonometrischen Reihen. *J. F. Math.* 155 (1926): 15-25.

34. Rademacher, H. Einige Sätze über Reihen von allgemeinen Orthogonal Funktionen. *Math. Ann.* 87 (1922): 112-138.

35. Riesz, M. Sur les fonctions conjuguées. *Math. Zeit.* 27 (1927): 218-244.

36. Spencer, D. C. A function theoretic identity. *Amer. J. of Math.* 65 (1943): 147-160.

37. Stein, E. M. Classes H^p, multiplicateurs, et fonctions de Littlewood-Paley. *C. R. Acad. Sci.* 263 (1966): 716-719, 780-781; 264 (1967): 107-108.

38. Stein, E. M. Maximal functions: spherical averages. *Proc. Nat. Acad. Sci. U.S.A.* 73 (1976): 2174-2175.

39. Stein, E. M. On the functions of Littlewood-Paley, Lusin and Marcinkiewicz. *Trans. Amer. Math. Soc.* 88 (1958): 430-466.

40. Stein, E. M. On the theory of harmonic functions of several variables, II *Acta Math.* 106 (1961): 137-174.

41. Stein, E. M. *Topics in Harmonic Analysis Related to the Littlewood-Paley Theory.* Annals of Math. Study #63. Princeton, N.J.: Princeton Univ. Press, 1970. Also *Singular Integrals and Differentiability Properties of Functions.* Princeton, N.J.: Princeton Univ. Press, 1970.

42. Stein, E. M., and Wainger, S. Problems in harmonic analysis related to curvature. *Bull. Amer. Math. Soc.* 84 (1978): 1239-1295.

43. Stein, E. M., and Weiss, G. On the theory of harmonic functions of several variables, I. The theory of H^p spaces. *Acta Math.* 103 (1960): 25-62.

44. Stein, E. M., and Zygmund, A. On the differentiability of functions. *Studia Math.* 23 (1964): 247-283.

45. Stromberg, J. O. Weak estimates for maximal functions with rectangles in certain directions. *Ark. Mat.* 15 (1977): 229-240.

46. Sunouchi, G. Theorems on power series of the class H^p. *Tohoku Math. J.* 8 (1956): 125-146.

47. Weiss, M., and Zygmund, A. A note on smooth functions. *Indag. Math.* 62 (1959): 52-58.

48. Zygmund, A. On certain integrals. *Trans. Amer. Math. Soc.* 55 (1944): 170-204.

49. Zygmund, A. On the convergence and summability of power series on the circle of convergence. *Fund. Math.* 30 (1938): 170-196.

50. Zygmund, A. On the Littlewood-Paley function g^*. *Proc. Nat. Acad. Sci.* 42 (1956): 208-212.

51. Zygmund, A. Proof of a theorem of Littlewood and Paley. *Bull. Amer. Math. Soc.* 51 (1945): 439-446.

52. Zygmund, A. *Trigonometric Series*. 2d ed. 2 vols. Cambridge, England, 1959.

53. Zygmund, A. Une remarque sur un theoreme de M. Kaczmarz. *Math. Zeit.* 25 (1926): 297-298. Also Sur l'application de la premiere moyenne arithmetique dans la theorie des séries orthogonales. *Fund. Math.* 10 (1927): 356-362.

PART TWO

Trigonometric Series

CONVOLUTION INEQUALITIES ON THE CIRCLE

William Beckner
University of Chicago
Svante Janson
University of Uppsala, Sweden
David Jerison
Massachusetts Institute of Technology

INTRODUCTION

Our purpose is to present some remarks on when it is possible to improve Young's inequality $\|f * \mu\|_p \leq \|f\|_p$, where μ is a positive measure of mass 1 on the circle group $\mathbf{T} = \mathbf{R}/\mathbf{Z}$, $f \in L^p = L^p(\mathbf{T})$ and $f * \mu(x) = \int_0^1 f(x - t)\, d\mu(t)$. The main theorem is:

Theorem 1. If μ is a positive measure of mass 1 on $\mathbf{T}$ and there exist C, $p > 1$, $\varepsilon > 0$ such that

$$\|f * \mu\|_{p+\varepsilon} \leq C\|f\|_p \quad \textit{for all } f \in L^p, \tag{1}$$

then for any q, $1 < q < \infty$, there exists $\delta > 0$ depending on C, p, q, and ε such that

$$\|f * \mu\|_{q+\delta} \leq \|f\|_q \quad \textit{for all } f \in L^q. \tag{2}$$

Since μ has mass 1, the inequality (2) has the best constant. The best value of δ is known only for a few cases of μ[2, 5, 9]. Of course, by interpolation with Young's inequality for $p = 1$ and $p = \infty$, the theorem is true for every value of q as soon as it is true for one value of q.

Corollary 1. If $h(x) \geq 0$, $h \in L^r$ *for some* $r > 1$ *and* $\int_0^1 h(x)\,dx = 1$, *then for* $1 < p < \infty$, *there exists* $\delta > 0$ *depending on* p, r, *and* $\|h\|_r$ *such that* $\|f * h\|_{p+\delta} \leq \|f\|_p$ *for all* $f \in L^p$.

The corollary follows from the theorem and the general form of Young's inequality: $\|f * h\|_q \leq \|f\|_p \|h\|_r$; $\frac{1}{q} = \frac{1}{r} + \frac{1}{p} - 1$. An alternative proof giving an explicit bound for δ is discussed later.

E. M. Stein [8] pointed out that there are many singular measures that satisfy (1). Indeed, any measure satisfying $|\hat{\mu}(n)| = O(|n|^{-\alpha})$, $\alpha > 0$ will do. He posed the problem, still unsolved, of characterizing those positive measures satisfying (1). Recently, D. Oberlin [6] has shown that the Cantor-Lebesgue measure λ defined below satisfies (2). This measure is substantially different from the previous examples because $\hat{\lambda}(n)$ does not tend to zero. (Another example can be found in A. Bonami [2].) The first step in Oberlin's proof is to obtain an inequality like (2) on the three-point group $\mathbf{Z}/3\mathbf{Z}$. It is crucial that the constant be the best as in (2) because the second step is to iterate this inequality infinitely often. We will show below that a variant of Theorem 1 yields the elementary "three-point" inequality that is the first step in Oberlin's proof. Thus our theorem has some indirect bearing on the truth of its hypothesis.

We would like to thank Tom Wolff for several helpful conversations and Stylianos Pichorides for the proof of Lemma 4.

PROOF OF THE THEOREM

Lemma 1. With the hypotheses of Theorem 1, there exists $\gamma > 0$ *depending on* C, p, ε *such that* $|\hat{\mu}(n)| \leq 1 - \gamma$ *for* $n \neq 0$.

In fact, define $\varphi_n\colon \mathbf{T} \to \mathbf{T}$ by $\varphi_n(x) = nx \bmod 1$. The measure $\mu_n(E) = \mu(\varphi_n^{-1}(E))$ satisfies $\hat{\mu}_n(1) = \hat{\mu}(n)$. Furthermore, $(f * \mu_n) \circ \varphi_n = (f \circ \varphi_n) * \mu$ and $\|f \circ \varphi_n\|_p = \|f\|_p$. Thus we deduce from (1) that $\| f * \mu_n \|_{p+\varepsilon} \leq C \| f \|_p$, and it then suffices to prove the lemma for $n = 1$. The set $S = \{\mu : \mu$ is a positive measure of mass 1, $\| f * \mu \|_{p+\varepsilon} \leq C \| f \|_p\}$ is a closed (and hence compact) subset of the closed unit ball of measures in the weak star topology. Therefore, there exists ν in S such that $|\hat{\nu}(1)| = \sup\{|\hat{\mu}(1)| : \mu \in S\}$. But every positive measure of mass 1 satisfying $|\hat{\mu}(1)| = 1$ is a point mass. These measures do not map L^p to $L^{p+\varepsilon}$ by convolution and so do not belong to S. Thus $|\hat{\nu}(1)| < 1$, and the lemma is proved.

By interpolation with Young's inequality we can assume that $p = 2$ in (1) and we need only deduce the theorem for $q = 2$ in (2). The multiplier T given by $(Tf)\hat{}(n) = |\hat{\mu}(n)|\hat{f}(n)$ is the composition of the multiplier $\{\hat{\mu}(n)\}$ with an isometry of L^2. Hence, by (1) and Lemma 1, there exists $\varepsilon > 0$, $\gamma > 0$:

$$\| Tf \|_{2+\varepsilon} \leq C \| f \|_2 \qquad \text{for } f \in L^2 \tag{3}$$

$$\| Tg \|_2 \leq (1 - \gamma) \| g \|_2 \qquad \text{for } g \in L^2,\ g \perp 1. \tag{4}$$

(The notation $g \perp 1$ means g is perpendicular to 1: $\int_0^1 g(x)\,dx = 0$.)

The theorem now follows from:

Lemma 2. Let T be a positive operator on L^2 satisfying (3) *and* (4) *and such that $T1 = 1$. Then there exists $\delta > 0$ such that $\| Tf \|_{2+\delta} \leq \| f \|_2$ for $f \in L^2$.*

This lemma is a variant of a theorem of Glimm [3]. To prove it, consider complex powers T^s of the operator T, $\operatorname{Re} s \geq 0$. (In our case, T^s is given by the multiplier $\{|\hat{\mu}(n)|^s\}$.) The operators T^s form an analytic family and a semigroup: $T^{s_1} \circ T^{s_2} = T^{s_1+s_2}$. Also, $g \perp 1$ implies $T^s g \perp 1$. It then follows from (4) that $\| T^s g \|_2 \leq (1 - \gamma)^{\operatorname{Re} s} \| g \|_2$ for all $g \in L^2$, $g \perp 1$. Next, using (3) we obtain

$$\| T^{s+1} g \|_{2+\varepsilon} \leq C(1 - \gamma)^{\operatorname{Re} s} \| g \|_2 \qquad \text{for } g \perp 1. \tag{5}$$

Normalize f in L^2 so that $f = 1 + g$ where $g \perp 1$. Define r by $2r = 2+\varepsilon$. We may as well assume that $r \leq 2$. Noting also that $\operatorname{Re} T^s g \perp 1$,

$$\| T^s f \|_{2+\varepsilon}^2 = \left(\int_0^1 (|T^s f|^2)^r \right)^{1/r} = \left(\int_0^1 (1 + 2\operatorname{Re} T^s g + |T^s g|^2)^r \right)^{1/r}$$

$$\leq \left(\int_0^1 |1 + 2\operatorname{Re} T^s g|^r \right)^{1/r} + \left(\int_0^1 |T^s g|^{2+\varepsilon} \right)^{1/r}$$

$$\leq \left(\int_0^1 |1 + 2\operatorname{Re} T^s g|^2 \right)^{1/2} + \| T^s g \|_{2+\varepsilon}^2$$

$$= \left(1 + 4 \int_0^1 |\operatorname{Re} T^s g|^2 \right)^{1/2} + \| T^s g \|_{2+\varepsilon}^2$$

$$\leq (1 + 4 \| T^s g \|_{2+\varepsilon}^2)^{1/2} + \| T^s g \|_{2+\varepsilon}^2 \leq 1 + 3 \| T^s g \|_{2+\varepsilon}^2 .$$

By (5), this is dominated by $1 + 3C^2 (1 - \gamma)^{2(\operatorname{Re} s - 1)} \| g \|_2^2$, which is less than $1 + \| g \|_2^2 = \| f \|_2^2$ for $\operatorname{Re} s$ sufficiently large. On the other hand, when $\operatorname{Re} s = 0$, $\| T^s f \|_2 \leq \| f \|_2$. Hence, E. M. Stein's analytic interpolation theorem implies that there is $\delta > 0$ such that $\| Tf \|_{2+\delta} \leq \| f \|_2$ $(T = T^1)$.

Let G be any compact group and let μ be a positive measure on G of mass 1. Denote by $\| f \|_p$ the norm of f in $L^p(dm)$, where m is the normalized Haar measure on G. Convolution is given by $f * \mu(x) = \int_G f(xy^{-1})\, d\mu(y)$. Another consequence of Lemma 2 is:

Theorem 2. With the notations above, if there exist C, $p > 1$, $\varepsilon > 0$ such that

$$\| f * \mu \|_{p+\varepsilon} \leq C \| f \|_p \qquad \text{for } f \in L^p(dm) \tag{1a}$$

$$\| g * \mu \|_2 \leq (1 - \gamma) \| g \|_2 \qquad \text{for } g \in L^2(dm) \text{ satisfying } g \perp 1, \tag{1b}$$

then for any q, $1 < q < \infty$, there exists $\delta > 0$ such that $\| f * \mu \|_{q+\delta} \leq \| f \|_q$ for all $f \in L^q(dm)$.

The only difference in the proof is in the non-Abelian case. In that case, instead of passing from the multiplier $\{\hat{\mu}(n)\}$ to $\{|\hat{\mu}(n)|\}$, we pass from the operator $Mf = f * \mu$ on $L^2(dm)$ to $|M| = (MM^*)^{1/2}$ and use the polar decomposition (see p. 197 of [7]). This argument shows that Lemma 2 is valid for operators T that are not necessarily positive, provided they also satisfy $T^*1 = 1$.

In the special case of the circle, we saw that hypothesis (1b) was redundant. This is not true for a group with discrete characters. In the case of a finite group, hypothesis (1a) is valid for any measure μ, since all norms are equivalent on a finite-dimensional vector space. Therefore, we have:

Corollary 2. Let G *be a finite, Abelian group, and let* μ *be a positive measure of mass* 1 *on* G. *If* $|\mu(\chi)| < 1$ *for every character* $\chi \neq 1$, *then there exists* $\delta > 0$ *depending on* q, $\max\{|\mu(\chi)| : \chi \neq 1\}$ *and* #G *such that* $\| f * \mu \|_{q+\delta} \leq \| f \|_q$ *for* $f \in L^q(dm)$ $(1 < q < \infty)$.

Corollary 2 and its application stated in (a) below were obtained independently by David Ritter.†

FURTHER REMARKS

(a) For each $\theta > 2$, denote by $\lambda_{\theta,n}$ the measure on **R**/**Z** with mass 2^{-n} at each point of the set

$$\{(1 - \theta^{-1}) \sum_{k=0}^{n-1} \varepsilon_k \theta^{-k} : \varepsilon_k = 0 \text{ or } 1\}.$$

Define the measure λ_θ as the weak limit of $\lambda_{\theta,n}$ as $n \to \infty$ (cf. [10]). D. Oberlin [6] proved that the classical Cantor-Lebesgue measure $\lambda = \lambda_3$ satisfies $\| f * \lambda \|_{p+\varepsilon} \leq \| f \|_p$ for some $\varepsilon > 0$ depending on p, $1 < p < \infty$. His first step was to prove the inequality $\| f * \mu \|_2 \leq \| f \|_p$ for some $p < 2$, $f \in L^p(\mathbf{Z}/3\mathbf{Z})$, with μ

---

†Thesis, Louisiana State University.

defined by $\mu(\{0\}) = \mu(\{1\}) = 1/2$, $\mu(\{2\}) = 0$. An iteration, using cyclic groups of order 3^n yields the inequality $\| f * \lambda \|_2 \leq \| f \|_p$ for $f \in L^p(\mathbf{T})$, with the same exponent p as in the "three-point" inequality. Corollary 2 yields analogous elementary inequalities with 3 replaced by any integer $k \geq 3$. Let μ be defined on $\mathbf{Z}/k\mathbf{Z}$ by $\mu(\{0\}) = \mu(\{1\}) = 1/2$ and $\mu(\{j\}) = 0$ for $j = 2, \ldots, k-1$. Then there exists $\varepsilon > 0$ depending on p and k such that $\| f * \mu \|_{p+\varepsilon} \leq \| f \|_p$ for $f \in L^p(\mathbf{Z}/k\mathbf{Z})$. Oberlin's iteration method then shows that for any rational value of θ, $\| f * \lambda_\theta \|_{p+\varepsilon} \leq \| f \|_p$ for $f \in L^p(\mathbf{T})$, where the exponents p and $p+\varepsilon$ are the same as those that can be obtained for the inequality on $\mathbf{Z}/k\mathbf{Z}$ and k is the numerator of θ in lowest terms. It would be interesting to decide whether such an inequality is true for irrational values of θ.

(b) The sharp exponent in the three-point inequality for L^2 can be calculated. In fact, $\| f * \mu \|_2 \leq \| f \|_p$ for all $f \in L^p(\mathbf{Z}/3\mathbf{Z})$ if and only if $p \geq \log 4/\log 3$. The extremal function f is given by $f(0) = 1$, $f(1) = f(2) = 1/3$.

The case $k = 3$ is exceptional among the cyclic groups because the cube of the real part of the first character is not orthogonal to 1. In general, small perturbations of the constant function (see Janson [5], Theorem 3) show that if $\| f * \mu \|_q \leq \| f \|_p$ for all $f \in L^p(\mathbf{Z}/k\mathbf{Z})$, then $(p-1) \geq c_k^2(q-1)$, where c_k is the modulus of the first Fourier coefficient of μ, that is, $c_k = |\hat{\mu}(1)|$. It seems likely that the converse is also true, but we have only been able to prove this in the case $k = 4$, by the series of reductions below.

Define a Poisson kernel on $\mathbf{Z}/k\mathbf{Z}$ as follows. Let m be the greatest integer in $(k-1)/2$. Then P_r is the multiplier

$$\sum_{j=-m}^{m} a_j e^{2\pi i m\theta/k} \longmapsto \sum_{j=-m}^{m} r^{|j|} a_j e^{2\pi i m\theta/k}.$$

One calculates easily that the Fourier coefficients of μ are dominated by those of the Poisson kernel P_r with $r = c_k$. Using the positivity of P_r, it is easy to see that $L^p \to L^q$ inequality for P_r implies the corresponding one for μ. In the case $k = 4$ the inequality can be reduced to a product of two-point inequalities by means of a correspondence between $\mathbf{Z}/4\mathbf{Z}$ and $\mathbf{Z}/2\mathbf{Z} \oplus \mathbf{Z}/2\mathbf{Z}$. Let x and y be two characters on $\mathbf{Z}/2\mathbf{Z} \oplus \mathbf{Z}/2\mathbf{Z}$ given by

$$x(0,1) = x(0,0) = 1; \quad x(1,0) = x(1,1) = -1.$$
$$y(0,0) = y(1,0) = 1; \quad y(0,1) = y(1,1) = -1.$$

Let the points 0, 1, 2, 3 of $\mathbf{Z}/4\mathbf{Z}$ correspond to (0,0), (0,1), (1,1), (1,0), respectively, of $\mathbf{Z}/2\mathbf{Z} \oplus \mathbf{Z}/2\mathbf{Z}$. Then the functions $\varphi_0 \equiv 1$, $\varphi_1 = \frac{1}{2}(x+y)$, $\varphi_{-1} = \frac{1}{2}(x-y)$ and $\varphi_2 = xy$ define, up to constant multiples, the real and imaginary parts of the characters of $\mathbf{Z}/4\mathbf{Z}$. Notice that the degree of each polynomial in x and y is $\deg \varphi_j = |j|$ for $j = 0, \pm 1, 2$. Thus the multiplier $\sum a_j \varphi_j \longmapsto \sum r^{|j|} a_j \varphi_j$ corresponds to the product of Poisson kernels on $\mathbf{Z}/2\mathbf{Z} \oplus \mathbf{Z}/2\mathbf{Z}$. The desired inequality now follows from Lemmas 1 and 2 of Beckner [1].

(c) We will now discuss the dependence of δ in Corollary 1 on $\|h\|_r$ and γ of Lemma 1. Let γ be defined by

$$1 - \gamma = \sup\{|\hat{h}(n)| : n \neq 0,\ h(x) \geq 0,\ \|h\|_1 = 1,\ \|h\|_r = A\}.$$

By the argument of Lemma 1 and a translation in x,

$$1 - \gamma = \sup\left\{ \int_0^1 (\cos 2\pi x) h(x)\, dx :\ h \text{ as above} \right\}.$$

Choose h so that $\int_0^1 (\cos 2\pi x) h(x)\, dx \geq 1 - 2\gamma$. Denote $E = \{x \in (0,1) : \cos 2\pi x > 1 - 4\gamma\}$. Then $|E| \leq 2\gamma^{1/2}$. Also,

$$\int_E h(x)\, dx + (1 - 4\gamma) \int_{C_E} h(x)\, dx \geq \int_0^1 (\cos 2\pi x) h(x)\, dx \geq 1 - 2\gamma$$

implies $\int_E h(x)\,dx \geq 1/2$. By Hölder's inequality, $\frac{1}{2} \leq \int_E h(x)\,dx \leq |E|^{1/r'} A$, where $\frac{1}{r} + \frac{1}{r'} = 1$. Hence

$$\gamma \geq c_r A^{-2r'}. \tag{6}$$

Denote the convolution of h with itself n times by $h*^n$. As in the proof of Theorem 1, we have

$$\| g * h*^{n+1} \|_q \leq A(1 - \gamma)^n \| g \|_2 \quad \text{for} \quad g \in L^2,\ g \perp 1;\ \frac{1}{q} = \frac{1}{2} + \frac{1}{r} - 1.$$

The same method as before yields the conclusion of Corollary 1 for $\delta = c_r\gamma/(1 + \log A)$. The example

$$h(x) = \begin{cases} \frac{1}{2}\varepsilon^{-1} + \frac{1}{2} & 0 < x < \varepsilon \\ \frac{1}{2} & \varepsilon \leq x \leq 1 \end{cases}$$

shows that the value $\delta \simeq \gamma/\log A$ is in the nature of best possible when γ is fixed and $A \to \infty$. In fact, let $f = h$ and observe that $\| h \|_2 \simeq \varepsilon^{-1/2}$ and $\| h * h \|_q \simeq \varepsilon^{-1+1/q}$. The details are left to the reader. On the other hand, estimate (6) gives a value of δ greater than $c/A^{2r'} \log A$. This can be improved:

Theorem 3. Let $1 < r < \infty$. There is a constant c depending only on r such that if $h(x) \geq 0$ on $\mathbf{R/Z}$, $\| h \|_1 = 1$, $\| h \|_r = A$, then for all $\delta \leq cA^{-2r'}$

$$\| f * h \|_{2+\delta} < \| f \|_2, \quad \textit{for} \quad f \in L^2 \left(\frac{1}{r} + \frac{1}{r'} = 1 \right).$$

This theorem is in the nature of best possible in the sense that there exists h as above and $f \in L^2$ such that $\| f * h \|_{2+\delta_1} > \| f \|_2$ with $\delta_1 = 40A^{-2r'}$. In fact, let $h(x) = \begin{cases} \varepsilon^{-1} & 0 < x < \varepsilon. \\ 0 & \varepsilon \leq x \leq 1. \end{cases}$ Then, denoting $a = \int_0^1 (\cos 2\pi x) h(x)\,dx$, $a \geq 1 - 10\varepsilon^2$ and $\| h \|_r = \varepsilon^{1/r'} = A$. Let $f_t(x) = 1 + t \cos 2\pi x$. Using the binomial expansion (see [5], Theorem 3),

$$\| f_t \|_2 = (1 + t^2/2)^{1/2} = 1 + \frac{1}{4} t^2 + O(t^4) \qquad \text{as } t \to 0$$

and

$$\| f_t * h \|^q = 1 + \frac{1}{4} a^2 t^2 (q - 1) + O(t^3).$$

The inequality $\| f_t * h \|_q \leq \| f_t \|_2$ for all $t > 0$ implies $(q - 1)a^2 \leq 1$. This is violated for $q = 2 + \delta_1$, $\delta_1 = 40\varepsilon^2 = 40A^{-2r'}$.

Theorem 3 follows from:

Lemma 3. Let h be as in Theorem 3. There exists c_1 depending on r such that for $N > c_1 A^{2r'}$, $\| h^{*N} \|_\infty \leq C$, where C is an absolute constant.

The theorem follows because we can now apply Theorem 1 to h^{*N} and show that $\| f * h^{*N} \|_{2+\delta_2} \leq \| f \|_2$ for a constant $\delta_2 > 0$ depending only on C in Lemma 3. As in the proof of Theorem 1, an analytic interpolation gives the desired inequality for h with δ = constant times $A^{-2r'}$.

To prove Lemma 3 we need:

Lemma 4. There is an absolute constant C such that for $\gamma < 1/4$,

$$|\{x \in (0,1) : |f(x)| > (2M + 1)(1 - \gamma)\}| \leq C\gamma^{1/2}/M,$$

for every $f(x) = \sum_{j=-M}^{M} c_j e^{2\pi i n_j x}$ with $n_{-M}, \ldots, n_{-1}, n_0, \ldots, n_M$ denoting $2M + 1$ distinct integers and c_j complex numbers of modulus 1.

We owe the proof of this lemma to S. Pichorides. For any positive integer k, the inequality of Hardy, Littlewood, and Pólya [4] says

$$\int_0^1 |f(x)|^{2k}\, dx \leq \int_0^1 |D_M(x)|^{2k}\, dx,$$

where $D_M(x) = \sum_{j=-M}^{M} e^{2\pi ijx}$ is the Dirichlet kernel. One can calculate directly that the second integral is dominated by $C(2M+1)^{2k-1}/\sqrt{k}$. In particular, the weak-type inequality gives

$$|\{x \in (0,1) : |f(x)| > (2M+1)(1-\gamma)\}|$$
$$\leq ((2M+1)(1-\gamma))^{-2k} C \frac{(2M+1)^{2k-1}}{\sqrt{k}}$$
$$= C \frac{(1-\gamma)^{-2k}}{\sqrt{k}(2M+1)} \leq C\gamma^{1/2}/M$$

provided $k \simeq \gamma^{-1}$.

We can use Lemma 4 to estimate the number of frequencies in the set $U_\varepsilon = \{j : |\hat{h}(j)| > 1 - \varepsilon^2\}$. Suppose that $\#U_\varepsilon \geq 2M+1$, and enumerate these integers by $n_{-M}, \ldots, n_M$. Note that when f is as in Lemma 4 with $c_j = \hat{h}(n_j)/|\hat{h}(n_j)|$,

$$\int_0^1 h(x)f(x)\,dx = \sum_{j=-M}^{M} |\hat{h}(n_j)| \geq (1-\varepsilon^2)(2M+1).$$

Denote $E = \{x \in (0,1) : |f(x)| > (2M+1)(1-2\varepsilon^2)\}$. Then as in the proof of (6), $\frac{1}{2} \leq \int_E h(x)\,dx \leq A|E|^{1/r'}$. Hence, using Lemma 4, $(2A)^{-r'} \leq |E| \leq C\varepsilon/M$. Therefore, for $\varepsilon < 1/4$, $\#U_\varepsilon \leq C\varepsilon A^{r'}$. For $\varepsilon \geq 1/4$ we will use the straightforward estimate from the Hausdorff-Young inequality: $\sum |\hat{h}(j)|^p \leq A^p$, where $p = \max(2, r')$. This implies $\#\{j : |\hat{h}(j)| > s\} \leq A^p s^{-p}$. In all, $\hat{h}(0) = 1$ and

$$\sum_{j\neq 0} |\hat{h}(j)|^N \leq C \sum_{\ell=3}^{\infty} (1-2^{-2\ell})^N 2^{-\ell} A^{r'} + C \sum_{\ell=0}^{\infty} (\tfrac{15}{16} 2^{-\ell})^N A^p 2^{\ell p}$$
$$\leq CA^{r'} \int_0^1 e^{-x^2 N}\,dx + CA^p (\tfrac{15}{16})^N$$
$$\leq CA^{r'}/N^{1/2} \leq C$$

for $N > \max(A^{2r'}, p)$. This concludes the proof of Lemma 3. In fact, we have shown slightly more, namely,

$$\sup_x |h*^N(x) - 1| \leq CA^{r'}/N^{1/2} \quad \text{for} \quad N > C_r(1 + \log A). \tag{7}$$

(d) An analogous statement to Lemma 3 on the real line is as follows. For $1 < r < \infty$, there exists an absolute constant C such that for $N \geq r'A^{2r'}$,

$$\sup\{\|h*^N\|_\infty : \|h\|_1 = 1, \|h\|_r = A\} \leq C.$$

(Here, we take convolution and norm on R.)

This follows immediately from a variant of a theorem of W. Beckner [1] stated below. The method of proof is the same as in [1].

Theorem 4. Let $1 < p < \infty$. *The value of*

$$\sup\{\|h*^N\|_p : \|h\|_1 = 1, \|h\|_p = A\}$$

is less than a constant multiple (independent of N) of the corresponding value attained for Gaussian functions.

It would be useful to be able to make a comparison between the line and the circle, since the results on the line are more precise. By analogy with (7), these estimates yield a small improvement on standard estimates in the local central limit theorem.

(e) In closing, let us recall a theorem of A. Zygmund [10, vol. II, p. 127] deduced from a theorem of Paley.

Theorem 5. If $\{\varphi_n(x)\}_{n=1}^\infty$ *is an orthonormal basis for* $L^2(d\nu)$ *and* $|\varphi_n(x)| \leq C$ *for all x, n, then any multiplier* $T(\Sigma\, c_n\varphi_n) = \Sigma\, a_n c_n \varphi_n$ *satisfying* $a_n = O(n^{-\alpha})$ *is bounded from* L^2 *to* L^q, *where* $\alpha = \frac{1}{2} - \frac{1}{q}$.

From the method of Theorem 2, we can conclude that if in addition $\varphi_1(x) \equiv 1$, $|a_1| = 1$, and $\sup_{n>1}|a_n| < 1$, then the multiplier T is a contraction from L^2 to $L^{2+\varepsilon}$ for some $\varepsilon : \|Tf\|_{2+\varepsilon} \leq \|f\|_2$ for $f \in L^2(d\nu)$.

REFERENCES

1. Beckner, W. Inequalities in Fourier analysis. *Ann. Math.* 102 (1975): 159-182.

2. Bonami, A. Étude des coefficients de Fourier des fonctions de $L^p(G)$. *Ann. Inst. Fourier (Grenoble)* 20 (1970): 335-402.

3. Glimm, J. Boson fields with nonlinear self-interaction in two dimensions. *Comm. Math. Phys.* 8 (1968): 12-25.

4. Hardy, G. H.; Littlewood, J. E.; and Pólya, G. *Inequalities*. 2d ed. London: Cambridge Univ. Press, 1952.

5. Janson, S. On hypercontractive for multipliers on orthogonal polynomials. To appear in *Ark. Mat.*

6. Oberlin, D. M. A convolution property of the Cantor-Lebesgue measure. To appear in *Colloq. Math.*

7. Reed, M., and Simon, B. *Methods of Modern Mathematical Physics, I: Functional Analysis*. New York: Academic Press, 1972.

8. Stein, E. M. Harmonic analysis on R^n. *Studies in Harmonic Analysis*. M.A.A. Studies Series 13, ed. J. M. Ash, 1976, 97-135.

9. Weissler, F. B. Logarithmic Sobolev inequalities and hypercontractive estimates on the circle. *J. Func. Anal.* 37 (1980): 218-234.

10. Zygmund, A. *Trigonometric Series*. 2d ed. London: Cambridge Univ. Press, 1959.

A CONVOLUTION STRUCTURE AND POSITIVITY OF A GENERALIZED TRANSLATION OPERATOR FOR THE CONTINUOUS Q-JACOBI POLYNOMIALS

George Gasper
Northwestern University

INTRODUCTION

In 1970 the author [6, 7] used a three-term recurrence relation due to Hyllerass for the coefficients in the linearization of the product of two Jacobi polynomials,

$$P_n^{(\alpha,\beta)}(x)P_m^{(\alpha,\beta)}(x) = \sum_{k=|n-m|}^{n+m} c(k, m, n; \alpha, \beta)P_k^{(\alpha,\beta)}(x), \quad (1)$$

to determine the set of V of all (α,β) with $\alpha, \beta > -1$ for which these coefficients are nonnegative, and then used this result to construct a natural convolution structure associated with the linearization and to prove the positivity of a generalized translation operator. Eight years later Koornwinder [12] showed that the nonnegativity of the linearization coefficients can also be proved for the (proper) subset $\{(\alpha,\beta) : \alpha \geq \beta \geq -1/2\}$ of V by writing each $c(k, m, n; \alpha, \beta)$ as a sum (with nonnegative coefficients) of squares of integrals of triple products of Jacobi polynomials.

Supported in part by NSF Grant MCS-8002507.

Recently, Mizan Rahman [14] showed that the linearization coefficients can be expressed as multiples of ${}_9F_8$ hypergeometric series from which their nonnegativity is obvious for the larger subset $\{(\alpha,\beta) : \alpha \geq \beta > -1,\ \alpha + \beta + 1 \geq 0\}$ of the set V, and then he extended [13] this to ${}_{10}\phi_9$ basic hypergeometric series representations for the coefficients in the linearization of the product of two continuous q-Jacobi polynomials $P_n^{(\alpha,\beta)}(x;\ q)$, from which the nonnegativity of the coefficients is obvious in the case $\alpha \geq \beta > -1$, $\alpha + \beta + 1 \geq 0$, $0 < q < 1$.

Here we shall show how Rahman's ${}_{10}\phi_9$ series can be used to completely determine all $(\alpha,\ \beta,\ q)$ with $\alpha,\ \beta > -1$, $0 < q < 1$, for which the coefficients in the linearization of the product of two continuous q-Jacobi polynomials are nonnegative and point out three of the most important applications: a convolution structure associated with the linearization, a Wiener-Lévy-type theorem, and positivity of a generalized translation operator. Since $P_n^{(\alpha,\beta)}(x;\ q)$ tends to $P_n^{(\alpha,\beta)}(x)$ as $q \to 1-$, the limiting case $q \to 1-$ of our argument gives an alternate proof of the Jacobi polynomial linearization result that is substantially simpler than that in [7]. At the end of the paper we will point out some open problems, one of which is related to Professor Zygmund's work [16] on the Riemannian theory for orthogonal systems.

NOTATION AND MAIN RESULTS

Following Rahman [13], we define the continuous q-Jacobi polynomials by

$$P_n^{(\alpha,\beta)}(x;\ q) = \frac{(q^{\alpha+1};\ q)_n(-q^{\beta+1};\ q)_n}{(q;\ q)_n(-q;\ q)_n}$$

$$\cdot {}_4\phi_3\left[\begin{matrix} q^{-n},\ q^{n+\alpha+\beta+1},\ q^{\frac{1}{2}}e^{i\theta},\ q^{\frac{1}{2}}e^{-i\theta} \\ q^{\alpha+1},\quad -q^{\beta+1},\ -q \end{matrix};q,\ q\right],$$

where $x = \cos\theta$, $(a;\, q)_n = (1 - \alpha)(1 - aq) \ldots (1 - aq^{n-1})$, $(a;\, q)_0 = 1$, and a ${}_{r+1}\phi_r$ basic hypergeometric series is defined by

$$ {}_{r+1}\phi_r\left[\begin{matrix} a_1, \ldots, a_{r+1} \\ b_1, \ldots, b_r \end{matrix};\, q,\, z\right] = \sum_{k=0}^{\infty} \frac{(a_1;\, q)_k \cdots (a_{r+1};\, q)z^k}{(b_1;\, q)_k \cdots (b_r;\, q)_k (q;\, q)_k} $$

whenever the series converges.

For $\alpha, \beta > -1$, $0 < q < 1$, the continuous q-Jacobi polynomials satisfy the orthogonality relation [4, Sec. 2]

$$ \begin{aligned} &\int_{-1}^{1} P_n^{(\alpha,\beta)}(x;\, q)P_m^{(\alpha,\beta)}(x;\, q)w^{(\alpha,\beta)}(x;\, q)\, dx \\ &+ P_n^{(\alpha,\beta)}(x_\beta(q);\, q)P_m^{(\alpha,\beta)}(x_\beta(q);\, q)w^{(\alpha,\beta)}(q) \\ &+ P_n^{(\alpha,\beta)}(-x_\alpha(q);\, q)P_m^{(\alpha,\beta)}(-x_\alpha(q);\, q)w^{(\beta,\alpha)}(q) = \frac{\delta_{mn}}{h_n^{(\alpha,\beta)}(q)}, \end{aligned} \tag{2} $$

with $x_\beta(q) = -\frac{1}{2}\left(q^{\beta+\frac{1}{2}} + q^{-\beta-\frac{1}{2}}\right)$, $w^{(\alpha,\beta)}(q) = 0$ if $\beta \geq -\frac{1}{2}$,

$$ w^{(\alpha,\beta)}(q) = \frac{2\pi(q^{-2\beta-1};\, q)_\infty}{(q;\, q)_\infty(q^{2\beta+2};\, q^2)_\infty(q^{-2\beta};\, q^2)_\infty(-q^{\alpha-\beta};\, q)_\infty(-q^{\alpha+\beta+1};\, q)_\infty} $$

if $-1 < \beta < -\frac{1}{2}$,

$$ h_n^{(\alpha,\beta)}(q) $$

$$ = \frac{[(q^2;\, q^2)_n]^4(q^{2n+2};\, q^2)_\infty(q^{2n+2\alpha+2};\, q^2)_\infty(q^{2n+2\beta+2};\, q^2)_\infty(-q^{n+\alpha+\beta+1};\, q)_\infty}{2\pi q^n(q;\, q)_n(q;\, q)_n(q^{n+\alpha+\beta+1};\, q)_n(q^{2n+\alpha+\beta+2};\, q)_\infty}, $$

and

$$ w^{(\alpha,\beta)}(x;\, q) = \frac{h(2x^2 - 1,\, 1)(1 - x^2)^{-\frac{1}{2}}}{h(x,\, q^{\frac{1}{2}})h(x,\, q^{\alpha+\frac{1}{2}})h(x,\, -q^{\beta+\frac{1}{2}})h(x,\, -q^{\frac{1}{2}})}, $$

where $(a;\, q)_\infty = (1 - a)(1 - aq)(1 - aq^2) \ldots$ and

$$ h(x,\, a) = \prod_{n=0}^{\infty}(1 - 2axq^n + a^2q^{2n}). $$

Hence, for α, $\beta > -1$, the polynomials $P_n^{(\alpha,\beta)}(x;\, q)$ are orthogonal with respect to a positive measure that is absolutely continuous on $(-1, 1)$ and has a point mass at $x_\beta(q) < -1$ when $-1 < \beta < -\frac{1}{2}$ and a point mass at $-x_\alpha(q) > 1$ when $-1 < \alpha < -\frac{1}{2}$.

Defining the linearization coefficients $g(k, m, n;\, \alpha, \beta, q)$ by

$$P_n^{(\alpha,\beta)}(x;\, q)P_m^{(\alpha,\beta)}(x;\, q)$$
$$= \sum_{k=|n-m|}^{n+m} g(k, m, n;\, \alpha, \beta, q)P_k^{(\alpha,\beta)}(x;\, q) \tag{3}$$

we shall prove:

Theorem 1. Let $S = \{(\alpha, \beta, q) : 0 < q < 1$ *and either* $\alpha \geq \beta > -1$, $\alpha + \beta + 1 \geq 0$ *or* $-\frac{1}{2} > \beta > -1$, $g(2, 2, 2;\, \alpha, \beta, q) \geq 0\}$.
If $(\alpha, \beta, q) \in S$, *then*

$$g(k, m, n;\, \alpha, \beta, q) \geq 0, \quad k, m, n = 0, 1, \ldots . \tag{4}$$

However, if α, $\beta > -1$, $0 < q < 1$ *and* $(\alpha, \beta, q) \notin S$, *then there exist positive integers* k, m, n *such that* $g(k, m, n;\, \alpha, \beta, q) < 0$. *S is a proper subset of* $\{(\alpha, \beta, q) : 0 < q < 1, \alpha \geq \beta > -1, \alpha \geq -\frac{1}{2}\}$.

To construct the convolution structure that follows from Theorem 1, we first observe that, since the zeros [15, Sec. 3.3] of $P_n^{(\alpha,\beta)}(x;\, q)$ are on the left side of the point $x = 1$ when $\alpha \geq -\frac{1}{2}$, $\beta > -1$, $0 < q < 1$ and since the coefficient of x^n in this polynomial is positive, $P_n^{(\alpha,\beta)}(1;\, q) > 0$ and so in this case we can define

$$p_n^{(\alpha,\beta)}(x;\, q) = \frac{P_n^{(\alpha,\beta)}(x;\, q)}{P_n^{(\alpha,\beta)}(1;\, q)}.$$

Then $p_n^{(\alpha,\beta)}(1;\, q) = 1$ and, by Askey and Ismail [2, Sec. 3.19], for $-1 \leq x \leq 1$

$$|p_n^{(\alpha,\beta)}(x;\, q)| \leq 1, \quad 0 < q < 1, \quad n = 0, 1, \ldots \tag{5}$$

in the continuous q-ultraspherical case $\alpha = \beta \geq -\frac{1}{2}$. From Andrews and Askey [1, Theorem 11 and equation (3.7)], we find that

$$P_n^{(\alpha,\beta)}(x;\ q) = \sum_{k=0}^{n} c(k,\ n) P_k^{(\gamma,\beta)}(x;\ q) \tag{6}$$

with $c(k, n) \geq 0$, $0 \leq k \leq n$, when $\alpha \geq \gamma > -1$ and $\beta > -1$, from which it follows on setting $\gamma = \beta$ and using the $\alpha = \beta \geq -\frac{1}{2}$ case of (5) that (5) also holds for $-1 \leq x \leq 1$ when $\alpha \geq \beta \geq -\frac{1}{2}$.

Let

$$I_\beta(q) = \begin{cases} [-1,\ 1], \text{ if } \beta \geq -\frac{1}{2}, \\ [-1,\ 1] \cup \{x_\beta(q)\}, \text{ if } -1 < \beta < -\frac{1}{2}, \end{cases}$$

$B = \{(\alpha,\ \beta,\ q) : 0 < q < 1,\ \alpha \geq \beta > 1$ and (5) holds for $x \in I_\beta(q)\}$, and for $\alpha \geq -\frac{1}{2}$ let

$$\begin{aligned} \gamma(k,\ m,\ n) &= \gamma(k,\ m,\ n;\ \alpha,\ \beta,\ q) \\ &= \int_{-1}^{1} p_k^{(\alpha,\beta)}(x;\ q) p_m^{(\alpha,\beta)}(x;\ q) p_n^{(\alpha,\beta)}(x;\ q) w^{(\alpha,\beta)}(x;\ q)\, dx \\ &\quad + p_k^{(\alpha,\beta)}(x_\beta(q);\ q) p_m^{(\alpha,\beta)}(x_\beta(q);\ q) p_n^{(\alpha,\beta)}(x_\beta(q);\ q) w^{(\alpha,\beta)}(q) \end{aligned}$$

and

$$h(n) = h(n;\ \alpha,\ \beta,\ q) = [\gamma(0,\ n,\ n;\ \alpha,\ \beta,\ q)]^{-1}.$$

Then

$$p_n^{(\alpha,\beta)}(x;\ q) p_m^{(\alpha,\beta)}(x;\ q) = \sum_{k=|n-m|}^{n+m} \gamma(k,\ m,\ n) h(k) p_k^{(\alpha,\beta)}(x;\ q) \tag{7}$$

and, by setting $x = 1$ and using (3) and (4),

$$\sum_k |\gamma(k,\ m,\ n) h(k)| = \sum_k \gamma(k,\ m,\ n) h(k) = 1$$

when $(\alpha,\ \beta,\ q) \in S$.

We shall say that a sequence $F = \{F(n)\}_{n=0}^{\infty}$ belongs to the class $b(\alpha,\ \beta,\ q)$ if its norm defined by

$$\| F \| = \sum_{n=0}^{\infty} |F(n)| h(n)$$

is finite. Define the convolution $F_1 * F_2$ of two sequences $F_1, F_2 \in b(\alpha, \beta, q)$ by

$$(F_1 * F_2)(n) = \sum_{k=0}^{\infty} \sum_{m=0}^{\infty} F_1(k) F_2(m) \gamma(k, m, n) h(k) h(m)$$

and define the transform $F^\wedge$ of a sequence $F \in b(\alpha, \beta, q)$ for $(\alpha, \beta, q) \in S \cap B$ by the absolutely convergent series

$$F^\wedge(x) = \sum_{n=0}^{\infty} F(n) p_n^{(\alpha,\beta)}(x; q) h(n), \quad x \in I_\beta(q).$$

Then the inversion formula is

$$F(n) = \int_{-1}^{1} F^\wedge(x) p_n^{(\alpha,\beta)}(x; q) w^{(\alpha,\beta)}(x; q)\, dx$$
$$+ F^\wedge(x_\beta(q)) p_n^{(\alpha,\beta)}(x_\beta(q); q) w^{(\alpha,\beta)}(q)$$

and, as in Hirschman [9] for ultraspherical polynomials and in Askey and Wainger [3] and Gasper [7] for Jacobi polynomials, Theorem 1 yields the convolution structure in Theorem 2, and the usual Banach algebra proof of the Wiener-Lévy theorem yields Theorem 3.

Theorem 2. If $(\alpha, \beta, q) \in S \cap B$ *and* $F_1, F_2, F_3 \in b(\alpha, \beta, q)$, *then*

(i) $F_1 * F_2 \in b(\alpha, \beta, q)$,

(ii) $\| F_1 * F_2 \| \leq \| F_1 \| \| F_2 \|$,

(iii) $F_1 * F_2 = F_2 * F_1$,

(iv) $F_1 * (F_2 * F_3) = (F_1 * F_2) * F_3$,

(v) $(F_1 * F_2)^\wedge(x) = F_1^\wedge(x) F_2^\wedge(x)$, $x \in I_\beta(q)$.

Theorem 3. If $\alpha \geq \beta \geq -\frac{1}{2}$, $0 < q < 1$, $-1 \leq x \leq 1$,

$$f(x) = \sum_{n=0}^{\infty} a(n) p_n^{(\alpha,\beta)}(x;\, q), \quad \sum_{n=0}^{\infty} |a(n)| < \infty,$$

and ψ *is a function holomorphic on an open set containing the range of* f, *then*

$$\psi(f)(x) = \sum_{n=0}^{\infty} b(n) p_n^{(\alpha,\beta)}(x;\, q) \quad \textit{with} \quad \sum_{n=0}^{\infty} |b(n)| < \infty.$$

The previously mentioned positivity of a generalized operator that follows from Theorem 1 is contained in:

Theorem 4. Suppose that $f(x)$ *is integrable with respect to the measure in the orthogonality relation* (2) *and let*

$$F(n) = \int_{-1}^{1} f(x) P_n^{(\alpha,\beta)}(x;\, q) w^{(\alpha,\beta)}(x;\, q)\, dx + f(x_\beta(q)) P_n^{(\alpha,\beta)}(x_\beta(q);\, q) w^{(\alpha,\beta)}(q),$$

$$F(n,\, m) = \int_{-1}^{1} f(x) P_n^{(\alpha,\beta)}(x;\, q) P_m^{(\alpha,\beta)}(x;\, q)\, dx + f(x_\beta(q)) P_n^{(\alpha,\beta)}(x_\beta(q);\, q) P_m^{(\alpha,\beta)}(x_\beta(q);\, q) w^{(\alpha,\beta)}(q).$$

If $(\alpha, \beta, q) \in S$, *then the operator that takes* $F(n)$ *into* $F(n, m)$, *the generalized translate of* $F(n)$, *is a positive operator in the sense that if* $F(n) \geq 0$, $n = 0, 1, \ldots$, *then* $F(n, m) \geq 0$, $n, m = 0, 1, \ldots$.

PROOF OF THEOREM 1

Let $\alpha, \beta > -1$, $0 < q < 1$, $2a = \alpha + \beta + 1$, and $p = q^2$. Without loss of generality we may assume that $n \geq m \geq 1$ in (3). Setting $m = n - s$ and $k = s + j$, it suffices to determine when

$$g_j = g(s+j,\, n-s,\, n;\, \alpha,\, \beta,\, q) \geq 0, \quad 0 \leq j \leq 2n-2s, \quad n > s \geq 0. \tag{8}$$

From (3) and Rahman [13, Sec. 4],

$$g_j = A_j a_j d_j, \quad 0 \le j \le 2n - 2s, \tag{9}$$

where $A_j > 0$,

$$a_j = \frac{(q^{2a};\, q)_{2s+j}(q^{\alpha-\beta+1-j};\, q)_j}{(q^{2a+2n-2s-j};\, q)_j} \cdot \frac{1 - q^{2a+2s+2j}}{1 - q^{2a}}, \tag{10}$$

and, if $j = 2k$ is even,

$$d_{2k} = \frac{(p^{\alpha+s+3/2};\, p)_k(p^{s-n-\alpha};\, p)_k(p^{\beta+n+1};\, p)_k(p^{1-a-k};\, p)_k}{(p^{s+1};\, p)_k(p^{s-n+1/2};\, p)_k(p^{2a+n+1/2};\, p)_k(p^{1-k+(\alpha-\beta)/2};\, p)_k}$$

$$\cdot \frac{(p^{-a-k+1/2};\, p)_k(p^{-k+1/2};\, p)_k p^{(\alpha+1/2)k}}{(p^{-k+(1+\alpha-\beta)/2};\, p)_k(p^{-\alpha-k};\, p)_k} \tag{11}$$

$$\cdot\, {}_{10}\phi_9\left[\begin{matrix} p^{\alpha},\, p^{1+\alpha/2},\, -p^{1+\alpha/2},\, p^{\alpha+1/2},\, p^{(\alpha-\beta)/2},\, p^{(\alpha-\beta+1)/2}, \\ p^{\alpha/2},\, -p^{\alpha/2},\, p^{1/2},\, p^{a+1/2},\, p^{a}, \end{matrix}\right.$$

$$\left.\begin{matrix} p^{s-n+k},\, p^{2a+n+k},\, p^{-s-k},\, p^{-k} \\ p^{\alpha+n+1-s-k},\, p^{-\beta-n-k},\, p^{\alpha+s+1+k},\, p^{\alpha+1+k} \end{matrix};\, p,\, p\right],$$

while, if $j = 2k + 1$ is odd,

$$d_{2k+1} = \frac{(p^{\alpha+s+3/2};\, p)_k(p^{s-n-\alpha};\, p)_k(p^{\beta+n+1};\, p)_k(p^{-a-k+1/2};\, p)_k}{(p^{s+1};\, p)_k(p^{s-n+1/2};\, p)_k(p^{2a+n+1/2};\, p)_k(p^{(\alpha-\beta+1)/2-k};\, p)_k}$$

$$\cdot \frac{(p^{-a-k};\, p)_k(p^{-k-1/2};\, p)_k p^{(\alpha+1/2)k}}{(p^{(\alpha-\beta)/2-k};\, p)_k(p^{-\alpha-1-k};\, p)_k} \tag{12}$$

$$\cdot\, {}_{10}\phi_9\left[\begin{matrix} p^{\alpha+1},\, p^{(\alpha+3)/2},\, -p^{(\alpha+3)/2},\, p^{\alpha+1/2},\, p^{(\alpha-\beta+2)/2},\, p^{(\alpha-\beta+1)/2}, \\ p^{(\alpha+1)/2},\, -p^{(\alpha+1)/2},\, p^{3/2},\, p^{a+1/2},\, p^{a+1}, \end{matrix}\right.$$

$$\left.\begin{matrix} p^{s-n+1+k},\, p^{2a+1+n+k},\, p^{-s-k},\, p^{-k} \\ p^{\alpha+1+n-s-k},\, p^{-\beta-n-k},\, p^{\alpha+2+s+k},\, p^{\alpha+2+k} \end{matrix};\, p,\, p\right].$$

From (9), (10), and (12) it is obvious that $g(1, 2, 2; \alpha, \beta, q) \geq 0$ if and only if $\alpha \geq \beta$. If $\alpha \geq \beta > -1$, $\alpha \leq -\frac{1}{2}$, and $\alpha + \beta + 1 < 0$, then $g(2, 2, 2; \alpha, \beta, q) < 0$, since the ${}_{10}\Phi_9$ in (11) is positive in this case. Hence, since Rahman has already observed that the above formulas give (8) for $\alpha \geq \beta > -1$, $\alpha + \beta + 1 \geq 0$, it remains to consider when (8) holds in the case

$$\alpha > -\frac{1}{2}, \ \beta > -1, \ \alpha + \beta + 1 < 0. \tag{13}$$

Let α, β satisfy (13). If j is odd, then the ${}_{10}\Phi_9$ series in (12) is positive and it is clear that (8) holds in this case. It is also clear that (8) holds for $j = 0$ and $j = 2n - 2s$. However, $g(2, 2, 2; \alpha, \beta, q)$ assumes both positive and negative values when (13) is satisfied, and so to complete the proof we have to show that if (13) holds and $g(2, 2, 2; \alpha, \beta, q) \geq 0$, then

$$g(s + 2k, n - s, n; \alpha, \beta, q) \geq 0, \ 1 \leq k \leq n - s - 1, \ n \geq 3. \tag{14}$$

Since, under these conditions, $g(s + 2k, n - s, n; \alpha, \beta, q)$ is a negative multiple of the ${}_{10}\phi_9$ in (11), proving (14) is equivalent to proving that this ${}_{10}\phi_9$ is negative for $1 \leq k \leq n - s - 1$, $n \geq 3$ whenever it is negative for $n = 2$, $s = 0$, $k = 1$. In view of the fact that the only positive term in the series expansion of this ${}_{10}\phi_9$ is the first term (which equals 1), if we write ${}_{10}\phi_9 = 1 + A_1(k, n, s) + \cdots + A_k(k, n, s)$, then it suffices to prove for $1 \leq k \leq n - s - 1$, $n \geq 3$ that

$$1 + A_1(k, n, s) \leq 1 + A_1(1, 2, 0),$$

which is equivalent to proving that

$$\begin{aligned} f_k \equiv {} & p^{\alpha+1}(1 - p)^3(1 - p^{2a+3})(1 - p^{k+\alpha+1}) \\ & \cdot (1 - p^{k+s-n-\alpha-1})(1 - p^{k+s+\alpha+1})(1 - p^{k+n+\beta}) \\ & - (1 - p^{\alpha+2})^3(1 - p^{\beta+3})(1 - p^k)(1 - p^{k+s-n})(1 - p^{k+s})(1 - p^{k+n+2a}) \\ & \geq 0, \ 1 \leq k \leq n - s - 1, \ n \geq 3. \end{aligned} \tag{15}$$

Setting

$$x = 1 - p^k,\ A = \alpha + 1,\ B = \beta + 3,\ C = \alpha + \beta + 4 = 2a + 3$$

and using expressions of the form

$$1 - p^{k+c} = p^c(x + p^{-c} - 1)$$

gives that

$$p^{-2s-2a} f_k = c_4x^4 + c_3x^3 + c_2x^2 + c_1x + c_0 \tag{16}$$

with (recall that $m = n - s$)

$$c_4 = p^A(1 - p)^3(1 - p^C) - (1 - p^{A+1})^3(1 - p^B), \tag{17}$$

$$\begin{aligned} c_3 = {} & p^A(1 - p)^3(1 - p^C)[p^{-s-A} + p^{-A} + p^{-n-\beta} + p^{m+A} - 4] \\ & - (1 - p^{A+1})^3(1 - p^B)[p^{-n-2a} + p^{-s} + p^m - 3], \end{aligned}$$

$$\begin{aligned} c_2 = {} & p^A(1 - p)^3(1 - p^C)[(p^{m+A} - 1)(p^{-A} + p^{-s-A} + p^{-n-\beta} - 3) \\ & + (p^{-A} - 1)(p^{-s-A} - 1) + (p^{-A} - 1)(p^{-n-\beta} - 1) \\ & + (p^{-n-\beta} - 1)(p^{-s-A} - 1] \\ & - (1 - p^{A+1})^3(1 - p^B)[(p^{-s} - 1)(p^{-n-2a} - 1) \\ & + (p^m - 1)(p^{-n-2a} - 1) + (p^m - 1)(p^{-s} - 1)], \end{aligned}$$

$$\begin{aligned} c_1 = {} & p^A(1 - p)^3(1 - p^C)[(p^{-A} - 1)(p^{-s-A} - 1)(p^{-n-\beta} - 1) \\ & + (p^{m+A} - 1)(p^{-A} - 1)(p^{-s-A} - 1) \\ & + (p^{m+A} - 1)(p^{-A} - 1)(p^{-n-\beta} - 1) \\ & + (p^{m+A} - 1)(p^{-s-A} - 1)(p^{-n-\beta} - 1)] \\ & - (1 - p^{A+1})^3(1 - p^B)(p^{-s} - 1)(p^m - 1)(p^{-n-2a} - 1), \end{aligned}$$

$$c_0 = p^A(1 - p)^3(1 - p^C)(p^{-A} - 1)(p^{-s-A} - 1)(p^{-n-\beta} - 1)(p^{m+A} - 1).$$

By utilizing Descartes' rule of signs, we shall show that if $F(x)$ denotes the polynomial on the right-hand side of (16), then

$$F(1 - p^k) \geq 0,\ 1 \leq k \leq n - s - 1,\ n \geq 3, \tag{18}$$

which will give (15) and complete the proof.

Clearly $c_0 < 0$, since only its last factor is negative.

To show that $c_4 < 0$, we first observe that

$$(1 - p^{A+1})(1 - p^B) - (1 - p)(1 - p^C) = p(1 - p^A)(1 - p^{B-1}) > 0.$$

Since $1 - p^{A+1} > 1 - p$, it then follows that

$$\left(\frac{1 - p^{A+1}}{1 - p}\right)^2 \frac{(1 - p^{A+1})(1 - p^B)}{(1 - p)(1 - p^C)} > 1,$$

which gives

$$(1 - p)^3(1 - p^C) - (1 - p^{A+1})^3(1 - p^B) < 0 \tag{19}$$

and hence $c_4 < 0$, since $0 < p^A < 1$.

For c_3 we first observe that

$$\begin{aligned} c_3 &= [(1 - p)^3(1 - p^C) - (1 - p^{A+1})^3(1 - p^B)][p^{-n-2a} + p^{-s} + p^m - 3] \\ &\quad - (1 - p)^3(1 - p^C)[p^{-n-2a} + p^m + 4p^A - p^{-n+A-\beta} - p^{m+2A} - 4] \\ &= [(1 - p)^3(1 - p^C) - (1 - p^{A+1})^3(1 - p^B)][p^{-s} - 1 + p^{-m}(p^m - 1)^2 \\ &\quad + p^{-m}(p^{-s-2a} - 1)] - (1 - p)^3(1 - p^C)(1 - p^A)p^{-m}[(p^m - 1)^2 \\ &\quad + p^{-A}(p^{m+A} - 1)^2 + (1 + p^{-A})(p^{-s-\beta} - 1)] \end{aligned}$$

from which, by (19), it follows that $c_3 < 0$ when $s \geq 1$. Hence the coefficients of $F(x)$ have (exactly) two variations of sign when $s \geq 1$.

Let $s = 0$. Since it is now possible for c_3 to be positive, we shall prove that the coefficients of $F(x)$ also have two variations of sign in this case by showing that if $c_3 > 0$, then $c_1 < 0$. First observe that for $s = 0$ we can express c_1 and c_3 in the forms

$$\begin{aligned} c_1 &= p^A(1 - p)^3(1 - p^C)(p^{-A} - 1)[2(p^{n+A} - 1)(p^{-n-\beta} - 1) \\ &\quad + (p^{-A} - 1)(p^{-n-\beta} + p^{n+A} - 2)], \end{aligned} \tag{20}$$

$$\begin{aligned} c_3 &= 2(p^{-A} - 1)[p^A(1 - p)^3(1 - p^C) - (1 - p^{A+1})^3(1 - p^B)] \\ &\quad + p^{-A}(p^{-n-\beta} + p^{n+A} - 2)[p^{2A}(1 - p)^3(1 - p^C) \\ &\quad - (1 - p^{A+1})^3(1 - p^B)]. \end{aligned} \tag{21}$$

If $c_3 > 0$, then since the two expressions in brackets in (21) are negative by (19), $p^{n-\beta} + p^{n+A} - 2 < 0$ and, by (20), $c_1 < 0$.

Hence, the coefficients of $F(x)$ have two variations of sign for all $s \geq 0$, and so, by Descartes' rule and the fact that $c_4 < 0$, to prove (18) it suffices to prove it for $k = 1$ and $k = n - s - 1$ with $n \geq 3$.

Setting $t = n - s - 2$ gives

$$f_1 = p^{-t-1}(1 - p)(1 - p^{A+1}) \cdot [(1 - p^{A+1})^2(1 - p^B)(1 - p^{s+1})(1 - p^{t+1})(1 - p^{s+t+C}) - (1 - p)^2(1 - p^C)(1 - p^{s+A+1})(1 - p^{t+A+1})(1 - p^{s+t+B})], \tag{22}$$

which is positive if and only if

$$\frac{(1 - p^{A+1})^2(1 - p^B)(1 - p^{s+1})(1 - p^{t+1})(1 - p^{s+t+C})}{(1 - p)^2(1 - p^C)(1 - p^{s+A+1})(1 - p^{t+A+1})(1 - p^{s+t+B})} > 1. \tag{23}$$

Observe that the left-hand side of (23) equals 1 when $s = t = 0$ and that the ratio

$$\frac{(1 - p^{s+1})(1 - p^{t+1})(1 - p^{s+t+C})}{(1 - p^{s+A+1})(1 - p^{t+A+1})(1 - p^{s+t+B})}$$

is symmetric in s and t. Thus to prove (23) it suffices to show that

$$\frac{(1 - p^{s+1})(1 - p^{s+t+C})}{(1 - p^{s+A+1})(1 - p^{s+t+B})}$$

is a (strictly) increasing function of s whenever $t \geq 0$, or, equivalently, that

$$h_s \equiv (1 - p^{s+2})(1 - p^{s+A+1})(1 - p^{s+t+B})(1 - p^{s+t+C+1}) - (1 - p^{s+1})(1 - p^{s+A+2})(1 - p^{s+t+B+1})(1 - p^{s+t+C})$$

is positive. By elementary manipulations, it can be shown that

$$h_s = p^{s+1}(1 - p)(1 - p^A)(1 - p^{t+B-1})(1 - p^{2s+t+C+2}),$$

which is clearly positive. Consequently $f_1 > 0$ and (18) holds for $k = 1$.

For the remaining case $k = n - s - 1$, set $t = n - s - 2$, $N = n - 2$, and observe that

$$\begin{aligned} f_{n-s-1} &= p^{-1}(1 - p)(1 - p^{A+1}) \\ &\cdot [(1 - p^{A+1})^3(1 - p^B)(1 - p^{N+1})(1 - p^{t+1})(1 - p^{N+t+C}) \\ &- (1 - p)^2(1 - p^C)(1 - p^{N+A+1})(1 - p^{t+A+1})(1 - p^{N+t+B})]. \end{aligned} \tag{24}$$

Since the expression in brackets in (24) becomes that in (22) when N is replaced by s, it follows that $f_{n-s-1} > 0$, which gives (18) for $k = n - s - 1$ and hence completes the proof of Theorem 1.

OPEN PROBLEMS

(a) Recall that in the section "Notation and Main Results" we pointed out that if $\alpha \geq \beta \geq -\frac{1}{2}$ and $-1 \leq x \leq 1$, then

$$|p_n^{(\alpha,\beta)}(x;\ q)| \leq 1,\ 0 < q < 1,\ n = 0,\ 1,\ \ldots. \tag{25}$$

Since it is proved in Szegö [15, Theorem 7.32.1] that the Jacobi polynomial case $q = 1$ of (25) holds if and only if $\alpha \geq \beta > -1$, $\alpha \geq -1/2$, it is natural to conjecture that (25) holds for $x \in I_\beta(q)$ when $\alpha \geq \beta > -1$, $\alpha \geq -\frac{1}{2}$. Unfortunately, one cannot prove this by using a projection formula of the type (6) with, for example, (γ,β) on the right-hand side replaced by $(-\frac{1}{2}, -\frac{1}{2})$, since $x_\beta(q) < -1$ and the coefficients are not all nonnegative when $\alpha = -1/2$, $-\frac{1}{2} > \beta > -1$. What seems to be needed is a generalization of the argument of Szegö [15, Sec. 7.32]. Once (25) is proved for $\alpha \geq \beta > -1$, $\alpha \geq -\frac{1}{2}$, $x \in I_\beta(q)$, the restriction in Theorem 2 that $(\alpha,\ \beta,\ q)$ be in the set B can be dropped.

(b) In [8, 5] the author set $R_n^{(\alpha,\beta)}(x) = P_n^{(\alpha,\beta)}(x)/P_n^{(\alpha,\beta)}(1)$ and proved that there is a product formula of the form (a dual of (1))

$$R_n^{(\alpha,\beta)}(x)R_n^{(\alpha,\beta)}(y) = \int_{-1}^{1} R_n^{(\alpha,\beta)}(z)\, d\mu_{x,y}^{(\alpha,\beta)}(z) \tag{26}$$

$n = 0, 1, \ldots,$

with $d\mu_{x,y}^{(\alpha,\beta)}(z) \geq 0$, $-1 \leq x, y, z \leq 1$, if and only if $\alpha \geq \beta \geq -1/2$ or $-1/2 > \beta > -1$, $\alpha + \beta \geq 0$. This suggests the conjecture that if $x, y \in I_\beta(q)$, $0 < q < 1$ and either $\alpha \geq \beta \geq -1/2$ or $-1/2 > \beta > -1$, $\alpha + \beta \geq 0$, then the product formula (7) has a dual of the form

$$p_n^{(\alpha,\beta)}(x; q)p_n^{(\alpha,\beta)}(y; q) = \int_{-\infty}^{\infty} p_n^{(\alpha,\beta)}(z; q)\, d\mu_{x,y}^{(\alpha,\beta)}(z; q), \tag{27}$$

$n = 0, 1, \ldots,$

with $d\mu_{x,y}^{(\alpha,\beta)}(z; q) \geq 0$. In proving (26), the author considered the kernel

$$K(x, y, z; \alpha, \beta) = \sum_{n=0}^{\infty} h_n^{(\alpha,\beta)} R_n^{(\alpha,\beta)}(x)R_n^{(\alpha,\beta)}(y)R_n^{(\alpha,\beta)}(z)$$

with

$$h_n^{(\alpha,\beta)} = \left(\int_{-1}^{1} \left[R_n^{(\alpha,\beta)}(x)\right]^2 (1 - x)^\alpha (1 + x)^\beta\, dx \right)^{-1}$$

and employed Zygmund's Riemannian theory for Jacobi series [16, Chapter 2] to show that if $\alpha \geq \beta \geq -1/2$, $\alpha > -1/2$, and $-1 < x, y < 1$, then (26) holds with an absolutely continuous measure

$$d\mu_{x,y}^{(\alpha,\beta)}(z) = K(x, y, z; \alpha, \beta)(1 - z)^\alpha (1 + z)^\beta\, dz.$$

This raises the problem of extending Zygmund's Riemannian theory to the continuous q-Jacobi polynomials (which in turn will require the derivation of an analogue of Darboux's asymptotic formula [16, Chapter 2, (14)]). It should be mentioned that since (26) is equivalent to the evaluation of the first term in Koornwinder's addition formula for Jacobi polynomials [10, 11], (27) also raises the problem of finding an addition formula for the continuous q-Jacobi polynomials.

The author wishes to thank Professor R. Askey for bringing the contents of [4] to his attention.

REFERENCES

1. Andrews, G. E., and Askey, R. *Enumeration of Partitions: The Role of Eulerian Series and q-Orthogonal Polynomials, Higher Combinatorics*. Edited by M. Aigner. Boston: D. Reidel Publishing Company, 1977, pp. 3-26.

2. Askey, R., and Ismail, M.E.-H. A generalization of ultra-spherical polynomials. *Studies in Analysis*. To appear.

3. Askey, R., and Wainger, S. *A Dual Convolution Structure for Jacobi Polynomials, in Orthogonal Expansions and Their Continuous Analogues*. Southern Illinois Press, 1968, pp. 25-36.

4. Askey, R., and Wilson, J. Some basic hypergeometric orthogonal polynomials that generalize Jacobi polynomials. To appear.

5. Gasper, G. Banach algebras for Jacobi series and positivity of a kernel. *Ann. Math.* 95 (1972): 261-280.

6. Gasper, G. Linearization of the product of Jacobi polynomials I. *Canad. J. Math.* 22 (1970): 171-175.

7. Gasper, G. Linearization of the product of Jacobi polynomials II. *Canad. J. Math.* 22 (1970): 582-593.

8. Gasper, G. Positivity and the convolution structure for Jacobi series. *Ann. Math.* 93 (1971): 112-118.

9. Hirschman, Jr., I. I. Harmonic analysis and ultraspherical polynomials. *Proc. of the Conference on Harmonic Analysis*, Cornell, 1956.

10. Koornwinder, T. Jacobi polynomials II. An analytic proof of the product formula. *SIAM J. Math. Anal.* 5 (1974): 125-137.

11. Koornwinder, T. Jacobi polynomials III. An analytic proof of the addition formula. *SIAM J. Math. Anal.* 6 (1975): 533-543.

12. Koornwinder, T. Positivity proofs for linearization and connection coefficients of orthogonal polynomials satisfying an addition formula. *J. London Math. Soc.* (2) 18 (1978): 101-114.

13. Rahman, Mizan. The linearization of the product of continuous q-Jacobi polynomials. *Canad. J. Math.* 33 (1981): 961-987.

14. Rahman, Mizan. A non-negative representation of the linearization coefficients of the product of Jacobi polynomials. *Canad. J. Math.* 33 (1981): 915-928.

15. Szegö, G. *Orthogonal Polynomials.* 4th ed. Amer. Math. Soc. Colloq. Pub., vol. 23. Providence, R.I.: Amer. Math. Soc., 1975.

16. Zygmund, A. Sur la theorie riemannienne de certains systemes orthogonaux, II. *Prace Matematyczno-Fizyczne,* 39 (1932): 73-117.

BERNSTEIN'S INEQUALITY FOR FINITE INTERVALS

Gabor Halász
Hungarian Academy of Science
Hugh L. Montgomery
University of Michigan

1. STATEMENT OF RESULTS

Bernstein's inequality asserts that

$$\| F' \|_{L^\infty(\mathbb{R})} \leq T \| F \|_{L^\infty(\mathbb{R})} \tag{1}$$

for any entire function F that is of exponential type at most T. Our object is to establish a corresponding inequality for a finite interval, say,

$$\| F' \|_{L^\infty[0,1]} \leq \Delta \| F \|_{L^\infty[0,1]}, \tag{2}$$

for a suitable class of functions F. As may be seen from the example in Sec. 4, it is no longer possible to allow F to be as general as in (1); we restrict ourselves to almost-periodic polynomials

$$F(x) = \sum_{n=1}^{N} a_n e^{it_n x} \tag{3}$$

Research supported in part by NSF Grant No. MSC80-02559.

and consider the way in which Δ may depend on N and the size of the t_n.

Theorem 1. Suppose that $F(x)$ is given by (3) *and that* $-T \leq t_n \leq T$ *for all n. Then* (2) *holds with*

$$\Delta = C(N^2 + T \log 6N),$$

where C is a suitable absolute constant.

By making a linear change of variable, we may replace the interval [0,1] by an arbitrary interval $[a,b]$, and we find that

$$\| F' \|_{L^\infty[a,b]} \leq C\left(\frac{N^2}{b - a} + T \log 6N\right)\| F \|_{L^\infty[a,b]}. \tag{4}$$

To judge the precision of this result, we note from the example $F(x) = e^{iTx}$ that

$$\Delta \geq T. \tag{5}$$

In Sec. 4 we use Chebychev polynomials to show that

$$\Delta \geq 2(N - 1)^2, \tag{6}$$

even when T is arbitrarily small. Thus our theorem is not far from being best possible. We conjecture that

$$\Delta \leq C(N^2 + L) \tag{7}$$

for some suitable absolute constant $C > 0$; our method would give this if we were able to sharpen slightly Lemma 1 below.

In a different direction, recall that Markov showed (see Schaeffer [4]) that if $P(x)$ is a polynomial of degree at most D, then

$$\| P' \|_{L^\infty[-1,1]} \leq D^2 \| P \|_{L^\infty[-1,1]}.$$

Equality is achieved when $P(x) = T_D(x)$, the D-th Chebychev polynomial. We raise the question of whether this bound might be improved if we further restrict P to having at most N nonzero coefficients. For example, it might be that

$$\| P' \|_{L^\infty[-1,1]} \leq CDN \| P \|_{L^\infty[-1,1]}$$

for some suitable constant C. Nothing better could be true, as we see by considering $P(x) = T_N(x^a)$ with $a = [D/N]$.

2. BASIC LEMMAS

We now formulate the basic estimate from which our theorem follows by means of routine complex analysis.

Lemma 1. *If* $F(x) = \sum_{n=1}^{N} a_n e^{w_n x}$, *where* $\operatorname{Im} w_n \geq 0$ *for all* n, *then for any* $x \geq 0$,

$$|F(-x)| \leq 30N^2 e^{6N\sqrt{x}} \| F \|_{L^\infty[0,1]}. \tag{8}$$

This bound was established by Halász [1] when $0 \leq x \leq 1$. For $x \geq 1$, the above bound is implied by an inequality of Turán [5], which asserts that

$$|F(-x)| \leq (2e(x + 1))^N \| F \|_{L^\infty[0,1]}.$$

If we could replace (8) by a sharper bound of the form

$$|F(-x)| \leq Ae^{AN\sqrt{x}} \| F \|_{L^\infty[0,1]}$$

for some constant $A > 0$, then we would be able to establish the conjectured inequality (7).

We shall employ the following weak form of an inequality due to Milloux and Schmidt. See Heins [2, 3] for stronger results.

Lemma 2. *Suppose that* $F(z)$ *is regular in the disc* $|z| \leq R$ *and that* $|f(z)| \leq M$ *in this disc, where* $M \geq 1$. *Suppose also that* $|F(x)| \leq 1$ *for* $0 \leq x \leq R$. *Then*

$$|F(z)| \leq M^{\frac{4}{\pi}\arctan\sqrt{r/R}}$$

for $|z| \leq r \leq R$.

The above inequality is best possible if we require only that F be regular in slit disc

$$\mathscr{D}_R = \{z : 0 < |z| \leq R, \quad 0 < \arg z < 2\pi\},$$

as is seen by considering the function

$$G(z) = \exp\left(\frac{2i}{\pi}(\log M) \log \frac{1 - \sqrt{z/R}}{1 + \sqrt{z/R}}\right).$$

Indeed, it is easily verified that

$$|G(Re^{i\theta})| = M$$

for $0 < \theta < 2\pi$, that

$$\lim_{\theta \to 0^+} |G(re^{i\theta})| = 1, \qquad \lim_{\theta \to 2\pi^-} |G(re^{i\theta})| = 1$$

for $0 < r < R$, and that

$$\max_{\substack{0 < p \leq r \\ 0 < \theta < 2\pi}} |G(\rho e^{i\theta})| = G(-r) = M^{\frac{4}{\pi} \arctan \sqrt{r/R}}.$$

Then the lemma follows by applying the maximum-modulus principle to the function $F(z)/G(z)$ in the domain $\mathscr{D}_R$.

3. PROOF OF THE THEOREM

We assume that $\| F \|_{L^\infty[0,1]} = 1$. Then by Lemma 1,

$$|F(-x)| \leq 30N^2 e^{6N\sqrt{x}}$$

for $x \geq 0$, and

$$|F(x)| \leq 30N^2 e^{6N\sqrt{x - 1}}$$

for $x \geq 1$. For Im $z \geq 0$, let

$$K(z) = 30N^2 \exp\big(6N(\sqrt{z - 1} - i\sqrt{z}) - iTz\big),$$

where the branches of the square roots are chosen so that $|F(x)| \leq |K(x)|$ for all real x. Since $F(z)/K(z) \to 0$ as $z \to \infty$, uniformly for Im $z \geq 0$, we deduce by the maximum-modulus principle that $|F(z)| \leq |K(z)|$ for all z in this half-plane. Suppose that x is fixed, $0 \leq x \leq 1$, and let $\mathscr{R}$ be the portion of the disc $|z - x| \leq R$ for which Im $z \geq 0$. If $z \in \mathscr{R}$, then $\mathrm{Re}(\sqrt{z-1} \leq \sqrt{R}$, $\mathrm{Re}(-i\sqrt{z}) \leq \sqrt{R}$, and $\mathrm{Re}(-iTz) \leq TR$, so that

$$|K(z)| \leq 30N^2 \exp(12N\sqrt{R} + TR).$$

Taking

$$R = \min\left(\frac{(\log 6N)^2}{144N^2}, \frac{\log 6N}{T}\right),$$

we find that $|F(z)| \leq (6N)^4$ for $z \in \mathscr{R}$. A similar argument applies to the lower half-plane, and hence this bound holds throughout the disc $|z - x| \leq R$. As $R \leq \frac{1}{2}$, we can apply Lemma 2 with $r = R(\log 6N)^{-2}$, so that

$$|F(z)| \leq (6N)^{\frac{16}{\pi}\sqrt{r/R}} = e^{16/\pi} = C_1$$

for $|z - x| \leq r$. Thus by Cauchy's inequality,

$$|F'(x)| \leq \frac{C_1}{r} = C_1 \max(144N^2, T \log 6N).$$

This gives the stated bound.

4. A BASIC EXAMPLE

We now construct an $F(x)$ to demonstrate the inequality (6). Let T_K be the K-th Chebychev polynomial, defined by the characteristic property that $T_K(\cos\theta) = \cos K\theta$. Hence $|T_K(x)| \leq 1$ for $-1 \leq x \leq 1$, and $T'(\cos\theta) = K\dfrac{\sin K\theta}{\sin\theta}$, so that $T'_K(1) = K^2$.

Take $F(x) = T_K\left(\dfrac{\sin \varepsilon\ (2x - 1)}{\sin \varepsilon}\right)$ where $0 < \varepsilon < \pi/2$. Then $|F(x)| \leq 1$ for $0 \leq x \leq 1$, and

$$F'(1) = 2K^2 \frac{\varepsilon}{\tan \varepsilon} \to 2K^2$$

as $\varepsilon \to 0^+$. But T_K is an even function for even K, and an odd function for odd K, so we may write F in the form

$$F(x) = \sum_{k=0}^{K} b_k e^{2i\varepsilon(K-2k)x} .$$

Thus $T = 2K\varepsilon$ and $N \leq K + 1$, which gives (6) as desired.

REFERENCES

1. Halász, Gabor. On the first and second main theorems in Turán's theory of power sums. *Studies in Mathematics in Memory of Pál Turán*. Hungarian Academy of Science, 1982. To appear.

2. Heins, Maurice. The problem of Milloux for functions analytic throughout the interior of the unit circle. *Amer. J. of Math.* 67 (1945): 212-234.

3. Heins, Maurice. *Selected Topics in the Classical Theory of Functions of a Complex Variable*. New York: Holt, Rinehart & Winston, 1962, p. 109.

4. Schaeffer, A. C. Inequalities of A. Markoff and S. Bernstein for polynomials and related functions. *Bull. Amer. Math. Soc.* 47 (1941): 565-579.

5. Turán, P. *Über eine neue Methode der Analysis und ihre Anwendungen*. Budapest: Akadémiai Kaidó, 1953.

SLOW POINTS OF GAUSSIAN PROCESSES

Jean-Pierre Kahane
Université de Paris-Sud

INTRODUCTION

In their papers "On Some Series of Functions" [11], Paley and Zygmund initiated the study of trigonometric series with random coefficients

$$\sum_{0}^{\infty} \pm\, a_n \cos(nt + \varphi_n) \qquad \text{(R)}$$

$$\sum_{0}^{\infty} a_n \cos(nt + 2\pi\omega_n). \qquad \text{(S)}$$

In (R) (for Rademacher) the a_n and φ_n are given, and the signs are chosen randomly with the natural probability. In (S) (for Steinhaus) the a_n are given, and the ω_n are chosen randomly on [0,1] with the natural probability. Very soon afterward, Paley, Wiener, and Zygmund considered Gaussian series

$$\sum_{0}^{\infty} a_n(\xi_n \cos nt + \xi_n' \sin nt) \qquad \text{(G)}$$

where the a_n are given and the ξ_n, ξ_n' are independent Gaussian normalized random variables [12]. The motivation for (G) was the

Fourier series of the Brownian motion function (or Wiener process). If $t \in [0,1]$, the Brownian motion function is

$$W(t) = \xi_0 t + \sum_1^\infty \frac{1}{\pi n\sqrt{2}} \left(\xi_n(\cos 2\pi nt - 1) + \xi_n' \sin 2\pi nt\right). \tag{W}$$

The almost sure properties of (R), (S), (G) are closely related; in particular, uniform convergence has the same probability (zero or one) for the three of them (for (R) and (S), this is due to P. Billard [1,2]; for (R) and (G), to M. Marcus and G. Pisier [7]). The question considered by Paley, Wiener, and Zygmund was an implicit conjecture of Jean Perrin in "Les Atomes" (1913), namely, that the Brownian motion function is almost surely nowhere differentiable. Actually, they prove more:

$$\text{a.s.} \quad \forall\, t \quad \overline{\lim_{h\to 0}} \frac{|W(t+h) - W(t)|}{|h|^{1/2+\varepsilon}} = \infty \tag{PWZ}$$

whenever $\varepsilon > 0$.

Is (PWZ) still valid when $\varepsilon = 0$? The answer is no, but very close to yes. The "close to yes" part is

$$\text{a.s.} \quad \forall\, t \quad \overline{\lim_{h\to 0}} \frac{|W(t+h) - W(t)|}{|h|^{1/2}} > 0; \tag{D}$$

it is due to A. Dvoretzky [3]. "No" means

$$\text{a.s.} \quad \exists\, t \quad \overline{\lim_{h\to 0}} \frac{|W(t+h) - W(t)|}{|h|^{1/2}} < \infty \quad [5]. \tag{K}$$

When t is given, the law of the iterated logarithm of Khintchin reads

$$\text{a.s.} \qquad \overline{\lim_{h\to 0}} \frac{|W(t+h) - W(t)|}{\sqrt{2|h| \log\log \frac{1}{|h|}}} = 1. \tag{Kh}$$

Therefore (Fubini) this equality holds almost surely almost everywhere. Therefore, the set of t's occurring in (K), that is,

$$W(t+h) - W(t) = O\left(|h|^{1/2}\right) \qquad (h \to 0),$$

has a.s. measure zero. We call these points "slow points." The set of slow points is not too small: it has a.s. Hausdorff dimension 1, and its intersection with the set of zeros $W^{-1}(0)$ has dimension $\frac{1}{2}$ [5,6]. In the opposite direction, there are also rapid points; this is due to the Paul Lévy theorem on the modulus of continuity

$$\omega_W(\delta) = \sup_{|t-t'|\leq\delta} |W(t) - W(t')| \qquad (t,t' \in [0,1])$$

that is

$$\text{a.s.} \quad \overline{\lim_{\delta\to 0}} \frac{\omega_W(\delta)}{\sqrt{2\delta \log \frac{1}{\delta}}} = 1. \tag{L}$$

The set of rapid points t, defined by

$$W(t+h) - W(t) = \Omega\left(|h|^{1/2}\sqrt{\log \frac{1}{h}}\right) \qquad (h \to 0),$$

is also of dimension 1, as proved by S. Orey and J. Taylor [8,10].

In contrast with Brownian motion, there is nothing like slow points and rapid points for lacunary trigonometric series: all points are ordinary points. Precisely, if

$$\lambda_{n+1}/\lambda_n > q > 1$$

(Hadamard lacunarity condition) and if

$$f(t) = \sum_0^\infty \rho_n \cos(\lambda_n t + \varphi_n)$$

satisfies a Lipschitz condition of order $\alpha (0 < \alpha < 1)$ at one point t_0, that is,

$$f(t_0 + h) - f(t_0) = O(|h|^\alpha) \qquad (h \to 0),$$

it satisfies the same everywhere and uniformly [9].

Since (R), (S), (G) can be lacunary trigonometric series, there is no general result on these series saying that there are slow points or rapid points. What can be found in the literature or easily established [4] are analogues of (D), (Kh), and (L) involving

$$s_j = \left(\sum_{2^j \le n < 2^{j+1}} a_n^2 \right)^{1/2} .$$

If the sequence s_j is dominated by a decreasing summable sequence, it is known that (R), (S), (G) converge uniformly a.s.; in each case we write $F(t)$ for the sum of the series. Suppose $0 < \alpha < 1$. If

$$\overline{\lim_{j\to 0}}\, 2^{j\alpha} s_j > 0,$$

we have

$$\text{a.s.} \quad \forall\, t \ \overline{\lim_{h\to 0}} \frac{|F(t+h) - F(t)|}{|h|^{\alpha}} > 0 \qquad (D_\alpha)$$

for (R), (S), (G), and we have

$$\text{a.s.} \quad \forall\, t \ \overline{\lim_{h\to 0}} \frac{|F(t+h) - F(t)|}{|h|^{\alpha} \sqrt{\log \log \frac{1}{h}}} > 0 \qquad (Kh_\alpha)$$

only for (G). In the opposite direction, if

$$\overline{\lim_{j\to\infty}}\, 2^{j\alpha} s_j < \infty,$$

we have

$$\text{a.s.} \qquad \overline{\lim_{\delta\to 0}} \frac{\omega_F(\delta)}{\delta^{\alpha} \sqrt{\log \frac{1}{\delta}}} < \infty \qquad (L_\alpha)$$

for all series (R), (S), (G). Note that (D_α), (Kh_α), and (L_α) are only inequalities and that the superior limit in (Kh_α) and (L_α) is constant (because of the zero-one law, and the translation invariance of the Gaussian series (G)). If we suppose

$$a_n = n^{-\frac{1}{2}-\alpha}$$

and we restrict ourselves to series (G), it is not difficult to check that (Kh_α) and (L_α) are best possible: both superior limits are different from 0 and from ∞. We intend to prove that (D_α) is also best possible; in other words, we have the following analogue of (K) (for series (G)):

$$\text{a.s.} \quad \exists\, t \ \overline{\lim_{h \to 0}} \frac{|F(t+h) - F(t)|}{|h|^\alpha} < \infty. \qquad (K_\alpha)$$

Comparing (Kh_α) and (K_α), there are again ordinary points (with the law of the iterated logarithm) and slow points.

The purpose of the paper is to prove (K_α). To make the reading easier, we first give a proof of (D) and (K), based on the independence of increments for Brownian motion. Then we prove (K_α) for (G) under the assumption $a_n \leq n^{-\frac{1}{2}-\alpha}$, with an additional information of the Hausdorff dimension of slow points; in principle, this last part can be read independently.

THE CASE OF BROWNIAN MOTION

We use the following expansion of $W(t)$ on [0,1]:

$$W(t) = \xi_0 t + \Sigma\, \xi_{jn}\, \Delta_{jn}(t) \qquad (j = 0, 1, \ldots ; 0 \leq n < 2^j).$$

In this formula Δ_{jn} is the piecewise linear function carried by the interval

$$I_{jn} = [n2^{-j},\ (n+1)2^{-j}]$$

whose derivative is $2^{j/2}$ on the left part of I_{jn} and $-2^{j/2}$ on the right part of I_{jn}. The coefficients ξ_0, ξ_{jn} are independent Gaussian normalized random variables.

Given $\lambda > 0$, we write

$$p(\lambda) = P(|\xi_{jn}| > \lambda)$$

(it does not depend on (j,n)) and we say that I_{jn} is black if $|\xi_{jn}| > \lambda$. We divide the interval [0,1] dyadically, each dyadic interval generating two dyadic intervals at the following step, and we kill the black intervals at each step. This is a birth and death process, where the population at the time (= step) j consists of "white" intervals I_{jn} (white = living).

Suppose $p(\lambda) \geq \frac{1}{2}$. Then the population of white intervals vanishes a.s. when j is large. That means, each $t \in [0,1]$ belongs to infinitely many black I_{jn}, a.s. For such a black interval I_{jn} we have

$$|W(t_{jn}) - W(t'_{jn})| \geq |\xi_{jn}| \|\Delta_{jn}\|_{\infty} > \lambda 2^{-\frac{j}{2}-1},$$

where t_{jn} is the middle point and t'_{jn} one of the endpoints of I_{jn}; hence we have at least one of the inequalities

$$|W(t) - W(t_{jn})| > \lambda 2^{-\frac{j}{2}-2}$$

$$|W(t) - W(t'_{jn})| > \lambda 2^{-\frac{j}{2}-2}$$

together with

$$|t - t_{jn}| < 2^{-j-1}, \quad |t - t'_{jn}| < 2^{-j};$$

therefore,

$$\varlimsup_{h\to 0} \frac{|W(t+h) - W(t)|}{|h|^{1/2}} > \frac{\lambda}{4}.$$

This proves (D).

If $p(\lambda) < \frac{1}{2}$, the probability that the population of white intervals does not vanish is strictly positive; that is, the set E that remains white does not disappear. The complement of E consists of disjoint black intervals B, and it is easy to see that the function obtained from $W(t)$ by linear interpolation over the intervals B and keeping $W(t)$ on E belongs to the Lipschitz class $\Lambda_{1/2}$, with a constant depending only on λ. But this is not enough to prove (K).

Given $\lambda > 0$ and $\kappa = 0, 1, 2, \ldots$, let us write

$$p_{\kappa} = p_{\kappa}(\lambda) = P(2^{\kappa}\lambda < |\xi_{jn}| \leq 2^{\kappa+1}\lambda),$$

and let us say that I_{jn} is κ-black if $2^{\kappa}\lambda < |\xi_{jn}| \leq 2^{\kappa+1}\lambda$. If I_{jn} is a κ-black interval, we denote by $\bar{I}_{jn}$ the union of all I_{jm} with $|m-n| \leq 2^{3\kappa}$. We proceed with the dyadic construction of white intervals as before, except that we kill the $\bar{I}_{jn}$ at each step instead of the black I_{jn}. Consider

$$p = \sum_{\kappa=0}^{\infty} p_{\kappa}(1 + 2^{3\kappa+1})$$

and suppose λ small enough to have $p < \frac{1}{2}$. Then again the set E that remains white is not empty with a strictly positive probability. This can be seen in many ways; in particular, the method of the final part of this paper applies.

Suppose $t \in E$ and $s \in [0,1]$ such that

$$2^{-\nu} \leq |t-s| \leq 2^{-\nu+1}.$$

Let us write

$$F_j(.) = \sum_n \xi_{jn} \Delta_{jn}(.).$$

For $j \leq \nu$ we have

$$|F_j(t) - F_j(s)| \leq \lambda 2^{j/2}|t-s|;$$

therefore,

$$\sum_{j=0}^{\nu} |F_j(t) - F_j(s)| \leq C\lambda\, |t-s|^{1/2}$$

(C: absolute constant, not the same in different lines). For $j > \nu$ we have

$$|F_j(t)| \leq \lambda 2^{-j/2-1}$$

and

$$|F_j(s)| \leq \lambda 2^{\kappa} 2^{-j/2-1}$$

if κ is the largest integer such that $|s-t| \geq 2^{3\kappa}2^{-j}$. Since $|s-t| \leq 2^{-\nu+1}$, we have $\kappa \leq \frac{j+1-\nu}{3}$ and

$$|F_j(s)| \leq c\lambda 2^{-\frac{\nu}{2}} 2^{-\frac{j-\nu}{6}};$$

therefore,

$$\sum_{j=\nu+1}^{\infty} |F_j(t) - F_j(s)| \leq c\lambda\, |t-s|^{1/2}.$$

There exists t (namely, $t \in E$) such that

$$|W(t+h) - W(t)| \leq c\lambda\, |h|^{1/2}$$

with a strictly positive probability. Because of the zero-one law there exists a.s. a t such that

$$W(t+h) - W(t) = O(|h|^{1/2}) \qquad (h \to 0),$$

which proves (K).

SLOW POINTS FOR SERIES (G): THEOREM AND PROOF

It will be convenient to write series (G) in the complex form (the result will be slightly stronger) and to work on the circle $\mathbf{T} = \mathbf{R}/\mathbf{Z}$. From now on, ζ_n, η_n, ... are independent Gaussian complex normalized random variables ($n = 1, 2, \ldots$).

Theorem. If

$$0 \leq a_n \leq n^{-\frac{1}{2}-\alpha} \qquad (0 < \alpha < 1;\ n = 1, 2, \ldots),$$

the random function $F(t)$ defined by the series

$$\sum_{n=1}^{\infty} a_n \zeta_n \exp(2\pi i n t) \tag{G'}$$

satisfies a.s.

$$F(t+h) - F(t) = O(|h|^{\alpha}) \qquad (h \to 0)$$

for a (random) set of t's, of Hausdorff dimension 1.

Proof. When $2^j \leq n < 2^{j+1}$, we write

$$a_n = \alpha_j \cos \theta_n, \quad \alpha_j = 2^{-j(\alpha+\frac{1}{2})}.$$

Then (G') is the half sum of the series

$$\sum_{j=0}^{\infty} \alpha_j \sum_{2^j \leq n < 2^{j+1}} (\zeta_n \cos \theta_n + \eta_n \sin \theta_n) \exp(2\pi int) \qquad (G^+)$$

$$\sum_{j=0}^{\infty} \alpha_j \sum_{2^j \leq n < 2^{j+1}} (\zeta_n \cos \theta_n - \eta_n \sin \theta_n) \exp(2\pi int). \qquad (G^-)$$

Each of the series (G^+) and (G^-) is of the form

$$\sum_{j=0}^{\infty} \alpha_j S_j(t),$$

where $S_j(t)$ is a random trigonometric polynomial whose spectrum is the interval $[2^j, 2^{j+1} - 1]$ and coefficients are independent complex normalized Gaussian variables. When necessary, we write $S_j^+(t)$ and $S_j^-(t)$, corresponding to (G^+) and (G^-). We remark that the values of $S_j(t)$ on 2^j equidistant points on the circle $\mathbf{T}$ are independent complex Gaussian variables, with the same L^2-norm $2^{j/2}$.

Let us write

$$t_{jm} = m2^{-j-1} \qquad (m = 1, 2, \ldots, 2^{j+1})$$

$$Z_{jm} = 2^{-j/2} S_j(t_{jm}).$$

The Z_{jm} are normalized complex Gaussian variables, and jointly Gaussian. Two of them, say, Z_{jm} and $Z_{j'm'}$, are independent whenever either $j \neq j'$ or $m - m'$ is even. When necessary, we write Z_{jm}^+ and Z_{jm}^-, corresponding to S_j^+ and S_j^-.

We choose an integer $N > 1/\alpha$. Given $\lambda > 0$, we define a sequence of random sets B_j (respectively, B_j^+, B_j^-) associated with the Z_{jm}'s (respectively, Z_{jm}^+, Z_{jm}^-). For each j, B_j is the union of the $\bar{I}_{jm}$ defined in the following way. When $|Z_{jm}| < \lambda$, $\bar{I}_{jm}$ is empty. When

$$\lambda 2^{\eta} \le |Z_{jm}| < \lambda 2^{\kappa+1} \qquad (\kappa = 0, 1, \ldots),$$

$\bar{I}_{jm}$ is the interval

$$[t_{jm} - 2^{N\kappa}2^{-j-1},\ t_{jm} + 2^{N\kappa}2^{-j-1}]$$

considered on the circle $\mathbf{T}$. We write

$$p_{\kappa} = p_{\kappa}(\lambda) = P(\lambda 2^{\kappa} \le |Z_{jm}| < \lambda 2^{\kappa+1})$$

and we suppose λ large enough, so that

$$\sum_{\kappa=0}^{\infty} 2^{N\kappa+3} \sqrt{p_{\kappa}} < \varepsilon$$

$\left(0 < \varepsilon < \frac{1}{2}\text{ given, } \lambda > \lambda(\varepsilon)\right)$. Given $k = 0, 1, \ldots,$ we write

$$N_k = \bigcup_{j \ge k} (B_j^+ \cup B_j^-)$$

$$E_k = \mathbf{T} \backslash N_k$$

$$E = \lim E_k.$$

Our purpose is to prove that 1°) a.s. E is nonvoid, and more precisely, the Hausdorff dimension of E is as near 1 as we want (by choice of λ), and 2°) for each t in E we have

$$F(t+h) - F(t) = O(|h|^{\alpha}) \qquad (h \to 0).$$

Proof of 1°). Let us start from stage k. There are 2^{k+1} dyadic intervals w limited by the points t_{km}. For each $\kappa \le \frac{k}{N}$ (and each choice + or -), the number of intervals $\bar{I}_{km}$ (m even, respectively m odd) of length $2^{N\kappa}2^{-k}$ is a binomial random variable of distribution $\mathscr{B}(2^k, p_{\kappa})$. This number is dominated by

$$(p_{\kappa} + (p_{\kappa}(1 - p_{\kappa}))^{1/2})2^k$$

except on an event of probability 2^{-k} (Bienaymé-Tchebycheff). Therefore, the number of dyadic intervals w contained in $B_k^+ \cup B_k^-$ does not exceed

$$8.2^k \sum_{\kappa \leq k/N} (p_\kappa + p_\kappa(1 - p_\kappa)^{1/2})2^{N\kappa} < \varepsilon\, 2^{k+1}$$

except on an event of probability $\frac{1}{N} k\, 2^{-k}$. Killing these dyadic intervals, it remains at least $(1-\varepsilon)2^{k+1}$ intervals w. At stage $k+1$, each of them generates two dyadic intervals limited by the points $t_{k+1,m}$, and we get W_{k+1} such dyadic intervals of length 2^{-k-2}. At stage j, we have W_j dyadic intervals of length 2^{-j-1}, whose endpoints are taken among the points t_{jm}. The number of such intervals contained in $B_j^+ \cup B_j^-$ does not exceed εW_j except on an event of probability $\frac{1}{N} j\, 2^{-j}$. Killing these intervals, it remains at least $(1-\varepsilon)W_j$ intervals that generate by division intervals of the following stage. Therefore,

$$W_{k+1} \geq 2(1-\varepsilon)2^{k+1}, \ \ldots, \ W_{j+1} \geq 2(1-\varepsilon)W_j, \ \ldots$$

except on an event of probability

$$\sum_{j=k}^{\infty} \frac{1}{N} j\, 2^{-j} .$$

Therefore, $P(E_k = \emptyset)$ is as small as we want; therefore, $E \neq \emptyset$ a.s.

We now sketch the proof that the Hausdorff dimension of E is close to 1 if λ is large.

Suppose we are given a Cantor set K constructed on $[0,1]$ with ratio of dissection 2^{-M}, so that the Hausdorff dimension of K is $\frac{1}{M}$. Instead of all dyadic subintervals of $[0,1]$ we consider only the dyadic intervals containing a portion of K; there are 2^i such intervals of length 2^{-Mi}. The i-th stage consists now of W_i intervals of this type. If ε is small enough, the proportion of these W_i intervals killed by the $B_j^+ \cup B_j^-$ when j runs from $Mi - 1$ to $(M+1)i - 1$ does not exceed 1/3, except on an event of very small probability (depending on i). Each of the remaining intervals generates two intervals of length $2^{-(M+1)^i}$ intersecting K; therefore, $W_{i+1} > \frac{4}{3} W_i$ with a large probability, and finally

$$\text{a.s.} \quad K \cap E \neq \emptyset.$$

This result is interesting in itself and easy to generalize. Using a small modification, we shall derive

$$\text{a.s.} \quad \dim E \geq 1 - \frac{1}{M}.$$

Given a Steinhaus sequence $\omega_j (j = 0, 1, \ldots)$ (the ω_j's are independent random variables equally distributed on $\mathbf{T}$) independent from the ζ_n and η_n, we consider the algebraic sums

$$B_{j,\omega} = B_j + \left[2^{-j-1}(\omega_j - 1),\ 2^{-j-1}\omega_j\right]$$

and we write

$$E_\omega = \mathbf{T}\backslash\overline{\lim}\ (B^+_{j,\omega} \cup B^-_{j,\omega}).$$

E_ω is contained in E, but the same argument as before shows that given K, M, and λ large enough $\big(\lambda > \lambda(M)\big)$, we have

$$\text{a.s.} \quad K \cap E_\omega \neq \emptyset. \tag{*}$$

Now the distribution of the $B_{j,\omega}$ is invariant under translation on $\mathbf{T}$. The same is true for E_ω. Since $K = -K$, (*) reads

$$\text{a.s.} \quad \text{meas}(K + E_\omega) = 1. \tag{**}$$

Because of the special structure of K it is easy to check that

$$\dim(K + E_\omega) \leq \dim E_\omega + \dim K = \dim E_\omega + M^{-1}.$$

This, compared to (**), proves $\dim E_\omega + M^{-1} \geq 1$. Therefore

$$\text{a.s.} \quad \dim E \geq 1 - \frac{1}{M}.$$

This ends the proof of 1°).

Proof of 2°). Here we can suppose

$$F(t) = \sum_{j=0}^{\infty} \alpha_j S_j(t),\ \alpha_j = 2^{-j(\alpha+\frac{1}{2})}.$$

We use a representation formula of $S_j(t)$ from the values $S_{jm} = S(t_{jm})$ $(m = 1, 2, \ldots, 2^{j+1})$ (sampling device). We have

$$S_j(t) = \sum_{1\le m\le 2^{j+1}} s_{jm} W_j(t - t_{jm}) \tag{i}$$

when $W_j(t)$ is chosen in the following way:

$$W_j(t) = \sum_{n\in\mathbf{Z}} v\big(2^j(t-n)\big)$$

$$v(t) = \int_{\mathbf{R}} \exp(2\pi iut)\hat{v}(u)\,du$$

where $\hat{v}(u)$ is a C^∞ function with compact support and $\hat{v}(u) = 1$ on $[1,2]$. Let us define $\eta(t)$ as the smallest even function decreasing on $[0,\infty]$ that majorizes

$$\sup\Big(|v(t)|,\ \sup_{0<\delta<2} \frac{1}{|\delta|}|v(t+\delta) - v(t)|\Big).$$

We observe that

$$\eta(t) = O\big(|t|^{-A}\big) \qquad (t \to \infty) \tag{ii}$$

for all $A > 0$. We write

$$H_j(t) = \sum_{n\in\mathbf{Z}} \eta(t - 2^j n)$$

so that

$$\begin{cases} |W_j(t)| \le H_j(2^j t) \\ |W_j(t+h) - W_j(t)| \le 2^j\,|h|\,H_j(2^j t) \quad \text{if} \quad 2^j|h| < 2. \end{cases} \tag{iii}$$

Suppose $2^{-\nu} \le |h| < 2^{-\nu+1}$. We write

$$F(t+h) - F(t) = \sum_{j=0}^{\nu} + \sum_{j=\nu+1}^{\infty} \alpha_j\big(S_j(t+h) - S_j(t)\big).$$

Consider first $j \le \nu$. Since $2^j|h| < 2$, using (i) and (iii) we have

$$|S_j(t+h) - S_j(t)| \le 2^j|h| \sum_{1\le m\le 2^{j+1}} |s_{jm}| H_j\big(2^j(t - t_{jm})\big).$$

Suppose $t \in E_0$. We divise the last sum according to the values of $|s_{jm}|$. We denote by $\Sigma^{(\kappa)}$ the sum corresponding to

$$\lambda 2^{\kappa} \leq |z_{jm}| < \lambda 2^{\kappa+1} \qquad (z_{jm} = 2^{-j/2} s_{jm})$$

($\kappa = 0, 1, \ldots$) and by Σ' the sum corresponding to $|z_{jm}| < \lambda$. We have

$$\begin{aligned} |\Sigma'| &\leq 2^{3j/2} \lambda |h| \sum_{1 \leq m \leq 2^{j+1}} H_j(2^j(t - t_{jm})) \\ &= 2^{3j/2} \lambda |h| \sum_{m \in \mathbf{Z}} \eta(t - \tfrac{m}{2}) \\ &< C \lambda 2^{3j/2} |h|. \end{aligned}$$

Remember that C is a constant, not the same for different lines.

$$\begin{aligned} \Sigma^{(\kappa)} &\leq 2^{3j/2} \lambda |h| 2^{\kappa+1} \sum_{|t_{jm}-t| > 2^{N\kappa} 2^{-j-1}} H_j(2^j(t - t_{jm})) \\ &\leq 2^{3j/2} \lambda |h| 2^{\kappa+1} \sum_{|m| \geq 2^{N\kappa}} \eta(\tfrac{m}{2}) \\ &\leq C \lambda 2^{3j/2} 2^{-\kappa} |h| \end{aligned}$$

because of (ii). Therefore,

$$\alpha_j |S_j(t+h) - S_j(t)| \leq C \lambda 2^{j(1-\alpha)} |h|$$

and

$$\sum_{j=0}^{\nu} \alpha_j |S_j(t+h) - S_j(t)| \leq C \lambda |h|^{\alpha}.$$

Consider now $j > \nu$, and again $t \in E_0$. We have

$$|S_j(t)| \leq \sum_{1 \leq m \leq 2^{j+1}} |s_{jm}| H_j(2^j(t - t_{jm}))$$

$$|S_j(t+h)| \leq \sum_{1 \leq m \leq 2^{j+1}} |s_{jm}| H_j(2^j(t + h - t_{jm})).$$

We decompose again these sums according to the values of Z_{jm}, in the form

$$\sum' + \sum_{\kappa=0}^{\infty} \sum^{(\kappa)}.$$

For $S_j(t)$ we obtain easily, using (i), (ii), and (iii),

$$|S_j(t)| \le C\, 2^{j/2}\lambda + \sum_{\kappa=0}^{\infty} C\, 2^{j/2}\lambda 2^{-\kappa}$$

$$\le C\, \lambda 2^{j/2}.$$

For $S_j(t+h)$ some more care is needed. Now

$$|\Sigma^{(\kappa)}| \le 2^{j/2}\lambda 2^{\kappa+1} \sum_{|t_{jm}-t|>2^{N\kappa}2^{-j-1}} H_j(2^j(t+h-t_{jm})).$$

In this sum

$$2^j(t - t_{jm}) \ge 2^{N\kappa-1}$$

$$|2^j h| \le 2^{j-\nu+1}.$$

When $N\kappa > j - \nu + 3$,

$$|2^j(t+h-t_{jm})| \ge 2^{N\kappa-2};$$

therefore,

$$|\sum^{(\kappa)}| \le C\, 2^{j/2}\lambda 2^{\kappa}. \sum_{m\ge 2^{N\kappa-2}} \eta(\tfrac{m}{2})$$

$$\le C\, 2^{j/2}\lambda 2^{-\kappa},$$

the last inequality deriving from the property (ii) of $\eta(t)$. On the other hand, when $N\kappa \le j - \nu + 3$, we simply write

$$|\Sigma^{(\kappa)}| \le C\, 2^{j/2}\lambda 2^{\kappa}.$$

We obtain

$$|S_j(t+h)| \le C\, \lambda 2^{j/2}\left(\sum_{\kappa\le\frac{j-\nu+3}{N}} 2^{\kappa} + \sum_{\kappa\ge\frac{j-\nu+3}{N}} 2^{-\kappa}\right);$$

therefore,

$$\alpha_j |S_j(t+h)| \leq C\,\lambda 2^{-j\alpha} 2^{\frac{j-\nu}{N}}.$$

Since we supposed $N > \frac{1}{\alpha}$, we get

$$\sum_{j=\nu+1}^{\infty} \alpha_j |S_j(t+h)| \leq C\,\lambda 2^{-\nu\alpha}$$

as well as

$$\sum_{j=\nu+1}^{\infty} \alpha_j |S_j(t)| \leq C\,\lambda 2^{-\nu\alpha}.$$

This proves

$$|F(t+h) - F(t)| < C\,\lambda |h|^{\alpha}$$

for all h when $t \in E_o$. For $t \in E_k$ we have the same result, but only for h small enough. In any case for $t \in E$

$$F(t+h) - F(t) = O(|h|^{\alpha}),$$

which proves 2°) and finishes the proof of the theorem.

REFERENCES

1. Billard, P. Séries de Fourier aléatoirement bornées, continues, uniformément convergentes. *Studia Math.* 22 (1963): 309-329.

2. Billard, P. Séries de Fourier aléatoirement bornées, continues, uniformément convergentes. *Ann. Scient. Ec. Norm. Sup.* 82 (1965): 131-179.

3. Dvoretzky, A. On the oscillation of the Brownian motion process. *Israël J. Math.* 1 (1963): 212-214.

4. Kahane, J.-P. Some random series of functions. *Heath Mathematical Monographs* (1968), 184 pp.

5. Kahane, J.-P. Sur l'irrégularité locale du mouvement brownien. *C. R. Acad. Sci. Paris* 278 (1974): 331-333.

6. Kahane, J.-P. Sur les zéros et les instants de ralentissement du mouvement brownien. *C. R. Acad. Sci. Paris* 282 (1976): 431-433.

7. Marcus, M. B., and Pisier, G. *Random Fourier Series with Applications to Harmonic Analysis*. Princeton, N.J.: Princeton Univ. Press, 1981.

8. Orey, S., and Taylor, S. J. How often on a Brownian path does the law of iterated logarithm fail? *Proc. London Math. Soc.* 28 (1974): 174-192.

9. Satô-Izumi, M. Lacunary Fourier series. *I. Proc. Japan Acad.* 31 (1955): 402-405.

10. Taylor, S. J. Regularity and irregularities on a Brownian path. *Ann. Inst. Fourier (Grenoble)* 24 (1974): 195-204.

11. Paley, R.E.A.C., and Zygmund, A. On some series of functions (1)(2)(3). *Proc. Cambridge Phil. Soc.* 26 (1930): 337-357; 26 (1930): 458-474; 28 (1932): 190-205.

12. Paley, R.E.A.C.; Wiener, N.; and Zygmund, A. Notes on random functions. *Math. Zeit.* 37 (1932): 647-668.

ADDED IN PROOF. More information on Brownian slow points can be found in an unpublished paper of Burgess Davis.

NOTES ON TRIGONOMETRIC POLYNOMIALS

S. K. Pichorides
Nuclear Research Center "Demokritos," Athens

Some rather unrelated problems concerning trigonometric polynomials will be the subject of this article. A common feature of these problems is that they have been settled recently. More specifically, we propose to discuss (1) the conjecture of Littlewood on exponential sums (settled by Konjagin and MacGehee, Pigno, and Smith independently), (2) a conjecture of Hayman on real trigonometric polynomials having all their roots real (settled by Kristiansen), and (3) another conjecture of Littlewood on polynomials with unimodular coefficients (settled by Körner and Kahane).

We shall use the standard notation $\| \ \|_p$ for the L^p norm, the normalized Lebesgue measure $dx/2\pi$, and we shall omit the limits of integration when they are 0 and 2π. We shall write sp g and $\tilde{g}$ for the spectrum and the conjugate of a 2π-periodic function g. C will denote an absolute positive constant not necessarily the same in all its occurrences.

THE CONJECTURE OF LITTLEWOOD ON EXPONENTIAL SUMS AND ITS ORIGIN

It is now more than fifty years since Hardy and Littlewood, in their effort to characterize those functions of L^q, $q > 1$, that remain in L^q after changing the order and argument of their Fourier coefficients, obtained the following remarkable inequality:

$$\| 1 + 2 \cos n_1 x + \cdots + 2 \cos n_N x \|_q \leq \| 1 + 2 \cos x + \cdots + 2 \cos Nx \|_q , \qquad (1)$$

$q = 2, 4, \ldots$

where $n_1, \ldots, n_N$ are distinct positive integers. In fact, the argument they gave was not conclusive. A few years later, the above inequality, which is a special case of a more general one known as Hardy-Littlewood rearrangement theorem, was slightly extended and correctly proved by Gabriel. It appears that it is still unknown if (1) holds for all $q \geq 2$ (that is, not necessarily even integers but still with constant 1). What is of interest to us here is the possible extension of (1) for $q < 2$, and above all, for $q = 1$.

Let's write $f(x) = 1 + 2 \cos n_1 x + \cdots + 2 \cos n_N x$ and $g(x) = 1 + 2 \cos x + \cdots + 2 \cos N_x$. Then (1) becomes $\| f \|_q \leq \| g \|_q$, $q = 2, 4, \ldots$. If we plot the graphs of $\| f \|_q^q$, $\| g \|_q^q$, as functions of q, we obtain two logarithmically convex curves that intersect at $q = 2$, and the curve $\| g \|_q^q$ stays above the curve $\| f \|_q^q$ when $q = 2, 4, \ldots$.

The above argument, due to Hardy and Littlewood, suggests the conjectures $\| f \|_q \leq \| g \|_q$, $q \geq 2$, and $\| f \|_q \geq \| g \|_q$, $q \leq 2$; in particular, the inequality $\| f \|_1 \geq \| g \|_1$ appears to be plausible. A slightly weaker version of this inequality is known as Littlewood's conjecture. It appeared explicitly for the first time in a joint paper of Hardy and Littlewood in 1948. This conjecture reads as follows:

$$\| \exp(in_1 x) + \cdots + \exp(in_N x) \|_1 \geq C \log N. \qquad (2)$$

Until last summer the best known result was a lower bound of the order of $\log N/(\log \log N)^2$. The proof of this result was based on a dyadic-type partition of the spectrum $n_1, \ldots, n_N$ coupled with an analytic inequality for H^1 functions and results from the theory of lacunary trigonometric series. Katznelson observed that a "little" injection of probability in this argument leads to $\log N/\log \log N$ and Kahane went even further up to $\log N/(\log \log N)^{0.5+\varepsilon}$, $\varepsilon > 0$. However, $\log N/(\log \log N)^{0.5}$ appeared to be the limit of this method (mainly because of its unsatisfactory analytic part).

Last November I was informed that S. Konjagin had proved the conjecture. I had access to an incomplete version of his paper (thanks to a letter of Kructsef). It appears that Konjagin knew stronger analytic inequalities pertaining to the problem, which, when combined with the above-mentioned dyadic-type decomposition of the spectrum, lead to the proof of the conjecture. This work was due to appear in the April issue of *Izvestia* (Acad. Sc. USSR).

In January 1981 I had the happy surprise to receive a letter from B. Smith containing a beautiful short proof of the conjecture. This proof was due to appear in the *Annales* in a joint paper of McGehee, Pigno, and Smith. We devote the next section to this proof.

References up to 1976 can be found in [5]. For the more recent results, see the references in the coming article of McGehee, Pigno, and Smith.

We close this section with a question that arose in the dyadic-type decomposition approach.

Let $F(x) = \exp(in_1 x) + \cdots + \exp(in_N x)$, *and* $E \subset [0, 2\pi]$ *with* $|E| = \pi$. *Is it true that* $\int_E |F| \geq C > 0$?

The weaker result $\int_E |F| \geq G/\|F\|$ led to the lower bound $(\log N)^{0.5}$ for $\|F\|_1$ and an analogous inequality for any set $E \subset [0, 2\pi]$ led to $\log N/(\log \log N)^2$. These weaker results depend only on the fact that the first Fourier coefficient of F is 1.

THE CONJECTURE OF LITTLEWOOD (CONTINUED)

The MGPS (McGehee, Pigno, Smith) proof of (2) can be considered as a rather remote descendent of a method introduced by P. Cohen twenty years ago (this method gave the first positive result toward the solution of the problem). However, the new elements in the proof deserve, in my opinion, the characterization of ingenius.

MGPS derive (2) from the following stronger result: "There is a function bounded by 1 in absolute value with Fourier coefficients having real parts exceeding C/k at n_k, $k = 1, \ldots, N$."

If we aim at (2) only, and not at the construction of the above-mentioned bounded function, we obtain the following, apparently neither stronger nor weaker, result:

Theorem. Let $F(x) = a_0 + a_1 \exp(ix) + \cdots + a_N \exp(iNx)$ *and let* $f_1(x) = a_k \exp(ikx) + \cdots + a_N \exp(iNx)$, $0 \leq k \leq N$, *be a tail of* F *such that* $\| f_1 \|_2 > 0.9 \| F \|_2$. *Let* f_2 *be defined in the same way, with* F *replaced by* $F - f_1$, *and similarly for* $f_3, \ldots, f_r$ *until* F *is exhausted. Then*

$$\| F \|_1 \geq 0.1\{(\| f_1 \|_2^2 / \| f_1 \|_\infty) + \cdots + (\| f_r \|_2^2 / \| f_r \|_\infty)\}. \quad (3)$$

If F is the exponential sum in (1), then the above splitting is that of MGPS, and (3) reduces to (2) if we observe that in this case $\| f_k \|_2^2 = \| f_k \|_\infty$ and make the most economical choice of the f_k's. It should be instructive to compare (3) with the trivial inequality (valid for any g)

$$\| g \|_1 \geq \| g \|_2^2 / \| g \|_\infty .$$

This theorem is a corollary of the MGPS proof (in fact its proof is essentially the same with that of MGPS).

Proof. Let n_k be the highest frequency in f_k. We prove by induction the existence of functions g_k, $k = 1, \ldots, r$, having the properties:

(a) $\| g_k \| \leq 1$,

(b) $\mathrm{sp}\ g_k \subset (-\infty, n_{r+1-k}]$,

(c) $$\int g_k(x)F(-x)\,dx \geq 0.1 \sum_{\mathrm{sp}\, f_n \subset (-\infty, n_{r+1-k}]} \left(\frac{\| f_n \|_2^2}{\| f_n \|_\infty} \right).$$

We take $g_1 = f_r / \| f_r \|_\infty$, in which case (a), (b), and (c) are trivial, and define

$$g_{k+1}(x) = \{\exp[-(1/4)h(-x)]\}g_k(x) + \left(\frac{1}{5}\right) f_{r-k}(x)/\| f_{r-k} \|_\infty ,$$

where

$$h(x) = \{|f_{r-k}|/\| f_{r-k} \|_\infty\} + i\{|f_{r-k}|/\| f_{r-k} \|_\infty\}^{\sim}.$$

(a) This follows from the elementary inequality $\exp(-\frac{1}{4}x) + \frac{1}{5}x \leq 1$, $0 \leq x \leq 1$.

(b) This is trivial.

(c) This is one of the most important ingredients of the MGPS proof.

We have: $$\int g_{k+1}(x)F(-x)\,dx = \int g_k(x)F(-x)\,dx + \left(\frac{1}{5}\right)\frac{\| f_{r-k} \|_2^2}{\| f_{r-k} \|_\infty} + A,$$

where

$$A = \int [\exp(-\frac{1}{4}h(-x) - 1]g_k(x)F(-x)\,dx.$$

In the definition of A, we may obviously replace F by $f_{r+1-k} + \cdots + f_r$. Using Schwarz's inequality, we have

$$|A| \leq \| \exp[-\frac{1}{4}h(-x)] - 1\|_2 \, \| f_{r+1-k} + \cdots + f_r \|_2 \leq$$

$$\leq 0.1 \| \exp[-\frac{1}{4}h(-x)] - 1\|_2 \, \| f_{r-k}\|_2 .$$

Examining separately the cases $|h| > 2$ and $h \leq 2$, we see that

$$|\exp[-\frac{1}{4}h(-x)] - 1| \leq \frac{1}{4}|h(-x)|.$$

(The case $|h| > 2$ is trivial. If $|h| < 2$, use, say, the MacLaurin series of $\exp x$.)

Since $\|h\|_2 \leq \sqrt{2}\,(\|f_{r-k}\|_2 / \|f_{r-k}\|_\infty)$, we obtain

$$|A| \leq 0.1 \cdot \frac{\sqrt{2}}{4}\left(\frac{\|f_{r-k}\|_2}{\|f_{r-k}\|_\infty}\right)^2 \leq 0.1\left(\frac{\|f_{r-k}\|_2}{\|f_{r-k}\|_\infty}\right)^2 .$$

Collecting the above results, we obtain (c).

On observing that $\|F\|_1 \geq \int g_r(x)F(-x)\,dx$, we get (3).

A CONJECTURE OF HAYMAN ON POLYNOMIALS HAVING ALL THEIR ROOTS ON THE UNIT CIRCUMFERENCE

Let $P(z) = a_0 + a_1 z + \cdots + a_n z^n$, $a_0, \ldots, a_n$ complex numbers, and assume that all the roots of P lie on the unit circumference. We also normalize P so that $\|P\|_\infty = 1$ (= maximum of P on $|z| = 1$). Among the many interesting problems concerning this class of polynomials, we shall examine one, due to Hayman, that states that $|a_k| \leq 0.5$, $k = 1, \ldots, n$.

Saff and Sheil-Small proved the validity of this conjecture in all but one case [7] (see also [6]). The exceptional case was the one corresponding to the middle coefficient when n is even. This case has been settled recently by Kristiansen and, as we shall see in a moment, is in fact equivalent to the whole conjecture.[†]

To get a better (I hope) feeling of the problem, let's translate it to another one concerning real trigonometric polynomials.

[†] I am indebted to J. P. Kahane, who pointed out to me the work of Kristiansen.

We assume $n = 2m$ (this is the only interesting case). Expressing the coefficients of P in terms of its roots and using the fact that the latter have modulus 1, we see easily that

$$a_{n-k} = u\bar{a}_k, \text{ for some } u \text{ with } |u| = 1,\ k = 1, \ldots, n,$$

and hence

$$P(z) = u^{0.5} z^m Q(z),$$

where the function $q(x) = Q(\exp ix)$ is a real trigonometric polynomial of degree m having all its roots real.

The conjecture of Hayman states that

$$|\hat{q}(k)| \leq 0.5 (= 0.5 \| q \|_\infty),\ k = 0, \ldots, m, \tag{4}$$

and the case $k = 0$ is the exceptional one mentioned above.

That $|\hat{q}(0)| = |\int q|$ is considerably smaller than $\| q \|_\infty$ is intuitively evident. The graph of q lies between $y = 1$ and $y = -1$, and it is forced to touch the real axis the maximum number of times permitted by its degree. As a consequence, we expect a mean value of q less than 1. The trivial example $q(x) = 0.5(1 + \cos mx)$ shows that (4) is best possible.

That the exceptional case is the only interesting one has been shown by Saff and Sheil-Small as follows:

Assume that (4) is true when $k = 0$ and apply it to q^2. We obtain

$$\int q^2 = |\hat{q}(0)|^2 + 2\{|\hat{q}(1)|^2 + \cdots + |\hat{q}(m)|^2\} \leq 0.5, \tag{5}$$

from which (4) follows immediately for all $k \neq 0$.

The same argument shows that (5), if true, suffices to show the conjecture in all but the exceptional case. This was the key idea in the work of Saff and Sheil-Small, who proved, in fact, a more general inequality [7]. For an even stronger inequality, see [6]. Since our goal here is the presentation of Kristiansen's argument, we take (5) for granted and refer to the above-mentioned articles for its proof.

We finish this section with an outline of Kristiansen's proof.

By a simple continuity argument, we know that an optimal f, that is, an f with $|\int f|$ as small as possible, exists. If we show that this optimal f is nonnegative, the proof is complete. Indeed, a classical theorem of F. Riesz and Fejer asserts that in this case $f = |q|^2$, where q is a trigonometric polynomial that can be taken real and satisfying the same conditions as f. Since $\int f = \int |q|^2$, (5) yields the desired inequality.

Thus, the problem reduces to showing that the optimal f can be taken nonnegative, assuming, as we may, that $\int f > 0$.

To show this we assume $f \ngeq 0$ and we look for a polynomial h such that $f + \varepsilon h$ contradicts the optimality of f for sufficiently small ε.

At this point the choice $h = f''$, which has been tried unsuccessfully before, shows that f has at least one root of even multiplicity. This is not enough; we have to show that *all* the roots are of even multiplicity. Here is Kristiansen's choice.

We take $h = f'g$, where

$$g(x) = \sum A_j\left(\cot\frac{a}{2} + \cot\frac{x - r_j}{2}\right),\ 0 < a < 2\pi,\ A_j > 0.$$

Here r_j are those roots of f' where f is negative and a, A_j are so chosen that all the roots of g lie in $\{x : f(x) \geq 0\}$.

A moment's thought shows that the graph of g consists of branches decreasing from $+\infty$ to $-\infty$ in each interval (r_{j-1}, r_j) and the choice of a and A_j's guarantees that $f'g$, which is obviously a polynomial of the same degree as f, is positive when $f < 0$. Combining these remarks (and a very careful counting of roots), we see that the maximum of $f + \varepsilon h$ is $1 + o(\varepsilon)$, while its mean value exceeds that of f by at least

$$\varepsilon \int_{f\geq 0} f'g = \varepsilon \int_{f\geq 0} f(-g').$$

Since the last integral is positive and independent of ε, we have obtained the desired contradiction.

It remains to show, and this appears to be the heart of the proof, that the above-mentioned choice of a and A_j is possible. To this end we order the r_j's and choose any c_j such that $r_j < c_j < r_{j+1}$ and $f(c_j) > 0$. We have to show that there are A_j's and a such that

$$\sum_j A_j \cot \frac{c_k - r_j}{2} = -\left(\sum A_j\right)\cot \frac{a}{2}, \quad k = 1, 2, \ldots, \quad A_j > 0.$$

We write $b_{kj} = \cot \dfrac{c_k - r_j}{2}$ and observe that this amounts to showing that there is a number A such that the system

$$\sum_j b_{kj} A_j = A$$

has a solution in positive A_j's. It turns out that this is always possible (see [3'] for the inductive proof). We note here that the validity of this assertion depends only on the monotonicity of the function cotangent.

A PROBLEM ON POLYNOMIALS WITH UNIMODULAR COEFFICIENTS

We denote by F_n the class of polynomials f of the form

$$f(x) = a_0 + a_1 \exp(ix) + \cdots + a_n \exp(inx), \quad |a_0| = \ldots = |a_n| = 1.$$

There are many interesting problems related to this class. We isolate and comment on one of them posed by Littlewood fifteen years ago: Find a sequence $f_n \in F_n$ such that

$$C_1\sqrt{n} \leq |f_n(x)| \leq C_2\sqrt{n}, \quad 0 \leq x \leq 2\pi. \tag{6}$$

This kind of problem attracted the interest of mathematicians long ago. In one of their earliest papers, Hardy and Littlewood solved half of the above-mentioned problem. They found a sequence $f_n \in F_n$ satisfying only the second inequality of (6). We note here that the well-known Rudin-Shapiro polynomials provide the most elegant solution in this case.

Four years ago, Byrnes almost solved the whole problem. We say almost because a small number $(\sim O(n^{3/4}))$ of the first and the last coefficients of his polynomials were not unimodular (their moduli were 1/2).

We owe to T. Körner the affirmative answer to this conjecture. Körner had the happy idea to "randomize." More precisely, he observed, and very neatly proved, that by "unimodulizing," in a carefully chosen random way, the bad coefficients of Byrnes's polynomials, he could obtain a set of polynomials that certainly contained one satisfying (6).

It appears that J. P. Kahane was not satisfied with this result. (I do know that he greatly liked it; what I mean is, he wanted more.) So he strengthened it in a way that proved and disproved quite a number of conjectures (for example, one of Erdös's asserting the existence of an absolute constant $A > 1$ such that $\|f_n\|_\infty > A\sqrt{n}$, for all $f_n \in F_n$). His probabilistic proof borrowed some of Körner's ideas, but the rest was a "tour de force." Here is his result.

There is a sequence $f_n \in F_n$ *such that*

$$f_n(x) = \sqrt{n}(1 + g_n(x)), \text{ with } \|g_n\|_\infty \to 0 \text{ as } n \to \infty. \tag{7}$$

Proofs, related results, and more on the history of the subject can be found in [1] and the references cited there.

We conclude that F_n is now less saturated with open problems, and this is our excuse for adding another one: *How fast can* $\|g_n\|_\infty \to 0$ *in* (7)?

The proof of (7) shows that there is a sequence f_n such that $\|g_n\|_\infty = O(n^{-(1/17)}\log n)$. On the other hand, the simple fact that the last Fourier coefficient of $|f_n|^2$ has absolute value 1 shows that in any case $\|g_n\|_\infty \geq C/n$.

There is, admittedly very slim, evidence that the answer to our question is: Not faster than $n^{-(1/2)}$. To see this, we show that if $f_n \in F_n$ has real coefficients, then $\| g_n \|_\infty > C/\sqrt{n}$. Incidentally, it is not known if there is $\{f_n\}$ with real coefficients satisfying (7) or even (6). Körner and Kahane's construction leads to polynomials with complex coefficients.

We use the same argument as before. In the present case the Fourier coefficients of $|f_n|^2$ are sums of +1 or -1 and hence, depending on the parity of n, either the even or the odd coefficients are sums of an odd number of +1 or -1. We conclude that these coefficients are not less than 2 in modulus.

It follows that $\| |f_n|^2 - n \|_\infty \geq \| |f_n|^2 - n \|_2 \geq C\sqrt{n}$, which in turn implies that $\| g_n \|_\infty \geq C/\sqrt{n}$.

REFERENCES

1. Kahane, J. P. Sur les polynômes à coefficients unimodulaires. *B.L.M.S.* 12 (1980): 321-342.

2. КОНЯГИН, С. В.: "О ПРОБЛЕМЕ ЛИТТЛВУДА." To appear.

3. Kristiansen, G. K. Some inequalities for algebraic and trigonometric polynomials. *J.L.M.S.* 2 (1979): 300-314.

3'. Kristiansen, G. K. Corrigendum of [3], *J.L.M.S.* 2 (1980): 575-576.

4. McGehee, O. C.; Pigno, L.; and Smith, B. Hardy's inequality and L^1 norm of exponential sums. To appear.

5. Pichorides, S. K. L^p norms of exponential sums. *Publications Mathématiques d'Orsay*, 73-77.

6. Pichorides, S. K. Une remarque sur les polynômes trigonometriques réels dont toutes les racines sont réelles. *C. R. Acad. Sc. Paris*, t. 286, 1978, Série A, 17-19.

7. Saff, E., and Sheil-Small, T. Coefficients and integral mean estimates for algebraic and trigonometric polynomials with restricted zeros. *J.L.M.S.* (2), 9 (1974): 16-22.

CERTAIN FUNCTION SPACES CONNECTED WITH ALMOST EVERYWHERE CONVERGENCE OF FOURIER SERIES

Mitchell H. Taibleson and Guido Weiss
Washington University in St. Louis

INTRODUCTION

In a recent collaboration with E. M. Stein [4], the authors obtained certain weak-type results for some maximal operators acting on Hardy spaces $H^p(\mathbf{R}^n)$ and $H^p(\Sigma_{n-1})$, $0 < p < 1$, associated with n-dimensional Euclidean space $\mathbf{R}^n$ and the surface of the unit sphere Σ_{n-1}. Let us describe a very special case of these results associated with Σ_1. Throughout this paper, we denote Σ_1 by T and refer to this domain as the *torus*, which we identify with the interval $\left(-\frac{1}{2}, \frac{1}{2}\right]$ (or the additive group of reals modulo 1).

If $f \in L^1(T)$, let $\hat{f}(k) = \int_{-1/2}^{1/2} f(t)e^{-2\pi ikt}\,dt$, $s_n(f;\,x) = \sum_{k=-n}^{n} \hat{f}(n)e^{2\pi ikx}$ be the n-th partial sum of the Fourier series of f and $\sigma_n^\delta(f;\,x) = \hat{f}(0) + \sum_{k=1}^{n} \frac{A_{n-k}^\delta}{A_n^\delta}\left\{\hat{f}(k)e^{2\pi ikx} + \hat{f}(-k)e^{-2\pi ikx}\right\}$ the

---

This research was supported by NSF grant MCS 7903 122. The authors are also grateful to Professors Richard Rochberg and Lu Shan-zhen for their useful comments concerning this paper.

corresponding Cesàro means of order δ, where $A_n^\delta = \binom{n+\delta}{n}$ (see [3]). In this setting the maximal operator S is defined by

$$(Sf)(x) = \sup_{n \geq 0} |\sigma_n^\delta(f;\, x)| \tag{1}$$

and the result obtained in [4] is:

Theorem 1. If $0 < p < 1$ *and* $\delta = \frac{1}{p} - 1$, *then the operator* S *is defined on* $H^p(T)$ *and*

$$|\{x : (Sf)(x) > \lambda\}| \leq \left(\frac{A \| f \|_{H^p}}{\lambda}\right)^p, \; \lambda > 0, \tag{2}$$

where A *is independent of* $f \in H^p$.†

This result cannot be extended to the case $p = 1$, since, as was pointed out in [4], there exists $f \in H^1(T)$ whose Fourier series diverges a.e. (When $p = 1$ we have $\delta = 0$ and, hence, $\sigma_n^0 = s_n$. A weak-type result (2) would imply the a.e. convergence of $s_n(f;\, x)$ for all $f \in H^1(T)$.) One of the motivations for this study is to obtain an alternate positive result that can be regarded as an extension to the case $p = 1$. To explain what we mean by this, it is convenient to explain, briefly, the method used to obtain Theorem 1.

In [4] we made use of the atomic characterization of $H^p : f \in H^p(t)$ if and only if $f = \sum_{k=1}^\infty m_k a_k$, where $\Sigma\, |m_k|^p < \infty$ and each a_k is a p-atom.‡ The "norm" $\| f \|_{H^p}$ is the infimum of the expressions $(\Sigma\, |m_k|^p)^{1/p}$ over all such representations of f. The basic weak-type estimate (2) is obtained by putting together two

---

† $|E|$ denotes the Lebesgue measure of the measurable set $E \subset T$.

‡ f is so represented as a linear functional on an appropriate space of test functions. This feature plays no role in the present study.

lemmas. The first asserts that there exists a uniform weak-type estimate for p-atoms:

Lemma 1. If f is a p-atom, $0 < p < 1$, then

$$|\{x : (Sf)(x) > \lambda\}| \leq \left(\frac{A}{\lambda}\right)^p$$

for all $\lambda > 0$, where A is independent of f.

The second allows us to "add up" these estimates:

Lemma 2. If $\{f_k\}$ is a sequence of measurable functions satisfying $|\{x : |f_k(x)| > \lambda\}| \leq \lambda^{-p}$, $0 < p < 1$, $\lambda > 0$ and $\sum_{k=1}^{\infty} |m_k|^p \leq 1$, then $|\{x : |\sum_{k=1}^{\infty} m_k f_k(x)| > \lambda\}| \leq \frac{2-p}{1-p}\lambda^{-p}$.

Our original motivation for Lemma 2 was the following result of E. M. Stein and N. J. Weiss [5]:

Lemma 3. If $\{f_k\}$ is a sequence of measurable functions satisfying $|\{x : |f_k(x)| > \lambda\} \leq \lambda^{-1}$, $\lambda > 0$, and $m = \{m_k\}$ is a numerical sequence, then

$$|\{x : |\sum_{k=1}^{\infty} m_k f_k(x)| > \lambda\}| \leq 4N(m)\lambda^{-1},$$

where

$$N(m) = \sum_{k=1}^{\infty} |m_k| \left(1 + \log \frac{\sum_{\ell=1}^{\infty} |m_\ell|}{|m_k|}\right).$$

The k-th summand of $N(m)$ is interpreted to be 0 if $m_k = 0$. The expression $N(m)$ does not occur in the original version of Lemma 2 in [5]; however, the weak-type inequality above is an immediate consequence of the result of Stein and Weiss and the positive homogeneity of the functional N.

We shall show that the maximal partial sum operator S satisfies a uniform weak-type estimate $|\{x : (Sf)(x) > \lambda\}| \leq C\lambda^{-1}$,

$\lambda > 0$, when f is a 1-atom. By using Lemma 3, we can then "add up" these estimates and obtain the weak-type result

$$|\{x : (Sf)(x) > \lambda\}| \leq 4CN(m)/\lambda, \tag{3}$$

for all $\lambda > 0$, whenever $f = \Sigma\, m_k a_k$, with $N(\{m_k\}) = N(m) < \infty$, and each a_k a 1-atom. We shall see that this implies that the Fourier series of f converges (to $f(x)$) a.e. It is in this sense that Theorem 1 can be extended to the case $p = 1$.

This furnishes us a motivation for studying spaces of such functions f that are generated by atoms, as well as more general objects that we shall call *blocks*. We shall see that the notion of entropy and the theory of random series of functions are related to this study. Moreover, these spaces can be studied from several points of view. For example, one can develop some of their functional analysis properties and they can be considered in the context of interpolation theory. In this paper, in addition to establishing the convergence result just described, we shall develop some of the basic properties of these spaces and their relation to entropy. In collaboration with Yves Meyer, we have investigated some of these other questions. These results shall appear elsewhere.

BLOCKS AND THE SPACES C_q THEY GENERATE

The atomic space $H^1_q(T)$ is generated by $(1, q)$-*atoms*. These are functions $a(x)$ supported in an interval $I \subset T$ satisfying the size condition $\| a \|_q \leq |I|^{(1/q)-1}$ and a cancellation property.† It can

†Usually, this cancellation property is simply $\int a = 0$. There exists a "special atom" that does not have mean $0 : a(x) \equiv 1$. It is sometimes useful to impose the condition that all moments $\int_T a(x)x^i\, dx$ vanish, $i = 0, 1, \ldots, s$ (see [6]).

be shown [6] that $H^1_q(T) = \{f = \sum_{k=1}^{\infty} m_k a_k : \Sigma |m_k| < \infty$, a_k a $(1, q)$-atom$\}$ is independent of q, $1 < q \leq \infty$, and any two of the norms

$$\| f \|_{1,q} = \inf\{\Sigma |m_k| : f = \Sigma m_k a_k\}$$

are equivalent. It turns out that in our study, the cancellation property is not relevant; however, the spaces we shall introduce (as "substitutes" for H^1) do depend on the index q.

Definition. A *q-block*, $1 < q \leq \infty$, is a function b supported in an interval $I \subset T$ satisfying

$$\| b \|_q \leq |I|^{(1/q)-1}. \tag{4}$$

That is, a q-block has the same properties of a $(1, q)$-atom except for the cancellation property.

We begin by establishing the uniform weak-type estimate for the maximal partial sum operator S applied to q-blocks.

Theorem 2. There exists a constant C, independent of the q-block b, such that $|\{x : (Sb)(x) > \lambda\}| \leq C\lambda^{-1}$ for all $\lambda > 0$.

Proof. There is no loss in generality in assuming Supp $b = I = (-\delta, \delta)$ and (4) holds for this interval I. Suppose $\lambda \geq |I|^{-1}$. We then use the deep result that S is of weak-type (q, q) :

$$|\{x : (Sb)(x) > \lambda\}| \leq C \frac{\| b \|_q^q}{\lambda^q} \leq C \frac{|I|^{1-q}}{\lambda^{q-1}} \frac{1}{\lambda} \leq \frac{C}{\lambda}^{\dagger}$$ (see [2]). Now assume $\lambda < |I|^{-1}$. We have

---

†Here, and elsewhere, we use the letter C to denote a constant, changing from place to place, that is independent of the principal parameters under discussion.

$$|\{x : (Sb)(x) > \lambda\}| \leq |\{x : (Sb)(x) > \lambda\} \cap \{x : |x| \leq |I|\}| + |\{x : (Sb)(x) > \lambda\} \cap \{x : |x| > |I|\}|.$$

The first summand is dominated by $|\{x : |x| \leq |I|\}| = 2|I| \leq 2/\lambda$. To estimate the second summand, we use the elementary inequality $|D_n(t)| \leq C/|t|$, $|t| \leq 1/2$, satisfied by the Dirichlet kernels D_n. Thus, if $x > |I| = 2\delta$, then

$$|S_n(b;\ x)| = |\int_{-1/2}^{1/2} b(x - t)D_n(t)\ dt| \leq \|b\|_1 \sup_{x-\delta\leq t\leq 1/2} |D_n(t)|$$

$$\leq 1 \cdot \sup_{x-\delta\leq t\leq 1/2} \frac{C}{|t|} \leq \frac{C}{x - \delta} < \frac{2C}{x}.$$

If $x < -|I| = -2\delta$, then the inequality $x - \delta \leq t \leq 1/2$ is replaced by $-1/2 \leq t \leq x + \delta$. In either case we have $|S_n(b;\ x)| < \dfrac{2C}{|x|}$; thus, the second summand is dominated by $|\{x : \dfrac{2C}{|x|} > \lambda\}| = \dfrac{4C}{\lambda}$.

Theorem 2 follows by combining these estimates (observe that C depends on q but not on b).†

This proof was also obtained independently by one of our students, Leonardo Colzani.

Corollary 1. If $m = \{m_k\}$ is a numerical sequence such that $N(m) < \infty$ and $\{b_k\}$ is a sequence of q-blocks, then $f = \Sigma\, m_k b_k$ satisfies

$$|\{x : (Sf)(x) > \lambda\}| \leq 4CN(m)/\lambda \tag{5}$$

for all $\lambda > 0$, where C is the constant in Theorem 2. Moreover, $s_n(f;\ x)$ converges to $f(x)$ a.e.

---

†This argument is not valid for $q = \infty$. Since, however, $\left(\int_I |b|^{q_1} \dfrac{dx}{|I|}\right)^{1/q_1} \leq \left(\int_I |b|^{q_2} \dfrac{dx}{|I|}\right)^{1/q_2}$ if $q_1 < q_2 \leq \infty$, we see that a q_1-block is also a q_2-block. In particular, if b is an ∞-block, then it is a 2-block and we can use the fact that S is of weak-type (2,2) in the argument that was just given.

Proof. Since, for each n, $\Sigma_k m_k s_n(b_k; x) = s_n(f; x)$ is uniformly and absolutely convergent, $(Sf)(x) \leq \Sigma_k |m_k|(Sb_k)(x)$. Thus, (5) is an immediate consequence of Lemma 3 and Theorem 2.

To see that the Fourier series of f converges a.e., we write $f = \sum_{k=1}^{N} m_k b_k + \sum_{k=N+1}^{\infty} m_k b_k = g + r$. Since g is a finite linear combination of q-blocks, it belongs to $L^q(T)$, $1 < q$, and thus has an a.e. convergent Fourier series. We can also choose $\alpha = \sum_{N+1}^{\infty} |m_k|\left(1 + \log\left(\sum_{N+1}^{\infty} |m_\ell|\right)/|m_k|\right)$ to be as small as we wish. Thus, applying (5) to the function r, it follows that if $\varepsilon > 0$,

$$\begin{aligned}
&\left|\{x : \lim_{N\to\infty} \sup_{k,j\geq N} |s_k(f;\, x) - s_j(f;\, x)| > \varepsilon\}\right| \\
&\leq \left|\{x : \lim_{N\to\infty} \sup_{k,j\geq N} |s_k(g;\, x) - s_j(g;\, x)| > \tfrac{\varepsilon}{2}\}\right| \\
&+ \left|\{x : \lim_{N\to\infty} \sup_{k,j\geq N} |s_k(r;\, x) - s_j(r;\, x)| > \tfrac{\varepsilon}{2}\}\right| \leq 0 \\
&+ \left|\{x : (Sr)(x) > \tfrac{\varepsilon}{2}\}\right| \leq \frac{16C}{\varepsilon}\alpha.
\end{aligned}$$

Since α is arbitrarily small, it follows that for each $\varepsilon > 0$,

$$\left|\{x : \lim_{N\to\infty} \sup_{k,j\geq N} |s_k(f;\, x) - s_j(f;\, x)| > \varepsilon\}\right| = 0.$$

Hence, the Fourier series of f converges a.e. (to $f(x)$, since $f \in L^1(T)$ and thus its Cesàro means converge to $f(x)$ a.e.).

Definition. The function space C_q consists of all

$$f = \Sigma\, m_k b_k, \tag{6}$$

where the b_k are q-blocks and $m = \{m_k\}$ satisfies $N(m) < \infty$. We endow C_q with a "norm" N_q defined by letting $N_q(f)$ be the infimum of the expressions $N(m)$ for all representations (6) of f.

Strictly speaking, N_q is a quasi-norm; that is, it satisfies all the properties of a norm except for Minkowski's inequality, for which we have the substitute

$$N_q(f + g) \leq (1 + \log 2)(N_q(f) + N_q(g)). \tag{7}$$

The proof of (7) is easy. We first observe that $\varphi(x) = 1 - x \log x - (1 - x) \log(1 - x)$, $0 \leq x \leq 1$, has a maximum at $x = \frac{1}{2}$ and $\varphi(\frac{1}{2}) = 1 + \log 2$. Choose representations of f and g, $f = \Sigma\, m_k b_k$, $g = \Sigma\, n_j c_j$, such that $N_q(f) \geq N(m) - \varepsilon$ and $N_q(g) \geq N(n) - \varepsilon$. Let $A = \Sigma\, |m_k|$, $B = \Sigma\, |n_j|$, and $x = \dfrac{A}{A + B}$.

Then

$$\begin{aligned} N_q(f + g) &\leq (A + B)\varphi(x) + \Sigma\, |m_k| \log \frac{A}{|m_k|} + \Sigma\, |n_j| \log \frac{B}{|n_j|} \\ &\leq (1 + \log 2)\left(A + B + \Sigma\, |m_k| \log \frac{A}{|m_k|} + \Sigma\, n_j \log \frac{B}{|n_j|}\right) \\ &= (1 + \log 2)(N(m) + N(n)) \\ &\leq (1 + \log 2)(N_q(f) + N_q(g) + 2\varepsilon). \end{aligned}$$

Since $\varepsilon > 0$ is arbitrary, (7) follows. Later, we shall show that the constant $1 + \log 2$ is best possible.

It is sometimes useful to use the functional $M(m) = \Sigma\, |m_k| \left(1 + \log^+ \dfrac{1}{|m_k|}\right)$ on numerical sequences $m = \{m_k\}$ instead of N. We then let $M_q(f)$ be the infimum of all numbers $M(m)$ over all representations (6) of $f \in C_q$. M_q is not positive homogeneous, but it does satisfy Minkowski's inequality

$$M_q(f + g) \leq M_q(f) + M_q(g). \tag{8}$$

Consequently, $d_q(f, g) = M_q(f - g)$ is a metric on C_q. It is easy to see that $N(m) < \infty$ if and only if $M(m) < \infty$. Moreover, the topologies on C_q induced by the metric d_q and the quasi norm N_q are the same. To see this, one first observes that if $M_q(f) < 1$, then $N_q(f) \leq M_q(f)$; consequently, $\lim_{n\to\infty} M_q(f_n) = 0$ implies $\lim_{n\to\infty} N_q(f_n) = 0$. The opposite implication follows from

$$M_q(f) \leq N_q(f)\left\{1 + \log \frac{1}{N_q(f)}\right\} \tag{9}$$

whenever $N_q(f) < 1/e$. To obtain (9), choose a representation $f = \Sigma\, m_k b_k$ so that $N(m) \leq N_q(f) + \varepsilon < 1/e$ ($\varepsilon > 0$ fixed). Let $A = \Sigma\, |m_j|$ and $B = N_q(f) + \varepsilon$. Then $N(m) = M(m) + A \log A \geq M(m) + B \log B$ since $A \leq N(m) \leq B < 1/e$ and $\varphi(x) = x \log x$ is decreasing on $[0, 1/e]$. Thus, $M_q(f) \leq M(m) \leq N(m) + B \log(1/B) \leq B(1 + \log(1/B))$. Since $\varepsilon > 0$ can be chosen as close to 0 as we wish, (9) follows.

An immediate consequence of the fact that $\| b \|_1 \leq 1$ for any q-block b are the useful inequalities

$$\| f \|_1 \leq N_q(f), \; \| f \|_1 \leq M_q(f) \tag{10}$$

for any $f \in C_q$.

We shall now list a number of properties of the spaces C_q:

(i) *C_q is complete;*

(ii) *The series* (6) *converges in the topology of C_q, in the L^1-norm and almost everywhere;*

(iii) *If $f \in C_q$ and $|g| \leq |f|$, g measurable, then $N_q(g) \leq N_q(f)$. In particular, if $f \in C_q$, then $|f| \in C_q$; thus, $\sup\{f, g\}$ and $\inf\{f, g\}$ belong to C_q if $f, g \in C_q$.*

(iv) *$L^q(T) \subset C_q$. If $1 \leq p < q \leq \infty$, there exists $f \in L^p(T)$ such that $f \notin C_q$. In particular, $C_{q_1} \underset{\neq}{\supset} C_{q_2}$ if $1 < q_1 < q_2$.*

The proof of (i) is straightforward (it is similar to the usual proofs of the completeness of L^q). The C_q and L^1 convergence of (6) is obvious. Lemma 3 can be applied to $\Sigma\, |m_k||b_k(x)|$, since $\| b \|_1 \leq 1$ for any q-block b, to obtain the a.e. absolute convergence of $\Sigma\, m_k b_k(x)$ (to $f(x)$, since we have L^1 convergence to f). The proof of (iii) is also immediate: $|g| \leq |f|$ implies $g = rf$ with r measurable and $|r| \leq 1$. Thus, if $f = \Sigma\, m_k b_k$, then $g = \Sigma\, m_k(rb_k)$. But rb_k is also a q-block. Thus, $N_q(g) \leq N(m)$ for $m = \{m_k\}$; since the infimum of such $N(m)$ is $N_q(f)$, we have (iii).

The fact that $L^q(T) \subset C_q$ is obvious: $\|f\|_q^{-1} f$ is a q-block supported in T whenever $f \in L^q(T)$. The other assertions in (iv) are considerably deeper facts. We begin their proof by establishing:

Lemma 4. *Let* $f = \frac{1}{|I|} \sum_{i=1}^{n} \chi_{I_i}$, *where the* $\{I_i\}$ *are* n *mutually disjoint intervals, all of length* $|I|$ *that are equally spaced so that they fit exactly inside a larger interval of length at least* $n(1 + \log n)^{q'}|I|$, $1 < q \leq \infty$, $\frac{1}{q} + \frac{1}{q'} = 1$. *Then*

$$N_q(f) = n(1 + \log n).$$

Before proving this lemma, let us observe that it immediately implies that inequality (7) is best possible. Let $f = \frac{1}{|I|} \chi_{I_1}$ and $g = \frac{1}{|I|} \chi_{I_2}$, where I_1, I_2 are disjoint intervals of length $|I|$ that are at a distance from each other that is at least $2[(1 + \log 2)^{q'} - 1]|I|$. Then f (and g) is a q-block such that $1 \leq \|f\|_1 \leq N_q(f) \leq 1$; thus, $N_q(f) = 1 = N_q(g)$. Hence, using Lemma 4, we have $N_q(f + g) = 2(1 + \log 2) = (1 + \log 2)(N_q(f) + N_q(g))$.

We now pass to the proof of Lemma 4. We see that f is a linear combination of n q-blocks, with coefficients equal to 1. Thus, $N_q(f) \leq \sum_{k=1}^{n} 1(1 + \log \frac{n}{1}) = n(1 + \log n)$. We need to show that if $f = \sum m_k b_k$, where each b_k is a q-block and $N(m) = N(\{m_k\}) < \infty$, then $n(1 + \log n) \leq N(m)$ (it is trivial that $n \leq N(m)$).

By (ii), $\sum m_k \{\chi_{\bigcup_{i=1}^{n} I_i} b_k(x)\} = f(x)$ a.e. Since $\chi_{\cup I_i} b_k$ is also a block, we can assume that $J_k = \operatorname{Supp} b_k \subset \bigcup_{i=1}^{n} I_i$. Observe that if $I_\ell, I_{\ell+1}, \ldots, I_{\ell+p}$ are the (consecutive) p intervals, $1 \leq p \leq n$, that J_k intersects, we can further assume that J_k contains no part to the left of I_ℓ or to the right of $I_{\ell+p}$ (since $b_k(x) = 0$ to the left of I_ℓ and to the right of $I_{\ell+p}$).

To make our next reduction, we shall use the following result (which is of independent interest).

Lemma 5. If $m = \{m_k\}$ and $n = \{n_k\}$ are two numerical sequences such that $|m_k| \le |n_k|$ for all k, then $N(m) \le N(n)$.

Proof. We can assume $0 \le m_k \le n_k$ for all k. Let $m(\varepsilon)$ be the sequence whose first term is $m_1 + \varepsilon$ and otherwise has the same terms as m. Clearly, it suffices to show that $\varphi(\varepsilon) = N(m(\varepsilon))$ is an increasing function of $\varepsilon > 0$. Let $A(\varepsilon) = \varepsilon + \sum_{\ell=1}^{\infty} m_\ell$. Then

$$\varphi(\varepsilon) = (m_1 + \varepsilon)\left(1 + \log \frac{A(\varepsilon)}{m_1+\varepsilon}\right) + \sum_{k=2}^{\infty} m_k\left(1 + \log \frac{A(\varepsilon)}{m_k}\right). \text{ Thus,}$$

$$\varphi'(\varepsilon) = 1 + \log \frac{A(\varepsilon)}{m_1+\varepsilon} > 0,$$ which shows φ is increasing.

We now can make the further assumption that $\| b_k \|_q = |J_k|^{-1/q'}$ (where $1/q' = 1 - 1/q$), since if this were not the case, we could represent f so that $N(\{m_k\})$ is smaller by Lemma 5.

Finally, we shall reduce the problem to the case where each J_k is contained in one of the intervals I_i. Once this is done, the desired inequality $n(1 + \log n) \le N(m)$ is easy to show. For in this case, we may write $f = \sum_{i=1}^{n} \sum_k m_k^i b_k^i$, where b_k^i is a q-block supported in $J_k^i \subset I_i$. Thus, $\frac{1}{|I|} \chi_{I_i}(x) = \sum_k m_k^i b_k^i(x)$ (in L^1 and a.e.) and it follows that $B_i \equiv \sum_k |m_k^i| \ge 1$. Consequently, we can choose a sequence $\{\tilde{m}_k^i\}$ satisfying $0 \le |\tilde{m}_k^i| \le |m_k^i|$ and $\sum_k |\tilde{m}_k^i| = 1$ for $i = 1, 2, \ldots, n$. By Lemma 5, $N(\{\tilde{m}_k^i\}) = \sum_i \sum_k |\tilde{m}_k^i| \left(1 + \log \frac{n}{|\tilde{m}_k^i|}\right) \le N(\{m_k^i\})$. Since $|\tilde{m}_k^i| \le 1$, $\log \frac{n}{|\tilde{m}_k^i|} \ge \log n$. Hence, $N(\{m_k^i\}) \ge \sum_i \sum_k |\tilde{m}_k^i| (1 + \log n) = n(1 + \log n)$.

To make this last reduction (where each J_k is contained in one of I_i), we can therefore assume $f = \sum_{k=1}^{\infty} m_k b_k$, supp $b_k = J_k$, $\|b_k\|_q = |J_k|^{-1/q'}$, J_k meets p consecutive intervals, as above, without "sticking out" to the left or right. Because of the homogeneity of N, we can assume $m_1 = 1$. Thus,

$$f = b_1 + \sum_{k=2}^{\infty} m_k b_k$$

and $J_1 = \operatorname{Supp} b_1$ meets, say, $I_1, I_2, \ldots, I_\nu$, $2 \leq \nu \leq n$. Let $L_1 = J_1 \cap I_1$, $L_\nu = J_1 \cap I_\nu$, and $L_i = I_i$, $2 \leq i \leq \nu - 1$. Put $b_1^{(j)} = b_1|_{L_j}$, $\mu_j = |L_j|^{1/q'}\|b_1^{(j)}\|_q$, and $a_j = \mu_j^{-1} b_1^{(j)}$, $j = 1, 2, \ldots, \nu$, so that a_j is a q-block supported in L_j.

The distance between two intervals I_i is at least $|I|\frac{n}{n-1}[(1 + \log n)^{q'} - 1]$. This fact, together with the increasing property of $\theta(t) = \frac{t}{t-1}[(1 + \log t)^{q'} - 1]$ for $t > 1$ $(1 \leq q' < \infty)$, gives us the estimate

$$\begin{aligned} |J_1| &\geq \sum_{i=1}^{\nu} |L_i| + (\nu - 1)\frac{n}{n-1}[(1 + \log n)^{q'} - 1]|I| \\ &\geq \sum_{i=1}^{\nu} |L_i| + \nu[(1 + \log \nu)^{q'} - 1]|I| \\ &= \nu|I|(1 + \log \nu)^{q'} + |L_1| + |L_\nu| - 2|I|. \end{aligned}$$

Hence, using Hölder's inequality,

$$\begin{aligned} \sum_{j=1}^{\nu} \mu_j = \sum_{j=1}^{\nu} |L_j|^{1/q'}\|b_1^{(j)}\|_q &\leq \left(\sum_{j=1}^{\nu} |L_j|\right)^{1/q'} \left(\sum_{j=1}^{\nu} \|b_1^{(j)}\|_q^q\right)^{1/q} \\ &= \left(\sum_{j=1}^{\nu} |L_j|\right)^{1/q'} \|b_1\|_q = \left(\sum_{j=1}^{\nu} |L_j|\right)^{1/q'} |J_1|^{-1/q'} \\ &\leq \left\{\frac{\sum_{j=1}^{\nu} |L_j|}{\nu|I|(1 + \log \nu)^{q'} + |L_1| + |L_\nu| - 2|I|}\right\}^{1/q'} \\ &= \left\{\frac{1 - \alpha}{(1 + \log \nu)^{q'} - \alpha}\right\}^{1/q'}, \end{aligned}$$

where $\alpha = \dfrac{2|I| - |L_1| - |L_\nu|}{\nu|I|}$. But $\dfrac{1 - \alpha}{(1 + \log \nu)^{q'} - \alpha} \leq$ $\dfrac{1}{(1 + \log \nu)^{q'}}$, since $\alpha \geq 0$ and $(1 + \log \nu)^{q'} \geq 1$. This shows

$$\sum_{j=1}^{\nu} \mu_j \leq \frac{1}{1 + \log \nu}. \tag{11}$$

Thus, $f = \sum_{k=2}^{\infty} m_k b_k + \sum_{j=1}^{\nu} \mu_j a_j$ and we have to show

$$(1 + A)(1 + \log(1 + A)) - \sum_{k=2}^{\infty} |m_k| \log |m_k|$$

$$\geq \sum_{k=2}^{\infty} |m_k| (1 + \log \frac{A + B}{|m_k|}) + \sum_{j=1}^{\nu} \mu_j \left(1 + \log \frac{A + B}{\mu_j}\right),$$

where $A = \sum_{k=2}^{\infty} |m_k|$ and $B = \sum_{j=1}^{\nu} \mu_j$. That is, we have to show

$$(1 + A)(1 + \log(1 + A)) \geq (A + B)(1 + \log(A + B)) - \sum_{j=1}^{\nu} \mu_j \log \mu_j.$$

If we consider both sides of this inequality as functions of A, we see that the derivative of the left side is $2 + \log(1 + A)$ and the derivative of the right side is $2 + \log(B + A)$. Since $B \leq 1$ (by (11)), we have $\log(1 + x) \geq \log(B + x)$ for $x \geq 0$. Thus, it is sufficient to check the inequality for $A = 0$, that is, to show that

$$\sum_{j=1}^{\nu} \mu_j \left(1 + \log \frac{B}{\mu_j}\right) \leq 1.$$

But a Lagrange multiplier argument shows that the maximum of this expression is attained for $\mu_j = B/\nu$, $j = 1, 2, \ldots, \nu$ and the maximum is $B(1 + \log \nu)$. By (11), $B(1 + \log \nu) \leq 1$ and this completes the proof.

Let us now show how Lemma 4 can be used to establish the second part of property (iv). Lemma 4 can be reformulated as follows. If I is an interval in T and n a positive integer, there exists a subset $S \subset I$ (of n equally spaced subintervals) such that $|S| = |I|(1 + \log n)^{-q'}$ and $N_q(\chi_S) = |I|(1 + \log n)^{1-q'}$. Consider first the case $q = \infty$ (thus $q' = 1$). Choose mutually disjoint subintervals I_k of T with corresponding subsets $S_k \subset I_k$ such that $N_\infty(\chi_{S_k}) = |I_k|$ and $|S_k| = |I_k|(1 + \log n_k)^{-1}$. Let $v_k > 1$, $k = 1, 2, \ldots$. Put $f = \Sigma\, v_k \chi_{S_k}$. We then have $N_\infty(f) \geq v_k N_\infty(\chi_{S_k}) = v_k|I_k|$ (by property (iii)) and $\int e^f = \sum e^{v_k}|S_k| = \sum e^{v_k}|I_k|(1 + \log n_k)^{-1}$. If we select $|I_k| = 2^{-k}$ (so this can be done within T), $v_k = 2^{2k}$ (so that $N_\infty(f) \geq \lim_{k\to\infty} v_k|I_k| = \infty$), and $n_k \geq \exp(e^{v_k})$ (so that $\int e^f < \infty$), we have an example of an exponentially integrable function f (which, a fortiori, belongs to L^p for all $p < \infty$) such that $N_\infty(f) = \infty$.

The case $1 \leq p < q < \infty$ requires a little more care. We make the same construction as before except that $q' > 1$ and we want $f \in L^p$ but $N_q(f) = \infty$. Thus, in this case, $N_q(f) \geq v_k N_q(\chi_{S_k}) = v_k|I_k|(1 + \log n_k)^{1-q'}$ and $\|f\|_p^p = \sum v_k^p|I_k|(1 + \log n_k)^{-q'}$. We therefore want this last sum to be finite, $\Sigma\, |I_k| \leq 1$ and $v_k|I_k|(1 + \log n_k)^{q'-1} \to \infty$ as $k \to \infty$. This can be done as follows. Let $|I_k| = 2^{-k}$, $v_k = 2^{\beta k}$, and n_k satisfy $1 + \log n_k = 2^{\gamma k}$.† It is easy to check that the conditions we need are fulfilled provided $(p\beta - 1)/q' < \gamma < (\beta - 1)/(q' - 1)$ and $\beta > (q - 1)/(q - p)$. Since $\{(\beta - 1)/(q' - 1)\} - \{(p\beta - 1)/q'\} = \{\beta(q - p) - (q - 1)\}/q' > 0$, appropriate values of β and γ can be found. This establishes the second part of property (iv).

---

†This condition does not necessarily guarantee that n_k is an integer; however, replacing n_k by the largest positive integer not exceeding n_k, $[n_k]$, does not change our basic estimates.

For technical reasons (which will be immediately obvious), let us consider the interval $[0, 1/e] = I^*$, instead of T, as the domain of the functions we shall study (alternatively, we can renormalize the Lebesgue measure so that $|T| = 1/e$).

The *entropy* $E(S)$ of a set $S \subset [0, 1/e]$ is defined as the infimum of all expressions $\Sigma_k |I_k| \log 1/|I_k|$ taken over all sequences of intervals $I_k \subset I^*$ that cover S. R. Fefferman [1] showed that the notion of entropy has many applications to Fourier analysis; in particular, he writes, "It seems likely that entropy has an intimate connection to the pointwise convergence of Fourier series." We shall see, in this section, that he was indeed correct when he made this observation.

Let $J(F) = \int_0^\infty E(\{x : |f(x)| > \lambda\})\, d\lambda$ be the *entropy of a function* f (this notion parallels the concept of the L^1-norm). We begin by proving:

Theorem 3. If $J(f) < \infty$, *then* $f \in C_\infty$. *In particular,* $\lim_{n\to\infty} s_n(f; x) = f(x)$ *a.e.*

Proof. Let $S_k = \{x : 2^{k-1} < |f(x)| \leq 2^k\}$, $k = 0, \pm 1, \pm 2, \ldots$. An easy calculation shows that

$$\frac{1}{2} \sum_{k=-\infty}^{\infty} 2^k E(S_k) \leq J(f) \leq 2 \sum_{k=-\infty}^{\infty} 2^k E(S_k).^\dagger \tag{12}$$

---

†The set function E is not a measure (additivity fails). Fefferman shows, however, that it does have some of the properties of a measure: (1) if $S \subset T \subset I^*$, then $E(S) \leq E(T)$; (2) $E(\cup_k S_k) \leq \Sigma_k E(S_k)$; (3) if $I (\subset I^*)$ is an interval and $S \subset I^*$ satisfies $|I| = |S|$, then $E(I) \leq E(S)$; (4) if $S_k \subset S_{k+1} \subset I^*$, $k = 1, 2, 3, \ldots$, then $E(\cup_k S_k) = \lim_{k\to\infty} E(S_k)$. In this paper we need to use these properties to obtain some of our assertions.

For k fixed, choose a sequence $\{I_{k,\ell}\}$ of mutually disjoint intervals of positive measure satisfying $S_k \subset \cup_\ell I_{k,\ell}$ and $E(S_k) \le \Sigma_\ell |I_{k,\ell}| \log 1/|I_{k,\ell}| \le E(S_k) + \varepsilon_k/2^k$, where $\varepsilon_k > 0$ and $\Sigma_k \varepsilon_k = \varepsilon > 0$ (this can be done, since $t \log 1/t$ is increasing in $[0, 1/e]$). Observe that $\log 1/|I_{k,\ell}| \ge 1$, since $|I_{k,\ell}| \le 1/e$. Thus, $\sum_{k,\ell} 2^k|I_{k,\ell}| \le \sum_{k,\ell} 2^k|I_{k,\ell}| \log 1/|I_{k,\ell}| \le \sum_k 2^k E(S_k) + \varepsilon$. This and (12) give us

$$\sum_{k,\ell} 2^k|I_{k,\ell}| \le \sum_{k,\ell} 2^k|I_{k,\ell}| \log |I_{k,\ell}| \le 2J(f) + \varepsilon. \qquad (13)$$

Let $a_0(x) = f(x)$ if $|f(x)| \le 1$ and $a_0(x) = 0$ otherwise; for each $k > 0$ let $a_{k,\ell}(x) = f(x)\chi_{S_k} \cap I_{k,\ell}(x)/2^k|I_{k,\ell}|$. Then

$f(x) = \sum_{k>0} 2^k \sum_\ell |I_{k,\ell}|a_{k,\ell}(x) + a_0(x)$. Since each $a_{k,\ell}$ is an ∞-block supported in $I_{k,\ell}$ and $ea_0(x)$ is an ∞-block supported in I^*, we have

$$M_\infty(f) \le \sum_{\substack{k,\ell \\ k>0}} 2^k|I_{k,\ell}| + \sum_{\substack{k,\ell \\ k>0}} 2^k|I_{k,\ell}|\log^+ \frac{1}{2^k|I_{k,\ell}|} + \frac{1}{e}.$$

By (13) the first summand is majorized by $2J(f) + \frac{1}{e}$. The second summand equals

$$\begin{aligned}\sum_{k>0} c_k &\equiv \sum_{k>0} 2^k \sum_{\{\ell:2^k|I_{k,\ell}|<1\}} |I_{k,\ell}|\log \frac{2^{-k}}{|I_{k,\ell}|} \\ &= \sum_{k>0} 2^k \sum_{\{\ell:|I_{k,\ell}|2^k<1\}} |I_k| \log \frac{1}{|I_{k,\ell}|} \\ &\quad - (\log 2) \sum_{k>0} k2^k \sum_{\{\ell:|I_{k,\ell}|2^k<1\}} |I_{k,\ell}| \\ &\equiv \sum_{k>0} a_k - \sum_{k>0} b_k.\end{aligned}$$

Since $c_k = a_k - b_k > 0$, we have $a_k \ge b_k > 0$. Moreover, from (13) we have $\Sigma a_k \le 2J(f) + \varepsilon < \infty$; therefore, $\Sigma c_k \le 2J(f) + \varepsilon$. Putting these estimates together and letting $\varepsilon \to 0$, we obtain

$$M_\infty(f) \leq 4J(f) + 1/e. \tag{14}$$

Since $N_\infty(f) \leq M_\infty$ when $M_\infty(f) \leqq 1$, we obtain, by the usual homogeneity argument, $N_\infty(f) \leqq \left(\frac{4}{1 - C^{-1}}\right) J(f)$.

This shows that the collection of measurable functions having finite entropy is contained in C_∞. These two classes, however, are not equal.

Theorem 4. There exists $f \in C_\infty$ such that $J(f) = \infty$.

Proof. Choose mutually disjoint intervals $I_k \subset I^*$ such that $|I_k| = 1/(2^k k(\log k)^3)$, $k = 3, 4, \ldots$. Since $\sum_{k=3}^{\infty} |I_k| \leq \sum_{k=3}^{\infty} 2^{-k}$ $= \frac{1}{4} < \frac{1}{e}$, this is possible. We claim that $f = \sum_{k=3}^{\infty} 2^k \chi_{I_k}$ has the desired properties. It is easy to check that $E(I) = |I| \log \frac{1}{|I|}$ for any interval $I \subset I^*$. Thus, $E(I_k) = \frac{(\log 2)k + \log k + 3 \log\log k}{k2^k (\log k)^3}$.

Since $I_k = \{x : f(x) = 2^k\}$, (12) shows that

$$\begin{aligned} J(f) &\geq \frac{1}{2} \sum_{k=3}^{\infty} 2^k \frac{(\log 2)k + \log k + 3 \log \log k}{k2^k (\log k)^3} \\ &\geq \frac{1}{2} \sum_{k=3}^{\infty} (\log 2)(\log k)^{-3} = \infty. \end{aligned}$$

If we let $b_k = \frac{1}{|I_k|} \chi_{I_k}$, then f has the ∞-block representation $f = \sum_{k \geq 3} m_k b_k$, with $m_k = 2^k |I_k|$. Thus,

$$\begin{aligned} M_\infty(f) \leq M(m) &= \sum_{k\geq 3} m_k \left(1 + \log \frac{1}{m_k}\right) \\ &= \sum_{k \geq 3} \frac{1}{k(\log k)^3} (1 + \log k + 3 \log \log k) \\ &\leq \sum_{k\geq 3} \frac{1}{k(\log k)^3} + \sum_{k \geq 3} \frac{\log k + 3 \log \log k}{k(\log k)^3} < \infty. \end{aligned}$$

This proves Theorem 4.

Observe that the function f in this proof does not belong to $L \log L$:

$$\int_{I^*} f \log^+ f = \sum_{k \geq 3} 2^k (\log 2^k) |I_k| = (\log 2) \sum_{k \geq 3} (\log k)^{-3} = \infty.$$

Hence, C_∞ is not contained in $L \log^+ L$.† On the other hand, a simple modification of the proof of Theorem 3 shows that $J(f) < \infty$ does imply $f \in L \log L$ (a fact that was already observed by R. Fefferman in [1]).

We shall content ourselves with these two results as an illustration of the connection between the notion of entropy and these spaces generated by blocks. We end this article with a brief indication of the connection between the ideas developed here and the behavior of the Fourier series of certain measures. We are grateful to Jean-Pierre Kahane for pointing out to us the results in his book [3] that bear such a close relationship with ours. In Chapter 10 of his book, Kahane considers a sequence $\{\theta_j\}$ in T and constructs the measure $\mu = \Sigma\, m_j \delta_{\theta_j}$, where δ_{θ_j} is the Dirac δ-measure concentrated at θ_j. He proves two results concerning the Fourier-Stieltjes series of μ. The first is:

Theorem 5. If $M(m) < \infty$, the Fourier-Stieltjes series of μ is bounded a.e.‡

This theorem can be established easily with the techniques we developed here. If $s_n(\delta_{\theta_j}; x)$ denotes the n-th partial sum of the Fourier-Stieltjes series of δ_{θ_j}, then, clearly, $s_n(\delta_{\theta_j}; x) = D_n(x - \theta_j)$. But, as we observed earlier, the Dirichlet kernel

†This shows (together with property (iv)) that for all p, $1 < p < \infty$, $C_\infty \not\subset L^p(T)$, $L^p(T) \not\subset C_\infty$.

‡Actually, Theorem 5 is a slight improvement of the result announced by Kahane.

satisfies the uniform estimate $|D_n(x - \theta_j)| \leq \frac{c}{|x - \theta_j|}$ for $|x - \theta_j| \leq \frac{1}{2}$. Thus, $S(\delta_{\theta_j}; x) = \sup_{n \geq 0} |s_n(\delta_{\theta_j}; x)|$ satisfies a uniform weak-type (1,1) estimate (as in Theorem 2). We can therefore apply Lemma 3, as we did in the proof of Corollary 1, to obtain Theorem 5.

The second result announced by Kahane can be stated as follows.

Theorem 6. If $M(m) = \infty$, then there exists a sequence $\{\theta_j\}$ such that the Fourier-Stieltjes series of μ is unbounded at almost every point.

We shall not prove this result here. We merely state it as an indication of the sharpness of the results on convergence we have obtained.

REFERENCES

1. Fefferman, Robert A. A theory of entropy in Fourier analysis. *Adv. in Math.* 30 (1978): 171-201.

2. Hunt, Richard. On the convergence of Fourier series. *Orth. Exp. and Their Cont. Analogs Proc. of Conf. at So. Ill. Univ.*, 1967, D. T. Haimo editor, pp. 235-256.

3. Kahane, Jean-Pierre. *Some Random Series of Functions.* Heath Math. Monographs. Lexington, Mass.: D. C. Heath and Co., 1968.

4. Stein, Elias M.; Taibleson, Mitchell H.; and Weiss, Guido. Weak type estimates for maximal operators on certain H^p classes. To appear in *Suppl. Rendiconti Circ. Mat. Palermo,* n.1 (1981): pp. 81-97.

5. Stein, Elias M., and Weiss, Norman J. On the convergence of Poisson integrals. *Trans. Amer. Math. Soc.* 140 (1969): 35-54.

6. Taibleson, Mitchell H., and Weiss, Guido. The molecular characterization of certain Hardy spaces. *Soc. Math. France, Astérisque* 77 (1980): 68-149.

EXPONENTIAL ESTIMATES IN MULTIPLIER ALGEBRAS

Misha Zafran
University of Washington

In [6] and [7], the author studied questions related to the operational calculus in some algebras of multipliers. Crucial in these investigations was the introduction of certain measures. In this note, we will further exploit their very special properties and obtain exponential estimates for the multiplier algebras M_p. These estimates are, in a sense, uniform in p, $1 < p < \infty$, $p \neq 2$ (see Theorem 1), and extend previous results of Rudin [5] and Igari [3]. We begin with some notations and definitions.

Let G be a locally compact Abelian group (or LCA group), with dual group Γ. Denote by $M(G)$ the class of finite, regular, Borel measures on G. We let $\|\mu\|$ be the total variation norm of μ. For $\mu \in M(G)$, let T_μ be the operator defined by $T_\mu(f) = \mu * f$.

A bounded operator T on $L_p(G)$ is called a multiplier if there exists a (necessarily unique) function $\hat{T} \in L_\infty(\Gamma)$ so that $T(f)^\wedge = \hat{T}\hat{f}$, for all integrable simple functions f. It will facilitate matters to write $\|\mu\|_{M_p}$ in place of $\|T_\mu\|_{M_p}$. We also sometimes use the abusive notation $\|\hat{T}\|_{M_p(G)}$ rather than $\|T\|_{M_p(G)}$.

Throughout, $\mathbb{R}$ denotes the real numbers, $\mathbb{Z}$ denotes the integers, $\mathbb{N}$ denotes the positive integers, and $\mathbb{T}$ denotes the circle group, which we identify with the interval $[-\pi, \pi)$.

We now state our principal result.

Theorem 1. Let $1 \leq p < \infty$ *with* $p \neq 2$, *and let* n *be a positive integer. Let* G *denote one of the groups* $\mathbb{T}^n$, $\mathbb{R}^n$, *or* $\mathbb{Z}^n$. *Then for every* $j \in \mathbb{Z}$, we have

$$\sup \| e^{ij\mu} \|_{M_p(G)} \geq c\beta^{|j|},$$

the supremum taken over all measures $\mu \in M(G)$ *with* $\| \mu \|_{M_p(G)} \leq 1$, *and* $\hat{\mu}$ *real-valued. Here* $c > 0$ *and* $\beta > 1$ *are constants independent of* p. *Moreover, in the case* $G = \mathbb{T}^n$ *or* $\mathbb{R}^n$, *we may take* $\beta = 3^{1/6}$.

The inequality of Theorem 1 sharpens results of Rudin [5, Theorem 2], and Igari [3, Lemma 1]. In these works, an estimate of the form

$$\sup \| e^{ij\mu} \|_{M_p(G)} \geq K_p^j, \; j = 1, 2, \ldots,$$

is obtained. Here the supremum is taken over all $\mu \in M(G)$ with $\hat{\mu}$ real-valued, and $\| \mu \| \leq 1$; K_p is a constant *depending* on p. In particular, $K_p \to 1$ as $p \to 2$. Also, it is clear that $\| e^{ij\mu} \|_{M_2(G)} = 1$ whenever $\hat{\mu}$ is real-valued.

The interest of Theorem 1 is that we can obtain estimates that are uniform in $p \neq 2$. The proof of the result requires the use of a class of measures introduced by the author in [6] and [7]. We now recall the constructions involved and the basic properties of these measures.

Notation 1. Let $\{t_k\}$ be a positive sequence so that $\sum_{k=1}^{\infty} t_k \leq 1$ and $t_n > 10 \sum_{k=n+1}^{\infty} t_k$, for all n. Let $\{x_k^n\}_{k=1}^{2^n}$ denote the set $\left\{ \sum_{i=1}^{n} \varepsilon_i t_i \,\middle|\, \varepsilon_i = 0 \text{ or } 1, \; 1 \leq i \leq n \right\}$. We assume $x_1^n < x_2^n < \cdots < x_{2^n}^n$, and define $x_{-k}^n = -x_k^n$, $1 \leq k \leq 2^n$.

Lemma 1. Let n be a positive integer. There exists a sequence of $\pm$ *so that if* $\mu_n = \frac{1}{2^{n+1}} \sum_{\substack{k=-2^n \\ k\neq 0}}^{2^n} \pm \delta_{x_k^n}$, *then* $\hat{\mu}_n$ *is real-valued, and*

$\|\hat{\mu}_n\|_\infty \leq 2^{1/2}2^{-n/2}$.

For this result, we refer the reader to Lemma 2.7 of [6] or Lemma 3.4 of [7]. The $\pm$ are related to the Rudin-Shapiro polynomials.

Notation 2. Let j be a positive integer, and let $\{t_k\}$ be as in Notation 1. Define $t_{k,s} = t_{(k-1)j+s}$, $1 \leq s \leq j$, $k = 1, 2, \ldots$. Let $\{x_{k,s}^n\}_{k=1}^{2^n}$ denote the set $\left\{ \sum_{i=1}^{n} \varepsilon_i t_{i,s} \middle| \varepsilon_i = 0 \text{ or } 1,\ 1 \leq i \leq n \right\}$, let $x_{-k,s}^n = -x_{k,s}^n$, $1 \leq k \leq 2^n$, and let $\mu_{n,s}$ be the measure associated with $\{x_{k,s}^n\}$ by Lemma 1. Define $\nu_{n,j} = \frac{1}{j} \sum_{s=1}^{j} \mu_{n,s}$.

The measures $\{\nu_{n,j}\}$ play a critical role in this work, largely due to the following two combinatorial properties.

Lemma 2. Let $1 \leq p < 2$, *and define* p' *by* $\frac{1}{p} + \frac{1}{p'} = 1$. *Assume the notations above, let* $j \in \mathbb{N}$, *and let* $\varepsilon > 0$. *There exists* $n_1(j) \in \mathbb{N}$ *so that if* $k \geq j + 1$ *and* $n \geq n_1(j)$, *then*

$$2^{nk/p'} \|\nu_{n,j}^k\|_{M_p(\mathbb{T})} < \varepsilon^k.$$

This result may be found in [6, Lemma 2.11].

Lemma 3. Let $1 \leq p < 2$, *and define* p' *by* $\frac{1}{p} + \frac{1}{p'} = 1$. *Assume Notations 1 and 2, let* $j \in \mathbb{N}$, *and let* $\{a_k\}_{k=1}^{j}$ *be a finite sequence of complex numbers. Then there exists* $n_2(j) \in \mathbb{N}$ *so that for all* $n \geq n_2(j)$, *we have*

$$\left\| \sum_{k=1}^{j} a_k 2^{nk/p'} \nu_{n,j}^k \right\|_{M_p(\mathbb{T})} \geq \frac{1}{2} \frac{j!}{j^j} 2^{-j/p'} |a_j|.$$

For a proof of this fact, we refer the reader to [6, Lemma 2.12]. A slight variant of the above construction may be found in Chapter 10 of [2].

We now begin the proof of Theorem 1. We first study the circle group.

Theorem 2. *Let* $1 \leq p < \infty$ *with* $p \neq 2$. *Then for every* $j \in \mathbb{Z}$, *we have*

$$\sup \| e^{ij\mu} \|_{M_p(\mathbb{T})} \geq c(3^{1/6})^{|j|},$$

the supremum taken over all $\mu \in M(\mathbb{T})$ *with* $\| \mu \|_{M_p(\mathbb{T})} \leq 1$, *and* $\hat{\mu}$ *real-valued. Here* $c > 0$ *is an absolute constant.*

Proof. We may assume that $j \in \mathbb{N}$ and $1 < p < 2$. Let $\frac{1}{p} + \frac{1}{p'} = 1$, and let $\varepsilon = 1/j$ in Lemma 2. Take $\alpha = 6$, and let $a_k = \frac{i^k \alpha^k 2^{-k/p'} j^k}{k!}$, $1 \leq k \leq j$, in Lemma 3. Fix $n = n(j) \geq \max(n_1(j), n_2(j))$, where $n_1(j)$ and $n_2(j)$ are as in Lemma 2 and 3, respectively. Let $\nu_{n,j}$ be the corresponding measure given in Notation 2. Define $\lambda = 2^{-1/p'} 2^{n/p'} \nu_{n,j}$. By the Riesz convexity theorem, we see that $\| \lambda \|_{M_p} \leq 1$, where $M_p = M_p(\mathbb{T})$ (see also [6, Lemma 2.8]). Then, using the Taylor series expansion,

$$e^{i\alpha j\lambda} = \sum_{k=0}^{j} \frac{i^k \alpha^k j^k \lambda^k}{k!} + \sum_{k=j+1}^{\infty} \frac{i^k \alpha^k j^k \lambda^k}{k!}. \tag{1}$$

By Lemma 2,

$$\begin{aligned} \Big\| \sum_{k=j+1}^{\infty} \frac{i^k \alpha^k j^k \lambda^k}{k!} \Big\|_{M_p} &\leq \sum_{k=j+1}^{\infty} \frac{\alpha^k j^k 2^{-k/p'} 2^{nk/p'} \| \nu_{n,j}^k \|_{M_p}}{k!} \\ &\leq \sum_{k=j+1}^{\infty} \frac{\alpha^k j^k 2^{-k/p'} (1/j)^k}{k!} \leq C, \end{aligned} \tag{2}$$

where C may be taken as e^6.

By Lemma 3.

$$\left\| \sum_{k=0}^{j} \frac{i^k \alpha^k j^k \lambda^k}{k!} \right\|_{M_p} \geq \left\| \sum_{k=1}^{j} \frac{i^k \alpha^k j^k 2^{-k/p'} 2^{nk/p'} \nu_{n,j}^k}{k!} \right\|_{M_p} - 1$$

$$\geq \frac{1}{2} \frac{j!}{j^j} 2^{-j/p'} \left(\frac{\alpha^j j^j 2^{-j/p'}}{j!} \right) - 1 \geq \frac{1}{2}\left(\frac{\alpha}{2}\right)^j - 1. \tag{3}$$

Hence by (1), (2), and (3), we have

$$\| e^{ij\alpha\lambda} \|_{M_p} \geq \frac{1}{2}\left(\frac{\alpha}{2}\right)^j - 1 - C. \tag{4}$$

Recalling that $\|\lambda\|_{M_p} \leq 1$, $\hat{\lambda}$ is real-valued, and that $\alpha = 6$, it is easy to see that (4) implies the theorem.

Corollary 1. Let $1 \leq p < \infty$ *with* $p \neq 2$, *and let* n *be a positive integer. Then for every* $j \in \mathbb{Z}$, *we have*

$$\sup \| e^{ij\mu} \|_{M_p(\mathbb{T}^n)} \geq c(3^{1/6})^{|j|},$$

the supremum taken over all $\mu \in M(\mathbb{T}^n)$ *with* $\|\mu\|_{M_p(\mathbb{T}^n)} \leq 1$, *and* $\hat{\mu}$ *real-valued. Here* $c > 0$ *is an absolute constant.*

Proof. We may assume $j \in \mathbb{N}$. Let $\lambda \in M(\mathbb{T})$ with $\|\lambda\|_{M_p} \leq 1$, $\hat{\lambda}$ real-valued, and so that $\| e^{ij\lambda} \| \geq c(3^{1/6})^j$. Define $\tau \in M(\mathbb{T}^n)$ by $\hat{\tau}(m_1, m_2, \ldots, m_n) = \hat{\lambda}(m_1)$ if $m_2 = \ldots = m_n = 0$ and $\hat{\tau}(m_1, m_2, \ldots, m_n) = 0$ otherwise. Here $(m_1, \ldots, m_n) \in \mathbb{Z}^n$. Then $\|\tau\|_{M_p(\mathbb{T}^n)} = \|\lambda\|_{M_p(\mathbb{T})} \leq 1$, $\hat{\tau}$ is real-valued, and $(e^{ij\tau})^\wedge|_{\mathbb{Z}} = (e^{ij\lambda})^\wedge$. Hence $\| e^{ij\tau} \|_{M_p(\mathbb{T}^n)} \geq \| e^{ij\lambda} \|_{M_p(\mathbb{T})}$, and the result follows immediately.

Corollary 2. Let $1 \leq p < \infty$ *with* $p \neq 2$, *and let* n *be a positive integer. Then for every* $j \in \mathbb{Z}$, *we have*

$$\sup \| e^{ij\mu} \|_{M_p(\mathbb{R}^n)} \geq c(3^{1/6})^{|j|},$$

the supremum taken over all $\mu \in M(\mathbb{R}^n)$ *with* $\|\mu\|_{M_p(\mathbb{R}^n)} \leq 1$, *and* $\hat{\mu}$ *real-valued. Here* $c > 0$ *is an absolute constant.*

Proof. Fix $j \in \mathbb{N}$. Let τ be as in the proof of Corollary 1. Then the support of τ is contained in $\mathbb{T}^n$, and we may view τ as a measure in $M(\mathbb{R}^n)$. Moreover, by a restriction theorem of de Leeuw [1, Sec. 3], we have $\| e^{ij\tau} \|_{M_p(\mathbb{R}^n)} \geq \| e^{ij\tau} \|_{M_p(\mathbb{T}^n)}$. The conclusion is now a consequence of Corollary 1.

We observe that Corollaries 1 and 2 and a simple argument involving approximate identities imply the following fact.

Lemma 4. Let $1 \leq p < \infty$ *with* $p \neq 2$, *and let* n *be a positive integer. Let* $G = \mathbb{T}^n$ *or* $\mathbb{R}^n$. *Then for every* $j \in \mathbb{Z}$, *we have*

$$\sup \| e^{ijf} \|_{M_p(G)} \geq c(3^{1/6})^{|j|},$$

the supremum taken over all $f \in L_1(G)$ *with* $\|f\|_{M_p(G)} \leq 1$, $\hat{f}$ *real-valued, and with* supp $\hat{f}$ *compact. Here* $c > 0$ *is an absolute constant.*

Corollary 3. Let $1 \leq p < \infty$ *with* $p \neq 2$, *and let* n *be a positive integer. Then for every* $j \in \mathbb{Z}$, *we have*

$$\sup \| e^{ijf} \|_{M_p(\mathbb{Z}^n)} \geq c\beta^{|j|},$$

the supremum taken over all $f \in \ell_1(\mathbb{Z}^n)$ *with* $\|f\|_{M_p(\mathbb{Z}^n)} \leq 1$, *and* $\hat{f}$ *real-valued. Here* $c > 0$ *and* $\beta > 1$ *are constants independent of* p.

Proof. Throughout this proof, we will write $\|\hat{T}\|_{M_p(G)}$ to denote the norm of the multiplier T on $L_p(G)$. We may assume that $1 < p < 2$ and that $j \in \mathbb{N}$. Let h be a function in $L_1(\mathbb{R}^n)$ so that $\|\hat{h}\|_{M_p(\mathbb{R}^n)} \leq 1$, $\hat{h}$ is real-valued, supp $\hat{h}$ is compact, and

$$\| e^{ij\hat{h}} \|_{M_p(\mathbb{R}^n)} \geq c(3^{1/6})^j . \qquad (5)$$

Choose $N \in \mathbb{N}$ so large that the function $\hat{h}_N(x) = \hat{h}(Nx)$ has support in $[-1, 1]^n$. By the dilation invariance of the norm of $M_p(\mathbb{R}^n)$, we have $\| \hat{h}_N \|_{M_p(\mathbb{R}^n)} = \| \hat{h} \|_{M_p(\mathbb{R}^n)} \leq 1$. Let φ be the periodic extension of $\hat{h}_N$ to $\mathbb{R}^n$ (with period 2π in each variable). By Theorem 2.3 of [4], $\| \varphi \|_{M_p(\mathbb{Z}^n)} \leq A \| \hat{h}_N \|_{M_p(\mathbb{R}^n)} \leq A$, where A depends only on n. Let $f \in \ell_1(\mathbb{Z}^n)$ with $\hat{f} = \varphi/A$. Then $\| \hat{f} \|_{M_p(\mathbb{Z}^n)} \leq 1$, $\hat{f}$ is real-valued, and by (5), it is not difficult to verify that

$$\| e^{ijA\hat{f}} \|_{M_p(\mathbb{Z}^n)} \geq c(3^{1/6})^j, \qquad (6)$$

where $c > 0$ is an appropriate constant (see [1, Theorem 4.5]). The result follows easily by (6).

Theorem 1 follows by Corollaries 1, 2, and 3.

REFERENCES

1. de Leeuw, K. On L_p multipliers. *Ann. Math.* 81 (1965): 364-379.

2. Graham, C., and McGehee, O. C. *Essays in Commutative Harmonic Analysis*. New York: Springer-Verlag, 1979.

3. Igari, S. Functions of L_p multipliers. *Tôhoku Math. J.* 21 (1969): 304-320.

4. Jodeit, M. Restrictions and extensions of Fourier multipliers. *Studia Math.* 34 (1970): 214-226.

5. Rudin, W. A strong converse of the Wiener-Levy theorem. *Canad. J. Math.* 14 (1962): 694-701.

6. Zafran, M. The functions operating on multiplier algebras. *J. Funct. Anal.* 26 (1977): 289-314.

7. Zafran, M. The spectra of multiplier transformations on the L_p spaces. *Ann. Math.* 103 (1976): 355-374.

8. Zygmund, A. *Trigonometric Series*. 2d ed., Vols. 1 & 2. New York: Cambridge Univ. Press, 1968.

PART THREE

Fourier Analysis on R^n and Real Analysis

ON WEIGHTED FRACTIONAL INTEGRALS

Angel E. Gatto, Cristian E. Gutiérrez, Richard L. Wheeden
Rutgers University

INTRODUCTION

Let w be a positive function in $L^1_{loc}(R)$ and f an integrable function with respect to the measure $w(x)\,dx$. If we call $I_1 f(x)$ the integral of f with respect to $w(t)\,dt$ over (a,x) and $I_\alpha f(x)$ the integral of $I_{\alpha-1} f$ with respect to $w(t)\,dt$ over (a,x), $\alpha =$ 2, 3, ..., then as in the unweighted case, a computation shows that

$$I_\alpha f(x) = c_\alpha \int_a^x f(t) \left(\int_t^x w(s)\,ds\right)^{\alpha-1} w(t)\,dt.$$

This leads us to consider

$$I_{\alpha,w} f(x) = \int_{-\infty}^{\infty} \frac{f(t)}{\left|\int_x^t w(s)\,ds\right|^{1-\alpha}}\, w(t)\,dt, \qquad (1)$$

for α a positive real number. The study of this integral can be reduced by a change of variables to the classical case. Calling

Supported by Consejo Nacional de Investigaciones Científicas y Técnicas, República Argentina.

Supported in part by NSF No. MCS80-03098.

$W(t) = \int_0^t w(s)\,ds$, we have $I_{\alpha,w}f(x) = I_{\alpha,1}(f \circ W^{-1})(W(x))$, where W^{-1} is the inverse function of W, and the basic properties of $I_{\alpha,w}f$ then follow from their analogues for $I_{\alpha,1}$.

In this paper we shall deal with an n-dimensional version of the integral (1). We shall assume that w is a strictly positive function in $L^1_{loc}(R^n)$ and define

$$I_{\alpha,w}f(x) = \int_{R^n} \frac{f(y)}{\left[\int_{B(x,|x-y|)} w(s)\,ds\right]^{1-\alpha/n}} w(y)\,dy, \qquad (2)$$

where $B(x,|x - y|)$ denotes the ball with center x and radius $|x - y|$. Our purpose is to study the behavior of this operator in weighted L^p, weighted *BMO*, and weighted Lipschitz classes. The theorems that are valid for the classical fractional integral have analogues for $I_{\alpha,w}$ with some mild conditions on w. Except for the L^p case (Theorem 1), we shall assume that the induced measure $\mu = m_w$ satisfies the doubling condition

$$\mu(B(x, 2r)) \leq C\,\mu(B(x, r)). \qquad (3)$$

We shall show (Lemma 1) that a measure μ satisfies (3) if and only if

$$\exists\ \beta,\ C,\quad 0 \leq \beta < 1,\quad C > 0 \quad \text{such that} \qquad (4)$$
$$\mu(B(x,\ \ell + r)) - \mu(B(x,\ \ell)) \leq C\,\mu(B(x,\ r))^{1-\beta}\mu(B(x,\ \ell))^{\beta},$$
$$\ell \geq r > 0.$$

This property is a generalization of the case μ = Lebesgue measure, in which $\beta = 1 - 1/n$, and enables us to handle the kernel in (2) as in the classical case.

Notation and Definitions

For a positive locally integrable function w, we shall denote by m_w the induced measure $m_w(E) = \int_E w(x)\,dx$. $B(x, r)$ or $B_r(x)$ will denote the ball with center x and radius r in R^n, $B'(x, r)$ or

$B'_r(x)$ its complement, and $\lambda B(x, r) = B(x, \lambda r)$. D^d stands for the class of measures μ that satisfy the doubling condition

$$\mu(B(x, tr)) \leq c\, t^d \mu(B(x, r)),\ t \geq 1,\ d \geq 1,$$

and $D^\infty = \bigcup_{d\geq 1} D^d$. $L_{p,w}$ denotes the usual weighted L^p space and $\| \ \|_{p,w}$ the norm in $L_{p,w}$. BMO_w will be the class of functions defined on R^n such that

$$\frac{1}{m_w(B_r(x))} \int_{B_r(x)} |f(t) - a_{B_r(x)}| w(t)\, dt \leq C \quad \forall\ r > 0,\ x \in R^n,$$

where C is independent of r and x; $\| f \|_{BMO_w}$ will denote the infimum of such C. By $\text{Lip}_w(\gamma)$, $\gamma > 0$, we denote the functions on R^n such that

$$|f(x + h) - f(x)| \leq c\, m_w(B_{|h|}(x))^{\gamma/n},$$

where c is independent of x and h; $\| f \|_{\text{Lip}_w(\gamma)}$ is the infimum of such C. M_w will denote the weighted maximal function, namely,

$$M_w f(x) = \sup_{r>0} \frac{1}{m_w(B(x, r))} \int_{B(x,r)} |f(y)| w(y)\, dy\,.$$

We will need the following results about M_w:

(i) The strong type (p, p), namely,

$$\| M_w f \|_{p,w} \leq c\, \| f \|_{p,w}, \qquad 1 < p < \infty;$$

(ii) The weak type $(1, 1)$, namely,

$$m_w\{x : |M_w f(x)| > \lambda\} \leq \frac{c}{\lambda} \| f \|_{1,w}, \qquad \lambda > 0.$$

We define formally

$$I_{\alpha,w} f(x) = \int_{R^n} \frac{1}{m_w(B_{|x-y|}(x))^{1-\alpha/n}} f(y) w(y)\, dy$$

and

$$\tilde{I}_{\alpha,w}f(x) = \int_{R^n}\left[\frac{1}{m_w(B_{|x-y|}(x))^{1-\alpha/n}} - \frac{\chi_{|y|>1}(y)}{m_w(B_{|y|}(0))^{1-\alpha/n}}\right]f(y)w(y)\,dy,$$

where $\chi_{|y|>r}(y)$ denotes the characteristic function of the set $\{y \in R^n : |y| > r\}$. The letter C will denote a positive constant not necessarily the same in each occurrence.

STATEMENTS OF THE RESULTS

Theorem 1. Assume that $m_w(B) \neq 0$ *for every ball* B, $m_w(R^n) = +\infty$ *and* $0 < \alpha < n$.

(i) *For* $f \in L_{p,w}$, $1 < p < n/\alpha$, $I_{\alpha,w}f(x)$ *converges absolutely a.e. and if* $1/q = 1/p - \alpha/n$, *then*

$$\| I_{\alpha,w}f \|_{q,w} \leq C \| f \|_{p,w}.$$

(ii) *For* $f \in L_{1,w}$, $I_{\alpha,w}f(x)$ *converges absolutely a.e. and*

$$m_w\{x : |I_{\alpha,w}f(x)| > \lambda\} \leq \left(\frac{C}{\lambda}\| f \|_{1,w}\right)^{n/(n-\alpha)}, \qquad \lambda > 0.$$

Theorem 2. Let $m_w \in D^\infty$, $0 < \alpha < n$. *Then for* $f \in L_{n/\alpha,w}$, $\tilde{I}_{\alpha,w}f(x)$ *converges absolutely a.e. and*

$$\| \tilde{I}_{\alpha,w}f \|_{BMO_w} \leq C \| f \|_{n/\alpha,w}.$$

Theorem 3. Let $m_w \in D^\infty$, $0 < \alpha < n$, $0 < \gamma$, *and* $\alpha + \gamma < n(1 - \beta)$, *where* $0 \leq \beta < 1$ *as in* (4). *Then for* $f \in \mathrm{Lip}_w(\gamma)$, $\tilde{I}_{\alpha,w}f(x)$ *converges absolutely for all* $x \in R^n$ *and*

$$\| \tilde{I}_{\alpha,w}f \|_{\mathrm{Lip}_w(\alpha+\gamma)} \leq C \| f \|_{\mathrm{Lip}_w(\gamma)}.$$

SOME PRELIMINARY LEMMAS

Lemma 1. *Let* μ *be a positive Borel measure on* R^n. *The following conditions are equivalent:*

(i) $\mu \in D^\infty$.

(ii) *There exist* β *and* C, $0 \leq \beta < 1$, $C > 0$, *which depend only on* μ *such that*

$$\mu(B(x, \ell + r)) - \mu(B(x, \ell)) \leq C\, \mu(B(x, r))^{1-\beta} \mu(B(x, \ell))^{\beta}, \quad (5)$$

where $\ell \geq r > 0$.

Proof. (i) => (ii). We shall show first that there exist $\bar{d}$ and C, $0 < \bar{d} \leq 1$, $C > 0$, such that

$$\mu(B(x, \ell + tr)) - \mu(B(x, \ell)) \geq C\, t^{\bar{d}}[\mu(B(x, \ell + r)) - \mu(B(x, \ell))] \quad (6)$$

for all $t \geq 1$, $\ell > 0$, $r > 0$. To prove (6) it is enough to prove that there exists a, $1 < a \leq 2$, such that

$$\mu(B(x, \ell + 2r)) - \mu(B(x, \ell)) \geq a[\mu(B(x, \ell + r)) - \mu(B(x, \ell))]. \quad (7)$$

In fact, for $t \geq 1$, let k be such that $2^k \leq t < 2^{k+1}$. Then

$$\mu(B(x, \ell + tr)) - \mu(B(x, \ell)) \geq \mu(B(x, \ell + 2^k r)) - \mu(B(x, \ell))$$
$$\geq a^k[\mu(B(x, \ell + r)) - \mu(B(x, \ell)],$$

and $a^k \geq a^{-1} t^{\log_2 a}$. Taking $\bar{d} = \log_2 a$, we get (6).

Let us show (7). We call $C_1 = B(x, \ell + r) - B(x, \ell)$ and $C_2 = B(x, \ell + 2r) - B(x, \ell + r)$. Let F be the family of all the balls of radius $r/2$ contained in C_1, and consider $F_2 = \{B_j\}_{j=1}^N$ a maximal subfamily of F of disjoint balls, that is, $B_j \cap B_k = \emptyset$, $j \neq k$, and if $B \in F$, $\exists\, j$ such that $B \cap B_j \neq \emptyset$. Let x_j be the center of B_j and $\bar{B}_j$ the ball with center $\bar{x}_j = x + (x_j - x)(1 + r/|x - x_j|)$ and radius $r/2$. Observe that $B_j \subset 3\bar{B}_j$, which implies $\mu(B_j) \leq C\, 3^d\, \mu(\bar{B}_j)$ where d is such that $\mu \in D^d$ with constant C. We also have $C_1 \subset \bigcup_1^N 3B_j$, $\bar{B}_j \cap \bar{B}_k = \emptyset$, $j \neq k$, and $\bigcup_1^N \bar{B}_j \subset C_2$. Therefore,

$$\mu(C_1) \leq \mu\left(\bigcup_1^N 3B_j\right) \leq \sum_1^N \mu(3B_j) \leq C3^d \sum_1^N \mu(B_j) \leq C3^d \sum_1^N C3^d \mu(\bar{B}_j)$$

$$= C^2 9^d \mu\left(\bigcup_1^N \bar{B}_j\right) \leq C^2 9^d \mu(C_2).$$

Then $\mu(C_1) + \mu(C_2) \geq (1 + C^{-2}9^{-d})\mu(C_1)$, which is (7).

Finally, from (6) for $\ell \geq r$, with $t = \ell/r$, we have

$$\mu(B(x, \ell + r)) - \mu(B(x, \ell)) \leq C^{-1}(\ell/r)^{-\bar{d}}\mu(B(x, 2\ell))$$

$$= C^{-1}(r/\ell)^{\bar{d}}\mu(B(x, 2\ell))^{1-\beta}\mu(B(x, 2\ell))^{\beta}$$

$$\leq C(r/\ell)^{\bar{d}}(2\ell/r)^{d(1-\beta)}\mu(B(x, r))^{1-\beta}2^{d\beta}\mu(B(x, \ell))^{\beta}.$$

Now taking β such that $\bar{d} = d(1 - \beta)$, we get (5).

To show that (ii) implies (i), we take $\ell = r$, obtaining

$$\mu(B(x, 2r)) \leq (C + 1)\mu(B(x, r)),$$

which shows that $\mu \in D^\infty$.

Lemma 2. If m_w is the measure induced by a positive locally integrable function w and $m_w(R^n) = +\infty$, then

(i) $\displaystyle\int_{B_R(x)} m_w(B(x,|x - y|))^\lambda w(y)\, dy = \frac{1}{\lambda + 1} m_w(B_R(x))^{1+\lambda},$

$\lambda > -1.$

(ii) $\displaystyle\int_{B'_R(x)} m_w(B(x,|x - y|))^\lambda w(y)\, dy = \frac{-1}{\lambda + 1} m_w(B_R(x))^{1+\lambda},$

$\lambda < -1.$

Proof. We shall only prove (i) for $-1 < \lambda < 0$, since the cases $\lambda \geq 0$ and (ii) can be obtained in a similar manner. Letting I denote the integral in (i), we will show that the following inequalities hold:

$$\frac{a-1}{a^{1+\lambda}-1} a^{\lambda} m_w(B_R(x))^{1+\lambda} \leq I \leq \frac{a-1}{a^{1+\lambda}-1} m_w(B_R(x))^{1+\lambda},$$

$\forall a > 1,\ -1 < \lambda < 0.$

To see this, let $y_k \in R^n$, $k = 0, 1, 2, \ldots$, be such that $|x - y_0| = R$, $|x - y_k| \to 0$, and if $B_k = B(x, |x - y_k|)$, then $m_w(B_k) = a^{-k} m_w(B_R(x))$. We have

$$I = \sum_{k=0}^{\infty} \int_{B_k - B_{k+1}} m_w(B(x, |x - y|))^{\lambda} w(y)\, dy$$

$$\leq \sum_{k=0}^{\infty} m_w(B_{k+1})^{\lambda} [m_w(B_k) - m_w(B_{k+1})]$$

$$= m_w(B_R(x))^{1+\lambda} (1 - a^{-1}) a^{-\lambda} \sum_{k=0}^{\infty} a^{-(1+\lambda)k}$$

$$= m_w(B_R(x))^{1+\lambda} \frac{a-1}{a^{1+\lambda}-1}.$$

Analogously, we get the other inequality. Now taking limits as $a \to 1$, we obtain (i).

Lemma 3. *Let $m_w \in D^\infty$, $0 < \alpha < n$, and β be as in Lemma 1. Then*

$$\left| \frac{1}{m_w(B(z, |z - y|))^{1-\alpha/n}} - \frac{1}{m_w(B(x, |x - y|))^{1-\alpha/n}} \right|$$

$$\leq C m_w(B(x, |x - z|))^{1-\beta} m_w(B(x, |x - y|))^{\beta + \frac{\alpha}{n} - 2},$$

for all x, y, z in R^n such that $|x - y| \geq 6|x - z|$.

Proof. We have

$$\left|\frac{1}{m_w(B(z,|z-y|))^{1-\alpha/n}} - \frac{1}{m_w(B(x,|x-y|))^{1-\alpha/n}}\right|$$

$$= \frac{1}{2-\alpha/n}\left|\int_{m_w(B(x,|x-y|))}^{m_w(B(z,|z-y|))} t^{\frac{\alpha}{n}-2}\,dt\right|$$

$$\leq \frac{1}{2-\alpha/n}\int_{m_w(B(x,|x-y|-2|x-z|))}^{m_w(B(x,|x-y|+2|x-z|))} t^{\frac{\alpha}{n}-2}\,dt$$

$$\leq \frac{1}{2-\alpha/n}\, m_w(B(x,|x-y|-2|x-z|))^{\frac{\alpha}{n}-2}$$

$$[m_w(B(x,|x-y|+2|x-z|)) - m_w(B(x,|x-y|-2|x-z|))].$$

Now using Lemma 1 and the doubling condition, we get the lemma.

Lemma 4. Let $m_w \in D^\infty$ *and* $\alpha < (1-\beta)n$, *where* β *is as in Lemma 1. Then the following integral is absolutely convergent and*

$$\int_{R^n}\left[\frac{1}{m_w(B(x,|x-y|))^{1-\alpha/n}} - \frac{1}{m_w(B(x+h,|x+h-y|))^{1-\alpha/n}}\right] w(y)\,dy = 0 \tag{8}$$

for all $x,\ h \in R^n$.

Proof. To show the absolute convergence of (8), observe that the integrand is locally integrable by Lemma 2(i), and on the other hand by Lemma 3, for $|x-y| > R > 6|h|$, it is less than $C\, m_w(B(x,|x-y|))^{\beta+(\alpha/n)-2}$, $0 \leq \beta < 1$, which by Lemma 2(ii) is an integrable function on that set.

Let us show (8). Take $R > 3|h|$. Since $B(x, R-|h|) \subset B(x+h, R) \subset B(x, R+|h|)$, there exists R', $|R'-R| \leq |h|$, such that $m_w(B(x, R')) = m_w(B(x+h, R))$. Then calling

$$I_R(x) = \int_{B_R(x)} \frac{1}{m_w(B(x,|x-y|))^{1-\alpha/n}}\, w(y)\,dy,$$

by Lemma 2 we have $I_{R'}(x) = I_R(x+h)$. Hence,

$$\int_{|x+h-y|<R}\left[\frac{1}{m_w(B(x,|x-y|))^{1-\alpha/n}}-\frac{1}{m_w(B(x+h,|x+h-y|))^{1-\alpha/n}}\right]w(y)\,dy$$

$$= \int_{|x+h-y|<R}\frac{1}{m_w(B(x,|x-y|))^{1-\alpha/n}}\,w(y)\,dy - I_{R'}(x)$$

$$= \int_{\{|x+h-y|\le R\}-\{|x-y|\le R'\}}\frac{1}{m_w(B(x,|x-y|))^{1-\alpha/n}}\,w(y)\,dy$$

$$- \int_{\{|x-y|\le R'\}-\{|x+h-y|\le R\}}\frac{1}{m_w(B(x,|x-y|))^{1-\alpha/n}}\,w(y)\,dy.$$

Since $|R - R'| \le |h|$, the last difference is in absolute value less than

$$\int_{R-|h|\le|x-y|\le R+|h|}\frac{1}{m_w(B(x,|x-y|))^{1-\alpha/n}}\,w(y)\,dy$$

$$\le \frac{1}{m_w(B_{R-|h|}(x))^{1-\alpha/n}}[m_w(B(x,\ R+|h|)) - m_w(B(x,\ R-|h|))],$$

which is by Lemma 1 and the doubling condition less than

$$C\, m_w(B_{|h|}(x))^{1-\beta} m_w(B_R(x))^{\beta+(\alpha/n)-1}.$$

Now taking into account that $\beta + (\alpha/n) - 1 < 0$ and $m_w(R^n) = \infty$ since $m_w \in D^\infty$, we get the result by letting $R \to \infty$.

PROOFS OF THE RESULTS

Proof of Theorem 1. We shall first show the following estimate:

$$|I_{\alpha,w}f(x)| \le C\,\|f\|_{p,w}^{p\alpha/n}[M_w f(x)]^{1-p\alpha/n}, \qquad 1 \le p < n/\alpha, \tag{9}$$

where M_w is the maximal function defined earlier. For a positive δ we write

$$I_{\alpha,w}f(x) = \int_{B_\delta(x)} \frac{f(y)}{m_w(B(x,|x - y|))^{1-\alpha/n}} w(y)\,dy$$

$$+ \int_{B'_\delta(x)} \frac{f(y)}{m_w(B(x,|x - y|))^{1-\alpha/n}} w(y)\,dy = I_1 + I_2.$$

To estimate I_1 we select y_k, $k = 0, 1, 2, \ldots$, such that $|x - y_0| = \delta$, $|x - y_k| \to 0$, and if $B_k = B(x,|x - y_k|)$, then $m_w(B_k) = 2^{-k} m_w(B_\delta(x))$. Then

$$|I_1| \le \sum_{k=0}^{\infty} \int_{B_k - B_{k+1}} \frac{|f(y)|}{m_w(B(x,|x - y|))^{1-\alpha/n}} w(y)\,dy$$

$$\le \sum_{k=0}^{\infty} \frac{m_w(B_k)}{m_w(B_{k+1})^{1-\alpha/n}} \frac{1}{m_w(B_k)} \int_{B_k} |f(y)|\, w(y)\,dy$$

$$\le m_w(B_\delta(x))^{\alpha/n}\, M_w f(x)\; 2^{1-\alpha/n} \sum_{k=0}^{\infty} 2^{-k\alpha/n}$$

$$\le C\, m_w(B_\delta(x))^{\alpha/n}\, M_w f(x).$$

On the other hand, if $1 < p < n/\alpha$, by Hölder's inequality we get

$$|I_2| \le \|f\|_{p,w} \left(\int_{B'_\delta(x)} \frac{w(y)}{m_w(B(x,|x - y|))^{(1-\alpha/n)p'}}\, dy \right)^{1/p'},$$

where $1/p + 1/p' = 1$. Since $(1 - \alpha/n)p' > 1$, by Lemma 2(ii) we obtain

$$|I_2| \le C \|f\|_{p,w}\, m_w(B_\delta(x))^{\alpha/n-1/p}.$$

For $p = 1$, since $m_w(B(x,|x - y|)) \ge m_w(B_\delta(x))$, $y \in B'_\delta(x)$, we also get the last estimate. Now choosing δ such that $m_w(B_\delta(x)) = [\|f\|_{p,w}/M_w f(x)]^p$, we have (9).

The theorem follows from (9) by raising both sides to the q-th power, observing that $(1 - \alpha p/n)q = p$, and using the strong type (p, p) and weak type $(1, 1)$ of M_w.

Proof of Theorem 2. Given a ball $B(x_0, r)$, we define

$$a_B = \int_{R^n} \left[\frac{\chi_{|x_0-y|\geq 6r}(y)}{m_w(B(x_0,|x_0-y|))^{1-\alpha/n}} - \frac{\chi_{|y|>1}(y)}{m_w(B(0,|y|))^{1-\alpha/n}} \right] f(y)w(y)\,dy$$

and

$$I(x) = \int_{R^n} \left[\frac{1}{m_w(B(x,|x-y|))^{1-\alpha/n}} - \frac{\chi_{|x_0-y|\geq 6r}(y)}{m_w(B(x_0,|x_0-y|))^{1-\alpha/n}} \right] f(y)w(y)\,dy.$$

The finiteness of the first integral follows from Hölder's inequality by observing that the expression in brackets belongs to $L_{n/(n-\alpha),w}$, which is a consequence of Lemmas 3 and 2(ii) and the fact that it is locally bounded. For the second one we shall prove

$$\int_{B(x_0,r)} |I(x)|w(x)\,dx \leq C\, m_w(B(x_0, r))\, \|f\|_{n/\alpha,w}, \tag{10}$$

where C is a constant that depends only on m_w and α.

Let us assume for a moment that (10) is true. Then for almost every point in $B(x_0, r)$, we can write $\tilde{I}_{\alpha,w}f(x) = I(x) + a_B$, which shows that $\tilde{I}_{\alpha,w}f$ is well defined in $B(x_0, r)$ and

$$\frac{1}{m_w(B(x_0, r))} \int_{B(x_0,r)} |\tilde{I}_{\alpha,w}f(x) - a_B|w(x)\,dx \leq C\, \|f\|_{n/\alpha,w}.$$

Since the arguments are true for any ball, we get the theorem.

It remains to prove (10). Let us write

$$I(x) = \int_{|x_0-y|\leq 6r} \frac{1}{m_w(B(x,|x-y|))^{1-\alpha/n}} f(y)w(y)\,dy$$

$$+ \int_{|x_0-y|>6r} \left[\frac{1}{m_w(B(x,|x-y|))^{1-\alpha/n}} - \frac{1}{m_w(B(x_0,|x_0-y|))^{1-\alpha/n}} \right]$$

$$f(y)w(y)\,dy = I_1(x) + I_2(x).$$

For $I_1(x)$, by Fubini's theorem we have

$$\int_{|x_0-x|<r} |I_1(x)|w(x)\,dx$$

$$\leq \int_{|x_0-y|<6r} |f(y)|w(y) \int_{|y-x|<7r} \frac{w(x)}{m_w(B(x,|x-y|))^{1-\alpha/n}}\,dx\,dy.$$

Since $B(y,|x-y|) \subset B(x, 2|x-y|)$, using the doubling condition and Lemma 2, we get

$$\int_{|x_0-x|<r} |I_1(x)|w(x)\,dx$$

$$\leq C \int_{|x_0-y|<6r} |f(y)|w(y) \int_{|y-x|<7r} \frac{w(x)}{m_w(B(y,|x-y|))^{1-\alpha/n}}\,dx\,dy$$

$$\leq C \int_{|x_0-y|<6r} |f(y)|w(y)m_w(B(y, 7r))^{\alpha/n}\,dy$$

$$\leq C\, m_w(B(x_0, r))^{\alpha/n} \int_{|x_0-y|<6r} |f(y)|w(y)\,dy.$$

Finally, using Hölder's inequality we get (10) for I_1. For I_2, using Hölder's inequality and Lemma 3, we obtain for $|x - x_0| < r$ that

$$|I_2(x)| \leq C\, \|f\|_{n/\alpha,w}\, m_w(B(x_0, r))^{1-\beta}$$

$$\left[\int_{|y-x_0|>6r} m_w(B(x_0,|x_0-y|))^{(\beta+\frac{\alpha}{n}-2)\frac{n}{n-\alpha}} w(y)\,dy\right]^{\frac{n-\alpha}{\alpha}}.$$

Therefore, by Lemma 2,

$$|I_2(x)| \leq C\, \|f\|_{n/\alpha,w}, \qquad x \in B(x_0, r).$$

Integrating over the ball $B(x_0, r)$ completes the proof.

Proof of Theorem 3. We begin by observing that since $f \in \mathrm{Lip}_w(\gamma)$, f is locally bounded and $f(y) = O\big(m_w(B(x,|x-y|))^{\gamma/n}\big)$ as

$|x - y| \to \infty$. Then since $\alpha + \gamma < (1 - \beta)n$, by Lemmas 2 and 3 it follows that $\tilde{I}_{\alpha,w} f(x)$ exists as a Lebesgue integral for all $x \in R^n$. Hence, using Lemma 4, we can write

$$\tilde{I}_{\alpha,w} f(x+h) - \tilde{I}_{\alpha,w} f(x) = \int [f(y) - f(x)] \left[\frac{1}{m_w(B(x+h, |x+h-y|))^{1-\alpha/n}} - \frac{1}{m_w(B(x, |x-y|))^{1-\alpha/n}} \right] w(y)\, dy$$

$$= \left(\int_{|x-y| \le 6|h|} + \int_{|x-y| > 6|h|} \right) [f(y) - f(x)] \left[\frac{1}{m_w(B(x+h, |x+h-y|))^{1-\alpha/n}} - \frac{1}{m_w(B(x, |x-y|))^{1-\alpha/n}} \right] w(y)\, dy$$

$$= I_1 + I_2 .$$

For I_1, using the Lipschitz condition, Lemma 2(i), and the doubling condition, we get

$$|I_1| \le C \|f\|_{\mathrm{Lip}_w(\gamma)} m_w(B(x, |h|))^{\gamma/n+\alpha/n} .$$

For I_2, using Lemma 3, the Lipschitz condition, and Lemma 2(ii), we also get

$$|I_2| \le C \|f\|_{\mathrm{Lip}_w(\gamma)} m_w(B(x, |h|))^{\gamma/n+\alpha/n} .$$

This concludes the proof of the theorem.

Remarks. There is an operator closely related to $I_{\alpha,w}$, namely,

$$I^{(2)}_{\alpha,w} f(x) = \int_{R^n} \frac{1}{m_w(B_{|x-y|}(y))^{1-\alpha/n}} f(y) w(y)\, dy .$$

For doubling measures, $C^{-1} I_{\alpha,w}f(x) \leq I^{(2)}_{\alpha,w}f(x) \leq C\, I_{\alpha,w}f(x)$ for some $C > 0$. Therefore, Theorem 1 holds for $I^{(2)}_{\alpha,w}$. Also $I^{(2)}_{\alpha,w}$ is a bounded operator from the homogeneous weighted atomic space H^1_w (see [2] for definition) into $L_{n/(n-\alpha),w}$, which follows using Lemma 3. This result gives by duality Theorem 2.

Finally, we want to mention that the following inequality similar to that in Lemma 1(ii) holds:

$$m_w(B_{|x-x_0|+r}(x)) - m_w(B_{|x-x_0|}(x)) \leq C\, m_w(B_r(x_0))^{1-\beta} m_w(B_{|x-x_0|}(x))^{\beta},$$

for $|x - x_0| \geq r$, $0 \leq \beta < 1$. With this estimate one can also prove that $I_{\alpha,w}$ is a bounded operator from the homogeneous weighted atomic space H^1_w into $L_{n/(n-\alpha),w}$.

REFERENCES

1. Calderón, A. Notes of the course "Análisis Real Avanzado," given at the Univ. of Buenos Aires, 1979.

2. Coifman, R., and Weiss, G. Extensions of Hardy spaces and their use in analysis. *Bull. Amer. Math. Soc.* 83 (1977): 569-645.

3. Lotkowski, E., and Wheeden, R. The equivalence of various Lipschitz condition on the weighted mean oscillation of a function. *Proc. Amer. Math. Soc.* 61 (1976): 323-328.

4. Macías, R. Interpolation theorems on generalized Hardy spaces. Doctoral dissertation, Washington Univ., St. Louis, 1974.

5. Zygmund, A. *Trigonometric Series*. 2d ed. New York: Cambridge Univ. Press, 1968.

THE DENSITY OF THE AREA INTEGRAL

Richard R. Gundy
Rutgers University

Consider $A(u)$, the Lusin area integral of a harmonic function u in the half-space R_+^2,

$$A^2(u)(x_0) = \iint_{\Gamma(x_0)} |\nabla u|^2(x, y)\, dx\, dy,$$

where $\Gamma(x_0) = \{(x, y) : |x - x_0| \leq \alpha y\}$ is a cone with vertex at $x_0 \in R^1 = \partial R_+^2$. The area integral is related to the nontangential maximal function $N(u) = \sup_{(x,y)\in\Gamma(x_0)} |u(x, y)|$ by the inequalities

$$\| N(u) \|_p \cong \| A(u) \|_p, \quad 0 < p < \infty,$$

which are the foundation of contemporary H^p-space theory [4]. There is another functional, defined on harmonic functions, that is intermediate between A and N in the sense that it shares properties of both of these operators: we shall call this functional D, the (maximal) density of the area integral. The object of this paper is to demonstrate that D is equivalent to A and N in the sense that:

This research was supported in part by a grant from the National Science Foundation.

Theorem 1.

$$\|A\|_p \cong \|D\|_p \cong \|N\|_p \quad \text{for} \quad 0 < p < \infty.$$

That D is intermediate between A and N is illustrated by the following ratio inequality, a corollary of the theorem:

Corollary 1.

$$\|A^2/N\|_p \cong \|A\|_p \quad \text{for} \quad 0 < p < \infty.$$

The functional D is defined by writing the area integral

$$\begin{aligned} A^2(u)(x_0) &= \iint_{\Gamma(x_0)} |\nabla u|^2 \, dx\, dy \\ &= \iint_{\{(x,y)\in u^{-1}(r)\cap\Gamma(x_0)\}} |\nabla u| \qquad \sigma_r(dx\, dy)\, dr \\ &= \int D(u;\ x_0)(r)\, dr, \end{aligned}$$

where $\sigma_r(dx\, dy)$ is the arc-length of the level set $u^{-1}(r)$ in the cone $\Gamma(x_0)$. This change-of-variables formula is known as the *co-area formula* in geometric measure theory (see Federer [6, Sec. 3.2]). (In our case, however, verification is easy: most of the subtleties of that formula are absent because (i) u is analytic and (ii) its singularities may be neglected because of the presence of the square on $|\nabla u|$.) Now define

$$D(u)(x_0) = \operatorname*{ess\ sup}_r D(u;\ x_0)(r),$$

so that D is the maximal density of A^2.

Since the function $D(u;\ x_0)(r)$ is defined on the range of u, we have the following estimate:

$$\begin{aligned} A^2(u)(x_0) &= \int_{-N(u)}^{N(u)} D(u;\ x_0)(r)\, dr \\ &\leq 2N(u)(x_0)D(u)(x_0) \end{aligned}$$

so that

$$\frac{A^2(u)(x_0)}{N(u)(x_0)} \le 2D(u)(x_0).$$

From this estimate we see that at least one part of the equivalence indicated in the corollary follows from Theorem 1. We return to this point below.

The proof of the theorem depends on some recent, remarkable inequalities obtained by M. Barlow and M. Yor [2] for a certain functional of Brownian motion stochastic integrals. This functional, the "local time" (and the maximal local time), is a probabilistic analogue of the geometric functional $D(u; x_0)(r)$ (and $D(u)(x_0)$ just described). The inequalities of Barlow and Yor relate the martingale maximal function (analogous to $N(u)$) and the quadratic variation (analogous to $A^2(u)$) to the maximal local time (analogous to D). The relation is an L_p-norm equivalence of exactly the type given in the theorem. As might be expected, the strategy of proof for the principal theorem of this paper consists of relating the Brownian functional to the geometric functional.

The study of local time for Brownian motion originated in the work of Paul Lévy. As the era of the Ito integral developed, much attention has been given to the study of local time for the stochastic integral (see, for example, Azéma and Yor [1]). Rather than give a too-brief summary of the theory of local time, permit us to give a definition of this notion by drawing a parallel with the area integral. Corresponding to "harmonic function," we have the probability counterpart, "stochastic integral." This is a function (random variable) defined on $\Omega \times R^1_+$ in somewhat the same way our $u(x, y)$ is defined on $R^1 \times R^1_+$ (although the analogy between the time variable t and the y-coordinate is limited). For convenience, let us replace $R^1 \times R^1_+$ by the unit disc $U = \{(x, y) : x^2 + y^2 \le 1\}$. Now we can turn the harmonic function u into a stochastic process by substituting a two-dimensional Brownian motion $(x_t^{(\omega)}, y_t^{(\omega)})$, started at the origin, stopped the first time $\tau(\omega)$ it hits the unit circle. It turns out that this stochastic process $u(x_t, y_t)$ is also a (one-dimensional) stochastic integral:

$$u(x_t, y_t) = u(0, 0) + \int_0^t \frac{\partial u}{\partial x}(x_s, y_s)\, dx_s + \frac{\partial u}{\partial y}(x_s, y_s)\, dy_s.$$

As a stochastic integral, it has an "intrinsic time," defined by

$$\langle u \rangle = \int_0^t |\nabla u|^2 (x_s, y_s)\, ds,$$

and for historical reasons, we denote

$$\langle u \rangle^{\frac{1}{2}} = S(u).$$

The random variable $S(u)$ is the analogue of the area integral $A(u)$. Here is the point: *both* $A(u)$ *and* $S(u)$ *may be considered as measures on* R^1. For $A(u)$, think of u as a mapping

$$u : \Gamma(x_0) \to R^1$$

and as such, induces a map of measures on Γ into measures on R^1:

$$u_*(dm) = dm \cdot u^{-1}.$$

As we have pointed out above, if $dm = |\nabla u|^2 (x, y)\, dx\, dy$, then

$$u_*(|\nabla u|^2\, dx\, dy) = D(u, x_0)(r)\, dr,$$

that is, $u_*(dm)$ is absolutely continuous. The facts about local time are exactly analogous: if we think of the (random) interval $[0, \tau(\omega)]$ as the substitute for $\Gamma(x_0)$, then $u(x_0(\omega), y_0(\omega)) = u \cdot z_0$ is a mapping

$$u \cdot z : [0, \tau(\omega)] \to R^1$$

and carries the random measure $|\nabla u|^2 (x_t(\omega), y_t(\omega))\, dt$ onto a random measure on R^1. It turns out that this measure on R^1 is also absolutely continuous:

$$(u \cdot z)_*(|\nabla u|^2\, dt) = L(r)\, dr.$$

The density $L(r) = L(u;\ \omega)(r)$ is the "local time at r," first introduced by Paul Lévy in the special case $u(x, y) = x$. It remains for us to relate $D(u;\ x_0)(r)$ to $L(u;\ \omega)(r)$. Here is an outline of how these ideas fit together. Fix the cone $\Gamma = \Gamma(1)$ and consider the Brownian path $z_t(\omega) = \left(x_t(\omega),\ y_t(\omega)\right)$ conditioned to terminate at the vertex $z = 1$. The conditioned motion is, of course, different from the original Brownian motion in that the latter diffuses uniformly in all directions from the origin, while the former is pulled toward a fixed point $z = 1$. For the conditioned process, the cone is significant: while the conditioned traveler is in the cone, he hardly feels the pull toward the vertex. In the cone, he meanders uniformly in all directions. This translates to a comparison between the area integral $A(u)$ and the intrinsic time $S(u)$. If we turn off the clock while z_t wanders outside the cone, then the intrinsic time *inside* the cone is given by

$$S^2(u\chi_\Gamma) = \int_0^\tau (|\nabla u|^2 \chi_{\Gamma(1)})(z_t)\, dt,$$

where $\chi_{\Gamma(1)}$ is the characteristic function of the cone. Furthermore, the occupation time of any subset $B \subset \Gamma(1)$ is given as

$$S_B^2(u\chi_\Gamma) = \int_0^\tau (|\nabla u|^2 \chi_{\Gamma(1)} \chi_B)(z_t)\, dt.$$

Let us take the expectation of this random variable $S_B^2(u\chi_\Gamma)$ with respect to the *conditioned measure* supported on those paths terminating at $z = 1$. Because the conditioned paths wander uniformly in Γ, we obtain

$$E\left(S_B^2(u\chi_\Gamma) \,\|\, z_\tau = 1\right) \cong \iint_B (|\nabla u|^2 \chi_{\Gamma(1)})(x,\ y)\, dx\, dy = A_B^2(u\chi_\Gamma)(1).$$

This equivalence leads to the *local estimation*

$$D(u;\ x_0)(r) \cong E\left(L(u\chi_\Gamma;\ \omega)(r) \,\|\, z_\tau = 1\right).$$

Now we take the suprema of both sides to obtain

$$D(u)(x_0) \leq E\left(\sup_r (L\, u\chi_\Gamma)(r) \parallel z_\tau = 1\right).$$

With this inequality, we may apply the results of Barlow and Yor, for whom the basic object of study is $\sup_r L(u;\, \omega)(r)$. Our aim is to obtain distribution function estimates for D in terms of A (or N). As usual, such estimates come from some more basic (easier) inequalities, usually L^2 estimates. (Recall the proof of the equivalence between the maximal function N and the area integral A.) In this case, the basic estimate is given above, and the L^2 inequality on the probability side is due to Barlow and Yor. (It should be remarked that, contrary to what has just been said, Barlow and Yor's result is *not* an elementary fact, even for L^2!) With this sketch, we now give the proofs.

THE PROOFS

The notation used in this section follows the conventions introduced above. However, the setting is the half-plane R^2_+ rather than the disc. This introduces some awkwardness on the probability side of the discussion, but certain formulas are simpler. The Brownian motion, instead of emanating from the origin, comes from infinity—the "background radiation" process used in [4] and [8]. This is not a probability measure on paths from infinity; rather, we have a set function $\mathscr{P}$ that is a σ-finite measure on continuous paths (x_t, y_t), $-\infty < t \leq 0$ with $\lim_{t\to-\infty} y_t = \infty$ and $y_0 = 0$. The measure $\mathscr{P}$ is characterized by the following requirement: given a level $y = N$, we may consider the random variables $\tau_N = \inf\{t : y_t = N\}$ and the pair (x_{τ_N}, y_{τ_N}). We demand that the distribution of (x_{τ_N}, y_{τ_N}) coincide with Lebesgue measure on the line $y = N$, and that $(x_{\tau_N+t}, y_{\tau_N+t})$ be distributed as two-dimensional Brownian motion started from the line $y = N$, killed when y_{τ_N+t} first vanishes.

We will also use *conditioned background radiation*; heuristically, this is the original process conditioned to terminate at a fixed point x_0 on the boundary $\partial R^2_+ = R^1$. This construction was first introduced by Doob [5] in a general context. The conditioned process (like the unconditioned, original process) is characterized by its transitions after the stopping times τ_N, together with the distribution of the variable (x_{τ_N}, N). Let $K_{x_0}(z)$ be the Poisson kernel centered at $x_0 \in R^1$, $z = (x, y)$. The density of x_{τ_N} is taken to be $K_{x_0}(x, N)$ and the transitions

$$p_t^{x_0}(z_1, z_2) = p_t(z_1, z_2)K_{x_0}(z_2)/K_{x_0}(z_1), \tag{1}$$

where $p_t(z_1, z_2)$ is the density of the two-dimensional Brownian process killed at the boundary ∂R^2_+. A full discussion of this process may be extracted from an appendix to P. A. Meyer's article [9] and from Doob [5]. For our purposes, the most important fact is that for $z_1, z_2 \in \Gamma(x_0)$,

$$p_t^{x_0}(z_1, z_2) \cong p_t(z_1, z_2), \tag{2}$$

where the constants for this equivalence depend only on the opening of the cone.

Lemma 1. Let $\Gamma = \Gamma(x_0)$ be a cone with vertex x_0 and W a Borel subset of R^2_+. Then

$$E\left(\int_{-\infty}^{0} |\nabla u|^2 \chi_{W\cap\Gamma(x_0)}(z_t)\, dt \,\|\, x_0\right) \geq C\, A_W^2(u)(x_0)$$

$$= \iint |\nabla u|^2 \chi_{W\cap\Gamma(x_0)}(x, y)\, dx\, dy.$$

Proof. Divide the cone $\Gamma(x_0)$ into parallel strips of width 2^{n-1}, $n = 0, \pm 1, \pm 2, \ldots$, and call each strip Γ_n. (Precisely $\Gamma_n = \{(x, y) : (x, y) \in \Gamma \text{ and } 2^{n-1} \leq y \leq 2^n\}$.) Now consider another cone

$\Gamma' = \Gamma'(x_0)$ with opening slightly larger than the opening for $\Gamma(x_0)$. Now define Γ'_n as the strip surrounding Γ_n:

$$\Gamma'_n = \{(x, y) : (x, y) \in \Gamma' \text{ and } 2^{n-1} - 2^{n-3} \leq y \leq 2^n + 2^{n-3}\}.$$

Then in particular, $\Gamma_0 \subset \Gamma'_0$ so that the Green function for Γ'_0, $g_0(z_1, z_2) \geq C > 0$ uniformly for $z_1, z_2 \in \Gamma_0$. The same lower bound holds for the Green functions $g_n(z_1, z_2)$ corresponding to Γ'_n with $z_1, z_2 \in \Gamma_n$, since each pair (Γ_n, Γ'_n) is a dilation of (Γ_0, Γ'_0).

Define stopping times

$$\tau_n = \inf\{t : z_t \in \Gamma_n\}$$

$$\gamma_n = \inf\{t : t > \tau_n, z_t \in \partial\Gamma'_n\}.$$

Then

$$E\left(\int_{-\infty}^{0} |\nabla u|^2 \chi_{W\cap\Gamma(x_0)}(z_t)\, dt \,\Big\|\, x_0\right)$$

$$\geq \sum_{n=1}^{\infty} E\left(\int_{\tau_{2n}}^{\gamma_{2n}} |\nabla u|^2 \chi_{W\cap\Gamma(x_0)}(z_t)\, dt \,\Big\|\, x_0\right)$$

$$\cong \sum_{n=-\infty}^{\infty} E\left(\iint |\nabla u|^2 \chi_{W\cap\Gamma_{2n}}\, g_{2n}(z_{\tau_{2n}}, (x, y))\, dx\, dy \,\Big\|\, x_0\right)$$

$$\geq C \sum_{n=-\infty}^{\infty} \iint |\nabla u|^2 \chi_{W\cap\Gamma_{2n}}(x, y)\, dx\, dy\; P(\tau_{2n} < \infty \,\|\, x_0)$$

$$\geq C \sum_{n=-\infty}^{\infty} \iint |\nabla u|^2 \chi_{W\cap\Gamma_{2n}}(x, y)\, dx\, dy.$$

This string of inequalities may be justified as follows. The first inequality is due to the fact that the larger strips Γ'_{2n} are disjoint pieces of the larger cone $\Gamma'(x_0)$. The equivalence is due to the equivalence (2) and the fact that the Green function is the time integral of the transition density. The last inequality is a consequence of the lower boundedness of the Green function mentioned above, and the fact that $P(\tau_{2n} < \infty) \geq C > 0$ for all n. (This is a probability calculation that flows easily from the form of the conditional density given above.) Finally, we remark that the same argument holds when τ_{2n} and γ_{2n} are replaced by τ_{2n+1} and γ_{2n+1}, so that

$$E\left(\int_{-\infty}^{0} |\nabla u|^2 \chi_{W\cap\Gamma(x_0)}(z_t)\, dt \,\|\, x_0 \right) \geq 2C\, A_W^2(u)(x_0)$$

as required.

Now we give the principal estimate.

Lemma 2. Let x_0 *be such that* $A(u)(x_0) < \infty$. *Let* $D(u)(x_0)$ *and* $L(u\chi_{\Gamma(x_0)})(\omega)$ *be the maximal functions of* $D(u;\, x_0)(r)$ *and* $L(u\chi_{\Gamma(x_0)};\, \omega)(r)$, *respectively. Then we have*

$$D(u)(x_0) \leq CE\left(L(u\chi_{\Gamma(x_0)}) \,\|\, x_0\right).$$

Proof. We apply Lemma 1 with the set $W = \{(x, y) : (x, y) \in \Gamma(x_0)$ and $r - \varepsilon \leq u(x, y) \leq r + \varepsilon\}$. Then

$$\begin{aligned} A_W^2(u)(x_0) &= \int_{r-\varepsilon}^{r+\varepsilon} D(u;\, x_0)(r')\, dr' \\ &\leq CE\left(\int_{-\infty}^{0} |\nabla u|^2 \chi_{\Gamma(x_0)\cap W}(x_t, y_t)\, dt \,\|\, x_0 \right) \\ &= E\left(\int_{r-\varepsilon}^{r+\varepsilon} L(u\chi_{\Gamma(x_0)})(r')\, dr' \,\|\, x_0 \right), \end{aligned}$$

so that we may divide each side by 2ε. By the Lebesgue differentiation theorem, the left-hand side tends to $D(u;\, x_0)(r)$ for almost every r, while the right-hand side is majorized by $E\left(L(u\chi_{\Gamma(x_0)}) \,\|\, x_0\right)$. This proves the assertion.

Proof of Theorem 1. The proof is divided into several cases.

Case 1. $\| D \|_p \leq c_p \| A \|_p$, $p \geq 1$. We have

$$C \int_{-\infty}^{\infty} |D(u)(x)|^p \, dx \leq \int_{-\infty}^{\infty} |E(L(u\chi_{\Gamma(x)}) \| x)|^p \, dx$$

$$\leq \int_{-\infty}^{\infty} E(L^p(u) \| x) \, dx$$

$$= E(L^p(u))$$

$$\leq c_p \| A \|_p^p ,$$

where the final inequality follows from the results of Barlow and Yor [2] and the equivalence of Brownian motion H^p and geometric H^p (see [4]).

Case 2. $\| D \|_p \leq c_p \| A \|_p$, $0 < p < 1$. To establish this inequality, we prove the following distribution function inequality:

$$\lambda^2 m(D > \lambda) \leq c_0 \int_{\{A \leq \lambda\}} A^2(u) + c_1 \lambda^2 m(A(u) > \lambda). \tag{3}$$

The use of this kind of inequality is well known by now (see Lemma 2.2 of [3] and Fefferman and Stein [7]). We may integrate both sides of this inequality to obtain the required norm estimates for the range $0 < p < 2$.

To prove inequality (3), we let $G = \{x : A(u)(x) \leq \lambda\}$ and define $W = \{(x', y) : (x', y) \in \cup \Gamma(x), x \in G\}$. The region W is bounded by a saw-toothed boundary line rising above the real axis. Now define

$$D_W(u)(x) = D(u\chi_W)(x)$$

and notice that $D_W(u)(x) = D(u)(x)$ for $x \in G$. Therefore,

$$\lambda^2 m(D > \lambda) \leq \lambda^2 m(\{D > \lambda\} \cap G) + \lambda^2 m(G^c), \tag{4}$$

$$= \lambda^2 m(D_W(u) > \lambda) + \lambda^2 m(A > \lambda),$$

so that we have reduced the problem to an estimation involving D_W. For this, we have

$$\begin{aligned}\lambda^2 m(D_W > \lambda) &\leq \int |D_W(u)(x)|^2\,dx \leq C \int E(|L_W(u)|^2 \| x)\,dx \\ &= CE(|L_W|^2) \\ &\cong CE(S_W^2(u)) \\ &= C \iint |\nabla u|^2 \chi_W(x, y)\, y\,dx\,dy \\ &= C \int A_W^2(u)(x)\,dx \\ &= C\left[\int_G A_W^2(u)(x)\,dx + \int_{G^C} A_W^2(u)(x)\,dx\right].\end{aligned}$$

The set G^C is a union of disjoint, open intervals and for $x \in G^C$, we have

$$\begin{aligned}A_W^2(u)(x) &\leq A_W^2(u)(x_1) + A_W^2(u)(x_2) \\ &= 2\lambda^2,\end{aligned}$$

where x_i, $i = 1, 2$ are the left and right end points of the component interval containing the point x. Therefore, we may majorize the above inequalities to conclude that

$$\begin{aligned}\lambda^2 m(D_W > \lambda) &\leq C\left[\int_G A_W^2(u) + 2\lambda^2 m(G^C)\right] \\ &= C\left[\int_{\{A\leq\lambda\}} A^2(u) + 2\lambda^2 m(A > \lambda)\right].\end{aligned}$$

Together with (4), this inequality concludes the proof of Case 2.

Case 3. $\|A\|_p \leq C_p \|D\|_p$ for $0 < p < \infty$. This assertion follows from the Schwarz inequality and the fact that $A^2/N \leq 2D$:

$$\begin{aligned}\int A^p(u)\,dx &= \int \left(\frac{A}{N^{\frac{1}{2}}}\right)^p N^{p/2} \\ &\leq \left[\int \left(\frac{A^2}{N}\right)^p\right]^{\frac{1}{2}} \cdot \left[\int N^p\right]^{\frac{1}{2}}\end{aligned}$$

so that

$$\int A^p(u)\,dx \leq C \int D^p(u)\,dx$$

as required.

Epilogue. The arguments in this paper are limited to the space R^2_+ and represented the state of my understanding at the time of the conference. A discussion of the n-dimensional result, together with other inequalities, will appear in a subsequent paper with R. Fefferman, M. Silverstein, and E. M. Stein.

REFERENCES

1. Azéma, J., and Yor, M. Temps locaux. *Asterisques Soc. Math. de France* (1978).

2. Barlow, M., and Yor, M. (Semi) martingales and local times. *Z. Wahr. verw. Gebiete* 55 (1981): 237-254.

3. Burkholder, D. L., and Gundy, R. F. Extrapolation and interpolation of quasi-linear operators on martingales. *Acta Math.* 124 (1970): 249-304.

4. Burkholder, D. L.; Gundy, R. F.; and Silverstein, M. L. A maximal function characterization of the class H^p. *Trans. Amer. Math. Soc.* 157 (1972): 137-153.

5. Doob, J. L. Conditional Brownian motion and boundary limits of harmonic functions. *Bull. Soc. Math. France* 85 (1957): 431-458.

6. Federer, H. *Geometric measure theory.* New York: Springer-Verlag, 1969.

7. Fefferman, C., and Stein, E. M. H^p spaces of several variables. *Acta Math.* 129 (1972): 137-194.

8. Gundy, R. F., and Varopoulos, N. Th. Les transformations de Riesz et les intégrales stochastiques. *C. R. Acad. Sc. Paris,* t. 289 (2 juillet 1979).

9. Meyer, P. A. Le dual de $H^1(R^\nu)$: démonstrations probabilistes. *Séminaire de Prob.* 11 (1977): 132-195. New York: Springer-Verlag.

THE HARDY-LITTLEWOOD MAXIMAL FUNCTION ON $L(p,1)$

Richard A. Hunt, Douglas S. Kurtz
Purdue University

The Hardy-Littlewood maximal function of a locally integrable function, f, is

$$Mf(x) = \sup \frac{1}{|Q|} \int_Q |f(t)|\, dt,$$

the supremum taken over all cubes, Q, containing x. In a previous paper [1], we considered norm inequalities of the form

$$\| Mf \|_{p\bar{q}} \le C \| f \|_{pq}, \quad 1 < p < \infty, \quad 1 \le q \le \bar{q} \le \infty, \tag{1}$$

where

$$\| f \|_{pq} = \begin{cases} \left(q \int_0^\infty w(\{x \in \mathbf{R}^n : |f(x)| > y\})^{q/p} y^{q-1}\, dy \right)^{1/q}, & 1 \le q < \infty \\ \sup\limits_{y>0} y\, w(\{x \in \mathbf{R}^n : |f(x)| > y\})^{1/p}, & q = \infty \end{cases}$$

is a norm on the weighted Lorentz space $L(p, q) = L(p, q; w(x)\, dx)$ for $1 < p < \infty$, $1 \le q \le \infty$, and w nonnegative and locally integrable.

Research supported in part by NSF grant no. MCS-7900918. This article appears in Vol. 32 (1983) of the *Indiana University Mathematics Journal*. It is copyrighted by them and reprinted with their permission.

Recall that $w \in A_p$, $1 < p < \infty$, if there is a constant C so that

$$\left(\frac{1}{|Q|}\int_Q w(x)\,dx\right)\left(\frac{1}{|Q|}\int_Q w(x)^{1-p'}\,dx\right)^{p/p'} \leq C$$

for all cubes $Q \subset \mathbf{R}^n$, and $w \in A(p, 1)$ if for any cube Q and measurable set $E \subset Q$,

$$\frac{|E|}{|Q|} \leq C\left[\frac{w(E)}{w(Q)}\right]^{1/p}.$$

In [1], it was shown that for $q > 1$, (1) holds if and only if $w \in A_p$, and for $q = 1$ and $\bar{q} = \infty$, (1) is equivalent to $w \in A(p, 1)$. Although it was also proved that $w \in A_p$ implies (1) with $\bar{q} = q$, the strongest possible inequality, the equivalence when $q = 1$ and $\bar{q} < \infty$ was unresolved. It is interesting that for these cases the previous method would only yield (1) implies $w \in A(p, 1)$, which is not a sufficient condition on the weight. In this note we prove:

Theorem 1. *Let* $1 < p < \infty$ *and* $1 \leq q < \infty$. $\|Mf\|_{pq} \leq C\|f\|_{p1}$ *if and only if* $w \in A_p$.

By the previous remark, this is a consequence of:

Theorem 2. *Let* $1 < p < \infty$ *and* $1 \leq q < \infty$. *If* $\|Mf\|_{pq} \leq C\|f\|_{p1}$, *then there is an* r, $1 < r < p$, *and an* s *so that* $\|Mf\|_{rs} \leq C\|f\|_{r1}$.

Since the maximal function is strong type (∞, ∞), using Theorem 2 and the Marcinkiewicz interpolation theorem [3, p. 197], we get $\|Mf\|_{pq} \leq C\|f\|_{pq}$, which implies $w \in A_p$, proving Theorem 1.

Let E be a measurable set, χ_E the characteristic function of E, and $E_y = \{x \in \mathbf{R}^n : M\chi_E(x) > y\}$. We will use:

Lemma 1. *There is a constant* $d > 1$, *depending only on* n, *so that for any* $0 < y < 1$, $0 < \alpha < \frac{1}{d}$, *and measurable set* E,

$$w(\{x \in \mathbf{R}^n : M\chi_E(x) > d\alpha y\}) \leq w(\{x \in \mathbf{R}^n : M\chi_{E_y}(x) > \alpha\}). \qquad (2)$$

Proof. This is a consequence of the two facts that $M\chi_{E_y}(x) = 1 > \alpha$ for a.e. $x \in E_y$, so E_y is essentially contained in the set on the right, and $M\chi_E(x) \leq dy\, M\chi_{E_y}(x)$ for $x \notin E_y$, where d depends only on n. This last estimate follows by applying a Whitney decomposition to E_y and estimating the average of χ_E over a cube in terms of these Whitney cubes [2, p. 19]. In particular, let $E_y = \cup Q_j$, where the cubes Q_j are pairwise disjoint,

$$\frac{1}{|Q_j|}\int_{Q_j} \chi_E(t)\, dt \leq By \qquad (3)$$

and

$$\operatorname{diam}(Q_j) \leq C \quad \text{distance} \quad (Q_j, \mathbf{R}^n - E_y), \qquad (4)$$

where B and C depend only on n. Let $x \in \mathbf{R}^n - E_y$, Q be a cube containing x, and suppose $Q_j \cap Q \neq \phi$. By (4), there is a constant, C_n, depending only on n, so that if Q^* is the cube concentric with Q and having diameter C_n times as big, then $Q_j \subset Q^*$. Thus, since $E \subset E_y$ and by (3),

$$\frac{1}{|Q|}\int_Q \chi_E(t)\, dt = \frac{1}{|Q|}\sum_{Q_j\cap Q\neq\phi}\int_{Q_j\cap Q} \chi_E(t)\, dt \leq \frac{1}{|Q|}\sum_{Q_j\subset Q^*}\int_{Q_j} \chi_E(t)\, dt$$

$$\leq \frac{(C_n)^n}{|Q^*|}\sum_{Q_j\subset Q^*} By|Q_j| \leq (C_n)^n By\,\frac{1}{|Q^*|}\int_{Q^*} \chi_{E_y}(t)\, dt$$

$$\leq dy\, M\chi_{E_y}(x),$$

when $d = (C_n)^n B$. Taking the sup over all such cubes, Q, shows $M\chi_E(x) \leq dy\, M\chi_{E_y}(x)$, and the lemma is proved.

Proof of Theorem 2. Assume $\| Mf \|_{pq} \le C \| f \|_{p1}$, $1 < p < \infty$, $1 \le q < \infty$. We wish to show that $\| Mf \|_{rs} \le C \| f \|_{r1}$ for some r, $1 < r < p$. Since M is sublinear and $L(r, s)$ is a normed space, it is sufficient to show that $\| M\chi_E \|_{rs} \le C \| \chi_E \|_{r1}$ for all measurable sets (see [3, p. 195]). Since $E_y = \{x \in \mathbf{R}^n : M\chi_E(x) > y\}$ and $w(\{x \in \mathbf{R}^n : M\chi_E(x) > 1\}) = 0$, our hypothesis becomes

$$\int_0^1 w(E_\alpha)^{q/p}\alpha^{q-1}\, d\alpha \le Cw(E)^{q/p}, \tag{5}$$

for any measurable set E. Using induction, we first prove

$$\int_0^1 w(E_\alpha)^{q/p}\alpha^{q-1}\left(\ln\frac{1}{\alpha}\right)^j \frac{d\alpha}{j!} \le C(Cd^q)^j w(E)^{q/p}, \tag{I_j}$$

where C is the constant in (5), d is the constant in the lemma, and E is any measurable set. (I_j) reduces to (5) when $j = 0$, so we need only show that (I_j) implies (I_{j+1}).

Fix E and apply (I_j) to E_y. Since the integrand is nonnegative, we get

$$\int_0^{1/d} w(\{x : M\chi_{E_y}(x) > \alpha\})^{q/p}\alpha^{q-1}\left(\ln\frac{1}{\alpha}\right)^j \frac{d\alpha}{j!} \le C(Cd^q)^j w(E_y)^{q/p}.$$

Applying the lemma and changing variables, the left-hand side is greater than

$$\int_0^{1/d} w(E_{d\alpha y})^{q/p}\alpha^{q-1}\left(\ln\frac{1}{\alpha}\right)^j \frac{d\alpha}{j!} = \frac{1}{(dy)^q}\int_0^y w(E_\beta)^{q/p}\beta^{q-1}\left(\ln\left(\frac{dy}{\beta}\right)\right)^j \frac{d\beta}{j!}.$$

Since $d > 1$, we decrease the last integral by replacing $\ln\left(\frac{dy}{\beta}\right)$ by $\ln\frac{y}{\beta}$. Then, multiply by $d^q y^{q-1}$ and integrate in y from 0 to 1 to get

$$\int_0^1 \frac{1}{y}\int_0^y w(E_\beta)^{q/p}\beta^{q-1}\left(\ln\frac{y}{\beta}\right)^j \frac{d\beta}{j!}\, dy \le (Cd^q)^{j+1}\int_0^1 y^{q-1}w(E_y)^{q/p}\, dy.$$

By (5) and the definition of E_y, the right-hand side is majorized by $C(Cd^q)^{j+1}w(E)^{q/p}$. Changing the order of integration on the left, we get

$$C(Cd^q)^{j+1}w(E)^{q/p} \geq \int_0^1 w(E_\beta)^{q/p}\beta^{q-1}\left(\int_\beta^1 \frac{1}{y}\left(\ln\frac{y}{\beta}\right)^j dy\right)\frac{d\beta}{j!}$$
$$= \int_0^1 w(E_\beta)^{q/p}\beta^{q-1}\left(\ln\frac{1}{\beta}\right)^{j+1}\frac{d\beta}{(j+1)!},$$

proving (I_{j+1}).

Let $0 < R < 1$ and rewrite (I_j) as

$$\int_0^1 w(E_\alpha)^{q/p}\alpha^{q-1}\left[\frac{R}{Cd^q}\ln\frac{1}{\alpha}\right]^j\frac{d\alpha}{j!} \leq CR^j w(E)^{q/p}.$$

Let $\delta = \dfrac{R}{Cd^q} > 0$. Summing in j, we get

$$\int_0^1 w(E)^{q/p}\alpha^{q-1}\left[\frac{1}{\alpha}\right]^\delta d\alpha \leq \frac{C}{1-R}w(E)^{q/p}. \tag{6}$$

Set $s = q - \delta$, $r = p\dfrac{s}{q} < p$. Then (6) becomes

$$\int_0^1 w(E_\alpha)^{s/r}\alpha^{s-1}\,d\alpha \leq Cw(E)^{s/r}, \quad \text{or}$$

$$\|M\chi_E\|_{rs} \leq C\,\|\chi_E\|_{r1}. \tag{7}$$

For small enough R, $r > 1$, and the theorem is proved.

REFERENCES

1. Chung, H. -M.; Hunt, R. A.; and Kurtz, D. S. The Hardy-Littlewood maximal function on $L(p, q)$ spaces with weights. *Indiana U. Math. J.* 31 (1982): 109-120.

2. Stein, E. M. *Singular Integrals and Differentiability Properties of Functions*. Princeton, N.J.: Princeton Univ. Press, 1970.

3. Stein, E. M., and Weiss, G. *Introduction to Fourier Analysis on Euclidean Spaces*. Princeton, N.J.: Princeton Univ. Press, 1971.

A NOTE ON THE EQUIVALENCE OF A_p AND SAWYER'S CONDITION FOR EQUAL WEIGHTS

R. A. Hunt, D. S. Kurtz, C. J. Neugebauer
Purdue University

Let $Mf(x) = \sup_{x \in Q \text{ cube}} \frac{1}{|Q|} \int_Q |f(y)|\, dy$ denote the Hardy-Littlewood maximal function on $\mathbf{R}^n$. We are interested in the inequality

$$\int [Mf(x)]^p w(x)\, dx \leq C \int |f(x)|^p w(x)\, dx, \qquad (*)$$

where $1 < p < \infty$ and $w(x)$ is a nonnegative function on $\mathbf{R}^n$.

Muckenhoupt [1] has shown that (*) is equivalent to

$$\left(\frac{1}{|Q|} \int_Q w(x)\, dx\right)\left(\frac{1}{|Q|} \int_Q w(x)^{1-p'}\, dx\right)^{p-1} \leq C. \qquad (A_p)$$

A special case of a result of Eric Sawyer [2] shows that (*) is equivalent to

$$\int_Q [M(\chi_Q w^{1-p'})(x)]^p w(x)\, dx \leq C \int_Q w^{1-p'}(x)\, dx < \infty. \qquad (S_p)$$

It is easy to show that (S_p) implies (A_p). In this note we give a relatively direct proof that (A_p) implies (S_p). By combining this result with Sawyer's proof, one obtains a proof that (A_p) implies (*). This line of proof involves neither the weak-type (p, p) result nor the fact that (A_p) implies $(A_{p-\varepsilon})$ for some $\varepsilon > 0$.

Let us introduce the maximal function

$$M_w f(x) = \sup_{x \in Q \text{ cube}} \frac{\int_Q |f(y)| w(y)\, dy}{\int_Q w(y)\, dy}.$$

If $w(x)$ satisfies the doubling condition, an appropriate covering lemma gives the weak-type (1, 1) result that

$$\int_{[M_w f > y]} w(x)\, dx \leq \frac{C}{y} \int |f(x)| w(x)\, dx.$$

Since clearly $\| M_w f \|_\infty \leq \| f \|_\infty$, (a simple special case of) the Marcinkiewicz interpolation theorem gives

$$\int [M_w f(x)]^{p'} w(x)\, dx \leq C_{p'}^{p'} \int |f(x)|^{p'} w(x)\, dx, \quad 1 < p' < \infty. \quad (**)$$

It is now easy to show that (A_p) implies (S_p). For $x \in \bar{Q} \subset Q$, (A_p) implies

$$\begin{aligned}
\frac{1}{|\bar{Q}|} \int_{\bar{Q}} \chi_Q(y) w^{1-p'}(y)\, dy &= \frac{1}{|\bar{Q}|} \int_{\bar{Q}} w^{1-p'}(y)\, dy \\
&\leq \left[\frac{C|\bar{Q}|}{\int_{\bar{Q}} w(y)\, dy} \right]^{1(p-1)} \\
&= \left[\frac{C \int_{\bar{Q}} \chi_Q(y) w^{-1}(y) w(y)\, dy}{\int_{\bar{Q}} w(y)\, dy} \right]^{p'/p} \\
&\leq [CM_w(\chi_Q w^{-1})(x)]^{p'/p}.
\end{aligned}$$

Since $\operatorname{supp}(f) \subset Q$ implies $Mf(x) = \sup\limits_{\substack{x \in \bar{Q} \\ \bar{Q} \subset Q}} \frac{1}{|\bar{Q}|} \int_{\bar{Q}} |f(y)|\, dy$, $x \in Q$, we have

$$[M(\chi_Q w^{1-p'})(x)]^p \leq [CM_w(\chi_Q w^{-1})(x)]^{p'}.$$

(S_p) then follows from (**) with $f = \chi_Q w^{-1}$.

REFERENCES

1. Muckenhoupt, B. Weighted norm inequalities for the Hardy maximal function. *Trans. Amer. Math. Soc.* 165 (1972): 207-227.

2. Sawyer, Eric T. A characterization of a two-weight norm inequality for maximal operators.

ON THE DECOMPOSITION OF L_1^1 (R) FUNCTIONS INTO HUMPS

Max Jodeit, Jr.
University of Minnesota

1. INTRODUCTION

The functions $F \in L^1(\mathbf{R})$ with $f = F'$ also in L^1 compose the Sobolev space L_1^1, with norm $\| F \| = \| F \|_1 + \| f \|_1$ $\left(\| F \|_1 = \int_{-\infty}^{\infty} |F(x)| \, dx \right)$. The problem is, given $f \in L^1$, to find conditions on f that tell whether $F(x) = \int_{-\infty}^{x} f(t) \, dt$ is in L^1. That $\int f(x) \, dx = 0$ (often used in the form $F(x) = -\int_x^{\infty} f(t) \, dt$ without mention) is necessary. If in addition $xf(x)$ is integrable, so is F. That isn't necessary: $F(x) = \sin x/(1 + x^2)$ is a counterexample.

The substance of Sec. 2 is that each f in $(L_1^1)'$ (the space of derivatives of L_1^1-functions) may be expressed as a series of functions $\pm f_n$ with $xf_n(x)$ integrable, such that $\| f \|_1 = \Sigma \| f_n \|_1$ and $\| F \|_1 = \Sigma \| F_n \|_1$. Each F_n is a "hump," a function in L_1^1 that increases and then decreases; $\| F_n \|_1 = \int (c_n - x) f_n(x) \, dx$; c_n is a point where F_n is greatest.

Secs. 3 through 6 exploit the hump decomposition in connection with results on fractional integrals of $(L_1^1)'$ functions, duality, pointwise multipliers [9] and composition operators [6], respec-

---

Partial support by NSF Grant No. MCS 79-00985.

tively. Many details are omitted, for the emphasis is on the use of the decomposition into humps.

Sec. 7 is on the Fourier transform side. I have not been able to make effective use of humps there; only a counterexample is given, and a few comments about L_1^1 and the Bessel potential space $\mathscr{L}_1^1$.

The following test for $F \in L^1$ (presumably known) arose in a conversation with C. Unguriano. If $h \in L^1$ is a hump, $g \in L^1_{loc}$ satisfies $|\int_a^b g(t)\,dt| \leq M$ for all a, b, and $f = gh$ has zero integral, then F is integrable. For if $x \gg 0$, say, $|F(x)| \leq \int_x^\infty |G(x) - G(t)||dh(t)| \leq Mh(x)$. One application is to $f(x) = \sin x/(1 + x^2)$.

The results of Secs. 5 and 6 also give tests for $F \in L_1^1$, in terms of the form of f.

The following lemma has been useful. A corollary of it is that $L_1^1 = E * L^1$, where $E(x) = \chi_+(x)e^{-x}$. It could also be used in Sec. 4 to show that $K = b - B$, where $b \in L^\infty$ and B is a primitive of b. Here $\mathscr{S}'$ is the L. Schwartz space of tempered distributions on **R**.

Lemma 1. *If $u \in \mathscr{S}'$ and $f = u' + u \in L^p$ for some p, $1 \leq p \leq \infty$, then $\|u'\|_p + \|u\|_p \leq C\,\|f\|_p$.*

See Beckenbach and Bellman [2, pp. 167 and 175]. The proof comes down to noting that $e^{-x}\left(A + \int_0^x e^t f(t)\,dt\right)$ cannot belong to $\mathscr{S}'$ unless $A = \int_{-\infty}^0 e^t f(t)\,dt$. The example given by the negative part of $\sinh x$ shows that a "growth" condition such as $u \in \mathscr{S}'$ is needed.

Many people at the University of Minnesota have graciously helped me to search for references. A. M. Fink of Iowa State referred me to [2]. To Richard Shaw I offer thanks for helpful conversations about this work.

2. A DECOMPOSITION THEOREM

Although it is not a necessary condition, integrability of $xf(x)$ *is* necessary for that of F if $xf(x)$ has constant sign. If, say, $xf(x) \leq 0$, then $F(x) \geq 0$ and a calculation gives $\int_{-R}^{R} F(x)\,dx = \int \min(R, |t|)f(t)\,|\,dt \to \int (-t)f(t)\,dt$ by monotone convergence. The point at which the sign changes may be any point c; if $(-t)$ in the last integral is replaced by $(c - t)$, the integrand has the useful features of constant sign. Functions of this sort are the basic pieces used in the decomposition of functions in L_1^1.

Definition. F is a *hump* (function) if there is a point c such that $F(x)$ decreases to 0 as x moves away from c toward infinity in either direction. The monotonicity need not be strict. If the hump F is absolutely continuous, $f = F'$ is a *polarized* function with *midpoint* any point c at which the sign of f changes.

A useful kind of hump F is the *trapezoid* of height 1, having $F'(x) = 1/\delta$ in $(a - \delta, a)$, $F'(x) = -1/\delta$ in $(b, b + \delta)$, and $F' = 0$ everywhere else.

The norm in L_1^1 of a hump function, if it is absolutely continuous and integrable, is $\|F\|_1 + 2\|F\|_\infty = \int (c - x)f(x)\,dx + \|f\|_1$.

Theorem 1. If $F \in L^1(\mathbf{R})$ is real-valued, absolutely continuous, and $f = F' \in L^1(\mathbf{R})$, then there exists a sequence of absolutely continuous "hump" functions H_n and signs $\varepsilon_n = \pm 1$ such that $F = \sum_n \varepsilon_n H_n$ and, with $h_n = H_n'$,

$$\|F\|_1 = \sum_n \|H_n\|_1, \quad \|f\|_1 = \sum_n \|h_n\|_1.$$

At most two of the H_n fail to have compact support. If the supports of two of the H_n overlap, then one is contained in the interior of the other and the corresponding signs are the same. Finally, if $n \neq m$, then $h_n h_m = 0$ almost everywhere.

Proof. The positive and negative parts of F may be treated separately, a sign -1 being given to each hump associated with the negative part. Nothing is lost by assuming $F \geq 0$. If $F(x) > 0$ for all x, exactly one of the humps will have the whole line as support. If the set Z of zeros of F is bounded on one side only, just one of the H_n will have unbounded support; if Z is bounded and nonempty, then two will. In the remaining case each H_n will have compact support.

In any case $\{x : F(x) > 0\} = \cup(a_n, b_n)$ yields an initial decomposition $F = \Sigma_n F_n$, where $F_n = F$ in (a_n, b_n) and $F_n = 0$ elsewhere. There is no loss in assuming that $F(x) > 0$ in some interval (a, b) with $F = 0$ elsewhere, for each piece F_n is treated in the same manner, separately and independently, to yield a single hump function and a countable number of (smaller) pieces of the same sort as F. Of course, (a, b) may be unbounded.

The construction of the humps is done using the idea of the rising (and setting!) sun lemma. Let c denote the least x in (a, b) such that $F(x) = \max F$, and put

$$H(x) = \min\{F(t) : \ t \text{ lies between } x \text{ and } c, \text{ inclusive}\}.$$

It is immediate that H is a hump with maximum at c, $H(c) = \|F\|_\infty$. It will be enough to check the absolute continuity of H on $(-\infty, c]$. Given $\varepsilon > 0$, choose $\delta > 0$ such that whenever $(\alpha_k, \beta_k]$ are pairwise disjoint intervals in $(-\infty, c]$ and $\Sigma(\beta_k - \alpha_k) < \delta$, then $\Sigma|F(\beta_k) - F(\alpha_k)| < \varepsilon$. Consider the term $|H(\beta_k) - H(\alpha_k)|$ of the corresponding sum for H. There exist numbers $\alpha \geq \alpha_k$ such that $H(\alpha_k) = F(\alpha)$. Choose the one closest to α_k and call it α_k^1. If $\alpha_k^1 \geq \beta_k$, the term under consideration is zero. If $\alpha_k^1 < \beta_k$, then $|H(\beta_k) - H(\alpha_k)| = H(\beta_k) - F(\alpha_k^1) \leq F(\beta_k) - F(\alpha_k^1)$. Since $\Sigma(\beta_k - \alpha_k^1)_+ \leq \Sigma\,(\beta_k - \alpha_k) < \delta$, $\Sigma|H(\beta_k) - H(\alpha_k)| \leq \Sigma|F(\beta_k) - F(\alpha_k^1)| < \varepsilon$. It follows that H is absolutely continuous.

The set $o \equiv \{x : F(x) > H(x)\} = \{x : \exists y \text{ between } x \text{ and } c \text{ with } F(x) > F(y)\}$ is an open subset of (a, b). The hump H is positive in (a, b) and constant on the component intervals of o. No component of o can be unbounded, or else F would not tend to 0 at infinity. The assertions in the theorem about the number of humps with unbounded support follow.

The remainder $R = F - H$ may be expressed as $R = \Sigma F_m$, the F_m given by $F - H$ on the component intervals I_m of the open set o. The process carried out on F in $I = (a, b)$ is next carried out on each F_m, in I_m, to complete the next stage of the process, which is repeated indefinitely or until there is no remainder. The assertion about overlapping supports of the humps follows because each one is continuous and is constant and positive on each component of the set where it is exceeded by the piece of the remainder used to define it.

It remains to show that the sum of the humps so obtained is actually equal to F. Return to the first stage of the decomposition of F: $F = H + R$. Since H is constant on the component intervals I_m of o and coincides with F otherwise,

$$\| f \|_1 = \text{var}(F) = \text{var}(H) + \text{var}(R).$$

As the decomposition process continues stage by stage, the total variation of F is expressible as the sum of the variations of a countable collection of humps plus the variation of the remainder. Thus the variation of F is at least as large as the sum of the variations of the humps.

To complete the proof it is enough to show that the inequality of the preceding paragraph can be reversed. Let $\varepsilon > 0$ be given. A finite number of points $x_0 < x_1 < \cdots < x_N$ exist such that

$$\text{var}(F) < \sum_{j=1}^{N} |F(x_j) - F(x_{j-1})| + \varepsilon. \tag{1}$$

It only increases the right-hand side if (in case it is absent) the point c is added to the x_j. Some of the points x_j may not belong to o. Distribute the ones that do among the component intervals of o in which they lie and add to the x_j the endpoints of those intervals, noting that none of the new points belongs to o. Break up the (new and no smaller) sum on the right-hand side of (1) into corresponding parts, which gives (since $F = H$ outside o)

$$\text{var}(F) < \text{var}(H) + \sum_1^M \sum_{x_{j-1}, x_j \in \bar{I}_m} |F_m(x_j) - F_m(x_{j-1})| + \varepsilon. \qquad (2)$$

The process of constructing humps may now be done on each piece F_m that is present, to get $F_m = H_m + R_m = H_m + \Sigma_i F_{mi}$, each F_{mi} being positive on a component of $o_m = \{x : F_m(x) > H_m(x)\}$, and zero elsewhere.

As before, distribute the x_j lying in o_m among the component intervals I_m in which they lie, and augment the x_j by adding in the endpoints of these component intervals. At each new point, $F_m = H_m$ so the inner sum in (2) for F_m is dominated by

$$\text{var}(H_m) + \sum_{i=1}^{M_m} \sum_{x_{j-1}, x_j \in \bar{I}_{mi}} |F_{mi}(x_j) - F_{mi}(x_{j-1})|.$$

The sum in (2) is thus replaced by a sum of variations of humps and a triple sum that has no more than $2N$ terms altogether; any of the x_j originally in o but not in any o_m make one of their two contributions to the sum to the variation of an H_m. It is more likely that an original x_j persists in belonging to successive open sets $o_m \supseteq o_{mm_1} \supseteq o_{mm_1m_2} \supseteq \cdots$. Suppose $\bar{x}$ is such an x_j. To simplify notation, let $I_1 \supseteq I_2 \supseteq \cdots$ now be used to denote component intervals containing $\bar{x}$, each obtained at the first, second, ..., stages. If $|I_k| \to 0$, the absolute continuity of F implies that the variation of the remainder corresponding to I_k tends to zero. But $|I_k| \to 0$ need not be: for example, let $F(x) = 0$ for $x > 1$, $F(x) = (1 + x^2)^{-1}$ for $x \leq -1$; in $(-1, 0)$ let $F(x) = 1/2$ and in $[0, 1]$ put $F(x) = \frac{1}{2} - x^2 \sin(\pi/6x)$. The interval $[-1, 0]$ "persists." Therefore, suppose that $\bar{x} \in I_\omega = \cap I_k$ has positive measure. Each interval I_k contains a first point c_k at which the maximum value of R_k occurs, so for each x in $I_\omega \subseteq I_{k+1}$ there exists a point

$y \in I_k$, between x and c_k, such that $R_k(x) > R_k(y)$: the endpoint of I_{k+1} nearest to c_k is such a y. It follows that the oscillation of R_k on the interval between c_k and the endpoint of I_ω nearest to c_k exceeds the oscillation of R_k on I_ω. Since each hump formed before the k-th stage is constant on I_k, the oscillation of R_k on I_k is the same as that of F. Thus $\operatorname{osc}_{I_k \sim I_\omega}(F) > \operatorname{osc}_{I_\omega}(F)$. Since $|I_k \sim I_\omega| \to 0$ as $k \to \infty$ and F is continuous, $\operatorname{osc}_{I_\omega}(F) = 0$ and thus again by the absolute continuity of F, $\operatorname{var}(R_k) \to 0$. Now the inner sum on the right side of (2) corresponding to $\bar{x}$ is seen to be bounded at each stage by a sum of variations of humps plus $\operatorname{var}(R_k)$. Since only a finite number of terms is involved, it follows that $\operatorname{var}(F) < \sum_n \operatorname{var}(H_n) + 2\varepsilon$ if for each $\bar{x}$ the hump construction process is stopped as soon as $\operatorname{var}(R_k) < \varepsilon/N$. This completes the proof.

Seven Remarks

(i) The preceding argument applies even when F is not integrable but does tend to zero at infinity.

(ii) If f is integrable and $\int f(x)\,dx \neq 0$ (say > 0) the argument, applied to $F(x) - \min_{t \geq x}(\max(0, F(t)))$, shows that any real function f in L^1 may be expressed as the sum of a function f_0 having the same sign as $\int f(x)\,dx$ and of a sequence $f_1, f_2, \ldots$ of polarized functions such that for $n > m \geq 0$, $f_n f_m = 0$ a.e. The real and imaginary parts of a complex-valued function may be treated separately; I don't know how to make an exact decomposition of norms for both F and f in that case.

(iii) Parts of the argument work for continuous functions of bounded variation too. For the total variation function is continuous, and this replaces the two applications of absolute continuity used to show $\operatorname{var}(R_k) \to 0$. Of course, the humps needn't be absolutely continuous, only continuous.

(iv) If the humps are removed from a function of bounded variation in such a way that the area between its graph and the x-axis is decreased, a decomposition can be made for integrable functions of bounded variation, and so forth.

(v) If $\{H_n\}$ is a sequence of humps such that $A = \Sigma \| H_n \|_1$ and $B = \Sigma \| h_n \|_1$ are both finite, then $f = \Sigma \pm h_n \in (L_1^1)'$ and $\| f \| (\equiv \| F \|) \leq A + B$. It is from this viewpoint that an analogy with atomic decompositions of H^1 (see [3]) is visible. If the conditions on signs and supports of the H_n hold and if $h_n h_m \geq 0$ when $n \neq m$, the other conclusions of the theorem hold.

(vi) A similar theorem holds for sequences. Thus if $a = \{a_k\}$, $A_k = \sum_{j=-\infty}^{k} a_j$, and both a, A are absolutely convergent, there exists a sequence of hump sequences $A_{n\cdot} = \{A_{nk}\}$ such that $A_k = \pm\Sigma_n A_{nk}$, $a_k = \pm\Sigma_n a_{nk}$ $(a_{nk} = A_{nk} - A_{n,k-1})$, and $\| A \|_1 = \Sigma\Sigma |A_{nk}|$, $\| a \|_1 = \Sigma\Sigma |a_{nk}|$. The analogues of the other conclusions hold with one exception: $a_{nk}a_{mk} \geq 0$ replaces "$h_n h_m = 0$ for $n \neq m$."

To prove this let $F(x)$ be the piecewise linear function with $F(k) = A_k$. Then by Theorem 1, $F(x) = \Sigma F_n(x)$ (one of $\pm F_n$ is a hump). Since F is a piecewise linear and $f_n f_m = 0$ if $n \neq m$, $f_n(x)$ takes values 0 or a_k in $(k - 1, k)$. Set $A_{nk} = F_n(k)$. It follows that $a_{nk} = \int_{k-1}^{k} f_n(t)\, dt$ and a_{mk} have the same sign if neither one is zero.

(vii) A representation $F = H_1 - H_2$, each H in L_1^1, is possible for an F in L_1^1 if and only if $xf(x)$ is integrable. Necessity has already been noted and used. For sufficiency choose c so that $\int_{-\infty}^{c} |f|\, dx = \int_{c}^{\infty} |f|\, dx$ and let $h_1(x) = f^+(x)$ for $x < c$, $h_1(x) = -f^-(x)$ for $x > c$. A little arithmetic shows that h_1 has zero integral, and the integrability of $xf(x)$ shows that each H is integrable.

3. FRACTIONAL INTEGRALS

Suppose $f \in L^1(\mathbf{R})$ and $F \in L^1(\mathbf{R})$. Then with $\gamma_\alpha = (\Gamma(\alpha) \cos \frac{\pi}{2} \alpha)^{-1}$, $(0 < \alpha < 1)$, the Riesz potential

$$R_\alpha f(x) \equiv \gamma_\alpha \int |x - y|^{\alpha-1} f(y)\, dy$$

satisfies

$$\| R_\alpha f \|_1 \leq 2^{1-\alpha} \gamma_\alpha \cdot \frac{2}{\alpha} \| f \|_1^{1-\alpha} \| F \|_1^\alpha. \tag{3}$$

Moreover, if $1 < r < (1 - \alpha)^{-1}$, then $R_\alpha f \in L^r$.

Remark. I suspect that the factor $2^{1-\alpha}$ is not needed in (3).

Proof. Consider a polarized function f with midpoint c. Since f has zero integral,

$$|R_\alpha f(x)| \leq \gamma_\alpha \int |f(y)| \left| \frac{1}{|x - y|^{1-\alpha}} - \frac{1}{|x - c|^{1-\alpha}} \right| dy. \tag{4}$$

Because of cancellation at infinity, the difference appearing in the integrand is integrable in x. That integral can be evaluated; its value is $2^{2-\alpha}|c - y|^\alpha/\alpha$. For any $A > 0$ and any positive t, $t^\alpha \leq A + \alpha(1 - \alpha)^s A^{-s} t$, where $s = (1 - \alpha)/\alpha$. Thus with C_α denoting the constant on the right-hand side of (3),

$$\| R_\alpha f \|_1 \leq C_\alpha \Big(A \| f \|_1 + \alpha(1 - \alpha)^s A^{-s} \int |f(y)| |c - y| \, dy\Big).$$

The integral appearing on the right is $\| F \|_1$, since f is a polarized function. The inequality may be added over the polarized functions into which an arbitrary f in $(L_1^1)'$ is decomposed. The inequality (3) then follows by minimizing in A.

That $R_\alpha f \in L^r$ when $r < q = (1 - \alpha)^{-1}$ is a consequence of Minkowski's integral inequality applied to (4), followed by a summation.

For some $f \in (L_1^1)'$, $R_\alpha f \notin L^q$. Otherwise the closed graph theorem gives $\| R_\alpha f \|_q \leq C \| f \|$ ($\| f \| \equiv \| F \|$), so a trapezoid $F_n(x) = 1$ for $|x| < 1$ and 0 for $|x| > 1 + \frac{1}{n}$ leads by Fatou's lemma to a contradiction, since $|x + 1|^{\alpha-1} - |x - 1|^{\alpha-1} \notin L^q$.

4. THE DUAL SPACE OF L_1^1

A continuous linear functional k on this space induces one on $(L_1^1)'$ by $K(f) \equiv -k(F)$. Since integrable function f with support in a bounded interval I, having integral zero, is in the latter space,

$$|K(f)| \leq C(\| F \|_1 + \| f \|_1) \leq C\left(\tfrac{1}{2}|I| + 1\right) \| f \|_1,$$

so there exists a locally bounded function $K(x)$, unique modulo constants, such that $K(f) = \int K(x)f(x)\,dx$ whenever $f \in (L_1^1)'$ has compact support. Also, $\inf_{W\in C} \| K - W \|_{L^\infty(I)} \leq C\left(1 + \tfrac{1}{2}|I|\right)$. It follows that for a.e. x, y

$$|K(x) - K(y)| \leq 2C + C|x - y|. \tag{5}$$

This also follows using Lemma 1, which implies $L_1^1 = E * L^1$ ($E(x) = \chi_+(x)e^{-x}$).

Functions K that satisfy this condition admit representation as $K = b + L$, where b is bounded and L is Lipschitz (if $\varphi \in C^\infty$ has support in $[-1, 1]$ and $\int \varphi(x)\,dx = 1$, then $K - \varphi * K$ is bounded and $\varphi * K$ is Lipschitz).

Conversely, a function K that satisfies (5) gives a bounded linear functional on $(L_1^1)'$, the action taking a form like that of BMO on atomic H^1:

$$K(f) = \Sigma \left\{ \int b(x)f_n(x)\,dx + \int \left(L(x) - L(c_n)\right)f_n(x)\,dx\right\},$$

where $f = \Sigma f_n$ is a decomposition from Theorem 1. The general term in the series is dominated by $\| b \|_\infty \| f_n \|_1 + \| L' \|_\infty \| F_n \|_1$, so $|K(f)| \leq \max(\| b \|_\infty, \| L' \|_\infty) \| f \|$. See [1].

Example. If numbers x_j are given for $-\infty < j < \infty$, then $\Sigma_j F(x_j)$ converges (absolutely) for every $F \in L_1^1$ if and only if there are positive numbers A, M such that every interval of length A contains at most M numbers x_j (multiplicities included).

Necessity follows from the closed graph theorem and application to trapezoid functions F. The condition on $\{x_j\}$ allows the assumption that $x_j \leq x_{j+1}$. For $n > m$, it follows that $n - m \leq M + M(x_n - x_m)/A$, so the function K defined by setting $K(x) = n$ when $x_n \leq x < x_{n+1}$ satisfies $|K(x) - K(y)| \leq 1 + M + (M/A)|x - y|$. Then for a polarized function f with midpoint c,

$$\begin{aligned} \int K(x)f(x)\,dx &= \sum_n n\big(F(x_{n+1}) - F(x_n)\big) \\ &= \lim_{N\to\infty}\Big\{-\sum_{-N}^{N} F(x_n) + NF(x_N) + (N+1)F(x_{-N})\Big\} \\ &= -\sum_n F(x_n), \end{aligned}$$

since for large N, $NF(x_N) = -N \int_{x_N}^{\infty} f(t)\,dt$

$$= -N \int_{x_N}^{\infty} \frac{(c - t)f(t)}{(c - t)}\,dt;$$

for, when $x_N > 2|c| + 1$, this is dominated by $(2N/x_N) \int_{x_N}^{\infty} (c - t)f(t)\,dt$, which tends to 0 as $N \to \infty$ (similarly $(N + 1)F(x_{-N}) \to 0$).

Corollary 1. If $F \in L_1^1$, then $\Sigma_n \max_{n\leq x\leq n+1} |F(x)| \leq C \| F \|$ (see [10]).

Remark. In particular, $\int xf(x)\,dx$ has meaning for each $f \in (L_1^1)'$, since f may be expressed as a sum of polarized functions. The integrals $\int_{-R}^{R} xf(x)\,dx$ need not converge. For an example, let the graph of F consist of isosceles triangles of altitude $1/n^2$, and unit base sitting on the x-axis with vertex above 2^n, $n \geq 1$. Then $\int_{-2^n}^{2^n} xf(x)\,dx > (2^n - 1)n^{-2}$.

5. MULTIPLIERS

Under what conditions on $m(x)$ is mF in L_1^1 whenever F is in L_1^1? A necessary and sufficient condition is that m be bounded, locally absolutely continuous, and that $\int_a^b |m'(x)|\,dx \leq A + B(b - a)$ for some constants A, B and any $a < b$.

Proof. (a) *Necessity.* By the closed graph theorem, $T_m F = mF$ is a bounded operator. Then $n(x) = |m(x)|^2/\|T_m\|^2$ gives an operator with norm ≤ 1. If $n(x) \geq R > 1$ in (a, b) (m is necessarily continuous), take for F a trapezoid that is 1 in (a, b) and 0 outside $(a - \delta, b + \delta)$. Then $\|F\| = (b - a) + \delta + 2$, so $R(b - a) < \int_a^b n(x)F(x)\,dx \leq \|nF\|_1 \leq \|F\| = (b - a) + \delta + 2$. Hence, $(b - a) \leq 2(R - 1)^{-1}$. The same argument applies to $n(x)^k$ so $(b - a) \leq 2(R^k - 1)^{-1}$ for any integer $k \geq 1$. Hence, $n(x) \leq 1$. The same trapezoid function leads to $\int_a^b |m'(x)|\,dx \leq \|(mF)'\|_1 + \|mF'\|_1 \leq 4\|T_m\| + \|T_m\|(b - a)$.

(b) *Sufficiency.* It is enough to verify that $mF \in L_1^1$ whenever F is a hump in L_1^1, with appropriate estimates. Now $\|mF\|_1 \leq \|m\|_\infty \|F\|_1$, and $\|(mF)'\|_1 \leq \|m'F\|_1 + \|m\|_\infty \|F'\|_1$. Since $\{F > \lambda\}$ is an interval,

$$\begin{aligned}\int |m'F|\,dx &= \int_0^{\|F\|_\infty} \int_{F>\lambda} |m'|\,dx\,d\lambda \\ &\leq \int_0^{\|F\|_\infty} (A + B\,\text{meas}\{F > \lambda\})\,d\lambda \\ &= A\|F\|_\infty + B\|F\|_1 \leq \max\left(\tfrac{1}{2}A, B\right)\|F\|.\end{aligned}$$

Thus $\|T_m F\| \leq \left(\|m\|_\infty + \max\left(\tfrac{1}{2}A, B\right)\right)\|F\|$.

Remark. The multipliers m admit representation in terms of bounded functions and sequences of L^1 functions. Let b denote a bounded function, $\{g_k\}$ a sequence of L^1 functions with $\operatorname{supp} g_k \subseteq [k, k+1]$, $\int g_k(x)\,dx = 0$, and $\|g_k\|_1 \leq A$. Then, with $E(x) = \chi_+(x)e^{-x}$, $m(x) = \Sigma_k G_k(x) + E * b(x)$ is a multiplier of L_1^1, and any multiplier may be so expressed. See Strichartz's paper [9] for the idea of localization.

6. COMPOSITION OPERATORS

Necessary and sufficient conditions on φ, Φ for $F(\varphi(x))$ or $\Phi(F(x))$ to be absolutely continuous for all absolutely continuous F were given by Fichtenholz [6] in 1922. They are, for φ, that φ be absolutely continuous and that $\varphi^{-1}(\{x\})$ have a finite number of boundary points, bounded independent of x and, for Φ, that Φ be a Lipschitz function. Since F was defined on a bounded closed interval, a little more is needed in the present context.

The case $F \circ \varphi$. This case is treated in detail because a "preservation of constants" phenomenon occurs. Suppose $F(\varphi(x)) \in L_1^1$ whenever $F \in L_1^1$. Then φ is necessarily continuous. With $R = \max_{\|t\|\leq r} |\varphi(t)|$, use $F \in L_1^1$ such that $F(x) = x$ for $|x| \leq R$, to see that φ is absolutely continuous on $[-r, r]$ for each $r > 0$. The closed graph theorem shows that $\|F \circ \varphi\| \leq \text{const}\,\|F\|$; in particular, there are constants C, C' such that $\|F \circ \varphi\|_1 \leq C\,\|F\|$, $\|F \circ \varphi'\|_1 \leq C'\|F\|$. For a trapezoid $F \equiv 1$ on (a, b),

$$|\varphi^{-1}(a, b)| \leq \|F \circ \varphi\|_1 \leq C\big((b - a) + \delta + 2\big) = C\,\|F\|,$$

which gives the first (of two) conditions on φ:

(i) $|\varphi^{-1}(a, b)| \leq 2C + C(b - a)$ for any numbers $a < b$.

An immediate consequence: $|\varphi(x)|$ is unbounded at $\pm\infty$.

Example. The change of variables $\varphi(t) = t^{1/3}$ does not preserve L_1^1.

The other condition may be approached as follows: for two numbers $a < b$, consider $\varphi^{-1}(a, b) = \cup_k(\alpha_k, \beta_k)$, and suppose there are at least N intervals (α_k, β_k). It may happen that $\varphi(\alpha_k) = \varphi(\beta_k)$ or that $\varphi(\alpha_k) \neq \varphi(\beta_k)$. Let $0 < 2\delta$ be a number less than the least of the oscillations of φ on the N intervals, and less than $(b - a)/2$, as well. Let F be zero outside $(a, a + 2\delta)$, $(b - 2\delta, b)$ and have as graph over each interval the upper boundary of an isosceles triangle of altitude 1 and base 2δ.

To see what happens to $F(\varphi(t))$, consider an interval (α_k, β_k) with $\varphi(\alpha_k) = a = \varphi(\beta_k)$. Each equation $\varphi(t) = a + \delta$, $\varphi(t) = a + 2\delta$ has at least two solutions there, so on (α_k, β_k) the variation of $F \circ \varphi$ is at least 4. This lower bound holds for each (α_k, β_k), so $4N \leq \|(F \circ \varphi)'\|_1 \leq C'\|F\| = C'(2\delta + 4)$. Since δ can be as small as desired, $N \leq C'$. Since φ is unbounded at infinity, it follows that $|\varphi(x)| \to \infty$ as $|x| \to \infty$.

Consider the complement $\circ$ of $E = \varphi^{-1}(\{x\})$ (note that $|E| \leq 2C$). If $\{\varphi > x\}$ has at least N^+ bounded and $U^+ (\leq 2)$ unbounded components, then for $\delta > 0$ sufficiently small, $\varphi^{-1}(x, x + \delta)$ has $2N^+ + U^+ \leq C'$ components. Now let N^+, N^-, and so forth, stand for the actual number of *bdd* components, and so on. It follows that $2(N^+ + N^-) + (U^+ + U^-) = 2(N^+ + N^-) + 2 \leq 2C'$, so the number of components of $\circ$ is at most $N^+ + N^- + 2 \leq C' + 1$, which gives the second condition (called (A) by Fichtenholz):

(ii) For each x, the equation $\varphi(t) = x$ holds for some isolated points t or in some intervals of constancy of φ, the total number of such points and intervals being at most C', and possibly zero.

Now if φ is locally absolutely continuous and satisfies the two conditions, the second implies that $F \circ \varphi$ is absolutely continuous, for, as Fichtenholz argues, the intervals (a_k, b_k) determined by $\varphi(\alpha_k)$ and $\varphi(\beta_k)$ have bounded overlaps if the (α_k, β_k)

are nonoverlapping. The same condition may be used in Banach's formula (Saks [7, Chapter 9 (6.4), p. 280]) for the variation of a continuous function to get

$$\| (F \circ \varphi)' \|_1 = \mathrm{var}(F \circ \varphi) \leq C' \ \mathrm{var}(F),$$

since except for a countable set of values of y, the solutions of $F(\varphi(t)) = y$ are no more in number than C' times those of the equation $F(x) = y$.

To show that $F \circ \varphi$ is integrable, suppose F is an absolutely continuous hump. Since $\{F > \lambda\}$ is an interval, the first condition gives

$$\int F(\varphi(t))\, dt = \int_0^{\|F\|_\infty} |\varphi^{-1}(F > \lambda)|\, d\lambda \leq 2C \| F \|_\infty + C \| F \|_1 = C \| F \| .$$

This inequality, applied to each term in the decomposition given by Theorem 1, thus holds for general F in L_1^1. That the same constants appear in the conditions on φ and in the inequalities for $F \circ \varphi$ is the "preservation of constants" referred to before. This may be stated as a theorem.

Theorem 2. A locally absolutely continuous function $\varphi : \mathbf{R} \to \mathbf{R}$ satisfies conditions (i) and (ii) with constants C and C', respectively, if and only if for every real $F \in L_1^1$,

$$\| F \circ \varphi \|_1 \leq C \| F \| \quad \text{and} \quad \| (F \circ \varphi)' \|_1 \leq C' \| F \| .$$

Examples. For $\varphi(t) = t^3$, $C = \sqrt[3]{4}/3$ and $C' = 1$; for $\varphi(t) = t^2$, $C = 1/\sqrt{2}$ and $C' = 2$.

The case $\Phi \circ F$. Very little needs to be added to the argument of Fichtenholz [6] to prove the following theorem (see Dahlberg [4]).

Theorem 3. The function Φ operates on L_1^1 if and only if Φ is locally Lipschitz and $\Phi(0) = 0$.

Sketch of proof. If $|\Phi(x) - \Phi(y)| \leq A(r)|x - y|$ when $|x| \leq r$ and $|y| \leq r$, then $|\Phi(F(x))| \leq A(\| F \|_\infty)|F(x)|$ and $\mathrm{var}(\Phi \circ F) \leq$

$A(\|F\|_\infty)\operatorname{var}(F)$, so $\|\Phi \circ F\| \leq A(\frac{1}{2}\|f\|_1)\|F\|$. To show necessity, let $F(x) = x$ for $|x| \leq r$ and $F \in L_1^1$ to see that Φ is absolutely continuous on $[-r, r]$. If F is chosen to have compact support, the necessity of $\Phi(0) = 0$ follows. To prove by contradiction that Φ' is essentially bounded in $[-r, r]$, suppose infinitely many of the $E_n = \{2^n < \Phi'(y) \leq 2^{n+1}\}$, $n = 0, 1, 2, \ldots$ have positive measure. The following construction is essentially the same as that of Fichtenholz. Choose a strictly monotone sequence $\{y_k\}$ in $[-r, r]$, each belonging to a different one of the sets E_k, and pairwise disjoint intervals (c_k, d_k) having y_k as an endpoint, such that $\Phi(d_k) - \Phi(c_k) > 2^{n_k}(d_k - c_k) \geq 2^k(d_k - c_k)$. On every other interval $(2^{-m-1}, 2^{-m})$, starting with $m = 0$ construct an accordion-fold graph oscillating from c_k to d_k and back a whole number N_k of times; N_k will be chosen later. These pieces are joined by line segments, and the whole thing completed outside $(0, 1)$ by two line segments ending at -1 and 2 on the x-axis. The total variation of each oscillating piece of this, the graph of a function $F^\#$, is $2N_k(d_k - c_k)$ and the variation of the line segments used to join the pieces is at most $4r$. Thus if $\sum_{k=0}^{\infty} N_k(d_k - c_k) < \infty = \sum_{k=0}^{\infty} 2^k N_k(d_k - c_k)$, $F^\#$ is in L_1^1 and $\Phi \circ F^\#$ is not, for the variation of $\Phi \circ F^\#$ on each oscillating piece is at least $2^k N_k(d_k - c_k)$. More detail may be found in [6, Sec. 5].

7. FOURIER TRANSFORMS

For $F \in L_1^1$, $\hat{F}(\xi) = \hat{f}(\xi)/i\xi$ so $\hat{f}(0) = 0$ and $\hat{f}'(0)$ exists. More conditions are needed to characterize elements of $(L_1^1)'$ in terms of their Fourier transforms. I don't know what they are, but they depend only on the behavior of $\hat{f}(\xi)$ near $\xi = 0$. For if $\varphi \in C_0^\infty$ and $\varphi(0) = 1$, then $\psi(\xi) = (1 - \varphi(\xi))/\xi$ is in L^2 and so is $\psi'(\xi)$. Therefore, $\psi \in A_1 = (L^1)\hat{}$, as does φ (cf. Esseen [5, pp. 13 and

31]), so $\hat{g} = (1 - \varphi)\hat{f}$ and $\hat{G} = \psi\hat{f}$ are in A_1 if $\hat{f}$ is. Thus for $f \in L^1$ and any $\delta > 0$, there is a function g in $(L_1^1)'$ such that $\hat{g}(\xi) = \hat{f}(\xi)$ for $|\xi| \geq \delta$.

The following example gives a function f in $H^1(\mathbf{R})$ with $\hat{f}'(0)$ existing, which is not in $(L_1^1)'$. Starting with a sequence ε_k decreasing to zero, $\Sigma\, \varepsilon_k = +\infty$, let f be that odd function that is 0 for $0 < x < 1$ and that equals $(-1)^k \varepsilon_k x^{-2}$ in $[2^k, 2^{k+1})$, $k = 0, 1, 2, \ldots$. Since f is a sum of odd atoms times coefficients $O(\varepsilon_k 2^{-k})$, f is in H^1. A computation shows that

$$\hat{f}(\xi)/i\xi = 2 \int_1^2 \sum_{k=0}^{\infty} (-1)^k \varepsilon_k \frac{\sin 2^k \xi x}{2^k \xi x} \cdot \frac{dx}{x}.$$

The following lemma will be proved later.

Lemma 2. If ε_k decreases to zero, then $g(x)$, given by $\sum_{k=0}^{\infty} (-1)^k \varepsilon_k \frac{\sin 2^k x}{2^k x}$, is uniformly bounded and tends to $\sum_{k=0}^{\infty} (-1)^k \varepsilon_k$ as $x \to 0$.

Thus by this lemma, $f(\xi)/i\xi = 2 \int_1^2 g(\xi x)\frac{dx}{x}$ converges to the limit $2 \log 2 \sum_0^{\infty} (-1)^k \varepsilon_k$. It remains to show that $F \notin L^1$. In the interval $[2^m, 2^{m+1})$,

$$\begin{aligned} F(x) &= - \int_x^{\infty} f(t)\, dt \\ &= -\left((-1)^m \varepsilon_m \left(\frac{1}{x} - \frac{1}{2^{m+1}}\right) + \sum_{2^k > x} \cdot (-1)^k \varepsilon_k \int_{2^k}^{2^{k+1}} \frac{dt}{t^2} \right) \\ &= -\left((-1)^m \varepsilon_m \left(\frac{1}{x} - \frac{1}{2^{m+1}}\right) + \sum_{m+1}^{\infty} (-1)^k \varepsilon_k 2^{-k-1} \right), \end{aligned}$$

so $|F(x)| \geq \varepsilon_m \left(\frac{1}{x} - \frac{1}{2^{m+1}}\right) - \varepsilon_{m+1} 2^{-m-2} \geq \varepsilon_m \left(\frac{1}{x} - \frac{3}{2^{m+2}}\right) \geq \varepsilon_m / 8x$ if $2^m \leq x \leq (7/6) 2^m$. Thus

$$\int_{2^m}^{(7/6)2^m} |F(x)|\,dx \geq c\varepsilon_m,$$

so $\Sigma\, \varepsilon_k = +\infty$ implies $F \notin L^1$.

Remark. If $\Sigma\, \varepsilon_k < \infty$, then $F \in L^1$, for then $xf(x)$ is integrable.

Proof of Lemma 2. (Cf. Zygmund [11, Chapter 2, (4.7) - (4.10)].) If $|x| \geq \pi$, then $|g(x)| \leq 2\varepsilon_0/\pi$. Otherwise (for $x > 0$), define m by $2^m x < \pi \leq 2^{m+1}x$. Splitting the sum defining $g(x)$ into two parts, one where $\sin 2^k x/2^k x$ is positive and decreasing, gives

$$|g(x)| \leq \left| \sum_{2^k x \leq \pi} (-1)^k \varepsilon_k \frac{\sin 2^k x}{2^k x} \right| + \frac{1}{x} \sum_{2^k x > \pi} \varepsilon_k 2^{-k}$$

$$\leq \varepsilon_0 + \frac{1}{x}\, \varepsilon_{m+1} \cdot 2^{-m} \leq \varepsilon_0 + \frac{2\varepsilon_{m+1}}{\pi},$$

by definition of m. Only the monotonicity of ε_k is used for uniform boundedness. Since $\sin u/u = 1 + O(u^2)$,

$$g(x) = \sum_{k=0}^{m} (-1)^k \varepsilon_k + \sum_{k=0}^{m} (-1)^k \varepsilon_k \left(\frac{\sin 2^k x}{2^k x} - 1 \right) + \sum_{m+1}^{\infty} \cdots.$$

The second term is dominated by

$$C \sum_{k=0}^{[m/2]} \varepsilon_0 (2^k x)^2 + C\, \varepsilon_{[m/2]} x^2 \sum_{[m/2]+1}^{m} 2^{2k}$$

$$< C\varepsilon_0 x^2 \cdot 2^{m+1} + C\varepsilon_{[m/2]} x^2 2^{2m} \to 0$$

as $x \to 0$ (forcing $m \to \infty$). The third term is dominated by $2\varepsilon_{m+1}/\pi$ as seen before, so $g(x)$ converges to the desired limit.

There may not be any characterization of L_1^1 in terms of Fourier transforms, for Lemma 1 shows that $L_1^1 = E * L^1$, and thus $\hat{F}(\xi) = \hat{g}(\xi)(1 + i\xi)^{-1} \in (L_1^1)\hat{}$ for any $g \in L^1$. Thus prior knowledge that $\hat{f}(\xi)/\xi \in A_1$ may be the only general condition.

In Stein's book [8], it is stated among other things in Chapter 5 (6.6) that $L_1^1(\mathbf{R}^n)$ and the Bessel potential spaces $\mathscr{L}_1^1(\mathbf{R}^n)$ are not comparable. For $n = 1$, each is isomorphic to L^1. For $n > 1$, is $L_1^1(\mathbf{R}^n)$ isomorphic to $L^1(\mathbf{R}^n)$? By using the multipliers supplied by Stein in Chapter 5, Sec. 4, Lemma 2, the following observation may be made: for $n \geq 1$, $\mathscr{L}_1^1(\mathbf{R}^n) \cap L_1^1(\mathbf{R}^n)$ consists of those functions $F \in L^1(\mathbf{R}^n)$ with grad $F \in H^1(\mathbf{R}^n)$ (equivalently, with $\Lambda F \in H^1$), for $F \in \mathscr{L}_1^1$ means $\sqrt{1 + |\xi|^2}\,\hat{F}(\xi) \in A_1(\mathbf{R}^n)$, and since the cited lemma shows that $|\xi|(1 + |\xi|^2)^{-1}$ is the Fourier transform of a measure, $(\Lambda F)\hat{}\,(\xi) = |\xi|\hat{F}(\xi) \in A_1$. If also $F \in L_1^1$, then $\xi_j\hat{F}(\xi) = (\xi_j/|\xi|)(\Lambda F)\hat{}\,(\xi)$, so $\Lambda F \in H^1$. Since the singular integral operators R_j are bounded on H^1, it follows that grad $F \in H^1$. Conversely, if $F \in L^1$ and grad $F \in H^1$, the formula $|\xi| = \Sigma(\xi_j/|\xi|)\xi_j$ shows that $\Lambda F \in L^1$ and again the cited lemma shows that $\sqrt{1 + |\xi|^2}\,\hat{F}(\xi) \in A_1$.

REFERENCES

1. Adams, R. A. *Sobolev Spaces*. New York: Academic Press, 1975.

2. Beckenbach, E. F., and Bellman, R. *Inequalities*. New York: Springer, 1965.

3. Coifman, R. R., and Weiss, Guido. Extensions of Hardy space and their use in analysis. *Bull. Amer. Math. Soc.* 83 (1977): 569-645.

4. Dahlberg, B. E. J. A note on Sobolev spaces. *Proc. of Symposia in Pure Mathematics*, 35, part 1 (1979): 183-185.

5. Esseen, C. -G. Fourier analysis of distribution functions. A mathematical study of the Laplace-Gaussian law. *Acta. Math.* 77 (1945): 1-125.

6. Fichtenholz, G. Note sur les fonctions absolument continues. *Acad. royale de Belgique, Bull. Cl. Sci.* (5) 8 (1922): 430-442.

7. Saks, S. *Theory of the Integral*. Dover edition, 1964.

8. Stein, E. M. *Singular Integrals and the Differentiability Properties of Functions*. Princeton, N.J.: Princeton Univ. Press, 1970.

9. Strichartz, R. S. Multipliers on fractional Sobolev spaces. *J. Math. Mech.* 16 (1967): 1031-1060.

10. Wiener, N. *Selected Papers of Norbert Wiener*, p. 281; Cambridge, Mass.: SIAM and MIT Press, 1964.

11. Zygmund, A. *Trigonometric Series*. 2d ed. New York: Cambridge Univ. Press, 1968.

ON INEQUALITIES OF CARLESON AND HUNT

Benjamin Muckenhoupt
Rutgers University

1. INTRODUCTION

The original form of the inequality of Carleson discussed here is the following.

Theorem 1. If $1 < p < \infty$, $\{I_k\}$ *is a set of disjoint subintervals of an interval* I, x_k *is the center of* I_k,

$$f(x) = \sum \frac{|I_k|^p}{|I_k|^p + |x - x_k|^p},$$

and E_α *is the set where* $f(x) > \alpha$, *then* $|E_\alpha| \le c|I|e^{-\alpha/c}$, *where* $|E_\alpha|$ *denotes the Lebesgue measure of* E_α *and* c *depends only on* p.

This result with $p = 2$ was proved by Carleson as Lemma 5 of [1]. As stated there, the I_k's cover I; this is, of course, equivalent to Theorem 1, since reducing the set $\{I_k\}$ reduces $|E_\alpha|$ without changing the right side of the conclusion. Theorem 1 is essentially a fact about Marcinkiewicz integrals and is treated in

---

Supported in part by NSF Grant No. MCS80-03098.

this way by Zygmund in [5]. Theorem 1 was also proved by Hunt in [3].

Theorem 1 can be frustrating to the user because of the term $|I|$ on the right side of the conclusion. In fact, integrating term by term shows that

$$\int_{-\infty}^{\infty} f(x)\,dx = \sum \int_{-\infty}^{\infty} \frac{|I_k|^p}{|I_k|^p + |x - x_k|^p}\,dx = \sum |I_k| \int_{-\infty}^{\infty} \frac{dx}{1 + |x|^p},$$

and from this it follows that

$$|E_\alpha| \le \frac{1}{\alpha} \sum |I_k| \int_{-\infty}^{\infty} \frac{dx}{1 + |x|^p}. \tag{1}$$

For α large, Theorem 1 gives the better estimate. For $\alpha = 1$, however, (1) is better if $\Sigma\ |I_k|$ is small compared to $|I|$. The obvious procedure when estimating $|E_\alpha|$ is to use (1) for small α and Theorem 1 for large α; unfortunately, the value of α at which one should change from one estimate to the other is a complicated expression.

Professor Zygmund mentioned in [5] that Theorem 1 was "known, but probably not too well known." This seems also to be true for the following modification of Theorem 1.

Theorem 2. If $1 < p < \infty$, $\{I_k\}$ is a disjoint set of intervals, and for $\alpha > 0$, E_α is the set defined in Theorem 1, then $|E_\alpha| \le c\ \Sigma\ |I_k| e^{-\alpha/c}$, where c depends only on p.

This version eliminates the problems mentioned above, since (1) is included in it. Theorem 2 was proved by Fefferman and Stein in [2]. In their notation the function f is $\mathscr{T}_r$. Their statement on p. 108 that $\mathscr{T}_r(\cdot)$ belongs to the exponential class does not include the dependence of the constants on the I_k's; however, the last sentence in Sec. 3 shows that $|E_\alpha|$ has the bound of Theorem 2. Their proof is fairly long, since they prove a general result about maximal functions of vector-valued functions.

This paper consists of the proof of two theorems. In Sec. 2 a simple proof is given of Theorem 2 that resembles the proofs of Theorem 1 in [1] and [3]. In Sec. 3 Theorem 2 is used to prove a strengthened version of a theorem of R. Hunt in [3] that estimates the size of the set where the conjugate of a periodic function has absolute value greater than one number and the Hardy-Littlewood maximal function is less than another number. The proof is similar to the proof given by Hunt for his estimate except that Theorem 2 is used instead of Theorem 1 and a slightly better estimate is used for the distribution function of the conjugate of a bounded function.

Throughout this paper the letter c will be used for positive constants that are not necessarily the same at each occurrence.

2. PROOF OF THEOREM 2

For $\alpha < 1$, the result follows trivially from (1). Therefore, assume $\alpha \geq 1$ and start with the fact that

$$|E_\alpha| \leq \frac{1}{\alpha} \int_{-\infty}^{\infty} \left[\sum_k \frac{|I_k|^p}{|I_k|^p + |x - x_k|^p} \right] \chi_{E_\alpha}(x)\, dx.$$

The right side is bounded by

$$\frac{2^p}{\alpha} \int_{-\infty}^{\infty} \left[\sum_k \int_{I_k} \frac{|I_k|^{p-1}}{|I_k|^p + |x - t|^p}\, dt \right] \chi_{E_\alpha}(x)\, dx,$$

and reversing the sum and inner integral with the outer integral gives

$$|E_\alpha| \leq \frac{2^p}{\alpha} \sum_k \int_{I_k} \left[\int_{-\infty}^{\infty} \frac{|I_k|^{p-1} \chi_{E_\alpha}(x)}{|I_k|^p + |x - t|^p}\, dx \right] dt.$$

By Theorem 2, p. 62 of [4], the inner integral on the right is bounded by $c\chi^*_{E_\alpha}(t)$, where $\chi^*_{E_\alpha}$ is the Hardy-Littlewood maximal function of χ_{E_α}. Therefore,

$$|E_\alpha| \leq \frac{c}{\alpha} \int_{UI_k} \chi^*_{E_\alpha}(t)\, dt.$$

Since $\chi^*_{E_\alpha}(t) \leq 1$ for all t, we have

$$|E_\alpha| \leq \frac{c}{\alpha} \int_0^1 |\{t|\chi^*_{E_\alpha}(t) > \beta\} \cap UI_k|\, d\beta.$$

From this

$$|E_\alpha| \leq \frac{c}{\alpha} \int_0^1 \min(|\{t|\chi^*_{E_\alpha}(t) > \beta\}|, |UI_k|)\, d\beta.$$

Using the usual weak-type (1, 1) estimate for the maximal function, Theorem 1, p. 5 of [4], we obtain

$$|E_\alpha| \leq \frac{c}{\alpha} \int_0^1 \min\left(\frac{C}{\beta}|E_\alpha|, \Sigma|I_k|\right) d\beta.$$

Now let $b = |E_\alpha| / \left[\sum|I_k| \int_{-\infty}^{\infty} \frac{dx}{1 + |x|^p}\right]$. By (1) and the fact that $\alpha \geq 1$, we see that $b \leq 1$ and

$$|E_\alpha| \leq \frac{c}{\alpha} \int_b^1 \frac{|E_\alpha|}{\beta} d\beta + \frac{c}{\alpha} \int_0^b \Sigma|I_k|\, d\beta.$$

Performing the integrations gives

$$|E_\alpha| \leq \frac{c|E_\alpha|}{\alpha} \log\left[\frac{c\ \Sigma|I_k|}{|E_\alpha|}\right] + \frac{c|E_\alpha|}{\alpha}.$$

Dividing by $|E_\alpha|$ and rearranging terms lead to

$$\frac{c}{\alpha} \log\left[\frac{|E_\alpha|}{c\ \Sigma|I_k|}\right] \leq \frac{c}{\alpha} - 1,$$

which is equivalent to the conclusion.

3. HUNT'S INEQUALITY

For an integrable function f with period 2π, let

$$\tilde{f}(x) = \lim_{\varepsilon \to 0^+} \int_{\varepsilon < |t| < \pi} \frac{f(x - t)}{2\pi \tan t/2} \, dt$$

be the usual conjugate function and

$$f^*(x) = \sup_{y \neq x} \frac{\int_x^y |f(t)| \, dt}{y - x}$$

be the Hardy-Littlewood maximal function. In [3], R. Hunt used Theorem 1 to prove the following.

Theorem 3. If $y > 0$, $\lambda > 0$, f has period 2π and is integrable on $[-\pi, \pi]$, and E is the subset of $[-\pi, \pi]$ where $f^*(x) \leq y$ and $|\tilde{f}(x)| > \lambda y$, then $|E| \leq ce^{-\lambda/c}$, where c is independent of f, λ, and y.

Like Theorem 1, this has the obvious flaw that the conclusion is weaker than another known estimate for some values of the variables. Here the usual weak-type (1, 1) estimate shows that $|E| \leq c\|f\|_1/\lambda y$, which is better if y is large. The modified version to be proved here is the following.

Theorem 4. If $y > 0$, $\lambda > 0$, f has period 2π, and E is the subset of $[-\pi, \pi]$ where $f^*(x) \leq y$ and $|\tilde{f}(x)| > \lambda y$, then $|E| \leq c \dfrac{\|f\|_1}{y} \cdot e^{-\lambda/c}$, where c is independent of f, λ, and y.

The proof uses the following lemma.

Lemma 1. If $g(x)$ has period 2π and is bounded, $y \geq \|g\|_\infty$ and E is the subset of $[-\pi, \pi]$ where $|\tilde{g}(x)| > y$, then $|E| \leq c \dfrac{\|g\|_1}{\|g\|_\infty} \cdot e^{-y/(c\|g\|_\infty)}$, where c is independent of g and y.

To prove Lemma 1, use the fact that $e^x \leq 82(e^x - x - 1)$ for $x \geq 1/6$ to show that if D is the subset of $[-\pi, \pi]$ where $|\tilde{g}(x)| > \|g\|_\infty$, then

$$\int_D e^{|\tilde{g}(x)|/(6\|g\|_\infty)}\,dx \le 82 \sum_{n=2}^{\infty} \int_{-\pi}^{\pi} \frac{|\tilde{g}(x)|^n\,dx}{n!\,(6\,\|g\|_\infty)^n} . \tag{2}$$

By [6, vol. 1, p. 261], $\|\tilde{f}\|_p \le 2p\|f\|_p$ for $p \ge 2$ and we see that the right side of (2) is bounded by

$$82 \sum_{n=2}^{\infty} \frac{n^n}{3^n n!\,\|g\|_\infty^n} \int_{-\pi}^{\pi} |g(x)|^n\,dx \le \frac{82\|g\|_1}{\|g\|_\infty} \sum_{n=2}^{\infty} \frac{n^n}{3^n n!} ;$$

the last sum is finite by Stirling's formula. Therefore, if $y \ge \|g\|_\infty$, then since $D \supset E$, we have

$$|E| \le e^{-y/(6\|g\|_\infty)} \int_E e^{|\tilde{g}(x)|/(6\|g\|_\infty)}\,dx \le c\,\frac{\|g\|_1}{\|g\|_\infty}\, e^{-y/(6\|g\|_\infty)} .$$

This completes the proof of Lemma 1.

The proof of Theorem 4 is similar to that of Theorem 3 in [3] except that Theorem 2 is used in place of Theorem 1 and Lemma 1 is used instead of the more common estimate $|E| \le ce^{-y/(c\|g\|_\infty)}$. The proof given here, therefore, omits the details that are the same as in [3].

To prove Theorem 4, observe first that since $f^*(x) \ge \|f\|_1/2\pi$ for all x, $|E| = 0$ if $y < \|f\|_1/2\pi$. Therefore, we may assume that $y \ge \dfrac{\|f\|_1}{2\pi}$. We may also assume $\lambda \ge 2$, since for $0 \le \lambda < 2$ the result follows from the usual weak-type (1, 1) estimate for the conjugate function. By translating if necessary, we may assume that $f^*(\pi) \le y$. Then, as usual, let the subset of $[-\pi, \pi]$ where $f^*(x) > 4y$ be written as UI_k where the I_k's are disjoint open intervals; let $g(x) = f(x)$ for x not in UI_k, $g(x) = \dfrac{1}{|I_k|}\displaystyle\int_{I_k} f(t)\,dt$ for x in I_k, and $b(x) = f(x) - g(x)$. Then if $3I_k$ is the interval with the same center as I_k and three times as long, $f^*(x) > y$ on $U3I_k$. Therefore, E is a subset of the union of

$$E_1 = \{x : |\tilde{g}(x)| > \lambda y/2\} \cap [-\pi, \pi]$$

and

$$E_2 = \{x : |\tilde{b}(x)| > \lambda y/2\} \cap [-\pi, \pi]\backslash U3I_k.$$

By Lemma 1,

$$|E_1| \leq \frac{c\|g\|_1}{\|g\|_\infty} e^{-\lambda y/(c\|g\|_\infty)}.$$

Since $\|g\|_\infty \leq 4y$ and $\lambda \geq 2$, this implies

$$|E_1| \leq \frac{c\|g\|_1}{y} e^{-\lambda/c} \leq \frac{c\|f\|_1}{y} e^{-\lambda/c}.$$

To estimate $|E_2|$, use the fact that $|\tilde{b}(x)| \leq cy \sum \frac{|I_k|^2}{x^2 + |I_k|^2}$ for x in $[-\pi, \pi]\backslash U3I_k$. Therefore, by Theorem 2,

$$|E_2| \leq c \sum |I_k| e^{-\lambda/c} \leq c \frac{\|f\|_1}{y} e^{-\lambda/c}.$$

Since $|E| \leq |E_1| + |E_2|$, this completes the proof of Theorem 4.

REFERENCES

1. Carleson, L. On convergence and growth of partial sums of Fourier series. *Acta Math.* 116 (1966): 135-157.

2. Fefferman, C., and Stein, E. M. Some maximal inequalities. *Amer. J. of Math.* 93 (1971): 107-115.

3. Hunt, R. An estimate of the conjugate function. *Studia Math.* 44 (1972): 371-377.

4. Stein, E. M. *Singular Integrals and Differentiability Properties of Functions.* Princeton, N.J.: Princeton Univ. Press, 1970.

5. Zygmund, A. On certain lemmas of Marcinkiewicz and Carleson. *J. of Approximation Theory* 2 (1969): 249-257.

6. Zygmund, A. *Trigonometric Series.* 2d ed. New York: Cambridge Univ. Press, 1968.

ON THE RESTRICTION OF FOURIER TRANSFORMS TO CURVES

Alberto Ruiz
Universidad Complutense de Madrid
and
University of Minnesota

The restriction of the Fourier transform to curves has been treated by Zygmund [6] for the circumference on the plane and by Prestini [1] in the case of a curve whose curvature and torsion do not vanish. In this paper we obtain some similar result, that is, $L^p \to L^q$ boundedness, for more general curves, by using geometric methods. The case of curves in the plane were done by Sjolin [2], except that in the (p, q)-boundary line $q = \frac{p'}{k+1}$, we reach this line with the restriction $p' > 2k$, which is Zygmund's $p' > 4$ for the circumference. In $\mathbf{R}^3$ we get Prestini's result, and the restriction $p < \frac{15}{13}$ in there remains; we prove also $L^p \to L^q$ boundedness for curves whose curvature never vanishes. We state a more general conjecture for curves with finitely many points with torsion and curvature zero. The only remaining thing in order to prove this conjecture is to establish Lemma 1 in zero curvature cases.

We prove that the results are sharp with respect to the condition $q > \frac{p'}{k+1}$. In order to determine whether or not these results are the sharpest with respect to the conditions on p's that arise from the methods used here, one would have to study the Fourier transform of the uniform measures along curves (t, t^{k_1}, t^{k_2}).

I would like to thank Professor A. Córdoba for his ideas and advice concerning these problems.

Theorem 1. Let's consider a curve $y = \gamma(x)$ *on* $\mathbf{R}^2$ *of class* C^{k+2} *that has a contact of order* $k - 1$ *with its tangent at the origin; in a small arc containing the origin we have*

$$\| f \|_{L^q(\gamma)} \leq c \| f \|_{L^p(\mathbf{R}^2)} \tag{1}$$

for every $f \in S(\mathbf{R}^2)$ *and* $1 < q < \frac{p'}{k+1}$, $p' > 4$, *and* $p' > 2k$ *in the case* $q = \frac{p'}{k+1}$. p' *here is such that* $\frac{1}{p} + \frac{1}{p'} = 1$.

Proof. For simplicity we prove it for the curve (t, t^k); the geometric construction can be repeated in the general case and all the properties remain the same.

(1) is equivalent to

$$\left| \int \hat{f}(x) g(x)\, d\sigma(x) \right| \leq c \| f \|_p$$

where $d\sigma(x)$ is the uniform measure on γ and $\| g \|^{q'} = 1$; using a classical density argument we can suppose g to be the step function

$$g = \Sigma\, a_j \chi_{Ij},$$

where the I_j's are intervals on our curve γ. Furthermore, we may assume, by taking δ sufficiently small depending on g, that every interval I_j has a projection of size δ on the $0x$-axis.

Then

$$1 = \| g \|_{q'}^{q'} \leq c\delta\, \Sigma\, |a_j|^{q'}$$

$$\| g \|_{q'}^{q'} \geq c'\delta\, \Sigma\, |a_j|^{q'}.$$

It holds

$c_2 \le \delta \sum |a_j|^{q'} \le c_1$

$$\left| \int_\gamma \hat{f}(x) \sum_{j=1}^{c\delta^{-1}} a_j \chi_{I_j}(x)\, d\sigma(x) \right| \le \frac{c}{\delta^k} \left| \int_{\mathbf{R}^2} \hat{f}(x) \frac{1}{j^{k-2}} a_j \emptyset_j(x)\, dx \right|, \qquad (2)$$

where $\emptyset_j$ is the characteristic function of the rectangle whose direction is the tangent to the curve at $x = j\delta$ and dimensions $\delta x j^{k-2} \delta^k$ (see Figure 1).

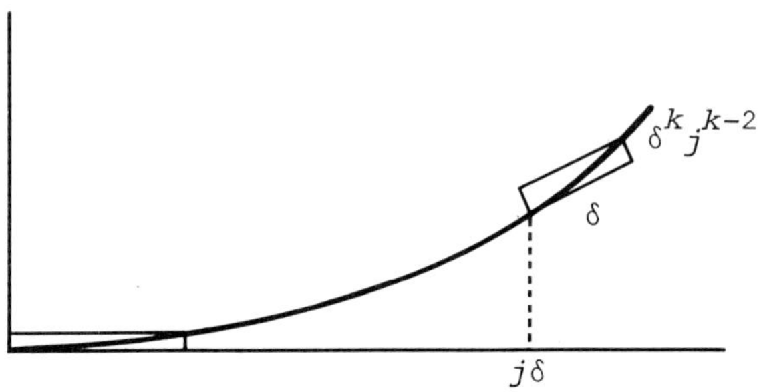

Figure 1

The last inequality follows from the continuity of $\hat{f}(x)$ and the compactness of γ; it holds for every $\delta < \delta_0$ with δ_0 small enough. (2) is bounded by

$$\frac{c}{\delta^k} \left| \int_{\mathbf{R}^2} \hat{f}(x) \sum \frac{a_j}{j^{k-2}} \emptyset_j(x) \right|$$

$$= \frac{1}{\delta^k} \left| \int f(x) \sum \frac{a_j}{j^{k-2}} \hat{\emptyset}_j(x)\, dx \right| \le \frac{1}{\delta^k} \| f \|_p \left\| \sum \frac{a_j}{j^{k-2}} \hat{\emptyset}_j \right\|_{p'}.$$

It suffices to prove that

$$\frac{1}{\delta^k} \left\| \sum \frac{a_j}{j^{k-2}} \hat{\emptyset}_j \right\|_{p'} \le c.$$

Assume $p' \geq 4$; we can use the Hausdorf-Young inequality and write

$$\left\| \sum_j \frac{a_j}{j^{k-2}} \hat{\phi}_j \right\|_{p'} = \left\| \left(\sum_j \frac{a_j}{j^{k-2}} \hat{\phi}_j \right)^2 \right\|_{p'/2}^{1/2} \tag{3}$$

$$= \left\| \sum_{n,m} \frac{a_n a_m}{n^{k-2} m^{k-2}} \hat{\phi}_n * \hat{\phi}_m \right\|_{p'/2}^{1/2}$$

$$\leq \left\| \sum_{n,m} \frac{a_n a_m}{n^{k-2} m^{k-2}} \phi_n * \phi_m \right\|_{s}^{1/2},$$

where $\frac{1}{s} + \frac{2}{p'} = 1$.

Let's consider $\phi_\ell * \phi_m$; we have the three following facts:

(a) $\left| \sum_{\ell,m} \chi_{\operatorname{supp}(\phi_\ell * \phi_m)}(x) \right| \leq C.$

It follows from the sizes of supp ϕ_ℓ and supp ϕ_m and from $\operatorname{supp}(\phi_\ell * \phi_m) \subset \operatorname{supp}(\phi_\ell) + \operatorname{supp} \phi_m$

(b) $|\operatorname{supp} \phi_\ell * \phi_m|$

$$|\operatorname{supp} \phi_\ell + \operatorname{supp} \phi_m| = \begin{vmatrix} \delta, & \delta^k k\ell^{k-1} \\ \delta, & \delta^k km^{k-1} \end{vmatrix} + \delta^{k+1}(\ell^{k-2} - m^{k-2})$$

$$= \{k|m^{k-1} - \ell^{k-1}| + \ell^{k-2} + m^{k-2}\}\delta^{k+1},$$

(c) $$|\phi_\ell * \phi_m(x)| \leq \int \phi_\ell(x-y)\phi_m(y)\,dy$$

$$\leq |\operatorname{supp} \phi_m(\cdot) \cap (\operatorname{supp} \phi_\ell - x)|$$

$$\leq |R_m \cap R_\ell| \leq \delta^k m^{k-2} \delta^k \ell^{k-2} \sin\alpha$$

$$= \frac{\ell^{k-2} m^{k-2} \delta^{2k}}{[|m^{k-1} - \ell^{k-1}|k + \ell^{k-2} + m^{k-2}]\delta^{k-1}}.$$

(See Figure 2 on the next page.)

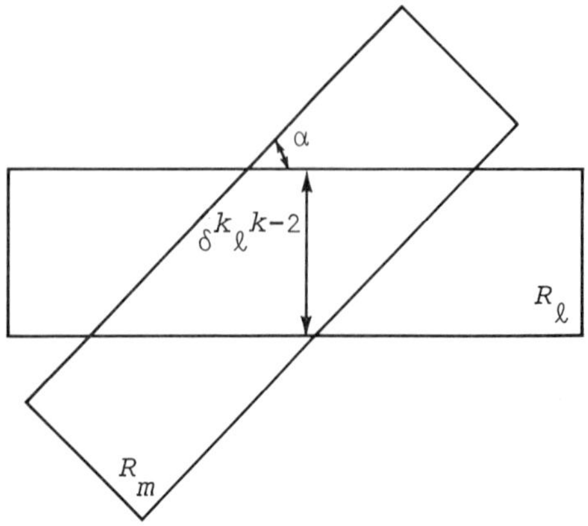

Figure 2

From (a) (3) is bounded by

$$\left[\int \left| \sum \frac{a_\ell a_m}{\ell^{k-2} m^{k-2}} \phi_\ell * \phi_m(x) \right|^s dx\right]^{1/2s}$$

$$\leq c \left| \sum \frac{|a_\ell| \, |a_m|^s}{\ell^{(k-2)s} m^{(k-2)s}} \int |\phi_\ell * \phi_m|^s ds\right]^{1/2s}$$

and from (b)

$$\leq \delta^{\frac{k+1}{2s}} \delta^{\frac{k+1}{2}} \left[c\delta^{-1} \sum_{m,\ell=1} \frac{|a_p|^s |a_m|^s}{[k|m^{k-1} - \ell^{k-1}| + \ell^{k-2} + m^{k-2}]^{s-1}} \right]^{1/2s} .$$

To get the result $q = \dfrac{p'}{k+1}$, $p' > 2k$, we bound the above by

$$\delta^{\frac{k+1}{2s}} \delta^{\frac{k+1}{2}} \left[\sum \frac{|a_\ell|^s |a_m|^s}{[|\ell - m|^{k-1} + 1]^{s-1}} \right]^{1/2s} .$$

Let us apply fractional integration $\ell^t \to \ell^s$ to $(|a_m|^s) \in \ell^t$ with the conditions: $\dfrac{1}{s} = \dfrac{1}{t} - \alpha$, $-\alpha = (s-1)(k-1) - 1$, $t = s'$; that is, $t = \dfrac{p'-2}{p'-k-1}$. Hence

$$\leq \delta^{\frac{k+1}{2s}} \delta^{\frac{k+1}{2}} \left[\sum |a_\ell|^{s \frac{p'-2}{p'-k-1}} \right]^{2 \cdot \frac{1}{2s} \frac{p'-k-1}{(p'-2)}}$$

$$= \delta^{\frac{k+1}{2s}} \delta^{\frac{k+1}{2}} \left[\sum^{c\delta^{-1}} |a_\ell|^{\frac{p'}{p'-k-1}} \right]^{\frac{p'-k-1}{p'}} .$$

By Hölder's inequality with $q' = \dfrac{p'}{p' - k - 1}$, that is, $q = \dfrac{p'}{k + 1}$, we obtain

$$\leq c\delta^{\frac{k+1}{2s}} \delta^{\frac{k+1}{2}} \delta^{-\frac{p'-k-1}{p'}} = c\delta^{k} .$$

This proves our statement; notice that the restriction $p' > 2k$ comes from the application of fractional integration.

To get the result $\frac{1}{q} > (k + 1)(1 - \frac{1}{p})$, let's bound (1, 4) by

$$\delta^{\frac{k+1}{2s}} \delta^{\frac{k+1}{2}} \left[\sum_m \frac{|a_m|^s}{\frac{(k-2)(s-1)}{2}} \sum_\ell \frac{|a_\ell|^s}{\frac{(k-2)}{2}(s-1)} \frac{1}{[(m - \ell) + 1]^{s-1}} \right]^{1/2s} .$$

Fractional integration allows us to write for $\frac{1}{q_1} = \frac{1}{p_1} - \alpha$, $\alpha = 2 - s$, and $q_1 = p_1'$, assuming $4 < p'$, the following:

$$\leq \delta^{\frac{k+1}{2s}} \delta^{\frac{k+1}{2}} \left[\sum_m \frac{|a_m|^{\frac{2s}{3-s}}}{\frac{(k+2)(s-1)}{2} \cdot \frac{2}{3-s}} \right]^{\frac{(3-s)}{2s}} .$$

Using Hölder's inequality for $t = q'\frac{p' - 3}{p'}$, we get

$$\leq \delta^{\frac{k+1}{2s} + \frac{k+1}{2}} \left[\sum |a_m|^{q'} \right]^{1/q'} \left[\sum^{c\delta^{-1}} m^{\frac{-q'(k-2)}{q'p'-sq'-p'}} \right]^{\frac{q'p'-3q'-p'}{p'q'}} .$$

If we assume $\dfrac{(k - 2)q'}{q'p' - 3q' - p'} < 1$, that is, $q < \dfrac{p'}{k + 1}$, we obtain the desired $c\delta^k$.

Remarks. 1. A homogeneity argument shows that we cannot improve the condition $q > \frac{p'}{k+1}$; in fact we can take

$$f(\xi_1, \xi_2) = \exp\left\{-\left(\frac{\xi_1}{\delta}\right)^2 - \left(\frac{\xi_2}{\delta^k}\right)^2\right\} \quad \text{for} \quad \delta > 0.$$

2. Let us consider $d\sigma$, the singular uniform measure along a neighborhood of the origin of the curve γ in Theorem 1; then $\widehat{d\sigma} \in L^p(\mathbf{R}^2)$ when $p > \max\{4, k+1\}$. In particular, if $d\sigma_1$ is the measure for (t, t^2) and $d\sigma_2$ is the one for (t, t^3), $\widehat{d\sigma_1}$ and $\widehat{d\sigma_2}$ are in $L^p(\mathbf{R}^2)$ for the same interval of p's. This fact follows by considering the dual of the "restriction operator."

Theorem 2. Let $\gamma(t)$ be an arc of a C^{k_2+1} curve around the origin 0 in $\mathbf{R}^3$. We assume that γ has nonzero curvature everywhere and its torsion is nonzero everywhere except at 0, and also that γ has a contact of order k_2 with its osculating plane at 0, where $k_2 > 2$. Then we have

$$\|\hat{f}\|_{L^q(\gamma)} \le c\,\|f\|_{L^p(\mathbf{R}^3)} \tag{4}$$

for every $f \in S(\mathbf{R}^3)$, when p and q satisfy

(a) $\frac{1}{q} > \frac{3+k_2}{p}$

$$p < \min\left(\frac{4k_2+3}{4k_2+1}, \frac{k_2+3}{k_2+2}, \frac{15}{13}\right).$$

(b) *For the following values of k_2,* (4) *is satisfied for the set of* (p, q) *considered in* (a) *and the additional ranges*

$$\frac{1}{q} > \frac{63}{5p'} - \frac{7}{5}, \quad p < \frac{23}{21}, \quad \text{when } k_2 = 5$$

$$\frac{1}{q} > \frac{36}{p'} - 3, \quad p < \frac{27}{25}, \quad \text{when } k_2 = 6$$

$$\frac{1}{q} > \frac{165}{p'} - 11, \quad p < \frac{31}{29}, \quad \text{for } k_2 = 7.$$

Remark. As we noticed in the introduction, this would be the $k_1 = 2$ case of the following conjecture:

Let $\gamma(t)$ be an arc of a C^{k_2+1} curve around the origin. Assume that γ has nonzero curvature and nonzero torsion at every point except possibly at the origin, where we assume it has a contact of order k_1 with the rectifying plane and a contact of order k_2 with its osculating plane for $2 < k_1 < k_2$. Then (4) holds when p and q satisfy:

(A) In case $(k_1, k_2) \neq (3, 4)$

$$\frac{1}{q} > \frac{k_1 + k_2 + 1}{p'}$$

$$p < \min\left(\frac{4k_2 - 2k_1 + 7}{4k_2 - 2k_1 + 5}, \frac{k_1 + k_2 + 1}{k_1 + k_2}, \frac{15}{13}\right)$$

if $k_1 > 2$. (4) holds on the line $\frac{1}{q} = \frac{k_1 + k_2}{p'}$ as well, with the same restrictions on p.

(B) The same as (b) in the statement of Theorem 2.

(C) If $(k_1, k_2) = (3, 4)$,

$$\frac{1}{q} > \frac{k_1 + k_2 + 1}{p'} \quad \text{and} \quad p < \frac{21}{19}.$$

To prove this conjecture Lemma 1 would need to be extended.

Proof of Theorem 2. Although $k_1 = 2$ in our theorem, we are going to keep the general notation as in the previous conjecture.

We do it again for the curve $\left(t, t^{k_1}, t^{k_2}\right)$ to avoid the difficult notation of the general case, whose proof is essentially the same.

We attempt to follow the steps of Theorem 1.

Let's take $f \in S(\mathbf{R}^3)$ such that $\hat{f}$ is continuous and exists everywhere along $\gamma(t)$.

$$\|\hat{f}\|_{L^q(\gamma)} = \sup_{\|g\|_{L^{q'}(\gamma)}=1} \left|\int_\gamma \hat{f}(x)g(x)\,d\sigma(x)\right|,$$

where $d\sigma(x)$ is the uniform measure along γ. We can suppose g to be $\sum a_k \emptyset_k$, where $\emptyset_k$ are characteristic functions of intervals on γ. Let us break down the support of $\emptyset_k$, taking δ small enough, so that

$$g = \sum_{k=1}^{c\delta^{-1}} a'_k \emptyset'_k$$

where the projections of supp $\emptyset_k$ on the x-axis are disjoint intervals of measure δ.

$$\|g\|^{q'}_{L^{q'}(\gamma)} = c\delta \sum |a_k|^{q'} = 1$$

$$\left|\int_\gamma \hat{f}(x)g(x)\,d\sigma(x)\right| = \left|\int_\gamma \hat{f}(x) \sum_{j=1}^{c\delta^{-1}} a_j \emptyset_j(x)\,d\sigma(x)\right| \tag{5}$$

$$\leq \frac{c}{\delta^{k_1+k_2}} \left|\int_{\mathbf{R}^3} \hat{f}(x) \sum_{j=1}^{c\delta^{-1}} \frac{a_j}{j^{k_1+k_2}} \bar{\emptyset}_j(x)\,dx\right|,$$

where $\bar{\emptyset}_j$ is the characteristic function of a parallelepiped whose edges are segments on the tangent, the normal, and the binormal to the curve at $x = j\delta$, of length δ, $\delta^{k_1} j^{k_1-2}$, and $\delta^{k_2} j^{k_2-3}$, respectively. The last inequality is valid for δ small enough depending on $\hat{f}$.

(5) is bounded by

$$\frac{1}{\delta^{k_1+k_2}} \left|\int_{\mathbf{R}^3} f(x) \Big(\sum_j \frac{a_j}{j^{k_2+k_1-5}} \hat{\bar{\emptyset}}_j(x)\,dx\right|$$

$$\leq \frac{1}{\delta^{k_1+k_2}} \|f\|_p \left\| \sum_j \frac{a_j}{j^{k_2+k_1-5}} \hat{\bar{\emptyset}}_j(\cdot) \right\|_{p'}.$$

It suffices to prove

$$\left\| \sum_{j=1}^{c\delta^{-1}} \frac{a_j}{j^{k_2+k_1-5}} \hat{\bar{\emptyset}}_j(\cdot) \right\|_{p'} \leq c\delta^{k_1+k_2}$$

with c independent of δ.

$$\left\| \sum_{j=1}^{c\delta^{-1}} \frac{a_j}{j^{k_2+k_1-5}} \hat{\bar{\emptyset}}_j(\cdot) \right\|_{p'} = \left\| \left(\sum_j \frac{a_j}{j^{k_1+k_2-5}} \hat{\emptyset}_j \right)^3 \right\|_{p'/3}^{1/3}$$

(let's denote $\emptyset_j$ instead of $\bar{\emptyset}_j$)

$$= \left\| \sum_{k,n,m} \frac{a_k a_n a_m}{(k.n.m)^{k_1+k_2-5}} \hat{\emptyset}_k \hat{\emptyset}_n \hat{\emptyset}_m \right\|_{p'/3}^{1/3}$$

$$= \left\| \sum_{k,n,m} \frac{a_k a_n a_m}{(k.n.m)^{k_1+k_2-5}} \emptyset_k * \emptyset_n * \emptyset_m \right\|_s^{1/3}$$

assuming $p'/3 \geq 2$ and $\frac{1}{s} + \frac{3}{p'} = 1$.

By Lemma 1 (recall that it is proved only when $k_1 = 2$)

$$\sum_{k,n,m} \chi_{\mathrm{supp}(\emptyset_k * \emptyset_n * \emptyset_m)}(x) \leq c.$$

Then we can write:

$$\left[\int \left| \sum \frac{a_k a_n a_m}{(k.n.m)^{k_1+k_2-5}} \emptyset_k * \emptyset_n * \emptyset_m \right|^s \right]^{1/3s}$$

$$\leq c_s \left| \sum \left| \frac{a_k a_n a_m}{(k.n.m)^{k_1+k_2-5}} \right|^s \int \left| \emptyset_k * \emptyset_n * \emptyset_m \right|^s dx \right]^{1/3s} . \qquad (6)$$

Let's consider $|\mathrm{supp}(\emptyset_k * \emptyset_n * \emptyset_m)|$ and $|\emptyset_k * \emptyset_n * \emptyset_m(x)|$.

The first number is a multiple of the volume of a trihedron whose edges are segments of length δ parallel to the tangents at $x = k\delta$, $x = n\delta$, $x = m\delta$, plus the volume due to the widening in the normal and binormal directions.

$$|\operatorname{supp} \emptyset_k * \emptyset_n * \emptyset_m| = \begin{vmatrix} \delta & \delta & \delta \\ k_1\delta^{k_1} k^{k_1-1} & k_1\delta^{k_1} n^{k_1-1} & k_1\delta^{k_1} m^{k_1-1} \\ k_2\delta^{k_2} k^{k_2-1} & k_2\delta^{k_2} n^{k_2-1} & k_2\delta^{k_2} m^{k_2-1} \end{vmatrix}$$

$$+ 4A_2 + \left(k^{k_1+k_2-5} + n^{k_1+k_2-5} + m^{k_1+k_2-5}\right)\delta^{k_1+k_2+1} \tag{7}$$

Let us write the first determinant $\Delta_1\delta^{k_1+k_2+1}$. A_2 is the volume of the trihedron shaped by two tangent vectors and one normal or binormal; hence A_2 is negligible compared to $\Delta_1\delta^{k_1+k_2+1}$.

We *claim* that there exists an $A \subset \operatorname{supp}(\emptyset_k * \emptyset_n * \emptyset_m)$ such that $|A| > \frac{1}{10^3}|\operatorname{supp} \emptyset_k * \emptyset_n * \emptyset_m|$ and $|\emptyset_k * \emptyset_n * \emptyset_m(x)| \geq c\,\|\emptyset_k * \emptyset_n * \emptyset_m\|_\infty$ for every $x \in A$.

In fact:

$$|\emptyset_k * \emptyset_n * \emptyset_m(x)|$$

$$= \iint \emptyset_k(y)\emptyset_n(z)\emptyset_m(x - y - z)\,dz\,dy$$

$$= \int_{\operatorname{supp}\emptyset_k} |\operatorname{supp} \emptyset_n \cap (x - y - \operatorname{supp} \emptyset_m)|\,dy$$

where

$$x - y - \operatorname{supp} \emptyset_m = \{z \in \mathbf{R}^3,\ z = x - y - x_m\ ,\ \text{when } x_m \in \operatorname{supp} \emptyset_m\}.$$

Pick x_0, y_0 in $\operatorname{supp} \emptyset_k$ such that the measure of $|\operatorname{supp} \emptyset_n \cap (x_0 - y_0 - \operatorname{supp} \emptyset_m)|$ is close to

$$\sup_{x,y \in \operatorname{supp}\emptyset_k} |\operatorname{supp} \emptyset_n \cap (x - y - \operatorname{supp} \emptyset_m)|.$$

We can choose x_0, y_0 so that there exist $t \in \operatorname{supp} \emptyset_m$ and $s \in \operatorname{supp} \emptyset_n$ satisfying $x_0 - y_0 - t = s$; and t is close to the center of $\operatorname{supp} \emptyset_m$ and s is close to the center of $\operatorname{supp} \emptyset_n$. Denote by A_i the "central part" of $\operatorname{supp} \emptyset_i$, that is, the most central body obtained when we divide supp $\emptyset_i$ by 12 planes, every four of which are equidistant parallel planes to each side. Let $A = A_k + A_n + A_m$; then

$$|A| > \frac{1}{10^3}\,|\text{supp } \emptyset_k * \emptyset_n * \emptyset_m|; \text{ suppose } \; x \in A$$

$$x = x_0 + (x_k - y_0) + (x_n - s) + (x_m - t).$$

If we denote $x_1 = x_0 + (x_n - s) + (x_m - t)$, then

$$|\text{supp } \emptyset_n \cap (x_1 - y_0 - \text{supp } \emptyset_m)| \geq \frac{1}{2}\,|\text{supp } \emptyset_n \cap (x_0 - y_0 - \text{supp } \emptyset_m)|.$$

Hence by taking $x_1 + (x_k - y_0)$, we can find $y \in \text{supp } \emptyset_k$ such that $x_k - y_0 = -(y - y_0)$ since we chose x_k in the "central part" of A_k.

Since the function $h(y_1, y_2, y_3) = |\text{supp } \emptyset_n \cap (x - y - \text{supp } \emptyset_m)|$, where $y = (y_1, y_2, y_3)$, is continuous and with a maximum as a function of every component y_i, we have for $x \in A$

$$\int_{\text{supp}\emptyset_k} |\text{supp } \emptyset_n \cap (x - y - \text{supp } \emptyset_m)|\, dy$$

$$\geq \frac{1}{4}\,|\text{supp } \emptyset_k|\,\frac{1}{2}|\text{supp } \emptyset_n \cap (x_0 - y_0 - \text{supp } \emptyset_m|$$

$$\geq \frac{1}{16}\,\|\emptyset_k * \emptyset_n * \emptyset_m\|_\infty,$$

which is our claim. Then we are able to estimate $\|\emptyset_k * \emptyset_n * \emptyset_m\|_\infty$:

$$\int_A \emptyset_k * \emptyset_n * \emptyset_m(x)\, dx \geq \frac{1}{10^3}|\text{supp } \emptyset_k * \emptyset_n * \emptyset_m|\frac{1}{16}\|\emptyset_k * \emptyset_\ell * \emptyset_m\|_\infty.$$

But

$$\int_A \emptyset_k * \emptyset_n * \emptyset_m(x)\, dx \leq \int_{\mathbf{R}^3} \emptyset_k * \emptyset_n * \emptyset_m(x)\, dx$$

$$= |\text{supp } \emptyset_k| \cdot |\text{supp } \emptyset_n| \cdot |\text{supp } \emptyset_m|$$

$$= \delta^{3(k_1+k_2+1)}\,(k \cdot n \cdot m)^{k_1+k_2-5}.$$

Then

$$\|\emptyset_k * \emptyset_n * \emptyset_m\| \leq C\left[\frac{\delta^{3(k_1+k_2+1)}(k\cdot n\cdot m)^{k_1+k_2-5}}{|\operatorname{supp}\,\emptyset_k * \emptyset_n * \emptyset_m|^{s-1}}\right]^{1/3s} . \tag{8}$$

The term $A_2 = \Delta_2\delta^{k_1+k_2-1}$ in (7) is important in the case $k = n \neq m$ or in symmetric situations,

$$|\operatorname{supp}\,\emptyset_k * \emptyset_k * \emptyset_m| \sim$$

$$\begin{vmatrix} -k^{2k_1-3}\delta^{2k_1-1} - \dfrac{k_2(k_2-1)}{k_1(k_1-1)}\delta^{2k_2-1}k^{2k_2-3} & , k^{k_1-2}\delta^{k_1} & , \dfrac{k_2(k_2-1)}{k_1(k_1-1)}k^{k_2-2}\delta^{k_2} \\ \delta & , k_1\delta^{k_1}k^{k_1-1} & , k_2\delta^{k_2}k^{k_2-1} \\ \delta & , k_1\delta^{k_1}m^{k_1-1} & , k_2\delta^{k_2}m^{k_2-1} \end{vmatrix} .$$

We may assume $k \geq n \geq m$. Let us consider two cases.

Case (a). $k = n > m$; write $M = k_1 + k_2$

$$c_s\delta^{\frac{2}{3}(M+1)}\delta^{\frac{M+1}{3s}}\left[\sum_{k>m}\frac{|a_k|^{2s}|a_m|^{s}}{\Delta_2^{s-1}}\right]^{\frac{1}{3s}} \tag{9}$$

where $\Delta_2 \geq |k - m|[k^{M-4} + k^{M-5}m + \cdots + m^{M-4}]$. (9) is less than

$$\delta^{\frac{2}{3}(M+1)}\delta^{\frac{M+1}{3s}}\left[\sum_{k>m}\frac{|a_k|^{3s}|a_m|^{s}}{\left[|k-m|k^{\frac{2(M-4)}{3}}m^{\frac{M-4}{3}}\right]^{s-1}}\right]^{\frac{1}{3s}} .$$

Taking $A_k = \dfrac{|a_k|^s}{k^{\frac{M-4}{3}(s-1)}}$, we have

$$\left[\sum_{k>m} \frac{A_k^2 A_m}{|k - m|^{s-1}}\right]^{\frac{1}{3s}} \leq [\sum_k |A_k|^{2t}]^{\frac{1}{3st}} \left[\sum \left[\frac{A_m}{|k - m|^{s-1}}\right]^{t'}\right]^{\frac{1}{3st}}.$$

Then by fractional integration $L^{2t} - L^{t'}$, with $t = \dfrac{3}{2(3 - s)}$,

$$(9) \qquad \leq \delta^{\frac{2}{3}(M+1)} \delta^{\frac{M+1}{3s}} [\sum |A_k|^{2t}]^{\frac{1}{2ts}}.$$

By Hölder:

$$\leq \delta^{\frac{2}{3}(M+1)+\frac{M+1}{3s}} [\sum |a_k|^{q'}]^{\frac{1}{q'}} \left[\sum_{k}^{\delta^{-1}} \frac{1}{k^{\frac{(M-4)(s-1)}{3-s}r'}}\right]^{\frac{3-s}{3sr'}}$$

where $\dfrac{1}{r} + \dfrac{1}{r'} = 1$, $q' = \dfrac{3s}{3 - s}r$, thus $r = \dfrac{q'(3 - s)}{3s} > 1$;

$r' = \dfrac{(3 - s)q'}{3q' - q's - 3s}$.

Consider two subcases:

(i) $\dfrac{(M - 4)(s - 1)}{3 - s} r' < 1$; that is, $q < \dfrac{3p'}{3M - 3 + p'}$.

Then the exponent of δ is

$$\frac{2}{3}(M + 1) + \frac{M + 1}{3s} - \frac{1}{q'} + \frac{(M - 4)(s - 1)q - 3q' + q's + 3s}{3sq'},$$

which is greater than M when $\dfrac{2}{3s} - \dfrac{1}{s} \geq 0$. This is true in the present case, since $s < 2$.

(ii) $\dfrac{(M - 4)(s - 1)}{3 - s} r' > 1$; thus $q > \dfrac{3p'}{3M - 3 + p'}$.

Then the exponent of δ is $-M + \dfrac{2}{3}(M + 1) + \dfrac{M + 1}{3s} - \dfrac{1}{q'} \geq 0$, equivalent to $q \leq \dfrac{p'}{M + 1}$; so if we suppose $p' \geq 6$, the only restriction on the range of p and q is $q \leq \dfrac{p'}{M + 1}$ (R1). By fractional integration $L^{2t} \to L^{t'}$, $\frac{1}{t} = \frac{1}{2t} - \alpha$, and $\alpha = 2 - s$ (here $p' > 6$), we can bound (9) by

$$\delta^{\frac{2}{3}(M+1)} \delta^{\frac{M+1}{3s}} \left[\sum |A_k|^{2t} \right]^{\frac{1}{2ts}}$$

and by Hölder's inequality

$$\leq \delta^{\frac{2}{3}(M+1)+\frac{M+1}{3s}} \left[\sum |a_k|^{q'} \right]^{\frac{1}{q'}} \left[\sum_k \frac{1}{k^{\frac{(M-4)(s-1)}{3s} r'}} \right]^{\frac{3-s}{3sr'}},$$

where $r = \frac{q'(3-s)}{3s} > 1$. Then the exponent of δ is positive if $q \leq \frac{p'}{m+1}$ (R1), assuming $p' > 6$.

Case (b). $k > n > m$.

Let's write

$$\delta^{\frac{2}{3}(M+1)+\frac{M+1}{3s}} \left[\sum_{k>n>m} \frac{|a_k|^s |a_n|^s |a_m|^s}{\Delta_1^{s-1}} \right]^{\frac{1}{3s}} \tag{10}$$

$$\Delta_1 = c \begin{vmatrix} 1 & 1 & 1 \\ k^{k_1-1} & n^{k_1-1} & m^{k_1-1} \\ k^{k_2-1} & n^{k_2-1} & m^{k_2-1} \end{vmatrix}$$

$= |k - n||n - m||m - k| P(k, n, m)$, where $p(k, n, m)$ is a homogeneous polynomial of degree $k_1 + k_2 - 5$ and positive coefficients, whose leading degree terms are

$$k^{k_2-3} n^{k_1-3} m \text{ and } k^{k_2-3} m^{k_1-3} n.$$

(10) is bounded above by

$$\delta^{\frac{2}{3}(M+1)+\frac{M+1}{3s}}\left[\sum_{k>n>m}\frac{|a_k|^s|a_n|^s|a_m|^s}{\left[|k-n||n-m||m-k|k^{k_2-3}n^{\frac{k_1-2}{2}}m^{\frac{k_1-2}{2}}\right]^{s-1}}\right]^{\frac{1}{3s}}$$

$$= \delta^{\text{Pow}}\left[\sum\frac{A_kA_nA_m}{\left(|k-n||n-m||m-k|k^{k_2-\frac{k_1}{2}-2}\right)^{s-1}}\right]^{\frac{1}{3s}},$$

where $A_k = \dfrac{|a_k|^s}{k^{\frac{(k_1-2)}{2}(s-1)}}$.

By Hölder's inequality:

$$= \delta^{\text{Pow}}\Big[\sum_k |A_k|^t\Big]^{\frac{1}{3st}}\left[\sum_k\Big|\sum_{u,v=1}^{k}\frac{A_{k-u}A_{k-v}}{k^{(k_2-\frac{k_1}{2}-2)(s-1)}u^{s-1}v^{s-1}|u-v|^{s-1}}\Big|^{t'}\right]^{\frac{1}{3st'}}$$

$$\leq \delta^{\text{Pow}}\Big[\sum_k |A_k|^t\Big]^{\frac{1}{3st}}$$

$$\left[\sum_k\frac{1}{k^{(k_2-\frac{k_1}{2}-2)(s-1)t'}}\left(\sum_{u=1}^{k}\frac{A_{k-u}}{u^{s-1}}\sum_{v=1}^{k}\frac{A_{k-v}}{v^{s-1}|u-v|^{s-1}}\right)^{t'}\right]^{\frac{1}{3st'}}$$

by Hölder's inequality

$$\leq \delta^{\text{Pow}}\Big(\sum_k |A_k|^t\Big)^{\frac{1}{t}}\left\{\sum_{k=1}^{c\delta^{-1}}\frac{1}{k^{(k_2-\frac{k_1}{2}-2)(s-1)t'}}\left(\sum_{u=1}^{k}\frac{A_{k-u}}{|u^{s-1}|}r\right)^{\frac{t'}{r}}\right..$$

$$\left.\cdot\left[\sum_{u=1}\left(\sum_{v=1}\frac{A_{k-v}}{v^{s-1}|u-v|^{s-1}}\right)^{r'}\right]^{\frac{t'}{r'}}\right\}^{\frac{1}{t'}}.$$

Fractional integration $\frac{1}{r'} = \frac{1}{r} - (2 - s)$ gives

$$\leq \delta^{\text{Pow}} \left(\sum |A_k|^t \right)^{\frac{1}{3st}}$$

$$\left[\sum_k^{\delta^{-1}} \frac{1}{k^{(k_2 - \frac{k_1}{2} - 2)(s-1)t'}} \left(\sum_{u=0}^{k} \left| \frac{A_{k-u}}{u^{s-1}} \right|^r \right)^{\frac{t'}{r}} \left(\sum_v \left| \frac{A_{k-v}}{v^{s-1}} \right|^r \right)^{\frac{t'}{r}} \right]^{\frac{1}{3st'}}$$

by Schwarz's inequality:

$$\leq \delta^{\text{Pow}} \left(\sum |A_k|^t \right)^{\frac{1}{3st}}$$

$$\left[\sum_k^{\delta^{-1}} \frac{1}{k^{(k_2 - \frac{k_1}{2} - 2)(s-1)t'}} \left(\sum_{u=0}^{k} |A_{k-u}|^{2r} \right)^{\frac{t'}{r}} \left(\sum_{u=0}^{k} \frac{1}{u^{(s-1)2r}} \right)^{\frac{t'}{r}} \right]^{\frac{1}{3st'}} .$$

If we take $t = 2r = \frac{4}{3 - s} \to t' = \frac{4}{s + 1}$,

$$\leq \delta^{\text{Pow}} \left[\sum_{k=1}^{\delta^{-1}} |A_k|^t \right]^{\frac{t}{st}} \left[\sum_{k=1}^{\delta^{-1}} \frac{1}{k^{(k_1 - \frac{k_2}{2} - 2)(s-1)t'}} \left(\sum_{u=0}^{k} \frac{1}{u^{(s-1)t}} \right)^{\frac{t'}{r}} \right]^{\frac{1}{3st'}}$$

providing $(s - 1) = \frac{4(s - 1)}{3 - s} < 1$, that is, $s < \frac{7}{5}$ (R2), and in case

$$(k_1 - \frac{k_2}{2} - 2)(s - 1)t' + \left(\frac{4(s - 1)}{3 - s} - 1\right)^{\frac{t'}{r}} < 1,$$

that is, $$s < \frac{4k_2 - 2k_1 + 7}{4k_2 - 2k_1 + 1} \quad \text{(R3)},$$

(10) is bounded by

$$\frac{2}{\delta^3}(M + 1) + \frac{1}{3s}(M + 1)^{\frac{4k_2 s - 2k_1 s + s - 7 - 4k_2 + 2k_1}{12s}} \left(\sum_R |A_k|^t \right)^{\frac{1}{st}}$$

$$\leq \delta^{Pow} \left[\sum_k \frac{|a_k|^{\frac{4s}{3-s}}}{k^{\frac{(k_1-2)(s-1)2}{3-s}}} \right]^{\frac{3-s}{4s}} .$$

Assuming $q' > \frac{4s}{3 - s}$, (R4)

by Hölder's inequality with $\frac{4s}{3 - s} t = q'$:

$$\leq \delta^{Pow} [\sum |a_k|^{q'}]^{\frac{1}{q'}} \left[\sum_{k=1}^{\delta^{-1}} \frac{1}{k^{\frac{2(k_1-2)(s-1)}{(3-s)} t'}} \right]^{\frac{3-s}{4st'}} ,$$

where $t' = \frac{q'(3 - s)}{3q' - sq' - 4s}$. And finally, in the case

$q' > \frac{4s}{3s + 2k_1 - 1 - 2k_2 s}$ (R5) (10) $\leq \delta^{k_1+k_2}$.

Let's summarize the above conditions (R1)(R2)(R3)(R4).

(I) If $7 \leq 2k_2 - k_1$

$$\frac{1}{p} > \frac{4k_2 - 2k_1 + 5}{4k_2 - 2k_1 + 7} \qquad \text{(R3)}$$

$$\frac{1}{q} > \frac{11}{4} - \frac{9}{4p} \qquad \text{(R4)}$$

$$\frac{1}{q} \geq k_1 + k_2 + 1 - \frac{k_1 + k_2 + 1}{p} \qquad \text{(R1)}.$$

(See Fig. 3 on p. 204.)

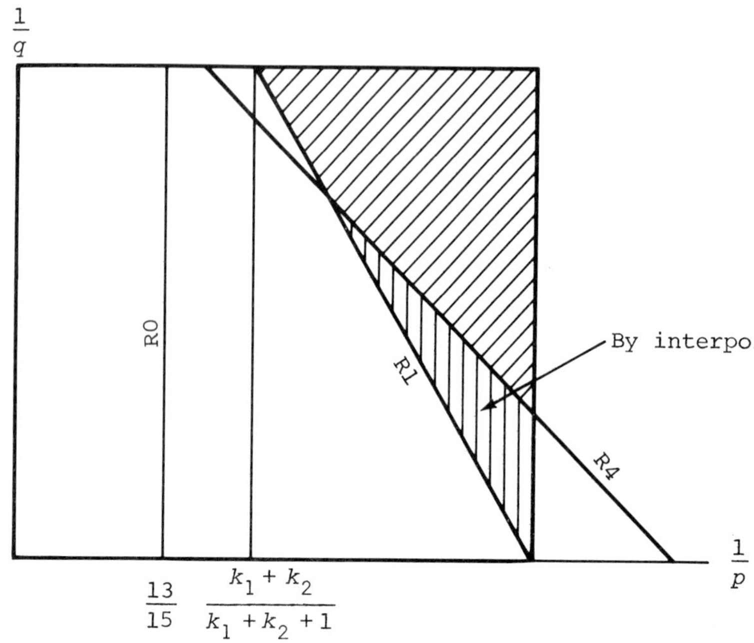

Figure 3

Using interpolation with the obvious case $(1, \infty)$, we get

$$\|\hat{f}\|_{L^q(\gamma)} \leq C \|f\|_p$$

when $\begin{cases} k_2 > 2k_1 - \frac{5}{2} \\ 7 \leq 2k_2 - k_1 \end{cases}$ for $\begin{cases} \frac{1}{q} > \frac{k_1 + k_2 + 1}{p} \\ p < \frac{4k_2 - 2k_1 + 7}{4k_2 - 2k_1 + 5} \end{cases}$

assuming $k_1 > 2$ also for $\frac{1}{q} = \frac{k_1 + k_2 + 1}{p}$

$$p < \frac{4k_2 - 2k_1 + 7}{4k_2 - 2k_1 + 5} .$$

When $k_2 > 2k_1 - \frac{5}{2}$, the inequality holds for

$$\frac{1}{q} \geq \frac{k_1 + k_2 + 1}{p}, \quad p < \frac{k_1 + k_2 + 1}{k_1 + k_2}.$$

(II) If $7 > 2k_2 - k$.

(IIa) $(t,\ t^2,\ t^3)(t,\ t^2,\ t^4)$. Figure 4 is an illustration of the above results.

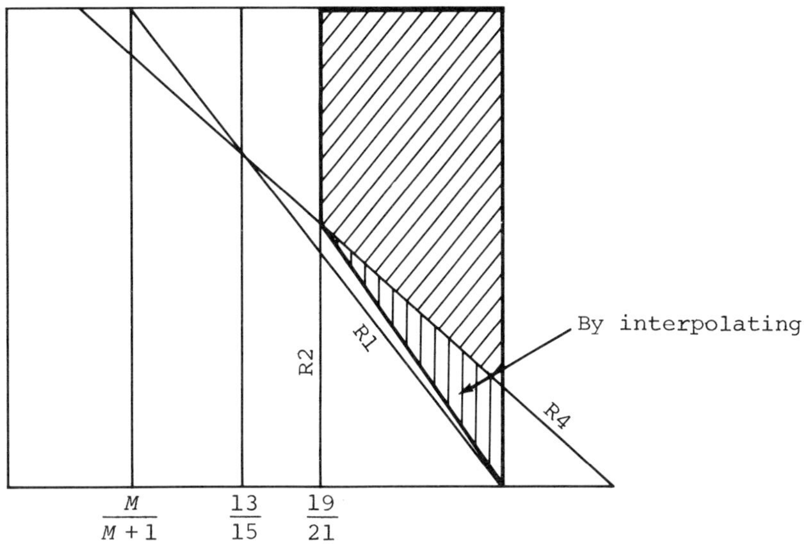

Figure 4

(IIb) $(t,\ t^3,\ t^4)$

$$\|\hat{f}\|_{L^q(\gamma)} \leq c\,\|f\|_p \quad \text{if} \quad \frac{1}{q} > \frac{8}{p'}$$
$$p < \frac{21}{19}.$$

To improve these results, let's estimate (10) in a different way:

$$\Delta_1 \geq c|k-n|\,|n-m|\,|m-k|\,(k\quad n\quad m)^{\frac{k_1+k_2-5}{3}}$$

$$(10) \leq c\delta^{\frac{2}{3}(k_1+k_2+1)+\frac{k_1+k_2+1}{3s}}\left[\sum_{k>n>m}^{\delta^{-1}} \frac{A_k A_n A_m}{(|n-m|\,|k-n|\,|m-k|)^{s-1}}\right]^{\frac{1}{3s}}$$

where $A_k = \dfrac{|a_k|^s}{k^{\frac{(k_1+k_2-5)(s-1)}{3}}}$

$$(10) \qquad \le [\sum_k A_k^t]^{\frac{1}{t}} \left[\sum_k \Big| \sum_{u,v=1}^{k} \frac{A_{k-u}A_{k-u-v}}{u^{s-1}v^{s-1}|u-v|^{s-1}}\Big|^{t'}\right]^{\frac{1}{t'}} .$$

From the Minkowski inequality (10) is bounded by

$$[\sum_k A_k^t]^{\frac{1}{t}} \sum_{n=1}^{k} \sum_{v<u} \left[\sum_{k>u}^{\delta^{-1}} \Big|\frac{A_{k-u}A_{k-u-v}}{u^{s-1}v^{s-1}|u-v|^{s-1}}\Big|^{t'}\right]^{\frac{1}{t'}}$$

$$= \sum_u \frac{1}{u^{s-1}} \sum_{v<u} \frac{1}{v^{s-1}|u-v|^{s-1}} \left[\sum_{k>u} |A_{k-u}A_{k-u-v}|^{t'}\right]^{\frac{1}{t'}} [\sum_k A_k^t]^{\frac{1}{t}}$$

$$\le [\sum_k A_k^t]^{\frac{1}{t}} \sum_u \sum_{v<u} \frac{1}{u^{s-1}v^{s-1}|u-v|^{s-1}}$$

$$\left[\sum_{k>u} |A_{k-u}|^{2t'}\right]^{\frac{1}{2t'}} \left[\sum_{k>u} |A_{k-u-v}|^{2t'}\right]^{\frac{1}{2t'}}$$

$$\le \sum_{u,v<u} \frac{1}{v^{s-1}u^{s-1}|u-v|^{s-1}} [\sum_k |A_k|^{2t'}]^{\frac{1}{t'}} [\sum_k |A_k|^t]^{\frac{1}{t}} .$$

Take $2t' = t$, that is, $t' = \frac{3}{2}$; then

$$(10) \le \delta^{\frac{2}{3}(M+1)+\frac{M+1}{3s}} (\sum |A_k|^2)^{\frac{1}{3s}} \delta^{\frac{3s-5}{3s}}$$

$$\le \delta^{\frac{2}{3}(k_1+k_2+1)+\frac{k_1+k_2+1}{3s}+\frac{3s-5}{3s}} [\sum |a_n|^{q'}]^{\frac{1}{q'}} \left[\sum_k^{\delta^{-1}} \frac{1}{k^{(k_1+k_2-5)(s-1)t}}\right]^{\frac{1}{3st}}$$

assuming $\frac{q'}{3s} > 1$ (R5)

$$\le \delta^{k_1+k_2} \text{ under the condition } \frac{(k_1+k_2-5)(s-1)q'}{q'-3s} < 1 \quad \text{(R6)}.$$

The result is true for

$$\frac{1}{q} > \frac{13}{15} \quad \text{(R0)}$$

$$\frac{1}{q} > \frac{3k_1 + 3k_2 - 10}{3} - \frac{k_1 + k_2 - 4}{p} \quad \text{(R6)}$$

$$\frac{1}{q} \geq k_1 + k_2 + 1 - \frac{k_1 + k_2 + 1}{p} \quad \text{(R1)}.$$

Therefore, the curves (t, t^2, t^5), (t, t^2, t^6), (t, t^2, t^7) satisfy the restriction theorem for Figure 5. For (t, t^2, t^3) and (t, t^2, t^4), we have the restriction theorem when

$$\frac{1}{q} > \frac{k_2 + 3}{p'}, \quad p > \frac{15}{13}.$$

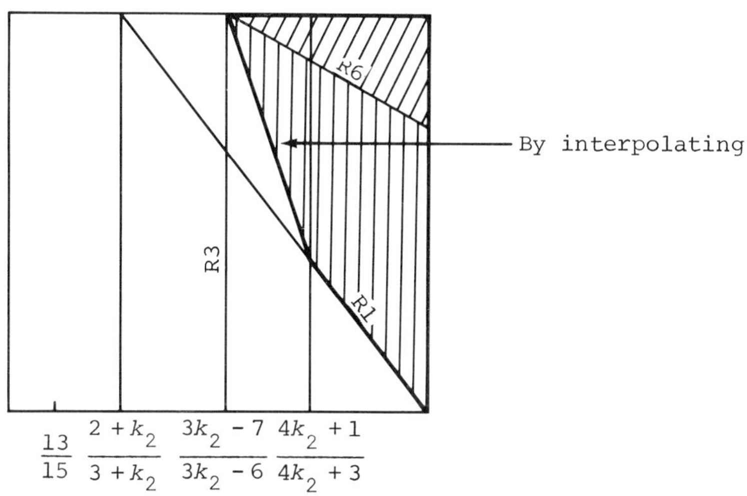

Figure 5

Lemma 1. Let us consider the curve (t, t^2, t^k), $k \geq 3$, *for* $0 \leq t \leq a$. *Let* R_i *be the Frenet parallelepiped at the point* $t = i\delta$, *for* i *an integer,* $i < a\delta^{-1}$, *whose dimensions are* δ *in the direction of the tangent,* δ^2 *in the direction of the normal, and* $\delta^k i^{k-3}$ *in the direction of the binormal. Then there exists a* $\delta_0 > 0$ *such that* $\Sigma\, \chi_{(R_h+R_i+R_j)}(x) < C_k$ *for every* $\delta < \delta_0$ *where* $R_h + R_i + R_j = \{x + y + z \in \mathbf{R}^3,\ x \in R_h,\ y \in R_i,\ z \in R_j\}$.

Proof. Let us denote by $V^M_{h,i,j}$ the set of vectors in $\mathbf{R}^3$, $v = (v_1, v_2, v_3)$, that can be written as

$$\begin{aligned} v_1 &= t_1 + t_2 + t_3 \\ v_2 &= t_1^2 + t_2^2 + t_3^2 \qquad (*) \\ v_3 &= t_1^k + t_2^k + t_3^k \end{aligned}$$

for $t_1 \in [(h - M)\delta, (h + M)\delta]$, $t_2 \in [(i - M)\delta, (i + M)\delta]$, $t_3 \in [(j - M)\delta, (j + M)\delta]$. Let θ be the number of different decompositions of any (v_1, v_2, v_3) in sums of the form (*). This θ is a finite number depending only on k and M, since for $M = 1$ it is bounded by the number of solutions of the algebraic system (*), which only depends on k. Thus

$$\sum_{i,j,h} \chi_{V^M_{hij}}(x) \le C_{M,k}.$$

We claim that there exists $M > 0$, independent of h, j, i, and δ, such that

$$R_h + R_i + R_j \subset V^M_{h,i,j} \quad \text{when} \quad h < i < j.$$

This claim proves Lemma 1, since when $h = i$ or $j = i$ it can be reduced to the two-dimensional case.

The first reduction is to substitute the arcs of curve building V^M_{hij} by the segments of tangents at $h\delta$, $i\delta$, $j\delta$ of length $4M$ and centered at the points $t = h\delta$, $i\delta$, $j\delta$ respectively.

We are going to prove that the set analogous to V^1_{hij} defined above is big enough to contain $R_h + R_i + R_j$.

Let us denote by τ_h, τ_i, τ_j the tangent vectors translated to 0, by η_j the translate of the normal plane, by ρ_j the translate of the rectifying plane. Let π be the plane spanned by τ_h and τ_i.

(a) Let $\pi \cap \rho_j$ be the intersection line of π and ρ_j. Let p be the intersection point of $\pi \cap \rho_j$ and the parallel line to τ_h through τ_i on π. Then $0p$ has projection on η_j of length $c\delta^2$, where c is independent of δ, h, i, and j.

(b) Let θ_j be the translate of the osculating plane at $t = j\delta$ and Q the intersection point of $\pi \cap \theta_j$ and the line parallel to τ_h through τ_i on π. Then $0Q$ has projection on η_j of length $C_j^{k-3}\delta^k$. If we assume (a) and (b), then by taking M large enough, depending on c, V_{hji}^M can absorb the widening of R_j in the normal and binormal directions; and so Lemma 1 follows. Now we are going to prove (a) and (b).

Consider Figure 6, where we drew the projection angles α and β. In the worst case, that is, $|h - i| = 1$, we have

$$0p = \frac{\delta \sin \delta}{\sin(\alpha + \delta)} .$$

Hence

$$0p \sin \beta = \frac{\delta \sin \beta \tan \delta}{\sin \alpha + \tan \delta \cos \alpha} \quad (11).$$

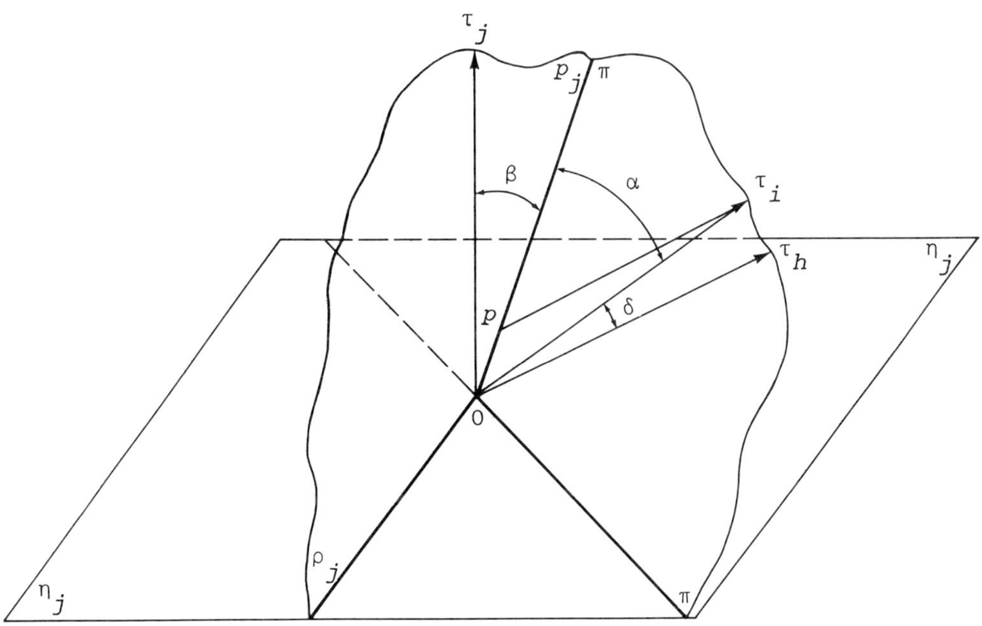

Figure 6

To estimate α and β, we look at the Gaussian map of (t, t^2, t^k) on the unit sphere (see Figure 7): α and β are of the order of the length of the segments $\overline{\tau_i A}$ and $\overline{A\tau_j}$, respectively (although we waste some length, we may substitute π and ρ_1 in the Gaussian map).

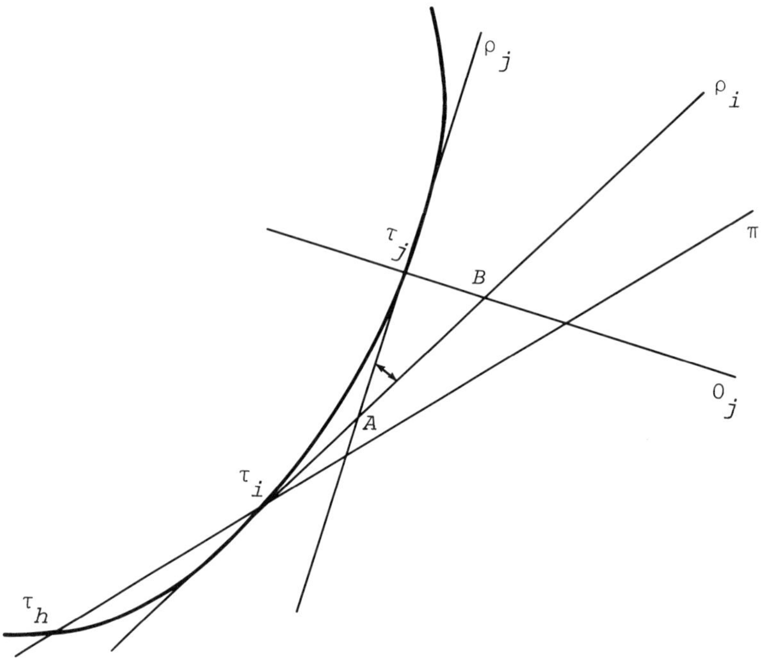

Figure 7

Since the curvature is positive, the arc-length of the indicatrix curve is of the same order as the arc-length of (t, t^2, t^k); hence the indicatrix is a curve (t, t^s) for some s depending on k (in a neighborhood of the origin).

In the worst case $(i = 0)$,

$$\overline{\tau_j A} \geq \frac{t}{s} = \frac{(j - i)\delta}{s}$$

$\overline{\tau_i A} < (j - i)\delta$, then Op sin. β

$$\geq \frac{\delta \sin \frac{(j - i)\delta}{s} \cdot \tan \delta}{\sin(j - i)\delta + \tan \delta \cos(j - i)\delta} \geq c_k \delta^s .$$

For the projection of $0Q$ considered in (b), we get a formula similar to (11)

$$0Q \sin \beta' = c \frac{\delta \sin \beta' \tan \delta}{\sin \alpha' + \tan \delta \cos \alpha'},$$

where β' is the segment $\overline{\tau_j B}$ and α' is $\overline{\tau_i B}$ in the indicatrix; hence

$$\overline{\tau_j B} = \overline{\tau_j A} \tan A \geq c_k (j - 1)\delta \tan A,$$

and

$$A = c \int_{\tau_i}^{\tau_j} \text{torsion}(t)\, dt = c \int_{i\delta}^{j\delta} t^{k-3}\, dt = c(j^{k-2} - i^{k-2})\delta^{k-2}.$$

Since $\alpha' < c(j - i)\delta$,

$$0Q \sin \beta' \geq \frac{c_k \delta \sin c(j - i)\delta \cdot (j^{k-2} - i^{k-2})\delta^{k-2} \tan \delta}{\sin c(j - i)\delta + \tan \delta \cos \alpha'} \geq c_k (j - i) p_{k-3} \delta^k,$$

where p_{k-3} is a polynomial of degree $k - 3$ and positive coefficients, whose term in j is j^{k-3}.

Remark. A homogeneity argument, similar to that in Remark 2, shows the impossibility of improving the condition $\frac{1}{q} > \frac{3 + k}{p}$ $\left(\text{that is, take } \hat{f}(\xi_1, \xi_2, \xi_3) = \exp\left\{-\left(\frac{\xi_1}{\delta}\right)^2 - \left(\frac{\xi_2}{\delta^2}\right)^2 - \left(\frac{\xi_3}{\delta^k}\right)^2\right\}\right)$.

REFERENCES

1. Prestini, E. A restriction theorem for space curves. *Proc. Amer. Math. Soc.* 70 (1978): 8-10.

2. Sjolin, P. Fourier multiplier and estimates of the Fourier transform of measures carried by smooth curves in $\mathbf{R}^2$. *Studia Math.* (1974): 169-182.

3. Tomas, P. A note on restriction. Preprint.

4. Tomas, P. A restriction theorem for the Fourier transform. *Bull. Amer. Math. Soc.* 81 (1975): 477-478.

5. Tomas, P. Restriction theorems for the Fourier transform. *Proc. Symp. Pure Math.* 35 (1979): 111-114.

6. Zygmund, A. On Fourier coefficients and transform of function of two variables. *Studia Math.* (1974): 189-202.

MINIMAL SMOOTHNESS FOR A BOUND ON THE FOURIER TRANSFORM OF A SURFACE MEASURE

T. Walsh
University of Florida

1

As is well known, the Fourier transform $\hat{\sigma}$ of surface measure σ on a compact hypersurface S in R^{n+1} with nonvanishing Gaussian curvature satisfies $|\hat{\sigma}(x)| \leq C|x|^{-n/2}$ provided S is sufficiently smooth (see [8]). For an outline of a proof, note that if Q is a nonsingular real symmetric $n \times n$ matrix, $\varphi \in \mathcal{D}(R^n)$ and $t > 0$,

$$\left| \int (\exp[it\, {}^t xQx])\varphi(x)\, dx \right| \tag{1}$$
$$= c_n |\det tQ|^{-1/2} \left| \int (\exp[-i\, {}^t xQ^{-1}x/t])\hat{\varphi}(x)\, dx \right| \leq C(Q)t^{-n/2}\|\hat{\varphi}\|_1 .$$

By Bernstein's theorem,

$$\|\hat{\varphi}\|_1 \leq \|\hat{\varphi} : {}^2L^1\| \leq C\|\varphi : B^{21}_{n/2}\|$$

(see [9]) where B^{pq}_{α} $(1 \leq p,\, q \leq \infty,\, \alpha > 0)$ is the homogeneous Besov space denoted Λ^{α}_{pq} in [9]. Suppose now $g \in C^{\alpha}(R^n)$, $\alpha > (n/2) + 3$, $g(0) = 0$, and g has a nondegenerate critical point at 0. Then by Morse's lemma there is a $C^{\alpha-2}$ diffeomorphism h^{-1} in a neighborhood U of 0 such that $g(h(x)) = {}^t xD^2 g(0)x$ (see [10]). It follows that if $spt\ \varphi \subset U$,

$$t^{n/2}\left|\int e^{itg(x)}\varphi(x)\,dx\right| \le C\|(\det Dh)\varphi\circ h : B^{21}_{n/2}\| \le C\|\varphi : B^{21}_{n/2}\|.$$

Note that Morse's lemma yields h of differentiability class 2 lower than g and an additional difference of 1 is needed because of the Jacobian det Dh. The purpose of the present paper is to obtain minimal smoothness conditions on the parametrized surface S and a measure carried by S. It will be shown that if M is an n-dimensional $B^{1}_{n/2+1}$ $(= B^{\infty 1}_{n/2+1})$ manifold, ζ is a $B^{1}_{n/2+1}$ immersion of M into R^{n+1}, $f \in B^{2\infty,1}_{n/2}(M)$ (L^2 is replaced by the weak L^p space $L^{2\infty}$), $\zeta(f)$ denotes the direct image of f under ζ, and f has compact support on which the Gaussian curvature of ζ is never 0, then $|\zeta(f)\hat{}\,(x)| \le C|x|^{-n/2}$ $(n \ge 3)$. The relation

$$\int_{-\infty}^{\infty} e^{it\zeta(u)f(u)}\,du = \int_{-\infty}^{\infty} e^{itv}[\zeta'(\zeta^{-1}(v))]^{-1}f(\zeta^{-1}(v))\,dv$$

seems to indicate that one higher order of differentiability for ζ compared with f may be necessary. The first half of the paper is concerned with the definition of various types of normed Besov spaces on not necessarily compact manifolds.

2

A Banach function space F on σ-finite measure space X, μ will be called nondegenerate if it is contained in L^1_{loc} and contains the indicator function 1_E of some set of positive measure E. F_+ will denote its subset of nonnegative elements. For α, $\beta \in R$ define the integral operator $H_{\alpha\beta}$ on nonnegative, measurable functions φ on R^*_+ by

$$H_{\alpha\beta}(\varphi)(t) = \int K_{\alpha\beta}(s/t)\varphi(s)s^{-1}\,ds,\quad K_{\alpha\beta}(u) = u^{-\alpha}\wedge u^{-\beta}.$$

C will denote a positive constant not necessarily the same at each occurrence, but the aim is to make each C depend only on C's defined earlier. For any function f on R^n, $t > 0$ put $f^t = f(t.)$, $f_t = t^{-n}f(t^{-1}.)$. For $f \in L^1(R^n)$ its Fourier transform $\hat{f}$ is defined by $\hat{f}(x) = \int e^{-ix.y}f(y)\,dy$, $x \in R^n$.

Definition. Suppose Ψ is a nondegenerate Banach function space on R_+^* in which H_{or} is bounded for some $r \in \mathbf{N}(= Z_+ - \{0\})$. Supposing F is a closed subset of R^n, define the space $B(F, \Psi)$ to consist of all bounded continuous complex functions f on F for which there exists a family

$$\mathscr{F} = \{f^{(\alpha)}(., \delta) : |\alpha| \leq r - 1, 0 < \delta \leq 1\}$$

of bounded continuous functions on F such that $f^0(., \delta) = f$ for all δ, and if ω on $(0, 1]$ is defined by

$$\omega(t) = \sup \left\{ \left|f^{(\alpha)}(y, \delta_1) - \sum_{|\beta| \leq r-|\alpha|-1} f^{(\alpha+\beta)}(x, \delta_2)(y - x)^\beta/\beta!\right| \right.$$
$$\left. \delta_1^{|\alpha|} : \delta_1 \leq \delta_2 \leq t, |y - x| \leq \delta_2, x, y \in F, |\alpha| \leq r \right\} \tag{2}$$

then $\omega \in \Psi$. For $M = \sup\{|f^{(\alpha)}(x, 1)| : |\alpha| \leq r - 1, x \in F\}$ put

$$\| f : B(F, \Psi) \| = \inf(\| \omega : \Psi \| + M)$$

extended over all possible families $\mathscr{F}$ (the norm thus depends on r, but as will be seen, different permissible r yield equivalent norms). In the special case $\Psi = \omega L^\infty$, $\omega \geq 0$, measurable and such that $H_{or}(\omega) \leq C\omega$, simply write $B(F, \omega)$.

3

Recall the definition of the Besov space $B(\Psi)$ on R^n. Suppose μ is a measure supported in $B(0, 1)$ such that for all $x \neq 0$, $\hat{\mu}(tx) \neq 0$ for some $t > 0$ and μ has vanishing moments of all orders $\leq r - 1$. Then $B(\Psi)$ consists of all bounded continuous functions f on R^n such that $1_{(0,1]} \| \mu. * f\|_\infty \in \Psi$ with norm defined by

$$\| 1_{(0,1]} \| \mu. * f\|_\infty : \Psi \| + \| f \|_\infty.$$

Different measures yield the same space with equivalent norms (see [3, 1]).

Lemma 1. *$B(F, \Psi)$ consists precisely of the restrictions to F of functions in $B(\Psi)$ and any norm of $B(F, \Psi)$ is equivalent to any norm defined by*

$$f \longrightarrow \inf\{ \| \tilde{f} : B(\Psi) \| : \tilde{f}|_F = f\}.$$

Proof. The fact that restriction to F is bounded from $B(\Psi)$ to $B(F, \Psi)$ follows from a well-known characterization of $B(\Psi)$ (included in the given definition if μ is allowed to be a measure with values in a function space): for $\varphi \in \mathcal{D}$ supported in $B(0, 1)$ such that $\int \varphi(x)x^{\alpha}\, dx = \delta_{\alpha 0}$, $|\alpha| \leq r - 1$, $f \in B(\Psi)$ put $f^{(\alpha)}(.,\delta) = D^{\alpha}\varphi_{\delta} * f$, $\alpha > 0$.

Similarly as for moduli of continuity (see [13, p. 177]) the converse direction follows by means of the Whitney extension operation. In fact, with the notation of [13, p. 177], put $\mathcal{E}_r(\mathcal{F})(x) = f^0(x)$, $x \in F$, $\mathcal{E}_r(\mathcal{F})(x) = \sum_i' P(x, p_i, \delta_i)\varphi_i^*(x)$, $x \in R^n - F$, where δ_i denotes the distance of Q_i from F and

$$P(x, y, \delta) = \sum_{|\alpha| \leq r-1} f^{(\alpha)}(y, \delta)(x - y)^{\alpha}/\alpha!, \quad x \in R^n, \ y \in F.$$

Observe that $\omega \in \Psi$ for ω as in (2) is equivalent to $1_{(0,\ 1)}\psi \in \Psi$ for

$$\psi(\delta_2) = \sup\{|P(x, b, \delta_1) - P(x, a, \delta_2)| : \\ |x - b| \leq \delta_1, \ \delta_1 \leq \delta_2, \ |b - a| \leq \delta_2, \ x, y \in F\}.$$

Hence it follows as in [13, p. 179] that for $a \in F$, $\delta \leq 1$,

$$\sup\{|\mathcal{E}_r(\mathcal{F})(x) - P(x, a, \delta)| : |x - a| \leq \delta, \ \mathrm{dist}(x, F) \leq C_1 \leq C_{nr}\omega(\delta).$$

This implies that for $\mathrm{dist}(a, F) \leq C\delta$, $\delta \leq 1$ and $\pi(a)$ denoting the point of F closest to a,

$$\sup\{|\mathcal{E}_r(\mathcal{F})(x) - P(x, \pi(a), (C + 1)\delta)| : \\ |x - a| \leq \delta, \ \mathrm{dist}(x, F) \leq C_1\} \leq C\omega(C\delta).$$

On the other hand, if $\operatorname{dist}(a, F) \geq C\delta$, then for $|x - a| < \delta$, $\operatorname{dist}(x, F) \leq C_1$,

$$\mathscr{E}_r(\mathscr{F})(x) = \sum P(x, p_i, \delta_i)\varphi_i^*(x),$$

$\sum \varphi_i^*(x) = 1$, and $\varphi_i^*(x) \neq 0$ for only a bounded number of indices i such that $|p_i - \pi(a)| \leq C\delta$, $C^{-1} \leq \delta_i/\delta(a) \leq C$; hence for $\Delta_i(x, a) = P(x, p_i, \delta_i) - P(x, \pi(a), \delta(a))$,

$$\sup_{|x-a|\leq C\delta} |(\partial/\partial x)^\ell \Delta_i(x, a)| \leq C\omega(C\delta(a))\delta(a)^{-\ell}.$$

Since $|D^\ell \varphi_i^*(x)| \leq C\delta(a)^{-\ell}$, it follows that

$$|D^r(\mathscr{E}_r(\mathscr{F})(x) - P(x, \pi(a), \delta(a)))| \leq C\delta(a)^{-r}\omega(C\delta(a));$$

hence for some polynomial $Q(., a, \delta)$ of degree $r - 1$ at most,

$$|\mathscr{E}_r(\mathscr{F})(x) - Q(x, a, \delta)| \leq C(\delta/\delta(a))^r\omega(C\delta(a)) \leq CH_{or}(\omega)(\delta).$$

Note also that for $x \in F$ $|f^{(\ell)}(x, \delta)|\delta^\ell/\ell! \leq C\delta^\ell \int_\delta^2 \omega(t)t^{-\ell-1}\,dt$; hence also $\|\mathscr{E}_r(\mathscr{F})\|_\infty \leq C\|f : B(F, \Psi)\|$.

4

The following technical result will be used repeatedly.

Lemma 2. Suppose φ_i, $0 \leq i \leq q$ are nonnegative measurable functions on R_+, $\alpha_i > 0$, $0 \leq i \leq q$, $|\alpha| = \sum_{i=0}^{q} \alpha_i$, $\varepsilon > 0$, $\varphi_0^* = \sup \varphi_0$. Then $\prod_{i=0}^{q} H_{-\infty\alpha_i}(\varphi_i) \leq C_{\alpha\varepsilon}[H_{-\infty|\alpha|-\varepsilon}(\varphi_0) + \varphi_0^* \sum_{i=1}^{q} H_{-\infty|\alpha|-\varepsilon}(\varphi_i^{|\alpha|/\alpha_i})]$.

Proof. For $\beta_i = \alpha_i/|\alpha|$ by the arithmetic-geometric mean inequality

$$\begin{aligned}
&\prod_{i=0}^{q} H_{-\infty\alpha_i}(\varphi_i)(t) \\
&= t^{|\alpha|}\int_t^\infty \cdots \int_t^\infty (\varphi_0(u_0)/\varphi_0^*)^{1-\beta_0}\varphi_0(u_0)^{\beta_0}u_0^{-\alpha_0+\varepsilon(1-\beta_0)\beta_0} \\
&\prod_{j\neq 0} u_j^{-\varepsilon\beta_0\beta_j}\prod_{i=1}^{q}(\varphi_0^*)^{\beta_i}\varphi_i(u_i)u_i^{-\alpha_i+\varepsilon(1-\beta_i)\beta_i} \\
&\prod_{j\neq i} u_j^{-\varepsilon\beta_i\beta_j}(u_0\cdots u_q)^{-1}\,du_0\cdots du_q \\
&\leq C_{\beta\varepsilon}t^{|\alpha|-\varepsilon(1-\beta_0)}\int_t^\infty \varphi_0(u)u^{-|\alpha|-\varepsilon(1-\beta_0)-1}\,du \\
&+ C_{\beta\varepsilon}\sum_{i=1}^{q} t^{|\alpha|-\varepsilon(1-\beta_i)}\varphi_0^*\int_t^\infty \varphi_i(u)^{1/\beta_i}u^{-|\alpha|+\varepsilon(1-\beta_i)-1}\,du.
\end{aligned}$$

5

The definition of $B((.)\Psi)$ manifolds requires the following chain rule.

Lemma 3. Suppose H_{or} is bounded in Ψ for some $r \in Z_+$. If g is a C^1 mapping of a closed subset F_1 of R^n into another F_2, $Dg \in B(F_1, \Psi)$ and $Df \in B(F_2, \Psi)$, then for some number $C = C(\|Dg : B(F_1, \Psi)\|)$,

$$\|D(f \circ g) : B(F_1, \Psi)\| \leq C\|Df : B(F_2, \Psi)\|.$$

Proof. By Lemma 1 it suffices to prove Lemma 3 for $F_2 = R^n$. By use of difference operators it is immediate that multiplication: $(g, h) \to gh$ is bounded from $B(\Psi) \times B(\Psi)$ into $B(\Psi)$; hence it suffices to show

$$\|Df \circ g : B(F_1, \Psi)\| \leq C\|Df : B(F_2, \Psi)\|.$$

By the intermediate space characterization of Besov spaces $\|h : B(\Psi)\| \leq C$ is equivalent to the existence of a function $\varphi \in \Psi_+$, nondecreasing in $(0, 2)$, vanishing in $(2, \infty)$, and such that

$\|\varphi : \Psi\| \leq 1$ and locally integrable mappings h^0, h^1 of the interval $(0, 1]$ into L^∞, C^r, respectively, such that

$$h = h^0(t) + h^1(t), \tag{3}$$

$$\sum_{j=0}^{r} t^j \|D^j h^0(t)\|_\infty / H_{-\infty j}(\varphi)(t) + \|h^1(t)\|_\infty / \varphi(t) \leq C, \ 0 < t \leq 1, \tag{4}$$

where the terms in (4) for $0 \leq j \leq r - 1$ may be omitted.

Suppose (3), (4) valid for $h = Df$. It may be supposed that $g = g^0(t) + g^1(t)$, where

$$\sum_{j=0}^{r} t^j \|D^{j+1} g^0(t)\|_\infty / H_{-\infty j}(\varphi)(t) + \|g^1(t)\|_\infty / (t\varphi(t)) \leq C, \ 0 < t \leq 1. \tag{5}$$

For $u = h \circ g$,

$$\begin{aligned} u &= h^0(t) \circ g^0(t) + [h^0(t) \circ g - h^0(t) \circ g^0(t)] + h^1(t) \circ g \quad (6)\\ &= u^0(t) + u^{10}(t) + u^{11}(t), \text{ say.} \end{aligned}$$

Clearly $\|u^{11}(t)\|_\infty \leq \|h^1(t)\|_\infty \leq C\varphi(t)$, and

$$\|u^{10}(t)\|_\infty \leq \|Dh^0(t)\| \ \|g^1(t)\|_\infty \leq C\varphi(t).$$

Thus it remains to show

$$t^r \|D^r u^0(t)\|_\infty \leq C\psi(t), \ 0 < t \leq 1, \ \|\psi : \Psi\| \leq 1. \tag{7}$$

For that purpose recall that for C^r functions F, G,

$$D^r(F \circ G) = \sum \left\{ c(\ell)(D^q F) \circ G \bigotimes_{j=1}^{q} D^{\ell_j} G : \sum_{j=1}^{q} \ell_j = r, \ \ell_j \geq 1, \ 1 \leq j \leq q \right\}. \tag{8}$$

Hence for $F = h^0(t)$, $G = g^0(t)$ by (4), (5),

$$t^r \|D^r u^0(t)\|_\infty \leq C \sum \left\{ H_{-\infty q}(\varphi)(t) \prod_{j=1}^{q} H_{-\infty \ell_j - 1}(\varphi)(t) : \sum_{j=1}^{q} \ell_j = r, \ell_j \geq 2 \right\}$$

(terms with $\ell_j = 1$ in (8) are bounded) which by Lemma 2 is bounded by $C_{r\varepsilon} H_{-\infty r-\varepsilon}(\varphi)(t)$. Since boundedness of $H_{-\infty r}$ in Ψ implies that of

$H_{-\infty r-\varepsilon}$ for some $\varepsilon > 0$ (see [2]), this yields the required estimate.

Remark. In case g is a diffeomorphism of F_1 onto F_2 and $\| Dg^{-1} \|_\infty \leq C$, it follows similarly that $\| Dg^{-1} : B(F_2, \Psi) \| \leq C$.

The following variant of Lemma 3 will be needed in Sec. 18.

Lemma 4. Suppose H_{or} is bounded in Ψ, $\| g : B(F_1, \Psi) \cap \Lambda^1(F_1) \| \leq 1$, $\| f : B(F_2, \psi) \cap \Lambda^1(F_2) \| \leq 1$ (Λ^1 is defined in Sec. 7). Then

$$\| f \circ g : B(F_1, \Psi) \| \leq C.$$

Proof. Use (6) with $f = h$ and note that the factors $Dg = DG$ in (8) still are bounded.

6

Lemmas 1 and 3 will be needed in slightly more general form. For some index set I suppose $\{F_{1\alpha} : \alpha \in I\}$ is a family of closed subsets of R^n and f_α is a continuous function on $F_{1\alpha}$, $\alpha \in I$. Write

$$|\!|\!| \{f_\alpha : \alpha \in I\} : B(\Psi) |\!|\!| \leq C \tag{9}$$

if in the definition of Sec. 2 a single function $\omega \in \Psi$, $\| \omega : \Psi \| \leq C$ and $M \leq C$ may be used. It is a consequence of Lemma 1 or follows from its proof that then each f_α may be extended to a function $\tilde{f}_\alpha$ on R^n such that

$$|\!|\!| \{\tilde{f}_\alpha : \alpha \in I\} : B(\Psi) |\!|\!| \leq C.$$

The latter relation is equivalent to $\| \rho : \Psi \| + \sup\{ |\tilde{f}_\alpha| : \alpha \in I\} \leq C$ for $\rho(t) = \sup\{ |\mu_t * \tilde{f}_\alpha| : \alpha \in I\}$, $t > 0$. The analogous generalization of Lemma 3 states that if $\{F_{2\alpha} : \alpha \in I\}$ is another family of closed subsets of R^n, g_α is C^1 from $F_{1\alpha}$ into $F_{2\alpha}$, and

$$\left\| \{Dg_\alpha : \alpha \in I\} : B(\Psi) \right\| \leq C,$$

then

$$\left\| \{D(f_\alpha \circ g_\alpha) : \alpha \in I\} : B(\Psi) \right\| \leq C \left\| \{Df_\alpha\} : B(\Psi) \right\| . \tag{10}$$

If $\Psi = \omega L^\infty$, then (9) is equivalent to $\sup\{ \| f_\alpha : B(\Psi, F_{1\alpha}) \| : \alpha \in I\} \leq C$.

7

For any subset E of R^n define $B(E, \Psi)$ to be the space of restrictions to E of functions in $B(\bar{E}, \Psi)$, $\bar{E} = \text{clos } E$. For $E \subset R^n$, $t > 0$, put $E^{-t} = \{x : B(x, t) \subset E\}$. Suppose M_0 is a C^1 manifold with atlas $\mathscr{A} = \{u_i : U_i \to R^n, i \in \mathscr{I}\}$, $\mathscr{I}$ countable, $u_i(U_i) = V_i \subset R^n$, $\mathscr{U} = \{U_i : i \in \mathscr{I}\}$, $\mathscr{V} = \{V_i : i \in \mathscr{I}\}$. Put $(.)\Psi = \{(.)\varphi : \varphi \in \Psi\}$ where $((.)\varphi)(t) = t\varphi(t)$. Call $\mathscr{A}$ a uniform $B((.)\Psi)$ atlas if

$$\left\| \{D(u_j \circ u_i^{-1}) : i, j \in \mathscr{I}\} : B(\Psi) \right\| \leq C. \tag{11}$$

Analogously, define a uniform $\Lambda^{k+\alpha}$ atlas, $k \in N$, $0 \leq \alpha < 1$, based on the definition of $\Lambda^{k-1+\alpha}(R^n)$ by

$$\| f : \Lambda^{k-1+\alpha}(R^n) \| = \sum_{\ell=0}^{k-1} \| D^\ell f \|_\infty + \sup\{|h|^{-\alpha} \| \Delta(h) D^{k-1} f \|_\infty : |h| \leq 1\}.$$

A uniform C^k atlas is a C^k atlas that is uniform Λ^k, $k \geq 1$. Two uniform $B((.)\psi)$ atlases will be called equivalent if their union is a uniform $B((.)\psi)$ atlas. $\mathscr{A}$ has Lebesgue number $\varepsilon > 0$ if $M_0 \subset \cup \mathscr{U}^{-\varepsilon}$. $\mathscr{U}^{-\varepsilon} = \{U_i^{-\varepsilon} = u_i^{-1}(V_i^{-\varepsilon}) : i \in \mathscr{I}\}$, in other words, $x \in M_0$ implies $B(u_i(x), \varepsilon) \subset V_i$ for some $i \in \mathscr{I}$. A $u\ell B((.)\Psi)$ atlas $\mathscr{A}$ is a uniform $B((.)\psi)$ atlas with a positive Lebesgue number. A $u\ell$ (uniform) $B((.)\Psi)$ manifold M is a C^1 manifold M_0 with an equivalence class of $u\ell$ (uniform) $B((.)\Psi)$ atlases. Define the following condition

$$B(x, \varepsilon) \subset V_i,\ i \in \mathscr{I} \Longrightarrow \#\{j : u_i^{-1}(B(x, \varepsilon)) \cap U_j \neq \emptyset\} \leq C. \quad (12)$$

Lemma 5. Given the uℓC^1 atlas $\mathscr{A}$ with Lebesgue number $\varepsilon > 0$, there is an open refinement $\mathscr{U}' = \{U'_j : j \in \mathscr{I}'\}$, $\kappa : \mathscr{I}' \to \mathscr{I}$, $U'_j \subset U_{\kappa(j)}$, $j \in \mathscr{I}'$ such that for $V'_j = u'_{\kappa(j)}(U'_j)$, $u'_j = u_{\kappa(j)}|U'_j$, $j \in \mathscr{I}'$, $\mathscr{A}' = \{u'_j : j \in \mathscr{I}'\}$, each V'_j is a ball of radius ε/C_1, $\mathscr{A}'$ has Lebesgue number ε/C_2, and for every $j \in \mathscr{I}'$ the number of $\ell \in \mathscr{I}'$ such that $U'_\ell \cap U'_j \neq \emptyset$ is bounded by C. Hence $\mathscr{U}'$ is the union of C disjointed subfamilies; hence (12) is satisfied for all $\varepsilon > 0$.

Proof. By assumption, the mappings $u_j \circ u_i^{-1}$ have a common Lipschitz constant C_0, say. It follows that if $B(x, r) \subset u_i(U_i \cap U_j)$, then $B\big(u_j \circ u_i^{-1}(x), r/C_0\big) \subset (u_j \circ u_i^{-1})B(x, r)$ and if $B(x, \eta) \subset V_j$, $u_i\big(u_j^{-1}B(x, \eta) \cap U_i\big) \cap V_i^{-2\varepsilon/3} \neq \emptyset$ and $\eta \leq \varepsilon/(3C_0)$, then $B(x, \eta)$ is contained in $u_j(U_i \cap U_j)$. Let $\mathscr{I}'$ denote a maximal family of pairs (x, i) such that $x \in V_i^{-\varepsilon}$, $i \in \mathscr{I}$ and the family $\{u_i^{-1}\big(B(x, \varepsilon(12C_0^3)^{-1})\big) : (x, i) \in \mathscr{I}'\}$ is disjointed. Then $\{u_i^{-1}\big(B(x, \varepsilon(6C_0)^{-1})\big) : (x, i) \in \mathscr{I}'\}$ is a covering of M_0. In fact, suppose $y \in U_i^{-\varepsilon}$ for some $i \in \mathscr{I}$; then $u_i(y)$ has distance less than $(12C_0^3)^{-1}$ from $u_i\big(u_j^{-1}B(x, \varepsilon(12C_0^3)^{-1} \cap U_i)\big)$ for some $(x, j) \in \mathscr{I}'$. It follows that $u_i(y) \in B\big(u_iu_j^{-1}(x), \varepsilon(6C_0^2)^{-1}\big)$, hence $u_j(y) \in B\big(x, \varepsilon(6C_0)^{-1}\big)$. Put $U'_{(x,i)} = u_i^{-1}\big(B(x, (3C_0)^{-1}\varepsilon)\big)$, $(x, i) \in \mathscr{I}'$, $\kappa: \mathscr{I}' \ni (x, i) \to i \in \mathscr{I}$. Finally, suppose $B\big(z, \varepsilon(3C_0)^{-1}\big) \subset V_i^{-\varepsilon+\varepsilon/(3C_0)}$ has nonempty intersection with $u_i\big(u_j^{-1}B(x, \varepsilon/(3C_0)) \cap U_i\big)$ for some $(x, j) \in \mathscr{I}'$. Then $B\big(x, \varepsilon/(3C_0)\big) \subset u_j(U_i \cap U_j)$ and $B(z, \varepsilon) \supset u_iu_j^{-1}$ $B\big(x, \varepsilon(3C_0)^{-1}\big) \supset u_iu_j^{-1}B\big(x, \varepsilon(12C_0^3)^{-1}\big) \supset B\big(u^iu_j^{-1}x, \varepsilon(12C_0^4)^{-1}\big)$.

By the last inclusion for distinct (x, j) the last written balls are disjoint. Since their measure is at least $C(C_0)^{-1}$

times the measure of the first term, the number of possible (x, j) is bounded by $C(C_0)$.

Corollary 1. A $u\ell B((.)\psi)$ manifold M has an equivalent $u\ell B((.)\Psi)$ atlas consisting of at most C charts (cf. [7]).

Proof. Suppose $\mathscr{I}' = \bigcup_{\ell=1}^{C} \mathscr{I}'_\ell$ such that $\mathscr{U}'_\ell = \{U'_j : j \in \mathscr{I}'_\ell\}$ is disjointed, $1 \leq \ell \leq C$. Put $U''_\ell = \bigcup \mathscr{U}'_\ell$ and define $u''_\ell : U''_\ell \to R^n$ by $u''_\ell|_{U'_j} = \tau(x_j) \circ u_{\kappa(j)}$, $j \in \mathscr{I}'$, $1 \leq \ell \leq C$, where $\tau(x)$ denotes translation by x, and $x_j \in R^n$ is chosen such that for distinct $j, j' \in \mathscr{I}'_\ell$, $\tau(x_p)u_{\kappa(p)}(U'_p)$, $p = j, j'$ have distance at least 1 from each other. Then $\mathscr{A}'' = \{u'' : 1 \leq \ell \leq C\}$ is an equivalent $u\ell B((.)\Psi)$ atlas.

8

Recall that every C^k manifold $(k \geq 1)$ admits an atlas with respect to which it is a C^∞ manifold [10, Sec. 4]. It can be shown similarly that any $u\ell B((.)\Psi)$ atlas is equivalent among $u\ell B((.)\psi)$ atlases to a $u\ell C^\infty$ atlas (for definition replace (11) by $\| Du_j u_i^{-1} : L^\infty_{k-1}(u_i(U_i \cap U_j)) \| \leq C_k$, $k = 1, 2, \ldots$). If C_0 has the meaning of Sec. 7 and $V'_i = V_i^{-\varepsilon/2}$, $U'_i = u_i^{-1}(V'_i)$, $i \in \mathscr{I}$, then for $t \leq \varepsilon/(4C_0)$, $i, j \in \mathscr{I}$,

$$x \in V'_i,\ B(x, t) \cap u_i\big(u_j^{-1}((V'_j)^{-\varepsilon/2}) \cap U'_i\big) \neq \emptyset \Longrightarrow \tag{13}$$

$$V'_i \cap B(x, t) \subset u_i(U'_i \cap U'_j).$$

Define $m^0 : M_0 \times M_0 \to [0, \infty]$ by

$$m^0(x, y) = \inf\{|u_i(x) - u_i(y)| : x, y \in U'_i,\ i \in \mathscr{I}\}$$

and the metric $m : M_0 \times M_0 \to R_+$ by

$$m(x, y) = \inf\left\{ \sum_{\ell=1}^{N} m^0(x_{\ell-1}, x_\ell) : \{x_\ell\}_{\ell=0}^{N} \subset M_0, x_0 = x, x_N = y, N \in \mathcal{N} \right\}$$

($m = m^0 \wedge (\varepsilon/(4C_0))$) would serve equally well). It follows that $x, y \in U_i' | u_i(x) - u_i(y) | \wedge m(x, y) < \varepsilon/(2C_0)$ implies

$$C_0^{-1} m(x, y) \leq |u_i(x) - u_i(y)| \leq C_0 m(x, y). \tag{14}$$

Replace $\mathcal{U}, \mathcal{V}$ by $\mathcal{U}', \mathcal{V}'$ but for simplicity retain the previous notation $\mathcal{U}, \mathcal{V}$. As normalization it may also be assumed that $\mathcal{A}$ has Lebesgue number 1 and $\varepsilon = 1$ in (12). If M is a $u\ell C^3$ manifold, then there is a C^2 Riemannian metric on M such that M is complete and m is equivalent to the geodesic distance.

9

Suppose $\mathcal{W} = \{W\kappa : \kappa \in K\}$ is an open covering of M_0 with countable index set K. For $\kappa, \kappa' \in K$ put $\kappa \sim \kappa'$ if $W_\kappa \cap W_{\kappa'} = \emptyset$. Suppose the number of elements in the sets $[\kappa] = \{\kappa' : \kappa' \sim \kappa\}$ is bounded, that is, $\#([\kappa]) \leq C$, $\kappa \in K$. Define the integral operator $H^1_{\alpha\beta}$ on functions f on $R_+^* \times K$ by

$$H^1_{\alpha\beta}(f)(t, \kappa) = H_{\alpha\beta}(f(., \kappa))(t), \quad \alpha, \beta \in R.$$

(Similar definition for functions on $R_+^* \times M$.) Suppose E_0 is a nondegenerate Banach function space on $R_+^* \times K$ (counting measure on K) such that H^1_{or} is bounded by C for some $r \in Z_+$. Define E_0^1 to consist of all functions f_1 on K such that

$$\| f_1 : E_0^1 \| = \| 1_{(1/2,1)} \otimes f_1 : E_0 \| < \infty$$

and then define E_0^0 to consist of all measurable functions f_0 on R_+^* such that

$$\| f_0 : E_0^0 \| = \sup\{ \| f_0 \otimes f_1 : E_0 \| : \| f_1 : E_0^1 \| = 1\} < \infty.$$

Clearly $\| 1_{(1/2,1)} : E_0^0 \| = 1$ and H_{or} is bounded in E_0^0. For

$$f_0^*(\kappa) = \sup\{|f(\kappa')| : \kappa' \sim \kappa\}, \; f^*(t, .) = \left[t^{-1} \int_{t/2}^{2t} f(s, .)\, ds \right]^*$$

suppose

$$\| f^* : E_0 \| \leq C \| f : E_0 \| . \tag{15}$$

Consequently,

$$\| f_1^* : E_0^1 \| \leq C \| f_1 : E_0^1 \| .$$

Given the $u\ell B((.)\psi)$ manifold $M = (M_0, \mathscr{A})$ satisfying (12), let $\sqcup \mathscr{V} = \sqcup\{V_i : i \in \mathscr{I}\}$ denote the union of the sets $V_i \times \{i\}$ and for functions f_i on V_i, $i \in \mathscr{I}$, $\sqcup\{f_i : i \in \mathscr{I}\}$ has the analogous meaning. If A is an r.i. (rearrangement-invariant) Banach function space on R with Lebesgue measure and (X, μ) is a σ-finite measure space, define $A(X)$ to consist of all functions f on X that are equimeasurable to functions f^* in A and put

$$\| f : A \| = \| f : A(X) \| = \| f^* : A \| .$$

Suppose the transformation $f \to f(./2)$ is bounded in A. Let Λ denote a measure on M_0 such that its direct image $u_i(\Lambda|_{U_i})$ on V_i equals $\Lambda_i \mathscr{L}^n | V_i$, where Λ_i is positive, continuous for $i \in \mathscr{I}$, and

$$\||\{\Lambda_i(x)^{-1} \Lambda_i|_{B(x,1/C)} : B(x, 1/C) \subset V_i,\; i \in \mathscr{I}\} : B(\Psi) \|| \leq C.$$

Observe that there exists a measure Λ^0 of this type such that the Λ_i^0 are uniformly equivalent to Lebesgue measure. Given Λ the space $A(\sqcup \mathscr{V})$ is well defined (with respect to the measure $\sqcup\{\Lambda_i \mathscr{L}^n : i \in \mathscr{I}\}$. If f is any (regular Borel) measure on M_0 absolutely continuous with respect to Λ, define the locally integrable function f^i on V_i by $u_i(f|_{U_i}) = f^i \Lambda_i$, $i \in \mathscr{I}$, and $f^{\mathscr{A}} = \{f^i : i \in \mathscr{I}\}$. The notation $\| \sqcup f^i : A(\cup \mathscr{V}) \|$ will be abbreviated to $\| \sqcup F^{\mathscr{A}} : A \|$. Let E denote the triple $(E_0, \mathscr{W}, \Lambda)$.

Suppose $\mathscr{W}$ has positive Lebesgue number with respect to the metric m; then $\mathscr{A}$ may and will be supposed to be subordinate to $\mathscr{W}$. Suppose μ is a distribution on R^n with support in $B(0, 1)$ and $f = \{f_i : i \in \mathscr{I}\}$ is a family of distributions on $\mathscr{V}$; then for $t > 0$,

$$\mu_t * f^{\mathscr{A}} = \{\mu_t * f^i : i \in \mathscr{I}\}$$

is well defined on $\mathscr{V}^{-t} = \{V_i^{-t} : i \in \mathscr{I}\}$. For $\kappa \in K$, $i \in \mathscr{I}$, let $\chi_{\kappa i}$ denote the indicator function in $\bar{R}_+ \times V_i$ of $\bigcup_{t=0} \{t\} \times V_i^{-t}$ or the empty set according as $U_i \subset W_\kappa$ or not so that $\chi_\kappa = \{\chi_{\kappa i} : i \in \mathscr{I}\}$. Also put $\xi_\kappa = \chi_\kappa(0, .)$. Finally if the measure f on M is absolutely continuous with respect to Λ, define

$$\| f : \mathscr{B}(\mathscr{A}, A, E, \mu) \| = \| \nu(f) : E \| , \tag{16}$$

where

$$\nu(f)(t, \kappa) = 1_{(0,1)}(t) \| \sqcup \chi_\kappa(t, .)\mu_t * f^{\mathscr{A}} : A \| + 1_{(1/2, 1)}(t) \| \sqcup \xi_\kappa f^{\mathscr{A}} : A \|$$

(if $f = \{f_i\}$, $g = \{g_i\}$, then fg denotes $\{f_i g_i\}$). Using $\mathscr{M}$ to denote the space of regular Borel measures on M as a simpler variant, let $\mathscr{B}(\mathscr{A}, \mathscr{M}, E, \mu)$ be defined in the same way except that convolution of measures $\mu_t * u_i(f|_{U_i})$, the measures $u_i(f|_{U_i})$ and $\mathscr{M}$ are used in appropriate places and $E = (E_0, \mathscr{W})$. It is easy to see that then $\mathscr{B}(\mathscr{A}, \mathscr{M}, E, \mu) = \mathscr{B}(\mathscr{A}, L^1, (E, \Lambda), \mu)$ (due to boundedness of $H^1_{0\infty}$ in E_0).

The space of functions $B(\mathscr{A}, A, (.) \otimes 1\, E, \mu)$ is defined analogously by use of $f \circ u_i^{-1}$ in place of $u_i(f|_{U_i})$ if f now denotes a locally integrable function on M.

10

Lemma 6. Suppose $\Psi = E_0^0$, *the* $u\ell B((.)\Psi)$ *atlas* $\mathscr{A}$ *is subordinate to* $\mathscr{W}$, *satisfies* (12), *and the conditions of Sec.* 8, μ *is a measure on* R^n,

$$\mu = \sum \{D^{\alpha}_{\mu-\alpha} : |\alpha| = r\},$$

where each $\mu_{-\alpha}$ *is supported in* $B(0, 1)$, *and for* $x \neq 0$ *there exists* $t \in R_+$ *such that* $\hat{\mu}(tx) \neq 0$. *Then the space* $\mathscr{B}(\mathscr{A}, A, E, \mu)$ *does not depend on* $\mathscr{A}$ *or* μ *and the norms defined by* (16) *are equivalent. Denote the resulting complete normable linear topological space by* $\mathscr{B}(M, A, E)$.

The same holds for the function space $B(\mathscr{A}, A, (.)\otimes 1E, \mu)$ *if* $\mu = \sum_{|\alpha|=r+1} D^{\alpha}_{\mu-\alpha}$, *in which case the notation* $B(M, A, (.)\otimes 1E)$ *is used.*

Proof. A proof will only be given for the spaces of distributions $\mathscr{B}(\mathscr{A}, A, E, \mu)$. For any function h on a subset E of R^n, put

$$h^*_t(x) = \sup\{|h(y)| : |y - x| < t, y \in E\}, x \in E.$$

Suppose $\|f : \mathscr{B}(\mathscr{A}, A, E, \mu)\| \leq 1$. Note that there are families $f^j(t) = \{f^j_i(t, .) : i \in \mathscr{I}\}$, $j = 0, 1$,

$$\varphi(t) = \{\varphi_i(t, .) : i \in \mathscr{I}\}, f^j_i, \varphi_i : R_+ \times V_i \to C, R_+$$

respectively such that $t \to f^0_i(t, .)$ is continuous from $(0, 1)$ into $C^r(V_i^{-t_0})$, φ_i is measurable, nonnegative, and nondecreasing with respect to its first variable in $(0, 1)$,

$$f^{\mathscr{A}} = f^0(t) + f^1(t), \ 0 < t \leq 1, \tag{17}$$

$$\sum_{j=0}^{r} t^j (D^j f^0_i(t, .))^*_t(x)/H_{-\infty j}(\varphi_i(., x))(t) \tag{18}$$

$$+ |f^1_i(t, x)|/\varphi_i(t, x) \leq C,$$

$$x \in V_i^{-t_0}, \ i \in \mathscr{I} \text{ and for } \varphi_A(t, \kappa) = \|\sqcup \xi_\kappa \varphi(t, .) : A\|$$

$$\|\varphi_A : E_0\| \leq 1, \tag{19}$$

for $0 < t \leq 1/4$, $t_0 = 3/4$, $\kappa \in K$.

To see that this is the case recall the integral representation

$$F = \lambda_t * F + \int_0^t \sigma_s * \mu_s * Fs^{-1}\, ds \tag{20}$$

for any $F \in \mathcal{D}'(R^n)$, where $\lambda, \sigma \in \mathcal{S}$ (see, for example, [4]). Choose $\omega \in \mathcal{D}$ such that $\omega \geq 0$, $\omega(x) = 1$ for $|x| \leq 1/2$, $\omega(x) = 0$ if $|x| \geq 1$. Define $\tilde{h}_i$ as the extension of h_i to R^n that equals 0 in $R^n \setminus V_i$, $i \in \mathcal{J}$. By (20) applied to $\tilde{f}^i$, it follows that in $V_i^{-1/2}$ for $t \leq 1/4$, $f^i = f_i^0(t) + f_i^1(t)$, where $f_i^0(t) = \lambda_{1/4} * \tilde{f}^i +$

$$\int_t^{1/4} (\omega^4\sigma_s) * \mu_s * f^i s^{-1}\, ds + \int_t^{1/4} [(1 - \omega^4)\sigma_s] * \mu_s * \tilde{f}^i s^{-1}\, ds,$$

$$f_i^1(t) = \int_0^t (\omega^4\sigma_s) * \mu_s * f^i s^{-1}\, ds + \int_0^t [(1 - \omega^4)\sigma_s] * \mu_s * \tilde{f}^i s^{-1}\, ds.$$

Observe that $\| (D^N(1 - \omega^4)\sigma_s)_s^* \|_1 \leq C_{MN} s^M$ for all $M, N \in Z_+$, $s \leq 1$. Then (18) and (19) follow from boundedness of H^1_{or} in E_0.

11

Lemma 7. Suppose g is a C^1 mapping of the open subset V into the open subset U of R^n, σ is a C^1 function on V, f is locally integrable on U, there exist $g^i(t)$, $\sigma^i(t)$, $f^i(t)$, $i = 0, 1$, such that $g^0(t) \in C^{r+1}(V)$, $\sigma^0(t) \in C^r(V)$, $f^0(t) \in C^r(U)$, $f^i(t)$ is supported in U^η ($\eta = 1/C$), $i = 0, 1$, $0 < t \leq 1$, $t \to \rho(t)$, $(t, x) \to \varphi(t, x)$ are nonnegative measurable on $R_+ \times M$, increasing (as function of the first variable) in $(0, 1]$, and

$$\sum_{j=0}^{r} t^j[(D^{j+1}g^0(t, .))_t^*(x) + (D^j\sigma^0(t, .))_t^*(x)]/H_{-\infty j}(\rho)(t) + (|Dg^1(t, x)| + |D\sigma^1(t, x)|)/\rho(t) \leq C, \tag{21}$$

$$\sum_{j=0}^{r} t^j(D^j f^0(t, x))_t^*/H^1_{-\infty j}(\varphi(., x))(t) \leq 1. \tag{22}$$

Then for $h = \sigma f \circ g$, $h = h^0(t) + h^1(t)$ *where*

$$t^r(D^r h^0(., x))^*_t + |h^1(t, x)| \le C_{r\varepsilon}[H_{or-\varepsilon}(\varphi(., g(x)))(t) + H_{or}(\varphi(g(x), .))(1)\, H_{or-\varepsilon}(\rho)(t)],\ 0 < t \le 1,\ t > 0.$$

Proof. First consider $H = f \circ g$. Define H^0, H^1, and so on, similarly as in (6):

$$H = H^0(t) + H^1(t) = H^0(t) + [H^{10}(t) + H^{11}(t)]$$
$$= f^0(t) \circ g^0(t) + [f^0(t) \circ g - f^0(t) \circ g^0(t)] + f^1(t) \circ g.$$

Observe that

$$|H^{10}(t, x)| \le Ct\rho(t) \sup\{|Df^0(t, y)| : |y - g(x)| \le Ct\rho(t)\};$$

hence for $t \le 1/C$,

$$|H^1(t, x)| \le C[\rho(t) H_{or}(\varphi(., g(x)))(1) + |f^1(t, g(x))|]. \quad (23)$$

Moreover for $k \ge 1$ by (8),

$$t^k(D^k H^0(t, .))^*_{t/C}(x) \le C \sum \{H_{-\infty\ell}(\varphi(., g(x)))(t) \prod_{i=1}^{\ell} H_{-\infty j(i)}(\rho)(t) : \ell + \sum_{i=1}^{\ell} j(i) = k,\ j(i) \ge 0\}.$$

Since $\rho \le C$, it may be supposed that $j(i) > 0$ in this summation. Hence by Lemma 2,

$$t^k(D^k H^0(t, .))^*_t(x) \le C_\varepsilon[H_{-\infty k-\varepsilon}(\varphi(., g(x)))(t) + \sup_t \varphi(t, g(x)) H_{-\infty k-\varepsilon}(\rho)(t)]. \quad (24)$$

Now put

$$h(t) = h^0(t) + h^1(t) = \sigma^0(t) H^0(t) + [\sigma^1(t) H^0(t) + \sigma H^1(t)], \quad (25)$$

say. Then by Leibniz's rule, (24), and Lemma 2,

$$t^r|D^r h^0(t, x)|$$

$$\leq C_{r\varepsilon}[H_{-\infty, r-2\varepsilon}(\varphi(., g(x)))(t) + \sup_s \varphi(s, g(x))H_{-\infty r-2\varepsilon}(\rho)(t)].$$

Also

$$|h^1(t, x)| \leq C[\rho(t)|H^0(t, x)| + |H^1(t, x)|].$$

By (23) this completes the proof of the lemma.

12

To complete the proof of Lemma 6, define a partition of unity on M as follows. Supposing $\varphi \in \mathscr{D}(B(0, 1))$, $\varphi \geq 1$ and $\int \varphi(x)\, dx = 1$, put $\mu_i = \varphi_{1/12} * 1_{V_i}^{-5/6}$, $\lambda_i^0 = \mu_i \circ u_i$ in U_i, $\lambda_i^0 = 0$ outside U_i,

$$\lambda_i = \lambda_i^0 / \sum_j \lambda_j^0, \quad \mu_i = \lambda_i \circ u_i^{-1}, \quad i \in \mathscr{I}.$$

Then by Lemma 3,

$$|||\{\mu_i : i \in \mathscr{I}\} : B((.)\Psi)||| \leq C. \tag{26}$$

Put $F = \{F_i : i \in \mathscr{I}\} = \{\mu_i f_i : i \in \mathscr{I}\}$. (17)-(19) clearly imply decompositions $F = F^0(t) + F^1(t)$, $0 < t \leq 1$ with analogous inequalities. Suppose (18) and (19) are valid for F in place of f.

Let $\mathscr{A}' = \{u_j : U_j' \to V_j' : j \in \mathscr{I}'\}$ be a $u\ell B((.)\Psi)$ atlas equivalent to $\mathscr{A}$, subordinate to $\mathscr{U}$ and likewise satisfying (12) and the condition of Sec. 8 and

$$g_{ij} = u_i \circ (u_j')^{-1} : u_j'(U_i \cap U_j') \to u_i(U_i \cap U_j'), \quad i \in \mathscr{I}, \ j \in \mathscr{I}'.$$

Let J_{ij} be the Jacobian determinant of g_{ij} and define

$$h_{ij} = J_{ij} F_i \circ g_{ij}, \quad i \in \mathscr{I}, \ j \in \mathscr{I}'.$$

For any function h on $u_j'(U_i \cap U_j')$, let $\tilde{h}$ denote its extension by 0 to all of V_j' (abuse of notation). Then for $f^{\mathscr{A}'} = \{f_j' : j \in \mathscr{I}'\}$,

$f'_j = \sum_{i \in \mathcal{I}} \tilde{h}_{ij}$. According to Lemma 7, with $u'_j(U_i \cap U'_j)$, V_i in place of U, V, suppose $h_{ij} = h^0_{ij}(t) + h^1_{ij}(t)$. Put

$$\bar{h}_{ij}(t) = t^r(D^r h^0_{ij}(t))^*_t + |h^1_{ij}(t)|, \ i \in \mathcal{I}, \ j \in \mathcal{I}'.$$

Note that for $t \leq 1/C$, $h^\ell_{ij}(t) \circ g_{ij}$, $\ell = 0, 1$, is supported in $g^{-1}_{ij}(V^{-\eta}_i)$. If $x \in V'_j$, $B(x, t) \cap g^{-1}_{ij}(V^{-\eta}_i) \neq \emptyset$ and $Ct < \eta$, then by (13),

$$V'_j \cap B(x, t) \subset g^{-1}_{ij}(u_i(U_i \cap U'_j)).$$

Hence for t sufficiently small, $\leq 1/C$, say,

$$((D^r h^0_{ij}(t))^{\sim})^*_{t/C} = [(D^r h^0_{ij}(t))^*_{t/C}]^{\sim}.$$

Put $f'^{\ell}_j(t) = \sum_{i \in \mathcal{I}} \tilde{h}^\ell_{ij}(t)$, $\ell = 0, 1$, and define

$$\bar{f}'_j(t) = t^r(D^r f'^0_j(t))^*_t + |f'^1_j(t)|.$$

By equivalence of $\mathcal{A}, \mathcal{A}'$ $|||\{g_{ij} : i \in \mathcal{I}, \ j \in \mathcal{I}'\} : B(\Psi)||| \leq C$, that is, there exists $\rho \in \Psi_+$, $\|\rho : \Psi\| \leq 1$ such that all g_{ij} in place of g hence J_{ij} in place of σ satisfy (21). Put

$$\bar{\varphi}_i(t) = H^1_{or-\varepsilon}(\varphi_i)(t) + H^1_{or}(\varphi_i)(1)H^1_{or-\varepsilon}(\rho)(t), \ i \in \mathcal{I},$$

$$\bar{\varphi}_A(t, \kappa) = \|\sqcup \xi_\kappa \bar{\varphi}(t, .) : A\|, \ \kappa \in K.$$

By boundedness of $H_{-\infty r-\varepsilon}$ in Ψ, $E^0_0 \otimes E^1_0 \subset E_0$, and (19),

$$\|\bar{\varphi}_A : E_0\| \leq C\|\varphi_A : E_0\| \leq C. \tag{27}$$

For any measurable function h on a measure space (X, m), let $\lambda_h = \lambda(h)$ denote its distribution function:

$$\lambda_h(s) = m(\{x : |h(x)| > s\}).$$

Define $\xi'_\kappa = \{\xi'_{\kappa j} : j \in \mathcal{J}'\}$, $\kappa \in K$ corresponding to $\mathcal{A}'$ analogously to the way ξ_κ, $\kappa \in K$ was defined for $\mathcal{A}$. It follows that

$$\lambda(\sqcup \xi'_\kappa \bar{f}'(t))(s) \le C\lambda(\sum_{\kappa'\sim\kappa} \xi_{\kappa'}\bar{\varphi}(t))(s/C) \le C \sum_{\kappa'\sim\kappa} \lambda(\sqcup \xi_{\kappa'}\bar{\varphi}(t))(s/C),$$

$\kappa \in K$, hence for $\bar{\varphi}'_A(t, \kappa) = \| \sqcup \xi'_\kappa \bar{f}'(t) : A \|$, $0 < t < 1/C$, $\kappa \in K$

$$\bar{\varphi}'_A(t, \kappa) \le C \sum_{\kappa'\sim\kappa} \| \sqcup \xi_{\kappa'}\bar{\varphi}(t) : A \| \le C \sum_{\kappa'\sim\kappa} \bar{\varphi}_A(t, \kappa')$$

and by (27) and (15), $\| \bar{\varphi}'_A : E_0 \| \le C$. But this implies

$$\| f : \mathcal{B}(\mathcal{A}', A, E, \mu') \| \le C \| f : \mathcal{B}(\mathcal{A}, A, E, \mu) \|$$

for any distribution μ' satisfying the hypotheses made on μ.

13

Suppose $M_0 = U$ is an open subset of R^n. It appears of interest to define Besov spaces in terms of $\mu_t * f(x)$, $0 < t < \operatorname{dist}(x, R^n - U)$, that is, with an upper bound on t depending on x. However, by choice of a suitable atlas, a possible definition is included in the present one. Suppose M_0 is a C^1 manifold and δ is a positive continuous function on M_0. Define the open covering $\mathcal{W}^C_\delta$ by $\mathcal{W}^C_\delta = \{\mathcal{W}^C_\delta(s) : s \in 2^Z\}$, $\mathcal{W}^C_\delta(s) = \{x \in M_0 : s/C < \delta(x) < Cs\}$. Call $\mathcal{A}$ a δ-uniform $B((.)\Psi)$ atlas if for $L(s) : R^n \to R^n$ defined by $L(s)x = sx$

$$\mathcal{A}_\delta = \{L(s) \circ u_i|_{U_i \cap W^2_\delta(s)} : i \in \mathcal{J}, s \in 2^Z\}$$

is a uniform $B((.)\Psi)$ atlas. If $M = (M_0, \mathcal{A})$, put $M_\delta = (M_0, \mathcal{A}_\delta)$ and say δ is a Lebesgue function for $\mathcal{A}$ if $\mathcal{A}_\delta$ has Lebesgue number $1/C$. Call M a $\delta u\ell B((.)\Psi)$ manifold if $M_\delta = (M, \mathcal{A}_\delta)$ is $u\ell B((.)\Psi)$. Supposing M to be a $\delta u\ell B((.)\Psi)$ manifold and that $\mathcal{A}_\delta$ satisfies (12) and Sec. 8, define $\chi^\delta_{si}(t, .)$ to be the indicator function of V^{-ts}_{is}, $V_{is} = u_i(U_i \cap W^2_\delta(s))$, and $\xi^\delta_s = \chi^\delta_s(0, .)$, $s > 0$, $i \in \mathcal{J}$. Suppose F_0 is a nondegenerate Banach function space on $R^*_+ \times R^*_+$ such that

$$\left\| \int_1^2 \int_1^2 f(s., t.)\, ds\, dt : F_0 \right\| \leq C \left\| f : F_0 \right\|. \tag{28}$$

Also suppose H^1_{or} is bounded in F_0. For $F = (F_0, \mathscr{H}, \Lambda)$, $f \in L^1_{loc}(M)\Lambda$, define $\| f : \mathscr{B}(\mathscr{A}, A, F) \|$ as $\| \nu(f) : F_0 \|$, where

$$\nu(f)(t, s) = \| \sqcup \chi_s^\delta \mu_{ts} * f\mathscr{A} : A \| + 1_{(1/2,1)}(t) \| \sqcup \xi_s f : A \|. \tag{29}$$

From F_0 on $R_+^* \times R_+^*$ obtain a related space $F_0^\#$ on $R_+^* \times 2^Z$ in the obvious way. For example, if f is a function on 2^Z, let $\tilde{f}$ denote its extension to R_+ which is constant on the intervals $(2^{j-1/2}, 2^{j+1/2}]$, $j \in Z$, for f on $R_+^* \times 2^Z$ put $\tilde{f}(t, s) = \big(f(t, .)\big)^\sim(s)$, t, $s \in R_+$ and finally $\| f : F_0^\# \| = \| \tilde{f} : F_0 \|$. Then the norm $\| . : \mathscr{B}(M_\delta, A, F^\#) \|$ is equivalent to the given one. To see this note that the image of $\Lambda_i \mathscr{L}^n$ under $L(s)^{-1}$ is $\Lambda_{il/s} \mathscr{L}^n$ and

$$\begin{aligned} &\| 1_s - 1_{V_{is}^{-t}} \mu_t * (f^i)^s : A(R^n, \Lambda_{il/s}\mathscr{L}^n) \| \\ &= \| 1_s - 1_{V_{is}^{-t}} (\mu_{ts} * f^i)^s : A(R^n, \Lambda_{il/s}\mathscr{L}^n) \| \\ &= \| 1_{V_{is}} - t\mu_{ts} * f^i : A(R^n, \Lambda_i\mathscr{L}^n) \|. \end{aligned}$$

14

Suppose F_0 is a nondegenerate Banach function space on $R_+^* \times R_+^*$ satisfying (28), h is a positive differentiable function on R_+ such that

$$0 < C_1^{-1} \leq th'(t)/h(t) \leq C_2. \tag{30}$$

Define the space F_0^h to consist of all locally integrable functions f on $R_+^* \times R_+^*$ such that for $f^h(t, s) = f\big(th(s), s\big)$,

$$\| f : F_0^h \| = \| f^{1/h} : F_0 \| < \infty.$$

As is easy to check, F_0^h is nondegenerate and boundedness of $H_{\alpha\beta}^1$ in F_0 implies its boundedness in F_0^h. Consider the following special case. Supposing X, Y are nondegenerate Banach function spaces on R_+^* such that H_{or} is bounded in X and $H_{\alpha\beta}$ in Y for some α, $\beta \in R$, let $F = X(Y)$. Put $\alpha_Y(t) = \| 1_{(t,\ 2t)} : Y \|$ for $t > 0$. Define $X^{\#}$ analogously to $F_0^{\#}$ by $\| f : X^{\#} \| = \| \tilde{f} : X \|$. Then for $E_0 = E_0^{h\#}$, it follows that $E_0^1 = F_0^{h1\#}$, where

$$\| f_1 : F_0^{h1} \| = \| (\alpha_Y f_1) \circ h^{-1} : X \|.$$

Hence for $T(f_0)f_1(t) = \| f_0(t/h)f_1 \circ h/\alpha_Y : Y \|$

$$\| f_0 : E_0^0 \| = \sup\{ \| T(f_0)f_1 : X \| : \| f_1 : X \| \leq 1\}.$$

Denote $L^p(R_+^*)$ by $\mathcal{L}^p$, $1 \leq p \leq \infty$. In the special case $X = \omega_0 \mathcal{L}^\infty$, $Y = \omega_1 \mathcal{L}^p$, $1 \leq p \leq \infty$ where ω_1 is a positive differentiable function on R_+^* satisfying (30), $i = 0, 1$ with $C_2 < r$ in the case of ω_0, it follows that $F_0^{h1} = \omega_1 \omega_0 \circ h \mathcal{L}^\infty$ and

$$\| f_0 : F_0^{h0} \| = \sup_{s>0} \left(\int_0^\infty [f_0(s/h(t))\omega_0(h(t))/\omega_0(s)]^p t^{-1}\, dt \right)^{1/p} \tag{31}$$

$$\leq C \sup_{s>0} \left(\int_0^\infty [f_0(t)(\omega_0(s/t)/\omega_0(s))]^p t^{-1} \right)^{1/p}.$$

In this case if M is a $\delta u\ell B((.)\omega_0)$ manifold, then $\mathcal{B}(M_\delta, L^p, \omega_0 \mathcal{L}^\infty (\omega_1 \mathcal{L}^p)^h)$ (suppose $M_\delta = (M_0, \mathcal{A}_\delta, \mathcal{W}_\delta^2)$) to be abbreviated to $\mathcal{B}(M_\delta, \omega_0 \mathcal{L}^\infty (\omega_1 \mathcal{L}^p)^h)$ consists of all measures on M_0 absolutely continuous with respect to Λ such that for $E_{ij} = \{x \in V_i : 2^{j-1} < \delta(u_i^{-1}(x)) < 2^{j+1}\}$, $E'_{ij}(t) = \{x \in E_{ij} : x + 2^j B(0, t/h(2^j)) \subset V_i\}$, $i \in \mathcal{I}$, $j \in Z$, $t > 0$

$$\varphi(t) = \sum_{i\in\mathscr{I}}\left[\sum_{h(2^j)>t}\int_{E'_{ij}(t)} |\omega_1(2^j)\mu_{2^j t/h(2^j)} * f^i(x)|^p \Lambda_i(x)\,dx + \sum_{h(2^j)<t}\int_{E_{ij}} |\omega_1(2^j) f^i(x)|^p \Lambda_i(x)\,dx\right]^{1/p} \tag{32}$$

$$\sup_{0<t<1} \varphi(t)/\omega_0(t) < \infty.$$

Note that $\underline{\omega}_0 \mathcal{L}^p \subset F_0^{h_0}$ for

$$\underline{\omega}_0(t) = \inf\{\omega_0(tu)/\omega_0(u) : u > 0\},\ t > 0. \tag{33}$$

15

Suppose $\mathscr{W}$ consists of the single set M and $E_0 = E_0^0 = \Psi$. Denote the corresponding space by $\mathscr{B}(M, A, \Psi)$. Then the measure f on M belongs to $\mathscr{B}(M, A, \Psi)$ if and only if for $\nu(t) = \| \sqcup \mu_t * f^{\mathscr{A}} : A \|$

$$\| 1_{(0,1)} \nu : \psi \| + \| \sqcup f^{\mathscr{A}} : A \| < \infty.$$

As regards function spaces $B(M, A, (.)\Psi)$ if $A = L^\infty$ write $B(M, (.)\Psi)$

Suppose X is a Banach function space on R_+ containing the indicator functions of compact subsets of R_+ such that $1/C \leq \| 1_{(1/2,1)} : X \| \leq C$, and for $t > 0$

$$\| 1_{(0,t)} : X \| \leq C \| 1_{(t/2,t)} : X \|. \tag{34}$$

For any subinterval (a, b) of R_+ put $X(a, b) = \{f : \| f : X(a, b) \| = \| 1_{(a,b)} f : X \| < \infty\}$. For any measurable subset E of R^n put $|E| = \mathscr{L}^n(E)$. Then define $\chi = |B(0, 1)|^{-1} 1_{B(0,1)}$ and for $f \in L^1_{\text{loc}}(R^n)$, $t > 0$

$$m_t(f) = \sup\{\chi_s * |f| : s \leq t\}.$$

Define the space $M(X)$ to consist of all $f \in L^1_{\text{loc}}(R^n)$ such that

$$\| f : M(X) \| = \| m(f) : X\| < \infty.$$

Although the space $M(X)$ need not be r.i. (see Lemma 9) under the hypotheses of Lemma 6 define $\mathscr{B}(\mathscr{A}, M(X), E, \mu)$ to consist of all $f \in L^1_{\text{loc}}(M)\Lambda$ such that for

$$\nu(f)(t, \kappa) = \sup\{\| 1_{V_i^{-\eta-t}}|\mu_t * f^i| + 1_{(1/2,1)}(t)1_{V_i^{-\eta}}|f^i| : M(X(0, 1))\| : U_i \in W_\kappa\},\ \eta$$

$$= 1/C,\ 0 < t < 1,\ \kappa \in K, \| f : \mathscr{B}(\mathscr{A}, M(X), E, \mu)\| = \| \nu(f) : E_0\| < \infty$$

(if $S \subset S' \subset R^n$ and f has domain S, assume 1_S, $f = 0$ outside S). Note that $M(X)$ has the following invariance property that permits extending the proof of Lemma 6 to the present spaces. If V_1, V_2 are open subsets of R^n, g is a C^1 diffeomorphism of V_1 onto V_2 such that $\| Dg\|_\infty, \| Dg^{-1}\|_\infty \leq C$, then

$$\| 1_{V_1} f \circ g : M(X(0, 1))\| \leq C\| 1_{V_2^{-1/C}} f : M(X(0, 1))\| .$$

This justifies the notation $\mathscr{B}(M, M(X), E)$ for the spaces just defined. To test whether f belongs to $\mathscr{B}(M, M(X), E)$, $m_s(\mu_t * f^i)$, $m_s(f^i)$ are needed only for $s \geq t$ or $s = 1$, respectively.

Lemma 8. *For* $f \in L^1_{\text{loc}}(M)\Lambda$ *define*

$$\nu'(f)(t, \kappa) = \sup\{\| 1_{V_i^{-\eta-t}} \mu_t * f^i : M(X(t, 1))\| + 1_{(1/2,1)}(t)m_1(1_{V_i^{-\eta}} f^i) : U_i \subset W_\kappa\}$$

for $0 < t < 1$, $\kappa \in K$. *Then*

$$\| f : B(M, M(X), E)\| \leq C\| \nu'(f) : E\| .$$

Proof. For definiteness, suppose $\mathscr{A}$ has Lebesgue number 1 and $\eta = 1/2$. For $s \leq t \leq 1$,

$$\chi_s * (1_{V_i^{-1/2-2t}} \chi_t * \mu_t * f^i)$$

$$\leq C(\chi_{t+s} * (1_{V_i^{-1/2-t}} \mu_t * f))$$

$$\leq Cm_t(1_{V_i^{-1/2-t}} \mu_t * f^i),$$

hence

$$m_s\left(1_{V_i^{-1/2-2t}}\ (\chi * \mu)_t * f^i\right) \leq Cm_t(1_{V_i^{-1/2-t}}\ \mu_t * f^i),$$

hence by (34) for $t \leq \eta = 1/C$,

$$\| 1_{V_i^{-1/2-\eta-t}}\ (\chi * \mu)_t * f^i : M(X(0,\ t)) \|$$

$$\leq C \| 1_{(0,t)} m_{t/2} (1_{V_i^{-1/2-t}}\ \mu_t * f^i) : X \|$$

$$\leq C \| 1_{(t/2,t)} m_{\bullet}\ (1_{V_i^{-1/2-\bullet}}\ \mu_t * f^i) : X \| \leq C\nu'(f)(t,\ \kappa).$$

In addition, by (20), in $V_i^{-1/2}$

$$f^i = \int_0^{1/4} \sigma_u * \mu_u * \tilde{f}^i u^{-1}\, du + \lambda_{1/4} * \tilde{f}^i;$$

hence for $s \leq 1/4$,

$$\chi_s * (1_{V_i^{-3/4}}\ |f^i|) \leq C(\sup_{r\leq 1} \chi_r * |\tilde{f}^i|$$

$$+ \int_0^{1/4} \chi_s * (1_{V_i^{-1/2}}\ [(\omega^4\sigma_u) * \mu_u * f^i])u^{-1}\, du$$

$$\leq C[m_1(\tilde{f}^i) + \int_0^{1/4} m_{s+u}(1_{V_i^{-1/2}}\ \mu_u * f^i)_u^{-1}\, du].$$

By (34) the mappings $\varphi \to 1_{(0,t)} \inf_{t/2\leq s\leq t} \varphi(s)$ are uniformly bounded in X_+, hence

$$\| 1_{(0,1)} m_{\bullet+u}(1_{V_i^{-3/4}}\ \mu_u * f^i) : X \| \leq C \| 1_{V_i^{-1/2}}\ \mu_u * f^i : M(X(0,\ 1)) \|.$$

Put $\psi(\kappa) = \| 1_{V_i^{-3/4}}\ f^i : M(X(0,\ 1)) \|$ so that

$$\psi(\kappa) \leq Cm_1(\tilde{f}^i) + C \int_0^{1/4} \nu'(f)(u,\ \kappa)u^{-1}\, du.$$

By the assumed boundedness of $H^1_{0\infty}$ in E_0 it follows that

$$\| \psi : E_0^1 \| \leq \| 1_{(1/2,1)} \otimes \psi : E_0 \| \leq C \| m_1(\tilde{f}^i) : E_0^1 \| + C \| \nu'(f) : E_0 \|.$$

Along with the estimate at the start of the proof if $\chi * \mu$ is used to define the norm of $\mathscr{B}(M, M(X), E)$, this shows that $\| f : \mathscr{B}(M, M(X), E) \| \leq c \| \nu'(f) : E \|$.

The spaces $M(X)$ are related to r.i. spaces defined as follows. If $f \in L^1_{\mathrm{loc}}$ on R put $f^{**}(t) = t^{-1} \int_0^t f_*(s)\, ds$ for $t > 0$ and $\| f : \Lambda(X) \| = \| f^{**} : X \|$. Recall that for $1 < p \leq \infty$, $\Lambda((.)^{-1/p} L^q)$ is the Lorentz space L^{pq}. As is easy to see, $M(X) \subset \Lambda(Y)(R^n)$ for $\| \varphi((.)^n) : Y \| = \| \varphi : X \|$, that is, $Y = X \circ (.)^n$. While for $X = (.)^{-1} \mathcal{L}^\infty$ or $\mathcal{L}^\infty$ these spaces agree (and with L^1, L^∞, respectively), this need not be the case.

Lemma 9. For any $\varepsilon > 0$, $1 < q < \infty$, *there exists a positive measurable function on* $(0, 1)$ *equimeasurable with* $(.)^{-1/q}$ *such that for any interval* I,

$$\int_I f(t)\, dt \leq c_{q\varepsilon} |I|^{1-\varepsilon}. \tag{35}$$

Proof. This follows by rearranging the complementary intervals of a Cantor set C in order of increasing length in $(0, 1)$. Define δ, $0 < \delta < 1/2$, by $2 = \delta^{\varepsilon q - 1}$ and suppose C is constructed by successive dissections of type $[2; 0, 1 - \delta; \delta]$ (see [15, p. 194]) of $[0, 1]$. Let the open "black" intervals removed at the k-th step be $J_{k\ell}$, $1 \leq \ell \leq 2^{k-1}$, leaving the disjoint closed white intervals $C_{k\ell}$, $1 \leq \ell \leq 2^k$, $k = 1, 2, \ldots$ such that $|J_{k\ell}| = \gamma\delta^{k-1}$ for $\gamma = 1 - 2\delta$ and the distance between two distinct $C_{k\ell}$, $1 \leq \ell \leq 2^k$ is at least $\gamma\delta^{k-1}$. On $J_k = \bigcup_{\ell=1}^{2^{k-1}} J_{k\ell}$ let f be a rearrangement of $(.)^{-1/q}$ restricted to $((2\delta)^k, (2\delta)^{k-1})$ such that the functions $f|_{J_{k\ell}}$, $1 \leq \ell \leq 2^k$ are equimeasurable. To see that f satisfies (35), suppose $\gamma\delta^k \leq |I| \leq \gamma\delta^{k-1}$. Then I intersects at most one interval $C_{k\ell}$, $1 \leq \ell \leq 2^k$. Hence,

$$\int_I f(t)\,dt = \int_{I\cap J_k} f(t)\,dt + \int_{I-J_k} f(t)\,dt$$

$$\leq (2\delta)^{-k/q}|I| + 2^{-k}\int_0^{(2\delta)^k} t^{-1/q}\,dt \leq C_{q\varepsilon}|I|^{1-\varepsilon}.$$

The tensor product of n functions as in (35) satisfies this inequality with the meaning of I generalized to an n-dimensional interval. It follows that if the function $\phi_Y : t \to \| 1_{(0,t)} : Y\|$ has the property that $t \to \phi_Y(t)t^{-\varepsilon}$, $\phi_Y(t)t^{-1+\varepsilon}$ are almost increasing or decreasing respectively for some $\varepsilon > 0$ (see [12, (1.11), (1.12)]), then $M(X(0, 1))$ is a proper subspace of $\Lambda(Y(0, 1))(R^n)$. As a result under this condition $B(R^n, \Lambda(Y), \Psi) = B(\Lambda(Y), \Psi) \approx \mathscr{B}(R^n, \Lambda(Y), \Psi)$ is a proper subspace of $B(R^n, M(X), \Psi) = B(M(X), \Psi)$. Thus suppose $\Lambda(Y(0, 1))(R^n) \neq M(X)$ $(Y = X \circ (.)^n)$ and for simplicity $n = 1$. By assumption for any $K > 0$, there exists f supported in $(0, 1)$ such that $\| f : \Lambda(X) \| \geq K$, $\| f : M(X) \| = 1$. It may be supposed that for some $N \in \mathbf{N}$ f is constant in the interval $((k - 1)/N, k/N)$, $k = 1, 2, \ldots, N$. Choose $\varphi \in \mathscr{D}$ supported in $[0, 1]$, $\varphi \neq 0$, with vanishing moments up to order $k + 1$. For $\varepsilon = 1/N$ put $f_1 = \sum_{k=1}^N f((k - 1/2)\varepsilon)\tau((k - 1)\varepsilon)\tilde{\varphi}^N$ $(\tilde{\varphi}(x) = \varphi(-x))$. Then for $\varepsilon(1 - 1/C(\varphi)) \leq t \leq \varepsilon(1 + 1/C(\varphi))$,

$$\| \varphi_t * f_1 : \Lambda(X) \| \geq C(\varphi)^{-1} \| f_1 : \Lambda(X) \| \geq C(\varphi)^{-1} K.$$

It follows that $\| f_1 : B(\Lambda(X), \Psi) \| \geq C(\varphi)^{-1} K \| 1_{(\varepsilon,2\varepsilon)} : \Psi \|$. On the other hand,

$$\| f_1 : B(M(X), \Psi) \| \leq C(\varphi) \| (./\varepsilon)^{k+2} \wedge 1 : \Psi \| \leq C(\varphi) \| 1_{(\varepsilon,2\varepsilon)} : \Psi \|.$$

16

For any vector field X on a $u\ell B((.)\Psi)$ manifold $M = (M_0, \mathscr{A})$ and $i \in \mathscr{I}$ let $X^i : V_i \to R^n$ denote the function corresponding canonically to the image vector field $T(u_i)X|_{U_i}$ on V_i. Then define $\| X : B(M, \Psi) \| = \| \nu(X) : \Psi \|$ for

$$\nu(X)(t) = \| \bigsqcup_i \chi_{\kappa i}(t, .)\mu_t * X^i \|_\infty + 1_{(1/2,1)}(t) \| \bigsqcup_i \xi_{\kappa i} X^i \|_\infty .$$

If $\Lambda^\alpha(R^n) \subset B(R^n, \Psi)$ $(\alpha \geq 0)$, it seems evident how to define a Λ^α $(\alpha \geq 0)$ vector field on M.

Supposing $\Lambda^1(R^n) \subset B(R^n, \Psi)$ for a Lipschitzian vector field X on M, its flow F_X is defined on a subset of $R \times M_0$. If X is bounded, the domain of F_X is all of $R \times M_0$. Let F_X^i denote the flow (with domain a subset of $R \times V_i$) determined by X^i on V_i, $i \in \mathcal{I}$. Suppose Λ is a measure on M_0 whose image $u_i(\Lambda|_{U_i})$ on V_i equals $\Lambda_i \mathcal{L}^n$ where Λ_i is positive continuous and $1/C \leq \Lambda_i(y)/\Lambda_i(x) \leq C$ if $|y - x| \leq 1$. For A as in Sec. 9, let $A(M_0)$ denote the Banach function space on M_0 with respect to Λ of rearrangements of functions in A. $\mathcal{W}$, K, E_0 have the same meaning as in Sec. 9. For $g \in L^1_{loc}(M)$, let $\omega(g)$ denote the function $\kappa \to \| 1_{W\kappa} g : A \|$, and correspondingly for $g \in L^1_{loc}(R \times M_0)$, put $\omega(g)(t, .) = \omega(g(t, .))$. Define the spaces $E(A)$, $E^1(A)$ by

$$\| g : E^{(1)}(A) \| = \| \omega(g) : E_0^{(1)} \| \quad (E^{(1)} = E_0, E_0^1).$$

Put $h_t = F_X(t, .)$. Recall that $DF_X^i(., x) (= (\partial/\partial x_j F^i(., x) : 1 \leq j \leq n))$, $x \in V_i$, satisfies the linear differential equation

$$(d/dt)Y = DX^i(F_X^i(t, x))Y, \; Y(0) = I, \; x \in V_i, \; t \in R \tag{36}$$

(cf[5, p. 22]). If for $x \in U_i$, $\rho(t, x) = \det DF^i(t, u_i(x))$, then $\rho(t, .)$ is the Radon-Nikodym derivative of $h_{-t}(\Lambda)$ with respect to Λ (for any compact subset K of M, $h_{-t}(\Lambda)(K) = \Lambda(h_t(K))$). Since $|DX^i| \leq C$, it follows that

$$C^{-1} \leq \rho(t, .) = \exp \int_0^t \operatorname{tr} DX^i(F_X^i(u, .))\, du \leq C, \; |t| \leq 1. \tag{37}$$

F_X gives rise to a continuous action $(t, \varphi) \to \varphi \circ h_{-t} = L_X(t)\varphi$, say, of R, + on the space $\mathscr{K}(M_0)$ of continuous functions of compact support on M_0. The action of L_X on the space of measures $\mathscr{M}(M_0)$ is defined by duality:

$$\langle L_X(-t)f, \varphi \rangle = \langle f, L_X(t)\varphi \rangle.$$

If $g \in L^1_{\text{loc}}(M)$, note that

$$L_X(t)(g\Lambda) = g \circ h_{-t}\rho(t)\Lambda = (L^*_X(t)g)\Lambda,$$

say. By (37), L^*_X is locally (that is, for $|t| \leq 1$) bounded in $L^1_{\text{loc}}(M)$. Let ν be a measure on R supported in (-1, 1) with vanishing moments up to order r such that neither of $\hat{\nu}(t)$, $\hat{\nu}(-t)$ is zero for all $t > 0$. Put $L_X(\nu)g(t, .) = L_X(\nu_t)g = \int L_X(s)g\, d\nu_t(s)$ for any $g \in L^1_{\text{loc}}(M)$ and define $L^*_X(\nu)$ similarly. The function space $B(X, A, E)$ consists of all locally integrable functions φ on M such that

$$\| \varphi : B(X, A, E) \| = \| L_X(\nu)\varphi : E(A) \| + \| \varphi : E^1(A) \| < \infty.$$

Define the space of distributions $\mathscr{B}(X, A, E)$ to consist of all measures $f = g\Lambda$, $g \in L^1_{\text{loc}}(M)$ such that

$$\| f : \mathscr{B}(X, A, E) \| = \| L^*_X(\nu)g : E(A) \| + \| g : E^1(A) \| < \infty. \tag{38}$$

Again without using the measure Λ on M, the space $\mathscr{B}(X, \mathscr{M}, E)$ of measures can be defined similarly as before using $L_X(\nu_t)f$ instead. The definition does not depend on ν; different ν yield equivalent norms. This follows from the integral representation (20). Note that σ hence λ can be chosen in $\mathscr{D}(R)$, that is, with compact support. In fact, choose $g \in \mathscr{D}(R)$ with integral 0 such that

$$\int_0^\infty \hat{g}(s)\hat{\nu}(s)s^{-1}\, ds = -\int_{-\infty}^0 \hat{g}(s)\hat{\nu}(s)s^{-1}\, ds = 1.$$

Measures Λ^0, Λ^1 give rise to the same space $B(X, A, E)$ if $C^{-1} \leq d\Lambda^1/d\Lambda^0 \leq C$ and to the same space $\mathscr{B}(X, A, E)$ if in addition $d\Lambda^1/d\Lambda^0 \in B(X, L^\infty, E^0_0)$.

Lemma 10. Suppose $f = g\Lambda$, $g \in L^1_{\text{loc}}(M)$; *then* $\| f : \mathscr{B}(X, A, E) \| \leq C$ *if and only if there is a locally integrable function* g^0 *on* $\{(t, x, s) : |t| \leq s \leq 1, x \in M_0\}$ *continuous with respect to its first and last variables and* $\varphi \in L^1_{\text{loc}}(R_+ \times M)$, $\| \varphi : E(A) \| \leq 1$ *such that for* $g^1(t, ., s) = L^*_X(t)g - g^0(t, ., s)$, $|t| \leq s \leq 1$,

$$\sum_{j=0}^{r} s^j |(d/dt)^j g^0(t, ., s)| / H^1_{-\infty j}(\varphi)(t, .) + |g^1(t, ., s)| / \varphi(t, .) \leq C$$

(the terms for $0 \leq j \leq r - 1$ *may be omitted). Analogous characterization of* $B(X, A, E)$.

Proof. Necessity follows from the integral representation (20).

17

Lemma 11. Suppose ϕ *is a bounded continuous function from the cube* $\{(t, x) : t \in R, x \in R^n, |t|, |x| \leq 1/C\}$ *into* R^n *such that for* $0 < s \leq 1$, $\phi = \phi^0(., s) + \phi^1(., s)$, *where* $|\phi^0| \leq C$, ϕ^0, ϕ^1 *have Lipschitz constant* C *with respect to the second variable* (x), ψ *is positive, measurable and bounded by* C *on* R_+, *and for* $0 < s \leq 1$

$$\sum_{j=0}^{r} s^j |D^j \phi^0(., s)| / H_{-\infty j}(\psi)(s) + \phi^1(., s)/\psi(s) \leq 1. \tag{39}$$

Let $t \to F(t)$ *denote the solution of the initial value problem*

$$d/dt\, F(t) = \phi(t, F(t)),\ F(0) = 0 \tag{40}$$

defined for $|t| \leq 1/C$. *Then for* $|t| \leq 1/C$, $0 < s \leq 1$, $F(t) = F^0(t, s) + F^1(t, s)$, *where*

$$\sum_{j=0}^{r} s^j |(d/dt)^{j+1} F^0(t, s)| / H_{-\infty j(1-\varepsilon)}(\psi)(s) \tag{41}$$

$$+ |F^1(t, s)| / (t\psi(s)) \leq C_\varepsilon .$$

Proof. Let $F^0(., s)$ denote the solution of the initial value problem

$$d/dt\, F^0(t, s) = \phi^0\big(t, F^0(t, s), s\big),\ F^0(t, s) = 0,\ |t| \leq 1/C. \quad (42)$$

By (39), $F^0(., s)$ is a $\psi(s)$-approximate solution of (40). By a well-known comparison theorem (see [5, p. 8]) for $|t| \leq 1/C$,

$$|F^1(t, s)| = |F(t) - F^0(t, s)| \leq C\psi(s)C^{-1}(e^{C|t|} - 1) \leq C_1\psi(s)|t|.$$

To complete the proof of (41) suppose

$$|(d/dt)^{\ell+1}F^0(t, s)| \leq CH_{-\infty\ell(1-\varepsilon)}(\psi)(s) \quad (43)$$

valid for $0 \leq \ell \leq j$. By (42) and (8),

$$|(d/dt)^{j+1}F^0(t, s)| \leq C \sum \left\{ \|D^q\phi(., s)\|_\infty \prod_{k=1}^{q} |(d/dt)^{\ell_k+1} F^0(t, s)| : \sum_{k=1}^{q} \ell_k + q = j \right\}.$$

Hence for $\ell = j = 0$, (43) is implied by (39), while for $\ell = j > 0$, the terms with $\ell_k = 0$ may be dropped; hence (43) follows by Lemma 2 and the induction assumption (cf. Sec. 5, 11).

18

Lemma 12. Suppose M is a $u\ell B\big((.)\Psi\big)$ *manifold, X is a* $B(\Psi)$ *vector field on M,* H_{or} *is bounded in* Ψ, λ *is a real-valued function on M such that* $\|\lambda : B(X, \Psi)\| \leq C$ *(valid if* $\|\lambda : B(M, \Psi)\| \leq C$ *and*

$$H_{1\infty}(\Psi_+) \subset E_0^0,\ H_{1\infty}(E_{0+}^0)(1) \subset R \quad (44)$$

$\big(H_{or}^1(E) \subset E\big)$ *(consequently M is a* $u\ell C^2$ *manifold). Then*

$$\mathscr{B}(\lambda X, A, E) \subset \mathscr{B}(X, A, E) \subset \mathscr{B}(M, A, E). \quad (45)$$

If also $|\lambda| \geq C^{-1}$, *then* $\mathscr{B}(X, A, E) = \mathscr{B}(\lambda X, A, E)$. *An equivalent norm on* $\mathscr{B}(X, A, E)$ *is given by*

$$L^1_{loc}(M)\Lambda \ni g\Lambda \longrightarrow \|L_X(\nu)g : E(A)\| + \|g : E^1(A)\| \quad (46)$$

(*cf.* (38)). *Analogous conclusions to* (45) *hold for the function spaces* $B(\lambda X, A, E)$, *and so forth, under the assumption that* $\Psi = E_0^0$ *and* (1b) *is satisfied.*

Proof. Recall that ρ is the solution of the initial value problem (36). The assumption $\| X : B(M, \Psi) \| \leq C$ implies that (39) is valid with a single $\psi = (.)^{-1} H_{1\infty}(\psi_0)$, $\| \psi_0 : \Psi \| \leq C$, ψ_0 nonnegative, nondecreasing in (0, 2), vanishing outside (0, 2) and all φ_X^i, $\phi_X^i(t, y) = DX^i(F^i(t, x))y$, $i \in \mathcal{I}$, $x \in V_i^{-\eta}$. Hence by Lemma 11, $\rho(t, .) = \rho^0(t, ., s) + \rho^1(t, ., s)$ for $|t| \leq s \leq 1$

$$s^{r+1} |(d/dt)^{r+1} \rho^0(t, ., s)| + |\rho^1(t, ., s)| \leq H_{1\infty}(\psi_0)(s). \tag{47}$$

Therefore, by the characterization of $\mathcal{B}(X, A, E)$ of Lemma 10, (46) is stronger than the original norm of $\mathcal{B}(X, A, E)$ (cf. (25)). By the argument for Lemma 4 and (37), the analogous estimate to (47) is valid for a splitting of $1/\rho$ likewise (or use $1 = \rho(t, x)\rho(-t, h_{-t}(x))$, $t \in R$, $x \in M$).

By Lemma 11 with $\| \psi : E_0^0 \| \leq 1$, $F_X^i = F^{i0}(t, ., s) + F^{i1}(t, ., s)$,

$$s^{r+1} |(d/dt)^{r+1} F^{i0}(t, ., s)| + |F^{i1}(t, ., s)| \leq Cs\psi(s).$$

Suppose $\| f : \mathcal{B}(M, A, E) \| \leq 1$, $f = g\Lambda$. By the intermediate space characterization of $\mathcal{B}(M, A, E)$ and the customary splitting of the composition $t \to g(F(t, .))$ (see Sec. 11), the norm of f defined by (46) is bounded by C. For the first inclusion of (45) note that $F_{\lambda X}(t, x) = F_X(\varphi(t, x), x)$, $t \in R$, $x \in M$, where

$$d/dt\varphi(t, x) = \lambda(F(\varphi(t, x), x)), \ \varphi(0, x) = 0.$$

By the customary splitting of $t \to g(F(t, x))$, it follows that the condition of Lemma 10 is satisfied with respect to the vector field λX. If also $|\lambda| \geq C^{-1}$, then by Lemma 4, $\lambda^{-1} \in B(M, L^\infty, \Psi)$.

Remark. If M is a $u\ell C^2$ manifold, X is a Λ^1 vector field, H_{01} is bounded in E_0^0 (hence $\Lambda^1(R^n) \subset B(R^n, E_0^0)$), and $\lambda \in \Lambda^1(M)$, then (45) and (46) are valid also.

19

Suppose M is a uniform C^2 manifold with Lipschitzian vector field X. For $x \in M$, let $\sigma(x)$ denote the sup of the set of nonnegative numbers t such that $F_X(., x)$ exists in $[-t, t]$. Suppose h is a positive differentiable function on R_+ satisfying (30) and for $\gamma(t) = h(t)/t$

$$x \in M_0, \ |X^i(u_i(x))| \leq 1/\gamma(t) \Longrightarrow$$

$$\exists\, j \in \mathcal{I}\big(B(u_j(x), t) \subset V_j\big) \text{ and } \big(y \in B(u_j(x), t) \Longrightarrow |X^i(y)| \leq 2/\gamma(t)\big). \tag{48}$$

By the Picard existence theorem $\sigma(x) \geq C^{-1}h(t)$ and $F_X(s, x) \in u_j^{-1}\big(B(u_j(x), t)\big)$ for $|s| \leq C^{-1}h(t)$. Choose a continuous function $\delta : M \to R_+$ such that $\delta(x) = \gamma^{-1}(|X^i(x)|^{-1})$ for some i such that $x \in U_i$ for all $x \in M$. By uniformity of $\mathcal{A}$ different choices lead to equivalent functions. It follows that (see Sec. 13)

$$h_t(W_\delta^{C_1}(s)) \subset W_\delta^{C_2C_1}(s), \ |t| \leq h(s)/C. \tag{49}$$

Suppose in addition M_δ, X as in Lemma 12 and put $\pi_s = 1_{(s/2,2s)} \circ \delta$, $s > 0$. For $f = g\Lambda$, $g \in L^1_{\text{loc}}(M)$ put $\varphi(t, s) = \| \pi_s L_X(\nu_{th(s)})g : A \|$ $0 < t < 1/C$, $\psi(t, s) = 1_{(1/2,1)}(t) \| \pi_s g : A \|$ and for F_0 as in Sec. 13 define the space $\mathcal{B}(X, h, A, F_0)$ to consist of all f such that

$$\| f : \mathcal{B}(X, h, A, F) \| = \| (1_{(0,1/C)} \otimes 1)\varphi + \psi : F \| < \infty. \tag{50}$$

It can be shown as before that up to equivalence of norms this definition does not depend on ν. By means of a $B(M_\delta, \Psi)$ partition of unity subordinate to $\mathcal{W}_\delta^2$ on M_δ obtain a vector field

$\tilde{X}$ such that on each $W_\delta^C(s)$, $\tilde{X} = \lambda_s h(s)X$ with $\| \lambda_s : B(M_\delta, \Psi) \| \leq C$, $C^{-1} \leq |\lambda_s| \leq C$, $s > 0$. As a consequence of Lemma 12, the norm determined by (50) is that of $B(\tilde{X}, A, F^\#)$. Define $R_{is} = (1/s)\{x \in V_i : s/2 < \delta(u_i^{-1}(x)) < 2s\} = (1/s)u_i^{-1}(W_\delta^2(s) \cap U_i)$, $i \in \mathscr{I}$, $s > 0$.

Lemma 13. Suppose M is a uniform C^2 manifold with Lipschitzian vector field X satisfying (48), *M is also a $\delta u \ell B(\Psi)$ manifold* (Ψ) *as in Lemma* 12, $E = F^\#$), *and*

$$|||\{h(s)s^{-1}X^i \circ L(s) : i \in \mathscr{I},\ s > 0\} : B(\Psi)||| \leq C. \tag{51}$$

Then $\mathscr{B}(M_\delta, A, F) \subset \mathscr{B}(X, h, A, F)$.

Proof. $\tilde{X}$ is a $B(\Psi)$ vector field on M_δ; hence the assertion results from Lemma 12.

20

Supposing M is a uniform $B((.)\Psi)$ manifold, S is a subset of M_0, say $\mathscr{A}$ has Lebesgue number $\varepsilon > 0$ on S if for each $x \in S$ there exists $i \in \mathscr{I}$ such that $x \in U_i$ and $B(u_i(x), \varepsilon) \subset V_i$. If the distance m is defined as in Sec. 8 for $S^\eta = \cup\{B_m(x, \eta) : x \in S\}$, $\mathscr{A}$ then also has Lebesgue number ε/C on $S^{\varepsilon/C}$. Modify the requirement (12) on $\mathscr{A}$ to

$$B(x, \varepsilon) \subset V_i,\ x \in u_i(S \cap U_i) \Rightarrow \#\{j : u_i^{-1}(B(x, \varepsilon)) \cap U_j \neq \emptyset\} \leq C. \tag{52}$$

Under this condition (which is satisfied by an equivalent atlas at any rate), there exists a family $\{\lambda_i : i \in \mathscr{I}\}$ of nonnegative functions on M such that $spt\lambda_i \subset U_i \cap S^{1/C}$, $\Sigma\, \lambda_i = 1$ on S^{1/C^2}, for any $x \in M$ the number of $i \in \mathscr{I}$ such that λ_i does not vanish on all of $B_m(x, 1/C)$ is bounded by C and

$$|||\{\lambda_i \circ u_i^{-1} : i \in \mathscr{I}\} : B(\Psi)||| \leq C$$

(cf. (26)). The definition of the spaces $B(M, A, E)$, $\mathcal{B}(M, A, E)$, and so on, and the results concerning them extend to spaces of functions and measures supported in a fixed set S ($B(M, S, A, E)$, $\mathcal{B}(M, S, A, E)$, and so on) if $\mathcal{A}$ has positive Lebesgue number $1/C$ on S. In fact, the previous definition of the norm is still valid (see (16)). The ambient manifold, however, may as well be replaced by its open submanifold $S^{1/C}$.

21

Lemma 14. Suppose M is a uniform C^2 manifold, ζ is a mapping of M into R^n, $\xi \in R^n$, $|\xi| = 1$, X is a Lipschitzian vector field such that $X(\xi.\zeta) = 1$ on M, and $f = g\Lambda$, $g \in L^1(M, \Lambda)$. Let $\zeta(f)$ denote the direct image of f in R^n. Then for $t > 0$, $\nu^k = \sum_{\ell=0}^{k} (-1)^{\ell}\binom{k}{\ell}$ $\tau(\ell)\delta$

$$|\zeta(f)^{\wedge}(t\xi)| \leq 2\int_{\sigma(y)\leq 3k\pi/t} d|f|(y) \qquad (53)$$

$$+ 2^{-k}\int_{\sigma(y)>2k\pi/t} d|L_X(\nu^k_{-\pi/t})f|(y)$$

$$= 2\int_{\sigma(y)\leq 3k\pi/t} |g(y)|d\Lambda(y) + 2^{-k}\int_{\sigma(y)>2k\pi/t} |L_X^*(\nu^k_{-\pi/t})g|(y)d\Lambda(y).$$

Proof. Observe that for $\beta = \xi.\zeta$, $(d/dt)\beta \circ h_t = \langle d\beta \circ h_t, X \circ h_t\rangle = 1$, hence for $\sigma(y) > k\pi/t$,

$$\exp(-it\beta(y)) = 2^{-k}\Delta^k(-\pi/t)\exp(-it\beta \circ h\,.\,(y))(0)$$

and

$$\int_{\sigma(y)>2k\pi/t} e^{-it\xi.\zeta(y)} g(y)d\Lambda(y)$$

$$= 2^{-k}\sum_{\ell=0}^{k} (-1)^{\ell}\binom{k}{\ell}\int_{\sigma(F_X(\ell\pi/t,y))>2k\pi/t} e^{-it\beta(y)} g(F_X(\ell\pi/t,y))$$

$$d(h_{-\ell\pi/t}\Lambda)(y).$$

Therefore,

$$\Big|\int_{\sigma(y)>2k\pi/t} e^{-it\beta(y)} g(y)\, d\Lambda(y) - 2^{-k}\int_{\sigma(y)>2k\pi/t} e^{-it\beta(y)}\, \Delta^{k}(\pi/t)$$

$$\big(g(F_X(.,\ y))\rho(.,\ y)\big)(0)\, d\Lambda(y)\Big| \le \int_{k\pi/t<\sigma(y)\le 3k\pi/t} |g(y)\, d\Lambda(y),$$

which implies (53).

22

It seems convenient to write $B^p(\Psi)$ for $B(\psi L^p)$ under appropriate conditions on the positive measurable function ψ on R_+.

Lemma 15. Suppose $M = (M_0, \mathscr{A})$ *is a uniform* C^2 *manifold,* ζ *is a* C^2 *mapping of* M *into* R^n*, the positive differentiable function* h *on* R_+ *satisfies* (30),

$$|D^2\beta_i| \le C\gamma'(\gamma^{-1}(|D\beta_i|)),\ \beta = \xi\ .\ \zeta,\ \beta_i = \beta\circ u_i^{-1},\ i\in\mathscr{I} \qquad (54)$$

(*see* Sec. 19). *Put* $M_0' = \{x \in M_0,\ d\beta(x) \ne 0\}$. *Let* $\delta = \delta_\xi : M_0' \to R_+$ *be a positive continuous function such that*

$$1/2 \le \delta(x)/\gamma^{-1}\big(|D\beta_i(u_i(x))|\big) \le 2$$

for some i *for every* $x \in M_0'$. *Suppose* ω *is a positive continuous function on* R_+ *such that for some* $\varepsilon > 0$, $t \to t^{-\varepsilon}\omega(t)$, $t^{-r+\varepsilon}\omega(t)$ *are increasing or decreasing respectively (equivalently* ω *satisfies* (30) *with* $C_2 = r - \varepsilon$), *define* $\underline{\omega}$ *according to* (33) *and* $\underline{\omega}_1$ *by* $1/\underline{\omega}_1 = H_{-\infty,-1}(1/\underline{\omega})$. *Suppose* $M_\delta' = (M_0', \mathscr{A}_\delta')$ (*see* Sec. 13) *is a uniform* $B^1((.)\underline{\omega}_1)$ *manifold with Lebesgue function* δ/C *on the support of a measure* f *on* M_0' *and for some* $\psi \in \underline{\omega}_1\mathscr{L}^1$ *such that* $\psi \ge 1$ *in* (1, 2)

$$\|\psi/\underline{\omega}^1 : \mathscr{L}^1\| \le 1, \qquad (55)$$

$N \geq r$ and if $x \in S_i = \{x \in V_i : \mathrm{dist}(x, u_i(U_i \cap \mathrm{spt}\, f)) < \delta(x)/C\}$, $\delta(x) = s$, $|z| \leq t \leq 1$, $B(u_i(x), (N+1)t) \subset V_i$, then

$$|\Delta^N(sz)(D\beta_i)(u_i(x))| \leq C\gamma(s)\psi(t). \tag{56}$$

Then

$$|\zeta(f)^\wedge(t\xi)| \leq Cb(\xi)\omega(t^{-1}),\ t \geq 1 \tag{57}$$

where for some distribution μ supported in $B(0, 1)$ as in Lemma 6

$$b(\xi) = \sup_{0<t<1} \omega(t)^{-1}\Bigg[\sum_{i\in\mathcal{I}} \int_{x\in V_i, h(\delta(u_i^{-1}(x)))<t} d|u_i(f)|(x)$$

$$+ \sum_{h(2^j)>t} \int_{E'_{ij}(t)} d|\mu_{2^j t/(Ch(2^j))} * u_i(f)|(x)\Bigg].$$

Proof. Choose the uniform $B^1(\underline{\omega}_1)$ atlas $\mathcal{A}'_\delta = \{u'_i : U'_i \to R^n, i \in \mathcal{I}'\}$ for M'_0 and the partition of unity $\{\lambda_i : i \in \mathcal{I}'\}$ on a neighborhood $S^{1/C}$ according to Sec. 20 and subordinate to $\mathcal{W}^2_\delta$ (according to Sec. 8 also). Define the vector field Y_i on $V'_i (= u'_i(U'_i \cap M'_0))$ by $Y_i = |D\beta_i|^{-2} D\beta_i$ and X on M'_0 by $X = \sum_{i\in\mathcal{I}'} \lambda_i T(u'^{-1}_i) Y_i$ where $T(u'^{-1}_i)Y_i$ is considered to be 0 in $M'_0 \setminus U'_i$. By (54) $|D\gamma^{-1}(|D\beta_i|)| \leq C$; hence if $|D\beta_i(x)| \geq \gamma(t)$ and $B(x, C^{-1}t) \subset V_i$, then $|y - x| \leq C^{-1}t \Rightarrow |D\beta_i(y)| \geq \gamma(t)/2$. Hence if h is replaced by $C^{-1}h(C.)$, then the vector field X satisfies (48) with the condition "$x \in S^{1/C}$" added to the hypothesis.

By Sec. 14 for $F_0 = \omega\mathcal{L}^\infty(\mathcal{L}^1)$, $\underline{\omega}\mathcal{L}^1 \subset F_0^{h0}$, hence $B(M'_\delta, S, \mathcal{M}, F^h)$ $(M'_\delta = (M'_0, \mathcal{A}'_\delta))$ is well defined on the uniform $B^1((.)\underline{\omega}_1)$ manifold M'_δ with positive Lebesgue number $1/C$ on S. By Lemma 14 (recall (49) and that h satisfies (30)), relation (57) holds with $b(\xi) = \| f : \mathcal{B}(X, S, h, \mathcal{M}, F^h)\|$. By Sec. 14 (see (32)), it remains to show

$$\mathcal{B}(M_\delta, S, \mathcal{M}, F^h) \subset \mathcal{B}(X, S, h, \mathcal{M}, F^h).$$

For $S_{si} = u_i'^{-1}(S \cap U_i' \cap W_\delta^2(s))$ by Lemma 13 it is enough to show

$$\left\|\left\{\gamma(s)Y^i \circ L(s)\right|_{L(s^{-1})(S_{si}^{s/C} \cap V_i^{-s/C})} : i \in \mathscr{I}',\ s > 0\right\} : B^1(\underline{\omega}_1)\right\| \leq C \tag{58}$$

$(S_{si}^{s/C} = \cup\{B(x, s/C) : x \in S_{si}\})$. For that purpose observe that in $S_{is}^{s/C} \cap V_i^{-s/C}$ for $|z| \leq t \leq 1/(NC)$

$$|\Delta^N(sz)(|D\beta_i|^{-2}D\beta_i)| \leq C\sum\left\{|D\beta_i|^{-m-2}\prod_{\ell=0}^{m}|\Delta^{k_\ell}(sz)\tau(k_\ell' sz)\ D\beta_i| : \sum_{\ell=0}^{m} k_\ell = N,\ k_\ell + k_\ell' \leq N\right\}. \tag{59}$$

By definition of δ, $C^{-1} \leq |D\beta_i|/\gamma(s) \leq C$, hence terms with $k_\ell = 0$ may be omitted from the product. By the identity

$$2^{mk}\Delta^k(y)f(x) = \Delta^k(2^m y)f(x) - \sum_{j=0}^{k-1}\sum_{i=0}^{m-1} 2^{(m-i)k}a_j\Delta^{k+1}(2^i y)f(x + 2^i jy)$$

(see [9, p. 321]) for any real-valued function f on R^n, $y \in R^n$, (56), and $|D\beta_i| \leq C\gamma(s)$, it follows by induction that for $t < 1/(Ck)$, $k = N, N-1, \ldots, 0$

$$|\Delta^k(sz)D\beta_i| \leq \gamma(s)H_{-\infty k}(\psi)(s)$$

in $S_{is}^{s/C} \cap V_i^{-s/C}$. Substitution in (59) and Lemma 2 yield

$$|\Delta^N(sz)(|D\beta_i|^{-2}D\beta_i)| \leq C\gamma(s)^{-1}H_{-\infty N-\varepsilon}(\psi)(s)$$

under the present restriction; hence (58).

23

For $1 \leq p, q, r \leq \infty$, $\alpha > 0$ let $\tilde{B}_\alpha^{pq,r}(R^n)$ denote the space $B(R^n, M((.)^{-n/p}L^q), (.)\not{L}^r)$ defined in Sec. 15 ($\Lambda = \mathscr{L}^n$). Thus for

$$\Omega_f(t) = \sup_{0<s<1} s^{n/p} \| \chi_s * \mu_t * f \|_\infty$$

$$\| f : \tilde{B}_\alpha^{p\infty,1} \| = \int_0^1 t^{-\alpha-1} \Omega_f(t)\, dt + \| \chi * f \|_\infty$$

(see Lemma 8). If $m_s(g) = \| \chi_s * g \|$ ($g = \mu_t * f$, or f) is replaced by g^{**}, denotes the corresponding space by $B_\alpha^{pq,r} (\subset \tilde{B}_\alpha^{pq,r},\ p > 1)$ (omit q if $p = q$). Note that the relation $B_{n/p}^{p\infty,1} \subset L^\infty$ $(1 < p \leq \infty)$ extends to

$$\tilde{B}_{n/p}^{p\infty,1} \subset L^\infty,\ 1 \leq p \leq \infty. \tag{60}$$

Like the former relation, this follows from the integral representation (20), since $\| (\sigma)_1^* \|_1 < \infty$. An atlas $\mathscr{A} = \{u_i : \in \mathscr{I}\}$, dom $u_i = U_i$, $i \in \mathscr{I}$, of a C^1 manifold M_0 will be called $B((.)\Psi)$ near a compact subset S of M_0 if there is an open neighborhood $\tilde{S}$ of S such that

$$D(u_j \circ u_i^{-1})|_{u_i(\tilde{S} \cap U_i \cap U_j)} \in B(\Psi),\ i,\ j \in \mathscr{I}.$$

Define the Banach function space Ψ_2 on R_+^* by

$$\| \varphi : \Psi_2 \| = \int_0^\infty \sup_{0<s\leq 1} s^{-1} |\varphi(st)| (1 + |\log t|) t^{-2}\, dt$$

and put $\Psi_n = (.)^{n/2} L^1$ if $n \geq 3$.

Proposition 1. Suppose M_0 is an n-dimensional C^1 manifold ($n \geq 1$) with C^2 atlas $\mathscr{A}$, $M = (M_0, \mathscr{A})$, ζ is a C^2 immersion of M into R^{n+1}, the Gaussian curvature of ζ does not vanish on a compact subset S of M, and if $n \geq 2$ then $\mathscr{A}$ and ζ are $B((.)\Psi_n)$ (more precisely $D\zeta$ is $B(\Psi_n)$) near S. Then

$$|\zeta(f)^\wedge(x)| \leq C(M,\ S,\ \zeta)\ \| f : \tilde{\mathscr{B}}_{n/2}^{2\infty,1}(M,\ S) \|\ |x|^{-n/2},\ x \in R^{n+1}. \tag{61}$$

Proof. By compactness of S it suffices to prove (61) under the assumption that M is an open subset of R^n with normal map N which is a diffeomorphism of M onto an open subset of S^n. Suppose $\xi \in R^{n+1}$, $|\xi| = 1$. For $x \in M$ let $A(x)$ denote the symmetric positive definite square root of

$$({}^t D\zeta(x)\, D\zeta(x))^{-1}, \quad (D\zeta = \partial(e_i . \zeta)/\partial x_j : 1 \leq i \leq n+1,\ 1 \leq j \leq n).$$

If ${}^t\xi$ denotes the transpose of the column vector ξ, the projection of ξ onto the tangent space to ζ has transpose ${}^t\xi D\zeta A$, hence $|{}^t\xi D\zeta A|^2 = 1 - (\xi\ .\ N)^2$. It follows that

$$C^{-1} \leq |D\xi\ .\ \zeta|/(|N - \xi| \wedge |N + \xi|) \leq C$$

and then that for $0 < t < 1$, $\xi \in S^n$ the set $\{x \in S : |D\xi\ .\ \zeta| < t\}$ is contained in $C = C(M, S, \zeta)$ balls of radius t (since this is true for ξ in an open neighborhood of $N(S)$).

Suppose $n \geq 2$; the case $n = 1$ is similar (see Remark in Sec. 18) and is well known (see [15, p. 197]). In Lemma 15 let $h(t) = t^2$, hence $\gamma'(t) = 1$ so that (54) is obviously satisfied and δ may be taken to be $|D\xi\ .\ \zeta|$. Regarding (56) for $n \geq 3$, note that if $\| D\zeta : B^1_{n/2} \| \leq C$, then for $x \in S^{1/C}$, $s \leq 1$, $|\Delta^N(sz)\zeta(x)| \leq C\psi(s|z|)$ where ψ is nondecreasing and $\| \psi : (.)^{n/2} L^1 \| \leq 1$, which implies

$$|\Delta^N(sz)\zeta(x)| \leq Cs \sup_{0<\sigma<1} \sigma^{-1}\psi(\sigma|z|) \leq CsH_{1\infty}\psi(|z|),$$

$$\| H_{1\infty}\psi : (.)^{n/2} L^1 \| \leq C \| \psi : (.)^{n/2} L^1 \|.$$

For $\| f : \tilde{\mathscr{B}}^{2\infty,1}_{n/2} \| \leq 1$, it remains to show

$$\int_{|D\xi.\zeta|<\sqrt{t}} d|f|(x) + \sum_{2^j \geq \sqrt{t}} \int_{2^j \leq |D\xi.\zeta| < 2^{j+1}} d|\mu_{t/2^j} * f|(x) \leq Ct^{n/2},\quad 0 < t \leq 1. \tag{62}$$

By (60) the first term on the left of (62) is bounded by $Ct^{n/2}$. The second is at most

$$C \sum_{2^j \geq \sqrt{t}} 2^{jn/2} \Omega_f(t/2^j) \leq Ct^{n/2} \int_0^{\sqrt{t}} \Omega_f(u) u^{-(n+2)/2} \, du \leq Ct^{n/2},$$

which completes the proof of (62).

Proposition 1 is best possible in the sense that in $\tilde{\mathscr{B}}^{2\infty,1}_{n/2}$ the index 1 cannot be raised nor can 2 or $n/2$ be lowered. Suppose $\lambda \in C^{\infty}$, $0 \leq \lambda \leq 1$, $\lambda = 1$ in $B(0, 1/2)$, $\lambda = 0$ outside $B(0, 3/4)$. Let $M = R^n$, $\zeta : x \to (x, |x|^2) \in R^{n+1}$. For the first claim suppose $1 < q \leq \infty$ and consider $f = \lambda |\log|.||^{\varepsilon} \mathscr{L}^n$, $0 < \varepsilon < 1/q'$. It is easy to see that $f \in \mathscr{B}^{pq}_{n/p} \subset \tilde{\mathscr{B}}^{pq}_{n/p}$, $1 < p \leq \infty$, but for $t \geq C_n$

$$|\zeta(f)^{\wedge}(te_{n+1})| \geq C^{-1} \int_{\sqrt{t}|x| \leq 1} |\log|x||^{\varepsilon} \, dx -$$

$$C \int_{1 \leq \sqrt{t}|x| \leq \sqrt{t}/2} (t/|x|)^N (\log|x|)^{\varepsilon-1} |x|^{-N} \, dx \geq C^{-1} (\log t)^{\varepsilon} t^{-n/2}.$$

For the second assertion put $f_b = \lambda e^{ib|.|^2} \mathscr{L}^n$, $b > 1$. Observe that

$$\| |.|^{n/2} \zeta(f_b)^{\wedge} \|_{\infty} \geq b^{n/2} |\zeta(f_b)^{\wedge}(be_{n+1})| = \|\lambda\|_1 \, b^{n/2}.$$

On the other hand, suppose φ is a positive continuous function such that $\lim_{t \to 0} t^{-n/2} \varphi(t) = \infty$, $t^{-\varepsilon}\varphi(t)$, $t^{-r+\varepsilon}\varphi(t)$ are increasing or decreasing respectively for some $r \in Z_+$. Then

$$\int_0^1 \varphi(t)^{-1} \| \mu_t * f_b \|_{\infty} t^{-1} \, dt$$

$$\leq C \int_0^1 ((tb)^r \wedge 1) \varphi(t)^{-1} t^{-1} \, dt \leq C\varphi(b^{-1})^{-1}.$$

It follows that

$$\| f_b : B^1(\varphi) \| \leq C\varphi(b^{-1})^{-1} = o(b^{n/2}), \text{ as } b \to \infty.$$

Thus (61) is not valid if $(.)^{n/2} \mathscr{L}^1$ is replaced by any larger space $\varphi \mathscr{L}^1$ even if A is taken to be L^{∞} instead of $M((.)^{-n/2} L^{\infty}) \subset L^{2\infty}$.

24 ESTIMATE OF $C(M, S, \zeta)$

Suppose $\{U_i : i \in \mathscr{I}\}$ is an open covering of M, π_i is an orthogonal projection p_i of R^{n+1} onto an n-dimensional subspace followed by an isometry of $p_i(R^{n+1})$ onto R^n, $\pi_i \circ \zeta$ is 1-1 on U_i, for $\zeta_i = (\pi_i|_{\zeta U_i})^{-1}$, $u_i = \pi_i \circ \zeta|_{U_i}$. $V_i = u_i(U_i)$, $i \in \mathscr{I}$, $\| D^j\zeta \|_\infty = \sup\{\| D^j\zeta_i \|_\infty : i \in \mathscr{I}\}$, $j = 1, 2$, $k^* = \| D^2\zeta\|_\infty$, $\| D\zeta \|_\infty \leq C$ $\mathscr{A} = \{u_i : i \in \mathscr{I}\}$ has Lebesgue number $\| D\zeta \|_\infty/k^* = \eta$, say, on S (M may be replaced by its submanifold $S^{1/(C\eta)}$), $\mathscr{A}$ satisfies (12) and is obtained by uniform shrinking from an atlas of larger charts as in Sec. 8, with $\varepsilon = 1/(C\eta)$,

$$\left|\!\left|\!\left| \{Du_j \circ u_i^{-1} : i, j \in \mathscr{I}\} : B(\Psi_n) \right|\!\right|\!\right| \leq C, \qquad (n \geq 2)$$

(Λ^1 in place of $B(\Psi_n)$ if $n = 1$), for $k = [n/2] + 1$

$$|\Delta^k(z)\zeta_i(x)| \leq Ck^*\psi(|z|), \; x \in V_i^{-1/(C\eta)}, \; |z| \leq (Ck\eta)^{-1}, \| \psi : \Psi_n \| \leq 1.$$

Supposing the measure Λ on M is uniformly equivalent to $\mathscr{L}^n$ on $\{V_i : i \in \mathscr{I}\}$ ($\Lambda = \Lambda^0$ as in Sec. 9, or $\zeta(\Lambda|_{U_i})$ is n-dimensional Hausdorff measure on $\zeta(U_i)$, $i \in \mathscr{I}$) let $\mathscr{B}(S^n)$ denote the collection of Borel subsets of S^n and σ_n normalized rotation-invariant measure on S^n and define

$$1/K_* = \sup\{\Lambda(N^{-1}(E))/\sigma_n(E) : E \in \mathscr{B}(S^n), \; \sigma_n(E) > 0\}. \tag{63}$$

Also suppose the norm $\| f : \tilde{\mathscr{B}}^{2\infty,1}_{n/2} \|$ is computed according to Sec. 15 with respect to $\mathscr{A}$, Λ. Then (61) is valid with

$$C(M, S, \zeta) = C(k^*)^{n/2}/K_* \tag{64}$$

(without assuming S to be compact but still $\Lambda(M) < \infty$ to avoid complications in defining the Fourier transform).

To see this, take $h(t) = k^*t^2$, hence $\gamma(t) = k^*t$,

$$C^{-1} \leq \delta(x)k^*/|D\beta_i(u_i(x))| \leq C.$$

$h(\delta(u_i^{-1}(x))) < t$ means $|D\beta_i(x)| \leq (k^*t)^{1/2}$ and if F_{ij} is defined by

$$F_{ij} = \{x \in V_i^{-1/(C\eta)},\ 2^{j-1} < \delta(u_i^{-1}(x)) < 2^{j+1}\},$$

$$F_{ij}^{2^j/C} \subset \{x \in V_i : N(x) \in \bigcup_{j=0}^{1} \left(B((-1)^j\xi, C2^jk^*)\setminus B((-1)^j\xi, 2^jk^*/C)\right)\} \tag{65}$$

for $i \in \mathscr{I}$, $2^j \leq (C\eta)^{-1}$. By a basic covering lemma in R^n, F_{ij} is contained in the union of a family of balls of radius $2^j/C$ with centers in F_{ij} any C_n of which have empty intersection, $\{B_{ij} : 1 \leq j \leq \#_{ij}\}$, say. By (12), (63), and (65), $\Sigma\{\#_{ij} : i \in \mathscr{I}\} \leq C(k^*)^n/K_*$, $2^j \leq (C\eta)^{-1}$. Supposing $\| f:\tilde{\mathscr{B}}^{2\infty,1}_{n/2}\| \leq 1$, it follows that (see Lemma 15) for some nondecreasing $\Omega \in (.)^{n/2}\mathcal{L}^1_+$, $\| \Omega : (.)^{n/2}\mathcal{L}^1 \| \leq 1$

$$\begin{aligned}|b(\xi)| &\leq Ct^{-n/2} \sum_{i\in\mathscr{I}} \int_{|D\beta_i(x)|\leq(k^*t)^{1/2}} 1_{V_i}(x)\, d\mathscr{L}^n(x) \\ &\quad + C(k^*)^n/K_* \sum \{\Omega(t/(2^jk^*))(2^j/t)^{n/2} : 2^j \geq (t/k^*)^{1/2}\} \\ &\leq C(k^*)^{n/2}/K_*.\end{aligned}$$

As S shrinks to a point $x \in M$ at which the principal curvature vector of ζ is $k(x)$ and the Gaussian curvature is $K(x)$, it follows that

$$\begin{aligned}C^{-1}|k(x)|^{n/2}/|K(x)| &\leq \liminf_{S\to\{x\}} C(M, S, \zeta) \\ &\leq \limsup_{S\to\{x\}} C(M, S, \zeta) \leq C|k(x)|^{n/2}/|K(x)|.\end{aligned}$$

In view of (1), which yields the well-known asymptotic estimate $|K(x)|^{-1/2}$ (under a stronger smoothness assumption on M, ζ and with a stronger norm on f), the usefulness of (64) seems to be limited to the case when the principal curvatures k_i, $1 \leq i \leq n$, of ζ satisfy $|k_i|/|k_j| \leq C$, $1 \leq i, j \leq n$.

REFERENCES

1. Besov, O. V.; Il'n, V. P.; and Nikolskii, S. M. *Integral Representations of Functions and Imbedding Theorems*, vol. 2. New York: Halsted Press, 1979.

2. Boyd, D. The spectral radius of averaging operators. *Pac. J. Math.* 24 (1968): 19-28.

3. Calderón, A. P. Intermediate spaces and interpolation, the complex method. *Studia Math.* 24 (1964): 113-190.

4. Calderón, A. P., and Torchinsky, A Parabolic maximal functions associated with a distribution. *Adv. in Math.* 16 (1977): 1-63.

5. Coddington, E. A., and Levinson, N. *Theory of Ordinary Differential Equations*. New York: McGraw-Hill, 1955.

6. Greub, W.; Halperin, S.; and Vanstone, R. *Connections, Curvature and Cohomology*, vol. 1. New York: Academic Press, 1972.

7. de Guzmán, M. Differentiation of integrals in R^n. *Springer Lecture Notes* 481 (1975).

8. Herz, C. S. Fourier transforms related to convex sets. *Ann. Math.* 75 (1962): 81-92.

9. Herz, C. S. Lipschitz spaces and Bernstein's theorem on absolutely convergent Fourier transforms. *J. Math. Mech.* 18 (1968): 283-323.

10. Milnor, J. *Morse Theory*. Princeton: Princeton Univ. Press, 1963.

11. Munkres, J. R. *Elementary Differential Topology*. Princeton: Princeton Univ. Press, 1963.

12. Sharpley, R. Spaces $\Lambda_\alpha(X)$ and interpolation. *J. Func. Anal.* 11 (1972): 479-513.

13. Stein, E. M. *Singular Integrals and Differentiability Properties of Functions*. Princeton: Princeton Univ. Press, 1970.

14. Thorpe, J. A. *Elementary Topics in Differential Geometry*. New York: Springer, 1979.

15. Zygmund. A. *Trigonometric Series*, vol. 1. New York: Cambridge Univ. Press, 1959.

PART FOUR

Singular Integrals and Pseudodifferential Operators

A GENERALIZED HERGLOTZ-BOCHNER THEOREM AND L^2-WEIGHTED INEQUALITIES WITH FINITE MEASURES

Rodrigo Arocena
Mischa Cotlar
Universidad Central de Venezuela

INTRODUCTION

In previous papers [7, 5, 4, 3], a generalized Herglotz theorem for finite measures in T and a (partially) generalized Herglotz-Schwartz theorem for nonfinite tempered measures in **R** were studied, and it was shown that these theorems furnish a unified approach to different prediction, interpolation, and L^2-weighted problems. However the important case of finite measures in **R** (the Bochner case), which is essential in some L^2-weighted problems related to the deep work of P. Koosis [10, 8], was not considered there. Also, the proofs given there, based on analytic functions, are not well-suited for some further developments. Here we establish the generalized Herglotz-Bochner theorem for finite measures, in T and in **R**, with a direct proof that doesn't use analytic functions and provides a better unicity condition. Then we show that this theorem and unicity condition lead to natural proofs of Koosis' theorems. The above results can be considered as generalizations of some results of Nehari and Adamian, Arov, and Krein [1] (and also can be interpreted in terms of "generalized stationary stochastic processes," see [6]). These considerations make possible some further developments. In particular they apply to distribution kernels and to vector-valued measures, leading to extensions of the dilation and commutation theorems of Nagy and Naimark and Sarason, Nagy, and Fojas, and to generalized Weyl relations and vectorial prediction. The lack of space doesn't allow us to include here these questions, which are detailed in [2].

EXTENSION OF THE HERGLOTZ-BOCHNER THEOREM

Consider first the case of the circle T and fix the following notations. Let $e_n(t) = \exp(int)$, $\hat{\mu}(n) = \mu(\bar{e}_n)$ (if μ is a measure in T), P = {the trigonometric polynomials $f(t) = \Sigma\, \hat{f}(n)e_n(t)$, $\hat{f}$ of finite support}, $P_1 = \{f \in P : \hat{f}(n) = 0\ \forall n < 0\}$, $P_2 = \{f \in P : \hat{f}(n) = 0\ \forall n \geq 0\}$ so that $P = P_1 + P_2$, $\mathbf{Z}_1 = \{m \in \mathbf{Z} : m \geq 0\}$, $\mathbf{Z}_2 = \{m \in \mathbf{Z} : m < 0\}$, and let

$$Ve_n = e_{n+1}, \qquad n \in \mathbf{Z}, \tag{1}$$

be the shift operator that induces an algebraic isomorphism of P onto P. Let $K(j, n)$ be a hermitean kernel in $\mathbf{Z} \times \mathbf{Z}$, that is, a function in $(j, n) \in \mathbf{Z} \times \mathbf{Z}$ with $K(j, n) = \overline{K(n, j)}$. If K is positive definite, p.d., that is, if $\Sigma_{jn}\, K(j, n)\, c_j \bar{c}_n \geq 0$ whenever (c_j) is of finite support, then we may associate with K a (eventually degenerated) scalar product in P and a corresponding Hilbert space $H \cong \bar{P}$, by setting $(e_j, e_n) = K(j, n)$. (2)

K is said to be a Toeplitz kernel if $K(j + 1, n + 1) = K(j, n)$, (3)

$\forall (j, n) \in \mathbf{Z} \times \mathbf{Z}$, or equivalently if there is a sequence $K(n)$ such that $K(j, n) = K(j - n)$, and such $K(n)$ is unique. If K is a p.d. Toeplitz kernel, then by (3), the shift V induces a unitary operator in H.

We say that K is a *generalized Toeplitz kernel*, GTK, if (3) holds only if $j \neq -1$, $n \neq -1$, or equivalently if there are four sequences $K_{\alpha\beta}(n)$, $\alpha, \beta = 1, 2$, such that $K(j, n) = K_{\alpha\beta}(j - n)$ if $j \in Z_\alpha$, $n \in Z_\beta$. (4)

In this case we write $K(j, n) \sim (K_{\alpha\beta}(n))$, and now only K_{11} and K_{22} are well determined while $K_{12}(n) = \overline{K_{21}(-n)}$ is determined only for $n > 0$. Now we cannot say (if K is p.d.) that V is unitary in H, but V still induces an isometric operator $V : H_{-1} \to H_0$, with domain H_{-1} = {the subspace spanned by the e_n, $n \neq -1$}, and range H_0 = {the subspace spanned by the e_n, $n \neq 0$}. (5)

Most GTK occurring in applications have the additional property $K_{\alpha\beta}(n) = \hat{\mu}_{\alpha\beta}(n)$, $\forall n \in \mathbf{Z}$, $\alpha, \beta = 1/2$, where $\mu_{\alpha\beta}$ are finite complex measures in $\mathbf{T}$. In this case we write $K \sim (\mu_{\alpha\beta})\hat{}$. (6)

If $K \sim (\mu_{\alpha\beta})\hat{}$ and $K \sim (\nu_{\alpha\beta})\hat{}$, then we write $(\mu_{\alpha\beta}) \sim (\nu_{\alpha\beta})$. (6a)

By the above remark (6a) holds iff $\mu_{11} = \nu_{11}$, $\mu_{22} = \nu_{22}$, and $\hat{\mu}_{12}(n) = \hat{\nu}_{12}(n)$ for $n < 0$, so that $\nu_{12} = \mu_{12} + hdt$, $h \in H^1(\mathbf{T})$. (6b)

Clearly for any four measures $(\mu_{\alpha\beta})$ there is a GTK K satisfying (6). In this case, and if K is p.d., then $(f_1 + f_2, f_1 + f_2) = \mu(f_1, f_2) \equiv \Sigma_{\alpha\beta} \int f_\alpha \bar{f}_\beta \, d\mu_{\alpha\beta}$, (7)

whenever $(f_1, f_2) \in P_1 \times P_2$, but not for arbitrary $(f_1, f_2) \in P \times P$. Thus, if (6) holds, then K is p.d. iff the sum in [7] is ≥ 0, $\forall$ $(f_1, f_2) \in P_1 \times P_2$. In this case we say that the matrix $(\mu_{\alpha\beta})$ is *weakly positive*, $(\mu_{\alpha\beta}) \rangle 0$. Instead $(\mu_{\alpha\beta})$ is said to be positive, $(\mu_{\alpha\beta}) \geq 0$, if for every $A \subset T$ is $(\mu_{\alpha\beta}(A))$ a p.d. numerical matrix, or equivalently if $\mu(f_1, f_2) \geq 0$ for any $(f_1, f_2) \in P \times P$. (7a)

Many GTK occurring in applications satisfy $K_{11} = K_{22}$, where $K \sim (K_{\alpha\beta})$. In this case K is p.d. iff

$$\sigma = \inf\{K(0, 0) + \sum_{j,n\neq 0} c_j \bar{c}_n K(j, n) - 2Re \sum_{j\neq 0} c_j K(j, 0)\} \geq 0, \quad (8)$$

where the inf is taken over all sequences $\{c_j\}$, $j \in Z - \{0\}$, of finite support, and $\sigma > 0$ iff $e_0 \notin H_0$. Moreover in this case, setting $T(ae_j) = \bar{a}\, e_{-j-1}$, T induces an antilinear isometry of H_0 onto H_{-1} with $Te_0 = e_{-1}$; hence $\text{dist}(e_{-1}, H_{-1}) = \text{dist}(e_0, H_0)$. (8a)

The classical Herglotz theorem says that a Toeplitz kernel K is p.d. iff it has a representation $K(j, n) = \hat{\mu}(j - n)$,

$\forall (j, n) \in Z \times Z$, where $\mu \geq 0$ is a positive measure in T, and such μ is unique. The corresponding theorem for the GTK is the following.

Theorem 1. Let $K(j, n)$ be a GTK. Then K is p.d. iff $K \sim (\mu_{\alpha\beta})\hat{}$ for some positive matrix measure $(\mu_{\alpha\beta}) \geq 0$.

Proof. Let K be p.d. and let H be the associated Hilbert space and $V : H_{-1} \to H_0$ the isometry defined in (5). Then there is a Hilbert space $H' \supset H$ and a unitary operator U in H' such that $Uf = Vf$ if $f \in H_{-1}$ and $U^{-1}f = V^{-1}f$ if $f \in H_0$. Each of the following sequences: $K_{11}(n) = (U^n e_0, e_0)$, $K_{22}(n) = (U^n e_{-1}, e_{-1})$, $F(n) = (U^n(e_0 + e_{-1}), (e_0 + e_{-1}))$, $G(n) = (U^n(e_0 + ie_{-1}), (e_0 + ie_{-1}))$, defines a p.d. Toeplitz kernel; hence by the classical Herglotz theorem there exist positive measures μ_{11}, μ_{22}, λ, η such that $\hat{\mu}_{11}(-n) = K_{11}(n)$, $\hat{\nu}_{22}(-n) = K_{22}(n)$, $\hat{\lambda}(-n) = F(n)$, $\hat{\eta}(-n) = G(n)$, and there is a complex measure μ such that $\hat{\mu}(-n) = (U^n e_0, e_1)$. Setting $d\mu_{12}(t) = e^{-it}d\mu(t)$, $\mu_{21} = \bar{\mu}_{12}$, we shall have $K(j, n) \sim (\mu_{\alpha\beta})\hat{}$, since $K(j, n) = (e_j, e_n) = (U^j e_0, U^n e_0) = (U^{j-n}e_0, e_0) = K_{11}(j - n)$ if $j \in Z_1$, $n \in Z_1$, and similarly in the remaining cases. Moreover, by (7a), $(\mu_{\alpha\beta}) \geq 0$. In fact, if $f_1(t) = \Sigma\, c_j e_j(t) \in P$, $f_2 = \Sigma\, d_j e_j(t) \in P$, then $\Sigma_{\alpha\beta} \int f_\alpha \bar{f}_\beta\, d\mu_{\alpha\beta} = \Sigma_{jn}[c_j\bar{c}_n(U^{j-n}e_0, e_0) + c_j\bar{d}_n(U^{j-n}e_0, Ue_{-1}) + d_j\bar{c}_n(U^{j-n}Ue_{-1}, e_0) + d_j\bar{d}_n(U^{j-n}Ue_{-1}, Ue_{-1})] = \| \Sigma\, c_j U^j e_0 + \Sigma\, d_j U^j U e_{-1} \|^2 \geq 0$. Conversely, if $K \sim (\mu_{\alpha\beta})\hat{}$ with $(\mu_{\alpha\beta}) \geq 0$, then by (7), K is p.d.

QED

If K is a p.d. GTK, then the construction in the above proof gives all the $(\mu_{\alpha\beta})$ satisfying $K \sim (\mu_{\alpha\beta})\hat{}$ with $(\mu_{\alpha\beta}) \geq 0$. (9)

In fact, $P \times P$ is generated by the elements $e_n^1 = \{e_n, 0\}$ and $e_m^2 = \{0, e_m\}$, and if (9) holds, then setting $\langle e_n^\alpha, e_m^\beta \rangle = \hat{\gamma}_{\alpha\beta}(n - m)$, $\alpha, \beta = 1, 2$, one obtains a "scalar product" in $P \times P$ and a corresponding Hilbert space $H' \approx \overline{P \times P}$, and the shift $Ue_n^{\alpha} = e_{n+1}^{\alpha}$ induces a unitary operator in H'. Identifying e_n with e_n^2 if $n < 0$, and with e_n^1 if $n \geq 0$, so that $(e_n, e_m) = K(n, m) = \hat{\mu}_{\alpha\beta}(n - m) = \langle e_n^{\alpha}, e_m^{\beta} \rangle$ if $e_n \in Z_{\alpha}$, $e_m \in Z_{\beta}$, we obtain that $H \subset H'$ and that $Uf = Vf$ for $f \in H_{-1}$.

Now if $e_{-1} \in H_{-1}$ (or if $e_0 \in H_0$), then the domain of V is H and the construction in the above proof gives that $\hat{\mu}_{12}(n) = (e_0, V^{n+1} e_{-1})$ is well determined for all $n \geq 0$ and, by (6b), $(\mu_{\alpha\beta})$ is well determined. Conversely, if $e_{-1} \notin H_{-1}$, $e_0 \notin H_0$, since $\hat{\mu}_{12}(0) = (e_0, U_{e-1})$, we shall obtain, for convenient choices of U, different values for $\hat{\mu}_{12}(0)$. Thus we obtain the following unicity condition.

Corollary 1. Let K be a p.d. GTK. Then there is only one matrix $(\mu_{\alpha\beta})$ satisfying (9) iff one of the following conditions is true: (i) $e_0 \in H_0$, or (ii) $e_{-1} \in H_{-1}$. If the kernel satisfies $K_{11} = K_{22}$, then (by (8a)) these two conditions are equivalent.

From (6b) it is clear that Theorem 1 can be restated as a "lifting theorem" as follows:

Corollary 2. If $(\mu_{\alpha\beta}) > 0$ is a weakly positive matrix measure in **T**, *then there exists an analytic function $h \in H^1(\mathbf{T})$ such that setting $\nu_{11} = \mu_{11}$, $\nu_{22} = \mu_{22}$, $\nu_{12} = \mu_{12} + hdt$ we shall have*

$$(\mu_{\alpha\beta}) \sim (\nu_{\alpha\beta}) \quad \text{and} \quad (\nu_{\alpha\beta}) \geq 0. \tag{10}$$

We now extend the above considerations to **R**. Let $R_1 = \{x \in \mathbf{R},\ x > 0\}$, $\mathbf{R}_2 = \{x < 0\}$, $R' = \mathbf{R}_1 \cup \mathbf{R}_2$, $\mathbf{R}_{\alpha\beta} = \mathbf{R}_{\alpha} \times \mathbf{R}_{\beta}$, $\alpha, \beta = 1, 2$. We say that $K(x, y)$ is a continuous generalized Toeplitz kernel, or simply a generalized Toeplitz kernel in **R**, GTK, if K is defined and continuous in $\mathbf{R}' \times \mathbf{R}'$ and if $K(x, y) = K(x + t, y + t)$ whenever (x, y) and $(x + t, y + t)$ belong to the same set $\mathbf{R}_{\alpha\beta}$, (11) or equivalently if there are four continuous functions $K_{\alpha\beta}(x)$, in **R**, with $K(x - y) = K_{\alpha\beta}(x - y)$, $\forall (x, y) \in \mathbf{R}_{\alpha\beta}$, $\alpha, \beta = 1, 2$. (11a)

In general, such a K cannot be continuous at the points $(0, x)$ or $(x, 0)$, but we may always assume that K is defined in $\mathbf{R} \times \mathbf{R}$ and continuous in $R' \times R'$ and let's say in $\bar{\mathbf{R}}_{11}$ and $\bar{\mathbf{R}}_{22} - \{0\}$. Such a K is then said to be positive definite, p.d., if $\Sigma_{x,y} K(x, y)c(x)\overline{c(y)} \geq 0$ whenever $c(x)$ has finite support.

Let K be a p.d. GTK, and for each positive integer N let K^N be the kernel in $\mathbf{Z} \times \mathbf{Z}$ defined by $K^N(m, n) = K(m/N, n/N)$. K^N is a p.d. GTK and by Theorem 1 $K^N_{\alpha\beta}(m, n) = \hat{\nu}^N_{\alpha\beta}(n - m)$, where $(\nu^N_{\alpha\beta}) \geq 0$, and also $\| \nu^N_{\alpha\beta} \| \leq (K_{11}(0)K_{22}(0))^{1/2}$. Hence, for every $(m, n) \in Z_\alpha \times Z_\beta$,

$$K_{\alpha\beta}(m/N, n/N) = \int_{-\Pi}^{\Pi} e^{-i(n-m)t} d\nu^N_{\alpha\beta}(t) = \int_{-N\Pi}^{N\Pi} e^{-i\frac{n-m}{N}y} d\lambda^N_{\alpha\beta}(Y),$$

where $(\lambda^N_{\alpha\beta}) \geq 0$ is a positive matrix measure with support in $(-N\Pi, N\Pi) \subset \mathbf{R}$ and with $\| \lambda^N_{\alpha\beta} \| \leq (K_{11}(0)K_{22}(0))^{1/2}$, $\forall N$. Therefore, there exists $(\nu_{\alpha\beta}) \geq 0$ such that $\nu_{\alpha\beta} = \lim_r \lambda^{Nr}_{\alpha\beta}$, in the weak topology, so that $K_{\alpha\beta}(t, s) = \hat{\nu}_{\alpha\beta}(s - t)$, for every pair of rational numbers $(t, s) \in R_\alpha \times \mathbf{R}_\beta$. Thus we have proved the following.

Theorem 2. (Generalized Bochner theorem). *If $K(x, y)$ is a positive definite continuous generalized Toeplitz kernel in* $\mathbf{R}$, *then there exists a positive matrix* $(\nu_{\alpha\beta}) \geq 0$ *of finite measures in* $\mathbf{R}$ *such that* $K \sim (\nu_{\alpha\beta})^\wedge$, *that is,* $K(x, y) = \hat{\nu}_{\alpha\beta}(y - x)$, $\forall (x, y) \in \mathbf{R}_{\alpha\beta}$, $\alpha, \beta = 1, 2$. (12)

Here, as before, $(\nu_{\alpha\beta}) \geq 0$ means $(\nu_{\alpha\beta}(A)) \geq 0$, $\forall A \subset R$. We say that the matrix $(\mu_{\alpha\beta})$ of finite measures is *weakly positive*,

$$(\mu_{\alpha\beta}) \succ 0, \text{ if } \sum_{\alpha,\beta=1,2} \int_R f_\alpha \bar{f}_\beta \, d\mu_{\alpha\beta} \geq 0 \tag{13}$$

for every $f_1 \in E_1$, $f_2 \in E_2$, where E_α is the vector space spanned by the functions $e_t(x) = e^{itx}$ with $t \in \mathbf{R}_\alpha$, $\alpha = 1, 2$. (14)

We write $(\nu_{\alpha\beta}) \sim (\mu_{\alpha\beta})$ if $\Sigma_{\alpha\beta} \int f_\alpha \bar{f}_\beta d\nu_{\alpha\beta} = \Sigma_{\alpha\beta} \int f_\alpha \bar{f}_\beta d\nu_{\alpha\beta}$ for every $(f_1, f_2) \in E_1 \times E_2$. Theorem 2 can then be restated as:

Corollary 3. If $(\mu_{\alpha\beta}) \rangle 0$ is a weakly positive matrix in **R**, *then there exists a function $h \in H^1(\mathbf{R})$ such that setting $\nu_{\alpha\alpha} = \mu_{\alpha\alpha}$ if $\alpha = 1, 2$, and $\nu_{12} = \mu_{12} + h\,dx$, $\nu_{21} = \mu_{21} + \bar{h}\,dx$, we have $(\nu_{\alpha\beta}) \geq 0$ and $(\nu_{\alpha\beta}) \sim (\mu_{\alpha\beta})$. Let* $e^{it} = (x - i)(x + i)^{-1}$ (15)

be the 1-1 correspondence between $\mathbf{R} = \{x, \ldots\}$ and $\mathbf{T} = \{e_1(t)\}$, which induces a correspondence between sets: $\mathbf{R}\cup\{\infty\} \supset A \leftrightarrow A' \subset T$, between measures in $\mathbf{R}\cup\{\infty\}$ and in **T**, *$\mu \leftrightarrow \mu'$, $\mu(A) = \mu'(A')$, and between functions: $w(x) = w'(e^{it})$.* (16)

Let K and $(\nu_{\alpha\beta}) \geq 0$ be as in Theorem 2; then $K' \sim (\nu'_{\alpha\beta})^\wedge$ is a p.d. GTK in $Z \times Z$ and

$$K'(m, n) = \int_T e^{-i(n-m)t} d\nu'_{\alpha\beta}(t) = \int_{\mathbf{R}} \left(\frac{x - i}{x + 1}\right)^{m-n} d\nu_{\alpha\beta}(x), \quad (17)$$

$\forall (m, n) \in Z_\alpha \times Z_\beta$. If also $K \sim (\mu_{\alpha\beta})^\wedge$, then $\mu_{\alpha\alpha} = \nu_{\alpha\alpha}$, $\mu_{12} = \nu_{12} + hdt$, $h \in H^1(\mathbf{R})$, and $\int \left((x - i)(x + i)^{-1}\right)^{m-n} h(x)\,dx = 0$. Hence from (17) it follows that the correspondence $K \leftrightarrow K'$ doesn't depend on the choice of $(\nu_{\alpha\beta})$. Thus, the problem of unicity of the representation (12), in **R**, *is reduced to the corresponding problem in* **T**; *hence:*

Corollary 4. Let K be a p.d. GTK in **R** *and K' defined by (17). Then the representation (12) is unique iff the corresponding representation for K' is unique, that is, iff K' satisfies the conditions of Corollary 1.*

Similarly is proved the following variant of Theorem 2, which can be interpreted in terms of "generalized stationary processes" [6]:

Corollary 5. Let $\sigma \geq 0$, $\mathbf{R}^\sigma_\alpha = \{x \in \mathbf{R}_\alpha : |x| > \sigma\}$, $\alpha = 1, 2$, and let K be a continuous kernel in **R** *such that $K(x, y) = K(x + t, y + t)$ whenever (x, y) and $(x + t, y + t)$ belong to the same set $\mathbf{R}^\sigma_\alpha \times \mathbf{R}^\sigma_\beta$.*

If K is p.d., then there exists a positive matrix $(\nu_{\alpha\beta})$ *in* **R** *such that* $K(x, y) = \hat{\nu}_{\alpha\beta}(y - x)$, $\forall (x, y) \in \mathbf{R}^{\sigma}_{\alpha} \times \mathbf{R}^{\sigma}_{\beta}$.

The proof is obtained by applying Theorem 2 to the kernel $L(x, y) = K(x + \sigma(sgx), y + \sigma(sgy))$.

APPLICATION TO L²-WEIGHTED INEQUALITIES

We now show that the generalized Herglotz-Bochner theorems, with their unicity conditions, lead to natural proofs of some results of P. Koosis. In applying Theorems 1 and 2, we shall use the following remark. If $d\mu_{\alpha\beta} = w_{\alpha\beta}(x)\,dx$, then $(\mu_{\alpha\beta}(A)) \geq 0$, $\forall A$, is equivalent to $(w_{\alpha\beta}(x)) \geq 0$, $\forall$ a.e. x, and hence to $w_1(x) \geq 0$, $w_{22}(x) \geq 0$ and $|w_{12}(x)|^2 \leq w_{11}(x)w_{22}(x)$. (18)

Koosis Theorem in **T**

Let Hf be the Hilbert transform: $He_n = -i(sgn)e_n$, $(sgn = (-1)^{\alpha}$ if $n \in Z_{\alpha})$, and let $w \in L^1(\mathbf{T})$, $w(t) \geq 0$, $w \not\equiv 0$. We say that w is a Koosis weight if there exists a nonzero function $q \in L^1(\mathbf{T})$, $q \geq 0$, such that $\int_T |Hf|^2\, qdt \leq \int_T |f|^2\, wdt$, $\forall f \in P$. (19)

Setting $\mu_{\alpha\beta} = (w - (-1)^{\alpha+\beta} q)\, dt$, (19) can be rewritten as $(\mu_{\alpha\beta}) \rangle\ 0$, $\alpha, \beta = 1, 2$. Hence by Corollary 2 and (18), (19) is equivalent to $w \geq q$ a.e., and there exists $h \in H^1(\mathbf{T})$, $h \not\equiv 0$, such that

$$(w - q)^2 - |w + q + h|^2 \geq 0 \text{ a.e.} \tag{20}$$

Remark 1. If we replace in (19) qdt and wdt by arbitrary positive measures ν and μ, then we shall obtain similarly that ν is absolutely continuous and that (20) holds with $q = d\nu/dt$, $w = d\mu/dt$. Letting $h = h_1 + ih_2$, $h_1 = Reh$, (20) is rewritten as

$$-2(h_1 + 2w)q - (|h|^2 + 2wh_1) \geq 0 \text{ a.e.} \tag{20a}$$

Let $A = \{h_1 + 2w \leq 0\}$, $q_1 = q1_{T-A} + w1_A$, so that $w \geq q_1 \geq q$, and (20a), or (20), holds if we replace q by q_1, and we get that $|h|^2 + 2wh_1 = 0$ in A and $|h|^2 + 2wh_1 < 0$ in $\mathbf{T} - A$. Hence

$$|h|^2 + 2wh_1 \leq 0 \text{ a.e.} \tag{20b}$$

Conversely, if (20b) holds with a nonzero h, then taking $0 \leq q \leq [-(|h|^2 + 2wh_1)/2(h_1 + 2w)] \wedge w$ (and replacing h by $h/2$, if necessary), we get (20a). Thus we have proved the following.

Theorem 3. $w \geq 0$ *is a Koosis weight iff there exists a nonzero function* $h \in H^1(T)$ *such that* (20b) *holds, or equivalently,*

$$|w + h| \leq w \text{ a.e.} \tag{21}$$

Letting $\mu_{\alpha\beta} = wdt$, $\alpha, \beta = 1, 2$, $\nu_{11} = \nu_{22} = wdt$, $\nu_{12} = (w + h)\,dt$, we have $(\mu_{\alpha\beta}) \sim (\nu_{\alpha\beta})$, $(\mu_{\alpha\beta}) \geq 0$, and if (21) holds, then also $(\nu_{\alpha\beta}) \geq 0$ (see (18)). Hence (21) means that the p.d. kernel $K(m, n) = \hat{w}(m - n)$ has two representations of the form (9), and by Corollary 1 there exists a nonzero element $g \in H \approx L^2(T, wdt)$ orthogonal to H_0; hence gw = constant. We thus obtain the following result of Koosis [9]:

Theorem 4. *Let* $0 \leq w \in L^1(T)$. *Then* w *is a Koosis weight iff* $w^{-1} \in L^1(T)$.

Let $W = \{w \geq 0 : w, w^{-1} \in L^1(T)\}$, so that $w \in W$ implies that $w = |\phi|$ with $\phi \in H^1$ an exterior function. Letting $h = -\phi\psi$, we get from (21) that:

Corollary 6. *Let* $0 \leq w \in L^1(\mathbf{T})$. *Then* $w \in W$ *iff* $\log w \in L^1$ and $\exists \psi \in H^\infty$, $\psi \neq 0$, *with* $\| \psi - |\phi|/\phi \|_\infty \leq 1$.

Let $V_w = \{h \in H^1;\ h \neq 0,\ |h + w| \leq w \text{ a.e.}\}$. The proof of Corollary 6 gives a bijection $\psi \to h = -\phi\psi$ between V_w and the set of best analytic approximations of $|\phi|/\phi$, which was parametrized in Adamian, Arov, and Krein [1]. On the other hand, if $Q_w = \{q \geq 0 : 0 > \int q\,dt < \infty$, and q satisfies (19)$\}$, then we saw that a nonzero q belongs to Qw iff

$$0 \leq q \leq [2w|h_1| - |h|^2][2(2w - |h_1|)]^{-1}, \; h \in V_w. \tag{22}$$

Thus the Adamian-Arov-Krein parametrization describes the set Q_w. If we set $Q = \cup\{Q_w : w \in W\}$ = {nonzero $q \geq 0$ satisfying (19) for some w}, then from the above considerations it follows that a nonzero $q \geq 0$ belongs to Q iff $\exists f \geq q$ with $|Hf|^2/|f| \in L^1$. (23)

Koosis Theorems in $\mathbf{R}$

Let us fix $\sigma > 0$, and let $0 \leq w \in L^1(\mathbf{R})$. We write $w \in W_\sigma$ if there exists a nonzero function $q \geq 0$ such that

$$\int_{\mathbf{R}} |Hf|^2 q \, dx \leq \int_{\mathbf{R}} |f|^2 w \, dx \tag{24}$$

for every trigonometric polynomial of the form $f = \Sigma_{|\lambda|>\sigma} c_\lambda e_\lambda$. In this case we write $q \in Q_{w,\sigma}$. As in the case of $\mathbf{T}$ we see that (24) is equivalent to $(\mu_{\alpha\beta}) \rangle 0$, with $\mu_{11} = \mu_{22} - (w - q)\, dx$, $\mu_{12} = e_{2\sigma}(w + q)\, dx$, and that Corollary 3 gives the following result: $w \in W_\sigma$ iff there exists a nonzero function $h \in H'(\mathbf{R})$ such that

$$|e_{2\sigma} w + h| + |w + e_{-2\sigma} h| \leq w \text{ a.e.} \tag{25}$$

In this case we write $e_{-2\sigma} h \in V_{w,\sigma}$. Moreover, a nonzero q satisfies

$$q \in Q_{w,\sigma} \text{ iff } 0 \leq q \leq \frac{2w|\mathrm{Re}\, g| - |g|^2}{2(2w - |\mathrm{Re}\, g|)}, \; g \in V_{w,\sigma}, \tag{26}$$

(if $2w - |\mathrm{Re}\, g| = 0$, then (26) is understood as $0 \leq q \leq w$). And since $g \in V_{w,\sigma}$ and $\rho \in (0, 1)$ implies $g_1 \equiv \rho g \in V_{w,\sigma}$ and $2w|\mathrm{Re}\, g_1| - |g_1|^2 \geq \rho(1 - \rho)|g|^2$, replacing g by g_1 in (26) we get a $q \in Q_{w,\sigma}$ with $[\log q(x)](1 + x^2)^{-1} \in L^1(\mathbf{R})$. We thus get the following result of P. Koosis [10].

Theorem 5. Let $0 \leq w \in L^1(\mathbf{R})$. *Then* $w \in W_\sigma$ *iff:* (i) $\log w(x)/(1 + x^2) \in L^1(\mathbf{R})$, *so that* $w = |\phi|$, $\phi \in H^1(\mathbf{R})$ *an exterior function, and* (ii) *there is a nonzero function* $\psi \in H(\mathbf{R})$ *such that* $|e_{2\sigma}|\phi|/\phi - \psi| \leq 1$ *a.e. Moreover,* q *can be chosen so that* $\frac{\log q(x)}{1 + x^2} \in L^1(R)$.

We now prove, by the same method, the second result of Koosis [10, 8].

Theorem 6. Let $0 \leq w \in L^1(\mathbf{R})$. *Then* $w \in W_\sigma$ *iff there is a nonzero entire function of exponential type* σ *such that*

$$\int_{\mathbf{R}} \frac{|\phi(x)|^2}{w(x)} \frac{dx}{1+x^2} < \infty. \tag{27}$$

Proof. To prove that (27) implies $w \in W$, we use, as Koosis does, an idea from [1]. (27) implies that $h_1 \equiv -e_{2\sigma}|\phi|^2 \in H^1(\mathbf{R}, dx/1+x^2)$ and that $|\phi|^2 = wf$ with $0 \leq f \in L^1(\mathbf{R}, (1+x^2)^{-1}\,dx)$; hence $\exists \psi \in H^\infty$ with $|1-\psi f| \leq 1$ a.e. Letting $h = \psi h_1$, we get $|h| \leq 2w$, $h \in H^1$, $h \not\equiv 0$, and $|e_{2\sigma}w + h| \leq w$ a.e., so that (25) holds.

Conversely, if $w \in W$, then we pass to T, as in (16). Then there exists $h' \in H^1(T)$, $h' \neq 0$, with $|e'_{2\sigma}w' + h'| \leq w'$ a.e. (with the notations of (16)). As in Theorem 4, the last inequality is interpreted as the nonunicity condition of Corollary 1 for the kernel

$$K \sim \begin{pmatrix} w' & e'_{2\sigma}w' \\ e'_{-2\sigma}w' & w' \end{pmatrix}^{\wedge}.$$

It is clear that the Hilbert space H associated with this K can be identified with a subspace of $L^2(w'dt)$, through the isomorphism U defined by $Ue'_n = e'_n\phi$ if $n \geq 0$ and $Ue'_n = e'_n\bar{\phi}$ if $n < 0$, where $\phi = e'_\sigma$. By the nonunicity there exists $v' \in H$, $v' \neq 0$, such that $\langle v', e'_n \rangle_k = 0$ for $n \neq 0$, so that $g = Uv' \in L^2(w'dt)$ satisfies $(g\bar{\phi}w')^\wedge(n) = 0$ for $n > 0$; hence $g\bar{\phi}w' = h_1$, $h_1 \in H^1$, and $g\phi w' = h_2 \in H^1(T)$. Let $G^1 = g + \bar{g}$. Then G' is real, $G' \neq 0$, and $w'G' = e'_{-\sigma}f'$ with $f' \in H^1(T)$. Passing back to $\mathbf{R}$, let $\phi = wG = e_{-\sigma}f$, $f \in H^1(\mathbf{R}, dx/1+x^2)$. It is clear that ϕ is an entire function and, being real with $e_\sigma\phi = f$, it is of exponential type ϕ. Moreover,

$$\int_{\mathbf{R}} \frac{|\phi(x)|^2}{w(x)} \frac{dx}{1+x^2} = \int_{\mathbf{R}} |G(x)|^2 w(x) \frac{dx}{1+x^2} = \int_{\mathbf{T}} |G'(t)|^2 w(t)\, dt < \infty.$$

QED

REFERENCES

1. Adamian, V.; Arov, D.; and Krein, M. Infinite Hankel matrices. *Funct. An. and Appl.* 2 (1968): 1-19.

2. Arocena, R., and Cotlar, M. Dilation of generalized Toeplitz kernels and L^2-weighted problems. To appear in *Proc. Symp. Harmonic Analysis*, held at the Univ. of Minnesota, 1981.

3. Arocena, R., and Cotlar, M. Generalized Toeplitz kernels and Adamian-Arov-Krein moment problems. To appear in *Integral Equations and Operator Theory*.

4. Arocena, R., and Cotlar, M. On a lifting theorem and some approximation problems. To appear in *Proc. International Seminar on Functional Analysis* held in Rio de Janeiro, 1980.

5. Arocena, R.; Cotlar, M.; and Sadosky, C. Lifting properties and L^2-weighted inequalities. To appear in *Adv. Math*, special volume dedicated to L. Schwartz.

6. Arocena, R., and León, J. On a generalization of the stationary processes. Preprint.

7. Cotlar, M., and Sadosky, C. On the Helson-Szegö theorem and modified Toeplitz kernels. *Proc. Symp. Pure Math. AMS* 35:I (1979): 383-407.

8. Koosis, P., Moyennes quadratiques de transformées de Hilbert et fonctions de type exponentiel. *C. R. Acad. Sc. Paris* 276 (1973): A-1201-1204.

9. Koosis, P. Moyennes quadratiques pondérées de fonctions périodiques et leurs conjuguées harmoniques. *C. R. Acad. Sc. Paris* 291 (1980): A-255-257.

10. Koosis, P. Weighted quadratic means of Hilbert transforms. *Duke Math. J.* 38 (1971): 609-634.

A GEOMETRIC CONDITION THAT IMPLIES THE EXISTENCE OF CERTAIN SINGULAR INTEGRALS OF BANACH-SPACE-VALUED FUNCTIONS

D.L. Burkholder
University of Illinois

Consider first the special case of the Hilbert transform

$$Hf(x) = \frac{1}{\pi}\int_{-\infty}^{\infty} \frac{f(t)}{x - t}\,dt.$$

If $1 \leq p < \infty$ and $f \in L^p(\mathbf{R})$, then for almost all x, the above integral exists in the principal value sense, satisfies the M. Riesz inequality

$$\| Hf \|_p \leq c_p \| f \|_p, \quad 1 < p < \infty, \tag{1}$$

and has many other remarkable properties.

On the other hand, much less is known, even though the question is quite old, about what happens if the scalar-valued function f is replaced by a function with values in a higher-dimensional Banach space. To be specific, let B be a real or complex Banach space with norm $|\cdot|$ and $L^p_B(\mathbf{R})$ the space of strongly measurable functions f on $\mathbf{R}$ with values in B such that $\| f \|_p^p = \int_{\mathbf{R}} |f(t)|^p\,dt$ is finite. Here $1 \leq p < \infty$ as above. If $f \in L^p_B(\mathbf{R})$ and

---

This research has been supported in part by a grant from the National Science Foundation.

$\varepsilon > 0$, then the truncated Hilbert transform of f, defined by

$$H_\varepsilon f(x) = \frac{1}{\pi} \int_{|t|>\varepsilon} \frac{f(x - t)}{t} \, dt, \tag{2}$$

is both strongly measurable and bounded. The question is: For what Banach spaces B does the limit

$$Hf = \lim_{\varepsilon \to 0} H_\varepsilon f$$

exist almost everywhere (or in some other reasonable sense) and satisfy the M. Riesz inequality? (Consider $B = \ell^1$ to see where some of the difficulties arise.)

The analogous question is particularly interesting for the more general singular integrals studied by Calderón and Zygmund [8] and underlies, at least implicitly, some of the work of Schwartz [15], Benedek, Calderón, and Panzone [2], Stein [16], Vági [17], and others.

Before pursuing this question, let us mention another that has also been of interest in recent years but otherwise may seem quite unrelated. Suppose that B is a real or complex Banach space and $1 < p < \infty$. Does there exist a positive real number $c_p(B)$ such that if $d = (d_1, d_2, \ldots)$ is a martingale difference sequence in $L_B^p(0, 1)$ and $\varepsilon_1, \varepsilon_2, \ldots$ are numbers in $\{-1, 1\}$, then

$$\| \varepsilon_1 d_1 + \cdots + \varepsilon_n d_n \|_p \le c_p(B) \| d_1 + \cdots + d_n \|_p \tag{3}$$

for all $n \ge 1$? (A sequence d in $L_B^p(0, 1)$ is a martingale difference sequence if it satisfies (5), an orthogonality condition.) We write $B \in$ UMD (the Banach space B has the unconditionality property for martingale differences) if such a number does exist. The class UMD is nonempty: $\mathbf{R} \in$ UMD [4]. In fact, if $L_B^p(0, 1)$ has an unconditional basis, then $B \in$ UMD (Aldous [1]). The class UMD does not depend on the choice of p (see [11] or Theorem 1 below). Therefore, $\ell^r \in$ UMD for $1 < r < \infty$, as can be seen by checking this for $p = r$ using term-by-term integration and the fact that $\mathbf{R} \in$ UMD.

Similarly, $L^r(0, 1) \in$ UMD for $1 < r < \infty$. Maurey and Pisier discovered a number of other properties of UMD. For example, if $B \in$ UMD, then B is superreflexive [11], but there do exist superreflexive spaces, hence uniformly convex spaces (Enflo [9]), that are not in UMD [12].

Is there a geometric characterization of Banach spaces B that belong to the class UMD? While studying this question, motivated partly by its possible relevance for singular integrals, the author discovered the following novel geometric condition on a Banach space.

A function $\zeta : B \times B \to \mathbf{R}$ is symmetric if $\zeta(x, y) = \zeta(y, x)$ and is biconvex if both $\zeta(., y)$ and $\zeta(x, .)$ are convex on B for all $x, y \in B$. We shall say that a Banach space B *satisfies the ζ-condition*, or is *ζ-convex*, if there is a symmetric biconvex function ζ on $B \times B$ such that $\zeta(0, 0) > 0$ and

$$\zeta(x, y) \leq |x + y| \quad \text{if} \quad |x| \leq 1 \leq |y|. \tag{4}$$

We note some properties of this condition:

(i) Any such function ζ determines the norm of B up to equivalence.

(ii) If B is ζ-convex, then any space isomorphic to B is ζ-convex.

(iii) The class of ζ-convex spaces contains ℓ^r and $L^r(0, 1)$, $1 < r < \infty$, as well as all finite-dimensional spaces.

(iv) A Banach space B is an inner product space if and only if there is a function ζ on $B \times B$, as above, with $\zeta(0, 0) = 1$. (To prove the necessity of this condition, consider

$$\zeta(x, y) = 1 + (x, y),$$

where (x, y) is the real part of the inner product of x and y. The sufficiency is much harder to prove.) For other properties and all the proofs, see [3 and 5].

How does the ζ-condition relate to the questions raised above about B-valued martingales? One of the main results of [3] can now be stated as follows: $B \in$ UMD if and only if B is ζ-convex.

Using this result, McConnell and the author [6] have been able to prove that if B is ζ-convex and $1 \leq p < \infty$, then

$$Hf = \lim_{\varepsilon \to 0} H_\varepsilon f$$

exists almost everywhere for all $f \in L_B^p(\mathbf{R})$ and satisfies the M. Riesz inequality if $1 < p < \infty$. In addition, strong-type and weak-type results hold for the maximal Hilbert transform $H^*f(x) = \sup_{\varepsilon > 0} |H_\varepsilon f(x)|$. The analogous results hold for more general singular integrals.

Our aim here is to give a brief description of some of the methods and results of [3, 5, and 6]. The original papers will have to be consulted for a number of other theorems and most of the proofs. However, the lemma that provides the transition from the martingale setting to the singular integral setting is proved in full.

It may be helpful to remark that no previous knowledge of martingale theory is needed to read this paper.

Constants will be denoted by c or C and dependence on B and p will be indicated. These symbols will often denote different positive real numbers from one use to the next.

THE MARTINGALE BACKGROUND

Let $f = (f_1, f_2, \ldots)$ be a B-valued martingale in $L_B^1(0, 1)$; that is, f is a sequence in $L_B^1(0, 1)$ such that its difference sequence $d = (d_1, d_2, \ldots)$, defined by $f_n = \sum_{k=1}^{n} d_k$, satisfies

$$d_{n+1} \perp \varphi(d_1, \ldots, d_n) \tag{5}$$

for all bounded continuous $\varphi : B \times \cdots \times B$ (n factors) $\to \mathbf{R}$ and all $n \geq 1$. The meaning of (5) is that the product of the scalar-valued function on the right-hand side and the B-valued function d_{n+1} has, over the unit interval, the integral 0 (the origin of B) with respect to Lebesgue measure.

As an example, consider an arbitrary function $F : [0, 1] \to B$ and, for each positive integer n, a partition of the interval $[0, 1)$ into a finite number of left-closed, right-open intervals. We assume that each interval in the n-th partition is the union of intervals in the $(n + 1)$-st. If $[a, b)$ is an interval in the n-th partition, let

$$f_n(t) = \frac{F(b) - F(a)}{b - a}, \quad t \in [a, b).$$

It is easy to check that this defines a sequence f in $L^1_B(0, 1)$ with a difference sequence d satisfying (5). In what follows, it would be sufficient to limit attention to such martingales.

If $\varepsilon = (\varepsilon_1, \varepsilon_2, \ldots)$ is a sequence of numbers in $\{-1, 1\}$ and f is a martingale with difference sequence d, we may consider the new martingale g defined by

$$g_n = \sum_{k=1}^{n} \varepsilon_k d_k, \quad n \geq 1.$$

This is the transform of f by ε and exhibits behavior somewhat similar to a conjugate harmonic function in case $B = \mathbf{R}$. Let $\|f\|_p = \sup_{n\geq 1} \|f_n\|_p$ and $g^*(t) = \sup_{n\geq 1} |g_n(t)|$. In the following, m denotes Lebesgue measure on the unit interval.

Theorem 1. Suppose that B is a ζ-convex Banach space and $1 < p < \infty$. Then, for all martingales f and g as above,

$$\|f\|_1 < \infty \Longrightarrow g \text{ converges a.e.}, \tag{6}$$

$$\lambda m(g^* > \lambda) \leq c(B) \|f\|_1, \quad \lambda > 0, \tag{7}$$

$$\|g\|_p \leq c_p(B) \|f\|_p. \tag{8}$$

Conversely, any of these three holding for all such f and g implies that B is ζ-convex.

Note that (8) is the same as the condition $B \in$ UMD, since $\|f_n\|_p$ is nondecreasing in n.

This theorem and the next remark extend some of the results of [4] to their most natural Banach-space settings. For other such extensions, see [3].

There are many other conditions relating martingales and their transforms that are equivalent to the ζ-condition. On the other hand, the square-function inequality

$$c_p(B)\,\|S(f)\|_p \leq \|f\|_p \leq C_p(B)\,\|S(f)\|_p \tag{9}$$

is valid only for Banach spaces in a much smaller class. Here $S(f)(t) = \left(\sum_{k=1}^{\infty} |d_k(t)|^2\right)^{1/2}$ and $1 < p < \infty$. Note that (9) implies (8) and if B is a Banach space in which the Khintchine inequality is valid, then (8) implies (9). Moreover, (9) gives the Khintchine inequality in the range $1 < p < \infty$. In this range, the Khintchine inequality is valid for B if and only if B is isomorphic to an inner product space (Kwapień [10]). By Theorem 1, inequality (8) does hold for such B. We can conclude that the square-function inequality (9) holds if and only if B is isomorphic to an inner product space, a result that can also be obtained from the work of Pisier [13].

We shall now describe a key part of the proof of Theorem 1, the part that connects the weak-type inequality (7) with the ζ-condition. For x and y in B, let $M(x, y)$ be the set of all B-valued martingales $f = (f_1, f_2, \ldots)$ on $[0, 1)$ such that $f_1 = x$ and, for some sequence $\varepsilon = (1, \varepsilon_2, \varepsilon_3, \ldots)$ in $\{-1, 1\}$, the transform g of f by ε satisfies

$$m(\{t : |g_n(t) - y| \geq 1 \quad \text{for some} \quad n \geq 1\}) = 1.$$

Let $\psi(x, y) = \inf\{\|f\|_1 : f \in M(x, y)\}$. The key step in the proof is to characterize the function ψ as follows: ψ is the greatest function $\varphi : B \times B \to \mathbf{R}$ such that

$$\varphi(x, y) = \varphi(x, 2x - y), \tag{10}$$

$$\varphi(\cdot, y) \quad \text{is convex,} \tag{11}$$

$$\varphi(x, y) \leq |x| \quad \text{if} \quad |y| \geq 1. \tag{12}$$

Now suppose that the weak-type inequality holds for all f and g as in the theorem. If $f \in M(0, 0)$, then $g^* \geq 1$ a.e. so, for $0 < \lambda < 1$,

$$\lambda = \lambda m(g^* > \lambda) \leq c(B)\|f\|_1.$$

This implies that $1/c(B) \leq \|f\|_1$, which in turn implies that $\psi(0, 0) > 0$. It is now easy to check that ζ, defined by

$$\zeta(x, y) = 2\psi\left(\frac{x + y}{2}, y\right),$$

is symmetric, biconvex, positive at the origin of $B \times B$, and satisfies

$$\zeta(x, y) \leq |x + y| \quad \text{if} \quad |y| \geq 1, \tag{13}$$

which is a little more than (4). Therefore, B is ζ-convex.

To go the other way, we may assume without loss of generality that there exists a symmetric biconvex function ζ on $B \times B$ with $\zeta(0, 0) > 0$ and such that (13) holds. Let

$$\varphi(x, y) = \zeta(2x - y, y)/2.$$

Then φ satisfies (10), (11), and (12), as is easily checked, so

$$\psi(x, y) \geq \varphi(x, y)$$

by the maximality of ψ. Thus $\psi(0, 0) > 0$, which can be used to prove the weak-type inequality (7) for the special case in which $g^* > 1$ a.e. But an additional argument gives the general case with

$$c(B) \leq 4/\zeta(0, 0) \tag{14}$$

for the optimal choice of the constant.

We note that the right-hand side of (14) may be replaced by $2/\zeta(0, 0)$ if $B = \ell^r$, $1 < r < \infty$, and this may be true for all ζ-convex spaces.

The following lemma contains all the information about B-valued martingales that we shall need. In fact, it is just a statement about the Rademacher sequence $r_0, r_1, \ldots,$ where r_k is the function on [0, 1) taking the values +1 and -1 alternately on $[0, 1/2^{k+1}), [1/2^{k+1}, 2/2^{k+1}), \ldots$.

Lemma 1. *Let B be a ζ-convex Banach space and $1 < p < \infty$. There is a positive real number $c_p(B)$ such that if ψ_k is a function on $\mathbf{R}^k$ with values in B and $v_k = \psi_k(r_0, \ldots, r_{k-1})$, then the norm of $f_n = \sum_{k=1}^{n} v_k r_k$ in $L_B^p(0, 1)$ satisfies*

$$c_p^{-1}(B)\| f_n \|_p^p \leq \int_0^1 \int_0^1 \left| \sum_{k=1}^{n} v_k(t) r_k(s) \right|^p ds\,dt \leq c_p(B)\| f_n \|_p^p \tag{15}$$

for all $n \geq 1$.

Proof. Because the Rademacher sequence is independent and satisfies $\int_0^1 r_k(t)\,dt = 0$, the sequence $v_1 r_1, v_2 r_2, \ldots$ satisfies the orthogonality property (5). Therefore, $f = (f_1, f_2, \ldots)$ is a B-valued martingale. Fix a number $s \in [0, 1)$, let $\varepsilon_k = r_k(s)$ so $\varepsilon_k \in \{-1, 1\}$, and let $g_n(t) = \sum_{k=1}^{n} \varepsilon_k v_k(t) r_k(t)$. Then $g = (g_1, g_2, \ldots)$ is the transform of f by $\varepsilon = (\varepsilon_1, \varepsilon_2, \ldots)$ and f is the transform of the martingale g by ε. Therefore, by Theorem 1, $\| g_n \|_p \approx \| f_n \|_p$. That is,

$$\int_0^1 \left| \sum_{k=1}^{n} v_k(t) r_k(t) r_k(s) \right|^p dt \approx \int_0^1 \left| \sum_{k=1}^{n} v_k(t) r_k(t) \right|^p dt. \tag{16}$$

For fixed t, the transformation $s \to (r_1(t)r_1(s), r_2(t)r_2(s), \ldots)$ has the same distribution as the transformation $s \to (r_1(s), r_2(s), \ldots)$. Therefore,

$$\int_0^1 |\sum_{k=1}^{n} v_k(t)r_k(t)r_k(s)|^p\,ds = \int_0^1 |\sum_{k=1}^{n} v_k(t)r_k(s)|^p\,ds$$

and integrating both sides of (16) with respect to s gives (15), the desired result.

Lemma 2. Let B be a ζ-convex Banach space and $1 < p < \infty$. There is a positive real number $c_p(B)$ such that if N is a positive integer and $a_1, \ldots, a_N$ and $b_1, \ldots, b_N$ belong to B, then for all $r \geq 0$,

$$\int_0^{2\pi} |v(re^{i\theta})|^p\,d\theta \leq c_p(B) \int_0^{2\pi} |u(re^{i\theta})|^p\,d\theta \tag{17}$$

where

$$u(re^{i\theta}) = \sum_{k=1}^{N} (a_k \cos k\theta + b_k \sin k\theta)r^k, \tag{18}$$

$$v(re^{i\theta}) = \sum_{k=1}^{N} (a_k \sin k\theta - b_k \cos k\theta)r^k. \tag{19}$$

Proof. Consider the Rademacher functions $r_1, r_2, \ldots$ as above and let $0 < \delta < 1/2$, $d_k = \delta r_{2k-1}$, $e_k = \delta r_{2k}$, $X_n = \sum_{k=1}^{n} d_k$, $Y_n = \sum_{k=1}^{n} e_k$, and $Z_n = X_n + iY_n$. Also, for convenience, let $Z_0 = 0$ and χ be the characteristic function of the open unit disc. Then

$$w_k = \prod_{j=0}^{k-1} \chi(Z_j)$$

is precisely the characteristic function of the set $\{t : \tau(t) \geq k\}$ where $\tau(t) = \inf\{n : |Z_n(t)| \geq 1\}$ with the usual convention that $\inf \phi = +\infty$. Thus, $w_1 = 1$ and for $k \geq 2$, w_k is a measurable function of $r_1, \ldots, r_{2k-2}$. Therefore, $w_1d_1, w_2d_2, \ldots$ satisfies the orthogonality property (5) and hence is a martingale difference sequence. The corresponding martingale, which we denote by f, has as its square function

$$S(f) = (\sum_{k=1}^{\infty} (w_k d_k)^2)^{1/2}$$
$$= \delta(\sum_{k=1}^{\infty} w_k)^{1/2}$$
$$= \delta\tau^{1/2}.$$

(Note that $w_k^2 = w_k$ and $|d_k| = \delta$.) If $n < \tau(t)$, then $|f_n(t)| = |X_n(t)| \leq |Z_n(t)| < 1$; if $n \geq \tau(t)$, then $|f_n(t)| = |X_{\tau(t)}(t)| \leq 1 + \delta \leq 2$. Therefore, $\|f\|_p \leq \|f\|_\infty \leq 2$ and, by the square-function inequality (9) for real-valued martingales,

$$\delta\|\tau^{1/2}\|_p = \|S(f)\|_p \leq c_p\|f\|_p \leq 2c_p. \tag{20}$$

In particular, $\tau = \sum_{k=1}^{\infty} w_k$ is finite almost everywhere so that

$$\lim_{n\to\infty} Z_{\tau\wedge n} = Z_\tau \text{ a.e.,} \tag{21}$$

$$|Z_{\tau\wedge n}| \leq 1 + 2\delta < 2, \tag{22}$$

$$1 \leq |Z_\tau| \leq 1 + 2\delta \text{ a.e.} \tag{23}$$

The absolute values of the third-order derivatives with respect to x and y of $r^k\cos k\theta$ and $r^k\sin k\theta$, the real and imaginary parts of $(x + iy)^k$, are majorized by $k^3 2^{k-3}$ in the disc $|z| < 2$. Let

$$\gamma = N^3 2^N \sum_{k=1}^{N} (|a_k| + |b_k|).$$

Then, by Taylor's theorem and the validity here of Laplace's equation $u_{xx} + u_{yy} = 0$, the function u defined in (18) must satisfy

$$|u(x+h, y+k) - u(x, y) - u_x(x, y)h - u_y(x, y)k - u_{xy}(x, y)hk| < \gamma\delta^3$$

if $|x + iy| < 1$ and $|h| = |k| = \delta$. If $w_k(t) = 1$, then $|Z_{k-1}(t)| < 1$ and $|Z_k(t)| < 2$. Therefore,

$$w_k|u(Z_k) - u(Z_{k-1}) - u_x(Z_{k-1})d_k - u_y(Z_{k-1})e_k - u_{xy}(Z_{k-1})d_k e_k| \le \gamma\delta^3 w_k.$$

Fix n and let

$$U = \sum_{k=1}^{n} w_k[u_x(Z_{k-1})d_k + u_y(Z_{k-1})e_k],$$
$$W = \sum_{k=1}^{n} w_k u_{xy}(Z_{k-1})d_k e_k,$$

and notice that $\tau \wedge n = \sum_{k=1}^{n} w_k$. Then, we have that

$$|u(Z_{\tau\wedge n}) - U - W| \le \gamma\delta^3(\tau \wedge n)$$

implying that

$$\| u(Z_{\tau\wedge n}) - U\|_p \le \| W\|_p + \gamma\delta^3 \| \tau \wedge n \|_p. \tag{24}$$

The last term can be estimated by replacing p by $2p$ in (20) to obtain

$$\| \tau \wedge n\|_p \le c_p \delta^{-2}. \tag{25}$$

To estimate the norm of W, write

$$W = \sum_{j=1}^{N} (a_j W^{2j-1} + b_j W^{2j}) \tag{26}$$

with W^j denoting the analogue of W for u^j where $u^{2j-1}(re^{i\theta}) = r^j \cos j\theta$ and $u^{2j}(re^{i\theta}) = r^j \sin j\theta$. The sequence

$$(w_1 u_{xy}^j(Z_0)d_1 e_1, \ldots, w_n u_{xy}^j(Z_{n-1})d_n e_n, 0, 0, \ldots)$$

satisfies (5), hence is a martingale difference sequence. The k-th term is bounded by $k^2\delta^2$. Therefore, by the square-function inequality (9) for real-valued martingales,

$$\| W^j \|_p \le c_p N^2 \delta^2 \left\| \left(\sum_{k=1}^{n} w_k^2 \right)^{1/2} \right\|_p$$

$$= c_p N^2 \delta^2 \| (\tau \wedge n)^{1/2} \|_p .$$

Using (20) and (26), we obtain $\| W \|_p \le c_p \gamma \delta$, and combining this with (24) and (25) gives

$$\| u(Z_{\tau \wedge n}) - U \|_p \le c_p \gamma \delta. \tag{27}$$

Similarly,

$$\| v(Z_{\tau \wedge n}) - V \|_p \le c_p \gamma \delta, \tag{28}$$

where

$$V = \sum_{k=1}^{n} w_k [v_x(Z_{k-1}) d_k + v_y(Z_{k-1}) e_k].$$

The next step is to show that

$$\| V \|_p \le c_p(B) \| U \|_p. \tag{29}$$

This is accomplished with the use of Lemma 1. Clearly, V may be expressed as a sum $\sum_{k=1}^{2n} v_k r_k$ of the type considered in that lemma; so by the conclusion of the lemma,

$$\| V \|_p^p \approx \int_0^1 \int_0^1 \left| \sum_{k=1}^{n} \delta w_k(t) \big[v_x\big(Z_{k-1}(t)\big) r_{2k-1}(s) + v_y\big(Z_{k-1}(t)\big) r_{2k}(s) \big] \right|^p ds\, dt.$$

Similarly,

$$\| U \|_p^p \approx \int_0^1 \int_0^1 \left| \sum_{k=1}^{n} \delta w_k(t) \big[u_x\big(Z_{k-1}(t)\big) r_{2k-1}(s) + u_y\big(Z_{k-1}(t)\big) r_{2k}(s) \big] \right|^p ds\, dt.$$

But the integrals with respect to s on the inside of these two double integrals are equal: $u_x = v_y$ and $u_y = -v_x$ and the two transformations $s \to (-r_1(s), r_2(s), -r_3(s), r_4(s), \ldots)$ and

$s \to (r_2(s), r_1(s), r_4(s), r_3(s), \ldots)$ have the same distribution. Thus, the two double integrals are equal and this implies (29).

We can now complete the proof of Lemma 2 as follows. Combining (27), (28), and (29), we have

$$\| v(Z_{\tau\wedge n}) \|_p \leq c_p(B) \| u(Z_{\tau\wedge n}) \|_p + c_p(B)\gamma\delta.$$

Using (21) and (22) and the continuity and boundedness of u and v on the closed disc $|z| \leq 2$, we obtain

$$\| v(Z_\tau) \|_p \leq c_p(B) \| u(Z_\tau) \|_p + c_p(B)\gamma\delta.$$

The same inequality holds for u^α and v^α where $u^\alpha(re^{i\theta}) = u(re^{i(\theta+\alpha)})$. Therefore,

$$\int_0^1 \int_0^{2\pi} |v(|Z_\tau|e^{i\alpha})|^p \, d\alpha \, dt \leq c_p(B) \int_0^1 \int_0^{2\pi} |u(|Z_\tau|e^{i\alpha})|^p \, d\alpha \, dt + c_p(B)\gamma^p\delta^p.$$

Let $\delta \to 0$. By (23) and the dominated convergence theorem, the desired inequality (17) follows for the special case $r = 1$. The case $r \geq 0$ is an immediate consequence. This completes the proof of Lemma 2.

For a somewhat shorter but less elementary proof in the same spirit, see [6].

Lemma 3. Suppose that B is a ζ-convex Banach space and $1 < p < \infty$. If $f \in L_B^p(-\pi, \pi)$ and

$$u(re^{i\theta}) = \frac{1}{2\pi} \int_{-\pi}^{\pi} \frac{1 - r^2}{1 - 2r\cos(\theta - t) + r^2} f(t)\, dt, \tag{30}$$

$$v(re^{i\theta}) = \frac{1}{2\pi} \int_{-\pi}^{\pi} \frac{2r\sin(\theta - t)}{1 - 2r\cos(\theta - t) + r^2} f(t)\, dt, \tag{31}$$

then for all $0 \leq r < 1$,

$$\int_{-\pi}^{\pi} |v(re^{i\theta})|^p \, d\theta \leq c_p(B) \int_{-\pi}^{\pi} |u(re^{i\theta})|^p \, d\theta. \tag{32}$$

Proof. Just as in the real-valued case, $u - u(0)$ and v are the uniform limits on the circle $|z| = r$ of functions of the form (18) and (19), respectively, where the a_k and b_k are the Fourier coefficients of f. Therefore, Lemma 2 implies that (32) holds for $u - u(0)$ and v and hence for u and v.

Now let

$$\tilde{f}_\varepsilon(\theta) = \frac{1}{2\pi} \int_{\varepsilon<|t|<\pi} f(\theta - t) \cot \frac{t}{2} \, dt,$$

where $f \in L_B^p(-\pi, \pi)$ is extended periodically, $1 < p < \infty$, and $0 < \varepsilon < 1$. Let $v(re^{i\theta})$ be as in (31) with $r = 1 - \varepsilon$. Then there is a nonnegative function K_r such that

$$|\tilde{f}_\varepsilon(\theta) - v(re^{i\theta})| \leq \frac{1}{2\pi} \int_{-\pi}^{\pi} |f(\theta - t)| K_r(t) \, dt$$

and $\int_{-\pi}^{\pi} K_r(t) \, dt/2\pi < 1$. Also, by (32),

$$\int_{-\pi}^{\pi} |v(re^{i\theta})|^p \, d\theta \leq c_p(B) \| f \|_p^p.$$

Consequently, by Young's inequality for convolution,

$$\| \tilde{f}_\varepsilon \|_p \leq c_p(B) \| f \|_p. \tag{33}$$

Returning to the truncated Hilbert transform defined in (2), we have the following parallel result for the nonperiodic case.

Lemma 4. Suppose that B is a ζ-convex Banach space and $1 < p < \infty$. If $f \in L_B^p(\mathbf{R})$, then for all $\varepsilon > 0$,

$$\| H_\varepsilon f \|_p \leq c_p(B) \| f \|_p. \tag{34}$$

This follows from (33) just as it does in the real case; see [14] or [18].

The method of rotation introduced by Calderón and Zygmund in [7] now becomes available and (34) implies a similar inequality for more general singular integrals. Here we shall limit our discussion to singular integrals with odd kernels and refer the reader to [6] for information about even kernels and related results. Let K be a scalar-valued function on $\mathbf{R}^n - \{0\}$ of the form

$$K(x) = \Omega(x)/|x|^n.$$

Assume that $\Omega(-x) = -\Omega(x)$, $\Omega(\alpha x) = \Omega(x)$ for all $\alpha > 0$, and

$$\int_0^1 \frac{\omega(\delta)}{\delta}\, d\delta < \infty,$$

where ω is the modulus of continuity of Ω restricted to the unit sphere. If $1 \leq p < \infty$, $f \in L^p_B(\mathbf{R}^n)$, and $\varepsilon > 0$, let

$$T_\varepsilon f(x) = \int_{|x-y|>\varepsilon} f(y)K(x - y)\, dy$$

and $T^*f(x) = \sup\limits_{\varepsilon>0} |T_\varepsilon f(x)|$. In the following theorem, Lebesgue measure on $\mathbf{R}^n$ is denoted by m.

Theorem 2. Suppose that B is a ζ-convex Banach space. Then for $1 \leq p < \infty$ and $f \in L^p_B(\mathbf{R}^n)$, the limit $Tf = \lim\limits_{\varepsilon \to 0} T_\varepsilon f$ exists almost everywhere,

$$\| T^*f \|_p \leq c_p(B) \| f \|_p, \quad 1 < p < \infty, \tag{35}$$

$$\lambda m(T^* f > \lambda) \leq c(B) \| f \|_1, \quad \lambda > 0, \tag{36}$$

so Tf satisfies the same inequalities and, for $1 < p < \infty$, exists also as a limit of the $T_\varepsilon f$ in $L^p_B(\mathbf{R}^n)$.

So we see that some of the classical results for singular integrals of scalar-valued functions, due to Calderón, Zygmund, and others, hold also for singular integrals of functions with values in a ζ-convex Banach space.

It is possible that the ζ-convexity of B is a necessary condition, as well as a sufficient condition, for (35) or (36) to hold, even in the special case of the Hilbert transform, but this has not yet been settled.

REFERENCES

1. Aldous, D. J. Unconditional bases and martingales in $L_p(F)$. *Math. Proc. Cambridge Phil. Soc.* 85 (1979): 117-123.

2. Benedek, A.; Calderón, A. P.; and Panzone, R. Convolution operators on Banach space valued functions. *Proc. Nat. Acad. Sci.* 48 (1962): 356-365.

3. Burkholder, D. L. A geometrical characterization of Banach spaces in which martingale difference sequences are unconditional. *Ann. Probability* 9 (1981): 997-1011.

4. Burkholder, D. L. Martingale transforms. *Ann. Math. Statistics* 37 (1966): 1494-1504.

5. Burkholder, D. L. Martingale transforms and the geometry of Banach spaces. *Proc. of the Third International Conference on Probability in Banach Spaces*, Tufts Univ., 1980. *Lecture Notes in Mathematics* 860 (1981): 35-50.

6. Burkholder, D. L., and McConnell, T. R. Manuscript in preparation.

7. Calderón, A. P., and Zygmund, A. On singular integrals. *Amer. J. of Math.* 78 (1956): 289-309.

8. Calderón, A. P., and Zygmund, A. On the existence of certain singular integrals. *Acta Math.* 88 (1952): 85-139.

9. Enflo, Per. Banach spaces which can be given an equivalent uniformly convex norm. *Israel J. Math.* 13 (1972): 281-288.

10. Kwapień, S. Isomorphic characterizations of inner product spaces by orthogonal series with vector valued coefficients. *Studia Math.* 44 (1972): 583-595.

11. Maurey, B. Système de Haar. *Seminaire Maurey-Schwartz*, 1974-1975, Ecole Polytechnique, Paris, 1975.

12. Pisier, G. Un exemple concernant la super-réflexivité. *Seminaire Maurey-Schwartz*, 1974-1975, Ecole Polytechnique, Paris, 1975.

13. Pisier, G. Martingales with values in uniformly convex spaces. *Israel J. Math.* 20 (1975): 326-350.

14. Riesz, M. Sur les fonctions conjuguées. *Math. Z.* 27 (1927): 218-244.

15. Schwartz, J. A remark on inequalities of Calderón-Zygmund type for vector-valued functions. *Comm. Pure Appl. Math.* 14 (1961): 785-799.

16. Stein, Elias M. *Singular Integrals and Differentiability Properties of Functions*. Princeton, N.J.: Princeton Univ. Press, 1970.

17. Vági, Stephen. A remark on Plancherel's theorem for Banach space valued functions. *Ann. Scuola Norm. Sup. Pisa Cl. Sci.* 23 (1969): 305-315.

18. Zygmund, A. *Trigonometric Series*, 2d ed. vols. I and II. New York: Cambridge Univ. Press, 1959.

ADDED IN PROOF: In response to the possibility raised in the final paragraph here, J. Bourgain has shown recently that the boundedness of the Hilbert transform in $L_B^p(\mathbf{R})$ implies that B is in *UMD*. Therefore, the ζ-convexity of B is both a necessary and a sufficient condition for the existence and boundedness of the Hilbert transform on $L_B^p(\mathbf{R})$.

ESTIMATIONS L^2 POUR LES NOYAUX SINGULIERS

R.R. Coifman
Yale University
A. McIntosh
MacQuarie University, Australia
Y. Meyer
Centre de Mathematiques de l'Ecole Polytechnique

Commençons par rappeler quelques définitions.

Soient $\Delta \subset \mathbf{R}^n \times \mathbf{R}^n$ la diagonale (définie par $y = x$, $x \in \mathbf{R}^n$, $y \in \mathbf{R}^n$) et $\Omega \subset \mathbf{R}^n \times \mathbf{R}^n$ le complémentaire de Δ.

Désignons par $K : \Omega \to \mathbf{C}$ une fonction possédant les deux propriétés suivantes:

$$|K(x, y)| \leq \frac{C}{|x - y|^n} \tag{1}$$

K est localement lipschitzienne sur Ω et l'on a

$$|\nabla_x K(x, y)| \leq \frac{C}{|x - y|^{n+1}} \quad et \quad |\nabla_y K(x, y)| \leq \frac{C}{|x - y|^{n+1}} . \tag{2}$$

A l'aide d'un tel noyau $K(x, y)$ on définit, pour tout $\varepsilon > 0$, l'opérateur tronqué T_ε par

$$T_\varepsilon f(x) = \int_{|y-x| \geq \varepsilon} K(x, y) f(y)\, dy . \tag{3}$$

Le problème fondamental de la théorie est alors de décider si T_ε envoie $L^2(\mathbf{R}^n)$ dans lui-même et si la norme, notée $\| T_\varepsilon \|_{op}$, de l'opérateur $T_\varepsilon : L^2(\mathbf{R}^n) \to L^2(\mathbf{R}^n)$ est bornée uniformément par rapport à $\varepsilon > 0$.

Lorsque, c'est le cas, nous dirons que $K(x, y)$ est un noyau de Calderón-Zygmund.

Si $K(x, y)$ est un noyau de Calderón-Zygmund, il existe un opérateur continu $T : L^2(\mathbf{R}^n) \to L^2(\mathbf{R}^n)$ tel que pour toute fonction $f \in C_0^\infty(\mathbf{R}^n)$ et tout x n'appartenant pas au support de f on ait

$$Tf(x) = \int_{\mathbf{R}^n} K(x, y) f(y)\, dy. \tag{4}$$

Naturellement, T n'est pas unique et un choix possible de T est la limite faible d'une sous-suite T_{ε_j}, $\varepsilon_j \to 0$.

On définit également l'opérateur maximal T_* par

$$T_* f(x) = \sup_{\varepsilon > 0} |T_\varepsilon f(x)|. \tag{5}$$

Avec ces notations, on a le théorème suivant (A. P. Calderón, M. Cotlar, et A. Zygmund).

Théorème 1. Pour tout noyau $K(x, y)$ vérifiant (1) *et* (2), *les deux conditions suivantes sont équivalentes*:

(a) *Il existe une constante $C > 0$ telle que, pour toute fonction $f \in C_0^\infty(\mathbf{R}^n)$ et tout $\varepsilon > 0$, on ait $\| T_\varepsilon(f) \|_2 \le C \| f \|_2$.*

(b) *Il existe un opérateur continu $T : L^2(\mathbf{R}^n) \to L^2(\mathbf{R}^n)$ tel que $K(x, y)$ soit la restriction au complémentaire de la diagonale du noyau-distribution $S(x, y)$ de T. Si l'une de ces deux conditions est satisfaite, alors l'opérateur maximal T_* se prolonge à tout $L^p(\mathbf{R}^n)$ pour $1 < p < +\infty$: $\| T_* f \|_p \le C_p \| f \|_p$ où C_p ne dépend que de p, n, $\| T \|_{op}$ et de la constante C figurant dans* (1) *et* (2).

On pourra se rapporter, par exemple, à [2, Chapitre 4]. Donnons des exemples de tels noyaux $K(x, y)$ en commençant par les exemples historiques qui ont motivé la théorie.

Si $K(x, y) = A(x, x - y)$ où $A(x, u)$, $x \in \mathbf{R}^n$, $u \in \mathbf{R}^n \setminus \{0\}$, vérifie les conditions

$$A(x, \lambda u) = \lambda^{-n} A(x, u) \quad \text{pour tout} \quad \lambda > 0 \tag{6}$$

$$\int_{|u|=1} A(x, u)\, d\sigma(u) = 0 \quad \text{pour tout} \quad x \in \mathbf{R}^n \tag{7}$$

$$|\partial_x^\alpha \partial_u^\beta A(x, u)| \le C_{\alpha,\beta} \quad \text{pour tout} \quad x \in \mathbf{R}^n \quad \text{et} \quad |u| = 1, \tag{8}$$

alors $K(x, y)$ est un noyau de Calderón-Zygmund (et certains auteurs appellent "noyaux de Calderón-Zygmund" les noyaux vérifiant (6), (7), et (8)).

L'opérateur T associé est alors défini par

$$Tf(x) = \text{v.p.} \int_{\mathbf{R}^n} K(x, y) f(y)\, dy = \lim_{\varepsilon \downarrow 0} \int_{|y-x| \ge \varepsilon} K(x, y) f(y)\, dy \tag{9}$$

et, pour toute fonction $f \in L^2(\mathbf{R}^n)$, l'intégrale singulière (9) existe presque partout.

Un second exemple est celui des opérateurs pseudodifférentiels classiques d'ordre 0. C'est à dire que l'on part d'un symbole $\sigma(x, \xi) \in C^\infty(\mathbf{R}^n \times \mathbf{R}^n)$ vérifiant les conditions de Hörmander

$$|\partial_\xi^\alpha \partial_x^\beta \sigma(x, \xi)| \le C_{\alpha,\beta}(1 + |\xi|)^{-|\alpha|} \tag{10}$$

et l'on définit $T = \sigma(x, D)$ par le formalisme usuel

$$\sigma(x, D) f(x) = \frac{1}{(2\pi)^n} \int_{\mathbf{R}^n} e^{ix.\xi} \sigma(x, \xi) \hat{f}(\xi)\, d\xi. \tag{11}$$

Alors le noyau distribution de $\sigma(x, D)$ s'écrit $S(x, x - y)$ et la distribution $S(x, .) \in \mathscr{S}'(\mathbf{R}^n)$ vérifie les propriétés suivantes

$$\sigma(x, \xi) = \int_{\mathbf{R}^n} e^{-i\xi.u} S(x, u)\, du \in L^\infty(\mathbf{R}^n \times \mathbf{R}^n) \tag{12}$$

et plus généralement $|\partial_x^\beta \sigma(x, \xi)| \le C_\beta$ si $\beta \in \mathbf{N}^n$. D'autre part

$$S(x, u) \in C^\infty(\mathbf{R}^n \times \mathbf{R}^n \setminus \{0\}) \tag{13}$$

est satisfait aux inégalités $|\partial_x^\alpha \partial_u^\beta S(x, u)| \le C_{\alpha,\beta} |u|^{-n-|\beta|}$ si $0 < |u| \le 1$ tandis que $|\partial_x^\alpha \partial_u^\beta S(x, u)| \le C_N |u|^{-N}$ pour tout $N \ge 1$ si $|u| \ge 1$, $\alpha \in \mathbf{N}^n$, $\beta \in \mathbf{N}^n$.

Les inégalités vérifiées par le noyau distribution sont bien plus précises que (1) et (2) et pour montrer que $\sigma(x, D)$ est borné sur $L^2(\mathbf{R}^n)$, on écrit, grâce à une partition C^∞ de l'identité, $S(x, u) = S_1(x, u) + S_2(x, u)$ où S_1 a les mêmes propriétés que S mais est, de plus nulle si $|u| \geq 1$ et où S_2 est nulle si $|u| \leq 1/2$. Alors $|S_2(x, x - y)| \leq \dfrac{C}{1 + |x - y|^{n+1}}$ et l'action de l'opérateur correspondant est triviale.

En ce qui concerne S_1, on est ramené à prouver une estimation locale. On définit, si $f \in L^2(\mathbf{R}^n)$, $v \in \mathbf{R}^n$, $x \in \mathbf{R}^n$, $F(v, x) = \int_{\mathbf{R}^n} S(v, x - y) f(y)\, dy$. Alors on a

$$\| F(v, x) \|_{L^2(dx)} \leq \| \sigma \|_\infty \| f \|_2$$

et, plus généralement,

$$\| \partial_v^\beta F(v, x) \|_{L^2(dx)} \leq \| \partial_x^\beta \sigma \|_\infty \| f \|_2$$

pour $0 \leq |\beta| \leq m$.

Puisque $m > \frac{n}{2}$, le théorème de trace de Sobolev s'applique et permet de restreindre $F(v, x)$ à la diagonale $v = x$ et d'obtenir une fonction $F(x, x)$ dans L^2_{loc}. Ceci termine la démonstration de la continuité de $\sigma(x, D) : L^2(\mathbf{R}^n) \to L^2(\mathbf{R}^n)$.

L'exemple suivant est de nature tout à fait différente. On désigne par $A : \mathbf{R} \to \mathbf{C}$ une fonction lipschitzienne: il existe une constante $C \geq 0$ telle que pour tout couple (x, y) de deux nombres réels on ait

$$|A(x) - A(y)| \leq C|x - y|. \tag{14}$$

Le noyau que l'on considère est $K(x, y) = \dfrac{A(x) - A(y)}{(x - y)^2}$ et A. P. Calderón a démontré en 1965 [1] que, pour toute fonction $f \in L^2(\mathbf{R})$, l'intégrale singulière v.p. $\int K(x, y) f(y)\, dy = g(x)$ existe presque partout et que $\| g \|_2 \leq C \| A' \|_\infty \| f \|_2$ où C est une constante absolue. L'opérateur T défini par $T(f) = g$ peut être considéré comme un opérateur pseudodifférentiel. Son symbole est alors, en posant $a(x) = A'(x)$, donné par la formule $\sigma(x, \xi) = \int e^{i\alpha x} \dfrac{|\xi + \alpha| - |\xi|}{\alpha}$ $\hat{a}(\alpha)\, d\alpha$ et n'appartient pas à $L^\infty(\mathbf{R} \times \mathbf{R})$. Or aucune des classes existantes d'opérateurs pseudodifférentiels ne permet de donner des estimations $L^2(\mathbf{R})$ pour des symboles non bornés.

L'opérateur de Calderón se situe donc "au delà des opérateurs pseudodifférentiels." En 1974 puis en 1976 deux des auteurs [2] généralisaient le théorème de Calderón aux noyaux $\frac{(A(x) - A(y))^k}{(x - y)^{k+1}}$ où $A : \mathbf{R} \to \mathbf{C}$ est lipschitzienne et $k \in \mathbf{N}$.

En 1977, A. P. Calderón démontrait [1] qu'en appelant L_k l'opérateur défini par le noyau ci-dessus, on a $\| L_k \|_{op} \leq C^k \| A' \|_\infty^k$ où $C \geq 1$ est une constante absolue dont la valeur n'était pas connue.

En présentant ses résultats au Congrès International d'Helsinki, A. P. Calderón pose le problème de montrer que $\| L_k \|_{op} \leq C_\varepsilon (1 + \varepsilon)^k \| A' \|_\infty^k$ pour tout $\varepsilon > 0$.

Nous allons maintenant donner une réponse positive à cette conjecture et, ce faisait, établir une liste de nouveaux noyaux de Calderón-Zygmund, c'est-à-dire de noyaux décrits par le Théorème 1.

Théorème 2. Soient K une partie compacte et convexe du plan complexe, U un ouvert contenant K et $F : U \to \mathbf{C}$ une fonction holomorphe dans U. Désignons par $A : \mathbf{R} \to \mathbf{C}$ une fonction telle que pour tout $x \in \mathbf{R}$, et tout $y \in \mathbf{R}$, on ait $\frac{A(x) - A(y)}{x - y} \in K$. Alors le noyau $N(x, y) = F\left(\frac{A(x) - A(y)}{x - y}\right) \frac{1}{x - y}$ définit un opérateur borné sur $L^2(\mathbf{R})$.

En employant la méthode de transférence de R. Coifman et G. Weiss, on peut donner à ce résultat une forme plus abstraite.

Soient $(\Omega, \mathscr{C}, \mu)$ un espace muni d'une tribu et d'une mesure σ-finie $\mu \geq 0$ sur $\mathscr{C}$. Désignons par U_t un groupe d'automorphismes de $(\Omega, \mathscr{C}, \mu)$ indexé par $\mathbf{R}$, préservant la mesure μ et tel que si $E \in \mathscr{C}$ et $\mu(E) < +\infty$, la mesure de la différence symétrique entre $U_t(E)$ et E tende vers 0 avec t.

Nous dirons qu'une fonction $A : \Omega \to \mathbf{C}$ est lipschitzienne relativement à U_t s'il existe une constante $C \geq 0$ telle que

$$\left\| \frac{A(U_s(x)) - A(x)}{s} \right\|_{L^\infty(dx)} \leq C.$$

Soient, de nouveau, K une partie compacte et convexe du plan complexe, U un ouvert contenant K et $F : U \to \mathbf{C}$ une fonction holomorphe dans U. Soit $A : \Omega \to \mathbf{C}$ une fonction lipschitzienne telle que, pour tout $s \in \mathbf{R}$, on ait

$$\frac{A(U_s(x)) - A(x)}{s} \in K.$$

Alors l'opérateur T défini par

$$Tf(x) = \text{v.p.} \int_{-\infty}^{+\infty} F\left(\frac{A(U_t(x)) - A(x)}{t}\right) f(U_t x) \frac{dt}{t}$$

est borné sur $L^2(\Omega, \mathscr{C}, d\mu)$.

Un cas particulier du Théorème 2 a une signification géométrique si particulière qu'il convient de le regarder comme un énoncé indépendant.

On désigne par $\varphi : \mathbf{R} \to \mathbf{R}$ une fonction lipschitzienne ($|\varphi(x) - \varphi(y)| \leq M|x - y|$ pour tout x réel et tout y réel) et par $\Gamma = \{(x, y) \in \mathbf{R}^2;\ y = \varphi(x)\}$ le graphe de la fonction φ. Soient Ω_1 et Ω_2 les ouverts limités par Γ et définis par $y < \varphi(x)$ et $y > \varphi(x)$. Suivant C. Kenig [4] on leur associe deux espaces de Hardy généralisés $H^2(\Omega_1)$ et $H^2(\Omega_2)$ définis comme suit: $F \in H^2(\Omega_1)$ signifie que $F : \Omega_1 \to \mathbf{C}$ est holomorphe et que

$$\sup_{\varepsilon > 0} \int_{\mathbf{R}} \left| F(x + i\varphi(x) - i\varepsilon) \right|^2 dx < +\infty$$

et de même pour $H^2(\Omega_2)$. Naturellement, si $\varphi = 0$, on retrouve les espaces de Hardy usuels.

On peut définir [4] un opérateur de trace $\theta_1 : H^2(\Omega_1) \to L^2(\Gamma)$ grâce auquel $H^2(\Omega_1)$ s'identifie à un sous-espace vectoriel fermé, noté $H^2(\Omega_1)|_\Gamma$ de $L^2(\Gamma)$ et l'on définit, de même, le sous-espace vectoriel fermé

$$H^2(\Omega_2)|_\Gamma \subset L^2(\Gamma) = L^2(\Gamma, ds) \quad \text{où} \quad ds = \sqrt{1 + \varphi'^2(x)}\, dx.$$

Théorème 3. L'espace $L^2(\Gamma)$ est la somme directe des deux sous-espaces fermés $H^2(\Omega_1)|_\Gamma$ et $H^2(\Omega_2)|_\Gamma$.

La somme directe n'est orthogonale que si $\varphi(x) = ax + b$. Cet énoncé avait été obtenu par A. P. Calderón en 1977 [1] sous l'hypothèse restrictive $\|\varphi'\|_\infty \le \delta$ ($\delta > 0$ venant de la méthode de perturbation employée). Notre contribution fut donc de nous affranchir de δ. En outre la méthode de A. P. Calderón reposait sur des techniques de variable complexe et, en particulier, sur l'usage de la représentation conforme. Nos démonstrations constituent un retour à l'analyse réelle et il y est fait un usage important des "mesures de Carleson" et des inégalités qui s'y rattachent.

Un dernier résultat intéressant est l'extension des théorèmes précédents à $\mathbf{R}^n$. Elle est obtenue par la méthode des rotations de Calderón.

Théorème 4. Supposons que $A : \mathbf{R}^n \to \mathbf{R}$ et $\varphi : \mathbf{R}^n \to \mathbf{R}$ soient deux fonctions lipschitziennes. C'est à dire qu'il existe une constante $C \ge 0$ telle que l'on ait $|\varphi(x) - \varphi(y)| \le C|x - y|$ pour tout $x \in \mathbf{R}^n$ et tout $y \in \mathbf{R}^n$ (et de même pour A). Alors le noyau $K(x, y) = \dfrac{A(x) - A(y)}{\left[|x-y|^2 + (\varphi(x) - \varphi(y))^2\right]^{\frac{n+1}{2}}}$ définit un opérateur borné sur $L^2(\mathbf{R}^n)$.

Très curieusement, le point de départ de ce travail fut une conjecture de Kato que nous avons pu résoudre par les mêmes méthodes d'analyse réelle.

Voilà ce dont il s'agit.

En retournant à la dimension 1, on désigne par D l'opérateur $-i\dfrac{d}{dx}$ et par $a \in L^\infty(\mathbf{R})$ une fonction essentiellement bornée telle que $\mathcal{R}e\, a(x) \ge 1$ presque-partout.

Suivant Kato [3] on peut définir la racine carrée $\sqrt{T}$ de l'opérateur accrétif $T = DM_a D$ comme un opérateur non borné dont le domaine est un sous-espace dense dans $L^2(\mathbf{R})$.

Plus précisément, $\sqrt{T} = \frac{2}{\pi} T \int_0^\infty \frac{d\lambda}{T + \lambda^2 I}$. Le problème posé par Kato est de montrer que le domaine de l'opérateur $\sqrt{T}$ est l'espace de Sobolev $H^1(\mathbf{R})$.

Théorème 5. Soit $H^1(\mathbf{R}) \subset L^2(\mathbf{R})$ l'espace de Sobolev usuel. Alors, avec les notations précédentes, $\sqrt{T} : H^1(\mathbf{R}) \to L^2(\mathbf{R})$ est continu.

Les démonstrations de ces résultats paraîtront ultérieurement.

REFERENCES

1. Calderón, A. P. Commutators, singular integrals on Lipschitz curves and applications. *Proc. of the International Congress of Mathematicians,* 1 (1978): 85-96.

2. Coifman, R. R., et Meyer, Y. Au delà des opérateurs pseudo-différentiels. *Astérisque* 57, *Soc. Math. France* (1978).

3. Kato, T. *Perturbation Theory for Linear Operators.* New York: Springer-Verlag, 1966.

4. Kenig, C. Weighted H^p spaces on Lipschitz domains. *Amer. J. Math.* 102 (1980): 129-163.

VECTOR VALUED INEQUALITIES FOR MULTIPLIERS

A. Córdoba
Universidad Autonoma de Madrid

In this paper we shall consider several inequalities of the following types:

$$\| (\sum |T_j f_j|^r)^{1/r} \|_p \leq C_{p,r} \| (\sum |f_j|^r)^{1/r} \|_p \qquad (*)$$

$$\| (\sum |S_j f|^2)^{1/2} \|_p \leq C_p \| f \|_p \qquad (**)$$

$1 \leq p,\ r \leq \infty$, where the involved operators are Fourier multipliers in R^n, that is, they are given as convolution with a tempered distribution, the kernel, whose Fourier transform, the multiplier, is bounded. In case of inequality (**), we shall assume, in addition, that if S_j is given by the formula $\widehat{S_j f} = m_j \hat{f}$, then

$$\sum |m_j|^2 \in L^\infty(\mathbf{R}^n).$$

We shall discuss some examples that, although different, will present certain common features: the role of weighted inequalities and the use of adequate square and maximal functions to decompose the operators and to control them.

1

If the T_j's are Calderón-Zygmund operators with uniform bounds, that is, $T_j f = K_j * f$ and each kernel K_j coincides with a smooth function outside the origin satisfying: $|K_j(x)| \leq C|x|^{-n}$, $|\nabla K_j(x)| \leq C|x|^{-n-1}$, uniformly on j, then inequality (*) holds if $1 < p, r < \infty$. This is a result of Benedek, Calderón, and Panzone [1]. In this paper we want to emphasize a different approach [3] based on the following inequality:

$$\int_{\mathbf{R}^n} |Tf(x)|^r \omega(x)\, dx \leq B(c;\ r;\ s) \int_{\mathbf{R}^n} |f(x)|^r A_s \omega(x)\, dx,$$

where $A_s \omega = [(\omega^s)^*]^{1/s}$ *-denotes the Hardy-Littlewood maximal function, f, ω are arbitrary functions in $C_0(R^n)$, and $B(c;\ r;\ s)$ depends only on the numbers r, s, $1 < r, s < \infty$ and the bound, C, of the Calderón-Zygmund operator T.

This is the simpler example of a vector-valued inequality implied by a weighted inequality. Both kinds of estimates are equivalent in a certain sense, and in several cases, one can proceed in the opposite direction. It should be noted that when $r = \infty$ and we restrict to the case $f_j = f$ for every j, inequality (*) is still true in some outstanding cases, for example, in R^1, if m_j is the characteristic function of the interval $(-j, +j)$, and $1 < p < \infty$.

If $\{\Delta_j\}$ is a dyadic decomposition of R^n and S_j is the multiplier operator given by the function $\chi_{\Delta j}$ then we have the Littlewood-Paley theory, which gives us the equivalence of norms:

$$\| (\sum |S_j f_j|^2)^{1/2} \|_p \sim \| f \|_p \quad \text{if} \quad 1 < p < \infty.$$

The "dyadic" character is crucial in the previous statement and, for example, if one substitutes dyadic by congruent, then it is not very hard to see that the norm-equivalence is destroyed. Nevertheless, we have the following result:

In R^n let us consider a cubic lattice $\mathscr{L} = \{Q_\nu\}_{\nu \in Z^n}$, that is, the Q_ν's are congruent cubes with disjoint interiors and such that $\mathbf{R}^n = \cup Q_\nu$.

Define, for each ν, the operator $\widehat{P_\nu f} = \chi_{Q\nu} \cdot \hat{f}$ and the square function $Gf(x) = (\sum |P_\nu f(x)|^2)^{1/2}$.

Theorem 1. For each $s > 1$ there exists a finite constant C_s such that for every f, $\omega \in C_0(R^n)$ we have

$$\int_{\mathbf{R}^n} |Gf(x)|^2 \omega(x)\, dx \le C_s \int_{\mathbf{R}^n} |f(x)|^2 A_s \omega(x)\, dx.$$

Proof. We may assume, without losing generality, that $\mathscr{L}$ is the unit lattice, that is, Q_ν is centered at the point $\nu \in Z^n$ and has volume equal to 1. Let φ be a smooth function with compact support and equal to the constant 1 on Q_0.

For every $y \in Q_0$ let us consider the Fourier multiplier

$$m_y(z) = \sum_\nu e^{2\pi i \nu \cdot y} \varphi(z - \nu), \; z \in \mathbf{R}^n.$$

Then the kernel $\mu_y = \hat{m}_y$ is a measure of finite total variation uniformly in $y \in Q_0$. More concretely, $\mu_y = \sum_\nu \hat{\varphi}(y + \nu)\delta_{y+\nu}$. Therefore,

$$|\mu_y * f(x)|^2 \le C \sum_\nu (1 + |\nu|)^{2n} |\hat{\varphi}(y + \nu)|^2 |f(x - y - \nu)|^2.$$

And since $\hat{\varphi}$ is rapidly decreasing, we have

$$\int_{\mathbf{R}^n} |\mu_y * f(x)|^2 \omega(x)\, dx \le C_N \sum_\nu (1 + |\nu|)^{2n-N} \int_{\mathbf{R}^n} |f(x)|^2 \omega(x + y + \nu)\, dx.$$

We may assume that $\omega \ge 0$.

Thus, taking $N \ge 4n + 1$, we have

$$\int_{Q_0} \int_{\mathbf{R}^n} |\mu_y * f(x)|^2 \omega(x)\, dx \le C_N \sum_\nu (1 + |\nu|)^{2n-N}$$

$$\int_{\mathbf{R}^n} |f(x)|^2 \left[\int_{Q_0} \omega(x + y + \nu)\, dy \right] dx \le C \int_{\mathbf{R}^n} |f(x)|^2 \omega^*(x)\, dx.$$

On the other hand, if $\widehat{T_\nu f}(\xi) = \varphi(\xi - \nu)\hat{f}(\xi)$, it follows

$$\mu_y * f(x) = \sum_\nu e^{2\pi i\nu\cdot y} T_\nu f(x)$$

and therefore

$$\int_{Q_0}\int_{\mathbf{R}^n} |\mu_y * f(x)|^2 \omega(x)\, dx\, dy = \sum_\nu \int_{\mathbf{R}^n} |T_\nu f(x)|^2 \omega(x)\, dx.$$

To finish we observe that $P_\nu f = P_\nu T_\nu f$ and we can apply the weighted inequality of [3].

Let us consider for each $k \in Z$ a decomposition of the strip $2^k \le x_n \le 2^{k+1}$ into congruent disjoint parallelepipeds whose sides are parallel to the coordinate axis, $\{Q_\nu^k\}$, and define

$$Sf(x) = \left(\sum_{k,\nu} |P_\nu^k f(x)|^2\right)^{1/2}.$$

Combining the Littlewood-Paley theorem with the previous result, we obtain:

Corollary 1. For each p, $2 \le p < \infty$, there exists a finite constant C_p, so that $\|Sf\|_p \le C_p \|f\|_p$ for every $f \in C_0(R^n)$.

Theorem 1 also allows us to make the following remark about trigonometric polynomials with gaps:

Corollary 2. For every trigonometric polynomial of the form

$$f(x) = \sum_N^{N+[\sqrt{N}]} a_k e^{2\pi i k^2 x} \quad \textit{we have} \quad \|f\|_4 \le 2(\Sigma |a_k|^2)^{1/2}.$$

Proof. Let us consider for each $\theta \in [0, 1]$ the polynomial

$$f_\theta(x) = \sum_N^{N+[\sqrt{N}]} a_k e^{2\pi i k\theta} e^{2\pi i k^2 x}.$$

Then

$$\int_0^1\int_0^1 |f_\theta(x)|^4\, dx\, d\theta = \sum_{\substack{k+j=p+q \\ k^2+j^2=p^2+q^2}} a_k a_j \bar{a}_p \bar{a}_q \leq 4(\Sigma\ |a_k|^2)^2.$$

Therefore, there exists $\theta \in [0, 1]$ so that

$$\| \sum a_k\, e^{2\pi i k\theta} e^{2\pi i k^2 x} \|_4 \leq 2^{1/2}(\Sigma\ |a_k|^2)^{1/2}.$$

The problem is that in general, $\{m_k(\theta)\} = \{e^{2\pi i k^{1/2}\theta}\}$, $k \in Z$, is not a multiplier of $L^4(T)$. However, let us consider $I_k =$ $(N^2 + 2kN,\ N^2 + 2(k+1)N) \subset R$ and let φ_k be a smooth function adapted to I_k as in Theorem 1.

Since

$$N^2 + 2kN < (N+k)^2 < N^2 + 2(k+1)N$$

$$N^2 + 2(k+1)N - (N^2 + 2kn) = 2N,$$

we have that

$$\sum_{N}^{N+[\sqrt{N}]} e^{2\pi i k\theta} \varphi_k(\xi)$$

is a Fourier multiplier of $L^p(R)$, $1 \leq p \leq \infty$, and we may use the de Leew's theorem to finish the proof of the corollary.

2

In R^2 let us consider a lacunary sequence of directions $\{\omega_j\}$, $j = 1, 2, \ldots$, $\omega_j/\omega_{j-1} \leq \rho < 1$ and let us denote by H_j the Hilbert transform in the direction ω_j, that is, $\widehat{H_j f} = m_j \hat{f}$, where $m_j(\xi) =$ $\text{sign}(\omega_j \cdot \xi)$. Then we have:

$$\| (\sum |H_j f_j|^r)^{1/r} \|_p \leq C_{p,r} \| (\sum |f_j|^r)^{1/r} \|_p,$$

$1 < p < \infty$ (see [4 and 6]).

Also if we define $\Delta_j = \{\xi : \omega_j \geq \arg(\xi) \geq \omega_{j+1}\}$,

$\Delta_0 = \mathbf{R}^2 - \bigcup_{j=1}^{\infty} \Delta_j$, and $\widehat{S_k f}(\xi) = \chi_{\Delta_k}(\xi)\hat{f}(\xi)$, $Sf = (\sum |S_k f|^2)^{1/2}$,

we have

$$\| Sf \|_p \simeq \| f \|_p, \quad 1 < p < \infty.$$

Similarly to the preceding section one may ask what happens to these inequalities when the lacunarity condition is substituted by a uniformly distributed sequence of angles: $\omega_j = \frac{2\pi j}{N}$, $j = 0, 1, \ldots, N-1$.

H_j is the Hilbert transform in the direction ω_j and S_j denotes the projection, at the Fourier transform side, over the angle

$$\frac{2\pi j}{N} \leq \arg(\xi) \leq 2\pi \frac{j+1}{N}.$$

Theorem 2. There exist constants $0 < a, C < \infty$ so that

$$\text{(i)} \quad \Big\| \Big(\sum_{j=1}^{N} |H_j f_j|^2 \Big)^{1/2} \Big\|_4 \leq C[\log N]^a \Big\| \Big(\sum |f_j|^2\Big)^{1/2} \Big\|_4.$$

$$\text{(ii)} \quad \Big\| \Big(\sum_{j=1}^{N} |S_j f|^2 \Big)^{1/2} \Big\|_4 \leq C[\log N]^a \| f \|_4.$$

A proof of statement (i) can be found in [2]. Here we shall sketch the proof of (ii).

First we may assume, without losing generality, that $1 \leq j \leq \frac{N}{8}$ so that $\frac{2\pi j}{N} \leq \frac{\pi}{4}$. Let us define $\widehat{P_j f} = \chi_{\Delta_j} \cdot \hat{f}$, where

$$\Delta_j = \Big\{\xi = \xi_1 + i\xi_2 : 1 \leq \xi_1 \leq 2, \frac{2\pi j}{N} \leq \arg(\xi) \leq 2\pi \frac{j+1}{N}\Big\}.$$

We want to compute

$$\sum_{j,k} \int_{\mathbf{R}^2} |P_j f(x) P_k f(x)|^2 \, dx = \sum_{|j-k|<N^{1/2}} \int_{\mathbf{R}^2} |P_j f(x) P_k f(x)|^2 \, dx + \sum_{|j-k|\geq N^{1/2}} \int_{\mathbf{R}^2} |P_j f(x) P_k f(x)|^2 \, dx = \text{I} + \text{II}.$$

Let us decompose each sector Δ_j into $[N^{1/2}]$ subsectors $\{\Delta_j^\alpha\}$, where $\Delta_j^\alpha = \{\xi = \xi_1 + i\xi_2 \in \Delta_j : \alpha N^{-1/2} \leq \xi_1 - 1 \leq (\alpha + 1)N^{-1/2}\}$.

It happens that if $|j - k| > N^{1/2}$, the overlapping of the sets $\Delta_j^\alpha + \Delta_k^\beta$, α, $\beta = 1, \ldots, N^{1/2}$, is finite, uniformly on N. Therefore,

$$\text{II} \leq \sum_{|j-k|\geq N^{1/2}} \sum_{\alpha,\beta} \int_{\mathbf{R}^2} |P_j^\alpha f(x) P_k^\beta f(x)|^2 \, dx$$
$$\leq \| (\sum_{j,\alpha} |P_j^\alpha f|^2)^{1/2} \|_4^4,$$

where the operators P_j^α have the natural definition $\widehat{P_j^\alpha f} = \chi_{\Delta_j^\alpha} \hat{f}$. We claim that $\| (\sum_{j,\alpha} |P_j^\alpha f|^2)^{1/2} \|_4 \leq C(\log N)^{1/4} \| f \|_4$ for some universal constant C.

To see this we take $\omega \in L^2(R^n)$, $\omega \geq 0$, and we consider

$$\sum_{j,\alpha} \int_{\mathbf{R}^2} |P_j^\alpha f(x)|^2 \omega(x) \, dx$$
$$= \sum_{\ell=1}^{N^{1/2}/8} \sum_{\alpha=1}^{N^{1/2}} \sum_{j=\ell N^{1/2}}^{(\ell+1)N^{1/2}} \int_{\mathbf{R}^2} |P_j^\alpha f(x)|^2 \omega(x) \, dx$$
$$\leq C_s \sum_{\ell=1}^{N^{1/2}/8} \sum_{\alpha=1}^{N^{1/2}} \int_{\mathbf{R}^2} |Q_\ell^\alpha f(x)|^2 M_s \omega(x) \, dx,$$

where $Q_\ell^\alpha f$ is given, at the Fourier transform side, as multiplication by the characteristic function of a rectangle, with sides parallel to the coordinate axes, and dimensions $N^{-1/2} \times 2N^{-1/2}$; $M_s\omega = (M\omega^s)^{1/s}$, $1 < s < \infty$, and M denotes the maximal function associated to the base of rectangles with directions in the set $\frac{2\pi j}{N}$, $j = 0, 1, \ldots, \frac{N}{8}$.

In establishing the last estimate, we have made a repeated use of Theorem 1. Using Hölder's inequality together with the known results for M (see [5]), we get

$$\text{II} \leq C[\log N] \| f \|_4^4 .$$

To finish, we estimate I in the following manner:

$$I = \sum_{\nu=0}^{1/2\log N} \sum_{2^{-\nu}N^{1/2}\le|j-k|<2^{-\nu+1}N^{1/2}} \int_{\mathbf{R}^2} |P_j f(x) P_k f(x)|^2\, dx + \sum_j \| P_j f \|_4^4 .$$

Since we always have $\sum_{j=1}^{N} \| P_j f \|_4^4 \le C \| f \|_4^4$ and we are looking for an estimate with a factor of $(\log N)^a$, we may estimate each block of the preceding sum independently: For each ν we decompose each sector Δ_j into subsectors

$$\Delta_j^\alpha = \{\xi = \xi_1 + i\xi_2 \in \Delta_j : \alpha 2^\nu N^{-1/2} \le \xi_1 - 1 \le (\alpha + 1)2^\nu N^{-1/2}\},$$

and we repeat the same arguments used in the estimation of II.

Finally we observe that by homogeneity we have proved the following: For each $n \in Z$ let us consider the family of sets

$$\Delta_{j,n} = \left\{\xi = \xi_1 + i\xi_2 : 2^n \le \xi_1 \le 2^{n+1},\ \frac{2\pi j}{N} \le \arg(\xi) \le \frac{2\pi(j+1)}{N}\right\}$$

and operators $\widehat{P_{jn} f} = \chi_{\Delta_{j,n}} \cdot \hat{f}$.

Then $\| (\sum_j |P_{jn} f|^2)^{1/2} \|_4 \le C(\log N)^{1/2} \| f \|_4$ uniformly on $n \in Z$.

We decompose $\Delta_j = \bigcup_{\ell=1}^{\log N} \bigcup_{n\equiv\ell (\mathrm{mod}[\log N])} \Delta_{j,n}$, which gives us the decomposition

$$(\sum_j |P_j f|^2)^{1/2} \le \sum_{\ell=1}^{\log N} (\sum_j |P_j^\ell f(x)|^2)^{1/2} \cdot [\log N],$$

where P_j^ℓ is given by the multiplier $\bigcup_{n\equiv\ell(\mathrm{mod}[\log N])} \Delta_{j,n}$.

The point is that if $n_1 \equiv n_2$ (mod[log N]) and if $n_1 > n_2$, then it follows that $2^{n_1} > N2^{n_2}$; that is, the smaller side of the rectangles corresponding to Δ_{j,n_1}, $j = 1, 2, \ldots, N$, is bigger than the diameter of the set $\bigcup_{\substack{n_2\equiv n_1\\ n_2<n_1}} \Delta_{j,n_2}$.

Furthermore, if we decompose each $\Delta_{j,n}$ into N "squares" $\{\Delta^{\alpha}_{j,n}\}$ of side $\simeq 2^n N^{-1}$ and if we define the associated multipliers operators $\widehat{P^{\alpha}_{j,n} f} = \chi_{\Delta^{\alpha}_{j,n}}(\xi)\hat{f}(\xi)$, then it is also true that the dimension of these "squares" is bigger than the diameter of the set considered before.

In the following we shall keep ℓ fixed and we shall assume that the index n ranges in the set of integers congruent to ℓ mod([log N]). We have:

$$\begin{aligned}
\Big\| \Big(\sum_j |P^{\ell}_j f|^2\Big)^{1/2} \Big\|_4^4 &\sim \Big\| \Big(\sum_{j,n} |P_{j,n} f|^2\Big)^{1/2} \Big\|_4^4 \\
&= \sum_n \Big\| \Big(\sum_j |P_{jn} f|^2\Big)^{1/2} \Big\|_4^4 + 2 \sum_{\substack{j,k\\ n_1>n_2}} \int_{\mathbf{R}^2} |P_{j,n_1} f(x) P_{k,n_2} f(x)|^2 \, dx \\
&\le C[\log N]^2 \|f\|_4^4 + 2 \sum_{\substack{j,k\\ n_1>n_2}} \int_{\mathbf{R}^2} \Big|\sum_{\alpha} P^{\alpha}_{j,n_1} f(x) P_{k,n_2} f(x)\Big|^2 \, dx \\
&= C[\log N]^2 \|f\|_4^4 + 2 \sum_{\substack{j,k\\ n_1>n_2}} \int_{\mathbf{R}^2} \Big|\sum_{\alpha} \widehat{P^{\alpha}_{j,n_1} f} * \widehat{P_{k,n_2} f}(\xi)\Big|^2 \, d\xi \\
&\le C[\log N]^2 \|f\|_4^4 + 8 \sum_{\substack{j,k\\ n_1>n_2}} \sum_{\alpha} \int_{\mathbf{R}^2} |P^{\alpha}_{j,n_1} f(x)|^2 |P_{k,n_2} f(x)|^2 \, dx \\
&\le C[\log N]^2 \|f\|_4 + 8 \Big\| \Big(\sum_{j,n,\alpha} |P^{\alpha}_{j,n} f|^2\Big)^{1/2} \Big\|_4^2 \cdot \Big\| \Big(\sum |P_{k,n} f|^2\Big)^{1/2} \Big\|_4^2 .
\end{aligned}$$

And since

$$\Big\| \Big(\sum_{j,n,\alpha} |P^{\alpha}_{j,n} f|^2\Big)^{1/2} \Big\|_4 \le C\|f\|_4 ,$$

the desired estimate now follows very easily.

3

Vector-valued spherical summation in R^2.

With $\alpha > 0$, $\mathbf{R} > 0$, let us define

$$\widehat{T_\alpha^R f}(\xi) = \left(1 - \frac{|\xi|^2}{R^2}\right)_+^\alpha \cdot \hat{f}(\xi).$$

Theorem 3. *Given any sequence of real numbers* $\{R_j\}$, *we have*

$$\|(\sum_j |T_\alpha^{R_j} f_j|^2)^{1/2}\|_{L^p(\mathbf{R}^2)} \leq C_{p,\alpha} \|(\sum_j |f_j|^2)^{1/2}\|_{L^p(\mathbf{R}^2)},$$

$$\frac{4}{3 + 2\alpha} < p < \frac{4}{1 - 2\alpha}.$$

Corollary 3. *Let* $R_j \nearrow \infty$ *be lacunary: then* $T_\alpha^{R_j} f \to f$ *a.e. if* $f \in L^p(\mathbf{R}^2)$, $4/(3 + 2\alpha) < p < 4/(1 - 2\alpha)$.

A proof of these results appears in [5]. Here we shall follow it very closely in sketching a proof of Theorem 3 based on Theorem 2.

Given $\delta > 0$ and $\varphi \in C_0^\infty(0, 1)$, let us define $\widehat{S_j^\delta f}(\xi) = \varphi\left(\frac{R_j - |\xi|}{\delta R_j}\right)\hat{f}(\xi)$. The theorem is then a consequence of the following lemma:

Lemma 1. $\|(\sum |S_j^\delta f_j|^2)^{1/2}\|_4 \leq C|\log \delta|^a \|(\sum |f_j|^2)^{1/2}\|_4$ *for some universal constants* C *and* a.

Let us consider a C^∞ partition of unity in S^1 of the form $1 = \sum_1^N \psi_n$, where ψ_n is supported in the set $\left|\frac{N}{2n}\theta - n\right| \leq 1$, where $N = [\delta^{-1/2}]$. We define $\widehat{T_j^n f}(\xi) = \psi_n(\arg(\xi))\widehat{S_j^\delta f}(\xi)$. The proof of the lemma contains two steps:

(1) Using geometric arguments, we can show the following estimate.

$$\|(\sum_j |S_j^\delta f_j|^2)^{1/2}\|_4 \leq C|\log \delta|^a \|(\sum_{j,n} |T_j^n f_j|^2)^{1/2})\|_4.$$

(2) We observe that the kernel K_j^n can be majorized in the form

$$|K_j^n(x)| \leq \sum_{\nu=0}^{\infty} 2^{-\nu} \frac{1}{|R_{j,\nu}^n|} \chi_{R_{j,\nu}^n}(x),$$

where $R_{j,0}^n$ is a rectangle with dimensions $\delta^{-1/2} \times \delta^{-1}$ and direction $2\pi j\delta^{1/2}$ and for each ν, $R_{j,\nu}^n$ is the result of expanding $R_{j,0}^n$ by the factor 2^{ν}.

With the notation of Theorem 2, we have $|T_j^n f| = |T_j^n S_j f| \leq M_N(S_j f)$.

Therefore, we are in the position of using the results of the preceding section together with the following substitute of the Hardy-Littlewood maximal theorem: $\| M_N \omega \|_2 \leq C(\log N) \| \omega \|_2$ (where M_N is the maximal function associated to the base of rectangles of R^2 with fixed eccentricity $= N$), to prove

$$\| (\sum_{j,n} |T_j^n f_j|^2)^{1/2} \|_4 \leq C|\log \delta|^a \| (\sum_j |f_j|^2)^{1/2} \|_4 .$$

REFERENCES

1. Benedek, A.; Calderón, A.; and Panzone, R. Convolution operators on Banach spaces valued functions. *Proc. Nat. Acad. Sci.* 18 (1962): 356-365.

2. Córdoba, A. The multiplier problem for the polygon. *Ann. Math.* 105 (1977): 581-588.

3. Córdoba, A., and Fefferman, C. Weighted norm inequalities for singular integrals. *Studia Math.* 57 (1976): 97-101.

4. Córdoba, A., and Fefferman, R. On the equivalence between the boundedness of certain classes of maximal and multipliers operators in Fourier Analysis. *Proc. Nat. Acad. Sci.* 74 (1977): 423-425.

5. Córdoba, A., and Lopez-Melero, B. Spherical summation: A problem of E. M. Stein. *Annales I. Fourier* 31 (1981): 147-157.

6. Nagel, A.; Stein, E.; and Wainger, S. Differentiation in lacunary directions. *Proc. Nat. Acad. Sci.* 80 (1978): 137-139.

ON SOME L^p VERSIONS OF THE HELSON-SZEGÖ THEOREM

M. Cotlar
Universidad Central de Venezuela
C. Sadosky
Howard University

INTRODUCTION

In previous papers [4, 2, and 1] a lifting theorem for quadratic forms and its applications to the Helson-Szegö, Koosis, and other weighted problems in L^2 were studied. Here we shall give a variant of that lifting theorem and some L^p versions and generalizations of the theorems of Helson-Szegö and Koosis, with self-contained proofs. Only the one-dimensional periodic case will be considered here; further applications to L^p spaces will be given elsewhere.

The generalization given here is based on the following remark. Let $\mu \geq 0$ and $\rho \geq 0$ be two finite measures in the unit circle T, M a fixed constant, and write $(\rho, \mu) \in R_M^{pq}$ if

$$(\int_T |Hf|^p \, d\rho)^{1/p} \leq M(\int_T |f|^q \, d\mu)^{1/q}, \quad \forall f \in C(T), \tag{1}$$

where Hf is the Hilbert transform of f, and write $\mu \in R^{pq}$ if $(\mu, \mu) \in R_M^{pq}$ for some M. Helson and Szegö [5] gave an explicit expression of all the $\mu \in R^{22}$, and in [4] their theorem was given the following more precise form: $(\mu, \mu) \in R_M^{22}$ iff $d\mu = w(t)\, dt$ and

$$\exists h \in H^1(T) \text{ with } w(t) \leq \operatorname{Re} h(t) \text{ and } |h(t)| \leq c_M w(t) \text{ a.e.} \tag{2}$$

Setting $\gamma(t) = (\text{Re } h(t))/w(t)$ and $T_w f = w^{-1}C(wf)$, where $Cf = f + i Hf$, (2) can be rewritten as

$$1 \leq \gamma(t) \leq c_M \quad \text{and} \quad \| T_w \gamma \|_\infty \leq c_M. \tag{2a}$$

Let us say that a linear transformation A is *u-bounded* in L^p, with u-norm $\leq c$, if $\forall f$, $\| f \|_p \leq 1$, there is a $g \in L^p$ with $|f| \leq g$ and $\| Ag \|_p \leq c$. Then, since $\| f \|_\infty \leq 1$ is equivalent to $|f(t)| \leq 1$, (2a) says that

$$\mu \in R^{22} \text{ iff } d\mu = wdt \text{ and } T_w \text{ is } u\text{-bounded in } L^\infty. \tag{2b}$$

Thus the essential result in Helson-Szegö's theorem is that it reduces the problem about the boundedness of the operator $w^{1/2} Hw^{-1/2}$ in L^2 to the easier problem of the u-boundedness of the operator $w^{-1}Cw$ in L^∞. In this form the theorem generalizes for $L^p : w \in R^{pp}$ iff $T_{w^{2/p}} = w^{-2/p}C(w^{2/p}f)$ is u-bounded in $L^{p/(p-2)}$. More precise results are given for the class R^{2p}, $2 \leq p < \infty$, which lead to a R^{2p} generalization of a theorem of Koosis [6]. More complete generalizations of Koosis' theorem were obtained by J. Rubio de Francia, and by L. Carleson and P. Jones [7, 3] (cf. also [8]). Finally let us observe that "boundedness" obviously implies "u-boundedness," and as is well known, the condition $T_{w^{2/p}}$ bounded in $L^{p/(p-2)}$ is equivalent to $w \in R^{p/2,p/2}$ and implies $w \in R^{pp}$, but not conversely. Now the above L^p version of Helson-Szegö's theorem says that this converse is true in the weaker "u-sense."

For simplicity we shall consider only the case of measures of the form $d\mu = wdt$, $d\rho = \rho(t)\, dt$, $0 \leq w$, $\rho \in L^1(T, dt)$.

We thank R. Arocena for a suggestion in the proof of Theorem 2a.

A LIFTING THEOREM

Let $T \sim [0, 2\Pi]$ be the unit circle, dt its Lebesgue measure, $e_n(t) = \exp(int)$, $n = 0, \pm 1, \pm 2, \ldots$, $P = \{$the trigonometric polynomials $f(t) = \Sigma\, c_n e_n(t) : c_n = \hat{f}(n) = 0$ for $|n| >$ some $n_0\}$, $P_+ = \{$the analytic polynomials $f \varepsilon P : \hat{f}(n) = 0$ for $n < 0\}$, $P_- = \{f \varepsilon P, \hat{f}(n) = 0$ for $n \geq 0\}$, so that every $f \varepsilon P$ has a unique representation $f = f_1 + f_2$, $f_1 \varepsilon P_+$, $f_2 \varepsilon P_-$, and let us write informally $f(t)$, $t \varepsilon T$, instead of $f(e_1(t))$.

We shall deal with 2×2 matrices $W = (w_{\alpha\beta})$, $\alpha, \beta = 1, 2$, whose elements are functions $w_{\alpha\beta}(t)$, $t \varepsilon T$, such that

$$w_{\alpha\beta} \varepsilon L^1(T),\ w_{21}(t) = \overline{w_{12}}(t), \text{ and } w_{11} = w_{22} \geq 0, \tag{3}$$

and will associate with W a quadratic hermitean form $W(F)$ in $C(T) \times C(T)$ that assigns to each $F = (f_1, f_2) \varepsilon C(T) \times C(T)$ the number

$$W(F) = W(f_1, f_2) = \textstyle\sum_{\alpha,\beta=1,2} \int f_\alpha \bar{f}_\beta(t) w_{\alpha\beta}(t)\, dt. \tag{3a}$$

If $V = (v_{\alpha\beta})$ is another such matrix, then we write $W \sim V$ if one of the following equivalent conditions (4), (4a) is satisfied:

$$W(f_1, f_2) = V(f_1, f_2) \text{ whenever } (f_1, f_2) \varepsilon P_+ \times P_-, \tag{4}$$

$$v_{11} = w_{11} \text{ and } v_{12} = w_{12} + h \text{ with } h \varepsilon H^1(T). \tag{4a}$$

We fix a seminorm $p(f)$ in $C(T)$, such that

$$p(f) = p(|f|),\ |f| \leq |g| \text{ implies } p(f) \leq p(g),\ p(f) \leq c\|f\|_\infty, \tag{5}$$

and set $p(F) = p(|f_1 + f_2|2)$ for $F = (f_1, f_2) \varepsilon C(T) \times C(T)$.

From (4) it is clear that if $W \sim V$ and $V(F) \leq p(F)$, $\forall F \varepsilon P \times P$, then $W(F) \leq p(F)$ for $F \varepsilon P_+ \times P_-$. The following theorem, which is a variant of an abstract lifting theorem given in [2], says that "within a fixed constant (=3)," the converse is also true.

Theorem 1. If W satisfies (3) and $W(F) \leq p(F)$, $\forall F = (f_1, f_2) \varepsilon P_+ \times P_-$, then there exists V such that $V \sim W$ and $V(F) \leq 3p(F)$, $\forall F \varepsilon P \times P$. In other words, there exists a function $h \varepsilon H^1(T)$ such that

$$V(F) \equiv W(f_1, f_2) + 2\mathrm{Re} \int_T f_1 \bar{f}_2 h(t)\, dt \leq 3p(|f_1 + f_2|^2), \tag{6}$$

for all (f_1, f_2) *of the larger space* $P \times P$. *Moreover, if* w_{12} *is real and* $w_{12} \leq w_{11}$, *then* $\text{Re } h \geq 0$. (6a)

Proof. By hypothesis we have, for $(f_1, f_2) \in P_+ \times P_-$,

$$\int_T (|f_1|^2 + |f_2|^2) w_{11}\, dt + 2\text{Re} \int_T f_1 \bar{f}_2 w_{12}\, dt \leq p(|f_1 + f_2|^2), \quad (6b)$$

or equivalently,

$$\int |f_1 + f_2|^2 w_{11} + 2\text{Re} \int f_1 \bar{f}_2 (w_{12} - w_{11})\, dt \leq p(|f_1 + f_2|^2). \quad (6c)$$

Letting $f_2 = 0$ in (6b), we get $\int |f_1|^2 w_{11} \leq p(|f_1|^2)$, and since $0 \leq f \in P$ implies $f = |f_1|^2$ with $f_1 \in P_+$, the last inequality and (6c) give

$$\int |f_1 + f_2|^2 w_{11}\, dt \leq p(|f_1 + f_2|^2), \ \forall (f_1, f_2) \in P \times P, \quad (7)$$

and $2\text{Re} \int f_1 \bar{f}_2 (w_{12} - w_{11})\, dt \leq p(|f_1 + f_2|^2)$. (7a)

Since every $f \in P_+$ can be written as $f = f_1 \bar{f}_2$ with $|f_1|^2 = |f_2|^2 = |f|$, $(f_1, f_2) \in \bar{P}_+ \times \bar{P}_-$ (closure in $C(T)$), (7a) gives that

$$\mathscr{C}(f) \equiv 2\text{Re} \int f(w_{12} - w_{11})\, dt \leq p(2|f| + 2\text{Re } f), \ \forall f \in P_+. \quad (7b)$$

Since $q(f) = p(2|f| + 2\text{Re } f)$ is a seminorm, $\mathscr{C}(f)$ can be extended to P so that $\mathscr{C}(f) \leq q(f)$ holds in P, and since $p(f) \leq c\| f \|_\infty$, $\mathscr{C}(f)$ is given by a measure that coincides with $w_{12} - w_{11}$ on P_+. Hence

$$\exists h \in H^1 \text{ such that } \mathscr{C}(f)$$

$$= 2\text{Re} \int f(w_{12} - w_{11})\, dt + \text{Re} \int fh\, dt \leq 2p(|f| + \text{Re } f), \quad (7c)$$

for all $f \in P$. Letting $f = f_1 \bar{f}_2$, $f_1, f_2 \in P$, and using the inequality $2|f| + 2\text{Re } f = 2|f_1 f_2| + 2\text{Re } f_1 \bar{f}_2 \leq |f_1|^2 + |f_2|^2 + 2\text{Re } f_1 \bar{f}_2 = |f_1 + f_2|^2$, we get from (7c) and (7) that

$$\int|f_1+f_2|^2w_{11}+2\mathrm{Re}\int f_1\bar{f}_2(w_{12}-w_{11})\,dt+\mathrm{Re}\int f_1\bar{f}_2h\leq 3p(|f_1+f_2|^2),$$

which is equivalent to (6). Letting $f_1 = -f_2$ in the last inequality, we see that $w_{12} - w_{11} \leq 0$ implies $\mathrm{Re}\, h \geq 0$. QED

Theorem 1a. Let $0 \leq \rho \in L^1(T)$. *The following conditions are equivalent "within a fixed constant."*

(a) $\int_T |f_1 - f_2|^2\rho(t)\,dt \leq p(|f_1 + f_2|^2),\ \forall(f_1, f_2)\in P_+ \times P_-$. (8)

(b) *There exists* $h \in H^1(T)$ *such that* $\mathrm{Re}\, h = h_1 \geq 2\rho \geq 0$ *such that, for every* $(g_1, g_2) \in P \times P$,

$$\int|g_2|^2\rho\,dt+\frac{1}{2}\int|g_1|^2\mathrm{Re}\,h-\frac{1}{2}\int|g_2|^2\mathrm{Re}\,h \tag{8a}$$
$$+\int\mathrm{Im}(g_1\bar{g}_2)\ \mathrm{Im}\,h\leq 3p(|g_1|^2).$$

(c) *There exists* $h = h_1 + ih_2 \in H^1$, *such that* $h_1 \geq 2\rho$ *and such that, for every* $(g_1, g_2) \in P \times P$,

$$\left|\int|g_1g_2|\,|\mathrm{Im}\,h|\,dt\right|^2$$
$$\leq\left[3p(|g_1|^2)-\frac{1}{2}\int|g_1|^2h_1\right]\left[\int|g_2|^2\left(\frac{1}{2}h_1-\rho\right)\right]. \tag{8b}$$

Proof. (*a*) *is equivalent to* (*b*). Letting $w_{11} = w_{22} = \rho$, $w_{12} = -\rho$, (8) becomes $W(f_1, f_2) \leq p(|f_1 + f_2|^2)$, $\forall(f_1, f_2) \in P_+ \times P_-$; hence by Theorem 1 there is $h = h_1 + ih_2 \in H^1$ such that $\forall(f_1, f_2) \in P \times P$,

$$\int|f_1-f_2|^2\rho\,dt+2\mathrm{Re}\int f_1\bar{f}_2h\,dt\leq 3p(|f_1+f_2|^2). \tag{8c}$$

Setting $g_1 = f_1 + f_2$, $g_2 = f_1 - f_2$, and taking in account that $2\mathrm{Re}[f_1\bar{f}_2h] = 2\mathrm{Re}(f_1\bar{f}_2)\mathrm{Re}\,h - 2\mathrm{Im}(f_1\bar{f}_2)\ \mathrm{Im}\,h = \frac{1}{2}(|g_1|^2 - |g_2|^2)\mathrm{Re}\,h + \mathrm{Im}(g_1\bar{g}_2)\ \mathrm{Im}\,h$, (8c) can be rewritten as (8a). Letting $g_1 = 0$, we get (as in the proof of (7)) that $\mathrm{Re}\,h \geq \rho$.

(b) is equivalent to (c). Substituting g_1 by $\lambda_1 g_1$, and g_2 by $\lambda_2 g_2$ in (8a), we get a positive hermitean form in λ_1, λ_2, and (8b) expresses that the determinant of this form is positive. QED

CHARACTERIZATION OF R^{2p}

Let $0 \leq \rho, w \in L^1(T)$, $2 \leq p < \infty$, and write $(\rho, w) \in R_M^{pq}$ instead of $(\rho\, dt, w dt) \in R_M^{pq}$. Since the Hilbert transform satisfies: $Hf = -if$ if $f \in P_+$ and $Hf = if$ if $f \in P_-$, the condition $(\rho, w) \in R_M^{2p}$ is equivalent to

$$\int |f_1 - f_2|^2 \rho\, dt \leq M^2 \left(\int |f_1 + f_2|^p w dt\right)^{2/p}, \ \forall (f_1, f_2) \in P_+ \times P_-. \qquad (9)$$

Theorem 2. $(\rho, \omega) \in R_m^{2p}$ *if and only if ("within a fixed constant")*

$$w(t)^{2/p} = \frac{|h_2(t)|^2}{a(t)\left(h_1(t) - 2\rho(t)\right)} \text{ and } w^{2/p} = \frac{|h|^2}{a_1 h_1}, \qquad (9a)$$

where $h = h_1 + ih_2 \in H^1(T)$ *with* $\rho \leq \frac{1}{2} h$, *and* $a, a_1 \in L^{p/(p-2)}$ *with*

$$\| a \|, \| a_1 \| \leq cM^2,\ a,\ a_1 \neq 0 \ a.e.\ (2 \leq p < \infty). \qquad (9b)$$

Remark 1. Theorem 2 gives a procedure to obtain all the pairs $(\rho, w) \in R_m^{2p}$ (assuming that $(\rho, w) \in R_m^{2p}$ if the right side of (9) is infinite). In fact, the two conditions (9a) are equivalent to

$$w^{2/p} = \frac{h_2^2}{a(h_1 - 2\rho)} \quad \text{and} \quad \frac{|h|^2}{h_2^2} \frac{h_1 - 2\rho}{h_1} a = a_1; \qquad (9c)$$

hence one takes any $h = h_1 + ih_2 \in H^1$ with $h_1 \geq 0$ and $a \in L^{p/p-2}$, then chooses $\rho \leq \frac{1}{2} h_1$ so that the second equality of (9c) is satisfied, and w so that the first equality of (9c) is satisfied.

Remark 2. If $\rho \leq \frac{1}{4} h_1$, then the second condition of (9a) alone is sufficient.

Proof. Taking $p(f) = M^2(\int |f|^{p/2} w\, dt)^{2/p}$ Theorem 1a, (c), gives that $(\rho, w) \in R_m^{2p}$ is equivalent to:

(a) There exists $h = h_1 + ih_2 \in H^1$ such that $h_1 - 2\rho \geq 0$ and such that, for every $(g_1, g_2) \in P \times P$,

$$|\int |g_1 g_2| \, |h_2| dt|^2$$
$$\leq [3M^2(\int |g_1|^p w)^{2/p} - \tfrac{1}{2} \int |g_1|^2 h_1][\int |g_2|^2 (\tfrac{1}{2} h_1 - \rho)]. \qquad (9d)$$

(9d) implies

$$\int |g_1 g_2| \, |h_2| dt \leq 3M(\int |g_1|^p w)^{1/p} \cdot [\int |g_2|^2 (\tfrac{1}{2} h_1 - \rho)]^{1/2}. \qquad (9e)$$

Setting $G_1 = g_1 w^{1/p}$, $G_2 = g_2(\frac{1}{2} h_1 - \rho)^{1/2}$, $b = |h_2|/ [w^{1/p}(\frac{1}{2} h_1 - \rho)^{1/2}]$, (9e) becomes

$$|\int G_1 G_2 b| \leq 3M \| G_1 \|_p \| G_2 \|_2, \qquad (9f)$$

$\forall G_1 \in L^p$, $G_2 \in L^2$. Taking $G_1^{p/2} = G_2 \in L^2$, (9f) gives

$|\int G_1^{\frac{p+2}{p}} b| \leq 3M$, $\forall \| G_1 \|_p = 1$, $G_1^{(p+2)/2} \in L^{2p/(p+2)}$. Hence

$b \in \left(L^{\frac{2p}{p+2}}\right)' = L^{2p/(p-2)}$, so that $|h_2|^2/[w^{2/p}(\frac{1}{2} h_1 - \rho)] = b^2$ $= a \in L^{p/(p-2)}$, which proves the first equality (9a). Now from (9d) it follows that $\int |g_1|^2 w^{2/p} \cdot [h_1/w^{2/p}]\, dt \leq 6M^2 (\int |g_1|^p w\, dt)^{2/p}$; hence $h_1/w^{2/p} = a' \in L^{p/(p-2)}$, $h_1^2 = w^{2/p} a' h_1$, which together with $h_2^2 = w^{2/p} a(h_1 - 2\rho) = w^{2/p} a'' h_1$ gives $|h|^2 = w^{2/p} a_1 h_1$. Thus, conditions (9a) are necessary. Conversely, assume that (9a) is satisfied. Then $|h_2|^2/[w^{2/p}(\frac{1}{2} h_1 - \rho)] = b^2$, $b \in L^{2p/(p-2)}$, and for every $G_1 \in L^p$, $G_2 \in L^2$, we have $|\int G_1 G_2 b| \leq \| G_1 \|_p [\int |G_2 b|^{p/(p-1)}]^{(p-1)/p} \leq \| G_1 \|_p [\, \| G_2 \|_2^{p/(p-1)} (\int |b|^{2p/(p-2)})^{(p-2)/(2p-2)}]^{(p-1)/p}$ $\leq c \| G_1 \|_p \| G_2 \|_2$. Hence (9f) and therefore (9e) hold. Since from the second inequality of (9a) it follows that $\int |g_1|^2 h_1\, dt \leq cM^2 (\int |g_1|^p w)^{2/p}$, $\forall g_1 \in P$, replacing M by $2M$, the last inequality and (9e) give (9d), which is equivalent to $(\rho, w) \in R_M^{2p}$. QED

Remark 3. We saw $(\rho, w) \in R_M^{2p}$ is equivalent to condition (9d), which in the case $p = 2$ takes the more transparent form:

$$|h_2(t)|^2 \leq [3M^2 w(t) - \frac{1}{2} h_1(t)][\frac{1}{2} h_1(t) - \rho(t)], \text{ a.e.} \qquad (9g)$$

For $p = 2$ and $\rho = w$, Theorem 2 contains the Helson-Szegö theorem.

The following theorem reduces for $p = 2$ to a (slightly more precise) theorem of P. Koosis [6].

Theorem 2a. Let $2 \leq p < \infty$, $0 \leq w \in L^1$. The following conditions are equivalent.

(a) *There exists* $\rho \in L^1$, $0 \leq \rho \neq 0$, *with* $(\rho, w) \in R_1^{2p}$.

(b) $w^{2/p} = |h(t)|^2/(a(t)h_1(t))$, *where* $h = h_1 + ih_2 \in H^1$, $h_1 \geq 0$, $a \in L^{p/(p-2)}$, $h_1 \neq 0$.

(c) $(1/w)^{1/(p-1)} \in L^1(T)$. *And* (c) *implies*

(d) *There exists* $h \in H^1$ *with* $|w^{1/(p-1)} - h| \leq w^{1/(p-1)}$, $h \neq 0$.

Proof. (*a*) *implies* (*b*). If $(\rho, w) \in R_1^{2p}$, $\rho \neq 0$, then also $(\frac{1}{2}\rho, w) \in R_1^{2p}$, and from Theorem 2 and Remark 2, (b) follows.

(*b*) *implies* (*c*). We have $1/w = [a(h_1/|h|^2)]^{p/2}$ where $h_1/|h|^2 = \text{Re}(1/h)$ extends to a positive harmonic function in the disc, since $h \in H^1$ and $h_1 \geq 0$ imply that $h(z) \neq 0$ in $|z| < 1$. Hence

$$\int (\frac{1}{w})^{1/p-1} dt = \int (a \text{ Re } \frac{1}{h})^{p/2(p-1)}$$
$$\leq (\int a^{p/(p-2)})^{\frac{p-2}{2p-2}} (\int \text{Re } \frac{1}{h})^{p/(2p-2)} < \infty.$$

(*c*) *implies* (*d*). Since $0 \leq w^{-1/(p-1)} \in L^1$, there exists an analytic function $f(z) = f_1(z) + if_2(z)$ such that $f_1(t) = w(t)^{-1/(p-1)}$

≥ 0 for $t \in T$, and therefore, $f_1(z) \geq 0$. Hence $1 + f(z) \neq 0$ for $|z| < 1$ and $(1 + f(z))^{-1} = h(z) \in H^\infty \subset H^1$. Besides

$$\left|\left(\frac{1}{w}\right)^{1/(p-1)} - h\right| = \left|w^{1/(p-1)} - \frac{1}{1 + f_1 + if_2}\right|$$

$$= \left|\frac{w^{1/(p-1)}(1 + if_2)}{1 + f_1 + if_2}\right| \leq w^{1/(p-1)} \text{ (in } T\text{)}.$$

Hence $\exists h \in H^1$ with $|w^{1/(p-1)} - h| \leq w^{1/(p-1)}$.

(c) implies (b). We just saw that (c) implies the inequality $|w^{1/(p-1)} - h|^2 \leq w^{2/(p-1)}$; hence $|h|^2 \leq 2w^{1/(p-1)}h_1 = 2w^{2/p}h_1 a$, where $a = w^{(2-p)/p(p-1)} \in L^{p/(p-2)}$.

(b) implies (a). Follows from Theorem 2 and Remark 2. QED

CHARACTERIZATION OF R^{pp}

The following theorem reduces for $p = 2$ to the Helson-Szegö theorem (in the more precise form given in [1]).

Theorem 3. Let $2 \leq p < \infty$ and $0 \leq w \in L^1(T)$. Then the following conditions are equivalent ("*within a constant factor*").

(a) $(w, w) \in R_M^{pp}$.

(b) *The operator T_w is u-bounded in $L^{p/(p-2)}$ with u-norm $\leq cM$, where $w^{2/p}T_w f = C(w^{2/p}f)$, $Cf = f + iHf$.*

(c) *For every $0 \leq \phi \in L^{p/(p-2)}$, $\|\phi\| \leq 1$, there exists $h = h_\phi \in H^1$, with* Re $h \geq 0$, *and $a = a_\phi \in L^{p/(p-2)}$, $\|a\|_{p/(p-2)} \leq cM^2$, $a \geq 0$, such that $2\phi w^{2/p} \leq \operatorname{Re} h \leq |h| = aw^{2/p}$.* (10)

Remark 4. Theorem 3 contains as particular case the ordinary M. Riesz inequalities. In fact, if $w \equiv 1$, then $T_w f = Cf$ is bounded in L^2 (by Bessel inequality) and by Theorem 3, T_w is bounded in L^4, that is, $(w, w) \in R_M^{44}$. Hence T_w is bounded in $L^{4/3}$, and by Theorem 3, it is bounded in L^8, and so forth.

Proof. (*a*) *implies* (*c*). By hypothesis,

$$\left(\int |Hf|^p w\, dt\right)^{1/p} \leq M\left(\int |f|^p w\, dt\right)^{1/p}, \ \forall f \in P. \tag{11}$$

For every $0 \leq \phi \in L^{p/(p-2)}$, $\|\phi\| \leq 1$, (11) implies that

$$\left(\int |Hf|^2 \phi w^{2/p}\, dt\right)^{1/2} \leq M\left(\int |f|^p w\, dt\right)^{1/p}, \tag{11a}$$

hence $(\phi w, w) \in R_M^{2p}$, and by Theorem 2, and the second equality of its condition (9a), (10) follows.

(*c*) *equivalent to* (*b*). Clear from the definition of u-boundedness.

(*c*) *implies* (*a*). By hypothesis, given $0 \leq \phi \in L^{p/(p-2)}$, $\|\phi\| \leq 1$, there exist $h \in H^1$, $\operatorname{Re} h \geq 0$, and $a \geq 0$, $\|a\|_{p/(p-2)} \leq cM^2$, such that

$$2\phi w^{2/p} \leq \operatorname{Re} h \leq |h| = a w^{2/p}, \ h = h_\phi, \ a = a_\phi, \tag{11b}$$

and in particular $\int |g|^2 |h| \leq cM^2 \left(\int |g|^p w\right)^{2/p}$, $\forall g \in P$.

Let's prove that (11) holds for every $f \in P$ (within a constant factor). Given $f \in P$, there exists $0 \leq \phi \in L^{p/(p-2)}$, $\|\phi\| = 1$, such that

$$\left(\int |Hf|^p w\, dt\right)^{2/p} = \int |Hf|^2 w^{2/p} \phi\, dt, \tag{11c}$$

and to this ϕ corresponds a $h \in H^1$ such that (11b) holds. Let

$$\varepsilon = \varepsilon_M < 1/(2M^2), \text{ and set}$$

$$E = \{t : \operatorname{Re} h(t) \leq \varepsilon\, |h(t)|\}, \ E' = T - E. \tag{11d}$$

From (11b) and (11c), we get

$$\left(\int |Hf|^p w\right)^{2/p} \leq \varepsilon \int_E |Hf|^2 |h| + \int_{E'} |Hf|^2 w^{2/p} \phi$$
$$\leq \left(\tfrac{1}{2} \int |Hf|^p w\right)^{2/p} + \underline{\int_{E'} |Hf|^2 w^{2/p} \phi}$$

$$\text{hence } \left(\int |Hf|^p w\right)^{2/p} \leq 2 \int_{E'} |Hf|^2 w^{2/p} \phi \leq 2 \int_{E'} |Hf|^2 |h|. \tag{11e}$$

Now in E' we have $\operatorname{Re} h(t) > \varepsilon |h(t)|$, and if we take $\varepsilon \sim 4/(M+1)$, then there exists $\lambda = \lambda_M$ such that for $t \in E'$ we have

$$|M + 1 - \lambda h(t)/|h(t)|| \leq M - 1, \text{ so that}$$
$$(M + 1)|h| - \lambda h = \eta(M - 1)|h|, \ |\eta(t)| \leq 1, \ \lambda = cst, \ \forall t \in E'.$$

This means that for every $t \in E'$ is positive definite the matrix

$$\begin{pmatrix} (M-1)|h| & (M+1)|h| - \lambda h \\ (M+1)|h| - \lambda\bar{h} & (M-1)|h| \end{pmatrix}$$

and therefore, taking $f_1 = f + iHf \in P_+$, $f_2 = \bar{f}_1 \in P_-$, we shall have

$$\int_{E'} |Hf|^2|h| + 2\lambda \operatorname{Re} \int_{E'} f_1\bar{f}_2 h \leq M \int_{E'} |f|^2|h|. \tag{11f}$$

From (11e) and (11f), and taking into account (11b), it follows that $\{\int|Hf|^p w\}^{2/p} \leq 2cM^2(\int|f|^p w)^{2/p} + 2\lambda \operatorname{Re} \int_E f_1\bar{f}_2 h$ (11g)

(we changed E' by E, since $f_1\bar{f}_2 h \in H^1$ and the last integral is zero if taken over $E \cup E' = T$), where $f_1\bar{f}_2 = |f_1|^2$. But since $\operatorname{Re} h \geq 0$, $2\lambda \operatorname{Re} \int_E f_1\bar{f}_2 h = 2\lambda \operatorname{Re} \int_E f_1^2 h = 2\lambda \int_E (\operatorname{Re} f_1^2)\operatorname{Re} h -$ $(\operatorname{Im} f_1^2)(\operatorname{Im} h) = 2\lambda \int_E [(\operatorname{Re} f_1)^2 - (\operatorname{Im} f_1)^2]\operatorname{Re} h - 2\lambda \int_E 2(\operatorname{Re} f_1)$ $(\operatorname{Im} f_2)\operatorname{Im} h \leq 4\lambda \int_E |f|^2 \operatorname{Re} h + 4\lambda \int|f|\,|Hf|\,|h| \leq 4\lambda cM^2(\int|f|^p w)^{2/p} +$ $4\lambda(\int|f|^2 h)^{1/2}(\int|Hf|^2 h)^{1/2} \leq 4\lambda cM^2(\int|f|^p w)^{2/p} + 4\lambda cM^2(\int|f|^p w)^{1/p}$ $(\int|Hf|^p w)^{1/p}$. (11h)

Setting $b^p = \int|f|^p w\,dt$, $a^p = \int|Hf|^p w\,dt$, we get from (11g) and (11h) that $a^2 \leq c_M b^2 + c_M ab$, hence $a \leq c'_M b$, which is equivalent to (11). QED

Theorem 3a. Let $2 \leq p < \infty$, $0 \leq \rho$, $w \in L^1$. *Then* $(\rho, w) \in R_M^{pp}$ *iff for every* $0 \leq \phi \in L^{p/(p-2)}$, $\|\phi\| \leq 1$, *there exist* $h = h_\phi \in H^1$, $\operatorname{Re} h \geq 0$, *and* $0 \leq a = a_\phi \in L^{p/(p-2)}$, $\|a\| \leq cM^2$, *such that* $\phi\rho^{2/p} \leq \frac{1}{2} \operatorname{Re} h = h_1$, *and*

$$w^{2/p} = h_2^2/[a(h_1 - 2\phi\rho^{2/p}] = |h|^2/(a_1 h_1),\ a_1 \in L^{p/(p-2)},\ \|a_1\| \leq cM^2.$$

Proof. As for (11a), one sees that the condition $(\rho, w) \in R_M^{pp}$ is equivalent to the following one: for each $0 \leq \phi \in L^{p/(p-2)}$, $\| \phi \| \leq 1$, $(\phi\rho^{2/p}, w) \in R_M^{2p}$ holds, and it is enough to apply Theorem 2. QED

Remark 5. The above arguments apply also to the case R_M^{pq}, $2 \leq p$, $q < \infty$ and to some other variants. The case $1 < p < 2$ is treated by duality.

REFERENCES

1. Arocena, R., and Cotlar, M. A generalized Herglotz-Bochner theorem and weighted inequalities with finite measures.

2. Arocena, R.; Sadosky, C.; and Cotlar, M. Weighted inequalities in L^2 and lifting properties. To appear in *Adv. Math.*, 1981.

3. Carleson, L., and Jones, P. Weighted norm inequalities and a theorem of Koosis. *Mittag-Leffler Report* No. 2, 1981.

4. Cotlar, M., and Sadosky, C. On the Helson-Szegö theorem and related class of Toeplitz kernels. *Proc. Symp. Pure Math. AMS* 35 (1979): 383-407.

5. Helson, H., and Szegö, G. A problem in prediction theory. *Ann. Math. Pura Appl.* 51 (1960): 107-138.

6. Koosis, P. Moyennes quadratiques pondérees de fonctions périodiques et de leurs conjugueés harmoniques. *C. R. Acad. Sc. Paris* 291 (1980): A 255-257.

7. Rubio de Francia, J. Boundedness of maximal functions and singular integrals in weighted L^p spaces.

8. Sawyer, E. T. Two weighted norm inequalities for maximal and integral operators. Italian-American Conference in Harmonic Analysis, Minneapolis, 1981.

SOME TOPICS IN CALDERÓN-ZYGMUND THEORY

R. Fefferman
University of Chicago

The classical theory of Calderón-Zygmund [1, 2, 16] deals with singular convolution operators on R^n. Its development proceeds by using the maximal operator of Hardy-Littlewood, introducing a way of decomposing a function in $L^1(R^n)$ into its "good" and "bad" parts, and then finally obtaining the desired estimates on the appropriate class of singular integrals. This class of singular integrals is invariant under the usual dilations of R^n, $x \to \delta x$, $\delta > 0$.

In this article we shall survey some of the recent results in the analogous theory that results when we consider the natural classes of convolution operators on $R^{n+m} = R^n \times R^m$ that are invariant under the two-parameter family of dilations $(x, y) \to (\delta_1 x, \delta_2 y)$, $\delta_1, \delta_2 > 0$ and $x \in R^n$, $y \in R^m$. In particular, we shall first consider the right maximal operator in this setting, the so-called strong maximal operator. Next we shall discuss a way of decomposing functions on $R^n \times R^m$ that play a role in the product setting similar to the Calderón-Zygmund decomposition or atomic decomposition in the classical case.

Last, we shall discuss the boundedness of a natural class of singular integral operators that are direct generalizations of the Calderón-Zygmund operators to the two-parameter setting.

To begin with, let us consider the maximal function, defined for functions f on $R^n \times R^m$ as follows: for $\delta_1, \delta_2 > 0$, let the dilation ρ_{δ_1,δ_2} be given by $\rho_{\delta_1,\delta_2}(x, y) = (\delta_1 x, \delta_2 y)$, and Q_1, Q_2 denote the unit cube in R^n, R^m, respectively.

Let $M_s(f)(x, y)$

$$= \sup_{\delta_1,\delta_2>0} \frac{1}{\delta_1^n \delta_2^m} \int_{(s,t)\in\rho_{\delta_1,\delta_2}(Q_1\times Q_2)} |f(x + s, y + t)\, ds\, dt.$$

Then the sharp results on this "strong maximal operator" are due to Jessen, Marcinkiewicz, and Zygmund in 1935 [14] and the result is that M_s maps $L \log^+ L(Q_1 \times Q_2)$ boundedly to weak $L^1(Q_1 \times Q_2)$:

$$m\{(x, y) \in Q_1 \times Q_2 | M_s(f)(x, y) > \alpha\} \leq \frac{C}{\alpha} \| f \|_{L\log^+L(Q_1\times Q_2)}.$$

Their proof is very elegant and proceeds by dominating M_s by a composition of two Hardy-Littlewood maximal operators acting in each of the factor spaces R^n and R^m. More recently, Cordoba and R. Fefferman [8] gave a different proof of the boundedness properties of M_s by proving a covering lemma, which we now state.

Covering Lemma: Given a collection, $\{R_k\}$ *of rectangles (here rectangle means a translate of a set of the form* $\rho_{\delta_1,\delta_2}(Q_1 \times Q_2)$*) in* $R^n \times R^m$ *contained inside* $Q_1 \times Q_2$*, there exists a subcollection* $\{\tilde{R}_i\}$ *such that*

$$|U\tilde{R}_i| > c|UR_k| \tag{1}$$

and

$$\| \Sigma\, \chi_{\tilde{R}_i} \|_{\exp(L)-1} \leq C|UR_k|. \tag{2}$$

There are advantages that come from the geometric approach to understanding the operator M_s. For example, the methods used in proving the covering lemma can be used to prove covering lemmas for other families of sets (see Strömberg [17], Cordoba and Fefferman [9], Cordoba [7], and so forth) where the corresponding maximal operator was not previously known to be bounded on the appropriate Orlicz spaces. The key feature of the covering lemma that allows us to attack more difficult maximal operators is the fact that we are able to dominate M_s, not by the composition of two one-parameter maximal functions, but by a single such maximal function. Thus simpler operators with fewer parameters involved control more complicated operators where direct iteration is impossible. Another application of the covering lemma is toward understanding the operator M_s better. It is natural to ask what happens when we replace Lebesgue measure by a positive Borel measure $d\mu$ on $R^n \times R^m$. In other words, we regard $d\mu$ as the only measure of interest for the moment and take all averages, L^p norms, and so on, with respect to $d\mu$.

Define accordingly M_s^μ by

$$M_s^\mu(f)(x, y) = \sup_{\delta_1,\delta_2>0} \frac{1}{\mu((x, y) + \rho_{\delta_1,\delta_2}(Q_1 \times Q_2))} \iint_{(x,y)+\rho_{\delta_1,\delta_2}(Q_1\times Q_2)} |f(s, t)|\, d\mu(s, t).$$

We ask for results like

$$\| M_s^\mu(f) \|_{L^p(d\mu)} < C_p \| f\|_{L^p(d\mu)}, \quad p > 1 \qquad (*)$$

(or the corresponding $L \log^+ L$ result).

There are measures (unlike in the one-parameter case where we have balls, not rectangles) $d\mu$ for which M_s^μ is unbounded on all the classes $L^p(d\mu)$ $p < \infty$. On the other hand, there must be some result that says that if the measure $d\mu$ looks enough like Lebesgue measure, then (*) holds. By use of a covering lemma argument we have the following: If $d\mu = \omega(x, y)\, dx\, dy$ and if $\omega(\cdot, y)$ as a function of x is uniformly in the Muckenhoupt class $A^\infty(R^n)$ and similarly $\omega(x, \cdot) \in A^\infty(R^m)$ uniformly, then $\| M_s^\mu f \|_{L^p(d)} \leq C \| f \|_{L^p(d\mu)}$, $p > 1$. The A^∞ assumption is somewhat similar in flavor to the doubling condition for a measure, so its appearance is perhaps not too surprising [11].

Next, let us consider an analogue in the product case, of the decomposition of a function into its "good" and "bad" parts. To be more specific, consider the atomic decomposition of a function in $H^1(R^n \times R^m)$. It is only rather recently that the space $H^1(R^n \times R^m)$ has been completely characterized in a satisfactory way. (See Gundy and Stein [13] and Merryfield [15].) In fact, we can regard $H^1(R^n \times R^m)$ to be the set of all $f \in L^1(R^n \times R^m)$ such that for $f*(x, y) = \sup_{\substack{|s-x|<\tau_1 \\ |t-y|<\tau_2}} |f*(P_{\tau_1} \cdot P_{\tau_2})(s, t)|$, $f^* \in L^1(R^n \times R^m)$.

$$\left(\text{here } P_{\tau_1} \cdot P_{\tau_2}(x, y) = \frac{c_n c_m \tau_1 \tau_2}{(|x|^2 + \tau_1^2)^{\frac{n+1}{2}}(|y|^2 + \tau_2^2)^{\frac{m+1}{2}}}\right).$$

Equivalently, $H^1(R^n \times R^m)$ could be defined via area integrals and, as expected, we have considerable choice, in the definition of the maximal function or area integral, as to which approximate identity is used.

Now, before giving a decomposition for H^1, let us consider the dual of $H^1(R^n \times R^m)$. From the one-parameter case of C. Fefferman [10], one might expect to be able to characterize those $\phi(x, y) \in (H^1)^*$ as follows: Let $\Phi(x, y, \tau_1, \tau_2) = \phi*(P_{\tau_1} \cdot P_{\tau_2})(x, y)$ be the biharmonic extension of ϕ to $R_+^{n+1} \times R_+^{m+1}$. Given cubes $Q^1 \subset R^n$ and $Q^2 \subset R^m$, define the Carleson region by $C(Q^1) = \{(x, \tau_1) \in R_+^{n+1} | x \in Q^1, 0 < \tau_1 < \text{side length of } Q^1\}$, define $C(Q^2)$ similarly, and set $C(Q^1 \times Q^2) = C(Q^1) \times C(Q^2)$. Then the guess would be that $\phi \in [H^1(R^n \times R^m)]^*$ iff

$$\iint_{C(Q^1 \times Q^2)} |\nabla_1 \nabla_2 \Phi|^2 \tau_1 \tau_2 \, dx \, d\tau_1 \, dy \, d\tau_2 \leq K|Q^1 \times Q^2|. \qquad (\sim)$$

Unfortunately, this guess is not correct (see Carleson [3]). However, somewhat later, in [4], Alice Chang showed that for $\phi \in (H^1)^*$ we have

$$\iint_{C(\Omega)} |\nabla_1\nabla_2\Phi|^2 \tau_1\tau_2 \, dx\, d\tau_1 \, dy\, d\tau_2 < K|\Omega| \tag{$\approx$}$$

for every open set $\Omega \subset R^n \times R^m$, where $C(\Omega)$ is defined as

$$\bigcup_{Q^1\times Q^2 \subset \Omega} C(Q^1 \times Q^2).$$

We want an atomic decomposition that gives the result that any function ϕ satisfying ($\approx$) acts continuously on $H^1(R^n \times R^m)$. In addition, as another application, it follows almost immediately from our decomposition that we can interpolate between $H^1(R^n \times R^m)$ and L^2 to get L^P, $1 < p < 2$. Now, to the decomposition. Corresponding to ($\approx$) there is a first (incorrect) guess at what an atom should be. We will say that a function $\alpha(x, y)$ on $R^n \times R^m$ supported on a rectangle $Q^1 \times Q^2$ is a pre-atom, provided

$$\text{for } x \in Q^1, \int_{Q^2} \alpha(x, y)\, dy = 0 \tag{3}$$

$$\text{for } y \in Q^2, \int_{Q^1} \alpha(x, y)\, dx = 0 \tag{4}$$

and

$$\alpha \in C^N(Q^1 \times Q^2), \tag{5}$$

where N is some integer depending only on n and m. We define the "norm" of a pre-atom α, $\|\alpha\|$ to be the C^N norm of $\tilde{\alpha}$, where $\tilde{\alpha}$ is the dilate of α that lives in a cube in R^{n+m} of side length 1.

Now, finally, we can give the definition of an atom corresponding to ($\approx$): A function $a(x, y)$ supported in an open set $\Omega \subset R^n \times R^m$ is an atom iff it can be written as a sum $a = \sum_R \alpha_R$, where the R are rectangles $\subset \Omega$ no one of which is contained in the double of any other, the α_R are pre-atoms in R, and $\Sigma|R|\, \|\alpha_R\|^2 \leq \frac{1}{|\Omega|}$. Since conditions (3) through (5) for pre-atoms force any two terms in the sum $\Sigma\, \alpha_R$ to be "almost orthogonal," the requirement that $\Sigma|R|\, \|\alpha_R\|^2 \leq \frac{1}{|\Omega|}$ just expresses, in terms of the α_R, the estimate $\|a\|^2_{L^2(\Omega)} \leq \frac{C}{|\Omega|}$. Then in [5] Chang and Fefferman show that every $f \in H^1(R^n \times R^m)$ can be written

$$f = \Sigma \lambda_k a_k, \quad \text{where} \quad \Sigma |\lambda_k| \leq c \|f\|_{H^1}$$

and a_k are atoms. Furthermore [6], each atom has H^1 norm essentially ≤ 1, so that, conversely, every sum $\Sigma \lambda_k a_k$ with $\Sigma|\lambda_k| < \infty$ and a_k atoms is in H^1.

We next wish to announce results on singular integral operators in the product setting. Recall that in [2], Calderón and Zygmund showed that for a kernel $K(x)$ on R^n that is C^1 away from 0 and satisfies

$$|K(x)| \leq \frac{C}{|x|^n} \tag{6}$$

$$\int_{\alpha<|x|<\beta} K(x)\, dx = 0 \quad \text{for all} \quad 0 < \alpha < \beta \tag{7}$$

and

$$|\nabla K(x)| \leq \frac{C}{|x|^{n+1}}, \tag{8}$$

the operators $T_\varepsilon f = f * (KX_{|x|>\varepsilon})$ are uniformly bounded on $L^p(R^n)$ for $1 < p < \infty$, and as $\varepsilon \to 0$ approach, in L^p norm a bounded operator. They also showed the existence a.e. of the pointwise limits $\lim_{\varepsilon\to 0} T_\varepsilon f(x)$, proving maximal inequalities such as

$$\left\| \sup_{\varepsilon>0} |T_\varepsilon f(x)| \right\|_p \leq C_p \|f\|_p \quad 1 < p < \infty.$$

Recently E. M. Stein and the author extended most of these results to the setting of $R^n \times R^m$. We shall not attempt here to give the results with the weakest possible assumptions on the kernels involved. Rather we shall just try to give the flavor of the theorems. Sharper results will be given in [12].

To begin with, we study the boundedness of the operators.

(i) Let $K(x, y)$ be a kernel on $R^n \times R^m$ that is C^2 away from $\{(x, y)|x = 0\} \cup \{(x, y)|y = 0\}$ and assume:

$$|K(x, y)| < \frac{C}{|x|^n |y|^m} \quad x, y \neq 0 \tag{9}$$

$$\int_{\alpha<|x|<\beta} K(x, y)\, dx = 0 \tag{10}$$

for all $y \in R^m$ and all $0 < \alpha < \beta$

$$\int_{\alpha<|x|<\beta} K(x, y)\, dy = 0$$

for all $x \in R^n$ and all $0 < \alpha < \beta$

and

$$|\nabla_x K(x, y)| \leq \frac{C}{|x|^{n+1}|y|^m}, \quad |\nabla_y K(x, y) \tag{11}$$

$$\leq \frac{C}{|x|^n |y|^{m+1}}, \ |\nabla_x \nabla_y K(x, y) \leq \frac{C}{|x|^{n+1}|y|^{m+1}}.$$

Then if $K_{\varepsilon,\delta}(x, y) = K(x, y)\chi_{|x|>\varepsilon}(x)\chi_{|y|>\delta}(y)$, we have $\| f*K_{\varepsilon,\delta} \|_p \leq C_p \| f \|_p$ for $1 < p < \infty$, where C_p does not depend on ε or δ, and the limit $\lim_{\varepsilon,\delta\to 0} f*K_{\varepsilon,\delta}$ exists in the L^p norm as ε, δ tend to zero independently.

(ii) If $w(x, y) \geq 0$ on $R^n \times R^m$ has the property that $w(\cdot, y) \in A^p(R^n)$ uniformly in y, $w(x, \cdot) \in A^p(R^m)$ uniformly in x, then under the same assumptions on K as in (i) we have

$$\| f * K \|_{L^p(w)} \leq C_p \| f \|_{L^p(w)} \qquad 1 < p < \infty.$$

(iii) If $K_{\varepsilon,\delta}$ is as above in (i), then

$$\| \sup_{\varepsilon,\delta>0} |f * K_{\varepsilon\delta}| \|_p \leq C_p \| f \|_p \text{ and } \lim_{\varepsilon,\delta\to 0} f * K_{\varepsilon\delta}(x, y)$$

exists a.e. for every $f \in L^p$, $1 < p < \infty$.

Finally, we refer the reader to [12] of R. Fefferman and E. M. Stein, where the $L \log^+ L$ versions of the theorems above, and with weaker assumptions on K, may be found.

REFERENCES

1. Calderón, A. P., and Zygmund, A. On singular integrals. *Amer. J. of Math.* 78 (1956): 289-309.

2. Calderón, A. P., and Zygmund, A. On the existence of certain singular integrals. *Acta Math.* 88 (1952): 85-139.

3. Carleson, L. A counterexample for measures bounded on H^p spaces for the bidisk. *Mittag Leffler Report* No. 7, 1974.

4. Chang, S. Y. Carleson measure on the bi disc. *Ann. Math.* 109 (1979): 613-620.

5. Chang, S. Y., and Fefferman, R. A continuous version of duality of H^1 with BMO on the bidisc. *Ann. Math.* 112 (1980): 179-201.

6. Chang, S. Y., and Fefferman, R. To appear in *Amer. J. of Math.*

7. Cordoba, A. Maximal functions, covering lemmas and Fourier multipliers. *Proc. Symp. in Pure Math. 35, Harmonic Anal. in Euclidean Spaces* (1979): 29-50.

8. Cordoba, A., and Fefferman, R. A geometric proof of the strong maximal theorem. *Ann. Math.* 102 (1975): 95-100.

9. Cordoba, A., and Fefferman, R. On differentiation of integrals. *Proc. Nat. Acad. Sci.* 74 (1977): 2211-2213.

10. Fefferman, C. Characterizations of bounded mean oscillation. *Bull. Amer. Math. Soc.* 77 (1971): 587-588.

11. Fefferman, R. Strong differentiation with respect to measures. *Amer. J. of Math.* 103 (1981): 33-40.

12. Fefferman, R., and Stein, E. M. Singular integrals on product domains. To appear.

13. Gundy, R., and Stein, E. M. H^p theory for the polydisk. *Proc. Nat. Acad. Sci.* 76 (1979): 1026-1029.

14. Jessen, B.; Marcinkiewicz, J.; and Zygmund, A. Note on the differentiability of multiple integrals. *Fund. Math.* 24 (1935): 217-234.

15. Merryfield, K. H^p spaces in poly-half spaces. Ph.D. thesis, Univ. of Chicago.

16. Stein, E. M. *Singular Integrals and Differentiability Properties of Functions*. Princeton, N.J.: Princeton Univ. Press, 1970.

17. Strömberg, J. O. Weak estimates on maximal functions with rectangles in certain directions. *Arkiv für Math.* 15 (1977): 229-240.

SINGULAR INTEGRALS WITH KERNELS OF MIXED HOMOGENEITIES

D.H. Phong
Columbia University
E.M. Stein
Princeton University

INTRODUCTION

The purpose of this article is to study two closely related classes of singular integral operators arising in noncoercive boundary value problems for elliptic differential equations. Consider, for example, a boundary value problem for the Laplace equation $\Delta u = f$, $b(x, D)u|_{b\Omega} = 0$ on a C^∞ domain Ω; there are then two ways of describing the solution u:

(a) either through a Green's function, that is, an integral formula of the type

$$u(x) = \int_\Omega \frac{2^{-1}\pi^{-n/2}\Gamma(n/2)}{|x - y|^{n-2}} f(y)\, dy + \int_\Omega K(x, y) f(y)\, dy,$$

where $1/|x - y|^{n-2}$ is the usual fundamental solution for the Laplacian and all the properties of u (such as regularity, non-coerciveness of the problem, and so on) will be determined by $K(x, y)$. Since $K(x, y)$ is harmonic, the integral operator H

This material is based on work supported by NSF Grant Nos. MCS78-27119 and MCS-80-03072.

with kernel $K(x, y)$ is smoothing in the interior. We are thus led to study the boundary behavior of operators with this last property, which we shall call here Hilbert integral operators for reasons that will be apparent later;

(b) or, following Calderón and Hörmander, observe that u is determined by $u_b = u|_{b\Omega}$ and that u_b satisfies a pseudodifferential equation on the boundary. Of importance now is the parametrix for this equation, which will be a classical pseudodifferential operator when the problem is coercive but may otherwise belong to a much more complicated class.

The setting we shall adopt here is that of the $\bar{\partial}$-Neumann problem on strongly pseudoconvex domains. It is expected, however, that similar considerations will apply as well to other subelliptic boundary value problems. Always present in these situations are two kinds of homogeneities: the usual isotropic one associated with the elliptic equation in Ω, and near $b\Omega$ a new parabolic one dictated by the noncoercive boundary conditions. The kernels arising thus are products of kernels homogeneous in the isotropic sense with kernels homogeneous in the parabolic sense. It will be seen below that there are many marked differences between them and the kernels encountered in classical examples.

SOME CLASSICAL EXAMPLES

In the classical examples the kernels that appear are homogeneous either in the isotropic or the parabolic sense but are not mixtures of both.

Isotropic kernels arise when the problem is coercive and can be given a rather complete description. Locally we may write $\Omega = \{(x, \rho) \in \mathbf{R}^n \times \mathbf{R}^+\}$, and parametrices in the boundary are simply classical pseudodifferential operators given by

$$Au(x) = \int e^{i<x,\xi>} a(x, \xi)\hat{u}(\xi)\, d\xi$$

$$|D_x^\alpha D_\xi^\beta a(x, \xi)| \le C(1 + |\xi|)^{-|\beta|}$$

which implies that

$$(Au)(x) = \int K(x, x - y)u(y)\, dy$$

$$|D_x^\alpha D_z^\beta K(x, z)| \le c|z|^{-n-|\beta|}. \tag{1}$$

The L^p and Schauder estimates for this class are well known. For Hilbert integral operators we also have at the same time a symbol representation

$$(Hu)(x, \rho) = \iint_0^\infty e^{i<x,\xi>} a(x, \rho, \mu, \xi)\hat{\mu}(\xi, \mu)\, d\mu\, d\xi$$

$$|\mu^\alpha \rho^\beta D_\mu^\gamma D_x^\delta D_\xi^\omega D_\rho^\varepsilon a|$$

$$\le C(1 + |\xi|)^{-|\omega|+(\gamma+\delta)-(\alpha+\beta)+1+\varepsilon}$$

and a kernel representation

$$(Hu)(x, \rho) = \iint_0^\infty K(x, \rho, \mu, x - y)u(y, \mu)\, d\mu\, dy \tag{2}$$

$$|D_x^\alpha D_\rho^\beta D_\mu^\gamma D_z^\delta K(x, \rho, \mu, z)| \le C(|z| + \rho + \mu)^{-(n+1)-(\beta+\gamma+|\delta|)}.$$

(In the above we have stated conditions only for operators of order 0. The statements for operators of any order can be easily reduced to this case.) For Schauder estimates it is important to note that $K(x, \rho, \mu, z)$ satisfies in addition to (2) a cancellation condition in the tangential variable z. We shall not discuss this type of estimate here but instead shall refer to [9]. From the point of view of L^p estimates, however, H is just the analogue in higher dimensions of the classical Hilbert integral (hence our terminology)

$$(Hv)(x) = \int_0^\infty \frac{v(y)\, dy}{x + y}$$

and bounds for H can simply be obtained as follows:

$$|(Hu)(x,\rho)|$$

$$\leq \iint_0^\infty |K(x,\rho,\mu,x-y)|^{\frac{1}{p}} |u(y,\mu)| \mu^{-\frac{\varepsilon}{q}} |K(x,\rho,\mu,x-y)|^{\frac{1}{q}} \mu^{\frac{\varepsilon}{q}}\, d\mu\, dy$$

$$\leq \left[\iint_0^\infty |K(x,\rho,\mu,x-y)|\,|u(y,\mu)|^p \mu^{-\frac{\varepsilon p}{q}}\, d\mu\, dy\right]^{\frac{1}{p}}$$

$$\times \left[\iint_0^\infty |K(x,\rho,\mu,x-y)| \mu^{\varepsilon}\, d\mu\, dy\right]^{\frac{1}{q}}$$

$$\leq C\rho^{\frac{\varepsilon}{q}} \left[\iint_0^\infty |K(x,\rho,\mu,x-y)|\,|u(y,\mu)|^p \mu^{-\frac{\varepsilon p}{q}}\, d\mu\, dy\right]^{\frac{1}{p}}.$$

Hence

$$\iint_0^\infty |(Hu)(x,\rho)|^p\, dx\, d\rho$$

$$\leq C' \iint_0^\infty \left\{\iint_0^\infty |K(x,\rho,\mu,x-y)| \rho^{\frac{\varepsilon p}{q}}\, d\rho\, dx\right\} \mu^{-\frac{\varepsilon p}{q}} |u(y,\mu)|^p\, d\mu\, dy$$

$$\leq C'' \iint_0^\infty |u(y,\mu)|^p\, d\mu\, dy,$$

establishing L^p boundedness. Note that throughout only the size of $K(x, \rho, \mu, z)$ mattered.

In the previous examples all directions are weighted equally. Analogues of classical singular integrals that are controlled by a different kind of homogeneity occur in the context of:

(a) the heat equation $\Sigma_{j=1}^n \partial^2/\partial x_j^2 - \partial/\partial t$ whose fundamental solution is $2^{-n}(\pi t)^{-n/2}\exp(-|x|^2/4t)$, a distribution homogeneous with respect to the dilation $\delta \cdot (x, t) = (\delta x, \delta^2 t)$;

(b) the boundary Laplacian $\Box_b$ on the Heisenberg group (Folland and Stein [3]). It is now well known that there is a close relationship between the complex structure of the Siegel domain of type II given by

$$U = \{(z, z_{n+1}) \in \mathbf{C}^{n+1};\quad \operatorname{Im} z_{n+1} - |z|^2 > 0\}$$

and the Lie group structure of the Heisenberg group:

$$H_n = \{(z, t) \in \mathbf{C}^n \times \mathbf{R}\}$$

$$(z, t) \cdot (z', t') = (z + z', t + t' + 2 \operatorname{Im} z \cdot \bar{z}').$$

In fact, H_n acts simply and transitively on $\bar{U}$ by

$$(z, t) \cdot (z', z_{n+1}) = (z + z', t + i|z|^2 + z_{n+1} + 2iz \cdot \bar{z}')$$

and hence can be identified with bU; the $\bar{\partial}_b$ operator can then be expressed in a simple manner in terms of the left-invariant vector fields on H_n. More precisely, with

$$X_j = \frac{\partial}{\partial x_j} + 2y_j \frac{\partial}{\partial t},\quad Y_j = \frac{\partial}{\partial y_j} - 2x_j \frac{\partial}{\partial t},\quad iT = \frac{\partial}{\partial t} \qquad 1 \leq j \leq n$$

and

$$\bar{\omega}^j = d\bar{z}^j \qquad 1 \leq j \leq n,$$

respectively, as bases for the Lie algebra of H_n and the (0, 1) forms on bU, a straightforward calculation shows that

$$\bar{\partial}_b u = \sum_{j=1}^{n} \frac{1}{2}\left(X_j + iY_j\right) u\bar{\omega}^j \equiv \sum_{j=1}^{n} (\bar{Z}_j u)\bar{\omega}^j.$$

Equip $\bar{U}$ with the metric defined by requiring that the $\bar{\omega}^j$, $1 \leq j \leq n$ together with $\bar{\omega}^{n+1} = 2^{1/2} \Sigma_{j=1}^{n} \bar{z}_j\, dz_j - i2^{-1/2} dz_{n+1}$ be orthonormal. The boundary Laplacian $\square_b \equiv \bar{\partial}_b^* \bar{\partial}_b + \bar{\partial}_b \bar{\partial}_b^*$ on (0, 1) forms with respect to the metric induced by bU then takes the form

$$\square_b\left(\sum_{j=1}^{n} \varphi_j \bar{\omega}^j\right) = \sum_{j=1}^{n} \left(-\frac{1}{2} \sum_{k=1}^{n} \left(Z_k \bar{Z}_k + \bar{Z}_k Z_k\right) - (n - 2)T\right)\varphi_j \bar{\omega}^j .$$

$\Box_b$ is not elliptic with the direction dual to T being characteristic at each point; nevertheless, for $n \geq 2$ L^p, Schauder estimates and regularity can still be derived from the fact that $\Box_b$ admits a fundamental solution

$$\psi = \frac{\Gamma(n-1)}{2^{2-2n}\Pi^{n+1}} \cdot (|z|^2 + it)^{-1}(|z|^2 - it)^{-n+1}, \tag{3}$$

which is a homogeneous distribution with respect to the dilation $\delta(z, t) = (\delta z, \delta^2 t)$;

(c) the Bergman kernel on strongly pseudoconvex domains (Fefferman [2]). This is the analogue of the operators (2) in the noncommutative and parabolic context of (b). For the Siegel upper half-space U parametrized as $H_n \times \mathbf{R}_+ = \{(z, t, \rho); \rho = \text{Im } z_{n+1} - |z|^2\}$, the Bergman projection is given by

$$(Hu)(z, t, \rho) = \iint_0^\infty K\big((z, t) \cdot (w, s)^{-1}; \rho + \mu\big)u(w, s; \mu)\, d\mu\, dw\, ds \tag{4}$$

with $K(z, t, \rho) = (2^{n-1}n!/\Pi^{n+1})(|z|^2 + \rho - it)^{-(n+1)}$. Observe that K is homogeneous with respect to the dilation $\delta(z, t, \rho) = (\delta z, \delta^2 t, \delta^2 \rho)$. Again L^p estimates depend only on the size of K and follow from a simple adaptation of the Hilbert integral arguments used earlier. The Bergman projection for strongly pseudoconvex domains in suitable coordinates is an asymptotic sum of terms of the type (4).

A NONCOERCIVE PROBLEM: THE $\bar{\partial}$-NEUMANN PROBLEM

The $\bar{\partial}$-Neumann problem is the problem of inverting the complex Laplacian $\Box = \bar{\partial}\bar{\partial}^* + \bar{\partial}^*\bar{\partial}$. From the inverse of $\Box$, usually called the Neumann operator and denoted N, one can easily derive, for example, the solution of $\bar{\partial}u = f$ which is orthogonal to holomorphic functions ($u = \bar{\partial}^*Nf$) and the Bergman projection ($H = I - \bar{\partial}^*N\bar{\partial}$). The $\bar{\partial}$-Neumann problem is in reality a boundary value problem, since the domains of $\bar{\partial}^*$ on C^∞ forms are defined by differential conditions at the boundary. The conditions are not coercive and L^2 estimates and regularity for strongly pseudoconvex domains were only obtained by Kohn [7] in the early 1960s. Later on a parametrix was constructed by Greiner and Stein [5], relying on earlier work of Folland and Stein [3] on $\square_b$. Here we shall be mainly interested in the kernel of N and the properties of the classes of operators defined by such kernels. For the sake of simplicity, we shall restrict ourselves to the case of the upper half-space U, in which $\square$ on (0, 1)-forms is given by (cf. [5])

$$\square\left(\sum_{j=1}^{n+1} \varphi_j \bar{\omega}^j\right) = \sum_{j=1}^{n} \left[-\frac{1}{2}\sum_{k=1}^{n}(Z_k\bar{Z}_k + \bar{Z}_k Z_k) - Z_{n+1}\bar{Z}_{n+1} - (n-2)T\right]\varphi_j\bar{\omega}^j$$
$$+\left[-\frac{1}{2}\sum_{k=1}^{n}(Z_k\bar{Z}_k + \bar{Z}_k Z_k) - Z_{n+1}\bar{Z}_{n+1} - nT\right]\varphi_{n+1}\bar{\omega}^{n+1}$$
$$\mathscr{D}(\square) = \left\{\sum_{j=1}^{n+1} \varphi_j\bar{\omega}^j,\ \varphi_{n+1}\big|_{bU} = 0,\ \bar{Z}_{n+1}\varphi_j\big|_{bU} = 0 \quad 1 \le j \le n\right\}.$$

Here $\bar{Z}_{n+1} = -i2^{1/2}\partial/\partial\bar{z}_{n+1}$. Only the equations for the tangential components are noncoercive, and they can be written under the scalar form

$$\left[-\frac{\partial^2}{\partial\rho^2} + (2\square_b + T^2)\right]\varphi = u,\ \left(T + \frac{\partial}{\partial\rho}\right)\varphi\big|_{bU} = 0. \tag{5}$$

Observe that $2\square_b + T^2$ is an elliptic differential operator on the boundary. If G denotes the Green's function with Dirichlet boundary conditions for $-\partial^2/\partial\rho^2 + (2\square_b + T^2)$ and φ_b the restriction of φ to bU, then

$$\varphi = Gu + e^{-\rho(2\square_b+T^2)^{1/2}}\varphi_b$$

and the boundary condition in (5) is equivalent to

$$[(2\square_b + T^2)^{1/2} - T]\varphi_b = \left(T + \frac{\partial}{\partial\rho}\right)Gu\big|_{bU}.$$

Greiner and Stein called the operator on the left $\square_+$. It is readily seen that $\square_+$ is not elliptic and that its characteristic variety of each point consists of a half-line. In contrast with $\Box_b$, whose fundamental solution is homogeneous in the parabolic sense, this suggests a mixture of both kinds of homogeneities. To invert $\Box_+$, introduce $\Box_- = (2\Box_b + T^2)^{1/2} + T$ and observe that

$$\Box_+\Box_- = 2\Box_b$$

and thus

$$\Box_+^{-1} = \frac{1}{2}\Box_-\Box_b^{-1},$$

which shows in particular that $\Box_+^{-1}$ is the composition of a classical isotropic pseudodifferential operator with a parabolic nonclassical one. Estimates for $\Box_+^{-1}$ can then be derived from estimates for $\Box_-$ and $\Box_b^{-1}$ separately. Applying the calculus of pseudodifferential operators with $\Box_b^{-1}$ viewed as an operator of type $S^{-1}_{1/2,1/2}$, we see readily that the main term on the symbol of $\Box_+^{-1}$ is a product of an isotropic symbol with a parabolic one; however, in general we do not know the asymptotic expansions of the Fourier transforms of such symbols. The same difficulties arise in finding the kernel of N. Answers to these questions are provided below by a different method.

MAIN THEOREMS

We list here the main theorems of this article, which provide answers to some of the questions raised in the previous sections. Set $\Phi(z, t, \rho) = 2|z|^2 + \rho^2 + t^2$, $\Psi(z, t, \rho) = |z|^2 + \rho - it$, $\Psi_0(z, t, \rho) = \rho - it$.

Theorem 1. [8, 10]. (a) *The Neumann kernel for* $U \subset \mathbf{C}^{n+1}$ $(n \geq 2)$ *up to isotropic terms* (*that is, of the type* (2)) *smoothing of order* 3 *and higher is given by*

$$(Nu)(z, t, \rho) \sim \frac{2^{n-1}\Gamma(n)}{\Pi^{n+1}}$$

$$\iiint_0^\infty \Phi^{-n}\big((z, t)\cdot(w, s)^{-1}, |\rho - \mu|\big)u(w, s, \mu)\, d\mu\, dw\, ds$$

$$-\iiint_0^\infty K_n\big((z, t)\cdot(w, s)^{-1}, \rho + \mu\big)u(w, s, \mu)\, d\mu\, dw\, ds$$

with

$$K_n(z, t, \rho) = |z|^{-2n+2}(\theta_n + \lambda_n),$$

where $\theta_n(z, t, \rho)$ and $\lambda_n(z, t, \rho)$ *are kernels with the following properties*:

(i) θ_n *and* λ_n *depend only on* $|z|^2$;

(ii) λ_n *is a rational function homogeneous of order* -4 *with respect to the parabolic dilations whose denominator is a power of* Ψ;

(iii) θ_n *is a sum of rational fractions homogeneous of orders* ≥ -2 *with respect to the isotropic dilation whose denominators are powers of* Φ;

(iv) $\lambda_n + \theta_n$ *vanish of order* $2n - 2$ *at* $z = 0$.

(b) *When* $n = 1$, $Z_1 N$ *up to isotropic kernels smoothing of order* 2 *and higher is given by*

$$(Z_1 Nu)(z, t, \rho) \sim -\frac{2}{\Pi^2}$$

$$\iiint_0^\infty (\bar{z}\Phi^{-2})\big((z, t)\cdot(w, s)^{-1}, |\rho - \mu|\big).$$

$$u(w, s, \mu)\, d\mu\, dw\, ds$$

$$+\frac{1}{\Pi^2}\iiint_0^\infty \left\{\frac{\bar{z}}{|z|^2}(\bar{\Psi}^2 - \bar{\Psi}_0^2\Phi^{-2})\right\}$$

$$\big((z, t)\cdot(w, s)^{-1}, \rho + \mu\big)u(w, s, \mu)\, d\mu\, dw\, ds.$$

(c) $\Box_+^{-1} v = -2^{-1/2} N\big(v \otimes \delta(\rho)\big)\big|_{bU}.$

Consequently, the kernel of $\square_+^{-1}$ *can be obtained from the kernel of* N *by letting* $\rho = \mu = 0$.

Remark 1. θ_n and λ_n above can be computed explicity, but their exact formulas are not important here. By clearing fractions, it is immediately seen that the Hilbert integral part of the kernel of N is of the product type referred to in the introduction, with singularities only on the diagonal of the boundary. In suitable coordinates, these kernels can be transferred to strongly pseudo-convex domains giving the top terms of the Neumann kernel.

Remark 2. There are other formulas for solutions of the $\bar{\partial}$ equation. Of particular interest in this connection are the recent ones for the ball of Charpentier [1] (for a kernel reducing to $\bar{\partial} * Nu$ when u is closed and which can be generalized to the case of weighted measures), and of Harvey and Polking [6], who gave an exact formula for $\bar{\partial} * N$ also exhibiting the product behavior. For another type of formula, see N. Stanton [11].

Boundedness for singular integrals on the boundary with mixed kernels is characterized by the following theorem.

Theorem 2. [10]. *Denote by* $E_k(z, t)$ *and* $H_\ell(z, t)$, *respectively, kernels homogeneous of degree* $-k$ *in the isotropic sense and* $-\ell$ *in the parabolic sense, and set*

$$K(z, t) = \begin{cases} K_k(z, t)H_\ell(z, t) & \text{for small } (z, t) \\ 0 & \text{outside a compact set.} \end{cases}$$

Then:

(a) *K is locally integrable if and only if* $k + \ell < 2n + 2$ *and* $k + \ell/2 < 2n + 1$.

(b) *Assume* $k + \ell = 2n + 2$, $k \leq 2n$. *Then K defines a distribution on* H_n *and* $Af = f * K$ *is bounded on* $L^2(H_n)$ *if and only if* $K_0(z, t) = E_k(z, 0)H_\ell(z, t)$ *has mean value 0 on the unit sphere.*

A then extends to a bounded operator on L^p, $1 < p < \infty$, *and is of weak type* (1, 1).

(c) *Assume* $k + \ell/2 = 2n + 1$, $\ell < 2$. *Then the exact analogue of* (b) *holds with the condition being now that* $K_0'(z, t) = E_k(z, t)H_\ell(z, 0)$ *have mean value* 0 *on the sphere.*

Remark 3. When $k = 2$, $\ell = 2$, and $E_{2n}(z, 0) \not\equiv 0$, $H_2(0, t) \not\equiv 0$, there is no distribution agreeing with K near 0 that gives rise to weak-type (1, 1) estimates. However, there is probably an L^p theory $(1 < p < \infty)$ if $E_{2n}(z, 0)$ has mean value 0 on the unit sphere in $\mathbf{C}^n$ and $H_2(0, t) = c/t$. $\square_+^{-1}$ belongs to the case $k + \ell = 2n + 2$, $k < 2$, and satisfies the condition in (b) of Theorem 2.

The Hilbert integrals coming from $Z_j Z_k N$, $Z_j \bar{Z}_k N$, $\bar{Z}_j \bar{Z}_k N$, $1 \leq j \leq n$, on the other hand, lead to kernels of the type $E_k(z, t, \rho)H_\ell(z, t, \rho)$ in the delicate case $k = 2n$, $\ell = 4$. In contrast with the classical Hilbert integrals, where only size played a role, the kernels have to satisfy a cancellation condition in the z-variables.

Theorem 3. [9]. *Consider* $K(z, t, \rho) = E_k(z, t, \rho)H_\ell(z, t, \rho)$ *and let* H *be the corresponding Hilbert integral operator. Then:*

(a) H *is bounded on* L^p $(1 < p < \infty)$ *when* $k + \ell < 2n + 4$, $k \leq 2n$, *or* $k + \ell/2 \leq 2n + 2$, $k > 2n$, *or* $k + \ell = 2n + 4$, $k < 2n$.

(b) *Assume* $k = 2n$, $\ell = 4$. *Then* H *is bounded on* L^2 *if*

$$\int_{|z|=\alpha, z\in \mathbf{C}^n} K(z, t, \rho)d(z) = 0 \quad \textit{for all} \quad \alpha > 0, t, \rho.$$

(c) *Assume* $K(z, t, \rho) = E_{2n}(z, t, \rho)M(|z|^2, t, \rho)$. *Then* H *is bounded on* L^p $(1 < p < \infty)$ *if* $E_{2n}(z, t, \rho)$ *has mean value* 0 *on the sphere* $|z| = \alpha$ *in* $\mathbf{C}^n$ *for each* $t, \rho, \alpha > 0$ *and* $M(\lambda, t, \rho)$ *satisfies*

$$|M(\lambda, t, \rho)| \leq \frac{C}{(|t| + \rho)^2}$$

$$\int_0^\infty \left|\frac{\partial M}{\partial \lambda}(\lambda, t, \rho)\right| d\lambda \leq \frac{C}{(|t| + \rho)^2}.$$

The proof of (c) in Theorem 3 relies heavily on recent results of Geller and Stein [4] on very singular convolution operators on the Heisenberg group.

We conclude by observing that the case of N and $\Box_+^{-1}$ indicates that composition of operators of two different types leads to operators having kernels that are products. It would be interesting to establish this directly and ultimately construct a class of pseudodifferential operators containing the ones considered here that would admit at the same time an approximate calculus and reasonably explicit kernel representation.

REFERENCES

1. Charpentier, P. Formules explicites pour les solutions minimales de l'equation $\bar{\partial}u = f$ dans la boule et dans le polydisque de $\mathbf{C}^n$. To appear.

2. Fefferman, C. The Bergman kernel and biholomorphic mappings of pseudo-convex domains. *Inventiones Math.* 26 (1974): 1-66.

3. Folland, G. B., and Stein, E. M. Estimates for the $\bar{\partial}_b$-complex and analysis on the Heisenberg group. *Comm. Pure Appl. Math.* 27 (1974): 429-522.

4. Geller, D., and Stein, E. M. Singular convolution operators on the Heisenberg group. *Bull. A.M.S.* 6 (1982): 99-103.

5. Greiner, P., and Stein, E. M. Estimates for the $\bar{\partial}$-Neumann problem. In *Princeton Math. Notes*. Princeton, N.J.: Princeton Univ. Press, 1977.

6. Harvey, R., and Polking, J. The $\bar{\partial}$-Neumann solution to the inhomogeneous Cauchy-Riemann equations in the ball in $\mathbb{C}^n$. To appear.

7. Kohn, J. J. Harmonic integrals on strongly pseudo-convex manifolds I and II. *Ann. Math.* 78 (1963): 112-148 and 79 (1964): 450-472.

8. Phong, D. H. On integral representations for the Neumann operator. *Proc. Nat. Acad. Sci.* 76 (1979): 1554-1558.

9. Phong, D. H., and Stein, E. M. On a class of operators of Hilbert integral type. To appear.

10. Phong, D. H., and Stein, E. M. Some further classes of pseudo-differential and singular integral operators arising in boundary value problems I. *Amer. J. of Math* 104 (1982): 141-172.

11. Stanton, N. The solution of the $\bar{\partial}$-Neumann problem on a strictly pseudo-convex Siegel domain. *Inventiones Math.* 66 (1982).

AN APPLICATION OF SINGULAR INTEGRALS TO A GROWTH PROBLEM FOR ENTIRE FUNCTIONS OF FINITE ORDER

Daniel F. Shea
Stephen Wainger
University of Wisconsin

Let $f(z)$ be an entire function of one complex variable z. We set

$$M(r,\ f) = \sup_{|z|=r} |f(z)|.$$

Then f is said to be of finite order ρ if

$$\limsup_{r\to\infty} \frac{\log \log M(r,\ f)}{\log r} = \rho < \infty.$$

For example, e^{z^n} is of order n, $\cos(z^{n/2})$ is of order $n/2$.

A basic problem of value distribution theory is to compare the rate of growth of various functionals of f. Some of these are $M(r,\ f)$,

$$m(r,\ f) = \inf_{|z|=r} |f(z)|,$$

$n(r) = n(r,\ 0;\ f) =$ the number of zeros of f in $|z| \leq r$,

$$N(r) = N(r,\ 0;\ f) = \int_0^r \frac{n(t)}{t}\, dt,$$

and

$$T(r,\ f) = \frac{1}{2\pi}\int_0^{2\pi} \log^+ |f(re^{i\theta})|\ d\theta.$$

The purpose of this paper is to point out how singular integral theory can be applied to such problems. It may seem strange that $n(r)$ appears on our list, since some functions of finite order, like e^{z^n}, have few or no zeros. However, functions of finite *nonintegral* order must have "many" zeros. (Thus, for instance, if f of order ρ has the same zeros as an entire function $\tilde{f}$ of order $\tilde{\rho} < \rho$, then $\log\{f(z)/\tilde{f}(z)\}$ is entire and of polynomial growth so that $f(z) = \tilde{f}(z)e^{P(z)}$ for a polynomial P and then ρ = degree of P.) This fact is made quite precise in the inequalities

$$\limsup_{r\to\infty} \frac{N(r)}{T(r,\ f)} \geq K(\rho), \tag{1}$$

$$\limsup_{r\to\infty} \frac{n(r)}{\log M(r,\ f)} \geq C(\rho), \tag{2}$$

where the constants depend only on ρ and vanish only when ρ = integer.

Here (1) is due to Nevanlinna [4] and (2) was proved by Valiron [8] and Pólya [5]. The problem of finding the sharp values of these constants is still open for $1 < \rho < \infty$; the presumed extremal function for both inequalities is

$$f_\rho(z) = \prod_{n=1}^{\infty} \left(1 + \frac{z}{n^\sigma}\right) \exp\left\{\sum_{k=1}^{q} \frac{1}{k}\left(\frac{z}{-n^\sigma}\right)^k\right\} \quad (\sigma = 1/\rho,\ q = [p]) \tag{3}$$

and would yield

$$C(\rho) = \frac{|\sin \pi\rho|}{\pi} \qquad (0 \leq \rho < \infty).$$

For $\rho \leq 1$, this is known [8, 5]; when $\rho > 1$, we have only

$$C(\rho) \geq \frac{|\sin \pi\rho|}{A \log(\rho + 1)|\sin \pi\rho| + \pi}$$

for an absolute constant A (see [5]). Thus it is not even known whether

$$\limsup_{r\to\infty} \frac{n(r)}{\log M(r,\ f)} \geq A|\sin \pi\rho| \tag{2'}$$

for some positive constant A.

(For the best known results for (1), see Edrei and Fuchs [1] and Miles and Shea [3].)

Attempts to prove inequalities like (2') are often preceded by attempts to prove the inequality for functions with all zeros on the negative real axis; observe that the conjectured extremal functions (3) belong to this class. Our goal here will be to sketch a proof of (2') for these f. For details, and a similar inequality involving $m(r, f)$, see [6].

Theorem 1. If $f(z)$ is entire of finite order ρ, and all the zeros of f lie on the negative real axis, then (2') *holds.*

The proof will use standard theorems giving L^p norm estimates for appropriate singular integrals. Since these estimates involve the usual gross constants, as in [7, Chapters 1 and 2], for instance, it is hard to imagine establishing by our method that $A = 1/\pi$ in (2').

The relevance of singular integral estimates to our problem comes from the representation

$$\log f(z) = (-1)^q z^{q+1} \int_0^\infty \frac{n(t)}{t^{q+1}} \frac{dt}{t+z} + P(z) \; (|\arg z| < \pi), \qquad (4)$$

where $q = [\rho]$ and P = polynomial. (If the zeros of f are $z_n = -t_n < 0$, we consider the Weierstrass product

$$\pi_q(z) = \prod_{n=1}^{\infty} \left(1 + \frac{z}{t_n}\right) \exp\left\{\sum_{k=1}^{q} \frac{1}{k}\left(\frac{z}{-t_n}\right)^k\right\}$$

and get (4) as a simple consequence of $f/\pi_q = e^P$ and

$$\log \pi_q(z) = \int_0^\infty \left\{\log\left(1 + \frac{z}{t}\right) + \sum_{k=1}^{q} \frac{1}{k}\left(\frac{z}{-t}\right)^k\right\} dn(t).)$$

By standard facts in function theory, when $\rho \neq$ integer we can ignore the $P(z)$ term in (4). Then from (4) we see, by using standard singular integral and maximal function theorems, that

$$\| \sup_\theta |\log f(re^{i\theta})|/r^{q+1} \|_{L^p(0,\infty;dr)} \leq A_p \left\{ \int_0^\infty \left(\frac{n(t)}{t^{q+1}} \right)^p dt \right\}^{1/p}. \tag{5}$$

Since $n(t)$ can be as big as $t^{\rho+o(1)}$, (5) requires

$$p > (q + 1 - \rho)^{-1}. \tag{6}$$

Of course, (5) does not by itself imply the theorem; using some easy tauberian arguments, (5) can be refined to

$$\int_{R_n}^\infty \left(\sup_\theta \frac{|\log f(re^{i\theta})|}{r^{q+1}} \right)^p dr \leq A_p^p \int_{R_n}^\infty \left(\frac{n(t)}{t^{q+1}} \right)^p dt \tag{7}$$

for a sequence $R_n \to +\infty$, provided $p = p_n$ satisfies (6) and

$$p_n \to p_* = (q + 1 - \rho)^{-1}. \tag{8}$$

Now (7) implies, in particular,

$$\limsup_{r\to\infty} \frac{n(r)}{\log M(r, f)} \geq 1/A_{p_*}. \tag{9}$$

By keeping track of the size of A_p, as it arises from our repeated use of standard theorems on maximal functions and singular integrals, we find

$$A_p \leq \tilde{A}(\frac{p}{p - 1} + p) \leq 2\pi\tilde{A} \csc(\frac{\pi}{p}) \tag{10}$$

for an absolute constant $\tilde{A}$. Combining (8) through (10) gives the theorem and shows that the decay of the factor $|\sin \pi\rho|$ in (2'), when $\rho \to$ integer, is just a reflection of the way A_p blows up near $p = 1$ and $p = \infty$ in the "standard theorems."

A discussion in [6] shows that (2') cannot be established by the "Pólya peak" or "proximate order" arguments that are usually sufficient for inequalities of this type (cf. [1, 3]).

One might ask whether our method might be used to prove (2') in general; after all, one might suppose that the zeros of f are carried on, say, j rays $re^{i\theta_\nu}$, $0 < r < \infty$, and try to obtain estimates independent of j.

However, one must consider that we have obtained a result stronger than the theorem; we actually proved

$$\underline{\lim_{r\to\infty}} \frac{n(r)}{\sup_\theta |\log f(re^{i\theta})|} > A|\sin \pi\rho|, \tag{11}$$

but examples due to Hayman [2, cf. 6], show (11) is not true in general.

One approach to (2') for general f might be to combine our technique with the subharmonicity of $\log|f|$. Thus one might start from

$$\log|f(z)| \leq \frac{1}{|c_z|} \int_{c_z} \log|f(\xi + i\eta)| \, d\xi \, d\eta = g(z),$$

where c_z is an appropriate disc, and try to use our technique on $g(z)$.

Another problem is to suppose f has all its zeros lying on j rays through 0, and find how the constant A_j in

$$\limsup_{r\to\infty} \frac{\log m(r, f)}{n(r)} \geq -A_j|\csc \pi\rho|$$

depends on j; a result in this direction is given in [6].

As a final problem that might be susceptible to our method, we mention that it would be interesting to show that

$$\limsup_{r\to\infty} \frac{\log m(r, f)}{\log M(r, f)} \geq -A_j$$

for f of finite order having all zeros on j rays through 0. For some counterexamples, and the relation of this problem to an ancient conjecture of Wiman, see Hayman [2].

REFERENCES

1. Edrei, A., and Fuchs, W. H. J. The deficiences of meromorphic functions of order less than one. *Duke Math. J.* 27 (1960): 233-249.

2. Hayman, W. K. The minimum modulus of large integral functions. *Proc. London Math. Soc.* (3) 2(1952): 469-512.

3. Miles, J., and Shea, D. F. On the growth of meromorphic functions having at least one deficient value. *Duke Math. J.* 43 (1976): 171-186.

4. Nevanlinna, R. *Le théorème de Picard-Borel et la théorie des fonctions méromorphes*. Paris: Gauthier-Villars, 1929.

5. Pólya, G. Bemerkungen über unendlichen Folge und ganzen Funktionen. *Math. Ann.* 88 (1923): 169-183.

6. Shea, D. F., and Wainger, S. Growth problems for a class of entire functions via singular integral estimates. *Illinois J. Math.* 25 (1981): 41-50.

7. Stein, E. M. *Singular Integrals and Differentiability Properties of Functions*. Princeton, N.J.: Princeton Univ. Press, 1970.

8. Valiron, G. Sur les fonctions entières d'ordre fini et d'order null, et en particular les fonctions à correspondence régulière. *Ann. Fac. Sci. Univ. Toulouse* (3) 5(1913): 117-257.

CONFERENCE
ON
HARMONIC ANALYSIS

THE WADSWORTH MATHEMATICS SERIES

W. Beckner, A. Calderón, R. Fefferman, P. Jones, *Conference on Harmonic Analysis in Honor of Antoni Zygmund*

M. Behzad, G. Chartrand, L. Lesniak-Foster, *Graphs and Digraphs*

J. Cochran, *Applied Mathematics: Principles, Techniques, and Applications*

A. Garsia, *Topics in Almost Everywhere Convergence*

K. Stromberg, *An Introduction to Classical Real Analysis*

CONFERENCE ON HARMONIC ANALYSIS IN HONOR OF ANTONI ZYGMUND

VOLUME II

Edited by

William Beckner
Alberto P. Calderón
Robert Fefferman
Peter W. Jones

University of Chicago

Wadsworth International Group
Belmont, California
A Division of Wadsworth, Inc.

Mathematics Editor: John Kimmel
Production Editor: Diane Sipes
Copy Editor: Janet Greenblatt
Typists: Elizabeth J. Ritter, Jo Ann Vine
Photographs: Patricia Evans

Printed in the United States of America

1 2 3 4 5 6 7 8 9 10---87 86 85 84 83

Library of Congress Cataloging in Publication Data

Conference on Harmonic Analysis (1981: Chicago, Ill.)
Conference on Harmonic Analysis in honor of Antoni Zygmund.

(Wadsworth international mathematics series)
Papers presented at the Chicago Conference on Harmonic Analysis, March 23-28, 1981; sponsored by the National Science Foundation and the University of Chicago.
1. Harmonic analysis—Addresses, essays, lectures. 2. Zygmund, Antoni, 1900- . I. Zygmund, Antoni, 1900- . II. Beckner, William. III. National Science Foundation (U.S.) IV. University of Chicago. V. Title. VI. Series.
QA403.C64 1981 515'.2433 82-11172
ISBN 0-534-98040-6 (v. 1)
ISBN 0-534-98041-4 (v. 2)
ISBN 0-534-98043-0 (set)

These proceedings are dedicated to Professor Antoni Zygmund in celebration of his eightieth birthday.

Antoni Zygmund

PREFACE

Between March 23 and 28, 1981, some two hundred mathematicians gathered at the University of Chicago to honor Professor Antoni Zygmund on the occasion of his eightieth birthday. It was a public tribute to the undisputed highest authority in Fourier series and one of the foremost analysts in the world.

Ever since the publication of its first edition in 1935, Professor Zygmund's famous *Trigonometric Series* has been one of the most influential books in the modern mathematical literature, making a permanent imprint on the mathematical work and outlook of many of the most distinguished present-day analysts.

His extensive research has covered many topics in Fourier series, real variables, and related fields. Many of his results—for example, those in the fine theory of differentiation, interpolation of linear operations, and singular integrals—now occupy a permanent and central place in analysis and have deeply influenced the modern development of other important areas of analysis, such as the theory of partial differential equations.

Professor Zygmund has also been an extraordinary teacher. He has had a large number of students—many of whom became top-ranking mathematicians themselves—and has had a profound influence in the development of analysis in several countries, including the United States, France, Argentina, Spain, and, of course, his native Poland. It is largely because of this influence that the United States has become the leader in the mathematical areas he cultivates.

These proceedings reflect the impact of Professor Zygmund's work and teaching. Most of the contributions to the conference are on topics he has either initiated or permanently shaped.

The Editors

CONTENTS

PART THREE: FOURIER ANALYSIS ON R^n AND REAL ANALYSIS 123

PART FOUR: SINGULAR INTEGRALS AND PSEUDODIFFERENTIAL OPERATORS 257

PART FIVE: HARDY SPACES 347

PART SIX: DIFFERENTIATION THEORY 507

PART SEVEN: PARTIAL DIFFERENTIAL EQUATIONS 555

PART EIGHT: OTHER TOPICS RELATED TO HARMONIC ANALYSIS 651

ANTONI ZYGMUND

Antoni Zygmund was born in Warsaw, Poland, on December 26, 1900, the son of Wincenty Zygmund and Antonina Perkowska Zygmund. After elementary school, he entered high school (gymnasium) in Warsaw in 1912. His early interests were in science, especially in astronomy, and he planned to become an astronomer.

At the outbreak of the First World War, his family was evacuated to Poltava, in the Ukraine, where it remained until autumn of 1918 before returning to (by then) independent Poland. In 1919 Zygmund entered the University of Warsaw, which had become Polish in 1915. The university was in its initial stages of development and the teaching of astronomy was still being organized. This situation, together with the interest in mathematical problems he had developed through independent study of astronomy, pushed Zygmund toward pure mathematics.

At that time there had emerged in Poland a group of talented young mathematicians who exerted a powerful attraction on university students. They were to develop what was later known as the Polish Mathematical School. Prominent among these young mathematicians were Waclaw Sierpinski (1882-1961), Stefan Mazurkiewicz (1888-1945), and Zygmunt Janiszewski (1888-1920). They founded a new mathematical periodical, *Fundamenta Mathematicae*, devoted primarily to the theory of sets and related topics, which had a strong impact on the subsequent development of mathematics in Poland. Zygmund was caught up in all this mathematical fervor. In addition, he was influenced by two of his older schoolmates, Stanislaw Saks and Zygmunt Zalcwasser. Saks was to play a very important role in the development of analysis in Poland. His primary interest was real variables, mainly differentiability and integrability of functions. His *Theory of the Integral*, first published in French in 1933, is a classic on the subject, and his *Analytic Functions*, written together with Zygmund, is still considered one of the best books on this topic. Alexander Rajchman, then an instructor and docent at the University of Warsaw, had made very important contributions to the theory of formal multiplication and uniqueness of trigonometric series, and Zygmund became particularly interested in the problems of this area. One may thus say that Zygmund was primarily a disciple of Saks and Rajchman.

Zygmund obtained his doctorate in 1923. A year earlier, he had been appointed instructor at the Polytechnical School in Warsaw, a position he kept until his departure to Wilno in 1930. In 1926 he was appointed docent at the University of Warsaw and also was awarded a Rockefeller Fellowship to spend a year in England with Hardy in Oxford and Littlewood in Cambridge. Unexpectedly, he encountered there one more mathematician who influenced him very deeply: R. E. A. C. Paley, a student of Littlewood, with whom Zygmund maintained a fruitful collaboration until Paley's early accidental death in 1933, a severe blow to Zygmund both personally and professionally.

In 1930, after his return from England, Zygmund was appointed professor at the University of Wilno. There he encountered Józef Marcinkiewicz, who became first his student and then his collaborator. Marcinkiewicz, whose name is familiar to everyone interested in functional analysis and Fourier series, was an extraordinary mathematician. His collaboration with Zygmund lasted almost ten years and produced a number of important results.

At the outbreak of the Second World War, in 1939, both Marcinkiewicz and Zygmund were mobilized as reserve officers in the Polish Army. Marcinkiewicz was taken prisoner in the East and disappeared without trace, most likely executed in captivity. Zygmund succeeded in returning to Wilno to his wife and son. Through the help of friends, mainly Professors Jacob Tamarkin and Jerzy Neyman, he managed to leave Poland and find his way to the United States, where he arrived early in 1940. The specter of impending mobilization and war weakened the need for university teachers and made jobs difficult to obtain. Nevertheless, Zygmund received an appointment as professor at Mount Holyoke College in Massachusetts.

In 1945 Zygmund moved to the University of Pennsylvania in Philadelphia. Two years later he was invited by Professor Marshall Stone to join the Department of Mathematics of the University of Chicago. A most impressive group of outstanding mathematicians was to gather there, A. A. Albert, S. S. Chern, S. MacLane, L. M. Graves, A. Weil, I. Kaplansky, P. Halmos, I. Segal, and E. Spanier among them. Zygmund went to the University of Chicago and remained there until he retired in 1980 as the Gustave and Ann Swift Distinguished Service Professor of Mathematics. At Chicago he had a large following and developed what has come to be known as the Chicago School of Analysis. The mathematical influence Zygmund exerted through his work, his teaching, and his students has spread across the United States and beyond, to South America and Europe.

Professor Antoni Zygmund has been awarded many honors and has held important positions in the mathematical community. He has honorary degrees from Washington University, the University of Toruń in Poland, the University of Paris, and the University of Uppsala in Sweden. The learned societies of which he is a member include the National Academy of Sciences of the United States; the Polish Academy of Sciences; the National Academy of Sciences of Buenos Aires, Argentina; the Royal Academy of Sciences of Madrid; and the National Academy of Science of Palermo, Italy.

Alberto P. Calderón

ACKNOWLEDGMENT

The Chicago Conference on Harmonic Analysis
March 23-28, 1981

was sponsored by

The National Science Foundation
grant number
MCS-8019742

and

The University of Chicago.

PART FIVE

HARDY SPACES

ON H^{∞} IN MULTIPLY CONNECTED DOMAINS

Lennart Carleson
Institut Mittag-Leffler, Djursholm, Sweden

1

In the paper [3] Forelli proved the following theorem. Let D be a finitely connected domain, such that no boundary component is a point. Let D^{∞} be the universal covering surface of D. Then there is a projection E from the space $H^{\infty}(D^{\infty})$ of bounded analytic functions on D^{∞} to $H^{\infty}(D)$ so that if $f \in H^{\infty}(D^{\infty})$ and $f_0 \in H^{\infty}(D)$, then

$$E(f_0 \cdot f) = f_0 E(f).$$

This immediately proves that the Corona theorem holds for D. Namely, if $F_0, G_0 \in H^{\infty}(D)$ are such that $|F_0| + |G_0| \geq \delta > 0$, then we have $F, G \in H^{\infty}(D^{\infty})$ so that

$$FF_0 + GG_0 \equiv 1 \text{ on } D^{\infty}.$$

If we apply E to this equation, we obtain solutions $E(F)$, $E(G)$ in $H^{\infty}(D)$.

If one could make the projection explicit, one would have an efficient method to study $H^{\infty}(D)$. The most obvious candidate for a projection is the following. We use this notation: points in some fixed copy $D^{\star}$ of D are denoted $z^{\star}$. Equivalent points are denoted z with notation $z \sim z^{\star}$. Assume that $P(z) \in H^{\infty}(D^{\infty})$ exists so that

$$\sum_{z \sim z^*} P(z) \equiv 1, \quad z^* \in D^*, \tag{1}$$

and

$$\sum_{z \sim z^*} |P(z)| \leq C, \tag{2}$$

where C is independent of $z^* \in D^*$. Then, clearly,

$$f_0(z^*) = E(f) = \sum_{z \sim z^*} P(z) f(z) \tag{3}$$

is a projection of the desired type.

We can use the explicit form of E to study the distribution of zeros of functions in $H^\infty(D^*)$. Let $g^*(z^*; c^*)$ be the Green's function for D^* with pole c^*. g^* always exists if $H^\infty(D^*)$ is non-trivial. $g(z; c)$ is the Green's function for D^∞. Then if $c \sim c^*$,

$$g^*(z^*; c^*) \equiv \sum_{z \sim z^*} g(z; c) \equiv \sum_{c \sim c^*} g(z; c). \tag{4}$$

Let $a_\nu^* \in D^*$ be a sequence of points that are zeros of a function $f_0(z^*) \in H^\infty(D^*)$. Then by Jensen's formula,

$$\sum_{\nu=1}^{\infty} g^*(a_\nu^*) < \infty. \tag{5}$$

Theorem 1. *If a partition function P exists and a_ν^* satisfies (5), then a_ν^* are zeros of some function in $H^\infty(D^*)$.*

Proof. The formula (4) shows that the sequence a_ν of all points $\sim a_\nu^*$, $\nu = 1, 2, \ldots,$ is a Blaschke sequence on (the disc) D^∞. Let $A(z)$ be the corresponding Blaschke product. Let $B(z)$ be the analogous Blaschke product for the sequence of points $c \sim c^*$. We note that

$$-\log|B(z)| = g^*(z^*; c^*), \quad z \sim z^*.$$

It follows, since $|A(c)|$ is independent of $c \sim c^*$ and $\neq 0$, that

$$|A(z)| + |B(z)| \geq \delta > 0.$$

Let

$$FA + GB \equiv 1$$

for $F, G \in H^\infty(D^\infty)$. Define

$$f(z^*) = \sum_{z \sim z^*} P(z)F(z)A(z).$$

Then clearly $f(a_\nu^*) = 0$ and $f(c^*) = 1$. Observe that this construction does not show that f has a_ν^* as exact zero set.

I am grateful to J. Garnett for many suggestions of improvements in the presentation.

2

Let us first show that P exists for a finitely connected domain D. It is convenient to represent D^∞ as the unit disc U modulo a group $\textcircled{G}$ of linear transformations. Let $\textcircled{F}$ be the fundamental domain of $\textcircled{G}$ that contains 0 and is bounded by $2s$ circles orthogonal to ∂U. We obtain new copies of $\textcircled{F}$ by reflecting in orthogonal circles. After n successive reflections, we have $2s \cdot (2s - 1)^{n-1} < (2s)^n$ orthogonal circles Γ_n. Then the length of each circle is $\geq a^n$ and the total projection on $|z| = 1 \leq 2\pi \cdot b^n$, where a, b are constants and $b < 1$. Furthermore, if $\gamma \in \Gamma_n$, then $\mathrm{dist}(\gamma, \Gamma_{n+1}) \geq c$ (length γ). We shall need these simple facts later.

Let $p_0 \in C^\infty(U)$ and assume that $p_0 \equiv 1$ in $\textcircled{F}$, $p_0 \geq 0$ and $p_0 \equiv 0$ outside the $2s$ copies of $\textcircled{F}$ that are adjacent to $\textcircled{F}$. Define

$$p(z) = p_0(z) \Big/ \sum_{S \in \textcircled{G}} p_0(Sz), \quad z \in U,$$

so that

$$\sum_{S \in \textcircled{G}} p(Sz) \equiv 1.$$

Observe that $p(z)$ has the same support as $p_0(z)$. We wish to construct $P(z)$ by solving the $\bar\partial$-problem for $p(z)$. Let therefore $f(z; \zeta)$, $z, \zeta \in U$, be functions in $H^\infty(U)$ for every $\zeta \in U$ and be uniformly bounded for $\zeta \in \mathrm{supp}(p)$ and finally satisfy $f(\zeta, \zeta) = 1$. Then

$$P(z) = p(z) - \frac{1}{\pi i}\iint_U \frac{f(z;\ \zeta)}{z-\zeta}\bar\partial p(\zeta)d\xi\ d\eta = p(z) - Q(z) \tag{6}$$

satisfies (1) and (2) if

$$\sum_{S \in \textcircled{G}} Q(Sz) \equiv 0, \tag{7}$$

$$\sum_{S \in \textcircled{G}} |Q(Sz)| \leq C. \tag{8}$$

Denote by $K(z;\ \zeta) = \dfrac{f(z;\ \zeta)}{z-\zeta}$ and assume that $f(z;\ \zeta)$ is defined for $\zeta \in \textcircled{F}$. We then define $K(z;\ \zeta)$ for $\zeta \in U$ so that the following identity holds:

$$S'(\zeta) \cdot K(Sz;\ S\zeta) = K(z;\ \zeta). \tag{9}$$

This defines $K(z;\ \zeta)$ uniquely and one can notice that $K(z;\ \zeta)$ has a pole with residue 1 for all $\zeta \in U$.

To prove (7) we note the following identity:

$$\pi i \sum_S Q(Sz) = \sum_S \iint_U K(Sz;\ \zeta)\bar\partial p(\zeta)d\xi\ d\eta$$
$$= \sum_S \iint_U K(z;\ S^{-1}(\zeta))\frac{1}{S'(S^{-1}(\zeta))}\bar\partial p(\zeta)d\xi\ d\eta.$$

Make here the substitution $\zeta = S(\zeta')$. The equality can be continued by

$$\sum_S \iint_U K(z;\ \zeta')\overline{S'(\zeta')}(\bar\partial p)(S\zeta')d\xi'\ d\eta'$$
$$= \iint_U K(z;\ \zeta')\bar\partial \sum_S p(S\zeta')\ d\xi'\ d\eta' = 0.$$

To prove (8) it is enough to prove

$$\sum_{S \in \textcircled{G}} |f(Sz;\ \zeta)| \leq C,$$

for $\zeta \in \textcircled{F}$. Choosing S corresponding to the $2s$ reflections, we obtain the same result for $\zeta \in \text{supp}(p)$.

We now recall the preliminary discussion of Ⓖ. Define

$$h(\theta) = (n + 1)^2 \text{ for } e^{i\theta} \in R_n = \operatorname{int}(\Gamma_n)\backslash\operatorname{int}(\Gamma_{n+1}).$$

Then

$$\int h(\theta)d\theta \leq 2\pi \cdot \sum(n + 1)^2 b^n.$$

Define

$$H(z) = \exp\left\{-\frac{1}{2\pi}\int_{-\pi}^{\pi}\frac{e^{i\theta} + z}{e^{i\theta} - z} h(\theta)d\theta\right\}.$$

Then for $z \in R_n$

$$|H(z)| \leq \exp\{-c_0 n^2\}$$

with $c_0 > 0$ independent of n. In Ⓕ, $|H(z)| \geq c_0$. We can therefore choose

$$f(z; \zeta) = H(z)/H(\zeta), \ \zeta \in Ⓕ,$$

and have

$$\sum |f(Sz; \zeta)| \leq \frac{1}{c_0} \sum_n (2s)^n e^{-c_0 n^2}.$$

3

This method can now be generalized to very general domains D. We shall here consider the case when $\partial D = E \subset \mathbf{R}$. It is necessary [1] that E has positive measure, to avoid $H^\infty(D)$ trivial. Unfortunately, I cannot prove the existence of P under this assumption alone. It is natural that the case when certain parts of E are very thin compared to Lebesgue measure are particularly difficult. We avoid this possibility by assuming that E is *homogeneous* in the following sense. There exists a constant c, only depending on E, so that for all $x_0 \in E$ and *all* $t > 0$,

$$m((x_0 - t, x_0 + t) \cap E) \geq ct. \tag{10}$$

Observe that (10) implies that $\infty \notin D$. The restriction in (10) as $t \to \infty$ is, however, not serious. If E is compact satisfying (10)

$t \leq 1$, we can obtain (10) for all t by making a suitable inversion $z \to (z - x_0)^{-1}$. From this observation it follows that all Cantor sets, even with a variable number of divisions in every step, satisfy (10). Our main result is then:

Theorem 2. A partition function P satisfying (1) *and* (2) *exists for homogeneous linear sets.*

The construction is best carried out in the given domain D and on its covering surface D^∞. We choose $p(z) \equiv 1$ in $D^\star$ and $\equiv 0$ on the rest of D^∞. $\bar{\partial} p \, d\xi \, d\eta$ will then be Lebesgue measure on the complementary intervals of E and formula (9) will simply be reversal of signs. The function $Q(z)$ is therefore

$$\int_{CE} K(z;\ \xi)\, d\xi, \tag{11}$$

where K has a pole with residue 1 at ξ. To deal with the singularity, we define

$$K(z;\ \xi) = \frac{f(z;\ \xi)}{z^\star - \xi} B(z;\ \xi) \cdot F(z;\ \xi) \tag{12}$$

with the following understanding:

(a) $f(z;\ \xi) \in H^\infty(D^\infty)$ and $f(\xi,\ \xi) = 1$ and $|f(z;\ \xi)|$ only depends on $z^\star \sim z$.

(b) $B(z;\ \xi)$ is the Blaschke product with zeros at all points $w \sim \xi$ on D^∞ *except* $w = \xi \in D^\star$.

(c) $F(z;\ \xi) \in H^\infty(D^\infty)$ and

$$F(\xi,\ \xi) = \frac{1}{B(\xi,\ \xi)}.$$

$K(z;\ \xi)$ is then a kernel with the correct pole and (7) holds if (8) holds.

(a′) To make $Q(z)$ bounded, we shall construct $f(z;\ \xi)$ so that

$$\int_{CE} \frac{|f(z^*;\ \xi)|}{|z^* - \xi|}\, d\xi \le C. \tag{13}$$

The conditions (a) and (a′) are the requirements on $f(z;\ \xi)$.

(b′) For the Blaschke product we need

$$|B(\xi,\ \xi)| \ge c_0 > 0. \tag{14}$$

This, however, is an immediate consequence of our assumption on E. Namely,

$$-\log B(z;\ \xi) = g^*(z^*;\ \xi) - g(z;\ \xi). \tag{15}$$

The right-hand side is independent of changes of scale: $z \to Mz$, $\xi \to M\xi$. We may therefore assume $\xi = 0$, $1 \in E$, $(-1,\ 1) \subset D$. Then

$$\left| g(z^*;\ 0) - \log \frac{1}{|z^*|} \right| \le C, \quad |z^*| = \frac{1}{2}$$

$$\left| g^*(z^*;\ 0) - \log \frac{1}{|z^*|} \right| \le C, \quad |z^*| = \frac{1}{2}$$

and by the maximum principle in (15), (14) holds.

Now observing $|B(z;\ \xi)| \le 1$, we can estimate (8) by

$$\sum_{z \sim z^*} \int \frac{|f(z^*;\ \xi)|}{|z^* - \xi|} |F(z;\ \xi)|\, d\xi \le \left(\sup_{\xi} \sum_{z \sim z^*} |F(z;\ \xi)| \right) \cdot C$$

by (13). Considering (b′), our requirement on $F(z;\ \xi)$ is therefore, by the maximum principle,

(c′) $$F(\xi,\ \xi) = 1, \qquad \xi \in \mathbf{R} \backslash E$$

$$\sum_{z \sim z^*} |F(z;\ \xi)| \le C, \quad z^* \in E.$$

To be able to construct $f(z;\ \xi)$ and $F(z;\ \xi)$, we need the estimates of what now corresponds to the Lebesgue measure in Sec. 2, that is, harmonic measure on D^∞. If $\sigma \subset \partial D^\infty$ with projection $\sigma^* \subset \partial D^*$, we obtain majorants of harmonic measure for (σ, D^∞) by considering (σ^*, D^*). Such estimates are the object of the following section.

4

We shall here study harmonic measure for the plane domain D^*. We omit in this section the $*$-notation, since all sets are schlicht. We denote the harmonic function in D whose boundary values are = 1 on F and 0 on $\partial D \backslash F$ by $u(z; F)$. For the method used, compare Benedicks [2].

Lemma 1. Let E be homogeneous, normalized so that $(-1, 1) \subset D$ and $1 \in E$. Then there are numbers $\theta > 0$ and $C < \infty$, only depending on the constant c in (10) so that

$$u\big(0;\, E \cap (|x| \geq R)\big) \leq C \cdot R^{-\theta} \tag{16}$$

for all $R \geq 1$.

Remark. P. Jones has pointed out to me that these types of inequalities are essentially Wiener criteria for the Dirichlet problem and can be found in Wiener's paper [5].

Proof. We first note that $\theta \leq \frac{1}{2}$, since this is the case if $E = (1, \infty)$. It does not seem likely that $\theta = \frac{1}{2}$ in the general case.

Let $A = c^{-1}$ and let $K(x, t)$ be the Poisson kernel for the complement of $(-\infty, -A) \cup (A, \infty)$. We assume also $R = A^n$, $n \geq 3$. Finally, denote by m_n the supremum of $u(x)$, $|x| < 1$, under the conditions made. We have for $-1 < x < 1$

$$u(x) \leq \int_{|t| \geq A} K(x;\, t)u(t)dt = \sum_{\nu=1}^{n-2} \int_{|t| = A^\nu}^{A^{\nu+1}} K(x;\, t)u(t)dt + O(A^{-n}).$$

When $A^{\nu} \leq |t| \leq A^{\nu+1}$, $\nu \leq n - 2$,

$$u(t) \leq m_{n-\nu-1}.$$

Furthermore,

$$\int_{\substack{|t|=A^{\nu} \\ t \notin E}}^{A^{\nu+1}} K(x;\, t)dt \leq q_{\nu}, \quad -1 < x < 1,$$

where

$$\Sigma q_{\nu} = q < 1$$

and q only depends on c, since $m\big(E \cap (A^{\nu} < |t| < A^{\nu+1})\big) \geq c \cdot A^{\nu}$. It follows that, for A large enough,

$$m_n \leq \sum_{\nu=1}^{n-2} q_{\nu} m_{n-\nu-1} + O(2^{-n}).$$

q_{ν} also satisfies

$$0 < q_{\nu} < c \cdot 2^{-\nu}. \tag{17}$$

If we introduce

$$m(z) = \sum_{0}^{\infty} m_n z^n$$

$$q(z) = \sum_{0}^{\infty} q_{\nu} z^{\nu},$$

the inequalities can be written

$$m(z)\big(1 - q(z)\big) << \Sigma 2^{-n} z^n.$$

Now since $q(1) = q < 1$ and q_{ν} satisfies (17), it follows that $(1 - q(z))^{-1}$ is holomorphic in $|z| < 1 + \delta$, $\delta = \delta(c)$. Furthermore,

$$m(z) << \left(\sum_{0}^{\infty} q^j(z)\right)(\Sigma 2^{-n} z^n)$$

and

$$m_n \leq (1 - \delta)^n \leq CR^{-\theta},$$

for some $\theta > 0$.

Lemma 1 takes care of the behavior of harmonic measure at large distances. We also need information on the local behavior.

Lemma 2. Let $I = (-1, 1)$ *be a complementary interval. Let* $F \subset E$ *have small measure,* $mF = \varepsilon$. *Then*

$$u(0; F) < C_0 \exp\left\{-c_0\left(\log \frac{1}{\varepsilon}\right)^{1/2}\right\}$$

where c_0, C_0 *only depend on* c.

Proof. Divide I into N equal intervals I_j of length $2/N$. N will be fixed later and will not depend on ε. Define

$$\psi_j(\varepsilon) = \sup \frac{N}{2}\int_{I_j} u(x; F)\,dx \tag{18}$$

for all configurations with c fixed. Consider some fixed configuration and some $x \in I_j$. Let $K(x; t)$ be the Poisson kernel for $(|t| \geq 1)$. Then

$$u(x; F) = \int_{|t|\geq 1} K(x; t)u(t)\,dt = \int_F + \int_S \tag{19}$$

where $S = \cup J_k$ and J_k are the complementary intervals of E. Let A be a (large) number to be defined later only depending on ε.

Denote by $J_k^\star$ the interval J_k expanded symmetrically around its midpoint in scale A. We divide $\{J_k\}$ into two groups:

I. $|J_k^\star \cap F| > A\varepsilon|J_k^\star|$.

II. The remaining intervals J_k.

There exists a disjoint subset of the $J_k^\star$'s of group I so that these expanded by the factor 3 cover $\bigcup_{(I)} J_k$. We denote the expanded intervals $J_k^\star$ without introducing a new indexing. Then corresponding to case I

$$\sum_{(I)} |J_k| \leq \sum_{(I)} m(S \cap J_k^\star) \leq \frac{3}{A\varepsilon}\sum_{(I)} m(J_k^\star \cap F) \leq \frac{3}{A}.$$

The contribution to the right-hand side of (19) of $J_k \subset S$ so that J_k belongs to group I has the estimate

$$\sum_{(I)} \int_{J_k} K(x;\ t)\,dt \le C \cdot A^{-1/2},$$

since $K(x, t)$ has the explicit form ($|t| > 1$)

$$K(x, t) = \frac{2}{\pi} \frac{\sqrt{1 - x^2}}{|1 - x| \cdot |t + 1| + |1 + x| \cdot |t - 1|} \cdot \frac{1}{\sqrt{t^2 - 1}}. \tag{20}$$

From this follows also that the first term in (19) is

$$O((\varepsilon(1 - |x|)^{-1})^{1/2}).$$

If J_k is in group II, then $J_k^{\star} \cap F$ behaves relatively to J_k as F does to $(-1, 1)$ except that ε has been changed to $A^2\varepsilon$. Write

$$J_k = \bigcup_{\nu=1}^{N} J_k(\nu), \quad |J_k(\nu)| = (1/N) \cdot |J_k|.$$

By the definition of ψ_j and by Lemma 1, we have

$$\frac{1}{|J_k(\nu)|} \int_{J_k(\nu)} u(t)\,dt \le \psi_\nu(A^2\varepsilon) + C \cdot A^{-\theta}.$$

Write

$$a_{j\nu} = \sum_k \frac{N}{2} \int_{I_j} K(x,\ t_\nu(J_k))\,dx \cdot |J_k|,$$

where t_ν is the endpoint of $J_k(\nu)$, closest to 0. Then

$$\psi_j(\varepsilon) \le C(N)(\varepsilon^{1/2} + A^{-1/2}) + \sum_\nu a_{j\nu}(\psi_\nu(A^2\varepsilon) + O(A^{-\theta})).$$

We shall show in a moment that if N is large (not depending on ε),

$$\sum_\nu a_{j\nu} \le q < 1, \tag{21}$$

where q only depends on c in (10). Accepting (21) and setting $h(\varepsilon) = \sup_j \psi_j(\varepsilon)$, we have

$$h(\varepsilon) \le C(\varepsilon^{1/2} + A^{-\theta}) + qh(A^2\varepsilon).$$

We take $A = \exp\left\{c'\left(\log \frac{1}{\varepsilon}\right)^{1/2}\right\}$ and iterate $\left[\left(\log \frac{1}{\varepsilon}\right)^{1/2}\right]$ times and obtain the desired estimate.

In proving (21), we may restrict ourselves to intervals J_k located in $(-A, A)$ and assume that there are finitely many intervals. Fix $\delta = \delta(c) > 0$.

Case 1. $J_k \subset \{1 \le |t| \le 1 + \delta\}$.

By (13), there are disjoint sets $E_k \subset E$ such that

$$\sup_{E_k} |t| \le \inf_{J_k} |t|$$

and

$$|E_k| \ge \lambda(c)\, |J_k|.$$

Then, since $K(x;\ t)$ is monotone for $t > 1$ and $t < -1$,

$$\sum_{\text{Case 1}} \left(\sum_{\nu} \frac{|J_k|}{N} K(x;\ t_\nu) - \int_{J_k(\nu)} K(x;\ t)\,dt \right) \le \sum \frac{1}{\lambda N} \int_{E_k} K(x;\ t)\,dt$$
$$\le \frac{1}{\lambda N} \int_{1 \le |t| \le 1+\delta} K(x;\ t)\,dt.$$

Case 2. J_k intersects $|t| \ge 1 + \delta$.

By (10), $J_k \subset \{|t| \ge 1 + \delta'\}$ for some $\delta'(c) > 0$, and

$$\sum_{\text{Case 2}} \left(\sum_{\nu} \frac{|J_k|}{N} K(x;\ t_\nu) - \int_{J_k(\nu)} K(x;\ t)\,dt \right) \le \sum \frac{|J_k|}{N} \sup_{J_k} |K'(x;\ t)|. \qquad (22)$$

Observe now the explicit expression (20) for K. The inequality (22) can be continued by

$$\frac{\text{Const.}}{N} \int_{|t| \ge 1+\delta'} |K'(x;\ t)|\,dt \le \frac{\text{Const.}}{N}$$

uniformly for $-1 < x < 1$. Consequently,

$$\sum_{\nu} a_{j\nu} \le \frac{N}{2} \int_{|t| \ge 1+\delta'} K(x;\ t)\,dt + O(N^{-1})$$
$$\le \frac{N}{2} \int_{I_j} dx \int_{|t| \ge 1+\delta'} K(x;\ t)\,dt + O(N^{-1}).$$

By (10), the first term is $\le q' < 1$ and hence (21) follows for N large enough.

Lemma 3. If in Lemma 2 we assume that $(-\infty, 1)$ *is a complementary arc, the conclusion of Lemma 2 is still valid.*

Proof. Let L be a large number and set $G = (-\infty, -L)$ and $E_1 = E \cup G$. Harmonic measures with respect to the outside of E are denoted u_1. Then by Lemma 1,

$$u_1(0;\, G) \leq C \cdot L^{-\theta},$$

and by Lemma 2,

$$u_1\left(-\frac{L}{2};\, F\right) \leq C_0 \exp\left\{-c_0\left(\log \frac{L}{\varepsilon}\right)^{1/2}\right\}.$$

By Harnack's inequality, there is a constant λ so that

$$u_1(0;\, F) \leq L^{\lambda} \cdot u_1\left(-\frac{L}{2};\, F\right).$$

Finally,

$$u(0;\, F) \leq u_1(0;\, F) + u_1(0;\, G) \leq L^{\lambda} C_0 \exp\left\{-c_0\left(\log \frac{L}{\varepsilon}\right)^{1/2}\right\} + C \cdot L^{-\theta}.$$

We choose $L = \exp\left\{c'\left(\log \frac{1}{\varepsilon}\right)^{1/2}\right\}$ and obtain the desired result.

Lemma 4. Let $(-1, 1) \subset D$ *and* $1 \in E$. *Let* $h_0(\sigma)$ *be harmonic measure at* 0 *of subsets* σ *of* E. *Then* h_0 *is absolutely continuous and*

$$\int_E h_0'(x) \exp\{c_1(\overset{+}{\log}\, h_0')^{1/2}\}dx \leq C_1$$

with c_1 *and* C_1 *only depending on* c.

Proof. By Lemma 3, h_0 is absolutely continuous. Let F_n be the set where $2^n \leq h_0' < 2^{n+1}$. Then $mF_n < 2^{-n}$ and by Lemma 3,

$$h_0(F_n) \leq C_0 \exp\{-c_0' n^{1/2}\}.$$

Hence

$$\{h_0' \exp\int c_1(\overset{+}{\log}\, h_0')^{1/2}\}dx \leq \sum \exp\{c_1' n^{1/2}\} h_0(F_n) \leq C_1$$

if c_1 is small enough.

5

We shall now start the program outlined in Sec. 3 and first construct our functions $f(z;\ \xi)$. The construction is essentially the same as used by P. Jones in [4].

We denote by $\varphi(x)$ the function $x^2/(1 + x^2)^2$. For $\xi \notin E$ let $\delta(\xi)$ be the distance from ξ to E. We solve the Dirichlet problem for the complement $D^* = D$ of E with boundary values, ξ fixed, $\xi \notin E$,

$$u(x;\ \xi) = \log \varphi\left(\frac{|x - \xi|}{\delta(\xi)}\right) - 2 \log\left\{1 + \int_{\substack{\delta(t) \leq \delta(\xi) \\ t \notin E}} \frac{\varphi\left(\frac{|x - t|}{\delta(t)}\right)}{|x - t|}\, dt\right\}. \tag{23}$$

Clearly $u \leq 0$. To show that this is indeed possible, we need to prove that $u(x;\ \xi)$ is summable with respect to harmonic measure h. If h_ξ denotes harmonic measure at ξ, we shall show

$$\int_E u(x;\ \xi)\, dh_\xi(x) \geq -C, \tag{24}$$

where C only depends on the constant of homogeneity.

We first estimate

$$I_1 = \int_E \log \varphi\left(\frac{|x - \xi|}{\delta(\xi)}\right) dh_\xi(x).$$

We can normalize so that $\xi = 0$ and belongs to a complementary interval $(-A,\ 1)$, $A \geq 1$. Then

$$-I_1 = -\int_{|x| \geq 1} \log \varphi(x)\, dh_0(x) = -\int_1^\infty h_0(|t| \leq x) \frac{\varphi'(x)}{\varphi(x)}\, dx$$

$$\leq -C \int_1^\infty x^{-\theta} \frac{\varphi'(x)}{\varphi(x)}\, dx < -C \int_1^\infty (\log \varphi(x)) x^{-\theta - 1} dx$$

by Lemma 1.

For the second term, I_2, in (23) we have the estimate, since $ab \leq e^{a-1} + b \log b$,

$$|I_2| \leq 2\int_E \log\left[1 + \int_{\delta(t)\leq 1}\right] \cdot (h_0'(x)|x|^{\theta+1})\frac{dx}{|x|^{\theta+1}} \tag{25}$$

$$\leq 2\int_E (h_0'|x|^{\theta+1})\log(h_0'|x|^{\theta+1})\frac{dx}{|x|^{\theta+1}} + 2\cdot e^{-1}\int_E\left(1 + \int_{\delta(t)\leq 1}\right)\frac{dx}{|x|^{\theta+1}}.$$

Here

$$\int_E h_0'(x)\log|x|\,dx \leq C\sum 2^{-n\theta}\cdot n$$

by Lemma 1 and

$$\int_E h_0'(x)\log\, h_0'(x)\,dx \leq C$$

by Lemma 4. For the last integral in (25), we have the estimate for $x \geq 1$

$$\int_1^\infty \frac{dx}{x^{\theta+1}}\int_{\delta(t)\leq 1}\frac{\varphi\left(\frac{|x-t|}{(t)}\right)}{\frac{|x-t|}{\delta(t)}}\frac{dt}{\delta(t)} \leq C\sum_{n=0}^\infty 2^{-n-n\theta}\int_{\delta(t)\leq 1}dt\int_{2^n}^{2^{n+1}}(\quad)\frac{dx}{\delta(t)} \tag{26}$$

and similarly for $x \leq -1$. Here

$$\int_{2^n}^{2^{n+1}}(\quad)\frac{dx}{\delta(t)} \leq \int_0^\infty \frac{\varphi(u)}{u}\,du\,. \tag{27}$$

Also, if $t \geq 2^{n+2}$, the left-hand side of (27) has the improved estimate [since $\delta(t) \leq 1$]

$$\int_{t/2}^\infty \frac{\varphi(u)}{u}\,du < C\cdot t^{-2}.$$

Hence (26) has the bound

$$C\sum_{n=0}^\infty 2^{-n-n\theta}\left(2^n + \int_{2^n}^\infty \frac{dt}{t^2}\right) \leq C.$$

This proves (24).

Let $v(z;\,\xi)$ be a conjugate function of $u(z;\,\xi)$ on D^∞ and define

$$f(z;\,\xi) = C(\xi)e^{u(z;\,\xi)+iv(z;\,\xi)}$$

where $C(\xi) \leq C$ is chosen so that $f(\xi,\,\xi) = 1$. We have just proved that $f(z;\,\xi)$ satisfies (a) of Sec. 3. It remains to prove (a′), that is, (13).

By subharmonicity, it is enough to prove (13) for $z^* = x$ real belonging to E. Then (13) equals

$$\int_{CE} \frac{e^{u(x;\,\xi)}\,d\xi}{|x - \xi|} = \int_{CE} \frac{\varphi\left(\dfrac{\dfrac{|x-\xi|}{\delta(\xi)}}{|x-\xi|}\right)}{\left(1 + \displaystyle\int_{\substack{\delta(t) < \delta(\xi) \\ t \in CE}}\right)^2}\, d\xi \leq \int_0^\infty \frac{du}{(1+u)^2} = 1,$$

simply by change of variables. The construction of $f(z;\,\xi)$ is therefore complete.

6

It remains to construct $F(z;\,\xi)$. Again we may normalize so that $\xi = 0$ and $\delta(0) = 1$. We now shall work on D^∞ and need some notation. A (half-)sheet of the covering surface is defined by a sequence $\Omega = (\omega_1, \omega_2, \ldots, \omega_n)$ of complementary intervals of E, $\omega_{i+1} \neq \omega_i$. Depending on if we first go up or down, Ω defines two (half-)sheets of D^∞. We are not going to introduce any separate notation to distinguish these two cases. n is the length $|\Omega|$ of Ω. There is also a natural partial ordering: $\Omega < \Omega'$, if $\Omega' = (\Omega, \Omega'')$. We are going to make the construction of $|F(x)|$, $x \in \partial D^\infty$, and then define $F(z)$ on D^∞ by solving the Dirichlet problem for $\log|F(x)|$. If h_z denotes harmonic measure on ∂D^∞ at z, the crucial estimate is

$$\int \log|F(x)|\,dh_0(x) \geq -C. \tag{28}$$

This will take care of the first condition in (c′) in Sec. 3.

The length of a complementary interval ω is denoted $|\omega|$ and its midpoint $\mu(\omega)$. To avoid special conventions later, let us assume that all $|\omega|$'s are different. Sometimes the notation ω will refer to a particular location of the slit ω in the covering surface D^∞; this will be clear from the context.

Let S be a collection of intervals ω in the ground sheet. Let Γ be the part of ∂D^∞ that is reached through these intervals:

$$\Gamma = \cup\, \partial\Omega, \ \Omega \supset \omega, \ \omega \in S.$$

Then the following analogues of Lemmas 1 and 3 of Sec. 3 hold.

Lemma 1'. *If all* $\omega \in S$ *are located in* $|x| \geq R$, *then*

$$h_0(\Gamma) \leq C \cdot R^{-\theta}.$$

Lemma 3'. *If* $\varepsilon = \sum |\omega|$, $\omega \in S$, *then*

$$h_0(\Gamma) \leq C' \exp\left\{-c'\left(\log \frac{1}{\varepsilon}\right)^{1/2}\right\}.$$

The lemmas follow immediately from Lemmas 1 and 3 if we replace E by $E \cup \bigcup_{\omega \in S} \omega$, and $h_0(\Gamma)$ by plane harmonic measure of $(\cup\, \omega)^-$.

Let t be real. Define

$$\psi(t;\, \omega) = \frac{|\omega|^2}{(t - \mu(\omega))^2 + |\omega|^2}$$

and

$$R(t;\, \omega) = \psi(t;\, \omega) \cdot \left(1 + \sum_{|\sigma| \leq |\omega|} \psi(t;\, \sigma)\right)^{-2}.$$

To describe $|F(x)|$, we define inductively functions $\varphi(x;\, \Omega)$ for x real.

We first set $\varphi(x;\, \emptyset) \equiv 1$. Let $\Omega = (\omega_1, \ldots, \omega_{n-1}, \omega_n) = (\Omega', \omega_n)$. We distinguish two cases.

A. If $(n \geq 2)$, $|\omega_n| \geq 2|\omega_{n-1}|$ and $|\mu(\omega_n) - \mu(\omega_{n-1})| \leq |\omega_n|$, we define

$$\varphi(x;\, \Omega) = R(x;\, \omega_{n-1}) \cdot \psi(\alpha_n;\, \omega_{n-1})\varphi(\mu_{n-1};\, \Omega')$$

where α_n is the endpoint of ω_n that is closest to μ_{n-1}.

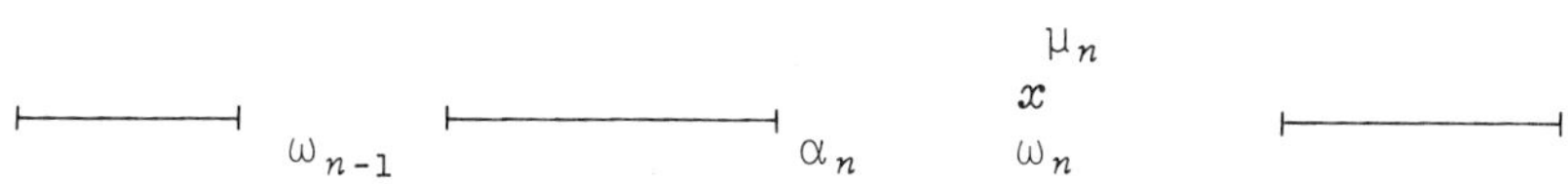

B. In all other cases, also for $n = 1$,

$$\varphi(x;\ \Omega) = R(x;\ \omega_n) \cdot \varphi(\mu(\omega);\ \Omega').$$

Let $\gamma > 0$ be a constant only depending on c to be defined shortly. On the boundary of the half-sheet Ω, we define

$$|F(x)| = \gamma^n \cdot \varphi(x^*;\ \Omega).$$

We now verify that this definition satisfies the second condition of (c′) in Sec. 3 if γ is small enough.

Denote by

$$S_n(x) = \sum_{|\Omega| = n} \varphi(x;\ \Omega).$$

Every Ω of length $n + 1$ can be written $\Omega = (\Omega^0,\ \omega_{n+1})$ with Ω^0 of length n or more generally ($p \geq 1$)

$$\Omega = (\Omega^{(p)},\ \omega_{n-p+1},\ \ldots,\ \omega_{n+1}).$$

For every p we consider the collection $\mathscr{C}(p)$ of Ω's so that case B occurs for $(\omega_{n-p},\ \omega_{n-p+1})$ with p minimal, $p > 0$. $p = n$ corresponds to no case B. We split the sum defining $S_{n+1}(x)$ according to $\mathscr{C}(p)$. If $|\Omega| = n + 1$ and $\Omega \in \mathscr{C}(p)$, we factorize $\varphi(x;\ \Omega)$ using the definitions:

$$\begin{aligned}
&\varphi(\mu_{n-p+1};\ \Omega^{(p)}) \cdot R(\mu_{n-p+1};\ \omega_{n-p+1}) \cdot \\
&\cdot R(\mu_{n-p+2};\ \omega_{n-p+1}) \cdot \psi(\alpha_{n-p+2};\ \omega_{n-p+1}) \\
&\quad\vdots \\
&R(\mu_n;\ \omega_{n-1}) \cdot \psi(\alpha_n;\ \omega_{n-1}) \\
&R(x;\ \omega_n) \cdot \psi(\alpha_{n+1};\ \omega_n).
\end{aligned}$$

We delete $R(\mu_{n-p+1};\ \omega_{n-p+1})$ and all ψ-factors except the last one. We then sum over the variables ω_i in the following order:

$$\Omega^{(p)},\ \omega_{n-p+1},\ \ldots,\ \omega_{n-1};\ \omega_{n+1},\ \omega_n$$

and observe the following inequalities:

$$\sum_{\Omega^{(p)}} \varphi(\mu_{n-p+1};\ \Omega^{(p)}) \leq \|S_{n-p+1}\|_\infty$$

$$\sum_{\omega_{n-j-1}} R(\mu_{n-j};\ \omega_{n-j-1}) \leq C$$

$$\sum_{\omega_{n+1}} \psi(\alpha_{n+1};\ \omega_n) \leq C$$

$$\sum_{\omega_n} R(x;\ \omega_n) \leq C.$$

The third inequality holds because the intervals ω_{n+1} that are in case A relatively some fixed ω_n have to grow exponentially. Using this subdivision $\mathscr{C}(p)$ we obtain

$$|S_{n+1}(x)| \leq \sum_{p=1}^{n+1} C^p \| S_{n-p+1} \|_\infty . \tag{29}$$

We now bound $\| S_n \|_\infty$ inductively. Clearly $\| S_0 \| = 1$. For $n = 1$, we always have definition B and therefore $\| S_1 \| \leq C$. Suppose now $\| S_\nu \|_\infty \leq (2C)^\nu$ for $\nu \leq n$. Then (29) implies

$$\| S_{n+1} \|_\infty \leq \sum_{p=1}^{n+1} C^p (2C)^{n-p+1} < (2C)^{n+1} \sum_1^\infty 2^{-\nu} = (2C)^{n+1}.$$

Hence if $\gamma < \gamma(c)$,

$$\sum \gamma^n \| S_n(x) \|_\infty \leq C,$$

which proves the second condition (c′) in Sec. 3.

7

The estimate (28) depends on the exponential decay of harmonic measure. In the group realization in the unit disc, this decay is due to the fact that the fundamental domains intersect $|z| = 1$ in sets of positive measure.

Lemma 5. Let $\Gamma_n = \bigcup_{|\Omega| \geq n} \partial\Omega$. *Then there exists* $\rho = \rho(c) < 1$ *so that*

$$h_0(\Gamma_n) \leq \rho^n.$$

Proof. Let $u(z)$ be the harmonic measure. Take Ω' of length $n - 1$. $\Omega' = (\omega_1, \ldots, \omega_{n-1})$. Let $v(z^*)$ be harmonic in the *plane* slit domain connected through ω_{n-1}, with boundary values $= 0$ on the two sides of E and $= 1$ on the two sides of ω, $\omega \neq \omega_{n-1}$. The homogeneity condition gives

$$v(z^*) \leq \rho < 1, \; z \in \omega_{n-1}.$$

Hence for $z \in \omega_{n-1}$ in Ω',

$$u(z) \leq v(z^*) \leq \rho.$$

Proceeding inductively, we find $u(z) \leq \rho^n$ in the ground sheets and have proved Lemma 5.

This lemma shows that the factor γ^n is of no importance for (28). In the sheets Ω of length n the boundary values have two components: one factor, R, depending on x and one constant, A_n, depending on the history of the sheet. Let us first analyze the x-factor and begin with case B.

Let $u(z)$ have boundary values $r(x) = -\log R(x;\, \omega_n)$ on $\partial\Omega$ and $= 0$ otherwise. If $H(x;\, \xi)$ is the plane harmonic measure discussed in Sec. 4, we have for $z \in \omega_n$, $z^* = \xi$,

$$u(z) \leq \int - \log R(x;\, \omega_n)\, dH(x;\, \xi) = \int r(x) H'(x)\, dx.$$

Let us normalize so that $\delta(\xi) = 1$, $L = |\omega_n|/\delta(\xi)$. Then by the lemmas in Sec. 4,

$$u(z) \leq C \int_1^\infty \psi(x)^{-\theta/4} \frac{dx}{x^{1+\theta}} + C \int_1^\infty \sum_{|\sigma| \leq L} \psi(x;\, \sigma) \frac{dx}{x^{1+\theta}}.$$

To estimate the last sum we need a bound for

$$I(\sigma) = \int_1^\infty \frac{|\sigma|^2}{(x - \mu(\sigma))^2 + |\sigma|^2} \frac{dx}{x^{1+\theta}}.$$

We first observe that if $\mu(\sigma) > 2L$ and $|\sigma| \leq L$, we have

$$I(\sigma) \leq \text{Const.} \frac{|\sigma|}{\mu(\sigma)^{1+\theta}}.$$

Hence $\sum I(\sigma)$ for this case is bounded. Let us now assume $2^{\ell-1} < \mu(\sigma) \leq 2^{\ell} < 2L$. Splitting the x-integral into $(1, 2^{\ell-2})$, $(2^{\ell-2}, 2^{\ell+1})$, $(2^{\ell+1}, \infty)$ we easily find

$$I(\sigma) \leq \text{Const.} \left(\frac{|\sigma|^2}{2^{2\ell}} + \frac{|\sigma|}{2^{\ell(1+\theta)}}\right).$$

For every $\ell = 1, 2, \ldots, [\log L] + 1$, we sum over σ and obtain a fixed bound:

$$\sum_{(\ell)} I(\sigma) \leq \text{Const.} \left(2^{-2\ell} \sum_{(\ell)} |\sigma|^2 + 2^{-\ell(1+\theta)} \sum_{(\ell)} |\sigma|\right)$$

$$\leq \text{Const.} \left(2^{-2\ell} \operatorname*{Max}_{(\ell)} |\sigma| + 2^{-\ell(1+\theta)}\right) \cdot 2^{\ell} \leq \text{Const.}$$

Together this implies $u(z) = O(\log L)$, that is, in our original situation

$$u(x) \leq \text{Const.} \log \frac{|\omega_n|}{\delta(x)} = \Delta(x; \omega_n), \quad x \in \omega_n. \tag{30}$$

Suppose now that (ω_{n-1}, ω_n) is in case A.

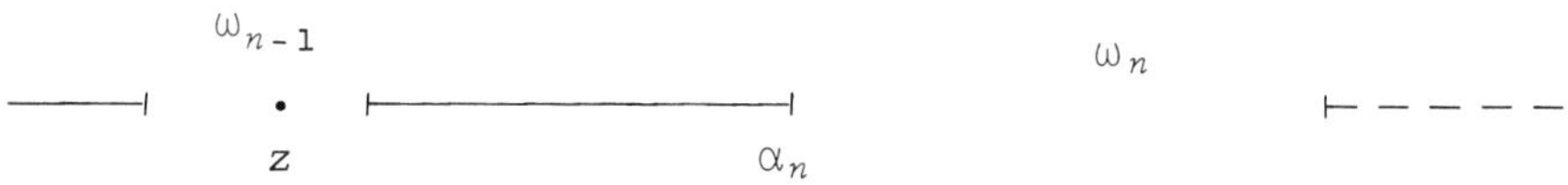

We proceed as before and consider $u(z)$ with boundary values $r(x) = -\log \psi(\alpha_n; \omega_{n-1}) - \log R(x; \omega_{n-1})$. Normalizing again so that $|\omega_{n-1}| = 1$, we find for $x \in \omega_n$

$$u(x) \leq \text{Const.} \left[\log\left(\frac{|\omega_{n-1}|}{\delta(x)} + 1\right) - \log \psi(\alpha_n; \omega_{n-1})\right] \tag{31}$$

$$= \Delta^{*}(x; \omega_{n-1}, \omega_n).$$

Observe here that $-\log \psi(\alpha_n; \omega_{n-1}) \leq - C \cdot \log(\text{dist.}(\omega_{n-1}, \omega_n))$.

Let us now keep ω_{n-1} fixed and consider all possible continuations $\omega_n \neq \omega_{n-1}$. For each such ω_n we have Ω with a variable boundary term $r(x) = r(x;\, \omega_n)$ as above. A majorant of the total contributions of these $r(x;\, \omega_n)$ to the value of $\log|F(z)|$ for $z \in \omega_{n-1}$ is obtained by solving the *plane* Dirichlet problem for $\mathbf{R}^2\backslash\omega_{n-1}$ with boundary values $\Delta(x;\, \omega_n)$ or $\Delta^*(x;\, \omega_{n-1},\, \omega_n)$ according to case B or A. Since everything used is scale invariant, we may assume $\delta(z) = 1$, and it follows that the total contribution is bounded on ω_{n-1}. The contribution to $\log|F(0)|$ from these variable factors in the boundary values can therefore be majorized by the harmonic measure $h_0(\Gamma)$ for $\Gamma = \bigcup_{\omega_n} \partial\Omega(\omega_1,\, \ldots,\, \omega_{n-1},\, \omega_n)$, since this harmonic measure is bounded from below on ω_{n-1}. The variable factors therefore satisfy (28).

We must now finally analyze the constant $A(\Omega) = \varphi(\mu;\, \Omega')$, $\mu = \mu_n$ or μ_{n-1} according to cases B and A. These constants are products of factors λ

$$\lambda = \begin{cases} \varphi(\alpha_j;\, \omega_{j-1}) \cdot R(\mu_j;\, \omega_{j-1}) & \text{(AA)} \\ \varphi(\alpha_j;\, \omega_{j-1}) \cdot R(\mu_{j+1};\, \omega_{j-1}) & \text{(AB)} \\ R(\mu_j;\, \omega_j) & \text{(BA)} \\ R(\mu_{j+1};\, \omega_j) & \text{(BB)} \end{cases}$$

where (XY) indicates that $(\omega_{j-1},\, \omega_j)$ is case X and $(\omega_j,\, \omega_{j+1})$ is case Y. If $\Omega_0 = (\omega_1,\, \ldots,\, \omega_{j-1},\, \omega_j,\, \omega_{j+1})$, then the factor λ is present in all constants $A(\Omega)$ with $\Omega \geq \Omega_0$. Since we are considering logarithms, this means that the contribution to (28) from each such factor is

$$(\log \lambda) \cdot h_0(\Gamma(\Omega_0)), \tag{32}$$

where

$$\Gamma(\Omega_0) = \bigcup_{\Omega \geq \Omega_0} \partial\Omega.$$

We must therefore prove that we can add (32) over all possible choices of Ω_0. The discussion is quite similar to the "variable" case treated above. We now base it on Lemmas 1′ and 3′ of Sec. 6. We observe that if μ is the center of some ω, then

$$C^{-1} \leq R(x;\ \sigma)/R(\mu;\ \sigma) \leq C \tag{33}$$

for $|x - \mu| \leq \dfrac{|\omega|}{3}$.

Case AA

We keep ω_{j-1}, ω_j fixed and consider the harmonic function on D^∞ with boundary values $= \log \lambda$ on $\bigcup_{\omega_{j+1}} \Gamma(\Omega_0)$ and $= 0$ otherwise. Clearly its values are $O(\log \lambda)$ on ω_j. Using Lemma 1, we see that the harmonic function with such boundary values on ω_j, as ω_j varies is uniformly bounded on ω_{j-1}. This case therefore follows as above.

Case AB

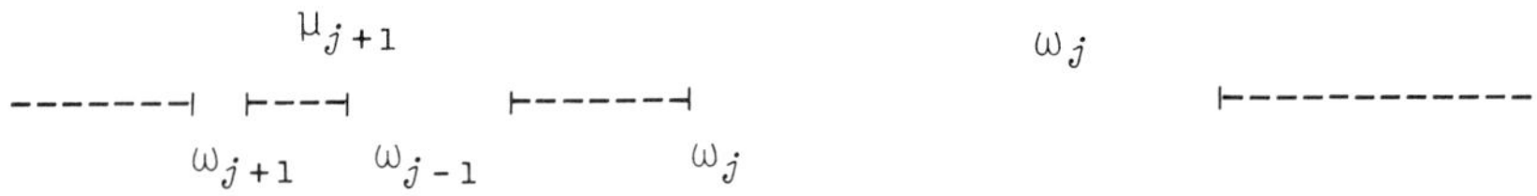

The factor $\varphi(\alpha_j;\ \omega_{j-1})$ is treated as in case AA. As above, the factor $-\log R(\mu_{j+1};\ \omega_j)$ contributes an estimate as in (31):

$$\Delta^*(x;\ \omega_{j-1},\ \omega_j) \text{ on } \omega_j.$$

As there, we can now let ω_j vary and obtain a bounded result on ω_{j-1}.

Case BA

Since $\log R(\mu_j;\ \omega_j)$ is bounded, this is a trivial case.

Case BB

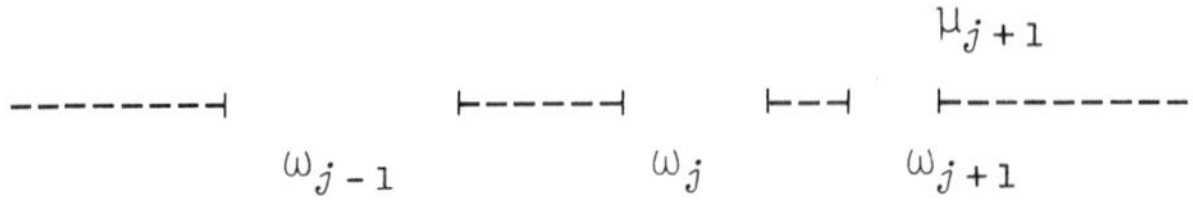

As ω_{j+1} varies, we obtain the estimate (30), $\Delta(x;\ \omega_j)$, on ω_j and then again a bounded estimate on ω_{j-1}.

This completes the proof.

REFERENCES

1. Ahlfors, L., and Beurling, A. Conformal invariants and function-theoretic nullsets. *Acta Math.* 83 (1950):101-129.

2. Benedicks, M. Positive harmonic functions vanishing on the boundary of certain domains in R^n. *Ark. Mat.* 18 (1980):53-72.

3. Forelli, F. Bounded holomorphic functions and projections. *Ill. J. Math.* 10 (1966):367-380.

4. Jones, P. L^∞-estimates for the $\bar{\partial}$-problem. To appear in *Acta Math.*

5. Wiener, N. The Dirichlet problem. *J. Math. Phys.* 3 (1924): 127-146.

TWO REMARKS ABOUT H^1 AND BMO ON THE BIDISC

Sun-Yung A. Chang
University of California, Los Angeles

1. INTRODUCTION

In [11], B. Maurey proved the existence of an unconditional basis for H^1 of the unit disc; his proof is based on Banach algebra techniques and is nonconstructive. In [3], L. Carleson constructed explicitly an unconditional basis for Re H^1, and in [13], P. Wojtaszczyk proved that the orthogonal Franklin system also is an unconditional basis. The proof in both [3] and [13] used explicit description of the dual space of H^1 of the disc, namely BMO. Although such explicit description for H^1 of the bidisc is still unknown, in Sec. 2 of this paper, we will use instead a Carleson measure characterization of the dual of H^1 of the bidisc (cf. [6]) to prove that the product Franklin system is an unconditional basis for Re H^1 of the bidisc. The existence of such basis for the bidisc case is also mentioned in Maurey.

In Sec. 3 of the paper, we show the existence of H^p-solutions $p < \infty$ for a special case of the corona problem of the bidisc. The proof generalizes T. Wolff's proof of the problem on the disc. We also use properties of Carleson measures on the

---

Research supported in part by NSF Grant and Mittag-Leffler Institute.

bidisc developed in [4] and [10]. A probabilistic approach to this result has been obtained by N. Varopoulous [12] (cf. also the related work of E. Amar [1]).

The author would like to thank Professor Carleson for many discussions about the subjects.

2. UNCONDITIONAL BASIS FOR $H^1(T^2)$

In this section, we establish the double Franklin system as unconditional basis for $H^1(T^2)$.

Let $\{\nu\}$ denote the collection of dyadic subintervals of $[0, 1]$, that is,

$$\nu = \left[\frac{k}{2^n}, \frac{k-1}{2^n}\right), \quad 0 \le k \le 2^n - 1, \; n = 0, 1, 2, \ldots,$$

and $\{f_\nu\}$ denote the Franklin system associated with $\{\nu\}$. For definitions and detailed properties of the Franklin system, cf. [7, 8]. For our purpose here, we will only use the following specific properties of the system, which are established in [7]:

(1) $\{f_\nu\}$ is an orthonormal set of functions which together with the constant function forms a basis for $L^2([0, 1])$.

(2) Each fixed f_ν satisfies

(a) $\int_0^1 f_\nu(t)\,dt = 0$, and there exists a constant C, and $q < 1$ independent of ν such that

(b) $$|f_\nu(t)| \le \frac{C}{|\nu|^{1/2}} q^{\frac{d(t,\nu)}{|\nu|}},$$

where d denotes the usual Lebesgue distance function and $|\nu|$ the Lebesgue measure of ν.

(c) $$|f_\nu(t_1) - f_\nu(t_2)| \le \frac{C|t_1 - t_2|}{|\nu|^{3/2}} q^{\frac{d([t_1, t_2],\nu)}{|\nu|}}.$$

Let ψ be a C^1 function defined on the real line supported on [-1, 1] with ψ even and

$$\int_{-1}^{1} \psi(t)dt = 0.$$

We will use the following notation: if $y > 0$, $\psi_y(t) = \frac{1}{y}\psi(t/y)$ and if $y = (y_1, y_2)$ and $t = (t_1, t_2) \in \mathbf{R}^2$, then

$$\psi_y(t) = \psi_{y_1}(t_1)\psi_{y_2}(t_2);$$

for each dyadic interval I, let I_+ denote the region

$$\left\{(t, y) : t \in I, \frac{|I|}{2} < y \leq |I|\right\};$$

for each dyadic rectangle $R = I \times J$, R_+ will denote the region $I_+ \times J_+$.

Although we do not know a specific description of the dual space of $H^1(T^2)$ [while on the unit circle T, Fefferman [9] characterized BMO (functions of bounded mean oscillation) as the dual of $H^1(T)$], one has [6] the following "Carleson measure" type of description of the space:

Lemma 1 ([6], Sec. 2). *Suppose $f \in L^1(T^2)$ and satisfies*

$$\int f(x_1, x_2)dx_1 = \int f(x_1, x_2)dx_2 = 0$$

for all $(x_1, x_2) \in T^2$. Then f is in the dual of $H^1(T^2)$ if and only if there exists some constant C such that

$$\sum_{\substack{R \subset \Omega \\ R \text{ dyadic Rectangle}}} \iint_{R_+} |f * \psi_y(t)|^2 \frac{dt\, dy}{y_1 y_2} \leq C|\Omega| \qquad (*)$$

for all open set $\Omega \subset T^2$.

Remark. One can think of the above result as a statement about the dual space of $H^1([0, 1] \times [0, 1])$ after the usual identification of T^2 as $[0, 1] \times [0, 1]$ with functions extending periodically at the endpoint 0, 1; that is, $f(0, x) = f(1, x)$ for $0 \leq x \leq 1$. The

Franklin basis [after deleting the function $f(x) = x$] all satisfy the periodic property.

The following result, which is a generalization of Lemmas 2, 3, and 4 in Carleson's paper [3] to the case of the product domain, is the main result of this section.

Theorem 1. If $f \in L^2(T^2)$ *satisfies*

$$\int f(x_1, x_2)\,dx_1 = \int f(x_1, x_2)\,dx_2 = 0,$$

and with expansion

$$f(x_1, x_2) = \Sigma\, C_{\mu\nu} f_\mu(x_1) f_\nu(x_2),$$

then condition (*) *in Lemma 1 is equivalent to*

$$\sum_{\mu \times \nu \subset \Omega} |C_{\mu\nu}|^2 \leq C|\Omega| \tag{**}$$

for all open set Ω *contained in* T^2.

Following the same logic as in [3], one obtains immediately:

Corollary 1. The double Franklin system $\left\{f_\mu \times f_\nu\right\}_{\text{all } \mu,\, \nu \text{ dyadic intervals}}$ *forms an unconditional basis of* $H^1(T^2)$.

With similar reasoning and proof as in [5], we obtain the following "unconditional" characterization of H^1.

Corollary 2. $H^1(T^2) = \{f \in L^1(T^2) : \Sigma\, C_{\mu\nu} f_\mu f_\nu$ *converges unconditional in* L^1, *where*

$$C_{\mu\nu} = \iint_{T^2} \{f(x_1, x_2) f_\mu(x_1) f_\nu(x_2)\,dx_1\, dx_2\}.$$

Carleson's [3] basis gives geometric insight about the structure of H^1. It remains open whether the product form of his basis also forms unconditional basis for $H^1(T^2)$. (The proof given below for Theorem 2 does not work for Carleson's basis.) On the other hand, it looks highly likely that recent work of P. Sjolin and J. O. Strömberg about characterization of $H^p(T)$, $0 < p < 1$, using Franklin basis and its higher splines, can be generalized to the product setting.

The following lemma lists elementary estimates that we will use in proving Theorem 1.

Lemma 2. Suppose μ, I *are dyadic intervals;* $\tilde{I}$ *the interval with same center as* I *and three times the length; then*

(1) *If* $|\mu| \leq |I|$, $\mu \subset \tilde{\tilde{I}}$, *then* $S_I^2(f_\mu) \leq C \dfrac{|\mu|^3}{|I|^3}$,

(2) *If* $|\mu| \leq |I|$, $\mu \cap \tilde{\tilde{I}} = \phi$, *then* $S_I^2(f_\mu) \leq C \dfrac{|\mu|}{|I|} q^{\frac{2d(\tilde{I},\, \mu)}{|\mu|}}$,

(3) *If* $|\mu| > |I|$, *then* $S_I^2(f_\mu) \leq C \dfrac{|I|^3}{|\mu|^3} q^{\frac{2d(\tilde{I},\, \mu)}{|\mu|}}$,

where C *is a constant depending only on* $\|\psi\|_\infty$, $\|\psi'\|_\infty$, *and*

$$S_I^2(f_\mu) = \iint_{I^+} |f_\mu * \psi_y(t)|^2 \frac{dt\, dy}{y}.$$

We will hence call a dyadic interval μ of type (i) with respect to I if condition (*) in Lemma 1 is satisfied $1 \leq i \leq 3$. For two dyadic rectangles $T = \mu \times \nu$ and $R = I \times J$, we say T is of type (i, j) with respect to R if μ is of type (i) w.r.t. I and ν is of type (j) w.r.t. J.

Proof of Lemma 2.

(1) If $|\mu| \leq |I|$ and $\mu \cap \tilde{\tilde{I}}$, then for $(t, y) \in I_+$,

$$\begin{aligned} |(f_\mu * \psi_y)(t)| &= \left|\int f_\mu(x)\psi_y(t - x)dx\right| \\ &= \left|\int f_\mu(x)(\psi_y(t - x) - \psi_y(t - C_\mu))dx\right| \\ &\leq \frac{\|\psi'\|_\infty}{|I|^2} \frac{1}{|\mu|^{1/2}} \int |x - C_\mu| q^{\frac{d(x,\, \mu)}{|\mu|}}\, dx \\ &\leq \frac{\|\psi'\|_\infty}{|I|^2} \frac{1}{|\mu|^{1/2}} \sum_{n=0}^{\infty} n|\mu|^2 q^{(n-1)} \leq C \frac{|\mu|^{3/2}}{|I|^2}, \end{aligned}$$

where C_μ is the center of the interval μ. In the second step above we have used property (a), in the third step property (b) of the Franklin function f_μ. The required estimate of $S_I^2(f_\mu)$ in the lemma follows directly from the estimate on $|f_\mu * \psi_y|$.

(2) Suppose $|\mu| \leq |I|$, $\mu \cap \tilde{\tilde{I}} = \phi$; notice that for each $(t, y) \in I_+$, $\psi_y(t - x)$ is supported on $\tilde{I}$ for each x; so if we divide $\tilde{I}$ into dyadic intervals I_j with $|I_j| = |\mu|$, then

$$|(f_\mu * \psi_y)(t)| = \left|\int f_\mu(x)\psi_y(t - x)\,dx\right|$$

$$\leq \sum_j \int_{I_j} |f_\mu(x)| \frac{\|\psi\|_\infty}{|I|}\,dx$$

$$\leq \frac{\|\psi\|_\infty}{|I|} \sum_{j=1}^{N} \frac{|I_j|}{|\mu|^{1/2}} q^{\frac{d(\tilde{I}, \mu)}{|\mu|} + j} \quad \text{where } N = \frac{|\tilde{I}|}{|\mu|}$$

$$\lesssim C \frac{1}{|I|} |\mu|^{1/2} q^{\frac{d(\tilde{I}, \mu)}{|\mu|}}.$$

The second to last step follows from property (b) of the Franklin basis f_μ.

(3) If $|\mu| > |I|$, then by property (c) of the Franklin function f_μ, we have, for $(t, y) \in I_+$,

$$|f_\mu * \psi_y(t)| = \left|\int_{\tilde{I}} (f_\mu(x) - f_\mu(C_I))\psi_y(t - x)\,dx\right|$$

$$\leq \frac{C}{|\mu|^{3/2}} \int_{\tilde{I}} |x - c_1| \frac{\|\psi\|_\infty}{|I|} q^{\frac{d([x, C_I], \mu)}{|\mu|}}\,dx$$

$$\leq C \frac{1}{|\mu|^{3/2}} |I| q^{\frac{d(\tilde{I}, \mu)}{|\mu|}},$$

where C_I is the center of the interval I.

Notation. For fixed dyadic rectangles $\mu \times \nu$, and $R = I \times J$. Let

$$S_R^2(f) = \iint_{R_+} |(f * \psi_y)(t)|^2 \frac{dt\, dy}{y_1 y_2}.$$

Notice

$$S_R(f_\mu f_\nu) = S_I(f_\mu) S_J(f_\nu).$$

Proof of Theorem 1. (*) ⇒ (**) Assume

$$\sum_{R \subset \Omega} S_R^2(f) \leq A|\Omega|$$

for all open set Ω. We will begin to estimate $\sum_{\mu \times \nu \subset \Omega} |C_{\mu\nu}|^2$, where

$$C_{\mu\nu} = \int_0^1 \int_0^1 f(x_1, x_2) f_\mu(x_1) f_\nu(x_2)\, dx_1\, dx_2.$$

For each fixed μ, ν

$$|C_{\mu\nu}| \leq \sum_R S_R(f) S_R(f_\mu f_\nu)$$

$$|C_{\mu\nu}|^2 \leq C\left(\sum_R S_R^2(f) S_R(f_\mu f_\nu)\right)\left(\sum_R S_R(f_\mu f_\nu)\right)$$

$$\leq C \sum_R S_R^2(f) S_R(f_\mu f_\nu)$$

The last step follows from the observation that

$$\sum_R S_R(f_\mu f_\nu) < \infty.$$

To see this, it suffices to verify the corresponding one-variable statement

$$\sum_I S_I(f_\mu) < \infty.$$

The latter follows easily from the estimates in Lemma 2 after we divide the intervals I into the three categories with the fixed interval μ of types (1), (2), (3) w.r.t. I accordingly. For example, let μ_N be the Nth interval adjacent to μ and of the same length $|\mu_N| = |\mu|$ ($\mu_0 = \mu$; $N > 0$ counting from the left, $N < 0$ to the right), then

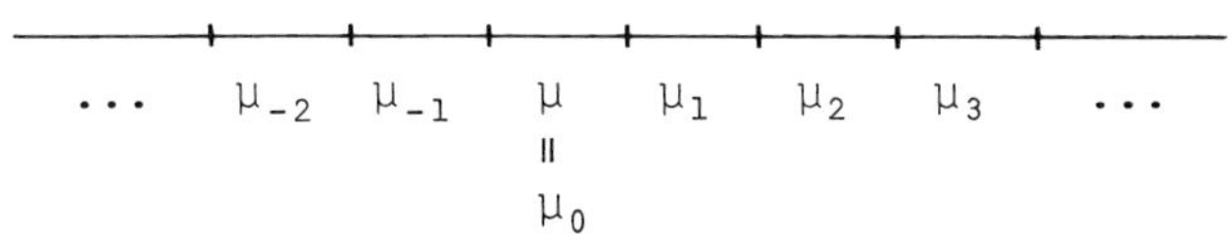

$$\sum_{\substack{I \\ \mu \text{ of type (2)} \\ \text{w.r.t } I}} S_I(f_\mu) \leq \sum_{N=-\infty}^{\infty} \sum_{I \subset \mu_N} \left|\frac{I}{\mu}\right|^{3/2} q^{\frac{d(\tilde{I},\, \mu)}{|\mu|}}$$

$$\leq 2 \sum_{N=0}^{\infty} \left(\sum_{n=1}^{\infty} \left(\frac{1}{2^n}\right)^{3/2} \cdot 2^n q^{(|N|-1)} \right) < \infty .$$

Types (1), (3) intervals could be handled similarly. Thus

$$\sum_{\mu \times \nu \subset \Omega} |C_{\mu\nu}|^2 \leq \sum_{\mu \times \nu \subset \Omega_0} \sum_R S_R^2(f) S_R(f_\mu f_\nu) = \sum_R S_R^2(f) \sum_{\mu \times \nu \subset \Omega_0} S_R(f_\mu f_\nu)$$

$$= \sum_{i,j=1}^{3} A_{ij},$$

where

$$A_{ij} = \sum_R S_R^2(f) \sum_{\substack{\mu \times \nu \subset \Omega \\ \mu \times \nu \text{ of type } (i,j) \text{ w.r.t. } R}} S_R(f_\mu f_\nu).$$

We will now estimate each term A_{ij} separately and indicate that each of them has a bound $\leq cA|\Omega|$. It is clear that the terms A_{12}, A_{21} can be handled the same way, and so can A_{13}, A_{31} and A_{23}, A_{32}. We will illustrate below ways to estimate terms A_{12}, A_{22}; the other terms A_{11}, A_{13}, A_{23}, A_{33} can be handled by similar methods.

Estimate of A_{12}

$$A_{12} = \sum_{R = I \times J} S_R^2(f) \sum_{\substack{\mu \times \nu \subset \Omega \\ |\mu| \leq |I|,\, \mu \subset \tilde{\tilde{I}} \\ |\nu| \leq |J|,\, \nu \cap \tilde{\tilde{J}} = \phi}} S_I(f_\mu) S_J(f_\nu).$$

For each integer N, let J_N be interval Nth adjacent to J as in the figure. Let

$$R_N^k = \left\{ I \times J : \frac{1}{2^k}|I \times J| < |\tilde{\tilde{I}} \times \tilde{\tilde{J}}_N \cap \Omega| \leq \frac{1}{2^{k-1}}|I \times J| \right\};$$

then

$$A_{12} \leq \sum_{R = I \times J} S_R^2(f) \sum_{|\mu| \leq |I|,\, \mu \subset \tilde{\tilde{I}}} S_I(f_\mu) \sum_{N=2}^{\infty} \sum_{\substack{\nu \subset J_N \\ \mu \times \nu \subset \Omega}} S_J(f_\nu)$$

$$\leq \sum_{N=2}^{\infty} \sum_{k=0}^{\infty} \sum_{\substack{R \in R_N^k \\ R = I \times J}} S_R^2(f) \sum_{\substack{|\mu| \leq |I| \\ \mu \subset \tilde{\tilde{I}}}} S_I(f_\mu) \sum_{\substack{\nu \subset J_N \\ \mu \times \nu \subset \Omega}} S_I(f_\nu)$$

$$= \sum_{|N|=2}^{\infty} \sum_{k=0}^{\infty} \sum_{R \in R_N^k} S_R^2(f) \Delta_N^k.$$

If $R = I \times J \in R_N^k$, $\mu \subset \tilde{\tilde{I}}$ with $|\mu| = \frac{1}{2^n}|I|$, $\nu \subset J_N$ with $|\nu| = \frac{1}{2^m}|J|$, then for each fixed m, $n \geq 0$ maximum number of such $\mu \times \nu \subset \Omega$ is $\leq 2^{m+n+1-k}$. Thus

$$\Delta_N^k \leq \sum_{\ell=0}^{\infty}\left(\sum_{n+m=k+\ell}\left(\frac{1}{2^n}\right)^{3/2}\left(\frac{1}{2^m}\right)^{1/2} q^{2^m|N|}\right)2^{\ell}$$

$$\leq C\left(\frac{1}{2^k}\right)^{3/2} q^{|N|}\left(\frac{1}{1-q^{|N|}}\right)^2.$$

And

$$A_{12} \leq C\sum_{|N|=2}^{\infty}\sum_{k=0}^{\infty}\sum_{R\in\mathcal{R}_N^k} S_R^2(f)\left(\frac{1}{2^k}\right)^{3/2} q^{|N|}\left(\frac{1}{1-q}\right)^2$$

$$\lesssim C\sum_{N=2}^{\infty}\sum_{k=0}^{\infty}\left(\frac{1}{2^k}\right)^{3/2} q^{|N|} A\left|\bigcup_{R\in\mathcal{R}_N^k} R\right| \quad \text{[by condition } (\star)]$$

$$\leq CA\sum_{N=2}^{\infty}\sum_{k=0}^{\infty}\left(\frac{1}{2^k}\right)^{3/2} q^{|N|} Nk\, 2^k \cdot |\Omega| \lesssim CA|\Omega|.$$

Estimate of A_{22}

$$A_{22} = \sum_{R=I\times J} S_R^2(f)\left(\sum_{\substack{\mu\times\nu\subset\Omega \\ |\mu|\leq|I|,\, \mu\cap\tilde{\tilde{I}}=\phi}} S_I(f_\mu)\right)\left(\sum_{\substack{|\nu|\leq|J| \\ \nu\cap\tilde{\tilde{J}}=\phi}} S_J(f_\nu)\right).$$

We will handle both the terms in μ and ν as the ν-term in the estimate of A_{12}. The only additional observation needed here is the following: if there exists $\mu \times \nu \subset \Omega$, with

$$|\mu| = \frac{|I|}{2^n},\ \mu \subset I_N\ |\nu| = \frac{|I|}{2^m},\ \nu \subset J_M \quad (n, m \geq 0,\ N, M = \pm 2, \pm 3, \ldots),$$

then $I \times J$ is contained in a set of measure $\leq (|N| + 1)2^n (|M| + 1)2^m |\mu \times \nu|$, and the collection of such $I \times J$ is contained in a set $\Omega_{2^n|N|,\, 2^m|M|}$ of measure $\leq 2^n(|N| + 1)2^m(|M| + 1)|\Omega|$. From this it follows that

$$A_{22} \leq \sum_{|N|=1}^{\infty}\sum_{|M|=1}^{\infty}\sum_{n=0}^{\infty}\sum_{m=0}^{\infty}\sum_{R\subset\Omega_{2^n|N|,2^m|M|}} (2^n)^{1/2}(2^m)^{1/2} q^{|N|2^n} q^{|M|2^m} S_R^2(f)$$

$$\leq \left(\sum_{N=1}^{\infty}\sum_{n=0}^{\infty}(2^n)^{1/2} q^{N2^n}\, 2^n(|N| + 1)\right)^2 A|\Omega_0| \leq CA|\Omega_0|.$$

(**) ⇒ (*) Assume there exists a constant A such that

$$\sum_{\mu \times \nu \subset \Omega} |C_{\mu\nu}|^2 \leq A|\Omega|$$

for all open set Ω. We will verify the same inequality for

$$\sum_{R \subset \Omega} S_R^2(f),$$

where $f = \sum C_{\mu\nu} f_\mu f_\nu$. Fix an open set Ω, let $\mathscr{R}_0 = \{$dyadic rectangles $R \subset \tilde{\tilde{\Omega}}\}$, let $f_1 = \sum_{\mu \times \nu \in \mathscr{R}_0} C_{\mu\nu} f_\mu f_\nu$; $f_2 = f - f_1$; then

$$(4) \quad \sum_{R \subset \Omega} S_R^2(f_1) \leq \sum_R S_R^2(f_1) \leqq \| f_1 \|_2^2 = \sum_{\mu \times \nu \in \mathscr{R}_0} |C_{\mu\nu}|^2 \leq CA|\Omega|.$$

$$(5) \quad \sum_{R \subset \Omega} S_R^2(f_2) \leq \sum_{R \subset \Omega} \Big(\sum_{\mu \times \nu \notin \mathscr{R}_0} S_R(f_\mu f_\nu) |C_{\mu\nu}| \Big)^2$$

$$\leq \sum_{R \subset \Omega} \Big(\sum_{\mu \times \nu \notin \mathscr{R}_0} S_R(f_\mu f_\nu) |C_{\mu\nu}|^2 \Big) \sum_{\mu \times \nu \notin \mathscr{R}_0} S_R(f_\mu f_\nu)$$

$$\leq C \sum_{R \subset \Omega} \sum_{\mu \times \nu \notin \mathscr{R}_0} S_R(f_\mu f_\nu) |C_{\mu\nu}|^2 = \sum_{i,j=1}^{3} S_{ij}$$

where

$$S_{ij} = \sum_{R \subset \Omega} \sum_{\substack{\mu \times \nu \notin \mathscr{R}_0 \\ \mu \times \nu \text{ of type } (ij) \text{ w.r.t. } R}} S_R(f_\mu f_\nu) |C_{\mu\nu}|^2 \qquad (S_{11} = 0).$$

The difference between the term A_{ij} and S_{ij} lies in the interchange of the role played by R and $\mu \times \nu$ in the formula. We again will only illustrate the estimate of the terms, say, S_{23}, S_{31}, to indicate the general pattern that each $S_{ij} \leq CA|\Omega|$.

<u>Estimate of S_{23}</u>

By Lemma 2, we have

$$S_{23} \leq \sum_{I \times J \subset \Omega} \sum_{\substack{|\mu| \leq |I|,\ \mu \cap \tilde{\tilde{I}} = \phi \\ |\nu| > |J|}} \left|\frac{\mu}{I}\right|^{1/2} q^{\frac{d(\tilde{I}, \mu)}{|\mu|}} \left|\frac{J}{\nu}\right|^{3/2} q^{\frac{d(\tilde{J}, \nu)}{|\nu|}} |C_{\mu\nu}|^2$$

$$= \sum_{I \times J \subset \Omega} \sum_{|N|=2}^{\infty} \sum_{M=0}^{\infty} \sum_{\mu \subset I_N} \sum_{M \leq \frac{d(\tilde{J}, \nu)}{|\nu|} \leq M+1} \left|\frac{\mu}{I}\right|^{1/2} q^{(N-1)\left|\frac{I}{\mu}\right| + \frac{d(I_N^c, \mu)}{|\mu|}} \left|\frac{J}{\nu}\right|^{3/2} q^M |C_{\mu\nu}|^2$$

$$\leq \sum_{|N|=2}^{\infty} \sum_{M=0}^{\infty} q^M \sum_{k=0}^{\infty} \sum_{\mu \times \nu \in \mathscr{R}_{N,M}^k} \Delta_{N,M}^k |C_{\mu\nu}|^2$$

where

$$\mathscr{R}_{N,M}^k = \left\{ \mu \times \nu : \frac{1}{2^k}|\tilde{\mu} \times \tilde{\nu}| \leq \left|(\tilde{\mu} \times \tilde{\nu}_{M+1}) \cap \left(\bigcup_{I \times J \subset \Omega} I_N \times J\right)\right| < \frac{1}{2^{k-1}}|\tilde{\mu} \times \tilde{\nu}| \right\}$$

and for each $\mu \times \nu \in \mathscr{R}_{N,M}^k$,

$$\Delta_{N,M}^k = \sum_{\substack{I \times J \subset \Omega, \\ \mu \subseteq I_N,\ M \leq \frac{d(\tilde{J},\nu)}{|\nu|} < M+1}} \left|\frac{\mu}{I}\right|^{1/2} \left|\frac{J}{\nu}\right|^{3/2} q^{(N-1)\left|\frac{I}{\mu}\right|}.$$

Observe that $\mu \subseteq I_N$, $M \leq \dfrac{d(\tilde{J},\ \nu)}{|\nu|} < M + 1$ imply that

$$|(\tilde{\mu} \times \tilde{\nu}_{M+1}) \cap (\tilde{I}_N \times \tilde{J}) = |\tilde{\mu} \times \tilde{J}|;$$

thus $\mu \times \nu \in \mathscr{R}_{N,M}^k$ implies that $|\tilde{J}| \leq \dfrac{1}{2^{k-1}}|\tilde{\nu}|$, and for each ℓ with $\left|\dfrac{J}{\nu}\right| = \dfrac{1}{2^{k+\ell}}$, the number of possible such J is $< 2^\ell$, so

$$\Delta_{N,M}^k \leq \sum_{\ell=0}^{\infty} \left(\frac{1}{2^{k+\ell}}\right)^{3/2} \cdot 2^\ell \sum_{m=0}^{\infty} \left(\frac{1}{2^m}\right)^{1/2} q^{(N-1)2^m} \leq Cq^N \left(\frac{1}{2^k}\right)^{3/2}.$$

We also have

$$\left|\bigcup_{\mu \times \nu \in \mathscr{R}_{N,M}^k} \mu \times \nu\right| \leq N(M + 1)k2^k|\Omega|.$$

Thus

$$S_{23} \leq C \sum_{|N|=2}^{\infty} \sum_{M=0}^{\infty} q^M \sum_{k=0}^{\infty} q^N \left(\frac{1}{2^k}\right)^{3/2} \sum_{\mu \times \nu \in \mathscr{R}_{N,M}^k} |C_{\mu\nu}|^2$$

$$\leq C \sum_{N=2}^{\infty} \sum_{M=0}^{\infty} q^{N+M} ANM \left(\sum_{k=0}^{\infty} \left(\frac{1}{2^k}\right)^{1/2} k\right) |\Omega| \lesssim CA|\Omega| \quad \text{[by condition (**)]}.$$

Estimate of S_{31}

Again by Lemma 2, we have

$$S_{31} \leq \sum_{I \times J \subset \Omega} \sum_{\substack{|\mu| > |I| \\ |\nu| \leq |J|,\ \nu \subset \tilde{\tilde{J}}}} \left|\frac{I}{\mu}\right|^{3/2} \left|\frac{\nu}{J}\right|^{3/2} q^{\frac{d(\tilde{I},\mu)}{|\mu|}} |C_{\mu\nu}|^2$$

$$= \sum_{I \times J \subset \Omega} \sum_{N=0}^{\infty} q^N \sum_{\substack{N \leq \frac{d(\tilde{I},\mu)}{|\mu|} < N+1 \\ |\nu| \leq |J|,\ \nu \subset \tilde{\tilde{J}}}} \left|\frac{I}{\mu}\right|^{3/2} \left|\frac{\nu}{J}\right|^{3/2} |C_{\mu\nu}|^2.$$

If we consider the set $\mathscr{L}_N^k = \left\{\mu \times \nu : \frac{1}{2^k}|\tilde{\mu} \times \tilde{\nu}| \leq |(\tilde{\mu}_{N+1} \times \nu) \cap \tilde{\tilde{\Omega}}| < \frac{1}{2^k}|\tilde{\mu} \times \hat{\nu}|\right\}$, then similar simple geometric arguments as before indicate that for each fixed $\mu \times \nu \in \mathscr{L}_N^k$, the number of $I \times J$ such that

$$N \leq \frac{d(\tilde{I}, \mu)}{|\mu|} < N + 1, \; \nu \subset \tilde{\tilde{J}},$$

is less than 2^ℓ when $\left|\frac{I}{\mu}\right| = \frac{1}{2^{k+\ell}}$ for each $\ell \geq 0$. Thus

$$S_{31} \leq \sum_{N=0} q^N \sum_{k=0}^{\infty} \sum_{\mu \times \nu \in \mathscr{L}_N^k} \left(\sum_{\ell=0}^{\infty}\left(\frac{1}{2^{k+\ell}}\right)^{3/2} \cdot 2^\ell \sum_{m=0}^{\infty}\left(\frac{1}{2^m}\right)^{3/2} \cdot 2\right)|C_{\mu\nu}|^2$$

$$\leq C \sum_{N=0}^{\infty} N q^N A |\Omega| \leq C|\Omega|.$$

Combining (4), (5), and the estimate on S_{ij}, we have established (**) from (*) and hence the proof of Theorem 2.

3. H^p-SOLUTION FOR CORONA PROBLEM ON THE BIDISC

In this section we will give another application of the characterization of the dual space of H^1 of the bidisc by a Carleson measure condition. Specifically, we will show that the Corona problem for two functions has an H^p-solution for all $p < \infty$. The proof we will give below is a direct generalization of T. Wolff's proof of the Corona problem on the disc. It does not depend on any explicit kernel formula for the solution of $\bar{\partial}$-equations. But the proof does not yield bounded solutions for the Corona problem as desired.

Theorem 2. Suppose f_1, f_2 are bounded analytic functions defined on the bidisc D^2, and satisfy

$$\sum_{j=1}^{2} |f_j(z)| \geq \delta$$

for all $z \in D^2$; then there exist $g_1, g_2 \in \bigcap_{p>\infty} H^p(D^2)$ with

$$\sum_{j=1}^{2} f_j(z) g_j(z) \equiv 1.$$

With the same reasoning as in T. Wolff's proof, Theorem 2 will be a consequence of Lemma 3 below. For any function g defined on D^2, we will use the notation $\partial_i g = \frac{\partial g}{\partial z_i}$, $\overline{\partial}_i g = \frac{\partial g}{\overline{\partial z_i}}$, $i = 1, 2$, $dA = \partial z_1 \overline{\partial} z_1 \partial z_2 \overline{\partial} z_2$. If μ is any Carleson measure on D or D^2, we will say $\mu \in C$ and denote its measure constant (cf. [2]) by $\|\mu\|_C$.

Lemma 3. Suppose g is a smooth function defined on D^2 and satisfies

(a) $|(\partial_1\partial_2 g)(z_1, z_2)| \log \frac{1}{|z_1|} \log \frac{1}{|z_2|} dA \in C$ on D^2,

(b) $\iint_{D^2} |\partial_1 g(z_1, z_2)| |\partial_2 h(z_1, z_2)| \log \frac{1}{|z_1|} \log \frac{1}{|z_2|} dA$
$\leq C\|h\|_{H^1}$ *for all* $h \in H^1$,

(c) *same condition as in (b) with* z_1, z_2 *interchanged,*

(d) $|g^2(z_1, z_2)| \log \frac{1}{|z_1|} \log \frac{1}{|z_2|} dA \in C$ on D^2.

Then there exists a solution v on D^2 with $\overline{\partial}_1\overline{\partial}_2 v = g$, the radial limit of v exists on T^2, and $\|v\|_{L^\infty(T^2)}$ is bounded by a constant depending only on the Carleson measure constant in (a)-(d).

One should remark that by Carleson measure on D^2 we mean a measure that satisfies the Carleson-type geometric condition w.r.t. open set as described in [4, 10]. Condition (b) [also (c)] is a mixed Carleson measure condition for one and two variables that is satisfied, for example, when

$$|\overline{\partial}_1 g(z_1, z_2)| \log \frac{1}{|z_1|} dA_1$$

is a Carleson measure in D with $\left\| \overline{\partial}_1 g(z_1, z_2) \log \frac{1}{|z_1|} \right\|_C$ uniform in z_2.

To prove Lemma 3, we will first establish some function theoretic facts of functions defined on T^2. Let $P_{i,j}$ denote the projection from $L^2(T^2)$ to the closed subspace of L^2 generated by $z_1^{in} z_2^{jm}$ $n \geq 0$, $m \geq 0$, i, $j = \pm 1$; $H_{ij} = \{h \in L^\infty(T^2),\ P_{ij}(h) = 0\}$.

Notice that if we let H denote the biharmonic in D^2 of $h \in L^2$, then $h \in H_{-1,-1}$ if and only if $\bar{\partial}_1\bar{\partial}_2 H = 0$. Furthermore, if we define $\mathrm{BMO}(T^2)$ as the dual of $H^1(T^2)$, and define the norm of $\varphi \in \mathrm{BMO}(T^2)$ as

$$\|\varphi\|_* = \inf_{\varphi = \sum_{i,j=\pm 1} P_{ij}(\varphi_{ij})} \sum_{i,j=\pm 1} \|\varphi_{ij}\|_\infty ,$$

then the following lemma is immediate:

Lemma 4. $d_{L^\infty(T^2)}(f, H_{i,j}) \approx \|P_{ij}(f)\|_*$ *for all* $f \in L^\infty(T^2)$ *for all* $i, j = \pm 1$.

If we use the notation $H^1(T^2) \equiv \{f \in L^1(T^2),\ P_{ij}(f) \in L^1$ for all $i, j = \pm 1\}$ and let $\mathscr{H}^1$ denote the analytic function in $H^1(T^2)$, that is, $\mathscr{H}^1 = \{f \in L^1(T^2),\ f = P_{11}(f)\}$, then the following lemma is also immediate:

Lemma 5. For all $f \in L^\infty(T^2)$,

$$\|P_{-1,-1}(f)\|_* = \sup_{\substack{h \in H^1 \\ \|h\|_{H^1} \le 1}} \left| \int_{T^2} fP_{11}(h) \right| = \sup_{\substack{h \in \mathscr{H}^1 \\ \|h\|_{\mathscr{H}^1} \le 1}} \left| \int_{T^2} fh \right| .$$

Proof of Lemma 3. Since g is smooth, there exists some $v_0 \in C^\infty(\bar{D}^2)$ that satisfies the equation $\bar{\partial}_1\bar{\partial}_2 v_0 = g$. To prove solution v exists with the desired bound on its norm, we need only to estimate $d_{L^\infty}(v_0, H_{-1,-1})$. Applying Lemmas 4 and 5, we have

$$d_{L^\infty}(v_0, H_{-1,-1}) \lesssim \sup_{h \in \mathscr{H}^1} \left| \int_{T^2} v_0 h \right| .$$

If we assume w.l.o.g. that

$$\int_T v_0(t_1, t_2)dt_2 = \int_T v_0(t_1, t_2)dt_1 = 0,$$

then by Green's theorem,

$$\int_{T^2} v_0 h = C \iint_{D^2} \Delta_1\Delta_2(v_0 h) \log \frac{1}{|z_1|} \log \frac{1}{|z_2|} \, dA .$$

Observe that if we take the biharmonic extension of h in D^2, then

$$\Delta_1\Delta_2(v_0 h) = \Delta_1\big((\Delta_2 v_0)h + \nabla_2 v_0 \cdot \nabla_2 h\big)$$

$$= (\Delta_1\Delta_2 v_0)h + \bar\partial_1(\Delta_2 v_0) \cdot \partial_1 h + \bar\partial_2(\Delta_1 v) \cdot \partial_2 h + \bar\partial_1\bar\partial_2 v_0 \cdot \partial_1\partial_2 h$$
$$= \partial_1\partial_2 g \cdot h + \partial_1 g \cdot \partial_2 h + \partial_2 g \cdot \partial_1 h + g\partial_1\partial_2 h.$$

We can then apply condition (a) to (b) listed to estimate each term independently (cf. [4, 10]).

To obtain Theorem 2 as a consequence of Lemma 3, we make the following observation, which is a direct consequence of Tom Wolff's theorem.

Lemma 6. Suppose h is smooth and satisfies

(a) $|h(z)|^2 \log \frac{1}{|z|}\, dz\, d\bar z \in C$ *in* D

(b) $|\partial h(z)| \log \frac{1}{|z|}\, dz\, d\bar z \in C$ *in* D

with constant $\| \ \|_C \le M$.

Then

(1) *there exists a function* v *with* $\bar\partial v = h$ *on* D *and* $\| v|_T \|_\infty \le CM$.

(2) *if* v_0 *is any smooth function* $\bar\partial v_0 = h$ *on* $\bar D$, *and* V_0 *is the harmonic extension into* D *of* $v_0|_T$, *then*

$$|\bar\partial V_0|^2 \log \frac{1}{|z|}\, dz\, d\bar z \in C$$

with measure constants $\le CM$.

Proof. (1) is Tom Wolff's result.

(2) Choose v as in (1); then $\| v|_T \|_\infty \le CM$ implies that

$$|\nabla V|^2 \log \frac{1}{|z|}\, dz\, d\bar z \in C$$

with measure constant $\le CM$ when V is the harmonic extension of v in D. If v_0 is another solution with $\bar\partial v = \bar\partial v_0 = h$, then $(v - v_0)|_T \in H^\infty(T)$; hence $\bar\partial V_0 = \bar\partial V$ in D when V_0 is the harmonic extension of $v_0|_T$.

Proof of Theorem 2. We will follow the same pattern of proof as in [2]. Let f_1, f_2 be two bounded analytic functions satisfying $|f_1(z)| + |f_2(z)| \geq 2\delta$ on D^2. Assume f_1, f_2 are smooth in $\overline{D^2}$, choose C^∞ function $u : \mathbf{C} \to \mathbf{C}$ with $u(z) = 0$ when $|z| < \frac{1}{2}$, $u(z) = 1$ when $|z| \geq 1$. Let

$$\psi_j(z) = \frac{1}{f_j(z)} \frac{u(f_j(z)/\delta)}{\sum_{j=1}^{2} u(f_j(z)/\delta)}, \quad j = 1, 2;$$

then $\sum \psi_j f_j = 1$. To modify ψ_j to analytic solutions ϕ_j, define $g_1(z) = \psi_1(z) + f_2(z)k(z)$, $g_2(z) = \psi_2(z) - f_1(z)k(z)$, then $g_1 f_1 + g_2 f_2 = 1$ and solve

$$(6) \quad \begin{cases} \bar{\partial}_j g_1 = \bar{\partial}_j(\psi_1) + f_2 \bar{\partial}_j k = 0 \\ \bar{\partial}_j g_2 = \bar{\partial}_j(\psi_2) - f_1 \bar{\partial}_j k = 0 \end{cases}, \quad j = 1, 2,$$

that is, we want to solve and obtain $\| k|_{T^2} \|_* \leq$ constant depends on δ, $\| f_j \|_\infty$, $j = 1, 2$.

$$(7) \quad \bar{\partial}_j k = -\frac{\bar{\partial}_j \psi_1}{f_2} = \frac{\bar{\partial}_j \psi_2}{f_1}, \quad j = 1, 2,$$

which is well defined, since $f_1\psi_1 + f_2\psi_2 = 1$. Let h_j denote

$$-\frac{\bar{\partial}_j \psi_1}{f_2} = \frac{\bar{\partial}_j \psi_2}{f_1}.$$

To solve $\bar{\partial}_j k = h_j$, $j = 1, 2$, with the desired bound, we apply Lemmas 3 and 6 in the following manner.

First notice that h_j, $j = 1, 2$, satisfies conditions (a) and (b) in Lemma 3 with the Carleson measure norm of

$$|h_1(z_1, z_2)|^2 \log \frac{1}{|z_1|} dA_1, \quad |\partial_1 h_1(z_1, z_2)| \log \frac{1}{|z_1|} dA_1,$$

$$|h_2(z_1, z_2)|^2 \log \frac{1}{|z_2|} dA_2, \quad |\partial_2 h_2(z_1, z_2)| \log \frac{1}{|z_2|} dA_2$$

depending only on δ and $\| f_j \|_\infty$, $j = 1, 2$, and uniform in $(z_1, z_2) \in D^2$; we will henceforth denote such norm by M. (The same proof is in T. Wolff's article.) Notice also that $\bar{\partial}_2 h_1 = \bar{\partial}_1 h_2$; applying (2) of Lemma 6, we obtain:

(8) There exists v_0 satisfying $\bar{\partial}_j v_0 = h_j$, $j = 1, 2$, and if v_0 is the biharmonic extension (that is, harmonic extension in z_1, z_2 separately) of $v_0|_{T^2}$, then

$$|\bar{\partial}_1 V_0(z_1, z_2)|^2 \log \frac{1}{|z_1|} dA_1$$

and

$$|\bar{\partial}_2 V_0(z_1, z_2)|^2 \log \frac{1}{|z_2|} dA_2$$

are both Carleson measures with constant M uniformly for $(z_1, z_2) \in D^2$.

Next, if we let $g = \bar{\partial}_2 h_1 = \bar{\partial}_1 h_2$, assume for a moment that g satisfies conditions (a)-(d) in Lemma 3 with measure constant $\leq CM$. Then Lemma 3 implies:

(9) There exists v with $\bar{\partial}_1 \bar{\partial}_2 v = g$ and $\| v|_{T^2} \|_\infty \leq CM$ and $v = v_0 + F$ in D^2 for some F biharmonic in D^2 with $\bar{\partial}_1 \bar{\partial}_2 F = 0$.

We now take the function v in (9) and try to solve

$$(10) \quad \begin{cases} \bar{\partial}_1 k_1 = h_1 - \bar{\partial}_1 v \\ \bar{\partial}_2 k_1 = 0 \end{cases} \quad \text{also} \quad \begin{cases} \bar{\partial}_1 k_2 = 0 \\ \bar{\partial}_2 k_2 = h_2 - \bar{\partial}_2 v. \end{cases}$$

We wish to obtain k_1, k_2 satisfy (10) and with $\| k_j|_{T^2} \|_\infty \leq CM$. (This would yield a bounded solution for the Corona problem.) But unfortunately, we are only able to obtain solutions with $\| k_j|_{T^2} \|_* \leq CM$, $j = 1, 2$, instead [see Remark (1) at the end]. To see this, first observe that from (8) and (9), $\bar{\partial}_2(h_1 - \bar{\partial}_1 v) = \bar{\partial}_1(h_2 - \bar{\partial}_2 v) = 0$, so both equations in (10) are well-set. We will now apply the one-variable technique in part (1) of Lemma 6 to solve (10). To be able to do so, we first check that both $h_1 - \bar{\partial}_1 v$

and $h_2 - \bar{\partial}_2 v$ satisfy the conditions (a) and (b) listed in Lemma 6, and with the measure constants uniform for all z D^2. Now if we denote V the biharmonic extension of v in D^2, then $\| v|_{T^2} \|_{L^\infty} \le CM$ implies that

$$\| \, |\bar{\partial}_j V(z_1, z_2)|^2 \log \frac{1}{|z_j|} \, dA \|_C \le CM, \; j = 1, 2.$$

Since $V = V_0 + F$ by (9), and

$$\| \, |\bar{\partial}_j V_0|^2 \log \frac{1}{|z_j|} \, dA \|_C \le CM, \; j = 1, 2,$$

by (8), we get the same property

$$\| \, |\bar{\partial}_j F|^2 \log \frac{1}{|z_j|} \, dA \|_C \le CM, \; j = 1, 2,$$

for F. Thus

$$h_j - \bar{\partial}_j v = h_j - \bar{\partial}_j (v_0 + F) = -\bar{\partial}_j F$$

satisfy (a), and

$$\partial_j (h \; - \bar{\partial}_j v) = \partial_j h_j - \partial_j \bar{\partial}_j v_0 - \partial_j \bar{\partial}_j F = 0$$

satisfy (b) trivially for $j = 1, 2$. If we apply Wolff's Lemma 6 to the first set of equations in (10), we obtain a solution u satisfying $\bar{\partial}_1 u = h_1 - \bar{\partial}_1 v$ with $\| u|_{T^2} \|_{L^\infty} \le CM$. Let k_1 be the orthogonal projection from $L^2(T)$ to $H^2(T)$ in z_2 of u; then k_1 satisfy both $\bar{\partial}_1 k_1 = h_1 - \bar{\partial}_1 v$ and $\bar{\partial}_2 k_1 = 0$ with $\| k_1 \|_* \le C \| u \|_{L^\infty}$ $\le CM$. Similarly, we obtain k_2 satisfy the end set equations in (10).

To finish the proof of Theorem 2, we need to verify the function $g = \bar{\partial}_1 h_2 = \bar{\partial}_2 h_1$ satisfies all the conditions (a)-(d) in Lemma 3. The verifications are lengthy but relatively routine. It only depends on the following facts about H^∞ functions f, g in D^2. (Cf. [4, 10].)

(1) $$\left| \frac{\partial f}{\partial z_j}(z_1, z_2) \right|^2 \log \frac{1}{|z_j|} \, dA_j \in C, \; j = 1, 2, \text{ uniform in } z \in D^2.$$

(2) $$\left|\frac{\partial f}{\partial z_j}(z_1, z_2)\right|^2 \left|\frac{\partial g}{\partial z_i}(z_1, z_2)\right|^2 \log\frac{1}{|z_j|}\log\frac{1}{|z_i|}\, dA \in C,$$

$$i, j = 1, 2;\ i \neq j.$$

(3) $$\left|\frac{\partial^2 f}{\partial z_1 \partial z_2}(z_1, z_2)\right|^2 \log\frac{1}{|z_1|}\log\frac{1}{|z_2|}\, dA \in C.$$

Furthermore, all the Carleson measure constants of above are bounded by constants depending only on $\|f\|_\infty$, $\|g\|_\infty$. We will leave the details to the reader.

Remarks

(1) Given h satisfy $\partial_1 h = \overline{\partial}_2 h = 0$ and

$$\|\, |h(z_1, z_2)|^2 \log\frac{1}{|z_1|}\, dA_1\|_C \leq M$$

uniformly in z_2, there may not exist a bounded solution for the equations

$$\begin{cases} \overline{\partial}_1 k = h \\ \overline{\partial}_2 k = 0. \end{cases}$$

This indicates that the BMO solution is about the best we can hope for for (10) under the given setting. I have learned the following example from Mok (private communication): Let $f = \log(1 - \overline{z}_1 z_2)$ and $h = \frac{\partial}{\partial \overline{z}_1} f$; then $\partial_1 h = \overline{\partial}_2 h = 0$, and

$$\|\, |h|^2 \log\frac{1}{|z_1|}\, dA_1\|_C \leq \|F(z_1, z_2)\|_{\mathrm{BMO}(z_1)} \leq C \cdot 2\pi$$

for all $z_2 \in D$. Suppose k is a solution with k bounded and $\overline{\partial}_1 k = h$, $\overline{\partial}_2 k = 0$; then $g = k - f$ is analytic in both z_1 and z_2. $g(0, 0) = k(0, 0)$ is finite, and when restricting to the subspace $z = z_1 = z_2$, and when $|z| = r$, we have $|g(z) + \log(1 - r^2)|$ bounded by $\|k\|_\infty$. Integrating over $|z| = r$, and letting $r \to 1$, yields a contradiction.

(2) Assuming only $f_1, f_2, \ldots, f_n$ exists in H^p $(p > 1)$ in Theorem 2 to start with, N. Varapoulous [12] proved that existence of some $g_1, \ldots, g_n \in H^q$ for some q depends on p with

$$\sum_{j=1}^{n} f_j g_j = 1.$$

It looks like methods used in this section also generalize to that setting (using area-function characterizations of H^p instead of Carleson measure properties of H^∞ in the formulation of Lemmas 3 and 6).

REFERENCES

1. Amar, E. Généralisation d'un théorème de Wolff à la boule de $\mathbf{C}^N$. Preprint.

2. Carleson, L. The Corona problem. *Proc. Fifteenth Scan. Congress* 968. *Lecture Notes in Math*, vol. 118, Springer-Verlag.

3. Carleson, L. An explicit unconditional basis in H^1. Institut Mittag-Leffler, Rap. No. 2 (1980).

4. Chang, S. Y. Carleson measure on the bi-disc. *Ann. Math.* 109 (1979):613-620.

5. Chang, S. Y., and Ciesielski, Z. Spline characterization of H^1. Preprint.

6. Chang, S. Y., and Fefferman, R. A continuous version of duality of H^1 and BMO on the bidisc. *Ann. Math.* (1980).

7. Ciesielski, Z. Properties of the orthonormal Franklin system. *Studia Math.* 73 (1963):141-157.

8. Ciesielski, Z. Properties of the orthonormal Franklin system II. *Studia Math.* 27 (1966):289-323.

9. Fefferman, C. Characterization of bounded mean oscillation. *Bull. Amer. Math. Soc.* 77 (1971):587-588.

10. Fefferman, R. Functions of bounded mean oscillation on the bi-disc. *Ann. Math.* (1979).

11. Maurey, B. Isomorphismes entre espaces H_1. *Acta Math.* 145 (1980):79-120.

12. Varapoulous, N. Probabilistic approach to some problems in complex analysis. Preprint.

13. Wojtaszczyk, P. The Franklin system is an unconditional basis in H^1. Preprint.

ON THE SOLUTION OF THE EQUATION $\Delta^m F = f$ For $f \in H^p$

Angel B. Gatto, Carlos Segovia
University of Buenos Aires

Julio R. Jiménez
University Central de Venezuela

DEFINITIONS AND NOTATION

Let $L^q(\text{loc})$, $1 < q < \infty$, be the space of the real-valued functions $f(x)$ on R^n that belong locally to L^q. We endow $L^q(\text{loc})$ with the topology generated by the seminorms

$$|f|_{q,B} = (|B|^{-1} \int_B |f(x)|^q \, dx)^{1/q},$$

where B is a ball in R^n and $|B|$ denotes its Lebesgue measure. For $f(x)$ in $L^q(\text{loc})$, we define a maximal function $n_{q,m}(f;\, x)$ as

$$n_{q,m}(f;\, x) = \sup_{\rho > 0} \rho^{-2m} |f|_{q,\, B(x,\, \rho)},$$

where m is a positive integer.

Let $\mathscr{P}_{2m-1}$ be the linear subspace of $L^q(\text{loc})$ formed by all the polynomials of degree at most $2m - 1$. This subspace is of finite dimension and therefore a closed subspace of $L^q(\text{loc})$. We denote by E^q_{2m-1} the quotient space of $L^q(\text{loc})$ by $\mathscr{P}_{2m-1}$. If $F \in E^q_{2m-1}$, we define the seminorm

$$\|F\|_{q,B} = \inf\{|f|_{q,B} : f \in F\}.$$

The family of all these seminorms induces on E^q_{2m-1} the quotient topology. This space E^q_{2m-1} is locally convex and complete.

Supported in part by Consejo Nacional de Investigaciones Cientificas y Técnicas, República Argentina.

For F in E^q_{2m-1}, we define a maximal function $N_{q,m}(F; x)$ as

$$N_{q,m}(F; x) = \inf\{n_{q,m}(f; x) : f \in F\}.$$

We say that an element F in E^q_{2m-1} belongs to $\mathscr{H}^p_{q,m}$, $0 < p \leq 1$, if the maximal function $N_{q,m}(F; x)$ belongs to $L^p(R^n)$. The "norm" of F in $\mathscr{H}^p_{q,m}$ is defined as $\|F\|_{\mathscr{H}^p_{q,m}} = \|N_{q,m}(F; \cdot)\|_{L^p}$.

To simplify the notation, we write $\mathscr{H}^p$ and $N(F; x)$ instead of $\mathscr{H}^p_{q,m}$ and $N_{q,m}(F; x)$, respectively.

Let us denote by $\mathcal{S}$ the set of all C^∞ functions ϕ satisfying that for any pair of nonnegative integers k and h

$$p_{k,h}(\phi) = \max_{|\alpha| \leq h} \sup_{x \in R^n} |\partial^\alpha \phi(x)| (1 + |x|)^k$$

is finite. The vector space $\mathcal{S}$ endowed with the family of norms $\{p_{k,h}\}$ becomes a locally convex and metrizable topological vector space. As usual, the dual space $\mathcal{S}'$ of $\mathcal{S}$ is called the space of tempered distributions. If $\phi(x)$ belongs to $\mathcal{S}$, let $\phi_t(x) = t^{-n}\phi(x/t)$.

We recall the definition of the Hardy spaces H^p on R^n. For $0 < p \leq 1$, H^p is the space of all tempered distributions f such that

$$f^\star(x) = \sup_{|x-y|<t} |(f \star \phi_t)(y)|$$

belongs to L^p, for some $\phi \in \mathcal{S}$ with $\int \phi(x)dx \neq 0$. The norm $\|f\|_{H^p}$ is defined as the norm of $f^\star(x)$ in L^p. A function $b(x)$ on R^n is called a p-atom if there is a ball B containing the support of $b(x)$ such that $\|b\|_\infty \leq |B|^{-1/p}$ and $\int b(x)x^\alpha dx = 0$ for all multi-indexes α, $|\alpha| \leq N$. The number N is assumed to be as large as necessary. It is well known that given f in H^p, there exist a numerical sequence $\{\lambda_i\}$ and a sequence $\{b_i(x)\}$ of p-atoms such that

$$f = \Sigma \lambda_i b_i$$

and $\sum |\lambda_i|^p \approx \|f^\star\|^p_{L^p}$. For more details, see [3].

We shall say that a class A E^q_{2m-1} is a p-atom in $\mathscr{H}^p_{q,m}$ if there exist a representative $a(x)$ of A and a ball B such that

$$\mathrm{supp}(a) \subset B$$

and

$$N(A;\, x) \leq |B|^{-1/p}.$$

Let $F \in E^q_{2m-1}$ and $f \in F$. Since f belongs to $L^q(\mathrm{loc})$, $\Delta^m f$ is defined in the sense of distributions. On the other hand, since any two representatives of F differ in a polynomial of degree smaller than $2m$, we get that $\Delta^m f$ is independent of the representative $f \in F$ chosen. Therefore, for $F \in E^q_{2m-1}$, we shall define $\Delta^m F$ as the distribution $\Delta^m f$, where f is any representative of F.

STATEMENT OF THE RESULTS

The main result of this paper is contained in the following.

Theorem 1. *Let $n(2m + n/q)^{-1} < p \leq 1$. The operator Δ^m defines a one-to-one application from $\mathscr{H}^p_{q,m}$ onto H^p. Moreover, there exist two positive and finite constants c and c' such that*

$$c\|F\|_{\mathscr{H}^p_{q,m}} \leq \|\Delta^m F\|_{H^p} \leq c'\|F\|_{\mathscr{H}^p_{q,m}} \tag{1}$$

hold for every $F \in \mathscr{H}^p_{q,m}$.

The case when the hypothesis $n(2m + n/q)^{-1} < p$ does not hold is trivial since:

Theorem 2. *If $n(2m + n/q)^{-1} \geq p$, $p \leq 1$, then $\mathscr{H}^p_{q,m} = \{0\}$.*

We also give a decomposition into atoms for elements of $\mathscr{H}^p_{q,m}$. The method used to obtain the decomposition was developed in [4].

Theorem 3. *Let F belong to $\mathscr{H}^p_{q,m}$, $n(2m + n/q)^{-1} < p \leq 1$. Then, there exist a numerical sequence $\{\mu_i\}$ and a sequence of p-atoms $\{A_i\}$ in $\mathscr{H}^p_{q,m}$ such that $F = \sum \mu_i A_i$ in E^q_{2m-1}. Moreover, there exist* *two positive constants* c_1 *and* c_2 *such that*

$$c_1\|F\|^p_{\mathscr{H}^p_{q,m}} \leq \sum|\mu_i|^p \leq c_2\|F\|^p_{\mathscr{H}^p_{q,m}}\,.$$

SOME PREVIOUS LEMMAS

Lemma 1. The maximal function $n(f;\, x)$ *associated to a function* f *in* $L^q(\mathrm{loc})$ *is lower semicontinuous.*

Proof. See Lemma 6 of [1].

Lemma 2. Let f_1 *and* f_2 *be two representatives of an element* $F \in E^q_{2m-1}$, *and* $P = f_1 - f_2$. *There exists a constant* c_α *such that*

$$|(\partial^\alpha P)(y)| \leq c_\alpha(n(f_1;\, x_1) + n(f_2;\, x_2))(|x_1 - y| + |x_2 - y|)^{2m-|\alpha|}$$

holds for every x_1, x_2, *and* y *in* R^n.

Proof. The proof is the same as Lemma 3 in [1].

Lemma 3. Let F *belong to* E^q_{2m-1} *with* $N(F;\, x_0) < \infty$. *Then:*

(i) There exists a unique f *in* F *such that* $n(f;\, x_0) < \infty$ *and, therefore,* $n(f;\, x_0) = N(F;\, x_0)$.

(ii) For any ball B, *there is a finite constant* c *depending on* x_0 *and* B *such that if* f *is the unique representative of* F *given in (i), then*

$$\|F\|_{q,B} \leq |f|_{q,B} \leq c\, n(f;\, x_0) = c\, N(F;\, x_0)\,.$$

The constant c can be chosen independently of x_0 provided that x_0 varies in a compact set.

Proof. Let us prove (i). Assume that f_1 and f_2 belong to F and both $n(f_1;\, x_0)$ and $n(f_2;\, x_0)$ are finite. Call P the polynomial $f_1 - f_2$ of degree at most $2m - 1$. Applying Lemma 2 with

$x_1 = x_2 = x_0 = y$ and varying α, $0 \leq |\alpha| \leq 2m - 1$, we get $\partial^\alpha P(x_0) = 0$, which shows that $P(x) = 0$.

Let us consider part (ii). We can assume $N(F;\, x_0) < \infty$, since, otherwise, there is nothing to prove. By part (i) already proved, let f be the unique representative of F satisfying $n(f;\, x_0) = N(F;\, x_0)$. For a given ball B, let ρ_0 be the least value of ρ for which $B \subset B(x_0,\, \rho)$. Then

$$\|F\|_{q,\,B} \leq |f|_{q,\,B} \leq \rho_0^{2m} |B|^{-1/q} |B(x_0,\, \rho_0)|^{1/q} \rho_0^{-2m} |f|_{q,\,B(x_0,\,\rho_0)}$$

$$\leq c_{x_0,B}\, n(f;\, x_0) = c_{x_0,B}\, N(F;\, x_0).$$

Corollary 1. If $\{F_k\}$ is a sequence of elements of E^q_{2m-1} converging to F in $\mathcal{H}^p$, $0 < p \leq 1$, then $\{F_k\}$ converges to F in E^q_{2m-1}.

Proof. For any ball B, by Lemma 3, we have

$$\|F_k - F\|^q_{q,\,B} \leq c^p |B|^{-1} \int_B N(F_k - F,\, x_0)^p\, dx_0 \leq c' \|F_k - F\|^p_{\mathcal{H}^p},$$

which proves the corollary.

Lemma 4. Let $\{F_k\}$ be a sequence in E^q_{2m-1} such that for a given point x_0, the series $\sum_k N(F_k;\, x_0)$ is finite. Then:

(i) The series $\sum_k F_k$ converges in E^q_{2m-1} to an element F and $N(F;\, x_0) \leq \sum_k N(F_k;\, x_0)$.

(ii) If f_k is the unique representative of F_k satisfying $n(f_k;\, x_0) = N(F_k;\, x_0)$, then $\sum_k f_k$ converges in L^q(loc) to a function f that is the unique representative of F satisfying $n(f;\, x_0) = N(F;\, x_0)$.

Proof. Since $\sum N(F_k;\, x_0)$ is finite, by Lemma 3, there exists a representative f_k of each F_k satisfying $n(f_k;\, x_0) = N(F_k;\, x_0) < \infty$. For any ball B and by Lemma 3, we have

$$\left| \sum_{k=1}^{K} f_k - \sum_{k=1}^{J} f_k \right|_{q,\,B} \leq \sum_{k=J+1}^{K} |f_k|_{q,\,B} \leq c \sum_{k=J+1}^{K} N(F_k;\, x_0).$$

This shows that $\sum_k f_k$ converges in L^q (loc) to a function f. Let us denote by F the class of f in E^q_{2m-1}. Since $n(f;\, x_0) \leq \sum_k n(f_k;\, x_0) = \sum_k N(F_k;\, x_0) < \infty$, we get that f is the unique representative of F satisfying $n(f;\, x_0) = N(F;\, x_0)$. On the other hand,

$$\left\| \sum_{k=1}^{K} F_k - F \right\|_{q,\,B} \leq \left| \sum_{k=1}^{K} f_k - f \right|_{q,\,B},$$

which shows the convergence in E^q_{2m-1} of the series $\sum_k F_k = F$.

Corollary 2. The space $\mathscr{H}^p$, $0 < p \leq 1$, *is complete.*

Proof. It is enough to show that if $\{F_k\}$ is a sequence such that $\sum_k \|F_k\|^p_{\mathscr{H}^p}$ is finite, then $\sum_k F_k$ converges in $\mathscr{H}^p$. Observe that for $0 < p \leq 1$,

$$\int \left(\sum_k N(F_k;\, x)\right)^p \, dx \leq \sum_k \int N(F_k;\, x)^p \, dx < \infty.$$

This implies that $\sum_k N(F_k;\, x)$ is finite almost everywhere. Then, by Lemma 4, $\sum_k F_k$ converges in E^q_{2m-1} to an element F. Now

$$N\left(\sum_{k=1}^{K} F_k - F;\, x\right) \leq \sum_{k=K+1}^{\infty} N(F_k;\, x).$$

Integrating the pth power of this inequality, we obtain

$$\left\| \sum_{k=1}^{K} F_k - F \right\|^p_{\mathscr{H}^p} \leq \sum_{k=K+1}^{\infty} \|F_k\|^p_{\mathscr{H}^p},$$

which shows that the series $\sum_k F_k$ converges in $\mathscr{H}^p$.

Lemma 5. If $g(x)$ *belongs to* L^q(loc) *and* $\phi \in \mathcal{S}$, *then for* $k > 2m+n$,

$$\int |g(y)|\,|\phi(y)|\,dy \leq c\, p_{k,0}(\phi)\; n(g;\, 0).$$

Proof. Since $|\phi(y)| \leq p_{k,0}(\phi)(1 + |y|)^{-k}$, then

$$\int |g(y)|\,|\phi(y)|\,dy \leq p_{k,0}(\phi) \int |g(y)|\,(1 + |y|)^{-k}\,dy.$$

Now,

$$\int_{B(0,\,1)} |g(y)|\,(1 + |y|)^{-k}\,dy \leq \int_{B(0,\,1)} |g(y)|\,dy \leq c\, n(g;\, 0).$$

Moreover,

$$\int_{B(0,\,2^{j+1})\sim B(0,\,2^j)} |g(y)|(1 + |y|)^{-k}\,dy \le 2^{-jk} \int_{B(0,\,2^{j+1})} |g(y)|dy \le c2^{(2m+n-k)j} n(g;\,0).$$

Since $2m + n - k < 0$, adding up these estimates, we get the lemma.

Corollary 3. If $g(x) \in L^q$(loc) and there is a point x_0 such that $n(g;\,x_0) < \infty$, then $g(x)$ defines a tempered distribution that satisfies

$$\left|\int g(y)\phi(y)dy\right| \le c\, p_{k,0}(\phi) n(g;\,x_0)(1 + |x_0|)^k$$

if $k > n + 2m$ and $\phi(y) \in \mathcal{S}$.

Proof. Using Lemma 5, we have

$$\int |g(y)||\phi(y)|dy = \int |g(x_0 + y)||\phi(x_0 + y)|dy$$
$$\le c\, p_{k,0}(\phi(x_0 + y))\, n(g(x_0 + y);\,0).$$

Since $p_{k,0}(\phi(x_0 + y)) \le (1 + |x_0|)^k p_{k,0}(\phi)$ and $n(g(x_0 + y);\,0) = n(g;\,x_0)$, we get the corollary.

Lemma 6. Let $g(x) \in L^q$(loc) $\cap\ \mathcal{S}'$ and $f = \Delta^m g$ in the sense of distributions. Then if $\phi \in \mathcal{S}$,

$$f^*(x) = \sup_{|x-y|<t} |(f^*\phi_t)(y)| \le c\, p_{k,2m}(\phi) n(g;\,x) \qquad (2)$$

holds for every x in R^n.

Proof. Let $|x - y| < t$. Since $f = \Delta^m g$ is a tempered distribution, we have

$$(f^*\phi_t)(y) = \langle f_z,\ \phi_t(y - z)\rangle = \langle \Delta_z^m g,\ \phi_t(y - z)\rangle$$
$$= t^{-2m}\langle g_z,\ (\Delta^m\phi)_t\,(y - z)\rangle.$$

Therefore,

$$(f^*\phi_t)(y) = t^{-2m}\int g(z)(\Delta^m\phi)_t\,(y - z)dz.$$

By a change of variables, we get

$$(f^*\phi_t)(y) = t^{-2m}\int g(x + tu)(\Delta^m\phi)(t^{-1}(y - x) - u)du.$$

Applying Lemma 5, we obtain

$$|(f^{*}\phi_t)(y)| \leq c\, n(g;\, x)\, p_{k,0}((\Delta^m\phi)(t^{-1}(y - x) - u)). \tag{3}$$

Since $|x - y| < t$, we have

$$1 + |u| \leq 1 + |u - t^{-1}(y - x)| + |t^{-1}(y - x)|$$
$$\leq 2(1 + |u - t^{-1}(y - x)|),$$

then

$$p_{k,0}((\Delta^m\phi)(t^{-1}(y - x) - u)) \leq 2^k\, p_{k,2m}(\phi).$$

This and (3) show that

$$|(f^{*}\phi_t)(y)| \leq c' p_{k,2m}(\phi) n(g;\, x),$$

for $|x - y| < t$, which implies (2).

Let $h(x)$ be the function defined as

$$\begin{aligned} h(x) &= |x|^{2m-n}\lg|x| \quad \text{if } n \text{ is even and } 2m - n \geq 0, \text{ and} \\ h(x) &= |x|^{2m-n} \quad \text{otherwise.} \end{aligned} \tag{4}$$

It is well known that if $b(x)$ is a bounded function with compact support, its potential $B(x)$, defined as

$$B(x) = \int h(x - y)b(y)dy,$$

is a locally bounded function and $\Delta^m B = b$ in the sense of distributions. For these potentials, we shall prove the following.

Lemma 7. Let $a(x)$ be a p-atom with vanishing moments up to the order $2m - 1$ and assume that $B(0, d)$ is the ball containing the support of $a(x)$ in the definition of p-atom. If

$$A(x) = \int h(x - y)a(y)dy,$$

then for $|x| \geq 2d$ and every α, there exists c_α such that

$$|(\partial^\alpha A)(x)| \leq c_\alpha\, d^{-n/p + 2m + n}\, |x|^{-n - |\alpha|}$$

holds.

Proof. Since $a(x)$ has vanishing moments up to the order $2m - 1$ and $|x| > 2d$, we have

$$(\partial^\alpha A)(x) = \int (\partial^\alpha h)(x - y)a(y)dy$$

$$= \int\left[(\partial^\alpha h)(x - y) - \sum_{|\beta| \le 2m-1} (\partial^{\alpha+\beta} h)(x)(-y)^\beta/\beta!\right]a(y)dy$$

$$= \int\left[\sum_{|\beta| = 2m} (\partial^{\alpha+\beta} h)(x - \theta y)(-y)^\beta/\beta!\right]a(y)dy,$$

where $0 < \theta < 1$. Taking into account that

$$|x - \theta y| \ge |x| - |y| \ge |x|/2,$$

and since for $|\beta| = 2m$, $(\partial^{\alpha+\beta}h)(x)$ is a homogeneous function of degree $-n - |\alpha|$, we obtain

$$|(\partial^\alpha A)(x)| \le c \int_{|y| \le d} |x - \theta y|^{-n-|\alpha|} dy\; d^{-n/p} d^{2m} \le c\, d^{-n/p + 2m + n}\, |x|^{-n-|\alpha|}.$$

Lemma 8. If $h(x)$ is the kernel defined in (4) *and* $|\alpha| = 2m$, *then*

(i) $(\partial^\alpha h)(x)$ *is a homogeneous function of degree* $-n$ *and* C^∞ *on* $R^n \sim \{0\}$.

(ii) $\int_{|x|=1} (\partial^\alpha h)(x)dx = 0.$

Proof. Part (i) follows immediately from the definition of $h(x)$.

To prove (ii), we define for $0 < \rho < 1$ and $|\theta| = 1$, the function $F(\rho\theta)$ as

$$F(\rho\theta) = |\Sigma_{n-1}|^{-1} \int_{|z|=1} h(z + \rho\theta)dz. \tag{5}$$

By a change of variables, and since h is a radial function, it follows that $F(\rho\theta) = f(\rho)$. Then, interchanging the order of integration, we get

$$f(\rho) = |\Sigma_{n-1}|^{-1} \int_{|\theta|=1} \left(|\Sigma_{n-1}|^{-1} \int_{|z|=1} h(z + \rho\theta)dz\right)d\theta$$

$$= |\Sigma_{n-1}|^{-1} \int_{|z|=1} \left(|\Sigma_{n-1}|^{-1} \int_{|\theta|=1} h(z + \rho\theta)\, d\theta \right) dx.$$

Now, since $\Delta^m h(x) = 0$ for $x \neq 0$, applying Pizetti's formula (see, for instance, [2]) to the innermost integral of the last term, we obtain

$$f(\rho) = |\Sigma_{n-1}|^{-1} \int_{|z|=1} \left(\sum_{k=0}^{m-1} c_k \rho^{2k} (\Delta^k h)(z) \right) dz,$$

which shows that $f(\rho)$ is a polynomial of degree less than $2m - 1$. Then, recalling that $f(\rho) = F(\rho\theta)$, if we take $2m$ derivatives with respect to ρ at $\rho = 0$ in (5), it follows that

$$0 = \int_{|z|=1} D_\rho^{2m} [h(z + \rho\theta)]_{\rho=0}\, dz = \sum_{|\alpha|=2m} \int_{|z|=1} (\partial^\alpha h)(z)\, dz\ \theta^\alpha (2m)!/\alpha!$$

for every θ. This implies (ii).

PROOF OF THE RESULTS

Proof of Theorem 1. Let $F \in \mathscr{H}^p$. Since $N(F; x)$ is finite a.e. on R^n, by Corollary 3, every representative of F is a function in $L^q(\text{loc}) \cap \mathcal{S}'$. Thus, if $\phi \in \mathcal{S}$ and $\int \phi(x)\, dx \neq 0$, by Lemma 6, we get

$$(\Delta^m F)^*(x) \leq c\ p_{k,2m}(\phi) N(F; x).$$

Taking the p-power and integrating, we obtain that $\Delta^m F$ belongs to H^p and $\|\Delta^m F\|_{H^p} \leq c\|F\|_{\mathscr{H}^p}$. This proves the second inequality in (1).

The fact that Δ^m is one to one on $\mathscr{H}^p$ follows from Lemma 9 of [1].

To prove that the application is onto, let us consider a p-atom $a(x)$ satisfying the hypotheses of Lemma 7. If $A(x) = \int h(x - y) a(y)\, dy$, we define

$$R(x, z) = A(x + z) - \sum_{|\alpha| \leq 2m-1} (\partial^\alpha h)(x - y) a(y)\, dy\ z^\alpha/\alpha!.$$

We shall prove first that for $|x| > 4d$, the following estimates hold: if $|z| < |x|/2$, then

$$|R(x, z)| \leq c\, d^{-n/p}(d/|x|)^{2m+n}|z|^{2m}, \tag{6}$$

if $|z| \geq |x|/2$ and $|x + z| > 2d$, then

$$|R(x, z)| \leq c\, d^{-n/p} d^{2m+n}|x + z|^{-n} + c\, d^{-n/p}(d/|x|)^{2m+n}|z|^{2m}, \tag{7}$$

and if $|z| \geq |x|/2$ and $|x + z| \leq 2d$, then

$$|R(x, z)| \leq c\, d^{-n/p} d^{2m}(d/|x|)^{-\mu} + c\, d^{-n/p}(d/|x|)^{2m+n}|z|^{2m}. \tag{8}$$

To prove (6), we observe that for $0 < \theta < 1$, $|x - \theta z| > |x| - |z| > |x|/2 > 2d$. Then, by the mean value theorem and Lemma 7, we get

$$|R(x, z)| \leq \sum_{|\alpha| = 2m} |(\partial^\alpha A)(x + \theta z)||z|^{|\alpha|}/\alpha! \leq c\, d^{-n/p}(d/|x|)^{2m+n}|z|^{2m}.$$

Now, let $|z| \geq |x|/2$. We have

$$|R(x, z)| \leq |A(x + z)| + \sum_{|\alpha| \leq 2m-1} |(\partial^\alpha A)(x)||z|^{|\alpha|}/\alpha!.$$

To estimate the right-hand side of this inequality, we shall consider first the terms under the summation sign. Since $|x| > 4d > 2d$, by Lemma 7, and observing that $|z|/|x| > 1/2$, we have

$$\begin{aligned} |(\partial^\alpha A)(x)||z|^{|\alpha|}/\alpha! &\leq c\, d^{-n/p} d^{2m+n}|x|^{-n+|\alpha|}|z|^{|\alpha|} \\ &\leq c\, d^{-n/p}(d/|x|)^{2m+n}|z|^{2m}. \end{aligned} \tag{9}$$

As for the other term, $|A(x + z)|$, we distinguish the cases $|x + z| > 2d$ and $|x + z| \leq 2d$. In the case $|x + z| > 2d$, we apply Lemma 7 for $\alpha = 0$, obtaining

$$|A(x + z)| \leq c\, d^{-n/p} d^{2m+n}|x + z|^{-n}.$$

This estimate together with (9) give (7). For $|x + z| \leq 2d$, we consider the cases n even and $2m - n \geq 0$, and n odd or $2m - n < 0$. In the first case, since $|x|^{2m-n}$ is a polynomial of degree smaller than $2m$, we have

$$A(x + z) = c\int |x + z - y|^{2m-n} \lg|x + z - y| \; a(y)dy$$
$$= c\int |x + z - y|^{2m-n}(\lg|x + z - y| - \lg|x|)a(y)dy.$$

Since $|x + z - y|/|x| \leq 3d/4d < 1$, we get

$$|A(x + z)| \leq c\, d^{-n/p} \int_{|y|\leq d} |x + z - y|^{2m-n}\lg(|x|/|x + z - y|)dy.$$

Recalling that $\lg t \leq \mu^{-1}t^{\mu}$ for any $0 < \mu$ and $t \geq 1$, and since $|y| \leq d$ implies $|x + z - y| \leq 3d$, we get

$$|A(x + z)| \leq c\, d^{-n/p}|x|^{\mu} \int_{|x+z-y|\leq 3d} |x + z - y|^{2m-n-\mu}\, dy$$
$$\leq c\, d^{-n/p}d^{2m}(d/|x|)^{-\mu}.$$

This estimate together with (9) give (8). For the case n odd or $2m - n < 0$, the proof is simpler since no logarithm appears, and the estimates obtained hold with $\mu = 0$.

Let us estimate

$$\rho^{-2m}\left(\rho^{-n}\int_{|z|\leq\rho} |R(x, z)|^{q} dz\right)^{1/q}, \quad \rho > 0.$$

According to the estimates obtained for $R(x, z)$ above, we split the domain of integration into three subsets:

$$D_1 = \{z : |z| \leq \rho, \ |z| < |x|/2\},$$
$$D_2 = \{z : |z| \leq \rho, \ |z| \geq |x|/2, \ |x + z| > 2d\},$$

and

$$D_3 = \{z : |z| \leq \rho, \ |z| \geq |x|/2, \ |x + z| \leq 2d\}.$$

By (6), we have

$$\rho^{-2m}\left(\rho^{-n}\int_{D_1} |R(x, z)|^{q} dz\right)^{1/q} \leq c\, d^{-n/p}(d/|x|)^{2m+n}. \qquad (10)$$

Let us assume $D_2 \neq \phi$. By (7), we get

$$\rho^{-2m}\left(\rho^{-n}\int_{D_2} |R(x, z)|^{q} dz\right)^{1/q}$$
$$\leq c\, d^{-n/p}d^{2m+n}\rho^{-2m}\left(\rho^{-n}\int_{|x+z|>2d} |x + z|^{-nq} dz\right)^{1/q} + c\, d^{-n/p}(d/|x|)^{2m+n}.$$

Since

$$\int_{|x+z|>2d} |x + z|^{-nq}\, dz \leq c\, d^{-nq+n},$$

and observing that $D_2 \neq \phi$ implies $\rho > |x|/2$, we get

$$\rho^{-2m}\left(\rho^{-n}\int_{D_2}|R(x,\ z)|^q\,dz\right)^{1/q} \leq c\,d^{-n/p}(d/|x|)^{2m+n}. \tag{11}$$

For the case $D_3 \neq \phi$, by (8), we have

$$\rho^{-2m}\left(\rho^{-n}\int_{D_3}|R(x,\ z)|^q\,dz\right)^{1/q}$$
$$\leq c\,d^{-n/p}d^{2m}(d/|x|)^{-\mu}\rho^{-2m}\left(\rho^{-n}\int_{|x+z|\leq 2d}dz\right)^{1/q} + c\,d^{-n/p}(d/|x|)^{2m+n}.$$

The assumption $D_3 \neq \phi$ implies $\rho > |x|/2$. Therefore,

$$\rho^{-2m}\left(\rho^{-n}\int_{D_3}|R(x,\ z)|^q\,dz\right)^{1/q}$$
$$\leq c\,d^{-n/p}(d/|x|)^{2m+n/q-\mu} + c\,d^{-n/p}(d/|x|)^{2m+n}. \tag{12}$$

The estimates (10), (11), and (12) and $|x|/d > 4$ imply

$$N(A;\ x) \leq c\,d^{-n/p}(d/|x|)^{2m+n/q-\mu}. \tag{13}$$

For $|x| \leq 4d$, we shall prove the following estimate for $R(x,\ z)$:

$$|R(x,\ z)| \leq c|z|^{2m}\left(d^{-n/p} + \sum_{|\alpha|=2m}K_\alpha^\star(a)\,(x)\right), \tag{14}$$

where $K_\alpha^\star$ is the maximal singular integral operator corresponding to the kernel $\partial^\alpha h(x)$; see Lemma 8. We have

$$R(x,\ z) = \int\left[h(x+z-y) - \sum_{|\alpha|\leq 2m-1}(\partial^\alpha h)(x-y)z^\alpha/\alpha!\right]a(y)\,dy$$
$$= \int_{|x-y|<2|z|} + \int_{|x-y|\geq 2|z|} = I_1 + I_2.$$

We can write, at least if y does not belong to the segment $[x,\ x+z]$,

$$U = h(x+z-y) - \sum_{|\alpha|\leq 2m-1}(\partial^\alpha h)(x-y)z^\alpha/\alpha!$$
$$= (2m-1)\sum_{|\alpha|=2m-1}(z^\alpha/\alpha!)\int_0^1\partial^\alpha h(x+tz-y)(1-t)^{2m-2}dt$$
$$+ \sum_{|\alpha|=2m-1}(\partial^\alpha h)(x-y)z^\alpha/\alpha!.$$

Since for $|\alpha| = 2m - 1$ the derivatives $\partial^{\alpha}h(x)$ are homogeneous functions of degree $-n + 1$, we get

$$|U| \leq c\left(\int_0^1 |x + tz - y|^{-n+1}(1 - t)^{2m-2}dt + |x - y|^{-n+1}\right)|z|^{2m-1}.$$

Therefore, since $|x - y| < 2|z|$ implies $|x + tz - y| \leq 3|z|$, we get the estimate

$$|I_1| \leq \int_{|x-y|<2|z|} |U|\,|a(y)|dy$$

$$\leq c\,d^{-n/p}\left[\int_0^1 (1 - t)^{2m-2}\left(\int_{|x+tz-y|\leq 3|z|} |x + tz - y|^{-n+1}dy\right)dt + \int_{|x-y|<2|x|} |x - y|^{-n+1}dy\right]|z|^{2m-1} \leq c\,d^{-n/p}|z|^{2m}.$$

To estimate I_2, we write

$$U = h(x + z - y) - \sum_{|\alpha|\leq 2m-1} (\partial^{\alpha}h)(x - y)z^{\alpha}/\alpha!$$

$$= \left[h(x + z - y) - \sum_{|\alpha|\leq 2m} (\partial^{\alpha}h)(x - y)z^{\alpha}/\alpha!\right] + \sum_{|\alpha|=2m} (\partial^{\alpha}h)(x - y)z^{\alpha}/\alpha!$$

$$= U_1 + U_2.$$

Let us estimate U_1. We have that for some $0 < s < 1$,

$$|U_1| \leq \sum_{|\alpha|=2m+1} |(\partial^{\alpha}h)(x + sz - y)|\,|z|^{2m+1}/\alpha!.$$

Since $|x - y| > 2|z|$ implies $|x + sz - y| > |x - y|/2$, and recalling that for $|\alpha| = 2m + 1$ the derivatives $\partial^{\alpha}h$ are homogeneous functions of degree $-n - 1$, we get $|U_1| \leq c|x - y|^{-(n+1)}|z|^{2m+1}$. Therefore,

$$|I_2| = \left|\int_{|x-y|>2|z|} U\,a(y)\,dy\right|$$

$$\leq \int_{|x-y|>2|z|} |U_1|\,|a(y)|dy + \left|\int_{|x-y|>2|z|} U_2\,a(y)\,dy\right|$$

$$\leq c\left(d^{-n/p} + \sum_{|\alpha|=2m} K_{\alpha}^{\star}(a)(x)\right)|z|^{2m}.$$

This ends the proof of (14).

Using (14), we get that for $|x| \leq 4d$,

$$N(A;\ x) \leq c\left(d^{-n/p} + \sum_{|\alpha| = 2m} K_\alpha^\star(a)(x)\right). \tag{15}$$

The estimates (13) and (15) for $N(A;\ x)$ allow us to write

$$\int N(A;\ x)^p\,dx = \int_{|x| > 4d} + \int_{|x| \leq 4d} \leq c\,d^{-n} \int_{|x| > 4d} (d/|x|)^{(2m+n/q-\mu)p}\,dx$$
$$+ c\,d^{-n} \int_{|x| \leq 4d} dx + c \sum_{|\alpha| = 2m} \int_{|x| \leq 4d} K_\alpha^\star(a)(x)^p\,dx,$$

which by Hölder inequality is bounded by

$$c\left(1 + \sum_{|\alpha| = 2m} d^{n(1-p/2)}\left(\int K_\alpha^\star(a)(x)^2\,dx\right)^{p/2}\right).$$

Since $K_\alpha^\star$ is of strong type 2-2, and since $\|a\|_2 \leq c\,d^{n(p-2)/2p}$, we get

$$\int N(A;\ x)^p\,dx \leq c. \tag{16}$$

Let ϕ belong to H^p. We know that there exist a numerical sequence $\{\lambda_i\}$ and a sequence $\{a_i(x)\}$ of p-atoms in H^p such that $\phi = \sum \lambda_i a_i$ and $\sum |\lambda_i|^p \leq c\|\phi\|_{H^p}$. If $A_i(x)$ are the potentials of the p-atoms $a_i(x)$, then from (16) we get $\sum |\lambda_i|\,\|A_1\|_{\mathscr{H}^p}^p < \infty$. By Corollary 2, there exists F in $\mathscr{H}^p$ such that $F = \sum \lambda_i A_i$. Recalling that we have already shown that Δ^m is a continuous operator from $\mathscr{H}^p$ into H^p, we get

$$\Delta^m F = \sum \lambda_i \Delta^m A_i = \sum \lambda_i a_i = \phi.$$

This shows that Δ^m is onto H^p. Moreover,

$$\|F\|_{\mathscr{H}^p}^p \leq \sum |\lambda_i|^p \|A_i\|_{\mathscr{H}^p}^p \leq c \sum |\lambda_i|^p \leq c' \|\phi\|_{H^p}^p = c' \|\Delta^m F\|_{H^p}^p.$$

This ends the proof of the theorem.

Proof of Theorem 2. Let $G \in \mathscr{H}_{q,m}^p$ and assume $G \neq 0$. Then there exists a representative $h(x)$ of G that is not a polynomial of degree less than or equal to $2m - 1$. It is easy to see that there exists a ball B with center at 0 and radius $r > 0$ such that for every $P \in \mathscr{P}_{2m-1}$,

$$\left(\int_B |h(y) - P(y)|^q\,dy\right)^{1/q} \geq a > 0.$$

Let x be a point such that $|x| > r$ and let $\rho = 2|x|$. Then $B \subset B(x, \rho)$. Let $g(x)$ be any representative of G. Then $g = h - P$ for some $P \in \mathscr{P}_{2m-1}$ and

$$\rho^{-2m}|g|_{q,\,B(x,\,\rho)} \geq |B(0,\ 1)|^{-1/q}\, 2^{-2m-n/q}|x|^{-2m-n/q}a = c|x|^{-2m-n/q},$$

showing that $N(F;\ x) \geq c|x|^{-2m-n/q}$, for $|x| > r$. Taking the pth power and integrating, we get a contradiction. Therefore, $G = 0$ and the theorem is proved.

To prove Theorem 3, a Calderón-Zygmund-type lemma is required. The statement of this lemma depends on a partition of unity given in the following:

Lemma 9 (a partition of unity). Let $\Omega \neq \phi$ be an open subset of R^n not coinciding with R^n. There exists a sequence $\{\phi_k\}_{k=1}^{\infty}$ of C_0^{∞} functions satisfying the following conditions

(i) $0 \leq \phi_k(x) \leq 1$ *and* $\Sigma\phi_k(x) = \chi_{\Omega}(x)$.

(ii) For each k, there is a ball $B_k = B(x_k,\ r_k) \subset \Omega$ such that $\operatorname{supp}(\phi_k) \subset B_k$*, and for every $z \in B_k$, $r_k \leq d(z,\ C\Omega) \leq c\, r_k$.*

(iii) For every k, the ball $B(x_k,\ 2r_k)$ is contained in Ω. Moreover, there exists an integer M such that the number of balls $B(x_j,\ 2r_j)$ that intersect $B(x_k,\ 2r_k)$ does not exceed M.

(iv) For every multiindex α, we have

$$|(\partial^{\alpha}\phi_k)(x)| \leq c_{\alpha} r_k^{-|\alpha|},$$

where c_α does not depend on k.

Proof. See [5].

Lemma 10 (Calderón-Zygmund-type lemma). Let F be an element of E^q_{2m-1} whose maximal function belongs to $L^p(R^n)$. Let

$n(2m + n/q)^{-1} < p \leq 1$. *Given* $t > 0$, *let* $\Omega = \Omega_t = \{x : N(F;\, x) > t\}$. *By Lemma* 1, *this is an open subset. Let* $\{\phi_k\}_{k=1}^{\infty}$ *be the partition of unity associated to* Ω *in Lemma* 9. *For each* k, *let* y_k *be a point of* $C\Omega$ *such that* $d(B(x_k, 2r_k), C\Omega) = d(B(x_k, 2r_k), y_k)$. *For a given representative* $f(x)$ *of* F, *by Lemma* 3, *there is a polynomial* $P(y_k, y)$ *of* $\mathscr{P}_{2m-1}$ *satisfying* $N(F, y_k) = n(f(y) - P(y_k, y);\, y_k)$. *For* $k = 1, 2, \ldots$, *we define the functions* $w_k(y) = \phi_k(y)(f(y) - P(y_k, y))$. *These functions* $w_k(y)$ *belong to* $L^q(\mathrm{loc})$. *Let us denote by* W_k *the class of* $w_k(x)$ *in* E^q_{2m-1}. *Then the following conditions are satisfied:*

(i) $N(W_k;\, x) \leq c\, N(F;\, x)$ *if* $x \in B(x_k, 2r_k)$.

(ii) $N(W_k;\, x) \leq ct(r_k/(|x - x_k| + r_k))^{2m + n/q}$ *if* $x \notin B(x_k, 2r_k)$.

(iii) *The series* $\sum_k N(W_k;\, x)$ *is pointwise convergent for almost every* x *in* R^n. *Moreover,*

$$\int \Big(\sum_k N(W_k;\, x)\Big)^p dx \leq \sum_k \int N(W_k;\, x)^p\, dx \leq c \int_{\Omega} N(F;\, x)^p\, dx.$$

(iv) *The series* $\sum_k W_k = W$ *converges in* E^q_{2m-1}, *and for almost every* x,

$$N(W;\, x) \leq \sum_k N(W_k;\, x).$$

(v) $\int N(W;\, x)^p\, dx \leq c \int_{\Omega} N(F;\, x)^p\, dx.$

(vi) *If* $G = F - W$, *there exists a representative of* G *having continuous derivatives up to the order* $2m - 1$, *and the derivatives of order* $2m - 1$ *satisfy a Lipschitz condition with constant not exceeding* ct. *Moreover,* $N(G;\, x) \leq ct$.

All the constants c appearing in the statement of the lemma are finite and independent of F and $t > 0$.

Proof. Let us prove (i). Let $x \in B(x_k, 2r_k)$. Assume $N(F;\, x) < \infty$. Otherwise, there is nothing to prove. Let f be a representative of F and $P(x, y)$ the polynomial of degree less than $2m$ satisfying

$n(f(y) - P(x, y); x) = N(F; x)$. We define the polynomial $Q_k(x, y)$ as

$$Q_k(x, y) = \sum_{|\alpha| \leq 2m-1} \partial_y^\alpha(\phi_k(y)P(x, y) - P(y_k, y))\Big|_{y=x}(y - x)^\alpha/\alpha!.$$

Let us estimate $\rho^{-2m}|w_k(y) - Q_k(x, y)|_{q, B(x, \rho)}$.

By the Leibnitz formula and reordering the sum, we can write

$$Q_k(x, y) = \sum_{|\alpha| \leq 2m-1}\left[\partial_y^\alpha(P(x, y) - P(y_k, y))\Big|_{y=x}(y - x)^\alpha/\alpha! \times \left(\sum_{|\gamma| = 2m-1-|\alpha|}(\partial^\gamma\phi_k)(x)(y - x)^\gamma/\gamma!\right)\right]. \tag{17}$$

From the hypothesis we get $|x_k - y_k| \leq 8r_k$. Therefore,

$$\begin{aligned}|y - x| + |y - y_k| &\leq \rho + |y - x| + |x - x_k| + |x_k - y_k| \\ &\leq 10(\rho + r_k).\end{aligned}$$

Then by Lemma 2 and since $t < N(F; x)$, we get

$$|\partial_y^\alpha(P(x, y) - P(y_k, y))| \leq c_\alpha N(F; x)(\rho + r_k)^{2m-|\alpha|}. \tag{18}$$

Assume first that $\rho \geq 2r_k$. Then

$$\begin{aligned}&|w_k(y) - Q_k(x, y)| \\ &\leq \phi_k(y)|f(y) - P(x, y)| + \phi_k(y)|P(x, y) - P(y_k, y)| + |Q_k(x, y)|. \end{aligned}\tag{19}$$

For the second term of the right-hand side of this inequality, taking $\alpha = 0$ in (18) and since $0 \leq \phi_k(x) \leq 1$, we obtain

$$\phi_k(y)|P(x, y) - P(y_k, y)| \leq c\, N(F; x)\rho^{2m}.$$

Since $|x - y| \leq 10\, r_k$, by Lemma 2, we have

$$|\partial_y^\alpha(P(x, y) - P(y_k, y))|_{y=x}| \leq c\, N(F; x) r_k^{2m-|\alpha|}.$$

Using this estimate, condition (iv) of Lemma 9, and (17), it follows that

$$|Q_k(x, y)| \leq \sum_{|\alpha| \leq 2m-1} c\, N(F; x) r_k^{2m-|\alpha|}\rho^{|\alpha|}\left(\sum_{|\gamma| \leq 2m-1-|\alpha|} c\, r_k^{-|\gamma|}\rho^{|\gamma|}\right).$$

Recalling that $\rho \geq 2r_k$, we obtain for the third term on the right-hand side of (19) the estimate

$$|Q_k(x, y)| \leq c\, N(F; x)\rho^{2m}.$$

Integrating (19) over $B(x, \rho)$ and using the estimates obtained, we get

$$\rho^{-2m}|w_k(y) - Q_k(x, y)|_{q, B(x, \rho)} \leq \rho^{-2m}|f(y) - P(x, y)|_{q, B(x, \rho)} + c\, N(F; x). \tag{20}$$

For the case when $\rho < 2r_k$, we rewrite $Q_k(x, y)$ as

$$Q_k(x, y) = \sum_{|\beta| \leq 2m-1} \left[(\partial^\beta \phi_k)(x)(y - x)^\beta/\beta! \times \left(\sum_{|\gamma| \leq 2m-1-|\beta|} \partial_y^\gamma (P(x, y) - P(y_k, y))\Big|_{y=x} (y - x)^\gamma/\gamma!\right)\right].$$

Adding and subtracting the expression

$$\phi_k(y)P(x, y) + \sum_{|\beta| \leq 2m-1} (\partial^\beta \phi_k)(x)(P(x, y) - P(y_k, y))(y - x)^\beta/\beta!$$

to $Q_k(x, y)$, we obtain

$$\begin{aligned} |w_k(y) - Q_k(x, y)| &\leq \phi_k(y)|f(y) - P(x, y)| \\ &+ \left|\left[\phi_k(y) - \sum_{|\beta| \leq 2m-1} (\partial^\beta \phi_k)(x)(y - x)^\beta/\beta!\right](P(x, y) - P(y_k, y))\right| \\ &+ \left|\sum_{|\beta| \leq 2m-1} \left[(\partial^\beta \phi_k)(x)(y - x)^\beta/\beta!\right]\right| \Big[P(x, y) - P(yk, y) \\ &- \sum_{|\gamma| \leq 2m-1-|\beta|} \partial_y^\gamma (P(x, y) - P(y_k, y))\Big|_{y=x} (y - x)^\gamma/\gamma!\Big] \\ &\leq |f(y) - P(x, y)| + A_1 + A_2. \end{aligned} \tag{21}$$

If $y \in B(x, \rho)$ and considering condition (iv) of Lemma 9 and (18) we get

$$\begin{aligned} A_1 &= \left|\sum_{|\beta| = 2m} \partial^\beta \phi_k(\xi)(y - x)^\beta/\beta!\right| |P(x, y) - P(y_k, y)| \\ &\leq c\, r_k^{-2m} \rho^{2m} N(F; x) r_k^{2m} \\ &= c\, N(F; x)\rho^{2m}. \end{aligned}$$

As for A_2, similar arguments show that

$$A_2 = \left| \sum_{|\beta| \leq 2m-1} \left[(\partial^\beta \phi_k)(x)(y-x)^\beta/\beta! \right.\right.$$
$$\left.\left. \times \left(\sum_{|\gamma| = 2m - |\beta|} \partial_y^\gamma (P(x, y) - P(y_k, y)) \Big|_{y=\xi} (y-x)^\gamma/\gamma! \right) \right] \right|$$
$$\leq c \sum_{|\beta| \leq 2m-1} r_k^{-|\beta|} \rho^{|\beta|} N(F; x)(\rho + r_k)^{|\beta|} \rho^{2m-|\beta|} \leq c N(F; x) \rho^{2m}.$$

Integrating (21) and taking into account the estimates obtained for A_1 and A_2, we get that (20) holds also for $\rho < 2r$. This shows condition (i).

Let us prove condition (ii). We shall estimate

$$\rho^{-2m} |w \ (y)|_{q, B(x, \rho)}.$$

This expression is equal to zero unless $B(x, \rho) \cap B(x_k, r_k) \neq \phi$. Since $x \notin B(x_k, 2r_k)$, we have $2r_k \leq |x - x_k| < \rho + r_k$. Then, $r_k < \rho$ and $|x - x_k| < 2\rho$. Thus, $|x - x_k| + r_k < 3\rho$. On the other hand, by condition (ii) of Lemma 9, $B(x_k, r_k) \subset B(y_k, 7r_k)$. Then

$$\int_{B(x, \rho)} |w_k(y)|^q \, dy \leq \int_{B(x_k, r_k)} |w_k(y)|^q \, dy$$
$$\leq \int_{B(y_k, 7r_k)} |f(y) - P(y_k, y)|^q \, dy.$$

Since $N(F; y_k) < t$, we get

$$\rho^{-2m} |w_k|_{q, B(x, \rho)} \leq c\, N(F; y_k)(r_k/\rho)^{2m+n/q}$$
$$\leq ct(r_k/(|x - x_k| + r_k))^{2m+n/q}, \qquad (22)$$

which implies condition (ii).

Let us prove condition (iii). Since $0 < p < 1$ and taking into account the estimates (i) and (ii) already obtained, we have

$$\int (\textstyle\sum_k N(W_k; x))^p \, dx \leq \sum_k \left\{ \displaystyle\int_{B(x_k, 2r_k)} N(W_k; x)^p \, dx + \int_{CB(x_k, 2r_k)} N(W_k; x)^p \, dx \right\}$$

$$\leq c\sum_k \int_{B(x_k,\, 2r_k)} N(F;\, x)^p\, dx + ct^p \int_{CB(x_k,\, 2r_k)} (r_k/(|x - x_k| + r_k))^{(2m+n/q)p}\, dx$$

$$\leq c\int_\Omega N(F;\, x)^p\, dx + ct^p \sum_k r_k^n \leq c\int_\Omega N(F;\, x)^p\, dx + ct^p|\Omega|$$

$$\leq c\int_\Omega N(F;\, x)^p\, dx.$$

Condition (iv) is a consequence of condition (iii) and Lemma 4. As for condition (v), it follows from conditions (iii) and (iv), already proved.

Let us prove condition (vi). From condition (iii) we get a point $x_0 \notin \Omega$ such that $\sum_k N(W_k;\, x_0) < \infty$. Since $x_0 \notin B(x_k,\, 2r_k)$ for every $k = 1, 2, \ldots,$ from (22), we have that $w_k(x)$ is the only representative of W_k satisfying $n(w_k;\, x_0) = N(W_k;\, x_0)$. Then by Lemma 4 the series $w = \sum_k w_k$ converges in $L^q(\text{loc})$ and supplies the representative of $W = \sum_k W_k$ satisfying $n(w;\, x_0) = N(W;\, x_0)$. Then the function $g(x) = f(x) - w(x)$ is a representative of $G = F - W$. Since $g(x)$ is equal to $f(x)$ on $C\Omega$ and $g(x) = \sum_k \phi_k(y) P(y_k,\, y)$ on Ω, this function $g(x)$ coincides with the function $F_2(y)$ defined by A. P. Calderón in the statement of Theorem 5 in [1]. Therefore, $N(G,\, x) \leq ct$ and $g(x)$ coincides almost everywhere with a function having continuous derivatives up to the order $2m - 1$. Moreover, the derivatives of order $2m - 1$ satisfy a Lipschitz condition with constant ct.

Proof of Theorem 3. The method we shall use to obtain the atomic decomposition is that developed in [4]. Following that method, as a consequence of Lemma 10, we can show that if H is an element of E^q_{2m-1} satisfying $N_{q,m}(H;\, x) \leq 1$ and $\int N(H;\, x)^r\, dx < \infty$, for some $n(2m + n/q)^{-1} < r < p \leq 1$, then there exist a numerical sequence $\{\lambda_j\}$ and a sequence of p-atoms $\{A_j\}$ of E^q_{2m-1} such that $H = \sum \lambda_j A_j$ in $\mathscr{H}^p_{q,m}$. Moreover, $\sum |\lambda_j|^p \leq c\int N_{q,m}(H;\, x)^r\, dx$. The proof of this result is essentially that of Lemma 4.2 of [4]. After this result is obtained, the proof of Theorem 3 can be obtained following the same lines of the proof of Theorem 4.3 of [4].

REFERENCES

1. Calderón, A. P. Estimates for singular integral operators in terms of maximal functions. *Studia Math.* 44 (1972):563-582.

2. Courant, R., and Hilbert, D. *Methods of Mathematical Physics*, vol. 2. New York: Interscience Publ., 1953.

3. Latter, R. H. A characterization of $H^p(R^n)$ in terms of atoms. *Studia Math.* 42 (1978):93-101.

4. Macías, R. A., and Segovia, C. A decomposition into atoms of distributions on spaces of homogeneous type. *Advances in Math.* 33 (1977):271-309.

5. Stein, E. M. *Singular Integrals and Differentiability Properties of Functions*. Princeton, N.J.: Princeton Univ. Press, 1970.

HIGHER GRADIENTS AND REPRESENTATIONS OF LIE GROUPS

J.E. Gilbert, P.A. Tomas
University of Texas, Austin

R.A. Kunze
University of California, Irvine

R.J. Stanton
Ohio State University

INTRODUCTION

To construct an H^p theory for $\mathbf{R}^n$, Stein and Weiss [13] showed that a solution ω of the system

$$\text{div } \omega = 0, \quad \text{curl } \omega = 0$$

automatically is harmonic and $|\omega|^p$ subharmonic when $p \geq (n-2)/n-1$. The existence of boundary values of H^p functions, $p \geq (n-2)/n-1$, can then be established, permitting the development of an H^p theory for solutions of the div-curl system [13]. Unlike the dimension 2 case, however, where boundary value results may also be based on factorization techniques, this H^p theory is available as yet only for such p. So-called "higher-gradient" systems of differential equations were introduced to provide an H^p theory in n dimensions for a full range of p. Stein and Weiss [13] in three dimensions and then later in general Calderón and Zygmund [3] and Stein and Weiss [12] established subharmonicity results for solutions to mth-order higher-gradient systems for $p \geq (n-2)/m+n-2$. Calderón

---

All authors were supported in part by NSF grants.

and Stein later observed that if d is any constant coefficient, first-order differential system whose solutions $df = 0$ are necessarily harmonic, then there exists some $p_0 < 1$ so that $|f|^p$ is subharmonic when $df = 0$ and $p \geq p_0$. This allows great freedom in the construction of H^1 theories via systems of differential operators and singular integrals.

The purpose of this note is to develop for noncompact Riemannian symmetric spaces an analogue of classical H^p theory.

In the theory envisaged, by contrast with the Euclidean theory, we shall see here that the only nontrivial H^2 spaces are those constructed through higher gradients (more precisely, operators with the same principal symbol). Riesz transforms provide solutions to Euclidean higher-gradient systems; we shall give group-theoretic interpretations and generalizations of this result. We shall also exhibit the analogy of higher-gradient systems with the Schmid $\mathfrak{D}$ operator [10]. The latter was used by Schmid to realize discrete series representations of semisimple Lie groups. We shall use the higher-gradient systems to realize another important class of irreducible unitary representations of semisimple Lie groups. Connections between the representation theory of a noncompact semisimple group G and the H^2 theory of the symmetric space G/K will be used frequently.

To orient the reader, we shall sketch the constituents of the H^p theory proposed:

(1) The theory should be based on elliptic systems of first-order differential operators that commute with the action of G. The kernel of the system should be contained in the kernel of a second-order elliptic differential operator.

(2) A theory of boundary values and of Poisson kernels should exhibit isomorphisms of spaces of functions on G/K with corresponding function spaces on the boundary.

(3) There should be a Riesz transform characterization of boundary values. This characterization should give a complete account of the overdetermined nature of the elliptic system.

(4) There should be a molecular decomposition of H^p functions.

(5) The differential equations, Riesz transforms, Poisson kernels, and boundary value operators should all commute with the action of G. In particular, H^2 should be invariant and one should obtain irreducible unitary representations from the action of G on H^2.

The construction of a full H^p theory in general is a formidable task, which we shall not accomplish in this note. Here we shall confine our discussion to the construction of nontrivial H^2 spaces for the Lorentz groups $SO(2k, 1)$. In the last section of the paper we shall discuss extensions of these results.

We wish to thank Professors B. Blank and A. W. Knapp for many valuable conversations on these topics.

NOTATION

In general, we shall follow the conventions of Knapp and Wallach [8].

Let G be a noncompact connected semisimple Lie group with finite center, Lie algebra $\mathfrak{g}$ and Cartan decomposition $\mathfrak{g} = \mathfrak{k} \oplus \mathfrak{p}$. Let K be the analytic subgroup corresponding to $\mathfrak{k}$, and we assume rank G = rank K. Thus if $\underset{\sim}{t}$ is a Cartan subalgebra, we may take $\underset{\sim}{t}$ compact and $\underset{\sim}{t} \subseteq \mathfrak{k}$. Let Δ_k be the compact roots, $\rho_{\mathfrak{k}}$ and $\rho_{\mathfrak{n}}$ are half the sum of the positive compact and noncompact roots.

If $\mathfrak{g} = \mathfrak{k} \oplus \alpha \oplus \mathfrak{n}$ is an Iwasawa decomposition, let A, N, and $V = \theta N$ be the corresponding analytic subgroups. Then

$g = \exp H(g)N(g)\kappa(g)$. Let ρ_+ be half the sum of the positive restricted roots, with multiplicites. Let M be the centralizer of α in K, M' the normalizer, and W the Weyl group M'/M.

If τ is an irreducible unitary representation of K on V, we let $C^\infty(G, \tau) = C^\infty(G, V) = \{f : G \to V | f(kg) = \tau(k)f(g) \text{ and } f \in C^\infty\}$.

EUCLIDEAN ANALOGUES

There is a simple Euclidean analogue of G/K, which clarifies the connection between Euclidean and non-Euclidean H^p theories.

We define affine motion groups through the semidirect product on $\mathbf{R}^n \times SO(n)$ defined by

$$(x, \theta) \cdot (y, \Psi) = (x + R_\theta y, \theta\Psi),$$

where R_θ is the rotation corresponding to θ.

The symmetric space $\mathbf{R}^n \times SO(n)/SO(n)$ is then isomorphic to $\mathbf{R}^n$, but the full motion group acts on the symmetric space by translations and by rotations. Differential operators invariant under the action of the group must therefore be constant coefficient and commute with rotations. It is not difficult to see that such first-order operators cannot be scalar valued, but must take values in a vector space on which rotations can be made to act. The classical systems of differential operators used to define H^p theory—div-curl, higher gradients, or the d, d^* system—all can be seen to possess rotational invariance, and therefore may be viewed as operators on a symmetric space.

The construction of such systems of operators was unified and generalized by Stein and Weiss [12]. They start with an irreducible unitary representation τ of $SO(n)$ on a vector space V. If $f : \bar{\mathbf{R}}^n \to V$ is a vector value function and we choose coordinates $\overline{x} = \Sigma x_i \overline{e}_i$, we may form the invariant operator

$$\nabla f = \sum \frac{\partial}{\partial x_i} f \otimes \overline{e}_i ;$$

∇ takes C^∞, V-valued functions to C^∞, $V \otimes \mathbf{C}^n$-valued functions. The group $SO(n)$ acts on $V \otimes \mathbf{C}^n$, by $\rho \cdot V \otimes z = \tau(\rho)V \otimes \rho \cdot z$, where $\rho \cdot z$ is the standard action of rotations on $\mathbf{C}^n$. There is a canonical subspace V_0 of $V \otimes \mathbf{C}^n$ on which $SO(n)$ acts irreducibly; it is called the Cartan composition of V and $\mathbf{C}^n$. Let P_0 denote the projection of $V \otimes \mathbf{C}^n$ onto $V_0^\perp$, and define the operator $\overline{\partial}$ on f by $\overline{\partial}f = (P_0\nabla)f$. Stein and Weiss [12] defined the $\overline{\partial}$ operator, showed that it reduced, for special choices of τ to div-curl, higher gradients, and so on, and showed that if $\overline{\partial}f = 0$, the components of f are harmonic. They also showed that solutions to the system $\overline{\partial}f = 0$ enjoy subharmonicity properties, so that one may construct an H^p theory for $\mathbf{R}^n+$ and obtain boundary values for $1 \leq p < \infty$. We shall sketch such a theory, for it indicates what we may expect in the non-Euclidean case.

We first define a boundary and a normal direction in a canonical way, by selecting a highest-weight vector $\overline{v}_0 + i\overline{v}_1$ for the standard representation of $SO(n)$ on $\mathbf{R}^n$. Then we define $\mathbf{R}^n+ = \{\overline{v} | \overline{v} \cdot \overline{v}_0 > 0\}$, and we define M to be the subgroup of $SO(n)$ that fixes v_0, and M_0 the subgroup that fixes $\overline{v}_0$ and $\overline{v}_1$. If $\xi \in \mathbf{R}^{n-1}$, let δ_ξ be the rotation that maps $\overline{v}_1$ to $\xi/|\xi|$ on the unit sphere. It is possible to construct a space $W \subset V$ on which $\tau|_{M_0}$ acts, and which characterizes the solutions to the τ system as follows.

If H^p is defined in the obvious way, then f is in H^p if and only if there is an $F : \mathbf{R}^{n-1} \to W$, $F \in L^p$, such that

$$\int_{\mathbf{R}^{n-1}} f(x + y)e^{ix\xi}\,dx = e^{-y|\xi|}\tau(\delta_\xi)\hat{F}(\xi).$$

The homogeneous multiplier $\tau(\delta_\xi)$ thus plays the role of the Riesz transforms. We have not been able to give a complete characterization of W, but if ϕ_τ is the highest-weight vector for (τ, V), it is known that $\tau(M_0)\phi_\tau \subset W$. In all known examples actual equality holds.

We shall need a further understanding of this theory to the special case that τ is the representation of SO(n) on symmetric m-tensors. Then $f : \mathbf{R}^n \to V$ satisfies the system $\overline{\partial} f = 0$ if and only if ∇f is a symmetric, traceless $m + 1$-tensor. This occurs if and only if f can be written as $f = \nabla^m u$ for a harmonic function u. If f is in H^p, then f has boundary values and $f|_{\mathbf{R}^{n-1}} = (-\Delta)^{-m/2}(\nabla_T)^m F$, where ∇_T is the boundary gradient operator and F is a complex-valued function. In this case, then, W is one-dimensional and the Riesz transforms may be given by iterates of the classical Riesz transforms, as in Calderón and Zygmund [3].

DIFFERENTIAL OPERATORS

To construct invariant differential operators on the non-Euclidean symmetric spaces G/K, we shall generalize the procedure of Stein and Weiss.

The maximal compact subgroup K plays the role of SO(n), and the space exp $\mathfrak{p}$ that of $\mathbf{R}^n$. But $\mathfrak{p}$ and K cannot be split apart, as $[\mathfrak{p}, \mathfrak{p}] \subset \mathfrak{k}$, and matters cannot be reduced to vector-valued functions on G/K. Moreover, solutions to any of the natural analogues of the Euclidean equation are eigenfunctions of the Casimir operator, which is not an elliptic operator. Both of these difficulties can be overcome in a standard way; our operators will act on function $f : G \to V$ with invariance properties. Technically speaking, our operators will act on sections of a homogeneous vector bundle over G/K, rather than on actual functions, but for the most part we shall not need this characterization.

We fix an order Δ_k^+ on the positive compact roots and choose an irreducible unitary representation τ of K on a space V, with highest-weight vector λ. The representation Ad of K acts on $\mathfrak{p}_{\mathbf{C}}$; we choose any order Δ_n^+ on the noncompact roots so that the highest-weight vector of Ad on $\mathfrak{p}_{\mathbf{C}}$ is a positive noncompact root. We let

V_0 be the complement of the irreducible subspace that contains the highest-weight vector of $\tau \otimes \mathrm{Ad}$ on $V \otimes \mathfrak{p}_{\mathbf{C}}$ and let P_0 be projection onto V_0.

Let

$$C^\infty(G, V) = \{f : G \to V | f(kg) = \tau(k)f(g),\ f \in C^\infty\}.$$

The gradient operator may be defined from $C^\infty(G, V)$ into $C^\infty(G, V \otimes \mathfrak{p}_{\mathbf{C}})$ by the formula $\nabla f = \Sigma X_i f \otimes \overline{X}_i$, where $\{X_i\}$ is any orthonormal basis of $\mathfrak{p}_{\mathbf{C}}$ and conjugation is with respect to $\mathfrak{g}$ in $\mathfrak{g}_{\mathbf{C}}$. We define $\mathcal{G}$ from $C^\infty(G, V)$ by

$$\mathcal{G}f = (P_0 \nabla)f.$$

The operator $\mathcal{G}$ commutes with the right action of G and is a generalization of the $\overline{\partial}$ system of Stein and Weiss. We have avoided the $\overline{\partial}$ notation because, in the case G/K is a Hermitian symmetric space, $\mathcal{G}$ is not the $\overline{\partial}$ operator on the Dolbeault complex.

The operator $\mathcal{G}$ is similar to the operator D of Schmid [10]. Note that

$$V \otimes \mathfrak{p}_{\mathbf{C}} = \sum_{\beta \in \Delta_n} m_\beta V_{\lambda + \beta};$$

define

$$V^- = \sum_{-\beta \in \Delta_n} m_\beta V_{\lambda + \beta}.$$

Let $P_- : V \otimes \mathfrak{p}_{\mathbf{C}} \to V_-$ be orthogonal projection, and $Df = P_- \circ \nabla f$.

Note $\ker \mathcal{G} \subset \ker D$. Knapp and Wallach [8] show that if $Df = 0$, then f is an eigenfunction of the Casimir, with eigenvalue $|\lambda + \rho - \rho_n|^2 - |\rho|^2$. Moreover, Schmid [10] shows that the principal symbol of D is $P_-(v \otimes \xi)$; that of $\mathcal{G}$ is $P_0(v \otimes \xi)$. Since $V_- \subset V_0$, $\ker P_0 \subset \ker P_-$ and the ellipticity of D implies that of $\mathcal{G}$. The Schmid operator D is elliptic if $\langle \lambda - 2\rho, \alpha \rangle \geq 0$ for $\alpha \in \Delta_k^+$. In general, D will not be elliptic, but the $\mathcal{G}$ operator will be. This will require separate proof.

In the case of motion groups, it is possible to show that $\mathcal{G}$ and the $\overline{\partial}$ system of Stein and Weiss are equivalent. We define

$T : C^\infty(\mathbf{R}^n \times SO(n, V)) \to C^\infty(\mathbf{R}^n) \otimes V$ by $(Tf)(x) = f(x, e)$. One readily computes that $T\mathscr{I}fT^{-1} = 2\bar{\partial}Tf$. The proof makes use of the fact that for motion groups $[\mathfrak{p}, \mathfrak{p}] = 0$. The $\mathscr{I}$ system is therefore equivalent to a system of first-order differential equations acting on vector-valued functions on $\mathbf{R}^n$. In the case of semisimple groups, $[\mathfrak{p}, \mathfrak{p}] \subset \mathfrak{k}$ and the $\mathscr{I}$ system does not reduce to a system of equations acting on valued functions on G/K, but must be viewed as an operator on G, acting on functions with transformation properties.

EXISTENCE RESULTS

We shall state our results for SO(4, 1), with the understanding that they are valid for SO(2k, 1). If J is the matrix defining the indefinite form

$$\sum_{J=1}^{4} x_j^2 - x_5^2,$$

then we define

$$SO(4, 1) = \{g \in GL(5) \,|\, gJg^T = J, \det g = 1, g_{55} \geq 1\}.$$

We choose a Cartan involution $\theta X = -X^T$. If E_{ij} is the matrix with a 1 in the (i, j) position and zeros elsewhere, we set $Y_j = E_{j5} + E_{5j}$. Then the root vectors of AdK on $\mathfrak{p}_{\mathbf{C}}$ are

$$E_{\alpha_1} = Y_1 + iY_2, \quad E_{\alpha_2} = Y_3 + iY_4;$$

and

$$\Delta_n = \{\pm\alpha_1, \pm\alpha_2\}, \ \Delta_k = \{\pm(\alpha_1 + \alpha_2), \pm(\alpha_1 - \alpha_2)\}.$$

We begin our construction of differential operators with a choice of positive roots for Δ_k; we shall choose $\Delta_k^+ = \{\alpha_1 + \alpha_2, \alpha_1 - \alpha_2\}$. We also choose orderings on Δ_n; if E_{α_1} is the highest weight for Ad on $\mathfrak{p}_{\mathbf{C}}$, we have two compatible orderings, $\Delta_1 = \{\alpha_1, \alpha_2\}$ and $\Delta_2 = \{\alpha_1, -\alpha_2\}$. If λ is a given dominant integral weight of K, there corresponds an irreducible unitary representation $\tau = \tau_\lambda$ of K. We may then construct the $\mathscr{I}$ operator, which is independent of Δ_n order, and we may construct, for each order Δ_i, a Schmid operator D_i.

Generally, ker $\mathscr{T}$ is contained in ker D_i but is not equal. Therefore, if λ is such that ker D_i leads to a discrete series or limit of discrete series in either order, then any invariant H^2 space must be a proper subspace of an irreducible representation and therefore trivial. In the Euclidean theory, Stein and Weiss [12] showed that any representation τ of K gives rise to an H^p theory on $\mathbf{R}_+^n$. In the semisimple case, only certain τ will lead to nontrivial H^2 theories. To be precise, we must have $\Lambda = \lambda + \delta_k - \delta_n$ not dominant with respect to Δ_i for either i, and if Λ is singular, we cannot have $\langle \Lambda, \alpha_2 \rangle = 0$. A simple computation shows that this is valid for $\lambda = m\alpha_1$, where $m \in \mathbf{Z}^+$. The representations of $K = \mathrm{SO}(4)$ that have highest weight $m\alpha_1$ are the representations of the rotation group on symmetric m-tensors. Therefore, the systems of differential equations that we obtain are algebraically the same as the higher-gradient systems, that is, they have the same principal symbol, and are easily seen to be elliptic.

For the remainder of the paper, we shall assume $\tau = \tau_\lambda$ for $\tau = m\alpha_1$, $m \in \mathbf{Z}^+$. The relationship with the Schmid operators may then be analyzed more precisely:

$$D_1 f = 0 \text{ iff } \nabla f \in \sum_{\alpha \in \Delta_1} m_\lambda \tau_{\lambda+\alpha} = (m+1)\alpha_1 \oplus (m\alpha_1 + \alpha_2),$$

$$D_2 f = 0 \text{ iff } \nabla f \in \sum_{\alpha \in \Delta_2} m_\lambda \tau_{\lambda+\alpha} = (m+1)\alpha_1 \oplus (m\alpha_1 - \alpha_2),$$

$$\mathscr{T} f = 0 \text{ iff } \nabla f \in (m+1)\alpha_1.$$

Lemma 1. ker $\mathscr{T}$ = ker $D_1 \cap$ ker D_2. *Moreover,* $\mathscr{T}$ *is elliptic and the kernel of* $\mathscr{T}$ *is contained in the eigenspace of the Casimir with eigenvalue* $m^2 + m - 2$.

This identifies the candidates for nontrivial H^2 spaces; the main result of this note is to prove that they are in fact nontrivial.

Theorem 1. There is a positive definite inner product $\langle\ ,\ \rangle$ *on* $C^\infty(G, V)$ *such that the closure of* $C^\infty \cap \ker \mathscr{T}$*, called* H^2*, is a nontrivial Hilbert space. Moreover, the right regular representation* $\pi(h)f(g) = f(gh)$ *forms an irreducible unitary representation of* G *on* H^2.

Remarks. The sketch of the proof will occupy most of the remainder of the paper. We shall therefore discuss the central ideas and difficulties.

Knapp and Wallach [8] constructed a Szegö map S for discrete series, which maps certain nonunitary principal series into the kernel of the Schmid operator. We verify that the parameters are independent of Δ_n order and therefore map into the kernel of $\mathscr{T}$. The Szegö map is nontrivial but has a large kernel; moreover, the Casimir has a highly degenerate principal symbol near the boundary, and in fact smooth solutions of the $\mathscr{T}$ system have zero boundary values. To identify precisely the kernel and range of the Szegö map, we use multiplicity formulas of Hotta and Parthasarathy [6], originally developed for discrete series.

To complete the proof, we need to find a positive definite inner product. In the case of discrete series, the inner product is that of $L^2(G, V)$, and the proof that the L^2 kernel of D is nontrivial requires extra work. Briefly, Atiyah and Schmid [1] observe the existence of a cocompact discrete subgroup Γ for which the Γ-index of D in $L^2(G, V)$ is related to the index of D on $L^2(\Gamma\backslash G/K, V)$. This latter may be computed by the L^2-index theorem; the computation relates the index of D to certain Chern classes.

It should be observed that these techniques are particularly adapted to L^2 theory and cannot be applied to nondiscrete series representations. The representations we realize are specifically nondiscrete series. A substitute is the theory of limits of complementary series, constructed by Knapp and Stein [7]. This requires us to relate the kernel of the intertwining operators to the kernel of the Szegö map. To accomplish this, we follow Blank [2] in defining a nonzero boundary value map, whose kernel can be analyzed precisely. We then obtain an inner product from the theory of complementary series.

The next section of the paper constructs Riesz transforms. We cannot readily use harmonic function theory to characterize the overdeterminedness of the $\mathfrak{J}$ system; the Casimir vanishes near the boundary. Instead we use an abstract embedding result of Langlands on the classification of irreducible unitary representations. The resulting Riesz transforms are like ∇^m but do not have the homogeneity of singular integrals; H^2 is not isomorphic to L^2.

In the final section of the paper we briefly discuss molecular decompositions. The existence of molecular decomposition for H^2 follows from an abstract result of Kunze [9]; we give a more concrete realization that relates the decomposition to the molecular decomposition for Bergman spaces given by Coifman and Rochberg [4].

SZEGÖ MAPPING

Since $\ker \mathscr{J} = \ker D_1 \cap \ker D_2$, we may use discrete series techniques to provide a mapping of boundary values into $\ker \mathscr{J}$. As in the Euclidean example we discussed earlier, the normal direction must be carefully chosen. We let $\boldsymbol{\alpha} = \mathbf{R}(E_{\alpha_2} + E_{-\alpha_2})$, A the corresponding analytic subgroup, $G = ANK$ the Iwasawa decomposition, ρ_+ half the sum of the positive restricted roots counted with multiplicity, and M the centralizer of α in K. Now $\pm\alpha_2$ is a fundamental sequence of Knapp and Wallach (since it is simple) and we may construct the corresponding Szegö map. Let σ_λ be the representation of M on W with highest weight $m\alpha_1$, and let $\nu = (1 - z)\rho_+$. The Szegö map $S_z : C^\infty(K, \sigma_\lambda) \to C^\infty(G, \tau_\lambda)$ is defined by

$$S_z F(g) = \int_K e^{\nu H(kg^{-1})} \tau_\lambda^{-1}(k(kg^{-1}))F(k)dk.$$

Proposition 1. If $z = 1/3$, S_z is a nontrivial map of $C^\infty(K, \sigma_\lambda)$ into $C^\infty(G, \tau_\lambda) \cap \ker \mathcal{T}$.

Proof. Knapp and Wallach identify parameters ν_i, σ_i for which S_{ν_i} maps $C^\infty(K, \sigma_\lambda)$ into $C^\infty \cap \ker D_i$. One easily checks that the parameters are independent of order Δ_i, and are $\nu = 2/3\rho_+$, $\sigma_\lambda \sim m\alpha_1$. To show S_z is nontrivial, Knapp and Wallach let $P_{\lambda\sigma}$ be the projection of V to W, ϕ_λ the highest vector of τ_λ, and $F_\lambda(k) = P_{\lambda\sigma}\tau_\lambda(k)\phi_\lambda$. Then

$$(S_\nu F_\lambda(e), \phi_\lambda) = \int_K |F|^2;$$

the latter is positive, since F is continuous and $F(e) = \phi_\lambda$.

Remark. Knapp and Wallach reformulate this result as follows. Let $\nu' = 2\rho_+ - \nu$; the nonunitary principal series $U(\sigma_\lambda, \nu')$ acts on the space

$$\{F : G \to W | F(\text{manx}) = e^{\nu' H(a)\sigma_\lambda}(m)F(x)\}$$

by the right regular representation. If F in $C^\infty(K, \sigma_\lambda)$ is appropriately extended, the Szegö map may be computed as

$$S_z F(x) = \int_K \tau_\lambda(k^{-1})F(kx)dk,$$

from which it is clear that S_z is equivariant.

Theorem 2. $S_{1/3}$ is a map onto $C^\infty(G, \tau_\lambda) \cap \ker \mathcal{T}$. Moreover, $\ker \mathcal{T}$ is irreducible.

Proof. We proceed through an identification of $\ker \mathcal{T}$ and again make use of discrete series techniques; in this case we compute precisely the multiplicities of the K-isotypic subspaces that occur in the decomposition of the right regular representation of G on $\ker D_i$. The relevant result is Theorem 1 of Hotta and Parthasarathy [6]. The result requires $\langle \lambda + \rho_k - \langle\phi\rangle, \alpha\rangle \geq 0$ for every

$\alpha \in \Delta_k^+$, $\phi \subset \Delta_i$, where $\langle\phi\rangle = \sum_{\alpha \in \phi} \alpha$. It is easily checked that the result applies if $m \geq 1$.

Let $m_\lambda(\mu)$ be the multiplicity of μ in the right regular representation of G in $\ker D_1 \cap \ker D_2$. Then $m_\lambda(\mu) \leq b_\lambda(\mu)$, where

$$b_\lambda(\mu) = \sum_{s \in W_k} \det(s)\phi[s(\mu + \rho_k) - (\lambda + \rho_k)],$$

where

$$\phi(\mu) = \sum_{k=0} \phi_k(\mu);$$

$\phi_0(0) = 1$, and $\phi_k(\mu) =$ number of distinct ways μ can be written as a sum of exactly k elements of Δ_i, $= 1, 2$. Moreover, equality $m_\lambda(\mu) = b_\lambda(\mu)$ holds if λ is far from the walls, which means, in the case $\lambda = m\alpha_1$, $m \geq 2$.

Since $\Delta_i = \{\alpha_1, \pm\alpha_2\}$, there can be no cancellation and $\phi_k(p\alpha_1 + q\alpha_2) = \delta_{k,p+q}$.

Lemma 2. Let $\lambda = m\alpha_1$, $m \geq 1$. *If* $\mu = a\alpha_1 + b\alpha_2$, *then* $m_\lambda(\mu) = 0$ *if* $b \neq 0$ *or* $a < m$; *if* $b = 0$, $a \geq m$, *then* $m_\lambda(\mu) \leq 1$. *If* $a \geq 2$, $m_\lambda(\mu) = 1$.

The lemma may be proved by straightforward computations. It is important to note that we need to use both the Δ_1 and Δ_2 orders to compute $b_\lambda(\mu)$, to show $m_\lambda(\mu) = 0$ if $b \neq 0$.

To complete the proof of the theorem, we make three observations:

(a) $m_\lambda(\lambda) = 1$.

(b) Let E_m be the sum of the K-isotypic subspaces of $C^\infty(G, \tau_\lambda)$ with highest weights $a\alpha_1$, $a \geq m$. Then E_m is irreducible under $\mathfrak{g}_{\mathbf{C}}$.

(c) The K-finite subspaces of $\ker \mathscr{T}$ and $\operatorname{Im} S$ are irreducible under the actions of $\mathfrak{g}_{\mathbf{C}}$. (Here $S \equiv S_{1/3}$.)

From (a), (b), and invariance, ker $\mathscr{T}$ is E_m. From the invariance of Im S and its nontriviality, it follows that the K-finite image of S is E_m. Hence S is onto, as was to be shown. To prove (a), note that $m_\lambda(\lambda) \leq 1$, so that we need only show $m_\lambda(\lambda) \neq 0$. Let χ_λ = trace τ_λ; it suffices to show $\chi_\lambda * \text{Im } S \neq 0$. But S commutes with the action of K, so that

$$\chi_\lambda * S(F_\lambda) = S(\chi_\lambda * F_\lambda) = S(F_\lambda) \neq 0.$$

The proof of (b) is immediate from the simple structure of E_m; the proof of (c) comes from the invariance of $\mathscr{T}$ and S. This completes the proof of the theorem.

BOUNDARY VALUES

To complete our analysis, we shall need a knowledge of the C^∞ kernel of S. We shall obtain such information from boundary values. Following Blank [2], we note that if $g = a_t \in A$ and w_0 is chosen in the Weyl group with $w_0^{-1} a_t w_0 = a_{-t}$, then

$$S_z F(a_t) = \int e^{\nu H(ka_{-t})} \tau_\lambda^{-1}(\kappa(ka_{-t}))F(k)\,dk$$

$$= \int e^{\nu H(ka_t w_0)} \sigma_\lambda^{-1}(\kappa(ka_t w_0))F(kw_0)\,dk$$

$$= \int e^{\nu H(ka_t)} \tau_\lambda^{-1}(\kappa(ka_t)w_0)F(kw_0)\,dk$$

$$= \iint_{M\times V} e^{\nu H(m\kappa(x)\alpha_t)} e^{2\rho_+ H(x)} \tau_\lambda^{-1}(\kappa(m\kappa(x)a_t)w_0)F(m\kappa(x)w_0)\,dx\,dm$$

but

$$H(m\kappa(x)\alpha_t) = H(\kappa(x)\alpha_t) = -H(x) + H(x\alpha_t)$$

and

$$\kappa(m\kappa(x)\alpha_t) = m\kappa(\kappa(x)\alpha_t) = m\kappa(x\alpha_t)$$

so that

$$S_z F(\alpha_t) = \iint e^{-\nu H(x)} e^{2\rho_+ H(x)} e^{\nu H(xa_t)} \tau_\lambda^{-1}(m\kappa(xa_t)w_0)F(m\kappa(x)w_0)\,dx\,dm$$

$$= \int_V e^{(2\rho_+ - \nu)H(x)} e^{\nu H(xa_t)} \tau_\lambda^{-1}(\kappa(xa_t)w_0)F(\kappa(x)w_0)dx.$$

Now letting $\nu' = 2\rho_+ - \nu$ and viewing F as an element of $U(\sigma_\lambda, \nu')$, with transformation properties under MAN, we obtain

$$S_z F(a_t) = \int e^{\nu H(xa_t)} \tau^{-1}((a_t^{-1}xa_t)w_0)F(xw_0)dx.$$

But

$$e^{\nu H(a_t^{-1}xa_t)} e^{\nu H(a_t)} = e^{\nu H(\delta_t x)} e^{\nu H(a_t)}$$

so that $S_z F(a_t)$ is

$$e^{\nu H(a_t)} \int_V e^{\nu H(\delta_t x)} \tau_\lambda^{-1}(\kappa(\delta_t x)w_0)F(xw_0)dx,$$

and

$$\lim_{t \to \infty} e^{\nu H(a_t)} S_z F(a_t) = \lim_{t \to \infty} \int e^{2\nu H(a_t)} e^{\nu H(\delta_t x)} \tau_\lambda^{-1}(\kappa(\delta_t x)w_0)F(xw_0)dx.$$

But $\kappa(\delta_t x)w_0$ converges almost everywhere to $m(xw_0)$, and $\tau^{-1}(\kappa(\delta_t x)w_0)$ boundedly to $\sigma_\lambda^{-1}(m(xw_0))$. Moreover,

$$e^{2\nu H(a_t)} e^{\nu H(\delta_t x)} F(xw_0)$$

converges dominatedly to $e^{\nu H(x)} F(wx_0)$ when $\nu = (1 - z)\rho_+$ satisfies Re $z > 0$. A comparison with Knapp and Stein [7] shows that

$$\lim_{t \to \infty} e^{\nu H(a_t)} S_z F(a_t) = A(w_0, \sigma_\lambda, z)F(e),$$

where A is the intertwining operator from $U(\sigma_\lambda, \nu')$ into $U(w_0\sigma_\lambda, w_0\nu')$, and $\nu = (1 - z)\rho_+$. If we define a boundary operator B from Im S_z to $U(w_0\sigma_\lambda, w_0\nu')$ by

$$BS_z F(g) = U(w_\sigma \sigma_\lambda, w_0\nu')(g) \lim_{t \to \infty} e^{\nu H(a_t)} S_z F(a_t),$$

it follows that:

Theorem 3. *If* Re $z > 0$, *then* $BS_z = A(w_0, \sigma_\lambda, z)$, *and the map* B *is equivariant.*

We note in particular that the above proof does not use special properties of S at $z = 1/3$, but is valid for $z > 0$.

CONSTRUCTION OF H^2

To construct H^2, we shall first construct a Hilbert space on a quotient of a nonunitary principal series and then use S to lift the structure to G/K.

Theorem 4 (Knapp and Stein, [7]). The map $A(w_0, \sigma_\lambda, z)$ *defines a positive definite norm on* $U(\sigma_\lambda, (1 - z)\rho_+)$ *if* $0 < z < 1/3$. *If* $z = 1/3$, *the norm is positive semidefinite, and there is a G-unitary Hilbert space structure on* $U(\sigma_\lambda, 1/3\rho_+)/\ker A(w_0, \sigma_\lambda, 1/3)$.

Remark. The result of Knapp and Stein was used to construct complementary series. If $z = 1/3$, then $\nu = (1 - z)\rho_+ = 2/3\rho_+$. This is the nonunitary principal series from which $S_{1/3}$ maps $U(\sigma_\lambda, 1/3\rho_+)$ onto $\ker \mathcal{T}$.

Corollary 1. $U(\sigma_\lambda, 1/3\rho_+)/\ker A(w_0, \sigma_\lambda, 1/3)$ *has a G-unitary structure,* $\| [F] \|_\sigma$.

Theorem 5. $\ker S_{1/3} = \ker A(w_0, \sigma_\lambda, 1/3)$.

Proof. We shall establish this result for K-isotypic subspaces. Since the nonunitary principal series are admissible, all the isotypic subspaces appear with finite multiplicity, and the results below follow from finite dimensional vector space theory.

Note first that $B \circ S = A(w_0, \sigma_\lambda, z)$ for $\mathrm{Re}\, z > 0$. In particular at $z = 1/3$, $\ker S \subset \ker A$. Therefore, $\mathrm{coker}\, A \subset \mathrm{coker}\, S$. But $\mathrm{coker}\, S \cong \mathrm{Im}\, S$ and at $z = 1/3$, $\mathrm{Im}\, S = \ker \mathcal{T}$ is irreducible. Since A commutes with G, $\ker A$ and $\mathrm{coker}\, A$ are invariant. Since $\mathrm{coker}\, A$ is not trivial, $\mathrm{coker}\, A = \mathrm{coker}\, S$ and $\ker A = \ker S$.

We may now define the H^2 norm. Take $f \in C^\infty(G, \tau_\lambda) \cap \ker \mathcal{T}$. There is a unique class $[F]$ in $U(\sigma_\lambda, 1/3\rho_+)/\ker A(w_0, \sigma_\lambda, 1/3)$ with $S[F] = f$. Define

$$\| f \|_\tau = \| [F] \|_\sigma .$$

This is a unitary structure on ker $\mathcal{T}$:

$$\begin{aligned} \|\pi(h)f(g)\|_\tau &= \|f(gh)\|_\tau = \|S[F](gh)\|_\tau \\ &= \|S(U(\sigma_\lambda, 1/3\rho_+))(h)[F]\|_\tau \\ &\equiv \|U(\sigma_\lambda, 1/3\rho_+)(h)[F]\|_\sigma = \|[F]\|_\sigma \\ &\equiv \|f\|_\tau . \end{aligned}$$

Theorem 6. *Let* H^2 = closure $C^\infty(G, \tau_\lambda) \cap \ker \mathcal{T}$. *Then* H^2 *is a non-trivial Hilbert space. The right regular representation* $\pi(h)f(g) = f(gh)$ *forms an irreducible unitary representation of* G *on* H^2.

Remark. This construction is analogous to defining $H^2(\mathbf{R}^n)$ as $L^2(\mathbf{R}^{n-1})$. It is a reasonable substitute for the index theorem, but clearly we cannot expect to do H^p theory. We discuss alternate characterizations in the last section of the paper.

RIESZ TRANSFORMS

In the Euclidean case of H^2 defined by higher gradients, every solution is determined by a single harmonic function. Since the degree of freedom of the $\mathcal{T}$ system is determined by the space W = Span $\tau_\lambda(M)\phi_\lambda$, one degree of freedom suggests an embedding of (π, H^2) into a nonunitary principal series $U(\sigma_0, \nu_0)$, where σ_0 is the trivial representation. Riesz transforms in the classical case take F into $(F, R_1F, \ldots, R_nF, \ldots, R_{i_1} \ \ldots \ R_{i_m}F)$, or, invariantly, $F \to (-\Delta)^{-m/2}\nabla^m F$. In the non-Euclidean case, the Riesz transforms will be given by the intertwining operator from $U(\sigma_0, \nu_0)$ to $U(\sigma_\lambda, 1/3\rho_+)$.

The action of the Weyl group on the infinitesimal character of one embedding yields all potential embeddings. We have obtained a quotient embedding at $\sigma_\lambda \sim m\alpha_1$, $\nu' = 1/3\rho_+$; the corresponding infinitesimal character is $(m + 1/2)\alpha_1 + \alpha_2$. After the action of the Weyl group, we obtain embeddings at $[\sigma_\lambda, i\mu]$, where $i\mu = \nu' - \rho_+$, at the following nonunitary principal series:

$$[m\alpha_1, 1/3\rho_+], \; [m\alpha_1, -1/3\rho_+],$$

and

$$[0\alpha_1, (m + \tfrac{1}{2})2/3\rho_+], \; [0\alpha_1, -(m + \tfrac{1}{2})2/3\rho_+].$$

Each of these pairs potentially gives (π, H^2) as a quotient and a subrepresentation of the nonunitary principal series. By the uniqueness in Langlands' classification, there is at most one quotient on any side of the unitary axis; by existence there is at least one. Since we are interested in obtaining (π, H^2) as the image of Riesz transforms, we wish to find $U(\sigma_0, \nu_0)$ for which (π, H^2) is embedded as a quotient. This occurs at

$$[m\alpha_1, 1/3\rho_+] \quad \text{and} \quad [0\alpha_1, -2/3(m + 1)\rho_+].$$

To obtain the Riesz transforms, we let $J'(\sigma_\lambda, i\mu)$ denote the irreducible quotient or subrepresentation realization of (π, H^2) at $U(\sigma_\lambda, \rho_+ + i\mu)$. We now know abstractly that $J'(0, -2/3(m + 1)\rho_+)$ is equivalent to $J'(m\alpha_1, 1/3\rho_+)$. To implement this isomorphism concretely, we let A_0 denote the intertwining map $A(w_0, m\alpha_1, 1/3)$ restricted to the quotient $J'(m\alpha_1, 1/3\rho_+)$; then A_0 is invertible, and A^{-1} gives the equivalence of $J'(m\alpha_1, -1/3\rho_+)$ with $J'(m\alpha_1, 1/3\rho_+)$. A_0 is a convolution operator on V, with kernel

$$J(\exp X) = \Omega\left(\frac{x}{\|x\|}\right)\Big/ \|x\|^{3(1-z)} = \Omega/\|x\|^2 .$$

Thus A_0^{-1} is formally a fractional integral operator composed with Euclidean Riesz transforms. To finish the characterization, we need an explicit equivalence of $J'(0\alpha_1, -2/3(m + 1)\rho_+)$ and $J'(m\alpha_1, -1/3\rho_+)$. From Knapp and Stein [7], it is known that this intertwining operator is given by convolution with a distribution at the origin.

If ∇ is the invariant gradient for V, one easily checks that ∇^m performs this intertwining. The Riesz transforms are then given by $A_0^{-1}\nabla^m$, where A_0 is a smoothing operator; this composition then plays the role of $(-\nabla)^{-m/2}\nabla^m$ in Euclidean H^2 theory, but it is not a singular integral operator on V, instead being the composition of a singular integral and a fractional integral.

MOLECULAR DECOMPOSITION

The representation (π, H^2) possesses a cyclic invariant subspace that transforms by τ, since the τ-isotypic subspace certainly occurs, and it must be cyclic by irreducibility. The results of Kunze [9] now imply that H^2 is a reproducing kernel space. There is an operator-valued kernel $R(g)$ on G, $R : V \to V$, satisfying

(i) $R(e) = I$.

(ii) $R(k_1 g k_2) = \tau(k_1) R(g) \tau(k_2)$.

(iii) R is positive definite.

(iv) (π, H^2) is unitarily equivalent to the right regular representation of G on the closure of

$$\left\{\sum R(x_i^{-1}x)v_i \,\middle|\, x_i \in G,\ v_i \in V\right\}$$

under the H^2 norm.

To give alternate characterizations of H^2, the precise form of R is important.

Lemma 3. $R(g)v = S_{1/3}\big(P_{\lambda\sigma}\,\tau(k^{-1})v\big)(g)$.

Proof. (i) and (ii) are immediate. The unitarity of (π, H^2) implies (iii), and the invariance and irreducibility of the relevant spaces implies (iv).

Remarks.

(1) The characterization of H^2 as a reproducing kernel space gives a molecular decomposition for H^2.

(2) In the special case $G = SL(2, \mathbf{R})$, the $\mathscr{T}$ operator is the Schmid D operator, and in appropriate local coordinates, the discrete series are realized in Bergman spaces:

$$\left\{f : \mathbf{R}_+^2 \to C \,\middle|\, \overline{\partial} f = 0,\ \iint |f(x + iy)|^2 y^{n-2} dx\, dy < \infty\right\}.$$

A molecular decomposition for such spaces was then given by Coifman and Rochberg [4]. The Coifman-Rochberg decomposition is more precise in that it relies on a fixed lattice of points $x_i \in G$. The decomposition given here may be regarded as a vector-valued analogue of classical molecular decompositions.

(3) The construction of reproducing kernels for unitary representations should lead to alternate characterizations of H^2, similar to the explicit descriptions of Bergman spaces given above. See Gross and Kunze [5] for a discussion. Such a characterization is essential to provide an H^p theory for $p \neq 2$. We shall take up this topic in a later paper.

REFERENCES

1. Atiyah, M., and Schmid, W. A geometric construction of the discrete series for semi-simple Lie groups. *Inv. Math.* 42 (1977):1-62.

2. Blank, B. E. Knapp-Wallach Szegö integrals and P-induced continuous series representations: The parabolic rank one case. Thesis, Cornell University, 1980.

3. Calderón, A., and Zygmund, A. On higher gradients of harmonic functions. *Studia Math.* 24 (1964):211-226.

4. Coifman, R. R., and Rochberg, R. Reparesntation theorems for holomorphic and harmonic functions in L^p. *Astérisque* 77 (1980).

5. Gross, K., and Kunze, R. A. Fourier-Bessel transforms and holomorphic discrete series. *Springer Lecture Notes in Math.* 266 (1972).

6. Hotta, R., and Parthasarathy, R. Multiplicity formulae for discrete series. *Inv. Math.* 26 (1974):133-178.

7. Knapp, A. W., and Stein, E. M. Intertwining operators for semi-simple groups. *Ann. Math.* 93 (1974):489-578.

8. Knapp, A. W., and Wallach, N. Szegö kernels associated with discrete series. *Inv. Math.* 34 (1976):163-200.

9. Kunze, R. A. Positive definite operator-valued kernels and unitary representations. *Proc. Conf. on Func. Anal.*, Irvine, Calif., 1966.

10. Schmid, W. On the realization of the discrete series of a semi-simple Lie group. *Rice Univ. Studies* 56 (1970):99-108.

11. Stein, E. M. Conjugate harmonic functions in several variables. *Proc. International Congress Math.* (1962):414-419.

12. Stein, E. M., and Weiss, G. Generalizations of the Cauchy-Riemann equations and representations of the rotation group. *Amer. J. Math.* 90 (1968):163-196.

13. Stein, E. M., and Weiss, G. On the theory of harmonic functions of several variables I. *Acta Math.* 103 (1960):26-62.

INTERPOLATION BETWEEN HARDY SPACES

Peter W. Jones
University of Chicago

1. INTRODUCTION

This paper is a survey of the results on interpolation between Hardy spaces. For $0 < p < \infty$ we denote by $H^p = H^p(\mathbf{R}^n)$ the real variables Hardy space consisting of functions f harmonic in $\mathbf{R}_+^{n+1}$ and satisfying $f^\star(x) \in L^p(\mathbf{R}^n)$. Here $f^\star$ is the nontangential maximal function defined by

$$f^\star(x) = \sup_{|x-y|<t} |f(y, t)|.$$

We always identify H^p functions with their distributional boundary values. When $n = 1$ we also consider for $0 < p < \infty$ the analytic Hardy spaces $H_a^p = H_a^p(\mathbf{R})$ consisting of functions f analytic on $\mathbf{R}_+^2$ and satisfying $f^\star \in L^p$. The norm of a function $f \in H^p$ is given by $\|f\|_{H^p} = \|f^\star\|_{L^p}$ with a similar definition for H_a^p. When $p = \infty$ there are several possible spaces that are natural "endpoint spaces" of the classes H^p and H_a^p. The first and most obvious is L^∞. When considering the classes H_a^p, we should consider as an endpoint space the class H_a^∞, the ring of bounded analytic functions on $\mathbf{R}_+^2$ endowed with the supremum norm. A third possibility is to consider BMO = BMO$(\mathbf{R}^n)$, the space of functions of bounded mean oscillation. We also consider the space BMO$_a$ = BMO$_a(\mathbf{R})$ with the obvious definition.

At first sight there seems to be no reason to distinguish between the spaces $H_a^p(\mathbf{R})$ and $H^p(\mathbf{R})$. When $0 < p < \infty$, the spaces $H_a^p(\mathbf{R})$ and $H^p(\mathbf{R})$ are isomorphic. (This follows from the Burkholder-Gundy-Silverstein theorem [2].) Since the Hilbert transform is bounded on $\mathrm{BMO}(\mathbf{R})$, it also follows that the spaces $\mathrm{BMO}_a(\mathbf{R})$ and $\mathrm{BMO}(\mathbf{R})$ are isomorphic. The real story is that $\mathrm{Re}\ (H_a^\infty) \neq \mathrm{Re}\ \big(L^\infty(\mathbf{R})\big)$ (the Hilbert transform is not bounded on L^∞); $H_a^\infty \subsetneq L^\infty(\mathbf{R})$ and it is a much smaller space. Consequently, it must be harder to interpolate between H_a^p and H_a^∞ than it is to interpolate between $H^p(\mathbf{R})$ and $L^\infty(\mathbf{R})$.

We will consider two methods of interpolation, namely the complex and real methods. The reader is referred to [1] for a full account of these methods. (The reader should be aware, however, that the results for interpolation between Hardy spaces that are stated in that book are not stated correctly!) Before proceeding to the statements of the known results, we point out that a good majority of these results are due to Professor Zygmund and his mathematical descendants.

2. THE COMPLEX METHOD

In this section we discuss the complex method of interpolation, due to Calderón [4]. In this method the spaces between X_0 and X_1 are denoted by $(X_0, X_1)_\theta$, where $0 < \theta < 1$. The complex method of interpolation can be viewed as a generalization of the celebrated Riesz-Thorin theorem. In his 1939 paper [25], Thorin proved the following result (modulo the modern notation).

Theorem 1 (Riesz-Thorin). If $1 \leq p_0 < p_1 \leq \infty$, *then* $(L^{p_0}, L^{p_1})_\theta = L^p$, $\dfrac{1}{p} = \dfrac{1-\theta}{p_0} + \dfrac{\theta}{p_1}$.

If we combine Thorin's result with the Marcel Riesz theorem on conjugate functions [21], proved in 1927, we obtain our first result:

$$(H_a^{p_0}, H_a^{p_1})_\theta = H_a^p,\ 1 < p_0 < p_1 < \infty,\ \frac{1}{p} = \frac{1-\theta}{p_0} + \frac{\theta}{p_1}. \tag{1}$$

The next result in this direction is due to Salem and Zygmund [23]. In that paper, the authors were looking for a simpler proof of a theorem due to Littlewood and Paley [19]:

$$\int_0^{2\pi} \frac{|S_{n(\theta)}f(\theta)|^p}{\log[n(\theta)+2]}\, d\theta \le C_p \int_0^{2\pi} |f(e^{i\theta})|^p\, d\theta,\ 1 < p \le 2.$$

Here

$$S_{n(\theta)}f(\theta) = \sum_{-n(\theta)}^{n(\theta)} a_k e^{ik\theta}$$

is a partial sum of

$$f(e^{i\theta}) = \sum_{-\infty}^{\infty} a_k e^{ik\theta}.$$

The Littlewood-Paley result is not terribly difficult for $p = 2$, or for $p = 1$ with f in H_a^1 of the circle. Salem and Zygmund had the idea to prove the L^p result, $1 < p < 2$, by interpolating H_a^1 and H_a^2 and then invoking the Marcel Riesz theorem on conjugate functions. [Notice that I am ignoring the difference between $H_a^p(\mathbf{R})$ and $H_a^p(\mathbf{T})$— almost any theorem for one space has an immediate analogue for the other.] Here is the result Salem and Zygmund proved to deduce the Littlewood-Paley theorem of above:

Theorem 2. Let

$$f(x) = \sum_0^{\infty} x_v z^v$$

be regular for $|z| < 1$ *and belong to the class* H^1/α $(\alpha > 0)$. *Let* $M(\alpha, \beta)$ *denote the maximum of the finite complex bilinear form*

$$\sum_{j=0}^{m} \sum_{h=0}^{n} a_{jh} x_j y_h$$

when the complex variables x_j, y_h *are subject to the conditions*

$$\sum_{0}^{n} |y_k|^{1/\beta} \leq 1 \qquad (\beta \geq 0)$$

and

$$\int_0^{2\pi} |f(re^{i\theta})|^{1/\alpha} d\theta \leq 1 \qquad (\alpha > 0),$$

the second condition holding for every $f(z)$ belonging to $H^{1/\alpha}$ and having $x_0, x_1, \ldots, x_m$ as its $m + 1$ first expansion coefficients, and for every nonnegative r inferior to 1. Then if the point (α, β) lies on the segment joining the points (α_1, β_1) and (α_2, β_2), that is to say, if $\alpha = t\alpha_1 + (1 - t)\alpha_2$ and $\beta = t\beta_1 + (1 - t)\beta_2$ $(0 < t < 1)$, we have

$$M(\alpha, \beta) \leq C(\alpha_1, \alpha_2) M^t(\alpha_1, \beta_1) M^{1-t}(\alpha_2, \beta_2),$$

the constant $C(\alpha_1, \alpha_2)$ depending on α_1 and α_2 only.

In 1950 Calderón and Zygmund [6] extended this result to (essentially) obtain

$$(H_a^{p_0}, H_a^{p_1})_\theta = H_a^p,\ 0 < p_0 < p_1 < \infty,\ \frac{1}{p} = \frac{1 - \theta}{p_0} + \frac{\theta}{p_1}. \tag{2}$$

Actually, they only showed that $(H_a^{p_0}, H_a^{p_1}) \supset H_a^p$, that being the only useful result. The opposite inclusion is true, but nobody bothered to prove it until much later (see [5]). The idea behind the proof of (2) goes back to Thorin's 1948 thesis [24]. One first selects an integer n such that $np_0 > 1$. For $f \in H_a^p$, one writes $f = BG = B(G^{1/n})^n$, where B is a Blaschke product and G is zero-free. By this trick and a clever argument (due as far as I can tell to Calderón and Zygmund), one reduces oneself to proving $(H_a^{np_0}, H_a^{np_1}) = H_a^{np}$, which follows from (1) since $1 < np_0 < np_1 < \infty$. Notice that in results (1) and (2), the endpoint $p = \infty$ is excluded. This is because the above proofs use the Marcel Riesz theorem, which fails for $p = \infty$.

After the paper of Calderón and Zygmund, there was a hiatus of some twenty years; it was not until 1972 that C. Fefferman and E. M. Stein [10] gave a new result. In their *Acta* paper they showed

$$(H^1, L^{p_1})_\theta = L^p, \; 1 < p_1 < \infty, \; \frac{1}{p} = 1 - \theta + \frac{\theta}{p_1} \tag{3}$$

and

$$(L^{p_0}, \text{BMO})_\theta = L^p, \; 1 < p_0 < \infty, \; \frac{1}{p} = \frac{1 - \theta}{p_0}. \tag{4}$$

Fefferman and Stein deduced their results from properties of their "sharp function," which is defined by

$$f^\#(x) = \sup_Q \frac{1}{|Q|} \int_Q \left| f(t) - \frac{1}{|Q|} \int_Q f(y)\,dy \right| dt,$$

where the above supremum is taken over all cubes Q containing x. Clearly $f^\# \in L^p$ whenever $f \in L^p$ and $1 < p \leq \infty$ and $f^\# \in L^\infty$ whenever $f \in$ BMO. Fefferman and Stein's theorem on the sharp function states that if $1 \leq p_0 \leq p < \infty$ and $f \in L^{p_0}$, then $\| f \|_{L^p} \leq C \| f^\# \|_{L^p}$. From this result it is not hard to deduce (4). Then (3) follows from (4) and H^1, BMO duality. As an exercise, the reader can verify that this approach does not allow one to compute the spaces (H^1, L^∞) or $(H^1, \text{BMO})_\theta$. It should be pointed out, however, that results (3) and (4) applied to L^2 are probably the most important results in the entire area. They give, for example, very traceable methods for proving L^p boundedness of convolution operators.

In 1977 Calderón and Torchinsky [5] extended the results of Fefferman and Stein to show

$$(H^{p_0}, H^{p_1})_\theta = H^p, \; 0 < p_0 < p_1 < \infty, \; \frac{1}{p} = \frac{1 - \theta}{p_0} + \frac{\theta}{p_1} \tag{5}$$

and

$$(H^{p_0}, L^\infty)_\theta \subset (H^{p_0}, \text{BMO})_\theta \subset H^p, \; 0 < p_0 < \infty, \; \frac{1}{p} = \frac{1 - \theta}{p_0}. \tag{6}$$

We will outline a proof of (5) later in this section—it is slightly different from the original proof of Calderon and Torchinsky. The result (6) was obtained by a careful examination of square functions.

This brings us to the most recent developments, which handle the case where one of the endpoints corresponds to $p = \infty$. In 1980 Jones [16] proved the following results:

$$(H_a^{p_0}, H_a^\infty)_\theta = H_a^p,\ 0 < p_0 < \infty,\ \frac{1}{p} = \frac{1-\theta}{p_0} \tag{7}$$

and

$$(H_a^{p_0}, \mathrm{BMO}_a)_\theta = H_a^p,\ 0 < p_0 < \infty,\ \frac{1}{p} = \frac{1-\theta}{p_0}. \tag{8}$$

Notice that (8) follows immediately from (6) and (7) because $(H_a^{p_0}, H_a^\infty)_\theta \subset (H_a^{p_0}, \mathrm{BMO}_a)_\theta$. As a direct corollary of (7), (8), and H^1, BMO duality, one obtains

$$(X_0, X_1)_\theta = L^p,\ \frac{1}{p} = 1 - \theta, \tag{9}$$

where X_0 equals either $H^1(\mathbf{R})$ or $L^1(\mathbf{R})$ and where X_1 equals either $H_a^\infty + \overline{H_a^\infty}$, $L^\infty(\mathbf{R})$, or BMO($\mathbf{R}$). Here is a rough idea of how one would go about proving (7) when $p_0 = 1$, $p = 2$, and $\theta = 1/2$. Take $f \in H_a^2$ with norm 1 and tile $\mathbf{R}_+^2$ into regions $\mathscr{R}_j$. The regions $\mathscr{R}_j$ should satisfy two properties. Firstly, there should be an integer $m(j)$ such that $|f(z)| \approx 2^{m(j)}$ on a large portion of $\mathscr{R}_j$. Secondly, $\sigma \equiv \Sigma\sigma_j$ should be a Carleson measure, where σ_j is arclength measure on $(\partial\mathscr{R}_j) \cap R_+^2$. Now put $f_j = f\chi_{\mathscr{R}_j}$ and notice that $f = \Sigma f_j$. We would like to solve the $\overline{\partial}$ problem $\overline{\partial}h_j = \overline{\partial}f_j$ with good L^∞ estimates on $\Sigma|h_j|$. To this end we invoke a magic formula. Let $\mu = \sigma/\|\sigma\|_C$, where $\|\sigma\|_C$ is the norm of the Carleson measure σ, and put

$$K_0(\mu, z, \zeta) = \frac{1}{z - \zeta}\exp\left\{\iint_{\mathrm{Im}\, w \le \mathrm{Im}\,\zeta}\left(\frac{-i}{z - \overline{w}} + \frac{i}{\zeta - \overline{w}}\right)d\mu(w)\right\},$$

$$K(\mu, z, \zeta) = \frac{2i}{\pi}\left(\frac{\mathrm{Im}\,\zeta}{z - \zeta}\right)K_0(\mu, z, \zeta).$$

Notice that $K(\mu, z, \zeta)$ is holomorphic (in z) on $R_+^2\backslash\zeta$ and has a simple pole at $z = \zeta$. Now put

$$h_j(z) = \iint_{R_+^2} K(\phi, z, \zeta)\overline{\partial}f_j(\zeta).$$

Then $\overline{\partial}h_j = \overline{\partial}f_j$ and by linearity, $\Sigma h_j \equiv 0$. Let $g_j = f_j - h_j$ and let $\alpha(\zeta) = 1 - \zeta$. Then $G_\zeta = \Sigma 2^{m(j)\alpha(\zeta)}g_\zeta$ is a holomorphic Banach space-valued function in $\{0 < \mathrm{Re}\,\zeta < 1\}$, and using the properties of $K(\mu, z, \zeta)$, one can show

$$\|G_{it}\|_{H_a^1} \le C,\ t \in R$$

and

$$\| G_{1+it} \|_{H^\infty_a} \leq C, \ t \in \mathbf{R}.$$

This does the job because

$$G_\theta = G_{1/2} = \Sigma(f_j - h_j) = \Sigma f_j = f.$$

The idea of using Carleson measures and solutions of $\overline{\partial}$ to attack problems like this has a long history. We say no more than the idea is a descendant of ideas coming from Carleson's solution of the Corona problem [7] and Hörmander's later modifications [12]. The reader is referred to [16] for a further discussion.

In 1980, after (7) and (8) had been obtained, Janson and Jones [15] completed the classification of the interpolation spaces in the complex method by showing

$$(H^{p_0}, L^\infty)_\theta = H^p, \ 0 < p_0 < \infty, \ \frac{1}{p} = \frac{1-\theta}{p_0} \tag{10}$$

and

$$(H^{p_0}, \mathrm{BMO})_\theta = H^p, \ 0 < p_0 < \infty, \ \frac{1}{p} = \frac{1-\theta}{p_0}. \tag{11}$$

As in the one-dimensional case, (11) follows from (6) and (10), and one can show

$$(X_0, X_1)_\theta = L^p, \ \frac{1}{p} = 1 - \theta, \tag{12}$$

where X_0 equals either H^1 or L^1 and where X_1 equals either L^∞ or BMO. The proof of (10) heavily uses the theory of H^p atoms. (As far as I know, the first person to use atoms explicitly in proving interpolation theorems was Peetre in his 1979 paper [20]. The idea was, however, lurking in the papers [8, 9, and 22].) A function a on $\mathbf{R}^n$ is an H^p atom if a is supported on a cube Q,

$$\| a \|_{L^\infty} \leq |Q|^{-1/p}$$

and

$$\int x^\alpha a \, dx = 0$$

for all monomials $x^\alpha = x_1^{\alpha_1} \ldots x_n^{\alpha_n}$ satisfying $|\alpha| \leq \alpha(p)$, where $\alpha(p)$ is a certain nonincreasing function of p. The fundamental result in this area, due to Coifman [8] when $n = 1$ and Latter [17] when $n > 1$, is:

Theorem 3 (on the atomic decomposition). Suppose $f \in H^p$, $0 < p < \infty$, *and suppose* $0 < p_0 < p$. *Then there are sequences* $\{a_j\}$ *of* H^p *atoms and* $\{\lambda_j\}$ *of complex numbers such that*

$$\int x^\alpha a_j \, dx = 0, \quad |\alpha| \leq \alpha(p_0)$$

and

$$\sum |\lambda_j|^p \leq C(p, p_0) \| f \|_{H^p}^p .$$

Other proofs of this theorem can be found in [4, 18, and 26]. (The reader should be aware, however, that none of these papers state their results as done above.) A combinatorial argument presented in [15] shows that one may refine the atomic decomposition to obtain the following result.

Theorem 4. Suppose $0 < p_0 < p < \infty$ *and suppose* $\| f \|_{H^p} = 1$. *Then there are dyadic cubes* Q_j *and functions* a_j *supported on the triples of* Q_j *such that* $f = \Sigma a_j$. *Furthermore, there are integers* $m(j)$ *such that*

(α) $\| a_j \|_{L^\infty} \leq C 2^{m(j)}$,

(β) $\Sigma 2^{pm(j)} |Q_j| \leq C$,

(γ) *if* $Q_k \subset Q_j$ *and* $k \neq j$ *then*

$$m(k) > m(j) \text{ and } Q_k \subsetneq Q_j,$$

(δ) $\displaystyle\sum_{Q_k \subset Q_j} |Q_k| \leq 2 |Q_j|$,

(ε) $\int x^\alpha a_j \, dx = 0, \quad |\alpha| \leq \alpha(p_0)$,

(ζ) *if* A *is any finite set of indices, then for every* x *there exists* $j(x) \in A$ *such that*

$$\sum_{j \in A} |a_j(x)| \leq C 2^{m(j(x))} \chi_{3Q_{j(x)}}(x) .$$

Now let $\alpha(\zeta) = \dfrac{p}{p_0}(1 - \zeta) - 1$ and let $G_\zeta = \Sigma 2^{m(j)\alpha(\zeta)} a_j$. Then the conclusions of the above theorem show $\| G_{it} \|_{H^{p_0}} \leq C$, $t \in \mathbf{R}$, and if $0 < \varphi < 1$, $\dfrac{1}{r} = \dfrac{1 - \varphi}{p_0}$, then $\| G_{\varphi + it} \|_{H^r} \leq C_r$, $t \in \mathbf{R}$. This yields the result of Calderón and Torchinsky (5). (In fact, the proof of (5) is much simpler once one has the theorem on the atomic decomposition. From the special *form* of the functions $\lambda_j a_j$ in [4], [17], or [18], one funds scalars β_j and a holomorphic function $a(\zeta)$ such that $G_\zeta = \Sigma\beta_j^{\alpha(\zeta)}\lambda_j a_j$ is an appropriate holomorphic function.) The point of the refinement of the atomic decomposition is that by (δ), a Carleson packing condition, the functions a_j are "almost disjoint." One can find functions $\tilde{a}_j$ such that $\| f - \Sigma\tilde{a}_j \|_{H^p} < 1/2$ and almost every point x is in the support of at most C functions $\tilde{a}_j$. Now let $\alpha(\zeta) = \frac{p}{p_0}(1 - \zeta) - 1$ and put $G_\zeta = \Sigma 2^{m(j)\alpha(\zeta)}\tilde{a}_j$. Then

$$\| G_{1+it} \|_{L^\infty} \leq C, \quad t \in \mathbf{R}.$$

The functions $\tilde{a}_j$ are also constructed so that

$$\| G_{it} \|_{H^{p_0}} \leq C, \quad t \in \mathbf{R}.$$

Since

$$\| f - G_\theta \|_{H^p} = \| f - \Sigma\tilde{a}_j \|_{H^p} < 1/2,$$

an iteration argument now shows that $H^p \subset (H^{p_0}, L^\infty)_\theta$.

Very recently, Tom Wolff [27] has shown that when $p_0 = 1$, (10) follows from (3) and

Theorem 5 (Wolff's theorem). Let A_j be Banach spaces, $1 \leq j \leq 4$, and suppose $A_1 \cap A_4 \subset A_2 \cap A_3$ and $A_1 \cap A_4$ is dense in A_2 and A_3. *Suppose also there exist $0 < \psi_2, \psi_3 < 1$ such that $(A_1, A_3)_{\psi_2} = A_2$ and $(A_2, A_4)_{\psi_3} = A_3$. Then there exist $0 < \theta_j < 1$ such that $(A_1, A_4)_{\theta_j} = A_j$, $j = 2, 3$.*

The result $(H^1, L^\infty)_\theta = L^p$ now follows from $(H^1, L^3)_\theta = L^p$, $[L^2, L^\infty)_\theta = L^p$, and Wolff's theorem. Notice, however, that this method *cannot* be used to give a quick proof of (7) because if (7) holds for one value of $p_0 < \infty$, then it follows immediately for all $0 < p_0 < \infty$ by factorization.

The results for the complex method are summarized in Table 1.

TABLE 1

X_0	X_1	$(X_0, X_1)_\theta$	Restrictions
$H_a^{p_0}$	$H_a^{p_1}$	H_a^p	$0 < p_0 < p_1 \leq \infty, \frac{1}{p} = \frac{1-\theta}{p_0} + \frac{\theta}{p_1}$
$H_a^{p_0}$	BMO_a	H_a^p	$0 < p_0 < \infty, \frac{1}{p} = \frac{1-\theta}{p_0}$
$H^1(\mathbf{R})$ or $L^1(\mathbf{R})$	$H_a^\infty + \overline{H_a^\infty}$	$L^p(\mathbf{R})$	$\frac{1}{p} = 1 - \theta$
H^{p_0}	H^{p_1}	H^p	$0 < p_0 < p_1 < \infty, \frac{1}{p} = \frac{1-\theta}{p_0} + \frac{\theta}{p_1}$
H^{p_0}	L^∞ or BMO	H^p	$0 < p_0 < \infty, \frac{1}{p} = \frac{1-\theta}{p_0}$
L^{p_0}	BMO	L^p	$1 \leq p_0 < \infty, \frac{1}{p} = \frac{1-\theta}{p_0}$

3. THE REAL METHOD

In this section we discuss the real method of interpolation. The reader is referred to [1] for basic properties and definitions. In the real method the spaces between X_0 and X_1 are denoted by $(X_0, X_1)_{\theta, q}$, where $0 < \theta < 1$ and $0 < q \leq \infty$. The first nontrivial result in this area came in 1963 and is due to Igari [13 and 14]. Unfortunately, his work seems to have been largely unnoticed. I, myself, was unaware of his work until the conference in honor of Professor Zygmund, at which time Guido Weiss kindly pointed it out to me. In [14], Igari shows that if $f \in L^p$ and $\alpha > 0$, then $f = g + h$ where $\|h\|_{L^\infty} \leq C\alpha$ and

$$\|g\|_{H^1} \leq C \int_{\{f^* > \alpha\}} f^*(x)\,dx$$

Igari proved this by using a Calderón-Zygmund decomposition. From this result, Igari deduced the following for sublinear operators T:

If $T : H^{p_j} \to L^{q_j}$, $1 \le p_j \le q_j < \infty$, then $T : H^p \to L^q$, where (13)

$$\frac{1}{p} = \frac{1-\theta}{p_0} + \frac{\theta}{p_1},\ \frac{1}{q} = \frac{1-\theta}{q_0} + \frac{\theta}{q_1},\ \text{and } 0 < \theta < 1,$$

and

If $T : H_a^{p_j} \to L^{q_j}$, $1 \le p_j \le q_j < \infty$, then $T : H_a^p \to L^q$, where (14)

$$\frac{1}{p} = \frac{1-\theta}{p_0} + \frac{\theta}{p_1},\ \frac{1}{q} = \frac{1-\theta}{q_0} + \frac{\theta}{q_1},\ \text{and } 0 < \theta < 1.$$

Notice that Igari did not state any results about interpolation between H^1 and L^∞. On the other hand, with a little bit of functional analysis plus Igari's result on decompositions of L_p functions, one obtains

$$(H^{p_0}, L^\infty)_{\theta, q} = H^{p,q},\ 1 \le p_0 < \infty,\ \frac{1}{p} = \frac{1-\theta}{p_0}. \tag{15}$$

Here $H^{p,q}$ denotes the class of harmonic functions f having the property that f^* is in the Lorentz space $L^{p,q}$. Result (15) was rediscovered by Riviere and Sagher in their 1973 paper [22]. In 1974, C. Fefferman, Riviere, and Sagher [9] extended result (15) to show

$$(H^{p_0}, H^{p_1})_{\theta, q} = H^{p,q},\ 0 < p_0 < p_1 < \infty,\ \frac{1}{p} = \frac{1-\theta}{p_0} + \frac{\theta}{p_1} \tag{16}$$

and

$$(H^{p_0}, L^\infty)_{\theta, q} = H^{p,q},\ 0 < p_0 < \infty,\ \frac{1}{p} = \frac{1-\theta}{p_0}. \tag{17}$$

The methods used by these three authors are closely related to those used to prove the atomic decomposition theorem (see Sec. 2). Indeed, Latter's proof in [17] of the atomic decomposition used some functions first appearing in [9] as a key ingredient. In turn, the form of the atomic decomposition as presented in [4, 17, 18, and 26] can be used to deduce (16) and (17). Suppose, for example, that $f \in H^p$, $0 < p < \infty$, $\|f\|_{H^p} = 1$ and let

$\Omega_k = \{f^* > 2^k\}$. Then for each $k \in \mathbf{Z}$ there are functions $a_{k,j}$ supported on Ω_k such that

$$\| a_{k,j}\|_{L^\infty} \le C2^k, \int x^\alpha a_{k,j}\, dx = 0$$

for $0 \le |\alpha| \le \alpha(p_0)$, and such that every point is in the support of at most C functions $a_{k,j}$. Furthermore, $\sum_{k,j} a_{k,j} = f$. If $\alpha > 0$, let

$$f_1^\alpha = \sum_{2^k > \alpha} \sum_j a_{k,j}$$

and let $f_2^\alpha = f - f_1^\alpha$. Using this splitting, one can rather easily verify that (17) holds for $0 < p_0 \le 1$ (for $1 < p_0 < \infty$, it is classical). Result (16) now follows from the reiteration theorem (see, for example, [1]).

The results of C. Fefferman, Rivière, and Sagher were extended by Hanks [11] in 1977 to include L^p and BMO as endpoints. His results are

$$(L^{p_0}, \mathrm{BMO})_{\theta,q} = L^{p,q}, \ 0 < p_0 < \infty, \ \frac{1}{p} = \frac{1-\theta}{p_0} \tag{18}$$

and

$$(H^{p_0}, \mathrm{BMO})_{\theta,q} = H^{p,q}, \ 0 < p_0 < \infty, \ \frac{1}{p} = \frac{1-\theta}{p_0}. \tag{19}$$

Hanks deduces (19) from (18). The idea is that (19) follows from (18) when $p > 1$. (There is a small slip in Hank's proof at this point—he asserts that $H^{p_0} \subset L^{p_0}$ when $p_0 < 1$, which is false. Hank's argument can easily be fixed by noticing that $p_\varepsilon * H^{p_0} \subset L^{p_0}$ and $p_\varepsilon * \mathrm{BMO} \subset \mathrm{BMO}$, where p_ε is the Poisson kernel of height ε.) The full result in (19) now follows from the reiteration theorem.

The classification of the intermediate spaces in the real method was completed by Jones [16] in 1980. The result is

$$(H_a^{p_0}, H_a^\infty)_{\theta,q} = (H_a^{p_0}, \mathrm{BMO}_a)_{\theta,q} = H_a^{p,q}, \ 0 < p_0 < \infty, \ \frac{1}{p} = \frac{1-\theta}{p_0}. \tag{20}$$

This result follows from Theorem 6 below and the reiteration theorem.

Theorem 6. Suppose $0 < p_0 < p < \infty$, $\alpha > 0$, *and suppose* $f \in H_a^p$. *Then* $f = F_\alpha + f_\alpha$, *where* $F_\alpha \in H_a^{p_0}$, $f_\alpha \in H_a^\infty$, *and*

$$\| F_\alpha \|_{H_a^{p_0}}^{p_0} \leq C_{p_0} \int_{\{f^* > a\}} |f^*(x)|^{p_0}\, dx,$$

$$\| f_\alpha \|_{H_a^\infty} \leq C\alpha.$$

We illustrate the proof of the theorem when $p_0 = 1$ and $p = 2$. Let $0_\alpha = \{x : f^\alpha(x) > \alpha\}$ and let T_α denote the union of the tents lying over 0_α, that is, the complement of the usual saw tooth region. Let $H_\alpha(z) = f(z)\chi_{T_\alpha}(z)$. Then $\overline{\partial} H_\alpha$ is a Carleson measure supported on the boundary of T_a and $\| \overline{\partial} H_\alpha \|_C \leq \sqrt{2}\alpha$. Let K be the kernel constructed in Sec. 2 and let

$$g_\alpha(z) = \iint_{\mathbf{R}_+^2} K(\overline{\partial} H_\alpha / \| \overline{\partial} H_\alpha \|_C,\ z,\ \zeta) \overline{\partial} H_\alpha(\zeta).$$

Then

$$f = H_\alpha - g_\alpha + (f - H_\alpha + g_\alpha) = F_\alpha + f_\alpha \in H_a^1 + H_a^\infty$$

and

$$\| f_\alpha \|_{H_a^\infty} \leq C\alpha, \quad \| F_\alpha \|_{H_a^1} \leq C \int_{0_\alpha} f^*(x)\, dx.$$

The results for the real method are summarized in Table 2. The index q can take any value $0 < q \leq \infty$.

TABLE 2

X_0	X_1	$(X_0, X_1)_{\theta, q}$	Restrictions
$H_a^{p_0}$	$H_a^{p_1}$	$H_a^{p,q}$	$0 < p_0 < p_1 \leq \infty$, $\frac{1}{p} = \frac{1-\theta}{p_0} + \frac{\theta}{p_1}$
$H_a^{p_0}$	BMO_a	$H_a^{p,q}$	$0 < p_0 < \infty$, $\frac{1}{p} = \frac{1-\theta}{p_0}$
H^{p_0}	H^{p_1}	$H^{p,q}$	$0 < p_0 < p_1 < \infty$, $\frac{1}{p} = \frac{1-\theta}{p_0} + \frac{\theta}{p_1}$
H^{p_0}	L^∞ or BMO	$H^{p,q}$	$0 < p_0 < \infty$, $\frac{1}{p} = \frac{1-\theta}{p_0}$
L^{p_0}	BMO	$L^{p,q}$	$0 < p_0 < \infty$, $\frac{1}{p} = \frac{1-\theta}{p_0}$

REFERENCES

1. Bergh, J., and Löfström, J. *Interpolation Spaces*. New York: Springer-Verlag, 1976.

2. Burkholder, D. L.; Gundy, R. F.; and Silverstein, M. L. A maximal function characterization of the class H^p. *Trans. Amer. Math. Soc.* 157 (1971):137-153.

3. Calderón, A. P. An atomic decomposition of distributions in parabolic H^p spaces. *Advances in Math.* 25 (1977):216-225.

4. Calderón, A. P. Intermediate spaces and interpolation, the complex method. *Studia Math.* 24 (1964):113-190.

5. Calderón, A. P., and Torchinsky, A. Parabolic maximal functions associated with a distribution, II. *Advances in Math.* 24 (1977):101-171.

6. Calderón, A. P., and Zygmund, A. On the theorem of Hausdorff-Young and its extensions. *Ann. Math.* study no. 25:166-188. Princeton University Press, 1950.

7. Carleson, L. Interpolation by bounded analytic functions and the Corona theorem. *Ann. Math.* 76 (1962):547-549.

8. Coifman, R. R. A real variable characterization of H^p. *Studia Math.* 51 (1974):269-274.

9. Fefferman, C.; Rivière, N. M.; and Sagher, Y. Interpolation between H^p spaces: The real method. *Trans. Amer. Math. Soc.* 191 (1974):75-81.

10. Fefferman, C., and Stein, E. M. H^p spaces of several variables. *Acta Math.* 129 (1972):137-193.

11. Hanks, R. Interpolation by the real method between BMO, L^α $(0 < \alpha < \infty)$ and H^α $(0 < \alpha < \infty)$. *Indiana Math. J.* 26 (1977):679-690.

12. Hörmander, L. Generators for some rings of analytic functions. *Bull. Amer. Math. Soc.* 73 (1967):943-949.

13. Igari, S. An extension of the interpolation theorem of Marcinkiewicz. *Proc. Japan Acad.* 38 (1962):731-734.

14. Igari, S. An extension of the interpolation theorem of Marcinkiewicz II. *Tôhoku Math. J.* 15 (1963):343-358.

15. Janson, S., and Jones, P. W. Interpolation between H^p spaces: The complex method. To appear in *J. Func. Anal.*

16. Jones, P. W. L^∞ estimates for the $\bar{\partial}$ problem in a half-plane. To appear in *Acta Math.*

17. Latter, R. H. A characterization of $H^p(\mathbf{R}^n)$ in terms of atoms. *Studia Math.* 62 (1978):93-101.

18. Latter, R. H., and Uchiyama, A. The atomic decomposition for parabolic H^p space. *Trans. Amer. Math. Soc.* 253 (1979):391-398.

19. Littlewood and Paley. Theorems on Fourier series and power series (III). *Proc. London Math. Soc.* 43 (1937):105-126.

20. Peetre, J. Two observations on a theorem of Coifman. *Studia Math.* 64 (1979):191-194.

21. Riesz, M. Sur les fonctions conjugées. *Math. Zeit.* 27 (1927):218-244.

22. Rivière, N. M., and Sagher, Y. Interpolation between L^∞ and H^1, the real method. *J. Func. Anal.* 14 (1973):401-409.

23. Salem, R., and Zygmund, A. A convexity theorem. *Proc. Nat. Acad. Sci.* 34 (1948):443-447.

24. Thorin, G. O. Convexity theorems. *Medd. Lunds Mat. Sem.* 9 (1948):1-57.

25. Thorin, G. O. An extension of a convexity theorem due to M. Riesz. *Medd. Lunds Univ. Mat. Sem.* 4 (1939).

26. Wilson, J. M. A simple proof of the atomic decomposition for $H^p(\mathbf{R}^n)$, $0 < p \leq 1$. To appear in *Studia Math.*

27. Wolff, T. H. A note on interpolation spaces. To appear in the proceedings of the 1981 conference in honor of N. M. Rivière, held at the University of Minnesota.

WEIGHTED H^1-BMO DUALITIES

Umberto Neri
University of Maryland

1. ATOMIC PREDUALS

The simplest examples of the dualities to be considered here occur between *atomic* h^1 and *weighted BMO* over spaces of homogeneous type. Thus, let $X = (X, d)$ be a space of homogeneous type ([2], Sec. 2) with respect to a pair of positive, mutually absolutely continuous Borel measures μ and ν. On X, we define the function space $\mathrm{BMO}_\mu(d\nu)$ to be the set of all $f \in L^1(d\nu)$ such that

$$\mu(B)^{-1} \int_B |f - f_{\nu(B)}| d\nu \leq C \tag{1}$$

uniformly for all balls $B \subset X$, where $f_{\nu(B)} = \nu(B)^{-1} \int_B f\, d\nu$.

If $\nu = \mu$, this space is the usual BMO. Otherwise, we shall call it *weighted BMO*, or BMV (for *bounded mixed variance*) as in [15].

To find the predual of $\mathrm{BMV} = \mathrm{BMO}_\mu(d\nu)$, we look for ν-measurable functions a on X such that

$$|\langle a, f\rangle| = \left| \int_X af\, d\nu \right| \leq \mu(B)^{-1} \int_B |f - f_{\nu(B)}| d\nu \tag{2}$$

The author was partially supported by the University of Minnesota and by a grant of the C.N.R. (National Research Council of Italy).

for each $f \in$ BMV. Clearly, any function a with

$$\int a\, d\nu = 0, \ \operatorname{supp}(a) \subset B, \text{ and } \|a\|_\infty \leq \mu(B)^{-1}$$

for some ball B in X will fulfill (2). We call these a *atoms of type* (μ, ν) and we denote by $h^1_\mu\,(d\nu)$ the space of all (infinite) linear combinations

$$\Sigma \lambda_j a_j, \text{ with } \Sigma |\lambda_j| < \infty$$

and a_j being such atoms. Note that for $\mu = \nu$, our atoms are just $(1, \infty)$ atoms in the terminology of [2].

Adapting some Coifman-Weiss arguments in [2], we have that if μ and ν are related by appropriate A_p conditions, then

$$[h^1_\mu(d\nu)]^* = \mathrm{BMO}_\mu(d\nu) \tag{3}$$

with action $\langle a, f\rangle$ as above on the atoms $a \in h^1_\mu(d\nu)$ (see [15]).

2. UNWEIGHTED H^1-BMO THEORY IN NONSMOOTH DOMAINS

Let us consider bounded, simply connected domains D in $\mathbf{R}^{n+1}$, $n \geq 2$, say, where the following general principle of harmonic functions (and other solutions of "good" P.D.E.'s) holds: "nontangential boundedness" implies the existence (a.e.) of "nontangential limits" on ∂D. Thus, in decreasing order of generality, we may consider

(A) N.T.A. domains (in the sense of Jerison and Kenig [12]),

(B) Lipschitz domains (for example, Hunt-Wheeden [11], Dahlberg [3]),

(C) C^1 domains (as in [6, 7, 8] and so forth).

On $X = \partial D$, we may have (Lebesgue) surface measure $d\sigma$ and/or harmonic measure $d\omega$, evaluated at some fixed $P_0 \in D$. The general problem is as follows: to find H^1-spaces of harmonic functions in D identifiable (via nontangential limits) with atomic h^1-spaces on ∂D, and hence yielding new preduals for BMO, or weighted BMO, on ∂D.

In case (A), if B is any "surface ball" (that is, intersection of a Euclidean ball, centered on ∂D, with ∂D itself), then $\sigma(B)$ may be infinite. So we must define BMO(∂D) as $\mathrm{BMO}_\omega(d\omega)$, taking $\mu = \nu = \omega$ in formula (1) of Sec. 1. The simplest way to introduce a Hardy H^1 space of harmonic functions is to solve a Dirichlet problem for the Laplacean with appropriate L^1 data. In short, introduce the class

$$H^1_{\mathscr{D}}(D,\ d\omega) = \{u : \Delta u = 0 \text{ in } D,\ u(P_0) = 0,\ (Nu) \in L^1(d\omega) \text{ on } \partial D\}$$

normed with $\|u\| = \|Nu\|_{L^1(d\omega)}$, where Nu denotes a nontangential maximal function of u (see [12]). Then Jerison and Kenig proved that this class of harmonic functions can be identified with the space $H^1_{\mathscr{D}}(\partial D,\ d\omega)$ of nontangential limits of such functions. Moreover, we have a duality

$$[H^1_{\mathscr{D}}(\partial D,\ d\omega)]^* = \mathrm{BMO}_\omega(d\omega) \tag{4}$$

and an atomic representation

$$H^1_{\mathscr{D}}(\partial D,\ d\omega) = h^1_\omega(d\omega) \tag{4$'$}$$

in terms of atoms of type $(\omega,\ \omega)$.

In the easier setting of C^1 domains, (4) and (4′) were first proved by Fabes, Kenig, and Neri in [8]. In the setting of N.T.A. domains, no other H^1-BMO dualities are known.

Before passing to cases (B) and (C), we must recall some definitions. A positive function w on ∂D is in $A_p(d\sigma)$, respectively $B_p(d\sigma)$, $1 < p < \infty$ if

$$(A_p) \qquad \left\{\sigma(B)^{-1}\int_B w\ d\sigma\right\}\left\{\sigma(B)^{-1}\int_B (1/w)^{1/(p-1)}\ d\sigma\right\}^{p-1} \le C_p,$$

respectively

$$(B_p) \qquad \left\{\sigma(B)^{-1}\int_B w^p\, d\sigma\right\}^{1/p} \leq C_p\left\{\sigma(B)^{-1}\int_B w\, d\sigma\right\},$$

for some constants $C_p > 0$ independent of the surface balls B. For brevity, let

$$A_\infty = \bigcup_{p>1} A_p, \qquad B_\infty = \bigcup_{p>1} B_p,$$

$$A_* = \bigcap_{p>1} A_p, \qquad B_* = \bigcap_{p>1} B_p.$$

As is well known, these conditions play a key role for the validity of weighted L^p estimates of maximal operators, singular integrals, and so forth. Now, a basic difference between C^1 and Litschitz domains is as follows:

> If D is Lipschitz, then $K = (d\omega/d\sigma) \in \{A_\infty(d\sigma) \cap B_2(d\sigma)\}$ (see [3, 5]);
>
> If D is C^1, then K and K^{-1} are in $A_*(d\sigma) \cap B_*(d\sigma)$ (cf. [8, Sec. 1]).

This difference reflects the fact that on the boundary of a C^1 domain we can use the L^p theory of *Cauchy-Calderón singular integrals* [1, 6], while this theory is still unavailable for Lipschitz domains. In 1981, Coifman, McIntosh, and Meyer (in press) proved the boundedness of these operators also in this case. As a result, the H^1-BMO theory of harmonic functions has been developed more fully for C^1 domains.

From now on, D shall always denote a C^1 domain in $\mathbf{R}^{n+1}$. Here, Fabes and Kenig [7] solved the Neumann problem

$$\Delta U = 0 \text{ in } D,\ \partial U/\partial N_Q = g \in h^1_\sigma(d\sigma) \text{ on } \partial D,$$

by means of single-layer potentials and found solutions U with gradients satisfying $N(|\nabla U|) \in L^1(d\sigma)$. In other words, the atomic space $h^1_\sigma(d\sigma)$ is contained in a class of Neumann data:

$$H^1_{\mathcal{N}}(\partial D, d\sigma) = \left\{f \in L^1(d\sigma),\ \int f\, d\sigma = 0,\ f = \partial U/\partial N_Q \right.$$
$$\left. \text{for some harmonic } U \text{ in } D \text{ with } N(|\nabla U|) \in L^1(d\sigma)\right\}.$$

Moreover, they proved the duality

$$[H^1_{\mathscr{N}}(\partial D,\ d\sigma)]^* = \mathrm{BMO}_\sigma(d\sigma) \tag{5}$$

along with the atomic representation

$$H^1_{\mathscr{N}}(\partial D,\ d\sigma) = h^1_\sigma(d\sigma). \tag{5'}$$

Since $K \in A_\infty(d\sigma)$ implies that $\mathrm{BMO}_\omega(d\omega) = \mathrm{BMO}_\sigma(d\sigma)$, comparing (4) and (5) we observe the existence of two distinct preduals of $\mathrm{BMO}(\partial D)$ for C^1 domains: the first being associated to a Dirichlet problem, the second to a Neumann problem.

3. WEIGHTED H^1-BMO DUALITIES

With $\mu = \sigma$ and $d\nu = d\omega$ on ∂D, we define $\mathrm{BMO}_\sigma(d\omega)$ as in Sec. 1 and let $h^1_\sigma(d\omega)$ denote the corresponding atomic predual. Thus, for any atom $a \in h^1_\sigma(d\omega)$, we have

$$\int a\ d\omega = 0,\ \mathrm{supp}(a) \subset B \text{ and } \|a\|_\infty \le \sigma(B)^{-1}.$$

Let us solve the Dirichlet problem

$$\Delta u = 0 \text{ in } D, \qquad \text{nontang. } \lim u = a \text{ on } \partial D$$

for an atom a in $h^1_\sigma(d\omega)$, by means of the Poisson integral $u = \mathscr{P}a$. That is, for each $X \in D$ and general f on ∂D,

$$[\mathscr{P}f](X) = \int_{\partial D} f\ d\omega^X = \int_{\partial D} f(y)k(X,\ y)d\omega(y),$$

where $k(X,\ y)$ is the *kernel function* for D normalized at $X = P_0$.

Now, letting

$$H^1_{\mathscr{D}}(D,\ d\sigma) = \{u : \Delta u = 0 \text{ in } D,\ u(P_0) = 0, \text{ and } (Nu) \in L^1(d\sigma)\},$$

we have the duality

$$[H^1_{\mathscr{D}}(D,\ d\sigma)]^* = \mathrm{BMO}_\sigma(d\omega) \tag{6}$$

and the atomic representation

$$H^1_{\mathscr{D}}(\partial D,\ d\sigma) = h^1_\sigma(d\omega) \tag{6'}$$

(see [10]). We note that (6) was announced earlier by Dahlberg [4], but his proofs have been unavailable to us.

An important step in the proof of (6) is given by the following:

Lemma. *There exists a geometric constant* $C > 0$ *such that*

$$\| N(\mathscr{P}a)\|_{L^1(d\sigma)} \le C, \text{ for all atoms } a \in h^1_\sigma(d\omega). \tag{*}$$

Proof. Let $a \in h^1_\sigma(d\omega)$ be an atom and fix a surface ball $B = B(x_0,\ r) \supset \operatorname{supp}(a)$. By a result of Jerison and Kenig [12, Theorem 7.1], there exists a constant $M > 1$ such that for all $j \in \mathscr{N} = \{1,\ 2,\ 3,\ \ldots\}$, all x_0 and x_1 in ∂D and all $X \in D$, if $|X - x_0| > M^j|x_1 - x_0|$, then

$$|k(X,\ x_1) - k(X,\ x_0)| \le M(1 - M^{-1})^j k(X,\ x_0). \tag{7}$$

Consider the M-adic dilations of B, that is, the surface balls $B_j = B(x_0,\ M^j r)$, $j \in \mathscr{N}$. On B_1, letting $d\sigma = K^{-1}\ d\omega$ and using Schwarz's inequality, we see that with $u = \mathscr{P}a$,

$$\begin{aligned}\int_{B_1} |Nu|\,d\sigma &\le \left\{\int_{B_1} (Nu)^2\ d\omega\right\}^{1/2}\left\{\int_{B_1} K^{-2}\ d\omega\right\}^{1/2}\\ &\le C_2\|Nu\|_{L^2(d\omega)}\,\sigma(B_1)^{1/2}[\sigma(B_1)/\omega(B_1)]^{1/2},\end{aligned}$$

since $K \in A_2(d\sigma)$. Hence, by the $L^2(d\omega)$ boundedness of $a \to Nu$ and the definition of a, it follows that

$$\int_{B_1} |Nu|\,d\sigma \le C_2'\left\{[\omega(B)/\sigma(B)]\,[\sigma(B_1)/\omega(B_1)]\right\}^{1/2} \le C.$$

On $\partial D\backslash B_1$ we apply (7), noting that $k(X,\ x_0)\int a\ d\omega = 0$. Thus, for any $x \in \partial D\backslash B_j$ and any $X \in \Gamma_x$ (truncated inner cone with vertex at x), we have

$$|u(X)| \le \int_B |k(X,\ y) - k(X,\ x_0)|\,|a(y)|\,d\omega(y)$$

(continued)

$$\le M(1-M^{-1})^j \int_B k(X, x_0)\sigma(B)^{-1}\, d\omega(y)$$

$$\le M(1-M^{-1})^j \sigma(B)^{-1}\omega(B)\omega(B_j)^{-1}.$$

Hence, with $\varepsilon = M^{-1}$, we see that for all $x \in \partial D\backslash B_j$,

$$[Nu](x) \le \varepsilon^{-1}(1-\varepsilon)^j \sigma(B)^{-1}\omega(B)/\omega(B_j). \tag{8}$$

Since D is a C^1 domain, $(d\sigma/d\omega) \in A_p(d\omega)$ for all $1 < p < \infty$. So a simple argument in [8, Lemma 2.6] shows that for any $\eta > 0$ there exists a constant $C_\eta > 0$ such that

$$[\omega(B)/\omega(B_j)][\sigma(B_j)/\sigma(B)] \le C_\eta M^{\eta j}.$$

Therefore,

$$\int_{\partial D\backslash B_1} |Nu|\, d\sigma \le \varepsilon^{-1}C_\eta \sum_{j\ge 2}[(1-\varepsilon)M^\eta]^j < \infty$$

if we fix $\eta > 0$ so small that $(1-\varepsilon)M^\eta < 1$. The proof is complete.

Alternatively, looking at the atomic space $h^1_\omega(d\sigma)$, we can also solve the Neumann problem by means of single-layer potentials in analogy to [7]. This approach leads to a duality of the form

$$\mathrm{BMO}_\omega(d\sigma) = [H^1_{\mathcal{N}}(\partial D, d\omega)]^* \tag{9}$$

and a corresponding atomic representation [9].

Let me close by mentioning an interesting open problem. The new results of Jerison and Kenig [13] suggest the possibility of solving the Neumann problem with L^p data, $1 < p < 2$, for a *Lipschitz* domain D. By [13], since $h^1_\sigma(d\sigma) \subset L^2(d\sigma)$, we know that a solution U of

$$\Delta U = 0 \text{ in } D, \qquad \partial U/\partial N_Q = a \in h^1_\sigma(d\sigma),$$

always exists. If an H^1-BMO duality of this type still holds, one would like to prove that, for some absolute constant $C > 0$,

$$\| N(|\nabla U|)\|_{L^1(d\sigma)} \le C$$

whenever a is an atom of type (σ, σ). At present, all this is unknown.

REFERENCES

1. Calderón, A. P. Cauchy integrals on Lipschitz curves and related operators. *Proc. Nat. Acad. Sci.* 74 (1977):1324-1327.

2. Coifman, R. R., and Weiss, G. Extensions of Hardy spaces and their use in analysis. *Bull. Amer. Math. Soc.* 83 (1977):569-645.

3. Dahlberg, B. E. J. Estimates of harmonic measure. *Archiv. for Rat. Mech. and Anal.* 65 (1977):272-288.

4. Dahlberg, B. E. J. A note on H^1 and BMO, "A tribute to Ake Pleijel." Uppsala University, 1980.

5. Dahlberg, B. E. J. On the Poisson integral for Lipschitz and C^1 domains. *Studia Math.* 66 (1979):13-24.

6. Fabes, E.; Jodeit, M.; and Rivière, N. Potential techniques for boundary value problems on C^1 domains. *Acta Math.* 141 (1978):165-186.

7. Fabes, E., and Kenig, C. On the Hardy space H^1 of a C^1 domain. *Arkiv. för Mat.* To appear.

8. Fabes, E.; Kenig, C.; and Neri, U. Carleson measures, H^1 duality and weighted BMO in nonsmooth domains. *Indiana Univ. Math. J.* 30 (1981):547-581.

9. Fabes, E.; Kenig, C.; and Neri, U. Neumann problem in weighted H^1 and preduals of BMV. Preprint, Univ. of Maryland, Feb. 1982.

10. Fabes, E.; Kenig, C.; and Neri, U. Weighted H_D^1-BMO dualities in C^1 domains. Preprint, Univ. of Maryland, July 1981.

11. Hunt, R. A., and Wheeden, R. L. On the boundary values of harmonic functions. *Trans. Amer. Math. Soc.* 132 (1968):307-322.

12. Jerison, D. S., and Kenig, C. E. Boundary behavior of harmonic functions in nontangentially accessible domains. *Advances in Math.* To appear.

13. Jerison, D. S., and Kenig, C. E. The Neumann problem on Lipschitz domains. *Bull. Amer. Math. Soc.* (N.S.) 4 (1981):203-207.

14. Muckenhoupt, B., and Wheeden, R. On the dual of weighted H^1 of the half-space. *Studia Math*. 63 (1978):57-79.

15. Neri, U. On functions with bounded mixed variance. *Studia Math*. To appear in vol. 75 (1982).

THE SHILOV AND BISHOP DECOMPOSITIONS OF $H^{\infty} + C$

Donald Sarason
University of California, Berkeley

1. INTRODUCTION

A compact Hausdorff space X admits several natural decompositions relative to any closed unital subalgebra A of $C(X)$. Two of these decompositions reflect, in some sense, the extent to which A is self-adjoint. The sets in the coarser of the two decompositions, the Shilov decomposition [12], are the maximal level sets of the largest C^*-subalgebra of A. Those in the finer of the decompositions, the Bishop decomposition [2, 5], are the maximal sets of antisymmetry of A. (A subset S of X is a set of antisymmetry of A if any function in A that is real valued on S must be constant on S.)

This paper is concerned with the case where $X = M(L^{\infty})$ and $A = H^{\infty} + C$. By L^{∞} is meant the C^*-algebra of essentially bounded, complex-valued, measurable functions with respect to Lebesgue measure on ∂D, the unit circle in the complex plane. This algebra will be identified, in the usual manner, with the algebra of continuous functions on $M(L^{\infty})$, its Gelfand space (space of multiplicative linear functionals). By H^{∞} is meant the space of boundary functions for bounded holomorphic functions in D, the unit disc, and by C is meant $C(\partial D)$. The largest C^*-subalgebra of $H^{\infty} + C$ will be denoted, as usual, by QC, and the sets in the Shilov decomposition of $M(L^\infty)$ associated with $H^\infty + C$ will be called QC-level sets. The subsets of $M(L^\infty)$ that are sets of antisymmetry for $H^\infty + C$ will be called simply sets of antisymmetry.

The following results will be established.

1. The Bishop decomposition of $M(L^\infty)$ associated with $H^\infty + C$ is strictly finer than the Shilov decomposition.

2. If f and g are functions in L^∞ and if, for each set of antisymmetry S, either $f|S$ or $g|S$ is in $H^\infty|S$, then the same conclusion holds for each QC-level set.

The first result will be proved in Sec. 2, where a function in H^∞ will be constructed whose imaginary part is continuous at the point $z = 1$ but whose restriction to $M_1(L^\infty)$, the fiber of $M(L^\infty)$ above the point 1, is not the restriction of any function in QC. Such a function is constant on each set of antisymmetry contained in $M_1(L^\infty)$, but it cannot be constant on every QC-level set contained in $M_1(L^\infty)$ (by the Stone-Weierstrass theorem).

The second result seems to say that even though the Shilov and Bishop decompositions of $M(L^\infty)$ due to $H^\infty + C$ are different, they are nevertheless somehow closely related. It would be nice to have a geometric picture of $M(H^\infty)$ that illuminates this relation, but that awaits further study. An example in Sec. 6 shows that the second result is not merely an instance of a general fact about uniform algebras.

The second result is closely related to and leads to a refinement of results on Toeplitz operators from [1]. Its proof depends on some of the latter results and on a theorem of T. Wolff. In Sec. 3 a preliminary lemma involving a duality relation is established. Sec. 4 contains what amounts to a concrete version of the second result, a theorem on the separation of the zeros of two

Blaschke products satisfying a certain disjointness condition. The second result will be derived from the separation theorem in Sec. 5.

T. Wolff [14] has found a more direct proof of the separation theorem using methods of J. Garnett and P. Jones. As will be explained in Sec. 5, this provides a new proof of the main result of [1]. That the separation theorem might be true was first suggested by J. Garnett.

Actually, the second result will be proved in a slightly stronger form than is stated above. It is known that each functional in $M(H^\infty + C)$ [$= M(H^\infty) - D$] has a unique representing measure on $M(L^\infty)$ [6, p. 182]. The support of any such representing measure will be called a support set. Support sets are easily seen to be sets of antisymmetry (although the precise relation between support sets and sets of antisymmetry for $H^\infty + C$ remains mysterious). The version of the second result to be established involves support sets instead of sets of antisymmetry.

Although the questions addressed in this paper are motivated in part by the general theory of function algebras, their resolution is largely a matter of classical analysis. It is hoped that the paper is therefore not out of place in a volume dedicated to Professor Zygmund.

The author is grateful to Sheldon Axler, Donald Marshall, and Thomas Wolff for their help.

In what follows, functions in L^∞ will be routinely identified with their harmonic extensions into D as well as with their Gelfand transforms.

2. PROOF OF RESULT 1

For c a point of ∂D, we let $M_c(L^\infty)$ denote the fiber of $M(L^\infty)$ above c. For f in L^∞, we let $\mathrm{dist}_c(f, QC)$ denote the distance of

$f|M_c(L^\infty)$ from $QC|M_c(L^\infty)$. As explained in Sec. 1, to establish the first result, it will be enough to construct a function f in H^∞ such that Im f is continuous at 1 but $\text{dist}_1(f, QC) > 0$. Since $\text{dist}_c(f, QC)$ is an upper semicontinuous function of c, we can guarantee the last condition by making sure $\text{dist}_c(f, QC)$ is bounded away from 0 along a sequence of points c tending to 1. The construction that follows is a substantial simplification due to D. Marshall of the author's original one (which involved Blaschke products).

Choose a sequence (c_n) of distinct points on ∂D converging to 1 and a sequence (r_n) of positive numbers tending to 0. Let u be any real-valued function on ∂D that is continuous except at the points of the sequence (c_n) and that has one-sided limits at c_n equal to r_n and $-r_n$. We can, for example, take u to be linear between successive points of the sequence (c_n). In particular, u is to be continuous at 1; it is automatic that $u(1) = 0$.

Let

$$h = e^{u+i\tilde{u}} + e^{-(u+i\tilde{u})},$$

where $\tilde{u}$ denotes the conjugate function of u. Then h is in H^∞, and the continuity of Im $h = (e^u - e^{-u})\sin\tilde{u}$ at 1 is clear. It only remains to show that $\text{dist}_{c_n}(h, QC)$ is bounded away from 0.

To simplify the notation, let c denote any one of the points c_n and r the corresponding r_n. As $z \to c$ radially, $u(z)$ tends to 0, and $\tilde{u}(z)$ tends either to $+\infty$ or to $-\infty$ [16, pp. 98, 108]. Hence $h(z) \to 0$ along some sequence tending to c. Let x be a cluster point in $M_c(H^\infty)$ of such a sequence, and let m be the representing measure of x on $M(L^\infty)$. The values of h on $M_c(L^\infty)$, and hence on supp m, lie in the ellipse E that is the image of the circle $|z| = e^r$ under the map $z \to z + 1/z$. If h(supp m) were not all of E, there would be a polynomial p having positive real part on h(supp m) but vanishing at 0. This would lead to the contradiction

$$0 = \text{Re}\; p(h(x)) = \int \text{Re}\; p(h)\,dm > 0,$$

so we can conclude that $h(\text{supp } m) = E$. The diameter of E exceeds 4, so, as every QC function is constant on supp m, we must have $\text{dist}_e(h, QC) \geqq 2$. This completes the proof.

3. A LEMMA

By BMO and VMO we mean the spaces of functions of bounded mean oscillation and vanishing mean oscillation on ∂D [8, 11]. Their relevance here comes through the equality $QC = \text{VMO} \cap L^\infty$. The subspaces of BMO and VMO consisting of the functions they contain whose harmonic extensions are holomorphic will be denoted by BMOA and VMOA.

It is a consequence of Fefferman's theorem that BMOA is the dual of the Hardy space H^1 under the pairing

$$\langle f, g \rangle = \frac{1}{2\pi} \int_{-\pi}^{\pi} f\bar{g}\, d\theta, \qquad f \in H^1, \qquad g \in \text{BMOA},$$

the integral on the right existing, for example, if one of the factors in the integrand is bounded (but not in general). Similarly, H^1 is the dual of VMOA under the analogous pairing. (For definiteness, we give H^1 its usual norm and BMOA the dual norm.)

Lemma. If b is a Blaschke product, the natural map of the quotient space VMOA/(bVMOA $\cap$ VMO) *into the quotient space* BMOA/(bBMOA $\cap$ BMO) *is an isometry.*

The following observation is needed: if the Blaschke product b divides the VMOA function f, then f/b is in VMOA. This follows from the characterization of VMOA as the class of functions f in H^2 that satisfy

$$\lim_{|a| \to 1} \left[\frac{1}{2\pi} \int_{-\pi}^{\pi} |f|^2 P_a \, d\theta - |f(a)|^2 \right] = 0$$

(P_a = Poisson kernel at a). It implies that the map in the above lemma is one-to-one. Similar reasoning establishes the analogous property of BMOA.

We can establish the lemma by showing that BMOA/(bBMOA ∩ BMO) is the second dual of VMOA/(bVMOA ∩ VMO) and that the natural map of the second space into the first one is the canonical embedding. Since the dual of VMOA/(bVMOA ∩ VMO) is the annihilator of bVMOA ∩ VMO in H^1, and BMOA/(bBMOA ∩ BMO) is the dual of the annihilator of bBMOA ∩ BMO in H^1, it will suffice to show that bVMOA ∩ VMO and bBMOA ∩ BMO have the same annihilator in H^1. We shall show that the annihilator of each is $(bH^\infty)^\perp$, the annihilator of bH^∞ in H^1.

Let $QA = QC \cap H^\infty$. The following result of T. Wolff [15] is needed.

Theorem 1. Any function in L^∞ can be multiplied into QC by an outer function in QA.

To see that the annihilator of bVMOA ∩ VMO in H^1 is contained in $(bH^\infty)^\perp$, consider any function g in H^1 that annihilates bVMOA ∩ VMO and any function f in H^∞. By Theorem 1, there is an outer function h in QA such that hbf is in QA. Then hf is in QA (by the observation following the statement of the lemma), so g annihilates hbf. Similarly, g annihilates $phbf$ for any polynomial p, and it follows by Beurling's theorem [8, p. 110] that g annihilates bf, as desired.

To complete the proof of the lemma we need only to show that $(bH^\infty)^\perp$ annihilates bBMOA ∩ BMO. For this, let b_n denote the nth partial product of b. Then $(b_nH^\infty)^\perp$ consists of bounded functions. As $(bH^\infty)^\perp$ is the norm closure of $\cup\,(b_nH^\infty)^\perp$ (by a standard duality argument), it is spanned by bounded functions. As any bounded function in $(bH^\infty)^\perp$ obviously annihilates bBMOA, the desired conclusion follows.

4. A SEPARATION THEOREM

The relevant facts about Toeplitz operators will now be stated; general background can be found in [10]. For f in L^∞, the Toeplitz operator on H^2 with symbol f will be denoted by T_f, and the closed subalgebra of L^∞ generated by H^∞ and f will be denoted by $H^\infty[f]$.

Theorem 2. For f and g in L^∞, the following conditions are equivalent:

(i) $T_{\bar f}T_g - T_{\bar f g}$ *is compact;*

(ii) $H^\infty[f] \cap H^\infty[g] \subset H^\infty + C$;

(iii) *For every support set S, either $f|S$ or $g|S$ is in $H^\infty|S$.*

Theorem 3. Let φ and ψ be Blaschke products with disjoint zero sequences (z_n) and (w_n). The following conditions are equivalent:

(i) $T_\varphi T_{\bar\psi} - T_{\varphi\bar\psi}$ *is compact;*

(ii) $\lim_{|z|\to 1} \max(|\varphi(z)|, |\psi(z)|) = 1$;

(iii) *For any positive number ε, there is a function h in H^∞ satisfying $\|h\|_\infty < 1 + \varepsilon$, $h(z_n) = 1$ except for finitely many n, and $h(w_n) = 0$ except for finitely many n.*

The implication from condition (i) of Theorem 2 to condition (ii) is due to A. L. Volberg [13]; the other implications can be found in [1].

The separation theorem states that the zeros of two Blaschke products satisfying the conditions of Theorem 3 can be separated by a function in QC.

Theorem 4. If φ and ψ satisfy the conditions of Theorem 3, there is a function u in QC such that $0 \leq u \leq 1$, $u(z_n) \to 1$, and $u(w_n) \to 0$.

The proof will be accomplished in three steps. To simplify slightly the explanation that follows, it will be assumed that φ and ψ have simple zeros; only a trivial rephrasing is needed to handle the general case.

Step 1. For any positive number δ, there is a function f in BMOA such that $f(z_n) = 1$ for all n, $f(w_n) = 0$ for all n, and $\text{dist}(f, \text{VMOA}) < \delta$.

To establish this we use condition (iii) of Theorem 3. Given $\varepsilon > 0$, there is, by that condition, a function h in H^∞ and a positive integer N such that $\|h\|_\infty \leq 1$, $h(w_n) = 0$ for $n > N$, and $h(z_n) = 1 - \varepsilon$ for $n > N$. Let $h_1 = \log(1 - h)/\log \varepsilon$. Then $h_1(z_n) = 1$ for $n > N$, $h_1(w_n) = 0$ for $n > N$, and h_1 is in BMOA, since its imaginary part is bounded. In fact, since

$$\| \text{Im } h_1 \|_\infty \leq \pi \Big/ \left(2 \log \frac{1}{\varepsilon}\right),$$

we have

$$\| h_1 - h_1(0) \|_{\text{BMO}} \leq \text{const.}/\log \frac{1}{\varepsilon}$$

for some absolute constant (proportional to the norm of the conjugation operator on BMO). Thus, if we choose ε sufficiently small, we get $\| h_1 - h_1(0) \|_{\text{BMO}} < \delta$. By Theorem 1, the Blaschke product with zero set $\{z_n, w_n : n > N\}$ can be multiplied into QA by an outer function. That yields a function h_2 in QA that vanishes at z_n and w_n for $n > N$ but is nonzero at z_n and w_n for $n \leq N$. Multiplying h_2 by a suitable polynomial, we can make it satisfy $h_2(z_n) = 1 - h_1(z_n)$ and $h_2(w_n) = -h_1(w_n)$ for $n \leq N$. The function $f = h_1 + h_2$ then has the required properties.

Step 2. There is a function g in VMOA such that $g(z_n) = 1$ for all n and $g(w_n) = 0$ for all n.

To see this, we take any f in BMOA such that $f(z_n) = 1$ for all n and $f(w_n) = 0$ for all n; step 1 guarantees that there is such an f. Let $b = \varphi\psi$. By step 1, the coset of f in BMOA/(bBMOA $\cap$ BMO) has distance 0 from the canonical image of VMOA/(bVMOA $\cap$ VMO). By the lemma of Sec. 3, that canonical image is closed, so the coset of f contains a function in VMO, which is the desired conclusion.

Step 3. The function u of the theorem exists.

It now follows by standard reasoning that if g is the function of step 2, then the function

$$u = \max(\min(\operatorname{Re} g, 1), 0)$$

has all of the required properties.

5. PROOF OF RESULT 2

As mentioned in Sec. 1, the result to be established is slightly stronger than the one stated there.

Theorem 5. *Let f and g be functions in L^∞ such that, for each support set S, either $f|S$ or $g|S$ is in $H^\infty|S$. Then, for each QC-level set S, either $f|S$ or $g|S$ is in $H^\infty|S$.*

We consider first a special case. Suppose φ and ψ are interpolating Blaschke products such that, for each support set S, either $\overline{\varphi}|S$ or $\overline{\psi}|S$ is in $H^\infty|S$. Then, by virtue of the unimodularity of φ and ψ on $M(L^\infty)$, one sees easily that $\max(|\varphi|, |\psi|)$ is identically 1 on $M(H^\infty + C)$, which implies condition (ii) of Theorem 3. The separation theorem therefore applies to φ and ψ. Because φ and ψ are interpolating Blaschke products, the zero set of each in $M(H^\infty)$ is the closure of its zero set in D. Thus, the separation theorem implies that no QC-level set in $M(H^\infty + C)$ contains a zero of both φ and ψ. Suppose S is a QC-level set in $M(L^\infty)$ and S' is the corresponding one in $M(H^\infty + C)$. Standard reasoning shows that

$S' = M(H^\infty|S)$, so at least one of $\varphi|S$ and $\psi|S$ is invertible in $H^\infty|S$; in other words, either $\overline{\varphi}|S$ or $\overline{\psi}|S$ is in $H^\infty|S$. This establishes the theorem for the special case under consideration. (Incidentally, for this case, one of $\varphi|S$ and $\psi|S$ is constant for any QC-level set S [3, Prop. 4].)

Now let f and g be any functions satisfying the hypotheses of the theorem. If φ and ψ are interpolating Blaschke products that are invertible in $H^\infty[f]$ and $H^\infty[g]$, respectively, then the preceding argument applies to φ and ψ, so that for each QC-level set S, either $\overline{\varphi}|S$ or $\overline{\psi}|S$ is in $H^\infty|S$. Hence, for each QC-level set S, either $\overline{\varphi}|S$ is in $H^\infty|S$ for every interpolating Blaschke product φ that is invertible in $H^\infty[f]$, or $\overline{\psi}|S$ is in $H^\infty|S$ for every interpolating Blaschke product ψ that is invertible in $H^\infty[g]$. By the Chang-Marshall theorem [9], $H^\infty[f]$ and $H^\infty[g]$ are generated, as closed subalgebras of L^∞, by H^∞ and the complex conjugates of interpolating Blaschke products. The desired conclusion about f and g is now immediate.

By virtue of what has just been proved, the following condition can be added to the equivalent conditions (i)-(iii) of Theorem 2:

(iv) *For every QC-level set S, either $f|S$ or $g|S$ is in $H^\infty|S$.*

It is quite simple to show that this condition implies the compactness of $T_{\bar{f}}T_g - T_{\bar{f}g}$; one can use either a straightforward adaptation of the argument in [11] or the localization technique of R. G. Douglas [4].

Wolff's proof of the separation theorem mentioned in Sec. 1 is a direct (albeit complicated) analytic argument that starts from condition (ii) of Theorem 3. Coupled with the argument just given, it shows that condition (iv) follows from condition (iii) of Theorem 2 without relying on the proof of the most difficult implication of that theorem, the implication from (iii) to (i). It thus provides a replacement for the hardest part of the proof of Theorem 2.

P. Jones [7] has proved a Corona-type theorem that has as a corollary the implication from condition (ii) to condition (iii) of Theorem 3. As explained in [1], this provides yet another proof of the difficult implication in Theorem 2. There are thus three quite distinct proofs of that implication, all, by the way, of comparable difficulty.

The author has been unable to decide whether the theorem proved in this section extends from two to three functions. Another problem of interest is that of finding a concrete structural condition on the pair f, g that is equivalent to the conditions of Theorem 2.

6. AN EXAMPLE

To show that the theorem in Sec. 5 does not generalize to arbitrary uniform algebras, we sketch here a simple example of a uniform algebra A on a compact Hausdorff space X and two functions, f and g, in $\mathbf{C}(X)$, with the following properties: (1) on each maximal set of antisymmetry S of A, either $f|S$ or $g|S$ is in $A|S$; (2) there is a level set S_0 of $A \cap \overline{A}$ such that neither $f|S_0$ nor $g|S_0$ belongs to $A|S_0$. The example is based on the standard one of an algebra whose Shilov and Bishop decompositions are distinct. We produce X by sewing together a square and a cylinder along an edge of the square and the axis of the cylinder. Namely, let X be the set of points (z_1, z_2) in $\mathbf{C}^2$ such that either $0 \leqq \text{Re } z_1 \leqq 1$, $0 \leqq \text{Im } z_1 \leqq 1$, $z_2 = 0$, or $0 \leqq \text{Re } z_1 \leqq 1$, $\text{Im } z_1 = 0$, $|z_2| \leqq 1$. Our square is

$$Q = \{(z_1, z_2) \in X : z_2 = 0\}$$

and our cylinder is

$$S_0 = \{(z_1, z_2) \in X : \text{Im } z_1 = 0\}.$$

For $n = 1, 2, \ldots,$ let

$$R_n = \{(z_1, 0) \in Q : 0 < \text{Re } z_1 < 1,\ 2^{-2n} < \text{Im } z_1 < 2^{-2n+1}\},$$

and for $0 \leq t \leq 1$, let

$$D_t = \{(t, z_2) \in S_0 : |z_2| < 1\}.$$

Let A be the algebra of functions in $C(X)$ that are holomorphic with respect to z_1 on each R_n and with respect to z_2 on each D_t. (The algebra $A|Q$ is the standard example referred to above.)

The level sets of $A \cap \overline{A}$ are $\overline{R}_n$ $(n = 1, 2, \ldots)$, S_0, and the remaining singletons. The maximal sets of antisymmetry of A are $\overline{R}_n$ $(n = 1, 2, \ldots)$, $\overline{D}_t$ $(0 \leq t \leq 1)$, and the remaining singletons.

Let the functions f and g in $C(X)$ be defined by

$$f(z_1, z_2) = \left(\frac{1}{2} - \text{Re } z_1\right)\overline{z}_2,\quad 0 \leq \text{Re } z_1 \leq \frac{1}{2},\ \text{Im } z_1 = 0$$
$$= 0,\ \text{otherwise},$$
$$g(z_1, z_2) = \left(\text{Re } z_1 - \frac{1}{2}\right)\overline{z}_2,\quad \frac{1}{2} \leq \text{Re } z_1 \leq 1,\ \text{Im } z_1 = 0$$
$$= 0,\ \text{otherwise}.$$

On each set of antisymmetry of A, one of these functions coincides with a function in A, namely, the zero function. However, neither f nor g coincides with a function in A on the level set S_0.

In this example, distinct nontrivial Gleason parts of A have disjoint closures. It appears likely that the geometric reason behind the theorem of Sec. 5 is some sort of intertwining of distinct Gleason parts of $H^\infty + C$.

In the example just given, $X = M(A)$, but a small modification produces an example in which the underlying space is the Shilov boundary. Namely, the Shilov boundary of A is the set Y consisting of all points of Q not belonging to any R_n and all points of S_0 not belonging to any D_t. If $B = A|Y$, then the level sets of $B \cap \overline{B}$ are the intersections with Y of the level sets of $A \cap \overline{A}$, and the maximal sets of antisymmetry of B are similarly related to those of A. Thus, on each set of antisymmetry of B, one of the two functions $f|Y$ and $g|Y$ coincides with a function in B, but neither $f|Y$ nor $g|Y$ coincides with a function in B on the level set $S_0 \cap Y$.

REFERENCES

1. Axler, S.; Chang, S.-Y. A.; and Sarason, D. Products of Toeplitz operators. *Integral Equations and Operator Theory* 1 (1978):285-309.

2. Bishop, E. A generalization of the Stone-Weierstrass theorem. *Pac. J. Math.* 11 (1961):777-783.

3. Clancey, K. F., and Gosselin, J. A. The local theory of Toeplitz operators. *Illinois J. Math.* 22 (1978):449-458.

4. Douglas, R. G. Banach algebra techniques in the theory of Toeplitz operators. *Regional Conference Series in Mathematics*, no. 15, Amer. Math. Soc., Providence, R.I., 1973.

5. Glicksberg, I. Measures orthogonal to algebras and sets of antisymmetry. *Trans. Amer. Math. Soc.* 105 (1962):415-435.

6. Hoffman, K. *Banach Spaces of Analytic Functions*. Englewood Cliffs, N.J.: Prentice-Hall, 1962.

7. Jones, P. W. Estimates for the Corona problem. *J. Func. Anal.* To appear.

8. Koosis, P. *Introduction to H_p Spaces*. New York: Cambridge Univ. Press, 1980.

9. Marshall, D. E. Subalgebras of L^∞ containing H^∞. *Acta Math.* 137 (1976):91-98.

10. Sarason, D. *Function Theory on the Unit Circle*. Blacksburg, Va.: Virginia Polytechnic Inst. and State Univ., 1978.

11. Sarason, D. On products of Toeplitz operators. *Acta Sci. Math.* (Szeged) 35 (1973):7-12.

12. Shilov, G. E. On rings of functions with uniform convergence. *Ukrain. Mat. Z.* 3 (1951):404-411.

13. Volberg, A. L. Two remarks concerning the theorem of S. Axler, S.-Y. A. Chang and D. Sarason. *J. Operator Theory*. To appear.

14. Wolff, T. Some theorems on vanishing mean oscillation. Doctoral Dissertation, Univ. of California, Berkeley, 1979.

15. Wolff, T. Two algebras of bounded functions. To appear.

16. Zygmund, A. *Trigonometric Series*, vol. I. New York: Cambridge Univ. Press, 1959.

A MODIFIED FRANKLIN SYSTEM AND HIGHER-ORDER SPLINE SYSTEMS ON R^n AS UNCONDITIONAL BASES FOR HARDY SPACES

Jan-Olov Strömberg
University of Stockholm
Princeton University

1. INTRODUCTION

Let $H^p(I)$ be the subspace of distributions in the Hardy space $H^p(\mathbf{R})$ that are supported in $I = [0, 1]$. The Franklin system ($m = 0$) and higher-order spline system ($m > 0$) have been studied and used as unconditional basis for $H^p(I)$, $p > 1/(m + 2)$ in [1, 4, 7, and 9-11]. The existence of an unconditional basis for H^1 was first shown by B. Maurey [8] and an explicit construction was made by L. Carleson [3]. The spline system as unconditional basis for the bi-Hardy space $H^1(I \times I)$ has been studied by S. Y. A. Chang.

The purpose of this paper is to define m-order spline systems on $\mathbf{R}$ and especially get a modified Franklin system ($m = 0$). We will also get spline systems on $\mathbf{R}^n$, $n > 1$. All these systems can also be made periodical to become spline systems on $\mathbf{T}^n$ by a simple summation. (The spline systems on I do not extend periodically to the circle $\mathbf{T} = \mathbf{R}/\mathbf{Z}$ and cannot be used as unconditional basis for the real Hardy spaces $H^p(\mathbf{T})$, $0 < p \leq 1$, without modification.)

We are especially interested to see how these spline systems on $\mathbf{R}^n$ can be used as unconditional basis for $H^p(\mathbf{R}^n)$, $n \geq 1$.

2. CONSTRUCTION OF TWO FUNCTIONS

Let $m \geq 0$. We will construct one function on **R** from which we get the m-order spline system on **R** and another function on **R** that we will use together with the first one to get the m-order spline system on $\mathbf{R}^n$.

Let $A_0 = \mathbf{Z}_+ \cup \{0\} \cup \frac{1}{2}\mathbf{Z}_-$ and $A_1 = A_0 \cup \left\{\frac{1}{2}\right\}$. A_0 splits **R** into intervals $\{I_\sigma\}_{\sigma \in A_0}$ (σ = left endpoint of I_σ). Let S_0^m be the subspace of functions f in $L^2(\mathbf{R})$ such that $f \in C^m(\mathbf{R})$ and is a real polynomial of degree $\leq m+1$ on each I_σ, $\sigma \in A_0$. Let S_1^m be the corresponding subspace of $L^2(\mathbf{R})$ with the set A_0 replaced by A_1. Note that $S_1^m = \{f;\ f(\cdot + 1) \in S_0^m\}$ and also that the set A_1 splits **R** into the same collection of intervals as A_0 did except for the interval $I_0 = [0, 1]$, which is split into two intervals $\left[0, \frac{1}{2}\right]$ and $\left[\frac{1}{2}, 1\right]$. It follows that S_0^m has codimension 1 in S_1^m, since the $m+1$ order derivative of a function f in S_1^m may have a jump at $\frac{1}{2}$ and if not, it is contained in S_0^m. Thus there is a function τ in $L^2(\mathbf{R})$ that is a uniquely defined modulo sign by

(i) $\tau \in S_1^m$,

(ii) $\tau \perp S_0^m$, that is, $\int_{\mathbf{R}} \tau(x) f(x)\,dx = 0$ for all $f \in S_0^m$,

(iii) $\|\tau\|_{L^2} = 1$.

Next let $\tilde{S}_0^m$ be the subspace of functions f in S_0^m that are supported in $[0, \infty)$ and let $\tilde{S}_1^m$ be the subspace of functions f in S_1^m that are supported in $[1, \infty)$. We observe that $\tilde{S}_1^m$ is a subspace of $\tilde{S}_0^m$ with codimension 1. Thus there is a function ρ in $L^2(\mathbf{R})$ that is a uniquely defined modulo sign by

(iv) $\rho \in \tilde{S}_0^m$,

(v) $\rho \perp \tilde{S}_1^m$,

(vi) $\|\rho\|_{L^2} = 1$.

By definition, both τ and ρ are piecewise polynomials of degree $\leq m + 1$ and are in $C^m(\mathbf{R})$; furthermore, the $(m+1)$-order derivatives of them are piecewise constant with discontinuities in A_1 resp. $\mathbf{Z}_+ \cup \{0\}$ and the following estimates hold with some constant r, $0 < r < 1$ (which may change from line to line):

$$|D^k \tau(x)| \leq Cr^{|x|}, \quad k = 0, \ldots, m + 1, \tag{1}$$

$$|D^k \rho(x)| \leq \begin{matrix} Cr^x, & x \geq 0 \\ 0, & x < 0 \end{matrix}, \quad k = 0, \ldots, m + 1. \tag{2}$$

The function τ also satisfies the moment conditions

$$\int_{\mathbf{R}} \tau(x) x^\alpha \, dx = 0, \quad \alpha = 0, \ldots, m + 1. \tag{3}$$

3. MAIN RESULTS

First we consider the one-dimensional case. Let $\nu = (j, k) \in \mathbf{Z} \times \mathbf{Z}$ and set

$$f_\nu(x) = 2^{j/2} \tau(2^j x - k).$$

Then we have:

Theorem 1

(a) $\{f_\nu\}_{\nu \in \mathbf{Z} \times \mathbf{Z}}$ *is a complete orthonormal system in* $L^2(\mathbf{R})$.

(b) *Let* $p > 1/(m + 2)$. *Then* f_ν *is in the dual space of* $H^p(\mathbf{R})$, $\nu \in \mathbf{Z} \times \mathbf{Z}$, *and if* $c_\nu = (f_\nu, f)$, *then the sum* $\sum_{\nu \in \mathbf{Z} \times \mathbf{Z}} c_\nu f_\nu$ *converges unconditionally to* f *in the* $H^p(\mathbf{R})$ *norm. Furthermore, the coefficients in a converging sum are uniquely defined; that is, if* $\sum c_\nu f_\nu$ *converges to a function* $f \in H^p(\mathbf{R})$ *in the norm, then* $c_\nu = (f_\nu, f)$.

(c) *Let* J_ν *be the interval* $[2^{-j}k, 2^{-j}(k + 1)]$ *and let* χ_ν *be its characteristic function. Then for any collections of coefficients* $\{c_\nu\}_\nu$ *such that only finitely many* c_ν *are non-vanishing, we have*

$$\left\| \sum_{\nu} c_{\nu} f_{\nu} \right\|_{H^p(\mathbf{R})} \leq C \left\| \left\{ \sum_{\nu} |c_{\nu}|^2 2^j \chi_{\nu} \right\}^{1/2} \right\|_{L^p(\mathbf{R})}, \quad p > 1/(m + 5/2). \quad (4)$$

On the other hand, if $f \in H^p(\mathbf{R})$, $p > 1/(m + 2)$ *and* $c_{\nu} = (f_{\nu}, f)$, *then*

$$\left\| \left\{ \sum_{\nu} |c_{\nu}|^2 2^j \chi_{\nu} \right\}^{1/2} \right\|_{L^p(\mathbf{R})} \leq C \| f \|_{H^p(\mathbf{R})}. \quad (5)$$

In higher dimensions we will use both the functions τ and ρ. We want to define a system of functions on $\mathbf{R}^n$, $n > 1$. Let Ω be the set of n-tuples of functions $\omega = (\omega_1, \ldots, \omega_n)$, where ω_i is either τ or ρ but $\omega_i = \tau$ for at least one i, $1 \leq i \leq n$. Thus Ω consists of $2^n - 1$ n-tuples. Now we define the function τ^{ω} on $\mathbf{R}^n$ by

$$\tau^{\omega}(x) = \tau^{\omega}(x_1, \ldots, x_n) = \prod_{i=1}^{n} \omega_i(x_i), \quad x \in \mathbf{R}^n, \ \omega \in \Omega.$$

For example, when $n = 2$, we get the three functions $\tau(x_1)\tau(x_2)$, $\tau(x_1)\rho(x_2)$, and $\rho(x_1)\tau(x_2)$. Let $\nu = (j, k_1, \ldots, k_n) \in \mathbf{Z} \times \mathbf{Z}^n$ and define the function f by

$$f_{\nu}^{\omega}(x) = 2^{nj/2} \tau^{\omega}\big(2^j x - (k_1, \ldots, k_n)\big), \quad x \in \mathbf{R}^n.$$

Then:

Theorem 2

(a) $\{f_{\nu}^{\omega}\}_{\Omega \times \mathbf{Z}^{n+1}}$ *is a complete orthonormal system in* $L^2(\mathbf{R}^n)$.

(b) $\{f_{\nu}^{\omega}\}_{\Omega \times \mathbf{Z}^{n+1}}$ *is an unconditional basis for* $H^p(\mathbf{R}^n)$ *when* $p > n/(n + m + 1)$; *that is, the corresponding statements of* Theorem 1(b) *hold with the sum taken over both* ν *and* ω.

(c) *The* n*-dimensional version of* (4) *holds when* $p > p(n, m)$, *and the* n*-dimensional version of* (5) *holds when* $p > n/(n + m + 1)$. (*The sum is taken over both* ν *and* ω *and* $\mathbf{R}$ *is replaced by* $\mathbf{R}^n$ *and* J_{ν} *is a cube.*)

Remark 1. With the methods that we give in detail when $n = 1$, one can get $p(n, m) = 1/\big(1/2 + (m + 2)/n\big)$.

Remark 2. $2^{(n)j}\chi_\nu$ may as well be replaced by $|f_\nu^{(\omega)}|^2$ in (4) and (5).

Remark 3. If we set $f_\nu^{\omega,T}(x) = \sum_{t \in \mathbf{Z}^n} f_\nu^{\omega}(x - t)$ when $\nu = (j, k_1, \ldots, k_n)$, $j \geq 0$, we get functions on $\mathbf{T}^n$, which together with the constant function 1 will be a complete orthonormal system. This system is an unconditional basis for the real Hardy space $H^p(\mathbf{T}^n)$ (defined with one dilation) when $p > n/(n + m + 1)$.

4. EXPLICIT DESCRIPTION OF τ AND ρ, m = 0

Since τ and ρ are linear on the intervals between the points $\sigma \in A_1$, it is enough to find their values at these points. We will in this section use the notation $\tau_\sigma = \tau(\sigma)$ and $\rho_\sigma = \rho(\sigma)$, $\sigma \in A_1$.

Definition. The *simple tent* Λ_σ in S_0^0, $\sigma \in A_0$, is the function in S_0^0 that is 1 at $x = \sigma$ but is 0 at all $\sigma_1 \in A_0$, $\sigma_1 \neq \sigma$.

Since τ is orthogonal to all simple tents in S_0^0 and ρ is orthogonal to all simple tents in S_0^0 with support in $[1, \infty)$, we will get the following equations after computing the integrals:

$$\tau_{\sigma-1} + 4\tau_\sigma + \tau_{\sigma+1} = 0, \qquad = 2, 3, 4, \ldots,$$
$$\tau_{\sigma-1/2} + 4\tau_\sigma + \tau_{\sigma+1/2} = 0, \qquad = -1/2, -1, -3/2, \ldots,$$
$$\tau_0 + 6\tau_{1/2} + 13\tau_1 + 4\tau_2 = 0,$$
$$2\tau_{-1/2} + 9\tau_0 + 6\tau_{1/2} + \tau_1 = 0,$$

and

$$\rho_{\sigma-1} + 4\rho_\sigma + \rho_{\sigma+1} = 0.$$

The characteristic equation $r^2 + 4r + 1 = 0$ has the two roots $r_1 = \sqrt{3} - 2$ and $r_2 = r_1^{-1} = -(2 + \sqrt{3})$. Since the functions must go to zero at infinity, we get

$$\tau_\sigma = \tau_1(\sqrt{3} - 2)^{\sigma-1}, \ \sigma = 1, 2, 3, \ldots,$$
$$\tau_\sigma = \tau_0(-\sqrt{3} - 2)^{2\sigma}, \ \sigma = 0, -1/2, -1, \ldots,$$

$$\rho_\sigma = \rho_1(\sqrt{3} - 2)^{\sigma-1}, \qquad = 1, 2, 3, \ldots,$$

and since ρ is linear on $[0, 1]$ and vanishes on the negative axis, we have $\rho_{1/2} = \frac{1}{2}\rho_1$ and $\rho_\sigma = 0$ for all $= 0, -1/2, -1, \ldots$. From the third and fourth rows of the equations above, we now get $\tau_0 = 2(\sqrt{3} - 1)\tau_1$ and $\tau_{1/2} = -(\sqrt{3} + 1/2)\tau_1$. Finally, by the normalization (iii) and (vi), one can get the values of $|\tau_1|$ and $|\rho_1|$. We leave the computation of these values to the interested readers.

It is now easy to check the estimates (1) and (2) when $k = 0, 1$ and $m = 0$. To see that the inequality (3) holds ($m = 0$), we can truncate the functions 1 and x outside a large compact set so that the truncated functions are in S_0^0. The corresponding part of the integrals will then be zero by orthogonality and the remainder of the integral can be made arbitrarily small if the compact set is chosen large enough, since τ is rapidly decreasing at infinity.

5. IDEAS FOR (1)–(3) WHEN m > 0

The derivatives $D^{m+2}\tau$ and $D^{m+2}\rho$ (taken in the sense of distributions) are measures supported in A_1, and in this section we use the notations τ_σ and ρ_σ, $\sigma \in A_1$ for the coefficients given by

$$D^{m+2}\tau = \sum_{\sigma \in A_1} \tau_\sigma \delta_\sigma,$$

$$D^{m+2}\rho = \sum_{\sigma \in A_1} \rho_\sigma \delta_\sigma$$

(δ_σ is the delta function at σ).

Let h be the function $h(x) = \Delta^{2m+4}(x_+^{2m+3})$, where $x_+^i = \max(x^i, 0)$ and Δ^i is defined by $\Delta^1 f(x) = f(x) - f(x - 1)$, $\Delta^i = \Delta^1\Delta^{i-1}$, $i = 2, 3, \ldots$, and set $h_k(x) = h(x - k)$, $k = 1, 2, \ldots$. Then $D^{m+2}h_k$ are functions in S_0^m, $k = 1, 2, \ldots$ (and also in $\tilde{S}_1^m$) and by orthogonality we have

$$\int D^{m+2} h_k \tau \, dx = 0,$$

which after integration by parts, gives us the equations

$$\sum_{\sigma \in A_1} h_k(\sigma)\tau_\sigma = 0, \; k = 1, 2, \ldots .$$

Observe that $h_k(\sigma)$ is zero for all $\sigma \in A_1$ except $\sigma = k + 1, \ldots,$ $k + 2m + 3$. We get the characteristic equation

$$\sum_j h(j) r^{j-1} \equiv Q_{2m+2}(r),$$

where Q_{2m+2} is a polynomial of degree $2m + 2$. By differentiation of geometric series, we can write

$$Q_j(r) = \frac{c_j(1 - r)^{j+2}}{r}\left(r\frac{d}{dr}\right)^{j+1}\frac{1}{1 - r}, \; j = 0, 1, \ldots .$$

Using a simple change of variable ($t = 1/r$), we see that if r_1 is a root to the equation $Q_j(r) = 0$, then $1/r_1$ is a root to the same equation. By this and an elementary mean value theorem, one can conclude by induction over j that the characteristic equation $Q_{2m+2}(r) = 0$ has $2m + 2$ distinct roots of which $m + 1$ roots $r_1, \ldots, r_{m+1}$ are in the open interval $(0, -1)$ and the remaining $m + 1$ roots are in the interval $(-\infty, -1)$. We conclude that

$$\tau_\sigma = C_1 r_1^\sigma + \cdots + C_{m+1} r_{m+1}^\sigma + C_{m+2} r_1^{-\sigma} + \cdots + C_{2m+2} r_{m+1}^{-\sigma}, \; \sigma = 2, 3, \ldots,$$

for some constants $C_1, \ldots, C_{2m+2}$, and since τ is in $L^2(\mathbf{R})$, $C_{m+2}, \ldots, C_{2m+2}$ are all equal to zero. We now get the estimate (1) for $x \geq 0$ by repeated integration. In the same way we also get (1) for $x < 0$ and (2). With the same truncation argument as we used in the case $m = 0$, we will get (3) also when $m > 0$.

6. ORTHOGONALITY

First we will see that $\{f_\nu\}_{\nu \in \mathbf{Z}^2}$ is an orthogonal system in $L^2(\mathbf{R})$. Let $A_\nu = \{x;\ 2^j x - k \in A_0\}$. The basic observation is that the collection of sets $\{A_\nu\}_{\mathbf{Z}^2}$ is totally ordered by inclusion; that is, if $\nu \neq \nu'$, then either $A_\nu \not\subseteq A_{\nu'}$ (we write then $\nu \prec \nu'$) or $A_{\nu'} \subsetneq A_\nu$ (we write then $\nu' \prec \nu$). Thus if $\nu = (j, k)$ and $\nu' = (j', k')$, then

$$\nu \prec \nu' \quad \text{is equivalent to} \quad \begin{cases} j < j' \\ j = j',\ k < k'. \end{cases}$$

Let $\nu \prec \nu'$ and consider the functions f_ν and $f_{\nu'}$. By a linear transformation, $f_{\nu'}$ can be transformed to (a multiple of) τ and, by the same linear transformation, f_ν will be transformed to a function in S_0^m, and since $\tau \perp S_0^m$, we conclude that $f_{\nu'} \perp f_\nu$. This proves that $\{f_\nu\}_{\mathbf{Z}^2}$ is an orthogonal system in $L^2(\mathbf{R})$.

Next we want to see that $\{f_\nu^\omega\}_{\Omega\times\mathbf{Z}^{n+1}}$ is an orthogonal system in $L^2(\mathbf{R}^n)$, $n > 1$. Let $\omega = (\omega_1, \ldots, \omega_n)$ and $\nu = (j, k_1, \ldots, k_n)$. Set

$$\omega_{i;j,k_i}(x_i) = 2^{j/2}\omega_i(2^j x_i - k_i),\ i = 1, \ldots, n.$$

Then

$$f_\nu^\omega(x) = \prod_{i=1}^n \omega_{i;j,k_i}(x_i),$$

and $\omega_{i;j,k_i}$ is of the form $2^{j/2}\tau(2^j x_i - k_i) = f_{j,k_i}(x_i)$ or $2^{j/2}\rho(2^j x_i - k_i)$. Let $\omega' = (\omega_1', \ldots, \omega_n')$ and $\nu' = (j', k_1', \ldots, k_n')$. Then we have

$$\int_{\mathbf{R}^n} f_\nu^\omega(x) f_{\nu'}^{\omega'}(x)\,dx = \prod_{i=1}^n \int_{\mathbf{R}} \omega_{i;j,k_i}(x_i)\omega'_{i;j',k_i'}(x_i)\,dx_i.$$

If $(\omega, \nu) \neq (\omega', \nu')$, we will see that there is an $i = i_0$ such that the integral in the x_i-direction vanishes in the product on the right-hand side. There are the following four cases:

Case 1. $j < j'$ (or similarly $j' < j$). Then we use that $\omega_i' = \tau$ for some $i = i_0$.

Case 2. $j = j'$, $\omega \neq \omega'$; then $\omega_i \neq \omega_i'$ for some $i = i_0$, say, $\omega_{i_0}' = \tau$, $\omega_{i_0} = \rho$ (or similarly, if $\omega_{i_0} = \tau$, $\omega_{i_0}' = \rho$).

Case 3. $j = j'$, $\omega = \omega'$, $(k_1, \ldots, k_n) \neq (k_1', \ldots, k_n')$; that is, $k_i \neq k_i'$ for some $i = i_0$, and $\underline{\omega_{i_0}' = \omega_{i_0} = \tau}$, say, $k_{i_0}' > k_{i_0}$ (or similarly $k_{i_0} > k_{i_0}'$).

Case 4. $j = j'$, $\omega = \omega'$, $(k_1, \ldots, k_n) \neq (k_1', \ldots, k_n')$; that is, $k_i \neq k_i'$ for some $i = i_0$ and $\underline{\omega_{i_0}' = \omega_{i_0} = \rho}$, say, $k_{i_0}' < k_{i_0}$ (or similarly $k_{i_0} < k_{i_0}'$).

By a linear transformation we can now transform the function $\omega_{i_0; j', k_{i_0}'}'$ to get the function τ (in cases 1-3) or the function ρ (in case 4), and by the same linear transformation $\omega_{i_0; j, k_{i_0}}$ will be transformed to a function in S_0^m (cases 1-3) resp. $\widetilde{S}_1^m$ (case 4). Since τ is orthogonal to S_0^m and ρ is orthogonal to $\widetilde{S}_1^m$, we conclude that the integral in the x_{i_0}-direction vanishes. This proves the orthogonality of the system $\left\{f_\nu^{\omega}\right\}_{\Omega \times \mathbf{Z}^{n+1}}$.

7. COMPLETENESS IN L^2

We will first show that $\left\{f_\nu\right\}_{\mathbf{Z}^2}$ is a complete system in $L^2(\mathbf{R})$.

Let $S_\nu = \{f;\ f(x) = g(2^j x - k) \text{ for some } g \in S_0^m\}$, $\nu = (j, k) \in \mathbf{Z}^2$, and let P_ν be the projection operator $P_\nu : L^2 \to S_\nu$. If $\nu' < \nu$, then $S_{\nu'} \not\subseteq S_\nu$ and $P_\nu P_{\nu'} f = P_{\nu'} P_\nu f = P_{\nu'} f$, $f \in L^2(\mathbf{R})$.

Let $f \in L^2(\mathbf{R})$ and let $c_\nu = \int_{\mathbf{R}} f\, f_\nu\, dx$; then

$$c_\nu f_\nu = P_{j, k+1} f - P_{j, k} f,$$
$$\nu = (j, k),$$

and

$$f - \sum_{|j| \le j_0} \sum_{|k| \le k_0} c_\nu f_\nu = f - \sum_{|j| \le j_0} (P_{j, k_0+1} f - P_{j, -k_0} f)$$
$$= f - P_{j_0, k_0+1} f - \sum_{j=-j_0}^{j_0 - 1} (P_{j+1, -k_0} f - P_{j, k_0+1} f) + P_{-j_0, -k_0} f.$$

The completeness of the system $\left\{f_\nu\right\}_{\mathbf{Z}^2}$ in $L^2(\mathbf{R})$ now follows easily if we can show that if $f \in L^2(\mathbf{R})$, then

(i) $\| P_{j,k} f - f \|_{L^2} \to 0$ as $j \to +\infty$ independent of k,

(ii) $\| P_{j,k} f \|_{L^2} \to 0$ as $j \to -\infty$ independent of k,

(iii) $\| P_{j+1,-k} f - P_{j,k+1} f \|_{L^2} \to 0$ as $k \to +\infty$ for each fixed j.

(i) follows by approximation in $L^2(\mathbf{R})$ by spline functions, which we leave to the reader.

For (ii) we use that $P_{j,k} f$ is piecewise a polynomial of degree $\leq m + 1$ to get

$$\| P_{j,k} f \|_\infty \leq C 2^{j/2} \| P_{j,k} f \|_{L^2} \leq C 2^{j/2} \| f \|_{L^2} .$$

Thus $f - P_{j,k} f$ converges uniformly to f as j tends to $-\infty$. We conclude that

$$\| f \|_{L^2} \geq \| f - P_{j,k} f \|_{L^2} \to \| f \|_{L^2}$$

and

$$\| P_{j,k} f \|_{L^2}^2 = \| f \|_{L^2}^2 - \| f - P_{j,k} f \|_{L^2}^2 \to 0 \text{ as } j \to -\infty.$$

To show (iii), we may assume that $f \in S_{j+1,0}$ and has compact support. Then

$$P_{j+1,-k} f = f + P_{j+1,-k} f - P_{j+1,0} f = f - \sum_{i=-k}^{-1} c_{j+1,i} f_{j+1,i}$$

and

$$\sum_{i=-k}^{-1} c_{j+1,i} f_{j+1,i}(x) = \int \left\{ \sum_{i=-k}^{-1} f_{j+1,i}(y) f_{j+1,i}(x) \right\} f(y)\,dy$$

and by (1) we get

$$\left| \sum_{i=-k}^{-1} f_{j+1,i}(y) f_{j+1,i}(x) \right| \leq C 2^{j/2} r^{2^{j+1}|x-y|}.$$

We conclude from this that $|P_{j+1,-k} f(x)| \leq C_f r^{2^j|x|}$ with the constant C_f independent of k. For any $\varepsilon > 0$, we can now find a large compact set K_ε independent of k such that we can write $P_{j+1,-k} f = g_k + g_k'$ with $g_k, g_k' \in S_{j+1,-k}$, $\| g_k \|_{L^2} < \varepsilon$, and g_k' supported in K_ε. We observe that $g_k' \in S_{j,k+1}$ if k is chosen large enough. Thus $P_{j,k+1} g_k' = g_k'$ and we get

$$\begin{aligned} P_{j+1,-k}f - P_{j,k+1}f &= P_{j+1,-k}f - P_{j,k+1}P_{j+1,-k}f \\ &= g_k + g_k' - P_{j,k+1}g_k - P_{j,k+1}g_k' \\ &= g_k - P_{j,k+1}g_k \end{aligned}$$

and

$$\| P_{j+1,-k}f - P_{j,k+1}f \|_{L^2} = \| g_k - P_{j,k+1}g_k \|_{L^2} \leq \| g_k \|_{L^2} \leq \varepsilon,$$

provided k is chosen large enough. This proves (iii) and finishes the proof of the completeness of the system $\{f_\nu\}_{\mathbf{Z}^2}$ in $L^2(\mathbf{R})$.

Next we will see that $\{f_\nu^\omega\}_{\Omega \times \mathbf{Z}^{n+1}}$ is a complete system in $L^2(\mathbf{R}^n)$. For simplicity we will only look at the case when $n = 2$. From the one-dimensional case, we conclude that the collection of products

$$\left\{f_{j_1,k_1}(x_1)f_{j_2,k_2}(x_2)\right\}_{\mathbf{Z}^2 \times \mathbf{Z}^2}$$

is a complete orthonormal system in $L^2(\mathbf{R}^2)$. However, it involves two independent dilations and that is the reason we use the system $\{f_\nu^\omega\}_{\Omega \times \mathbf{Z}^2}$, which involves only one dilation. We split the products into three cases:

Case 1. $j_1 = j_2$. These products are still in our system as the functions f_ν^ω with $\omega = (\tau, \tau)$.

Case 2. $j_1 > j_2$. We will see that these products are in the closure of the span of the functions f_ν^ω with $\omega = (\tau, \rho)$.

Case 3. $j_1 < j_2$. These products are in the closure of the span of the functions f_ν^ω with $\omega = (\rho, \tau)$.

We will only look at case 2, as case 3 can be treated in the same way. We claim that the product $f_{j_1,k_1}(x_1)f_{j_2,k_2}(x_2)$ is in the closure of the span of functions

$$\left\{f_{j_1,k_1}(x_1)2^{j/2}\rho(2^{j_1}x_2 - k)\right\}_{k \in \mathbf{Z}}$$

provided $j_1 > j_2$. This will again be a one-dimensional problem. By dilation argument we may assume that $j_1 = 0$ and $j_2 < 0$. It is enough to show that the function f_{j_2,k_2} that is in S_{j_2,k_2+1} can be approximated in $L^2(\mathbf{R})$ by linear combinations of the functions $\{\rho(\cdot - k)\}_{k \in \mathbf{Z}}$.

First we approximate f_{j_2,k_2} by a function f in S_{j_2,k_2+1} with compact support. Let k_0 be so large that the interval $[-k_0, k_0]$ contains the support of f. Set $\widetilde{S}_k = \{f : f(\cdot + k) \in \widetilde{S}_0^m\}$ and let $\widetilde{P}_k$ be the projection operator $\widetilde{P}_k : L^2(\mathbf{R}) \to \widetilde{S}_k$. Then $\widetilde{P}_{-k_0} f = f$ and $\widetilde{P}_{k_0} f = 0$, since $f \in \widetilde{S}_{-k_0}$ and since

$$\| f \|_{L^2}^2 - \| \widetilde{P}_{k_0} f \|_{L^2}^2 = \| f - \widetilde{P}_{k_0} f \|_{L^2}^2 = \| f \|_{L^2}^2 + \| \widetilde{P}_{k_0} f \|_{L^2}^2 .$$

The last inequality holds, since f and $\widetilde{P}_{k_0} f$ have disjoint supports. Thus

$$f = \widetilde{P}_{-k_0} f - \widetilde{P}_{k_0} f = \sum_{k=-k_0}^{k_0-1} \widetilde{P}_k f - \widetilde{P}_{k+1} f,$$

and $\widetilde{P}_k f - \widetilde{P}_{k+1} f$ is a multiple of $\rho(\cdot - k)$. This shows that f can be written as a linear combination of the functions $\{\rho(\cdot - k)\}_{k=-k_0}^{k_0-1}$ and hence f_{j_2,k_2} is in the closure of the span of the functions $\{\rho(\cdot - k)\}_{k \in \mathbf{Z}}$. This completes the proof of the completeness of the system $\{f_\nu^\omega\}_{\Omega \times \mathbf{Z}^3}$ in $L^2(\mathbf{R}^2)$.

8. PROOF OF THEOREM 1(a) AND (b)

First we observe that f is in Lip$(m + 1)$, that is, $D^m f$ is in Lip 1 and hence is in the dual space of $H^p(\mathbf{R})$ when $1/p \leq 2 + m$. We will first prove Theorem 1(c) by an area integral approach. The inequalities (4) and (5) will then be the basic ingredients in the proof of Theorem 1(b). At the end of this paper, we will show how (b) can be proved directly with atoms.

Proof of (4) and (5). It is enough to show (5) for f in a dense subspace of $H^p(\mathbf{R})$, since if f_j converges to f, so will the corresponding coefficients $c_{j;\nu}$ converge to c_ν by the duality, and hence

(5) will hold for a general f if the sum over ν is restricted to be finite. The estimate for an infinite sum then follows by monotone convergence. Thus we may assume that the Fourier transform of f is C^∞, compactly supported and vanishing near the origin.

Let $\tilde{\nu} = (2^{-j}k, 2^{-j}) \in \mathbf{R}^2_+$, $\nu = (j, k) \in \mathbf{Z}^2$, and let μ be the measure on $\mathbf{R}^2_+$ defined by $\mu = \sum_\nu \delta_{\tilde{\nu}}$, where $\delta_{\tilde{\nu}}$ is the Dirac measure at $\tilde{\nu}$. Set $\tau_t(x) = t^{-1}\tau(x/t)$, $\tilde{\tau}_t(x) = \tau_t(-x)$, and define the function F on $\mathbf{R}^2_+$ by

$$F(x, t) = \tilde{\tau}_t * f(x),$$

where f is the finite sum $\sum_\nu c_\nu f_\nu$ [in showing (4)] or the function given in (5). Observe that in both cases we have $F(\tilde{\nu}) = 2^{j/2}c_\nu$, $\nu \in \mathbf{Z}^2$. We define the functions

$$AF(z) = \left\{ \iint_{0 \leq z - x \leq t} |F(x, t)|^2 d\mu(x, t) \right\}^{1/2}$$

and

$$g^*_\lambda F(z) = \left\{ \iint_{\mathbf{R}^2_+} (1 + |x - z|/t)^{-2\lambda} |F(x, t)|^2 d_\mu(x, t) \right\}^{1/2}, \quad \lambda > 0.$$

Then AF is identical to the expression

$$\left\{ \sum_\nu |c_\nu|^2\, 2^j \chi_\nu \right\}^{1/2}.$$

Now we let ψ be a C^∞ function supported in [-1, 1] such that

$$\int \psi(x)x^\alpha\, dx = 0, \quad \alpha = 0, 1, \dots, m + 1 \tag{6}$$

and

$$\int_0^\infty |\hat{\psi}(s\xi)|^2 \frac{ds}{s} = 1 \text{ for } \xi \neq 0. \tag{7}$$

Let $\psi_s(x) = s^{-1}\psi(x/s)$ and define the function G on $\mathbf{R}^2_+$ by $G(y, s) = \psi_s * f(y)$ and set

$$AG(z) = \left\{ \iint_{|y-z|<s} |G(y, s)|^2 \frac{dy\, ds}{s^2} \right\}^{1/2}$$

and

$$g^*_\lambda G(z) = \left\{ \iint_{\mathbf{R}^2_+} (1 + |y - z|/s)^{-2\lambda} |G(y, x)|^2 \frac{dy\, ds}{s^2} \right\}^{1/2}, \quad \lambda > 0.$$

By A. P. Calderón and A. Torchinsky [2, Theorem 6.9, p. 56, and Theorem 3.5, p. 20], we get

$$\| AG \|_{L^p} \sim \| f \|_{H^p}, \quad 0 < p < \infty$$

and

$$\| g_\lambda^\star(\,\cdot\,) \|_{L^p} \leq C \| A(\,\cdot\,) \|_{L^p}, \quad 1/\lambda < \min(p, 2).$$

To obtain the inequalities (4) and (5), we have only to show the pointwise estimates

$$AG(z) \leq Cg_\lambda^\star F(z), \quad \lambda < m + 5/2, \tag{8}$$

$$AF(z) \leq Cg_\lambda^\star G(z), \quad \lambda < m + 2. \tag{9}$$

By (5) we have the identity

$$F(x, t) = \iint_{\mathbf{R}_+^2} \tau_t \star \psi_s(y - x) G(y, s) \frac{dy\, ds}{s},$$

and since $F(\tilde{\nu}) = c_\nu 2^{j/2}$, we also have the identity

$$G(y, s) = \iint \tau_t \star \psi_s(y - x) F(x, t) t\, d\mu(x, t).$$

We have to make estimates on the convolution $\tau_t \star \psi_s$. Let $D^{-1}g$ denote the primitive function of g, that is,

$$D^{-1}g(x) = \int_0^x g(y)dy.$$

By (3), $D^k\tau$ will satisfy (1) even for $k = -1, \ldots, -m - 2$; and similarly, by (6), $D^k\psi$ will be supported in $[-1, 1]$ even for $k = -1, \ldots, -m - 2$. By repeated integration by parts we get

$$|\tau_t \star \psi_s(y)| = |D^{m+1}\tau_t \star D^{-m-1}\psi_s(y)| \leq C(s/t)^{m+1}(1 + |y|/t)^{-N} t^{-1},$$

for any $N > 0$ when $s \leq t$, and

$$|\tau_t \star \psi_s(y)| = |D^{-m-2}\tau_t \star D^{m+2}\psi_s(y)| \leq C(t/s)^{m+2}(1 + |y|/s)^{-N} s^{-1},$$

for any $N > 0$ when $s \geq t$. If we use that $D^{m+2}\tau$ is a sum of point-measures, rapidly decreasing at infinity, we can also get the estimate

$$\| (1 + |\cdot|/t)^N(\tau_t * \psi_s)\|_{L^2} \le C(s/t)^{m+3/2}t^{-1/2},$$

for any $N > 0$ when $s \le t$.

To estimate $AG(z)$, we let $|y - z| < s$ and get from these inequalities

$$|\tau_t * \psi_s(y - x)| \le C(t/s)^\varepsilon (1 + |x - z|/t)^{-3-m+\varepsilon}t^{-1}, \text{ when } t \le s,$$

and

$$|\tau_t * \psi_s(y - x)| \le C(s/t)^\varepsilon (1 + |x - z|/t)^{-N}t^{-1}, \text{ when } t \ge s,$$

for any $N > 0$ and $\varepsilon > 0$. Plugging these estimates into the identity for $G(y, s)$ above, we get, after using the Cauchy-Schwarz inequality,

$$|G(y, s)|^2 \le C\iint_{\mathbf{R}^2_+} (\min(s/t, t/s))^\varepsilon$$
$$\times (1 + |x - z|/t)^{-5-2m+3\varepsilon}|F(x, t)|^2 d\mu(x, t).$$

Integrating over the cone $|y - z| < s$, we get, after changing the order of integration, the inequality (8) with $\lambda = m + 5/2 - 3\varepsilon/2$.

To estimate $AF(z)$, we let $|x - z| < t$ and get from the estimates of $\tau_t * \psi_s$

$$|\tau_t * \psi_s(y - x)| \le C(t/s)^\varepsilon (1 + |y - z|/s)^{-N}s^{-1}, \text{ when } s \ge t$$

and

$$\| (1 + |(\cdot - z)|/s)^{m+2-\varepsilon}(\tau_t * \psi_s(\cdot - x))\|_{L^2} \le C(s/t)^\varepsilon s^{-1/2}, \text{ when } s \le t,$$

for any $N > 0$ and $\varepsilon > 0$. Using these estimates in the identity for $F(x, t)$ together with the Cauchy-Schwarz inequality, we get

$$|F(x, t)|^2 \le C\iint_{\mathbf{R}^2_+} (\min(s/t, t/s))^\varepsilon$$
$$\times (1 + |y - z|/s)^{-4-2m+2\varepsilon}|G(y, s)|^2 \frac{dy\, ds}{s^2}.$$

Integrating over the cone $|x - z| < t$, we get, after changing the order of integration, the inequality (9) with $\lambda = m + 2 - \varepsilon$. This finishes the proof of Theorem 1(c).

Proof of Theorem 1(b). As a consequence of Theorem 1(c), the partial sums $\Sigma_\nu (f_\nu, f)f_\nu$ are uniformly bounded in the H^p norm by the H^p norm of f, $f \in H^p(\mathbf{R})$, $p > 1/(m+2)$. We will also get the following lemma.

Lemma. The set of finite linear combinations of the functions $\{f_\nu\}$ *are dense* $H^p(\mathbf{R})$, $p > 1/(m+2)$.

Proof of the Lemma. Define $S_N f$ by

$$S_N f = \sum_{\substack{|j|<N \\ |k|<N}} (f_\nu, f)f_\nu, \quad (\nu = (j, k)), \; f \in H^p(\mathbf{R}).$$

Since $L^2(\mathbf{R}) \cap H^p(\mathbf{R})$ is dense in $H^p(\mathbf{R})$, it is enough to show that

$$\| f - S_N f \|_{H^p} \to 0 \text{ as } N \to 0, \; f \in L^2(\mathbf{R}) \cap H^p(\mathbf{R}).$$

Using dominated convergence, we see that the left side of (5) goes to zero as N tends to infinity if the sum is only taken over $\nu = (j, k)$ with $|j|$ or $|k|$ larger than N. It follows that for each $\varepsilon > 0$ there is an N_ε such that $\| S_N f - S_{N_\varepsilon} f \|_{H^p}^p < \varepsilon$ for all $N \geq N_\varepsilon$. Let $M(\;)$ be the nontangential maximal function. Then we have

$$\| f - S_N f \|_{H^p}^p \leq \int_{|x|<R} |M(f - S_N f)|^p dx + C\int_{|x|>R} |M(f - S_{N_\varepsilon} f)|^p dx$$
$$+ C\int_{|x|>R} |M(S_N f - S_{N_\varepsilon} f)|^p dx.$$

The third integral is less than $C\varepsilon$ independent of R, and since $f - S_{N_\varepsilon}$ is in $H^p(\mathbf{R})$, the second integral will be less than ε if R is large enough; and finally, using Hölder's inequality, the first integral is bounded by $C_R \| M(f - S_N f) \|_{L^2}^p \leq C_R \| f - S_N f \|_{L^2}^p$, which goes to zero as N tends to infinity by the completeness of $\{f_\nu\}$ in $L^2(\mathbf{R})$. This finishes the proof of the lemma. (We have assumed that $p < 2$, which is the only case of interest.)

Now let $\{\nu_i\}_{i=1}^\infty$ be any enumeration of $\{\nu\}_{\mathbf{Z}^2}$ and let $S_K^P f$ denote the partial sum

$$\sum_{i=1}^{K} (f_{\nu_i}, f) f_{\nu_i}, \quad f \in H^p(\mathbf{R}).$$

Let $f \in H^p(\mathbf{R})$. For each $\varepsilon > 0$ we can find a finite linear combination f_ε of the functions $\{f_\nu\}$ such that $\| f - f_\varepsilon \|_{H^p} < \varepsilon$. Then $S_K^p f_\varepsilon = f_\varepsilon$ if K is large enough. Since the partial sums are uniformly bounded in the H^p norm, we get

$$\| f - S_K^p f \|_{H^p} \leq C \| f - f_\varepsilon \|_{H^p} + C \| S_K^p (f - f_\varepsilon) \|_{H^p} \leq C \| f - f \|_{H^p} < C\varepsilon,$$

for any $\varepsilon > 0$ provided K is large enough. This proves the unconditional convergence. The uniqueness of the coefficients is now easy. If $\sum_\nu c_\nu f_\nu$ converges to f in the H^p norm, we apply the partial sums to f_{ν_0}, which is in the dual space. On the one hand, we get the limit c_{ν_0}; on the other hand, this limit must be (f_{ν_0}, f), which shows that $c_\nu = (f_\nu, f)$. With this we have finished the proof of Theorem 1.

9. PROOFS OF THEOREM 2(b) AND (c)

The area integral approach will work in higher dimension in exactly the same way as in the one-dimensional case. It involves a summation over ω that is harmless, since Ω is a finite set. To estimate the convolutions $\tau_t^\omega * \psi_s$, we can no longer use integration by parts, but instead we can, for instance, make estimates on the Fourier transform side. With this method, we get (4) in Theorem 2(c) when $p > 1/(1/2 + (m + 3/2)/n)$ and (5), and consequently also Theorem 2(b) when $p > 1/(1/2 + (m + 2)/n)$.

Using atoms we obtain inequality (5) and Theorem 2(b) when $p > n/(n + m + 1)$. Since atoms have been used on Franklin systems on [0, 1] (see, for example, [5 and 10]), we will only sketch the proof briefly.

Let f be an atom supported in a cube Q with side s and center x_0 such that

$$\int fx^\alpha dx = 0, \ |\alpha| \le m, \ \| f \|_\infty \le s^{-n/p}$$

and let x_ν denote the point $(2^{-j}k_1, \ldots, 2^{-j}k_n) \in \mathbf{R}^n$ and $|\nu| = 2^{-j}$ for $\nu = (j, k_1, \ldots, k_n) \in \mathbf{Z}^{n+1}$. Then

$$|(f_\nu^\omega, f)| \le \| f_\nu^\omega \|_{L^1(Q)} \| f \|_\infty$$
$$\le Cs^{-n/p}|\nu|^{n/2} \min\left\{1, r^{(|x_0 - x_\nu| - ns)/|\nu|}\right\}, \text{ when } |\nu| \le s,$$

and using the mth-order Taylor expansion of f_ν^ω at x_0 and the moment conditions of f, we get

$$|(f_\nu^\omega, f)| \le Cs^{1+m+n-n/p}|\nu|^{-n/2-m-1} r^{|x_\nu - x_0|}|\nu|, \text{ when } |\nu| \ge s.$$

Let $M(\ )$ be the nontangential maximal function and let Sf be *any* partial sum

$$\sum_{\nu,\omega} (f_\nu^\omega, f) f_\nu^\omega.$$

Looking at the moment conditions of τ^ω, we find that $|M(\tau^\omega)(x)| \le C(1 + |x|)^{-m-n-2}$ and consequently

$$|M(f_\nu^\omega)(x)| \le C|\nu|^{-n/2}(1 + |x - x_\nu|/|\nu|)^{-m-n-2}.$$

By summation we now get for $\varepsilon > 0$ and $|x - x_0| \ge 2ns$

$$|M(Sf)(x)| \le \sum_{\nu,\omega} |(f_\nu^\omega, f)| |M(f_\nu^\omega)(x)| \le C(|x - x_0|/s)^{-m-n-1+\varepsilon} s^{-n/p}$$

and

$$\left\{\sum_{\nu,\omega} |(f_\nu^\omega, f)|^2 |\nu|^{-n} \chi_\nu(x)\right\}^{1/2} \le C(|x - x_0|/s)^{-m-n-1+\varepsilon} s^{-n/p}.$$

By these estimates and L^2 estimates when $|x - x_0| < 2ns$, we get

$$\| Sf \|_{H^p} \le \| M(Sf) \|_{L^p} \le C, \ p > n/(n + m + 1),$$

and

$$\left\| \left\{\sum_{\nu,\omega} |(f_\nu^\omega, f)|^2 |\nu|^n \chi_\nu\right\}^{1/2} \right\|_{L^p} \le C, \ p > n/(n + m + 1).$$

Taking limits with finite sums of atoms, we get inequality (5) and $\| Sf \|_{H^p} \le C\| f \|_{H^p}$ for all $f \in H^p(\mathbf{R}^n)$, $p > n/(n + m + 1)$.

To prove that finite linear combinations of $\{f_\nu^\omega\}$ are dense in $H^p(\mathbf{R}^n)$, $p > n/(n + m + 1)$, is now easy if we use the

L^2-completeness and that

$$|M(Sf - f)(x)| \leq C_f|x|^{-n-m-1+\varepsilon} \text{ for } |x| > c_f$$

uniformly for all partial sums Sf when f is a finite sum of atoms. The rest of the proof of Theorem 2(b) is the same as in the one-dimensional case.

An open problem is to find a better range of p for the n-dimensional version of inequality (4) when $n \geq 2$. (Added in proof:) A better range of p for the n-dimensional version of (4) has been found by Peter Sjögren, and the author of this paper is now aware of methods that will give the sharp range.

REFERENCES

1. Bockariev, S. V. Existence of basis in the space of analytic functions in the disc and some properties of the Franklin system. *Mat. Sbornik* 95 (1974):3-18 (in Russian).

2. Calderón, A. P., and Torchinsky, A. Parabolic maximal functions associated with a distribution. *Adv. Math.* 16 (1975):1-64.

3. Carleson, L. An explicit unconditional basis in H^1. Institut Mittag-Leffler, Report No. 2, 1980.

4. Ciesielski, Z. Equivalence, unconditionality and convergence a.e. of the spline bases in L_p spaces. Approximation theory, Banach Center Publications, vol. 4, pp. 55-68.

5. Ciesielski, Z. The Franklin orthogonal system as unconditional basis in Re H^1 and VMO. Preprint, 1980.

6. Ciesielski, Z., and Domsta, J. Estimates for the spline orthonormal functions and for their derivatives. *Studia Math.* 44 (1972):315-320.

7. Ciesielski, Z.; Simon, P.; and Sjölin, P. Equivalence of Haar and Franklin bases in L^p spaces. *Studia Math.* 60 (1976):195-211.

8. Maurey, B. Isomorphismes entre espace H^1. *Acta Math.* 145 (1980):79-120.

9. Schipp, F., and Simon, P. Investigation of Haar and Franklin series in the Hardy spaces. Preprint, 1980.

10. Sjölin, P., and Strömberg, J.-O. Basis properties of Hardy spaces. University of Stockholm, Report No. 19, 1981.

11. Wojtaszczyk, P. The Franklin system is an unconditional basis in H^1. To appear in *Arkiv för Matematik*.

A CONSTRUCTIVE PROOF OF THE FEFFERMAN-STEIN DECOMPOSITION OF BMO ON SIMPLE MARTINGALES

Akihito Uchiyama
University of Chicago

Let

$$\Omega = (0, 1].$$

Let $\mathfrak{F}$ be the σ-field of all Borel measurable sets in Ω. Then $(\Omega, \mathfrak{F}, dx)$ is a probability space. Let $d \geq 3$ be an integer. For each integer $n \geq 0$, let $\mathfrak{F}_n$ be the sub-σ-field of $\mathfrak{F}$ generated by

$$((k - 1)d^{-n}, kd^{-n}], \ k = 1, \dots, d^n.$$

Set

$$I_{k_1, k_2, \dots, k_n} = ((k_1 - 1)d^{-1} + (k_2 - 1)d^{-2} + \cdots + (k_n - 1)d^{-n}, (k_1 - 1)d^{-1} + (k_2 - 1)d^{-2} + \cdots + k_n d^{-n}]$$

for each $k_1, \dots, k_n \in \{1, \dots, d\}$. For $f \in L^1(\Omega)$, set

$$f_n = E[f \mid \mathfrak{F}_n]$$

and

$$\Delta f_n = f_n - f_{n-1}.$$

Let

$$V = \left\{ \mathfrak{x} = \{x_k\}_{k=1}^{d} \in \mathbf{C}^d : \sum_{k=1}^{d} x_k = 0 \right\},$$

and let A be a linear operator from V to V. For $f \in L^1(\Omega)$ and

---

This work was supported in part by Science Research Foundation of Japan (General Research © 1980).

$k_1, \ldots, k_{n-1} \in \{1, \ldots, d\}$, set

$$\mathfrak{r} = \left\{ \Delta f_n \left(I_{k_1, k_2, \ldots, k_{n-1}, k} \right) \right\}_{k=1}^{d} \in V.$$

Define

$$\left\{ \Delta g_n \left(I_{k_1, \ldots, k_{n-1}, k} \right) \right\}_{k=1}^{d} = A\mathfrak{r}.$$

Set

$$g = \sum_{n=1}^{\infty} \Delta g_n. \tag{1}$$

It can be easily shown that if $f \in L^2(\Omega)$, then the right-hand side of (1) converges in $L^2(\Omega)$ and

$$\| g \|_{L^2} \leq C \| f \|_{L^2}.$$

Set

$$Tf = g. \tag{2}$$

Definition. For $f \in L^2(\Omega)$, let

$$\| f \|_{\mathrm{BMO}} = \sup_n \left\| E\left[\sum_{k=n+1}^{\infty} |\Delta f_k|^2 \,|\, \mathfrak{F}_n \right] \right\|_{\infty}^{1/2}.$$

If $\| f \|_{\mathrm{BMO}} < \infty$, we say f belongs to BMO.

Assume that $A_1, \ldots, A_m$ are one or more linear transformations on V and let $T_1, \ldots, T_m$ be the corresponding operators defined by (2). Let $A_0 = I$ and T_0 be the identity operator. S. Janson [4] showed

Theorem

$$\mathrm{BMO} = \sum_{j=0}^{m} T_j L^{\infty}$$

if and only if

(i) $A_1^*, \ldots, A_m^*$ *do not have a common real eigenvector, where* A_j^* *is the adjoint operator of* A_j *from* V *to* V.

This result corresponds to the Fefferman-Stein decomposition of $\mathrm{BMO}(\mathbf{R}^m)$. In [4], Janson showed only the existence of $g_0, \ldots, g_m \in L^{\infty}$ such that for $f \in \mathrm{BMO}$,

$$f = \sum_{j=0}^{m} T_j g_j . \tag{3}$$

In this note, we give a construction of $g_0, \ldots, g_m$.

Notation. $\mathbf{a}$, $\mathbf{X}$, $\mathbf{Y}$, ... denote $(m+1)$-dimensional row vectors. $\mathfrak{x}$, $\mathfrak{y}$, $\mathfrak{z}$, ... denote d-dimensional column vectors. For

$$\mathfrak{x} = \begin{pmatrix} x_1 \\ \vdots \\ x_d \end{pmatrix} \quad \text{and} \quad \mathfrak{y} = \begin{pmatrix} y_1 \\ \vdots \\ y_d \end{pmatrix},$$

$$\langle \mathfrak{x}, \mathfrak{y} \rangle \text{ denotes } \sum_{k=1}^{d} x_k \overline{y}_k , \qquad |\mathfrak{x}| \text{ denotes } \left(\sum_{k=1}^{d} |x_k|^2 \right)^{1/2}$$

and similarly for $|\mathbf{a}|$. $\mathbf{X}_k$ and $\mathfrak{x}_j$ denote the kth row vector and the $(j+1)$th column vector of the same $d \times (m+1)$ matrix, respectively. $\mathbf{C}$ denotes the set of all complex numbers. The letter c denotes various positive constants.

Lemma 1. The condition (i) *is equivalent to the following:*

(ii) *For any*

$$(a_0, \ldots, a_m) \in \mathbf{C}^{m+1}$$

and any

$$\mathfrak{y} \in V,$$

there exist

$$\mathfrak{x}_0, \ldots, \mathfrak{x}_m \in V$$

such that

$$\sum_{j=0}^{m} A_j \mathfrak{x}_j = \mathfrak{y}, \tag{4}$$

$$\sum_{j=0}^{m} \operatorname{Re} a_j \cdot \operatorname{Re} \mathfrak{x}_j + \operatorname{Im} a_j \cdot \operatorname{Im} \mathfrak{x}_j = 0, \tag{5}$$

$$\sum_{j=0}^{m} |\mathfrak{x}_j| \leq C|\mathfrak{y}|, \tag{6}$$

where C depends on $A_1, \ldots, A_m$ but is independent of $\mathfrak{y}$ and $(a_0, \ldots, a_m)$.

Remark 1. (5) means that each row vector $\mathbf{X}_1, \ldots, \mathbf{X}_d$ of the $d \times (m+1)$ matrix

$$(\mathfrak{x}_0, \ldots, \mathfrak{x}_m)$$

is orthogonal to $(a_0, \ldots, a_m)$ in $\mathbf{R}^{2(m+1)}$. Similarly, for $\mathfrak{x}, \mathfrak{y} \in C^d$,

$$\operatorname{Re}\langle \mathfrak{x}, \mathfrak{y} \rangle$$

is equal to the inner product in $\mathbf{R}^{2d}$.

Proof. (ii) → (i). Suppose $\mathfrak{y}_0$ is a common real eigenvector of $A_1^*, \ldots, A_m^*$. Let $a_0, \ldots, a_m$ be its eigenvalues. If

$$\sum_{j=0}^{m} A_j \mathfrak{x}_j = \mathfrak{y}_0$$

and (5) holds, then

$$|\mathfrak{y}_0|^2 = \operatorname{Re}\left\langle \sum_{j=0}^{m} A_j \mathfrak{x}_j, \mathfrak{y}_0 \right\rangle = \operatorname{Re} \sum_{j=0}^{m} \left\langle \mathfrak{x}_j, A_j^* \mathfrak{y}_0 \right\rangle$$

$$= \operatorname{Re} \sum_{j=0}^{m} \left\langle \mathfrak{x}_j, a_j \mathfrak{y}_0 \right\rangle = \left\langle \operatorname{Re} \sum_{j=0}^{m} \bar{a}_j \mathfrak{x}_j, \mathfrak{y}_0 \right\rangle = 0,$$

and (ii) is false.

(i) → (ii). It suffices to show

$$\left\{ \sum_{j=0}^{m} A_j \mathfrak{x}_j : \mathfrak{x}_j \in V \text{ and (5) holds} \right\} = V \tag{7}$$

for any $(a_0, \ldots, a_m) \in C^{m+1}$. Then (6) follows from a compactness argument. If $a_0 = 0$, then (7) is clear.

The left-hand side of (7) is a subspace of V over $\mathbf{R}$. So, to show (7), it suffices to show that for any $\mathfrak{y} \in V$, there exists $\mathfrak{x}_0, \ldots, \mathfrak{x}_m \in V$ such that (5) and

$$\operatorname{Re}\left\langle \sum_{j=0}^{m} A_j \mathfrak{x}_j, \mathfrak{y} \right\rangle \neq 0 \tag{8}$$

hold. (Remember Remark 1.)

Case 1. $\mathfrak{y}$ *is not of the form* $\lambda\mathfrak{y}_0$, *where* $\lambda \in \mathbf{C}$ *and* $\mathfrak{y}_0$ *is a real vector, and* $a_0 \neq 0$. Set

$$\mathfrak{y} = \begin{pmatrix} y_1 \\ \vdots \\ y_d \end{pmatrix}.$$

Then, there exist two components, say, y_1 and y_2, that satisfy

$$\mathrm{Re}(ia_0 \cdot \overline{y}_1) > 0$$

and

$$\mathrm{Re}(ia_0 \cdot \overline{y}_2) < 0,$$

where $i = (-1)^{1/2}$. So, set

$$\mathfrak{x}_0 = \begin{pmatrix} ia_0 \\ -ia_0 \\ 0 \\ \vdots \\ 0 \end{pmatrix}, \mathfrak{x}_1 = \cdots = \mathfrak{x}_m = 0.$$

Then (5) and (8) hold.

Case 2. $\mathfrak{y}$ *is a constant multiple of a real vector and* $a_0 \neq 0$. By (i), $\mathfrak{y}$ is not a common eigenvector of $A_1^*, \ldots, A_m^*$. So at least one of $\{A_j^*\mathfrak{y}\}_{j=1,\ldots,m}$, say, $A_1^*\mathfrak{y}$, is independent of $\mathfrak{y}$ over $\mathbf{C}$. Take $\tilde{\mathfrak{x}} \in V$ such that

$$\langle \tilde{\mathfrak{x}}, A_1^*\mathfrak{y} \rangle = a_0,$$
$$\langle \tilde{\mathfrak{x}}, \mathfrak{y} \rangle = 0.$$

Set

$$\mathfrak{x}_0 = \overline{a}_1 \tilde{\mathfrak{x}},$$
$$\mathfrak{x}_1 = -\overline{a}_0 \tilde{\mathfrak{x}},$$
$$\mathfrak{x}_j = 0 \text{ for other } j.$$

Then

$$\mathrm{Re}\left\langle \sum_{j=0}^{m} A_j \mathfrak{x}_j, \mathfrak{y} \right\rangle = \mathrm{Re}\left\langle A_1 \mathfrak{x}_1, \mathfrak{y} \right\rangle = -|a_0|^2.$$

So (8) holds. Since

$$\sum_{j=0}^{m} a_j \overline{\mathfrak{x}}_j = a_0 \overline{\mathfrak{x}}_0 + a_1 \overline{\mathfrak{x}}_1 = 0,$$

(5) holds.

Main Lemma. Let $A_1, \ldots, A_m$ satisfy (i). *Let $R > 0$ be a sufficiently large number depending only on $A_1, \ldots, A_m$. Let*

$$\mathfrak{y} \in V,$$

$$\mathbf{a} = (a_0, \ldots, a_m) \in \mathbf{C}^{m+1},$$

$$b_1, \ldots, b_d \geq 0$$

be such that

$$\min(R - |\mathbf{a}|, 1) \geq \left\{\left(|\mathfrak{y}|^2 + \sum_{k=1}^{d} b_k^2\right)\Big/ d\right\}^{1/2}. \tag{9}$$

Then there exist

$$\mathfrak{x}_0, \ldots, \mathfrak{x}_m \in V \tag{10}$$

such that

$$\sum_{j=0}^{m} A_j \mathfrak{x}_j = \mathfrak{y} \tag{11}$$

and

$$R - |\mathbf{a} + \mathbf{X}_k| \geq b_k, \; 1 \leq k \leq d, \tag{12}$$

hold.

Proof. First we show this for three special cases and then for the general case.

Case 1. $R - |\mathbf{a}| \geq 2d^{1/2}$.

In this case, set

$$\mathfrak{x}_0 = \mathfrak{y}, \; \mathfrak{x}_1 = \cdots = \mathfrak{x}_m = 0.$$

Then (11) is trivial and

$$R - |\mathbf{a} + \mathbf{X}_k| \geq R - |\mathbf{a}| - |\mathbf{X}_k| \geq 2d^{1/2} - |\mathbf{X}_k| \geq d^{1/2} \geq b_k,$$

since

$$\left(|\mathfrak{y}|^2 + \sum_{k=1}^{d} b_k^2\right)\Big/ d \leq 1.$$

Case 2.

$$2d^{1/2} \geq R - |\mathbf{a}| \geq 1, \tag{13}$$

$$\left(|\mathfrak{y}|^2 + \sum_{k=1}^{d} b_k^2\right)\Big/ d = 1, \tag{14}$$

$$\max_{1 \le k \le d} b_k < 1/2. \tag{15}$$

By Lemma 1, there exist

$$\mathfrak{x}_0, \mathfrak{x}_1, \ldots, \mathfrak{x}_m \in V$$

such that (4)-(6) hold. (11) is clear from (4). By (5) and (6), $\mathbf{X}_1, \ldots, \mathbf{X}_d$ are orthogonal to $\mathbf{a}$ in $\mathbf{R}^{2(m+1)}$ (see Remark 1), and

$$|\mathbf{X}_k| \le c,\ k = 1, \ldots, d.$$

So,

$$|\mathbf{a} + \mathbf{X}_k| = (|\mathbf{a}|^2 + |\mathbf{X}_k|^2)^{1/2} \le |\mathbf{a}|(1 + c|\mathbf{X}_k|^2/|\mathbf{a}|^2)$$
$$\le R - 1 + c/R \le R - b_k,$$

by (13) and (15). Thus, (11) and (12) hold.

Case 3. (13), (14), *and*

$$\max_{1 \le k \le d} b_k \ge 1/2. \tag{16}$$

We may assume

$$b_1 \ge b_2 \ge \cdots \ge b_d.$$

Then by (16) and (14),

$$\sum_{k=1}^{d} b_k/d \le \left(\sum_{k=1}^{d} b_k^2/d\right)^{1/2} - c(b_1 - b_d)^2 \le 1 - c(|\mathfrak{y}|^2 \pm (b_1 - b_d)^2).$$

So we can take

$$\mathfrak{z} = \begin{pmatrix} z_1 \\ \vdots \\ z_d \end{pmatrix} \in V \cap \mathbf{R}^d$$

such that

$$|\mathfrak{z}| \le c(b_1 - b_d), \tag{17}$$

$$1 - z_k \ge b_k + c(|\mathfrak{y}|^2 + (b_1 - b_d)^2),\ 1 \le k \le d. \tag{18}$$

Set

$$\mathfrak{x}'_j = a_j \mathfrak{z}/|\mathbf{a}|,\ j = 0, \ldots, m,$$

$$\mathfrak{y}' = \mathfrak{y} - \sum_{j=0}^{m} A_j \mathfrak{x}'_j. \tag{19}$$

Then,

$$|\mathfrak{y}'| \le |\mathfrak{y}| + c(b_1 - b_d).$$

By Lemma 1, there exist

$$\mathfrak{x}_j'' \in V, \; j = 0, \ldots, m,$$

such that

$$\sum_{j=0}^{m} A_j \mathfrak{x}_j'' = \mathfrak{y}', \tag{20}$$

$$\sum_{j=0}^{m} \operatorname{Re} a_j \cdot \operatorname{Re} \mathfrak{x}_j'' + \operatorname{Im} a_j \cdot \operatorname{Im} \mathfrak{x}_j'' = 0, \tag{21}$$

$$\sum_{j=0}^{m} |\mathfrak{x}_j''| \le c(|\mathfrak{y}| + (b_1 - b_d)). \tag{22}$$

Set

$$\mathfrak{x}_j = \mathfrak{x}_j' + \mathfrak{x}_j''.$$

By (19) and (20),

$$\sum_{j=0}^{m} A_j \mathfrak{x}_j = \mathfrak{y}.$$

So (11) holds. Since $\mathbf{a} + \mathbf{X}_k'$ and $\mathbf{X}_k''$ are orthogonal in $\mathbf{R}^{2(m+1)}$ by (21), we get

$$|\mathbf{a} + \mathbf{X}_k| = |\mathbf{a} + \mathbf{X}_k' + \mathbf{X}_k''| \le |\mathbf{a} + \mathbf{X}_k'|(1 + c|\mathbf{X}_k''|^2/|\mathbf{a} + \mathbf{X}_k'|^2)$$

$$\le |\mathbf{a} + \mathbf{X}_k'| + c(|\mathfrak{y}|^2 + (b_1 - b_d)^2)/R$$

by (22). Since

$$|\mathbf{a} + \mathbf{X}_k'| = |\mathbf{a}| + z_k \le R - 1 + z_k \le R - b_k - c(|\mathfrak{y}|^2 + (b_1 - b_d)^2)$$

by (18), we get

$$|\mathbf{a} + \mathbf{X}_k| \le R - b_k - c(|\mathfrak{y}|^2 + (b_1 - b_d)^2) + c(|\mathfrak{y}|^2 + (b_1 - b_d)^2)/R.$$

So (12) holds if R is large enough.

General Case. Note that by the above three cases, the case

$$R - |\mathbf{a}| \ge \left\{\left(|\mathfrak{y}|^2 + \sum_{k=1}^{d} b_k^2\right)\Big/ d\right\}^{1/2} = 1 \tag{9'}$$

was completely shown. In the general case, set

$$\alpha = \left(\left(|\mathfrak{y}|^2 + \sum_{k=1}^{d} b_k^2\right)\Big/ d\right)^{1/2}.$$

By (9), $\alpha \le 1$ and

$$R/\alpha - |\mathbf{a}/\alpha| \geq \left\{\left(|\mathfrak{y}/\alpha|^2 + \sum_{k=1}^{d}(b_k/\alpha)^2\right)\Big/ d\right\}^{1/2} = 1. \tag{9''}$$

So

$$R/\alpha,\ \mathbf{a}/\alpha,\ \mathfrak{y}/\alpha,\ b_1/\alpha,\ \ldots,\ b_d/\alpha$$

satisfy (9′). So we get

$$\tilde{\mathfrak{x}}_j \in V,\ j = 0, \ldots, m,$$

such that

$$\sum_{j=0}^{m} A_j \tilde{\mathfrak{x}}_j = \mathfrak{y}/\alpha \tag{11'}$$

and

$$R/\alpha - |\mathbf{a}/\alpha + \tilde{\mathbf{X}}_k| \geq b_k/\alpha \tag{12'}$$

hold. Set

$$\mathfrak{x}_j = \alpha \tilde{\mathfrak{x}}_j .$$

Then (11) and (12) follow from (11′) and (12′).

The construction of $g_0, \ldots, g_m$ *of* (3). Let

$$\|f\|_{\text{BMO}} \leq 1,\ E[f] = 0.$$

Let

$$\mathbf{g}_0 \equiv (0, \ldots, 0) \in \mathbf{C}^{m+1}.$$

Let $R > 0$ be as in the main lemma. Of course,

$$R - |\mathbf{g}_0| > 1 \geq \left(E\left[\sum_{n=1}^{\infty} |\Delta f_n|^2 \middle| \mathfrak{F}_0\right]\right)^{1/2}.$$

Assume that we have constructed the $\mathfrak{F}_n$-measurable $\mathbf{C}^{m+1}$-valued function

$$\mathbf{g}_n = (g_{0n}, \ldots, g_{mn})$$

such that

$$R - |\mathbf{g}_n| \geq \left(E\left[\sum_{h=n+1}^{\infty} |\Delta f_h|^2 \middle| \mathfrak{F}_n\right]\right)^{1/2}, \tag{23}$$

$$\sum_{j=0}^{m} T_j g_{jn} = f_n . \tag{24}$$

Take any $I = I_{k_1, \ldots, k_n} \in \mathfrak{F}_n$. Set

$$\mathfrak{y} = \begin{pmatrix} \Delta f_{n+1}(I(1)) \\ \vdots \\ \Delta f_{n+1}(I(d)) \end{pmatrix} \in V, \tag{25}$$

$$\mathbf{a} = \mathbf{g}_n(I) \in \mathbf{C}^{m+1},$$

$$b_k = \left\{ E\left[\sum_{h=n+2}^{\infty} |\Delta f_h|^2 \,\Big|\, \mathfrak{F}_{n+1} \right](I(k)) \right\}^{1/2}, \tag{26}$$

where

$$I(k) = I_{k_1, \ldots, k_n k}.$$

Then

$$E\left[\sum_{h=n+1}^{\infty} |\Delta f_h|^2 \,\Big|\, \mathfrak{F}_n \right](I) = \left(|\mathfrak{y}|^2 + \sum_{k=1}^{d} b_k^2 \right) \Big/ d.$$

So by (23) and the main lemma, there exist $\mathbf{X}_1, \ldots, \mathbf{X}_d \in \mathbf{C}^{m+1}$ such that (10)-(12) hold. Set

$$\mathbf{g}_{n+1}(I(k)) = \mathbf{a} + \mathbf{X}_k, \; k = 1, \ldots, d.$$

In this way, we can define $\mathfrak{F}_{n+1}$-measurable $\mathbf{g}_{n+1}$ on Ω.

By (10),

$$E[\mathbf{g}_{n+1} | \mathfrak{F}_n] = \mathbf{g}_n .$$

By (12) and (26),

$$R - |\mathbf{g}_{n+1}| \geq \left(E\left[\sum_{h=n+2}^{\infty} |\Delta f_h|^2 \,\Big|\, \mathfrak{F}_{n+1} \right] \right)^{1/2}.$$

By (24), (25), and (11),

$$\begin{aligned} \left\{ \sum_{j=0}^{m} T_j g_{j,\,n+1}(I(k)) \right\}_{k=1}^{d} &= \left\{ \sum_{j=0}^{m} T_j g_{j,\,n}(I) \right\} + \sum_{j=0}^{m} A_j \mathfrak{x}_j \\ &= \left\{ f_n(I) + \Delta f_{n+1}(I(k)) \right\}_{k=1}^{d} \\ &= \left\{ f_{n+1}(I(k)) \right\}_{k=1}^{d}. \end{aligned}$$

So we can define $\{\mathbf{g}_n\}_{n=1}^{\infty}$ with (23) and (24). By (23),

$$\|\mathbf{g}_n\|_\infty \leq R.$$

Let

$$\mathbf{g} = \lim_{n\to\infty} \mathbf{g}_n \text{ in } L^2(\Omega).$$

Then this is the desired function.

REFERENCES

1. Chao, J.-A., and Taibleson, M. H. A sub-regularity inequality for conjugate systems on local fields. *Studia Math.* 46 (1973): 249-257.

2. Fefferman, C., and Stein, E. M. H^p spaces of several variables. *Acta Math.* 129 (1972):137-193.

3. Gundy, R. F. Inegalites pour martingales un et deux indices: L'espace H^p. *Springer Lecture Notes in Math.* 774 (1980):251-334.

4. Janson, S. Characterization of H^1 by singular integral transforms on martingales and $\mathbf{R}^n$. *Math. Scand.* 41 (1977):140-152.

5. Jones, P. W. Carleson measures and the Fefferman-Stein decomposition of BMO($\mathbf{R}$). *Ann. of Math.* 111 (1980):197-208.

6. Jones, P. W. L^∞ estimates for the $\bar{\partial}$ problem in a half-plane. To appear in *Acta Math.*

7. Taibleson, M. H. *Fourier Analysis on Local Fields.* Princeton, N.J.: Princeton Univ. Press, 1975.

PART SIX

DIFFERENTIATION THEORY

RELATIONS BETWEEN PEANO DERIVATIVES AND MARCINKIEWICZ INTEGRALS

Sagun Chanillo
Richard L. Wheeden
Rutgers University

The purpose of this paper is to describe the results obtained in our papers [5 and 6]. To make the presentation as short as possible, we do not prove any of the facts stated below; complete proofs can be found in [5 and 6].

In [5], the main results are distribution function inequalities relating Marcinkiewicz integrals and Peano derivatives. These inequalities imply as corollaries many local and global results, including some new facts, and thus provide a fairly unified approach to the study of Marcinkiewicz integrals. They have the advantage of also leading easily to weighted norm inequalities for A_∞ weight functions. The distribution function inequalities involve several other operators, such as nontangential maximal functions and area integrals related to differentiation. The purpose of [6] is to study pointwise estimates that connect the Peano maximal function of a given order with an appropriate nontangential maximal function. This helps in understanding the hierarchy among the operators related to Marcinkiewicz integrals and simplifies some of the corollaries of the distribution function inequalities.

The local results mentioned above are theorems of Marcinkiewicz [11] and Stein and Zygmund [14, 15, and 16] giving

---

Research partly supported by NSF Grant No. MCS80-03098.

necessary and sufficient conditions for the finiteness a.e. in a set of a Marcinkiewicz integral. The global results involve L^p and weak L^p relations between Peano maximal functions, nontangential maximal functions, and Marcinkiewicz integrals. Such results are ultimately connected to the question of when a function is the potential of an element of L^p or H^p. Global results have been proved by many authors: see, for example, [11, 20, 10, 18, 13, 8, 9, 19, and 7].

The most familiar of the Marcinkiewicz integrals is

$$\mathcal{M}_1(F)(x) = \left(\int_{\mathbf{R}^n} \frac{|F(x+u) + F(x-u) - 2F(x)|^2}{|u|^{n+2}}\, du\right)^{1/2}.$$

Frequently, in local considerations, one wishes to study the truncated integral

$$\mathcal{M}_1(F)(x, \delta) = \left(\int_{|u|<\delta} \frac{|F(x+u) + F(x-u) - 2F(x)|^2}{|u|^{n+2}}\, du\right)^{1/2}, \quad \delta > 0.$$

More generally, for $0 < \alpha < 2$, we consider

$$\mathcal{M}_\alpha(F)(x) = \left(\int_{\mathbf{R}^n} \frac{|F(x+u) + F(x-u) - 2F(x)|^2}{|u|^{n+2\alpha}}\, du\right)^{1/2}$$

and the truncated version $\mathcal{M}_\alpha(F)(x, \delta)$ in which the integration is only extended over $|u| < \delta$. For technical reasons, we also define

$$\mathcal{M}^*_\alpha(F)(x) = \sup_{s>0} \left(\int_{\mathbf{R}^n} \frac{|F(x+u, s) + F(x-u, s) - 2F(x, s)|^2}{|u|^{n+2\alpha}}\, du\right)^{1/2},$$

where $0 < \alpha < 2$ and $F(x, y)$ is an appropriate function in $\mathbf{R}^{n+1}_+$. We will always assume that $F(x, y)$ is harmonic, which in particular includes the case when $F(x, y)$ is the Poisson integral of a function on the boundary. Another convenient class of functions is the solutions of the heat equation, which would include Gauss-Weierstrass extensions of functions on the boundary. It can be proved that

$\mathcal{M}_\alpha^\star(F)(x)$ and $\mathcal{M}_\alpha(F)(x)$ are pointwise equivalent if $F(x, y)$ is the Poisson integral of $F(x)$.

We also consider the unsymmetric Marcinkiewicz integrals

$$\mathcal{D}_\alpha(F)(x) = \begin{cases} \left(\displaystyle\int_{\mathbf{R}^n} \frac{|F(x+u) - F(x)|^2}{|u|^{n+2\alpha}} \, du \right)^{1/2}, & 0 < \alpha < 1, \\ \left(\displaystyle\int_{\mathbf{R}^n} \frac{|F(x+u) - F(x) - u \cdot A(x)|^2}{|u|^{n+2\alpha}} \, du \right)^{1/2}, & 1 < \alpha < 2, \end{cases}$$

where $A(x)$ is a vector with n components. We define $\mathcal{D}_1(F) = \mathcal{M}_1(F)$. Similarly, for harmonic $F(x, y)$,

$$\mathcal{D}_\alpha^\star(F)(x) = \sup_{s>0} \begin{cases} \left(\displaystyle\int_{\mathbf{R}^n} \frac{|F(x+u, s) - F(x, s)|^2}{|u|^{n+2\alpha}} \, du \right)^{1/2}, & 0 < \alpha < 1, \\ \left(\displaystyle\int_{\mathbf{R}^n} \frac{|F(x+u, s) - F(x, s) - u \cdot \nabla_1 F(x, s)|^2}{|u|^{n+2\alpha}} \, du \right)^{1/2}, & 1 < \alpha < 2, \end{cases}$$

and $\mathcal{D}_1^\star(F) = \mathcal{M}_1^\star(F)$. Here ∇_1 denotes the gradient in the x-variables,

$$\nabla_1 = \left\langle \frac{\partial}{\partial x_1}, \ldots, \frac{\partial}{\partial x_n} \right\rangle,$$

as distinguished from the full gradient

$$\nabla = \left\langle \nabla_1, \frac{\partial}{\partial y} \right\rangle.$$

The truncations obtained by only integrating over $|u| < \delta$ are denoted $\mathcal{D}_\alpha(F)(x, \delta)$ and $\mathcal{D}_\alpha^\star(F)(x, \delta)$. We then clearly have $\mathcal{M}_\alpha(F)(x) \le 2\mathcal{D}_\alpha(F)(x)$ and $\mathcal{M}_\alpha(F)(x, \delta) \le 2\mathcal{D}_\alpha(F)(x, \delta)$, as well as the corresponding inequalities for $\mathcal{M}_\alpha^\star$ and $\mathcal{D}_\alpha^\star$. If $F(x, y)$ is the Poisson integral of $F(x)$, it can be shown that $\mathcal{D}_\alpha^\star(F)(x)$ and $\mathcal{D}_\alpha(F)(x)$ are pointwise equivalent, assuming when $1 < \alpha < 2$ that $\nabla_1 F(x, s)$ has a limit as $s \to 0$ and defining $A(x)$ to be this limit.

We now define the Peano maximal functions $\mathcal{P}_\alpha(F)(x)$. These are inherent in [4] and were explicitly studied in [3]. Let

$$\mathscr{P}_\alpha(F)(x) = \sup_{h>0} \begin{cases} h^{-\alpha}\left(h^{-n}\int_{|u|<h} |F(x+u) - F(x)|^2\, du\right)^{1/2}, & 0 < \alpha \le 1, \\ h^{-\alpha}\left(h^{-n}\int_{|u|<h} |F(x+u) - F(x) - u \cdot A(x)|^2\, du\right)^{1/2}, & 1 < \alpha \le 2, \end{cases}$$

where $A(x)$ is a vector with n-components. To eliminate problems at the boundary, we shall also consider for harmonic $F(x, y)$, the function

$$\mathscr{P}^\star_\alpha(F)(x) = \sup_{s,h>0} \begin{cases} h^{-\alpha}\left(h^{-n}\int_{|u|<h} |F(x+u, s) - F(x, s)|^2\, du\right)^{1/2}, & 0 < \alpha \le 1, \\ h^{-\alpha}\left(h^{-n}\int_{|u|<h} |F(x+u, s) - F(x, s) - u \cdot \nabla_1 F(x, s)|^2\, du\right)^{1/2}, & 1 < \alpha \le 2. \end{cases}$$

As usual, if $F(x, y)$ is the Poisson integral of its boundary values $F(x)$ and $\nabla_1 F(x, s)$ has a limit as $s \to 0$ and we take $A(x)$ in the definition of $\mathscr{P}_\alpha(F)(x)$ to be this limit, then $\mathscr{P}_\alpha(F)(x)$ and $\mathscr{P}^\star_\alpha(F)(x)$ are pointwise equivalent.

The function $\mathscr{P}^\star_\alpha(F)$ is closely related to a function $Z^\star_\alpha(F)$ whose study is somewhat easier. We set

$$Z^\star_\alpha(F)(x) = \sup_{s,h>0} \begin{cases} h^{-\alpha}\left(h^{-n}\int_{|u|<h} |F(x+u, |u|+s) - F(x+u, s)|^2\, du\right)^{1/2}, & 0 < \alpha \le 1, \\ h^{-\alpha}\left(h^{-n}\int_{|u|<h} \left|F(x+u, |u|+s) - F(x+u, s) - |u|\dfrac{\partial F}{\partial s}(x+u, |u|+s)\right|^2 du\right)^{1/2}, & 1 < \alpha \le 2. \end{cases}$$

$Z^\star_\alpha(F)$ is easier to localize than $\mathscr{P}^\star_\alpha(F)$, and it is possible to make pointwise estimates on $Z^\star_\alpha(F)$ that yield global results for $\mathscr{P}^\star_\alpha(F)$. To write down the relation between $\mathscr{P}^\star_\alpha(F)$ and $Z^\star_\alpha(F)$, we let

$$N^{(\alpha)}(F)(x) = \sup_{\Gamma(x)} \begin{cases} y^{1-\alpha}\left|\dfrac{\partial F}{\partial y}(u, y)\right|, & 0 < \alpha \le 1, \\ y^{2-\alpha}\left|\dfrac{\partial^2 F}{\partial y^2}(u, y)\right|, & 1 < \alpha \le 2, \end{cases}$$

where $\Gamma(x) = \{(u, y) : |x - u| < ay\}$, a nontangential cone with suitably large aperture a. We then have

$$\mathscr{P}_\alpha^\star(F)(x) \le c[Z_\alpha^\star(F)(x) + N^{(\alpha)}(F)(x)], \tag{1}$$

except that when $\alpha = 1$, $N^{(\alpha)}(F)(x)$ must be replaced by $N(|\nabla F|)(x)$, where N denotes the ordinary nontangential maximal function. Moreover, the roles of $\mathscr{P}_\alpha^\star(F)(x)$ and $Z_\alpha^\star(F)(x)$ may be interchanged in (1). The proof of (1) is essentially an application of the mean value theorem.

It is also possible to show that for an arbitrary harmonic function $F(x, y)$,

$$Z_\alpha^\star(F)(x) \le c \sup_{h>0} \left(\frac{1}{h^n} \int_{|u|<h} [N^{(\alpha)}(F)(x+u)]^{p_0}\, du \right)^{1/p_0}, \tag{2}$$

where $p_0 = 2n/(n + 2\alpha)$. This inequality together with its weighted version is the main result of [6] referred to earlier. The idea of its proof is to use the mean value property for harmonic functions to estimate $Z_\alpha^\star(F)(x)$ in terms of the product of two "box" maximal functions

$$T_{\lambda,r}\left(y^{k-\alpha} \frac{\partial^k F}{\partial y^k} \right)(x), \quad k - 1 < \alpha \le k, \quad k = 1, 2, \ldots,$$

for two choices of each of the parameters λ and r; here

$$T_{\lambda,r}(F)(x) = \sup_{h>0} \left(h^{-\lambda n} \iint_{\substack{|x-u|<h \\ 0<y<h}} y^{\lambda n - n - 1} |F(u, y)|^r\, du\, dy \right)^{1/r}.$$

From this estimate, (2) follows by [1] or [17].

Inequality (2) in conjunction with (1) easily implies global norm inequalities for $\mathscr{P}_\alpha^\star(F)$. For example, letting

$$\|f\|_p = \left(\int_{\mathbf{R}^n} |f(x)|^p\, dx \right)^{1/p}, \quad \|f\|_{p,\infty} = \sup_{t>0} t \left| \{x : |f(x)| > t\} \right|^{1/p},$$

$0 < p < \infty$, we have:

Theorem 1. Let $0 < \alpha < 2$, $F(x, y)$ *be harmonic in* $\mathbf{R}_+^{n+1}$, *and assume that* $\nabla_1 F(x, y) \to 0$ *as* $y \to \infty$ *if* $\alpha = 1$. *Then*

$$\|\mathscr{P}_\alpha^\star(F)\|_p \le c\|N^{(\alpha)}(F)\|_p, \quad 2n/(n+2\alpha) < p < \infty,$$

$$\|\mathscr{P}_\alpha^\star(F)\|_{p_0,\infty} \le c\|N^{(\alpha)}(F)\|_{p_0}, \quad p_0 = 2n/(n+2\alpha).$$

The assumption that $\nabla_1 F(x, y) \to 0$ as $y \to \infty$ if $\alpha = 1$ can be dropped if we replace $N^{(1)}(F)$ on the right by $N(|\nabla F|)$. For another estimate of the weak norm of $\mathscr{P}_\alpha^\star(F)$, see Theorem 12.

For $\alpha > 0$, we define the αth (harmonic) derivative $F^{(\alpha)}(x, y)$ of $F(x, y)$ by

$$F^{(\alpha)}(x, y) = \frac{e^{-i\beta\pi}}{\Gamma(\beta)}\int_0^\infty \frac{\partial^m F}{\partial y^m}(x, y+s)s^{\beta-1}\,ds,$$

where $m = [\alpha] + 1$ and $m = \alpha + \beta$. In this and all that follows, we assume that $F(x, y)$ satisfies a decay condition that we shall refer to as the "cone condition." This may be stated as the requirement that there exist $x_0 \in \mathbf{R}^n$ and $\eta > 0$ such that

$$\sup_{\Gamma(x_0),\, y>1} y^\eta |F(u, y)| < \infty.$$

The aperture a of the cone $\Gamma(x_0)$ is taken to be fixed and sufficiently large. The cone condition ensures, for example, that $F^{(\alpha)}(x, y)$ exists and is harmonic, and that $F^{(k)}(x, y) = \frac{\partial^k}{\partial y^k}F(x, y)$ for positive integers k and

$$\left(F^{(\alpha_1)}\right)^{(\alpha_2)}(x, y) = F^{(\alpha_1+\alpha_2)}(x, y).$$

We also need to introduce $N(F^{(\alpha)})(x)$, the ordinary nontangential maximal function of $F^{(\alpha)}$, as well as both the Lusin area integral

$$S(F^{(\alpha)})(x) = \left(\iint_{\Gamma(x)} |F^{(\alpha+1)}(u, y)|^2 \frac{du\,dy}{y^{n-1}}\right)^{1/2}$$

and the Littlewood-Paley-Zygmund function

$$g_\lambda^\star(F^{(\alpha)})(x) = \left(\iint_{\mathbf{R}_+^{n+1}} \left(\frac{y}{y + |x-u|}\right)^{\lambda n} |F^{(\alpha+1)}(u, y)|^2 \frac{du\,dy}{y^{n-1}}\right)^{1/2}.$$

It is not difficult to show that $N^{(\alpha)}(F)(x) \le cN(F^{(\alpha)})(x)$; if α is an integer, $N^{(\alpha)}(F)(x)$ and $N(F^{(\alpha)})(x)$ of course coincide.

SUFFICIENT CONDITIONS FOR THE FINITENESS OF MARCINKIEWICZ INTEGRALS AND RELATED GLOBAL RESULTS

Our aim now is to see how the Peano maximal function $\mathscr{P}^*_\alpha(F)$ controls $\mathscr{D}^*_\alpha(F)$. As a first step in doing this, one proves that for $F(x, y)$ satisfying the cone condition,

$$\mathscr{D}^*_\alpha(F)(x) \leq c[\mathscr{A}_\alpha(F)(x) + S(F^{(\alpha)})(x)],$$

$$\mathscr{A}_\alpha(F)(x) = \sup_{s>0}\left[\int_{\mathbf{R}^n}\left(\int_0^{|u|} yF_{yy}(x+u, y+s)\,dy\right)^2 \frac{du}{|u|^{n+2\alpha}}\right]^{1/2}. \tag{3}$$

Truncated versions of (3) are also valid, and these are helpful for local results. The term $\mathscr{A}_\alpha(F)(x)$ is a tangential component that can be estimated pointwise in terms of $g^*_\lambda(F^{(\alpha)})(x)$ for $\lambda < 1 + (2\alpha/n)$ (see [20, 10, and 13]. However, the restriction that λ be strictly less than $1 + (2\alpha/n)$ causes a slight loss of precision in theorems obtained by this method. Sharper results can be gotten by estimating $\mathscr{A}_\alpha(F)$ in terms of $Z^*_\alpha(F)$ through a distribution function inequality in the spirit of Burkholder and Gundy [2]. For technical reasons, we consider an approximation to $\mathscr{A}_\alpha(F)$ defined for $0 < \varepsilon$, $R < \infty$, by

$$\mathscr{A}_\alpha(F)(x, R, \varepsilon) = \sup_{s>\varepsilon}\left(\int_{|v|<R}\left|\int_0^{|x-v|} yF_{yy}(v, y+s)\,dy\right|^2 \frac{dv}{|x-v|^{n+2\alpha}}\right)^{1/2}.$$

$\mathscr{A}_\alpha(F)(x, R)$ is defined similarly with $\varepsilon = 0$. Note that $\mathscr{A}_\alpha(F)(x, R, \varepsilon) \uparrow \mathscr{A}_\alpha(F)(x)$ as $\varepsilon \downarrow 0$ and $R \uparrow \infty$. The first distribution function estimate is as follows.

Theorem 2. Let $F(x, y)$ be a harmonic function in $\mathbf{R}^{n+1}_+$ that satisfies the cone condition, and let $\beta >> 1$ and $0 < \gamma < 1$. Then for $0 < \alpha < 2$ and $t > 0$,

$$|\{x : \mathscr{A}_\alpha(F)(x, R, \varepsilon) > \beta t,\ Z^*_\alpha(F)(x) + S(F^{(\alpha)})(x) + N^{(\alpha)}(F)(x) \leq \gamma t\}|$$
$$\leq c(\gamma/\beta)^2|\{x : \mathscr{A}_\alpha(F)(x, R, \varepsilon) > t\}|,$$

where c depends only on α and n.

Now let us indicate the local and global results that follow from this.

Theorem 3. *Let $F(x, y)$ be a harmonic function in $\mathbf{R}_+^{n+1}$ that satisfies the cone condition and for which $N(F)(x) \in L^2_{\text{loc}}(\mathbf{R}^n)$. If $E \subset \mathbf{R}^n$ and $0 < \delta < \infty$, then*

(a) *$\mathscr{D}_\alpha^\star(F)(x, \delta)$ is finite a.e. in E if both $\mathscr{P}_\alpha^\star(F)(x)$ and $N(F^{(\alpha)})(x)$ are finite in E, $0 < \alpha < 2$.*

(b) *If, in addition, F satisfies the semigroup property $F(\cdot, y_1 + y_2) = F(\cdot, y_1) \star P(\cdot, y_2)$, $y_1, y_2 > 0$, where P denotes the Poisson kernel, then $\mathscr{M}_1^\star(F)(x, \delta)$ is finite a.e. in E if $\mathscr{P}_1^\star(F)(x)$ is finite in E.*

This result yields the following basic corollary.

Theorem 4. *Let $F(x) \in L^2_{\text{loc}}(\mathbf{R}^n)$, E be a subset of $\mathbf{R}^n$, and $0 < \delta < \infty$.*

(a) *If $\alpha = 1$ and $\mathscr{P}_1(F)(x, \delta)$ is finite in E, then $\mathscr{M}_1(F)(x, \delta)$ is finite a.e. in E.*

(b) *Let $0 < \alpha < 2$, $\alpha \neq 1$, and $m = [\alpha] + 1$. Assume that $F \in L^p(\mathbf{R}^n)$ for some p satisfying $1 \leq p < n/(m - \alpha)$, and let $I_{m-\alpha}(F)(x)$ be the fractional integral of F of order $m - \alpha$. If both $\mathscr{P}_\alpha(F)(x, \delta)$ and $\mathscr{P}_m(I_{m-\alpha}F)(x, \delta)$ are finite in E, then $\mathscr{D}_\alpha(F)(x, \delta)$ is finite a.e. in E.*

The assumption that $F \in L^p(\mathbf{R}^n)$ is made to guarantee the existence of $I_{m-\alpha}F$. Part (a) is due to Stein and Zygmund [15 and 14]; part (b) for $n = 1$ was also proved by Stein and Zygmund [16].

To give a rough idea of how Theorem 3 is deduced from Theorem 2, let $\beta \to \infty$ in Theorem 2 to get for any $k > 0$,

$$|\{x : \mathscr{A}_\alpha(F)(x, R, \varepsilon) = \infty, Z_\alpha^\star(F)(x) + S(F^{(\alpha)})(x) + N^{(\alpha)}(F)(x) \leq k\}| = 0.$$

This means essentially by letting $\varepsilon \to 0$ that the finiteness of $Z_\alpha^\star(F)$,

$S(F^{(\alpha)})$, and $N^{(\alpha)}(F)$ in a set E implies that $\mathscr{A}_\alpha(F)(x, R)$ is finite a.e. in E, and by the truncated form of (3), that $\mathscr{D}_\alpha^\star(F)(x, \delta)$ is finite a.e. in E. Now assume only that $\mathscr{P}_\alpha^\star(F)$ and $N(F^{(\alpha)})$ are finite in E. Using (1) with $\mathscr{P}_\alpha^\star$ and $Z_\alpha^\star$ interchanged, and theorems of Calderón and Stein (see [14]) relating the finiteness of non-tangential maximal functions and area integrals, it then follows that $Z_\alpha^\star(F)$ and $S(F^{(\alpha)})$ are finite a.e. in E. Hence, the finiteness of $N(F^{(\alpha)})$ and $\mathscr{P}_\alpha^\star(F)$ in E ensures that of $Z_\alpha^\star(F)$, $S(F^{(\alpha)})$, and $N^{(\alpha)}(F)$ a.e. in E. The finiteness of $\mathscr{D}_\alpha^\star(F)(x, \delta)$ a.e. in E then follows. For part (b) of Theorem 3, the semigroup condition allows us to show that $N(|\nabla_1 F|)(x) \leq c\mathscr{P}_1^\star(F)(x)$. Hence, by using the results of Calderón and Stein, we can conclude that if $\mathscr{P}_1^\star(F)$ is finite in E, then $Z_1^\star(F)$, $N^{(1)}(F)$, and $S(F^{(1)})$ are finite a.e. in E, from which part (b) of Theorem 3 follows. Theorem 4 is now almost immediate if we first note that since its conclusions are local, we may assume $F \in L^2(\mathbf{R}^n)$ and then use the pointwise equivalences of $\mathscr{P}_\alpha^\star$ and $\mathscr{P}_\alpha$ and of $\mathscr{D}_\alpha^\star$ and $\mathscr{D}_\alpha$.

It should be remarked that the results of Stein and Zygmund for $\alpha = 1$ were originally formulated as follows: if $F(x)$ has a first Peano derivative in E, that is, if there is a vector $A(x)$ such that

$$\left(h^{-n}\int_{|u|<h} |F(x+u) - F(x) - u \cdot A(x)|^2\, du\right)^{1/2} = o(h) \text{ as } h \to 0 \qquad (4)$$

for $x \in E$, then $\mathscr{M}_1(F)(x, \delta)$ is finite a.e. in E. In fact, using the Whitney extension theorem, Calderón and Zygmund show in [4] that (4) is equivalent a.e. to the analogous condition with o replaced by 0. Thus (4) is actually equivalent a.e. to

$$\left(h^{-n}\int_{|u|<h} |F(x+u) - F(x)|^2\, du\right)^{1/2} = O(h),\ h \to 0, \qquad (5)$$

which is the same as the condition that $\mathscr{P}_1(F)(x, \delta) < \infty$. We would also like to point out that the proofs of Theorem 2 and the rest of the results below do not use either the Whitney extension theorem or the harmonic derivative splitting theorem given in [14]. The role of the extension theorems thus appears to be extraneous to the mechanism by which $\mathcal{M}_\alpha^\star(F)$ and $\mathcal{P}_\alpha^\star(F)$ are related, but is a means of passing from 0 conditions to o conditions.

By a rather standard argument, one can deduce the following global result from Theorem 2.

Theorem 5. Let $F(x, y)$ *be harmonic and satisfy the cone condition. Then for* $0 < \alpha < 2$,

(a) $\|\mathcal{D}_\alpha^\star(F)\|_p \leq c\|N(F^{(\alpha)})\|_p$, $2n/(n + 2\alpha) < p < \infty$,

(b) $\|\mathcal{D}_\alpha^\star(F)\|_{p,\infty} \leq c(\|N(F^{(\alpha)})\|_{p,\infty} + \|\mathcal{P}_\alpha^\star(F)\|_{p,\infty})$, $2n/(n + 2\alpha) \leq p < \infty$.

Part (a) was essentially already known (see [14]) and can be deduced from the technique based on $g_\lambda^\star$ that was mentioned earlier. Note that the right side of (a) is the H^p norm of $F^{(\alpha)}$. To obtain the theorem from Theorem 2 is simple; (2) lets us omit the norm of $Z_\alpha^\star(F)$ in the strong-type result and [2] lets us omit the norm of $S(F^{(\alpha)})$.

In case the harmonic function $F(x, y)$ satisfies the semigroup property, it is possible to show that for $\alpha = 1$ the norms of $N(F^{(1)})$ are majorized by those of $\mathcal{P}_1^\star(F)$. Similarly, under suitable conditions, if

$$I_\beta F(x, y) = \int_0^\infty F(x, y + s)s^{\beta-1}ds, \quad \beta > 0, \tag{6}$$

then for $0 < \alpha < 2$, $\alpha \neq 1$, the norms of $N(F^{(\alpha)})$ can be majorized by those of $\mathcal{P}_m^\star(I_{m-\alpha}F)$, $m = [\alpha] + 1$. Thus, we obtain the following two results from the last theorem.

Theorem 6. Let $F(x, y)$ *be a harmonic function in* $\mathbf{R}_+^{n+1}$ *that satisfies the cone condition and the semigroup property. Then*

(a) $\|\mathcal{M}_1^\star(F)\|_p \leq c\|\mathcal{P}_1^\star(F)\|_p$, $2n/(n + 2) < p < \infty$,

(b) $\|\mathcal{M}_1^\star(F)\|_{p,\infty} \le c\|\mathcal{P}_1^\star(F)\|_{p,\infty}$, $2n/(n+2) \le p < \infty$.

Theorem 7. Let $0 < \alpha < 2$, $\alpha \ne 1$, *and* $m = [\alpha] + 1$. *If* $F(x, y)$ *is harmonic in* $\mathbf{R}_+^{n+1}$ *and satisfies the cone condition for some* $\eta > m - \alpha$ *and* $I_{m-\alpha}F(x, y)$ *satisfies the semigroup property, then*

(a) $\|\mathcal{D}_\alpha^\star(F)\|_p \le c\|\mathcal{P}_m^\star(I_{m-\alpha}F)\|_p$, $2n/(n+2\alpha) < p < \infty$,

(b) $\|\mathcal{D}_\alpha^\star(F)\|_{p,\infty} \le c(\|\mathcal{P}_m^\star(I_{m-\alpha}F)\|_{p,\infty} + \|\mathcal{P}_\alpha^\star(F)\|_{p,\infty})$, $2n/(n+2\alpha) \le p < \infty$.

NECESSARY CONDITIONS FOR THE FINITENESS OF MARCINKIEWICZ INTEGRALS AND RELATED GLOBAL RESULTS

In this section, the object is to use Marcinkiewicz integrals to obtain information about Peano maximal functions. To do so, we need to study in a quantitative form the "desymmetrization principle" of Stein and Zygmund [15]. The Marcinkiewicz integral in effect controls the function $\mathcal{P}_\alpha^\star(F)(x)$ in terms of two distribution function inequalities. To precisely write down these results, we first define the symmetric maximal function $m_\alpha^\star(F)(x)$ by

$$m_\alpha^\star(F)(x) = \sup_{h,s>0} h^{-\alpha}\left(h^{-n}\int_{|u|<h} |F(x+u, s) + F(x-u, s) - 2F(x, s)|^2\, du\right)^{1/2}, \quad 0 < \alpha < 2.$$

Clearly, $m_\alpha^\star(F)(x) \le \mathcal{M}_\alpha^\star(F)(x)$. Moreover, if $F(x, y)$ is the Poisson integral of $F(x)$, then $m_\alpha^\star(F)$ and $m_\alpha(F)$ are pointwise equivalent.

The first distribution function inequality shows that $\mathcal{M}_\alpha^\star(F)$ controls $N(F^{(\alpha)})$. Actually, to be more precise, $\mathcal{M}_\alpha^\star(F)$ controls $g_\lambda^\star(F^{(\alpha)})$ for $\lambda > 1 + (2\alpha/n)$. For technical reasons, we study an approximation to $g_\lambda^\star(F^{(\alpha)})$, namely, for $0 < \varepsilon < R < \infty$, the function

$$g_{k-\alpha,\lambda}^\star(F^{(\alpha)})(x, R, \varepsilon) = \left(\iint_{\substack{|u|<R\\ \varepsilon<y<R}} \frac{y^{\lambda n+2(k-\alpha-1)}}{(y+|x-u|)^{\lambda n}} |F^{(k)}(u, y)|^2 \frac{du\, dy}{y^{n-1}}\right)^{1/2}.$$

One can show (see [12]) that if we define

$$g^{\star}_{k-\alpha,\lambda}(F^{(\alpha)})(x) = \lim_{R\uparrow\infty,\,\varepsilon\downarrow 0} g^{\star}_{k-\alpha,\lambda}(F^{(\alpha)})(x, R, \varepsilon),$$

then $g^{\star}_{\lambda}(F^{(\alpha)})$ and $g^{\star}_{k-\alpha,\lambda}(F^{(\alpha)})$ are pointwise equivalent.

Theorem 8. *Let $F(x, y)$ be harmonic in $\mathbf{R}^{n+1}_{+}$ and satisfy*

$$\frac{\partial^k F}{\partial y^k}(x, y + s) = F(\cdot, s) \star \frac{\partial^k P}{\partial y^k}(\cdot, y),\ y,\ s > 0$$

for some $k \geq 2$. Given $0 < \alpha < 2$, let

$$M_t = \{x : \mathscr{M}^{\star}_{\alpha}(F)(x) > t\},\quad t > 0,$$

and let $\chi^{\star}_{M_t}$ denote the Hardy-Littlewood maximal function of χ_{M_t}. Then for $\lambda > 1 + (2\alpha/n)$, β and N suitably large, and $0 < \gamma < 1$,

$$\left|\left\{x : g^{\star}_{k-\alpha,\lambda}(F^{(\alpha)})(x, R, \varepsilon) > \beta t,\ \chi^{\star}_{M_{\gamma t}}(x) \leq \frac{1}{N}\right\}\right|$$
$$\leq c\left(\frac{\gamma}{\beta}\right)^2 |\{x : g_{k-\alpha,\lambda}(F^{(\alpha)})(x, R, \varepsilon) > t\}|,$$

where c depends only on k, α, λ, and n.

The second result describes how $Z^{\star}_{\alpha}(F)$ and $m^{\star}_{\alpha}(F)$ are related. To state it, we define a truncated version of $Z^{\star}_{\alpha}(F)$ by

$$Z^{\star}_{\alpha}(F)(x, R, \varepsilon) = \sup_{\substack{h>0\\ \varepsilon<s<R}} \begin{cases} h^{-\alpha}\left(h^{-n}\displaystyle\int_{\substack{|v|<R\\|v-x|<h}} |F(v, |v-x| + s) - F(v, s)|^2\, dv\right)^{1/2}, & 0 < \alpha \leq 1, \\ h^{-\alpha}\left(h^{-n}\displaystyle\int_{\substack{|v|<R\\|v-x|<h}} |F(v, |v-x| + s) - F(v, x) - |v-x|F_y(v, |v-x| + s)|^2\, dv\right)^{1/2}, & 1 < \alpha \leq 2. \end{cases}$$

Note that $Z^{\star}_{\alpha}(F)(x, R, \varepsilon) \uparrow Z^{\star}_{\alpha}(F)(x)$ as $R \uparrow \infty$ and $\varepsilon \downarrow 0$.

Theorem 9. *Let $F(x, y)$ be harmonic in $\mathbf{R}^{n+1}_{+}$ and satisfy the cone condition (even with $\eta = 0$). Given $0 < \alpha < 2$, let*

$$H_t = \{x : m^{\star}_{\alpha}(F)(x) + N^{(\alpha)}(F)(x) > t\},\quad t > 0,$$

except that when $\alpha = 1$, $N^{(\alpha)}(F)$ *is replaced by* $N(|\nabla F|)$. *Let* $\chi^{\star}_{H_t}$ *denote the Hardy-Littlewood maximal function of* χ_{H_t}. *Then for suitably large* β *and* N *and* $0 < \gamma < 1$,

$$\left|\left\{x : Z^{\star}_{\alpha}(F)(x, R, \varepsilon) > \beta t,\ \chi^{\star}_{H_{\gamma t}}(x) \le \frac{1}{N}\right\}\right|$$

$$\le c\left(\frac{\gamma}{\beta}\right)^2 |\{x : Z^{\star}_{\alpha}(F)(x, R, \varepsilon) > t\}|,$$

where c *depends only on* α *and* n.

In essence, Theorem 8 implies that the finiteness of $\mathcal{M}_{\alpha}(F)$ in a set E implies the finiteness of $N(F^{(\alpha)})$ a.e. in E, while Theorem 9 shows us that if $N^{(\alpha)}(F)$ and $m_{\alpha}(F)(x)$ are finite in E, then $Z_{\alpha}(F)$ is finite a.e. in E. From (1), we can then conclude that $\mathcal{P}_{\alpha}(F)(x)$ is finite a.e. in E. This strategy enables us to prove the following result.

Theorem 10. *Let* $F(x) \in L^2_{\text{loc}}(\mathbf{R}^n)$, $E \subset \mathbf{R}^n$, *and* $0 < \delta < \infty$.

(a) *If* $\alpha = 1$ *and* $\mathcal{M}_1(F)(x, \delta)$ *is finite in* E, *then* $\mathcal{P}_1(F)(x, \delta)$ *is finite a.e. in* E.

(b) *If* $0 < \alpha < 2$, $\alpha \ne 1$, *let* $m = [\alpha] + 1$. *Assume* $F \in L^p(\mathbf{R}^n)$ *for some* p *satisfying* $1 \le p < n/(m - \alpha)$ *and let* $I_{m-\alpha}(F)(x)$ *be the fractional integral of* F *of order* $m - \alpha$. *If* $\mathcal{M}_{\alpha}(F)(x, \delta)$ *is finite in* E, *then both* $\mathcal{P}_{\alpha}(F)(x, \delta)$ *and* $\mathcal{P}_m(I_{m-\alpha}F)(x, \delta)$ *are finite a.e. in* E.

Part (a) was proved for $n = 1$ in [15] and for $n > 1$ in [14]. For $n = 1$, under the stronger assumption that $\mathcal{D}_{\alpha}(F)(x, \delta)$ is finite in E, part (b) was proved in [16]. It follows by combining Theorems 4 and 10 that for $F \in L^2_{\text{loc}}$, the finiteness of $\mathcal{M}_{\alpha}(F)(x, \delta)$ is equivalent a.e. to the finiteness of $\mathcal{D}_{\alpha}(F)(x, \delta)$.

We now consider the global conclusions to be drawn from the two distribution function estimates in this section. From the first one, we get the following result.

Theorem 11. *Let* $F(x, y)$ *be harmonic in* $\mathbf{R}_+^{n+1}$ *and satisfy the cone condition and* $F^{(k)}(\cdot, y + s) = F(\cdot, y) * \frac{\partial^k P}{\partial s^k}(\cdot, s)$, $y, s > 0$ *for some* $k \geq 2$. *If* $0 < \alpha < 2$ *and* $\lambda > 1 + (2\alpha/n)$, *then*

(a) $\|g_\lambda^\star(F^{(\alpha)})\|_p \leq c\|\mathscr{M}_\alpha^\star(F)\|_p$, $2/\lambda < p < \infty$,

(b) $\|g_\lambda^\star(F^{(\alpha)})\|_{p,\infty} \leq c\|\mathscr{M}_\alpha^\star(F)\|_{p,\infty}$, $2/\lambda \leq p < \infty$.

Part (a) was essentially already known by combining norm inequalities for the Littlewood-Paley functions g and $g_\lambda^\star$.

Theorem 9 yields the following result.

Theorem 12. *Let* $F(x, y)$ *be harmonic in* $\mathbf{R}_+^{n+1}$ *and satisfy the cone condition (even for* $\eta = 0$). *Then for* $0 < \alpha < 2$,

$$\|\mathscr{P}_\alpha^\star(F)\|_{p,\infty} \leq c(\|m_\alpha^\star(F)\|_{p,\infty} + \|N^{(\alpha)}(F)\|_{p,\infty}),\ 2n/(n + 2\alpha) \leq p < \infty.$$

The corresponding strong-type statement for $2n/(n + 2\alpha) < p < \infty$ is omitted, since it gives no information not already included in Theorem 1.

We may combine the last two results with the fact that $m_\alpha^\star(F)(x) \leq \mathscr{M}_\alpha^\star(F)(x)$ to get:

Theorem 13. *Let* $F(x, y)$ *be harmonic in* $\mathbf{R}_+^{n+1}$ *and satisfy the cone condition and* $F^{(k)}(\cdot, y + s) = F(\cdot, y) * \frac{\partial^k P}{\partial s^k}(\cdot, s)$, $y, s > 0$, *for some* $k \geq 2$. *If* $0 < \alpha < 2$, *then*

(a) $\|N(F^{(\alpha)})\|_p + \|\mathscr{P}_\alpha^\star(F)\|_p \leq c\|\mathscr{M}_\alpha^\star(F)\|_p$, $2n/(n + 2\alpha) < p < \infty$,

(b) $\|N(F^{(\alpha)})\|_{p,\infty} + \|\mathscr{P}_\alpha^\star(F)\|_{p,\infty} \leq c\|\mathscr{M}_\alpha^\star(F)\|_{p,\infty}$,
$2n/(n + 2\alpha) \leq p < \infty$.

Part (a) actually contains no information not already given in Theorems 11 and 1 and is only included for completeness. Also, if one omits the norms of $\mathscr{P}_\alpha^\star(F)$, the statements are valid for $0 < p < \infty$.

We summarize the main inequalities listed in Theorems 5 and 13 in the following corollary.

Corollary 14. Let $F(x, y)$ *be harmonic in* $\mathbf{R}_+^{n+1}$ *and satisfy the cone condition and the semigroup property* $F(\cdot, y+s) = F(\cdot, y) * P(\cdot, s)$, $y, s > 0$. *Then for* $0 < \alpha < 2$,

(a) $\|\mathscr{D}_\alpha^\star(F)\|_p \approx \|\mathscr{M}_\alpha^\star(F)\|_p \approx \|N(F^{(\alpha)})\|_p$, $2n/(n+2\alpha) < p < \infty$,

$\|\mathscr{M}_1^\star(F)\|_p \approx \|\mathscr{P}_1^\star(F)\|_p$, $2n/(n+2) < p < \infty$,

(b) $\|\mathscr{D}_\alpha^\star(F)\|_{p,\infty} \approx \|\mathscr{M}_\alpha^\star(F)\|_{p,\infty} \approx \|N(F^{(\alpha)})\|_{p,\infty} + \|\mathscr{P}_\alpha^\star(F)\|_{p,\infty}$, $2n/(n+2\alpha) \le p < \infty$,

$\|\mathscr{M}_1^\star(F)\|_{p,\infty} \approx \|\mathscr{P}_1^\star(F)\|_{p,\infty}$, $2n/(n+2) \le p < \infty$.

If $\alpha \neq 1$, $F(x, y)$ satisfies the cone condition for some $\eta > m - \alpha$, where $m = [\alpha] + 1$, and $I_{m-\alpha}F$ satisfies the semigroup property, then by using Theorem 1 and the facts

$$\mathscr{P}_m^\star(I_{m-\alpha}F) \le c[\mathscr{P}_\alpha^\star(F) + N(|\nabla^m I_{m-\alpha}F|)]$$

and

$$N\left(\frac{\partial^m}{\partial y^m} I_{m-\alpha}F\right) = cN(F^{(\alpha)}),$$

the term $N(F^{(\alpha)})$ may be replaced by $\mathscr{P}_m^\star(I_{m-\alpha}F)$ in both (a) and (b) above.

We point out that if $F(x, y)$ is the Poisson integral of $F(x)$, then by the pointwise equivalences mentioned earlier, one can drop the $\star$'s in all the preceding global results. In particular, if $F = I_\alpha f$ in the sense of (6) for $f \in H^p$, $p < n/\alpha$, then $F^{(\alpha)} = f$ and we get

$$\|\mathscr{D}_\alpha(F)\|_p \approx \|\mathscr{M}_\alpha(F)\|_p \approx \|f\|_{H^p}, \quad 2n/(n+2\alpha) < p < n/\alpha,$$

and

$$\|\mathscr{D}_\alpha(F)\|_{p_0,\infty} \approx \|\mathscr{M}_\alpha(F)\|_{p_0,\infty} \le c\|f\|_{H^{p_0}}, \quad p_0 = 2n/(n+2\alpha).$$

For the last statement, we have also used Theorem 1. See also [19].

L^q-ANALOGUES AND WEIGHTS

It is not necessary to restrict attention to the L^2 metric in the results above. For $0 < \alpha < 2$ and $0 < q < \infty$, define

$$\mathscr{M}^{\star}_{\alpha,q}(F)(x) = \sup_{s>0}\left(\int_{\mathbf{R}^n} \frac{|F(x+u,s)+F(x-u,s)-2F(x,s)|^q}{|u|^{n+q\alpha}}\,du\right)^{1/q}.$$

Similarly, for $0 < \alpha \le 1$, let

$$\mathscr{P}^{\star}_{\alpha,q}(F)(x) = \sup_{s,h>0} h^{-\alpha}\left(h^{-n}\int_{|u|<h} |F(x+u,s)-F(x,s)|^q\,du\right)^{1/q},$$

$$Z^{\star}_{\alpha,q}(F)(x) = \sup_{s,h>0} h^{-\alpha}\left(h^{-n}\int_{|u|<h} |F(x+u,|u|+s) - F(x+u,s)|^q\,du\right)^{1/q}.$$

Analogous definitions can be given for $1 < \alpha \le 2$. It is then possible to derive versions of Theorem 2 for $2 < q < \infty$ and of Theorem 9 for $0 < q < \infty$. Also, inequality (2) holds for $0 < q < \infty$ if we replace $2n/(n + 2\alpha)$ by $qn/(n + q\alpha)$. The estimate (1) holds for $0 < q < \infty$ too. Local and global results can be obtained from these facts. For example, we have:

Theorem 15. Let $F(x, y)$ be harmonic in $\mathbf{R}^{n+1}_+$ and satisfy the cone condition and the semigroup property. If $0 < \alpha < 2$ and $2 < q < \infty$, then

(a) $\|\mathscr{D}^{\star}_{\alpha,q}(F)\|_p + \|\mathscr{D}^{\star}_{\alpha}(F)\|_p \approx \|\mathscr{M}^{\star}_{\alpha,q}(F)\|_p + \|\mathscr{M}^{\star}_{\alpha}(F)\|_p \approx \|N(F^{(\alpha)})\|_p$, $qn/(n + q\alpha) < p < \infty$,

(b) $\|\mathscr{D}^{\star}_{\alpha,q}(F)\|_{p,\infty} + \|\mathscr{D}^{\star}_{\alpha}(F)\|_{p,\infty} \approx \|\mathscr{M}^{\star}_{\alpha,q}(F)\|_{p,\infty} + \|\mathscr{M}^{\star}_{\alpha}(F)\|_{p,\infty} \approx \|N(F^{(\alpha)})\|_{p,\infty} + \|\mathscr{P}^{\star}_{\alpha,q}(F)\|_{p,\infty}$,

$qn/(n + q\alpha) \le p < \infty$.

The global results have weighted versions too if we replace $\|\cdot\|_p$ and $\|\cdot\|_{p,\infty}$ by their respective weighted analogues, namely,

$$\|f\|_{L^p_w} = \left(\int_{\mathbf{R}^n} |f(x)|^p w(x)\,dx\right)^{1/p}$$

and

$$\|f\|_{L^p_w,\infty} = \sup_{0<t<\infty} t^{-1}\left(\int_{\{x:|f(x)|>t\}} w(x)\,dx\right)^{1/p}.$$

The weight w is assumed to be in A_∞ and, for example in (a) of the preceding theorem, the restriction $p_0 < p < \infty$ [where $p_0 = qn/(n + q\alpha)$] is replaced by $\mu p_0 < q$ and $\mu p_0 < p$, where μ is the doubling order of w: that is, where

$$\int_{B_1} w(x)\,dx \leq c\left(\frac{|B_1|}{|B_2|}\right)^{\mu} \int_{B_2} w(x)\,dx$$

for all balls B_1 and B_2 with $B_2 \subset B_1$. In statement (b), which still has a term involving $\mathscr{P}^*_{\alpha,q}(F)$, the restriction $p_0 \leq p < \infty$ is replaced simply by

$$\sup_{0<t<1} t^p \int_{|x|<t^{-p_0/n}} w(x)\,dx < \infty.$$

REFERENCES

1. Barker, S. R. An inequality for measures on a half-space. *Math. Scand.* 44 (1979):92-102.

2. Burkholder, D. L., and Gundy, R. F. Distribution function inequalities for the area integral. *Studia Math.* 44 (1972): 527-544.

3. Calderón, A. P. Estimates for singular integral operators in terms of maximal functions. *Studia Math.* 44 (1972):167-186.

4. Calderón, A. P., and Zygmund, A. Local properties of solutions of elliptic partial differential equations. *Studia Math.* 20 (1961):171-225.

5. Chanillo, S., and Wheeden, R. L. Distribution function estimates for Marcinkiewicz integrals and differentiability. To appear in *Duke Math J.*

6. Chanillo, S., and Wheeden, R. L. Inequalities for Peano maximal functions and Marcinkiewicz integrals. To appear.

7. DeVore, R., and Sharpley, R. Maximal operators. To appear.

8. Fefferman, C. L. Inequalities for strongly singular convolution operators. *Acta. Math.* 124 (1970):9-36.

9. Gatto, A. B. E.; Jiménez, J. R.; and Segovia, C. On the solutions of the equation $\Delta^m F = f$ for $f \in H^p$. This volume, p. 394.

10. Hirschman, I. I., Jr. Fractional integration. *Amer. J. of Math.* 75 (1953):531-546.

11. Marcinkiewicz, J. Sur quelques integrals du type de Dini. *Ann. Soc. Polonaise Math.* 17 (1938):42-50.

12. Segovia, C., and Wheeden, R. L. On certain fractional area integrals. *J. Math. Mech.* (now *Indiana Univ. Math. J.*) 19 (1969):247-262.

13. Stein, E. M. The characterization of functions arising as potentials I. *Bull. Amer. Math. Soc.* 67 (1961):102-104.

14. Stein, E. M. *Singular Integrals and Differentiability Properties of Functions.* Princeton Math. Series No. 30. Princeton, N.J.: Princeton Univ. Press, 1970.

15. Stein, E. M., and Zygmund, A. On the differentiability of functions. *Studia Math.* 23 (1964):247-283.

16. Stein, E. M., and Zygmund, A. On the fractional derivatives of functions. *Proc. London Math. Soc.* 14A (1965):249-264.

17. Torchinsky, A. Weighted norm inequalities for the Littlewood-Paley function g_λ^*. *Proc. Symp. Pure Math.* 35, part 1 (1979): 125-131.

18. Waterman, D. On an integral of Marcinkiewicz. *Proc. Internat. Congress of Mathematicians* 2 (1954):185-186.

19. Weiss, G. Lecture notes.

20. Zygmund, A. On certain integrals. *Trans. Amer. Math. Soc.* 55 (1944):170-204.

SOME RESULTS AND OPEN PROBLEMS IN DIFFERENTIATION OF INTEGRALS

Miguel de Guzmán
Universidad Complutense de Madrid

The purpose of this brief note is to present in an expository way some developments in the theory of differentiation of integrals in which one can strongly feel the influence of A. Zygmund, in their origin, their style, and their motivation. For more technical details the reader is referred to the recent work of the author in 1981.

The theory of differentiation, born from the fundamental theorem of Lebesgue in 1910, has expanded into two main branches, one more abstract and global, in the direction of the Lebesgue-Radon-Nikodym theorem, and the other more local and concrete. It has been in the latter direction where the influence of Zygmund has been more strong and decisive. Through the problems he has raised, some solved by himself, some by others, he has indicated one direction of research that has proved extremely useful, both for the development of the theory in itself and for the links that have been established with other branches of the modern real analysis and in particular with the real variable developments of the last thirty years in Fourier analysis.

1

When one considers the classical theorem of Lebesgue about the differentiation of the integral of a function f in $L^1(\mathbf{R}^n)$ through cubic intervals or through spheres containing the point at which one differentiates, several questions arise in a natural way. Restricting ourselves to $\mathbf{R}^2$ for the sake of clarity, one can ask, for example: (1) What happens if instead of square intervals one considers rectangles?; (2) How about rectangles in one fixed direction for each point x of the plane, such directions varying from point to point?

These two simple questions, their surprising answers, in a good part given by Zygmund, and the subsequent questions and developments they have stimulated have strongly marked the ulterior research in the field.

2

In 1927, Nikodym constructed a set N of measure 1 in the unit square Q of $\mathbf{R}^2$ so that for each point $x \in N$ there is a straight line $\ell(x)$ such that $\ell(x) \cap N = \{x\}$. From this fact Zygmund, as Nikodym points out at the end of his paper, deduced in an easy way that the Lebesgue differentiation theorem is false when one takes, instead of square intervals, rectangles containing the point at which one differentiates, and this is so even if one imposes that $f \in L^\infty(\mathbf{R}^2)$; that is, the Lebesgue density theorem does not hold for arbitrary rectangles. Moreover, one observes from the same fact of the existence of the Nikodym set that even if one fixes for each point x of $\mathbf{R}^2$ a single direction $d(x)$ and takes the means of the integral $\int f$ for $f \in L^\infty(\mathbf{R}^2)$ over the rectangles containing x in that direction $d(x)$, the differentiation property may completely fail.

3

Some years after this result came the strong density theorm of Saks in 1933 and the more complete results of Zygmund [18] and Jessen, Marcinkiewicz, and Zygmund [4] telling that rectangular parallelepipeds in $\mathbf{R}^n$ are a good basis for the differentiation of $\int f$, $F \in L \times (\log^+ L)^{n-1}(\mathbf{R}^n)$, provided one fixes the direction of their axes.

The general philosophy underlying many of the questions proposed by Zygmund in differentiation has been inspired by these results. A differentiation basis in $\mathbf{R}^n$ can have too many sets to present good differentiation properties, not differentiating the integrals of a desirable class of functions, like, for instance, L^p. A restriction imposed on the geometry of the basis or a restriction on the space of functions whose integrals we would like to differentiate, or both, can give an acceptable theorem of differentiation.

4

An instance of the foregoing situation is the following. According to the Jessen-Marcinkiewicz-Zygmund theorem, if in $\mathbf{R}^n$ one considers the basis of intervals with *no restriction* on their sides, one obtains the differentiation of $\int f$, for $f \in L(\log^+ L)^{n-1}(\mathbf{R}^n)$. What will happen if we consider the basis of intervals in $\mathbf{R}^n$ with s ($s \leq n$) of their sides of the same length, the others being arbitrary? This was the question raised and answered by Zygmund in 1967. This basis differentiates $L(\log^+ L)^{n-s}(\mathbf{R}^n)$.

5

In this same spirit was raised by Zygmund the question about the role that a space of functions in $\mathbf{R}^2$ like $L(\log^+ L)^{1/2}(\mathbf{R}^2)$ can play in differentiation. Perhaps can one say that a basis of intervals of side lengths d, D, with $1 > D > d > D^2$, differentiates $L(\log^+ L)^{1/2}(\mathbf{R}^2)$? This question was answered by R. Moriyón in 1975. Such a basis does not differentiate any space worse in the appropriate sense than $L(\log^+ L)(\mathbf{R}^2)$.

However, the question about the possible role in differentiation of a space like $L(\log^+ L)^{1/2}(\mathbf{R}^2)$ remains still open. Can one construct a differentiation basis such that it differentiates $L(\log^+ L)^{1/2}(\mathbf{R}^2)$ and no worse space?

6

Another problem with the same flavor suggested by Zygmund about bases of parallelepipeds was the following: Let $\phi(s_1, s_2, \ldots, s_{n-1})$ be a function from $[0, \infty) \times [0, \infty) \times \cdots \times [0, \infty)$ to $[0, \infty)$, which is separately increasing in each variable with $\phi(0, 0, \ldots, 0) = 0$. Consider the basis of all intervals in $\mathbf{R}^n$ with side lengths s_1, $s_2, \ldots, s_{n-1}, \phi(s_1, \ldots, s_{n-1})$. What are the differentiation properties of this basis? One good guess, looking at the Jessen-Marcinkiewicz-Zygmund theorem, could be that it differentiates $L(\log^+ L)^{n-2}(\mathbf{R}^n)$. That for $n = 3$ it is so has been proved by A. Córdoba in 1978. The question for $n > 3$ is still open.

7

Another interesting question proposed by Zygmund was connected with the following result of Saks [13]: there exists a function f in $L^1(\mathbf{R}^2)$ such that the basis of intervals of $\mathbf{R}^2$ differentiates $\int f$ at almost no point x of $\mathbf{R}^2$ (the upper derivative of $\int f$ at almost each x is $+\infty$). The question, both natural and practical, proposed by Zygmund is now the following: given $g \in L^1(\mathbf{R}^2)$, is it possible to choose a direction so that the basis of all rectangles in such direction differentiates $\int g$ [that is, the derivative of $\int g$ at almost each $x \in \mathbf{R}^2$ is $g(x)$]? If the answer were positive, for a given g, a change of axes would ease many considerations. However, the answer, found by Marstrand in 1977, is negative. The situation is similar in some other more general cases studied by El Helou [2] and L. Melero [7].

8

The main method by which one has traditionally attacked problems in the local theory of differentiation of integrals has been to obtain a covering lemma in some form similar to the classical one of Vitali [16] by which Lebesgue obtained his theorem. Its connection with the Hardy-Littlewood maximal theorem and the further manipulation of this latter tool by an adequate process of iteration have given rise to most of the positive results in the field. Zygmund has been very deeply conscious of this strong dependence of the theory on geometric facts and sought himself and stimulated others to seek methods in differentiation not involving covering lemmas. Negative results such as the one of Nikodym quoted in Sec. 2, the one of Saks in Sec. 3, or that of Marstrand in Sec. 7 did not involve covering methods, but these tools proved rather ineffective to produce positive results and to attack some of the many interesting open problems in differentiation. For this reason, one can say that the recent methods involving the application of powerful techniques of Fourier analysis to solve problems that still seem far from being successfully tackled by other more classical geometric considerations come to fulfill Zygmund's expectations. Such methods were first introduced by Nagel, Rivière, and Wainger [9, 10, and 11] and were later perfected by Stein and Wainger [15].

It is now justifiably expected that such methods may give a complete answer to the question formulated in Sec. 1 (2): Let be given for each point x of $\mathbf{R}^2$ a direction $d(x)$. For each x consider the family $B(x)$ of all rectangles centered at x, one of whose sides has direction $d(x)$. What are the differentiation properties of this basis? The basis can be very bad, as the Nikodym set shows, but if the field of directions varies in a sufficiently smooth way, the basis is rather good. This problem and many others connected with it have been successfully explored by Stein and Wainger by means of their useful tools.

REFERENCES

1. Córdoba, A. $s \times t \times \phi(s, t)$. Mittag-Leffer Institute, Report No. 9 (1978).

2. El Helou, J. Recouvrement du tore T^q par des ouverts aléatoires. Preprint.

3. de Guzmán, M. Real variable methods in Fourier analysis. North-Holland Mathematics Studies 46, Amsterdam, 1981.

4. Jessen, B.; Marcinkiewicz, J.; and Zygmund, A. Note on the differentiability of multiple integrals. *Fund. Math.* 25 (1935): 217-234.

5. Lebesgue, H. Sur l'integration des fonctions discontinues. *Ann. Sci. Ecole Norm. Sup.* 27 (1910):361-450.

6. Marstrand, J. M. A counter-example in the theory of strong differentiation. *Bull. London Math. Soc.* 9 (1977):209-211.

7. Melero, B. L. A negative result in differentiation. To be published in *Studia Math.*

8. Moriyón, R. On the derivation properties of a class of bases, Appendix III. In M. de Guzmán, *Differentiation of Integrals in* $\mathbf{R}^n$. Berlin: Springer, 1975.

9. Nagel, A.; Rivière, N. M.; and Wainger, S. On Hilbert transforms along curves. *Bull. Amer. Math. Soc.* 80 (1974):106-108.

10. Nagel, A.; Rivière, N. M.; and Wainger, S. On Hilbert transforms along curves, II. *Amer. J. Math.* 98 (1976):395-403.

11. Nagel, A.; Rivière, N. M.; and Wainger, S. A maximal function associated to the curve (t, t^2). *Proc. Nat. Acad. Sci.* 73 (1976):1416-1417.

12. Nikodym, O. Sur la mesure des ensembles plans dont tous les points sont rectilinéairement accessibles. *Fund. Math.* 10 (1927):116-168.

13. Saks, S. On the strong derivatives of functions of an interval. *Fund. Math.* 25 (1935):235-252.

14. Saks, S. Théorie de l'integrale. Warszawa, 1933.

15. Stein, E. M., and Wainger, S. Problems in harmonic analysis related to curvature. *Bull. Amer. Math. Soc.* 84 (1978):1239-1295.

16. Vitali, G. Sui gruppi di punti e sulle funzioni di variabile reali. *Atti Accad. Sci. Torino* 43 (1908):75-92.

17. Zygmund, A. A note on the differentiability of multiple integrals. *Colloq. Math.* 16 (1967):199-204.

18. Zygmund, A. On the differentiability of multiple integrals. *Fund. Math.* 23 (1934):143-149.

TANGENTIAL BOUNDARY BEHAVIOR OF HARMONIC EXTENSIONS OF L^p POTENTIALS

Alexander Nagel, Walter Rudin
University of Wisconsin

Joel H. Shapiro
Michigan State University

1. INTRODUCTION

We describe, mostly without proofs, some results on the boundary behavior of harmonic functions in classes modeled on the space $\mathfrak{D}$ of functions harmonic in the open unit disc with finite Dirichlet integral. Detailed proofs will appear elsewhere [6].

For our purposes, $\mathfrak{D}$ is best regarded as the space of Poisson integrals of functions f square integrable on the unit circle T, with the additional restriction:

$$\sum_{-\infty}^{\infty} |n|\,|\hat{f}(n)|^2 < \infty, \tag{1}$$

where $\hat{f}(n)$ is the nth Fourier coefficient of f. Such f are somewhat more regular than "typical" L^2 functions, but still not necessarily continuous, or even bounded. We study how the additional regularity of f affects the boundary behavior of its harmonic extension u.

This type of problem was considered by Salem and Zygmund [7], who showed that the Fourier series of each $f \in L^2(T)$ satisfying (1) converges at each point of T, with the possible exception of

Research partially supported by the National Science Foundation.

a set of logarithmic capacity zero. This implies that each $u \in \mathfrak{D}$ has a radial limit at each point of T, with the possible exception of such a set. In fact, the same is true for nontangential limits.

Our contribution is to show that if larger classes of exceptional sets—intermediate between "log-capacity zero" and "measure zero"—are allowed, then the functions in $\mathfrak{D}$ will converge within regions that meet the unit circle *tangentially*. We describe the precise relationship between the curvature of the approach region and the size of the exceptional sets for convergence within these regions. In particular, we show that each $u \in \mathfrak{D}$ has at almost every $\zeta \in T$ a limit as $z \to \zeta$ through an approach region making *exponential* contact with T at ζ.

To make these matters precise, consider the following regions in the open unit disc U. If $c > 0$ and $\gamma \geq 1$, let

$$\mathcal{A}_{\gamma,c}(\varphi) = \left\{ re^{i\theta} \in U : 1 - r > c\left|\sin\left(\frac{\theta - \varphi}{2}\right)\right|^{\gamma} \right\},$$

while if $\gamma > 0$, let

$$\mathcal{E}_{\gamma,c}(\varphi) = \left\{ re^{i\theta} \in U : 1 - r > \exp\left[-c\left|\sin\left(\frac{\theta - \varphi}{2}\right)\right|^{-\gamma}\right] \right\}.$$

Thus $\mathcal{A}_{\gamma,c}(\varphi)$ has *order of contact* γ with T at $e^{i\varphi}$, while $\mathcal{E}_{\gamma,c}(\varphi)$ has *exponential contact*. Note that the regions $\mathcal{A}_{1,c}(\varphi)$ are the usual nontangential approach regions.

We say a function u defined in U has *$\mathcal{A}_\gamma$-limit L at $e^{i\varphi}$* if $u(z) \to L$ as $z \to e^{i\varphi}$ within $\mathcal{A}_{\gamma,c}(\varphi_0)$ for every $c > 0$. A similar definition applies to $\mathcal{E}_\gamma$-limits. We can state the classical results in this language: if $f \in L^2(T)$ and u is the Poisson integral of f, then u has an $\mathcal{A}_1$-limit at almost every point of T. If, in addition, f satisfies (1), then u has an $\mathcal{A}_1$-limit at every point of T, with the possible exception of a set of logarithmic capacity zero. Our results for the space $\mathfrak{D}$ can be stated as follows.

Theorem 1. Suppose $u \in \mathfrak{D}$. Then

(a) *U has an $\mathcal{E}_1$-limit at almost every point of T.*

(b) *If, in addition, $0 < \beta < 1/2$ and $\gamma = (1 - 2\beta)^{-1}$, then U has an $\mathfrak{A}_\gamma$-limit at every point of T, with the possible exception of a set of $C_{\beta,2}$-capacity zero.*

Here the capacity $C_{\beta,2}$ is the analogue for the unit circle of the corresponding Bessel capacity on R^n (see [5], for example). It coincides with the classical capacity $C_{1-2\beta}$ discussed by Kahane and Salem in [3; Chapter 3, p. 33]. In particular, $C_{1/2,2}$ is the logarithmic capacity, so when $\beta = 1/2$, our result coincides with the previously mentioned one of Salem and Zygmund.

The correct setting for this work is a more general one, motivated by the fact that $\mathfrak{D} = P[K * L^2]$, where P is the Poisson integral for the unit disc,

$$K(\theta) = \left|\sin \frac{\theta}{2}\right|^{-1/2} \sim \sum_{-\infty}^{\infty} (|n| + 1)^{-1/2} e^{in\theta},$$

and $*$ denotes the convolution on T. We prove a generalization of Theorem 1 valid for the classes $P[K * L^p]$ where $1 \leq p < \infty$ and K is a positive, integrable, even function on $[-\pi, \pi]$ that is decreasing on $(0, \pi]$. As a by-product of our work, we answer a question of H. S. Shapiro and A. L. Shields concerning the zeros of holomorphic functions in classes $P[K * L^2]$. These results are stated in detail in the next section.

As the reader has probably guessed, our tangential convergence theorems follow from weak-type estimates on maximal functions associated with our approach regions. For $p > 1$, we use Hansson's strong-type capacity inequality [2] to obtain strong-type maximal estimates, which are in turn crucial to the proof of part (b) of Theorem 1 and its generalizations. We state these results precisely in Secs. 3 and 4. In Sec. 5 we show how the strong-type maximal estimates enter into the proof of Theorem 1, part (b); and in Sec. 6 we discuss Carleson measures for the classes $P[K \star L^p]$.

2. TANGENTIAL CONVERGENCE THEOREMS

We prefer to work in the upper half-space R_+^{n+1} of $(n + 1)$-dimensional Euclidean space, instead of the unit disc; so R^n replaces the unit circle as the boundary. We write L^p for $L^p(R^n)$ and denote points of R_+^{n+1} by (x, y) with $x \in R^n$ and $y > 0$. P will denote the Poisson integral for R_+^{n+1}, with $\{P_y : y > 0\}$ the corresponding Poisson kernels on R^n. $\star$ will denote convolution on R^n.

Potential Spaces and Dirichlet-Type Spaces

Suppose K is a ***kernel*** on R^n; that is, K is positive, integrable, radially symmetric, and $K(x)$ decreases as $|x|$ increases. For $1 \leq p < \infty$, let

$$L_K^p = \{K \star F : F \in L^p\}$$

denote the space of L^p potentials associated with K, and let

$$h_K^p = P[f] : f \in L_K^p$$

be the corresponding space of harmonic extensions to R_+^{n+1}. Every h_K^p is thus an analogue of the Dirichlet space $\mathfrak{D}$ of Sec. 1. In particular, it follows from Plancherel's theorem that h_K^2 consists of the harmonic extensions of functions $f \in L^2$ satisfying the additional conditions

$$\int_{R^n} |\hat{K}(\lambda)|^{-2} |\hat{f}(\lambda)|^2 d\lambda < \infty \tag{2}$$

where ^ denotes the Fourier transform on L^2.

To avoid trivialities, we always assume that $K \notin L^q$, where $p^{-1} + q^{-1} = 1$. Thus L_K^p, and therefore h_K^p, always contains unbounded functions.

Approach Regions

For $(x, y) \in R_+^{n+1}$, let

$$K_y(x) = P[K](x, y) = P_y \star K(x)$$

and set

$$r(y) = r_{K,p}(y) = \| K_y \|_q^{-p/n}, \quad y > 0.$$

For $x_0 \in R^n$ and $\beta > 0$, define

$$\Omega(x_0) = \Omega_{K,\beta}^p(x_0) = \{(x, y) \in R_+^{n+1} : |x - x_0| < \beta r(y)\}.$$

Thus $\Omega(x_0)$ is the region in R_+^{n+1} with spherical cross section of radius $\beta r(y)$ at height y above R^n. Since $K \notin L^q$, we know that $r(y) \to 0$ as $y \to 0^+$, so $\Omega(x_0)$ approaches R^n only at the point x_0 (its "vertex"). Moreover, standard estimates of Poisson integrals show that $y^{-1}r(y) \to 0$ as $y \to 0^+$, so the boundary of $\Omega(x_0)$ actually approaches R^n *tangentially* at x_0.

We define Ω_K^p *limits* of functions defined on R_+^{n+1} exactly as in the last section.

Capacity

Following Meyers [5], if $1 < p < \infty$, we define the (K, p)-capacity of a subset E of R^n as follows. Let $T_{K,p}(E)$ denote those nonnegative $F \in L^p$ for which $K \star F \geq 1$ everywhere on E. Note that since K and F are both positive, the convolution $K \star F$ make sense (possibly $+\infty$) at every point of R^n.

The *capacity* of E is

$$C_{K,p}(E) = \inf\{\|F\|_p^p : F \in T_{K,p}(E)\},$$

where $\|\cdot\|_p$ denotes the L^p norm.

It is easy to check that $C_{K,p}$ is subadditive and monotone increasing on the subsets of R^n, and that a subset E has capacity zero if and only if $K \star F \equiv \infty$ on E for some nonnegative F in L^p [5, Sec. 2].

This last comment shows that the members of the potential space L_K^p are defined and finite at $C_{K,p}$-almost every point of R^n.

Main Results

We can now state our main results on tangential convergence. In what follows, $1 \leq p < \infty$, $p^{-1} + q^{-1} = 1$, $K \notin L^q$, $f \in L_K^p$, and $u = P[f]$.

Theorem 2. *There is a set E of Lebesgue measure zero such that u has Ω_K^p limit $f(x)$ at x for every $x \in R^n \backslash E$.*

Theorem 3. *If, in addition, $p > 1$ and $K = H * G$ where H and G are also kernels, then there is a set E with $C_{H,p}(E) = 0$ such that u has Ω_G^p limit $f(x)$ at x for each $x \in R^n \backslash E$.*

Thus Theorem 3 shows precisely how the degree of tangential convergence influences the size of the exceptional set.

Examples: Bessel Potentials

The most important class of kernels are the Bessel kernels $K = g_\alpha$ defined for $0 < \alpha \leq n$ by

$$\hat{g}_\alpha(\lambda) = (1 + |\lambda|^2)^{-\alpha/2}, \quad \lambda \in R^n.$$

The corresponding capacities $B_{\alpha,p}$ make sense for $\alpha p \leq n$ and are called *Bessel capacities*. Observe that $g_\alpha * g_\beta = g_{\alpha+\beta}$. Properties of these kernels are worked out in detail in (for example) Meyers [5, Sec. 7, p. 279]. The analogous kernels for the unit circle are the ones considered in Kahane and Salem [3, Chapter 3]:

$$\Phi_\alpha(\theta) = \left|\sin\frac{\theta}{2}\right|^{\alpha-1} \sim \sum (|n| + 1)^{-\alpha} e^{in\theta}$$

for $0 < \alpha < 1$, and

$$\Phi_1(\theta) = -\log\left|\sin\frac{\theta}{2}\right| \sim \sum (|n| + 1)^{-1} e^{in\theta}.$$

As $y \to 0^+$, we have the following asymptotic estimates on the radius $r_{g_\alpha, p}(y) = r_{\alpha,p}(y)$, defined in "Approach Regions" of Sec. 2:

$$r_{\alpha,p}(y) \sim \begin{cases} y^{1-(\alpha p/n)} & \text{if } \alpha p < n \\ (-\log y)^{-1/n(q-1)} & \text{if } \alpha p = n,\ p > 1 \\ (-\log y)^{-1/n} & \text{if } \alpha = n,\ p = 1 \end{cases} \tag{3}$$

where $p^{-1} + q^{-1} = 1$. By analogy with the work of Sec. 1, we define approach regions for $c > 0$: for $\gamma \geq 1$,

$$\mathfrak{A}_{\gamma,c} = \{(x, y) \in R_+^{n+1} : y > c|x - x_0|^\gamma\},$$

while if $\gamma > 0$, then

$$\mathcal{E}_{\gamma,c} = \{(x, y) \in R_+^{n+1} \quad y > \exp(-c|x - x_0|^{-\gamma})\}.$$

Then estimate (3) shows that the regions $\Omega^p_{g_{\alpha,\beta}}$ essentially coincide with the classes of regions:

$\mathfrak{A}_{\gamma,c}$ if $\alpha p < n$, where $\gamma = n/(n - \alpha p)$,

$\mathcal{E}_{\gamma,c}$ if $\alpha p = n$ and $p > 1$, where $\gamma = n(q - 1)$,

and

$\mathcal{E}_{n,c}$ if $\alpha = n$ and $p = 1$.

Thus Theorems 2 and 3 have the following corollaries, which generalize Theorem 1. Here $f \in L^p_{g_\alpha}$ where $\alpha p \leq n$, so $u = P[f] \in h^p_{g_\alpha}$; and "a.e." refers to Lebesgue measure on R^n.

Corollary 1.

(a) *If* $\alpha p < n$ *and* $p > 1$, *then* u *has* $\mathfrak{A}_{n/(n-\alpha p)}$*-limit* $f(x)$ *at a.e.* x *in* R^n.

(b) *If* $\alpha p = n$ *and* $p > 1$, *then* u *has* $\mathcal{E}_{n(q-1)}$*-limit* $f(x)$ *at a.e.* x *in* R^n.

(c) *If* $\alpha = n$ *and* $p = 1$, *then* U *has* $\mathcal{E}_n$*-limit* $f(x)$ *at a.e.* x *in* R^n.

Corollary 2. Suppose in addition to the hypotheses above, that $1 < p < \infty$, $\alpha p \leq n$, *and* $\alpha = \tau + \kappa$, *where* τ *and* κ *are positive* *numbers. Then u has $\mathfrak{A}_{n/(n-\tau p)}$-limit $f(x)$ at each $x \in R^n$ with the possible exception of a set of $B_{\kappa,p}$ capacity zero.*

Remark. If we return to the unit circle and the case $p = 2$, then as noted earlier, the capacity $B_{\alpha,2}$ corresponds to the classical capacity $C_{1-2\alpha}$ of Kahane and Salem [3, Chapter 3] for $0 < \alpha \leq 1/2$, with the understanding that C_0 is logarithmic capacity. Note that the index in Kahane and Salem's capacity refers to the exponent associated with the ***kernel***, while the index α of the Bessel capacity refers to the exponent associated with the ***Fourier transform*** of the kernel.

Application to Zeros of h_K^p Functions

The next result generalizes one proved by H. S. Shapiro and A. L. Shields in the case $n = 1$, $p = 2$ for special kernels and holomorphic functions [8, Theorem 3], and it answers a question posed by them [8, p. 224].

Theorem 4. *Suppose $1 \leq p < \infty$ and suppose $(y_j)_1^\infty$ is a sequence of positive numbers with $\sum r_{K,p}^n(y_j) = \infty$. Then there exists a sequence $(x_j)_1^\infty$ in R^n such that no nontrivial function in the class h_K^p vanishes at each point (x_j, y_j) of R_+^{n+1}.*

Proof. The hypothesis on (y_j) ensures that we can choose open balls B_j of radius $r_{K,p}(y_j)$ such that each point of R^n lies in infinitely many B_j. Let x_j be the center of B_j and set $z_j = (x_j, y_j) \in R_+^{n+1}$. If $x \in R^n$, then x belongs to some sequence $B_{j_1}, B_{j_2}, \ldots$ of balls; hence $z_{j_1}, z_{j_2}, \ldots$ all belong to $\Omega_{K,1}^p(x)$. So if $u \in h_K^p$ vanishes at each z_j, then it has Ω_K^p limit zero at each x for which it has an Ω_K^p limit. By Theorem 2 this happens for almost every x in R^n. Since u is the Poisson integral of its boundary function, this implies $u \equiv 0$, which completes the proof.

3. TANGENTIAL MAXIMAL FUNCTIONS: WEAK-TYPE INEQUALITIES

If u is a complex valued function defined on R_+^{n+1}, define for $1 \leq p < \infty$ and $\beta > 0$:

$$\mathfrak{M}_{K,p,\beta}u(x_0) = \sup\{|u(x, y)| : (x, y) \in \Omega_{K,\beta}^p(x_0)\}.$$

If $u = P[f]$ we will also write this as $\mathfrak{M}_{K,p,\beta}f(x_0)$. Our fundamental estimate on this tangential maximal function comes from a direct comparison with the following L^p-variant of the Hardy-Littlewood maximal function. For $f \in L^p$ and $x_0 \in R^n$, let

$$M_pF(x_0) = \sup_{r>0}\left\{\frac{1}{m(B_r(x_0))}\int_{B_r(x_0)}|F|^p\,dm\right\}^{1/p},$$

where $B_r(x_0)$ is the ball of radius r in R^n, centered at x_0, and m is Lebesgue measure on R^n. Clearly M_p is subadditive, and by the usual Hardy-Littlewood maximal theorem, it is of weak type (p, p). We have:

Lemma. Suppose $1 \leq p < \infty$ *and* $\beta > 0$. *Then there exists* $A = A(K, p, \beta) > 0$ *such that if* $f = K * F$ *for* $F \in L^p$, *then*

$$\mathfrak{M}_{K,p,\beta}f(x_0) \leq AM_pF(x_0)$$

for every $x_0 \in R^n$.

This lemma immediately gives the following weak-type estimate, which by standard arguments yields Theorem 2.

Theorem 5. For each $1 \leq p < \infty$ *and* $\beta > 0$ *there exists* $A = A(K, p, \beta) < \infty$ *such that for every* $f = K * F$ *with* $F \in L^p$:

$$m\{x \in R^n : \mathfrak{M}_{K,p,\beta}f(x) > \lambda\} \leq \left(\frac{A\|F\|_p}{\lambda}\right)^p$$

for every $\lambda > 0$.

Theorem 5 is "best possible" in the following sense. $\mathfrak{M}_{K,p,\beta}$ is not of weak type (r, r) for any $r < p$. Moreover, if Ω is any region in R_+^{n+1} that approaches R^n only at 0, and $\Omega(x_0) = \Omega + x_0$, then the Ω-maximal function

$$M_\Omega f(x_0) = \sup\{|u(x, y)| : (x, y) \in \Omega(x_0)\}$$

($u = P[f]$) is of weak type (p, p) on L_K^p only if there is a "slab" in R_+^{n+1}:

$$S = \{(x, y) \in R_+^{n+1} : y < y_0\},$$

and a $\beta > 0$ such that

$$\Omega \cap S \subset \Omega_{K,\beta}^p(0) \cap S.$$

We remark that the lemma is a consequence of the following convolution inequality, whose proof is fairly straightforward. If $1 \leq p < \infty$, then there exists $A = A(n, p) < \infty$ such that if $F \in L^p$ *and* K *is a nonnegative, radial decreasing function on* R^n, *then*

$$|K * F(x)| \leq AM_pF(x_0)\left[|x - x_0|^{n/p}\|K\|_q + \|K\|_1\right]$$

for all x *and* x_0 *in* R^n.

The lemma follows from this inequality upon replacing K by K_y and taking the supremum of both sides for $|x - x_0| < r_{K,p}(y)$ and $y > 0$.

4. STRONG-TYPE ESTIMATES

For $1 < p < \infty$ we have the following improvement of Theorem 5.

Theorem 6. *Suppose* $1 < p < \infty$ *and* $\beta > 0$. *Then there exists* $A = A(K, p, \beta) < \infty$ *such that if* $f = K * F$ *for* $F \in L^p$, *then* $\mathfrak{M}_{K,p,\beta}f \in L^p$ *and* $\|\mathfrak{M}_{K,p,\beta}f\| \leq A\|F\|_p$.

Since M_p is not of strong type (p, p), this result shows that $\mathfrak{M}_{K,p,\beta}$ is not "equivalent" to M_p if $1 < p < \infty$. As we will see in Sec. 5, Theorem 6 is the main step in the proof of Theorem 3.

Toward the Proof of Theorem 6

We need the following geometry. For $x_0 \in R^n$, let

$$\Gamma(x_0) = \{(x, y) \in R_+^{n+1} : |x - x_0| < y\}$$

denote the right circular cone in R_+^{n+1} with vertex x_0; and for u continuous on R_+^{n+1}, let

$$Nu(x_0) = \sup\{|u(x, y)| : (x, y) \in \Gamma(x_0)\}$$

be the usual nontangential maximal function. For E an open subset of R^n, let

$$S(E) = R_+^{n+1} \setminus \bigcup_{x \notin E} \Gamma(x),$$

the usual "Carleson set" over E, and for F an open subset of R_+^{n+1}, $1 \leq p < \infty$, and $\beta > 0$, let

$$J(F) = J_{K,\beta}^{p}(F) = \{x \in R^n : \Omega_{K,\beta}^{p}(x) \cap F \neq \emptyset\}.$$

So $J(F)$ is a sort of "shadow" of F on R^n. Then we have the following two results, the first of which is an interesting geometric lower bound for capacity. Assume $1 < p < \infty$ and $\beta > 0$.

Proposition 1. There exists $A = A(K, p, \beta) < \infty$ such that for every open set $E \subset R^n$:

$$m\{J_{K,\beta}^{p}(S(E))\} \leq AC_{K,p}(E).$$

Proposition 2. If u is continuous in R_+^{n+1} and $1 < p < \infty$, then for every $\lambda > 0$:

$$m\{\mathfrak{M}_{K,p,\beta}u > \lambda\} \leq AC_{K,p}\{Nu > \lambda\}$$

where $A = A(K, p, \beta) < \infty$.

Proposition 2 follows from Proposition 1 upon observing that if $E = \{Nu > \lambda\}$, then $J_{K,\beta}^{p}(S(E))$ is an open set containing $\{\mathfrak{M}_{K,p,\beta}u > \lambda\}$. Proposition 1 follows from the weak-type estimate of

Theorem 5 and the fact that if $F \in T_{K,p}(E)$, then (since $K * F \geq 1$ on E) $P[K * F] \geq \lambda_0$ on $S(E)$, where λ_0 depends only on the dimension n and not on E. We describe a similar argument in more detail in the next section.

Finally, we require:

Hansson's Strong-Type Estimate [2, Theorem 2.4]. *If* $1 < p < \infty$, *then there exists* $A = A(K, p) < \infty$ *such that for each* $F \in L^p$:

$$\int_0^\infty C_{K,p}\{|K * F| > \lambda\} d(\lambda^p) \leq A\|F\|_p^p.$$

Special cases of this result have previously been obtained by Adams [1] and Maz'ya [4]. Note that the estimate is a significant improvement over the trivial weak-type estimate

$$C_{K,p}\{|K * F| > \lambda\} \leq \left(\frac{\|F\|_p}{\lambda}\right)^p, \tag{4}$$

which is just a statement of the fact that for every $F \in L^p$ and $\lambda > 0$, the function $|F|/\lambda$ belongs to $T_{K,p}(\{|K * F| > \lambda\})$.

<u>Proof of Theorem 6</u>

Suppose $f = K * F$ for $F \in L^p$, and write $u = P[f]$, $v = P[F]$. Then it is easy to see that $Nu \leq K * Nv$, so

$$\begin{aligned}
\|\mathfrak{M}_{K,p,\beta}u\|_p^p &= \int_0^\infty m\{\mathfrak{M}_{K,p,\beta}u > \lambda\} d(\lambda^p) \\
&\leq A\int_0^\infty C_{K,p}\{Nu > \lambda\} d(\lambda^p) \qquad \text{(Prop. 2)} \\
&\leq A\int_0^\infty C_{K,p}\{K * Nv > \lambda\} d(\lambda^p) \\
&\leq A\|Nv\|_p^p \qquad \text{(Hansson's estimate)} \\
&\leq A\|F\|_p^p,
\end{aligned}$$

where the last inequality follows from the Hardy-Littlewood non-tangential maximal theorem.

5. CAPACITY VS. CURVATURE: PROOF OF THEOREM 3

Recall that Theorem 3 describes how the curvature of our approach regions influences the capacity of the exceptional sets for convergence of h_K^p functions within these regions. Theorem 3 follows by standard arguments from part (b) of the next result, which in turn follows from part (a) and Hansson's estimate.

Theorem 7. Suppose $1 < p < \infty$, $\beta > 0$, *and* $K = H \star G$, *where* H *and* G *are also kernels.*

(a) *Then there exists* $A = A(p, \beta, G, H) < \infty$ *such that*

$$C_{H,p}\{\mathfrak{M}_{G,p,\beta}u > \lambda\} \leq AC_{K,p}\{Nu > \lambda\}$$

for every u *continuous on* R_+^{n+1} *and every* $\lambda > 0$.

(b) *If, in addition,* $u = P[K \star F]$ *where* $F \in L^p$, *then*

$$\int_0^\infty C_{H,p}\{\mathfrak{M}_{G,p,\beta}u > \lambda\}d(\lambda^p) \leq A\|F\|_p^p$$

for A *as in part* (a).

Note that part (a) above is a capacitary analogue of Proposition 2 of the last section. It follows from a similar analogue of Proposition 1, which we state and prove in detail.

Proposition 3. Under the hypotheses of Theorem 7 we have for every open subset E *of* R^n:

$$C_{H,p}\{J_{G,\beta}(S(E))\} \leq AC_{K,p}(E),$$

where $A = A(p, \beta, G, H) < \infty$.

Proof. Write $M_G = M_{G,\beta,p}$ and $J_G = J_{G,\beta}^p$. Suppose $F \in T_{K,p}(E)$ and write $u = P[K \star F]$ and $v = P[G \star F]$. Since $K \star F \geq 1$ on E, we know that $u \geq \lambda_0$ on $S(E)$, where $\lambda_0 > 0$ is independent of E. Thus

$$J_G(S(E)) \subset \{\mathfrak{M}_G u > \lambda_0\}$$

and so

$$\begin{aligned} C_{H,p}\{J_G(S(E))\} &\le C_{H,p}\{\mathfrak{M}_G u > \lambda_0\} \\ &= C_{H,p}\{\mathfrak{M}_G P[H * G * F] > \lambda_0\} \\ &\le C_{H,p}\{H * \mathfrak{M}_G P[G * F] > \lambda_0\} \\ &\le \left(\frac{\|\mathfrak{M}_G v\|_p}{\lambda_0}\right)^p \qquad (v = P[G * F]) \\ &\le \left(\frac{A\|F\|_p}{\lambda_0}\right)^p, \end{aligned}$$

where the next-to-last inequality follows from the trivial weak-type estimate (4) of Sec. 4, and the last inequality follows from our strong-type estimate Theorem 6. Take the infimum of the right side of the above inequality as F ranges through $T_{K,p}(E)$. The proof is complete.

6. CARLESON MEASURES

We call a finite positive Borel measure μ on R_+^{n+1} a *Carleson measure for* h_K^p if

$$\int |u|^p \, d\mu < \infty$$

for every $u \in h_K^p$. Stengenga [9] has characterized the (K, p) Carleson measures for $1 < p < \infty$ as follows: they are precisely those μ for which there exists $A > 0$ such that

$$\mu(S(E)) \le AC_{K,p}(E) \tag{5}$$

for every open subset E of R^n. Stengenga also gives examples that show that if (5) holds merely for open balls E, then μ need not be a Carleson measure.

Using Theorem 6 and standard arguments, we can obtain a more geometric condition on μ that is *sufficient* for it to be a Carleson measure for h_K^p. To state our result, we define for every open subset E of R^n a sort of "(K, p, β)-Carleson region" over E:

$$Q^p_{K,\beta}(E) = R^{n+1}_+ \setminus \bigcup_{x \notin E} \Omega^p_{K,\beta}(x).$$

Theorem 8. Suppose $1 < p < \infty$, $\beta > 0$; *and suppose there exists* $A < \infty$ *such that*

$$\mu\{Q^p_{K,\beta}(B)\} \leq Am(B)$$

for every open ball B *in* R^{n+1}_+. *Then* μ *is a Carleson measure for* h^p_K.

However, this condition is *not* necessary. Using an idea communicated to us by Stengenga, we can show that for each kernel K and $1 < p < \infty$, there exists a Carleson measure μ for h^p_K that does *not* satisfy the hypothesis of the theorem.

REFERENCES

1. Adams, D. R. On the existence of capacitary strong type estimates in R^n. *Ark. Mat.* 14 (1976):125-140.

2. Hansson, K. Imbedding theorems of Sobolev type in potential theory. *Math. Scand.* 45 (1979):77-102.

3. Kahane, J. P., and Salem, R. Ensembles parfaits et séries trigonométriques. *Actualités Sci. Ind.* No. 1301. Paris: Hermann, 1963.

4. Maz'ya, V. G. On capacitary estimetes of strong type for fractional norms. In Russian. *Zap. Sem. LOMI Leningrad* 70 (1977): 161-168.

5. Meyers, N. G. A theory of capacities for potentials of functions in Lebesgue classes. *Math. Scand.* 26 (1970):255-292.

6. Nagel, A.; Rudin, W.; and Shapiro, J. H. Tangential boundary behavior of functions in Dirichlet-type spaces. *Annals of Math.* 115 (1982).

7. Salem, R., and Zygmund, A. Capacity of sets and Fourier series. *Trans. Amer. Math. Soc.* 59 (1946):23-41.

8. Shapiro, H. S., and Shields, A. L. On the zeros of functions with finite Dirichlet integral, and some related function spaces. *Math. Zeit.* 80 (1962):217-229.

9. Stengenga, D. A. Multipliers of the Dirichlet space. *Ill. J. Math.* 24 (1980):113-139.

STRONG DIFFERENTIALS IN L^p

G.V. Welland
University of Missouri, St. Louis

In the paper [1], A. Córdoba has shown that a conjecture of A. Zygmund is true. This conjecture was made in 1935 and roughly states that for the purpose of differentiation of the integral, the differentiation basis B_ϕ, which consists of rectangles oriented with sides parallel to a fixed set of coordinate axes and having side lengths that vary with one of s, t, and $\phi(s, t)$, behaves in the same way as B_2, the differentiation basis that allows rectangles oriented as in B_ϕ with at most two distinct side lengths. The function ϕ increases monotonically from 0 in both the variables, s and t. In [3], M. Weiss investigated strong differentiation in L^p. Among the facts that she established is the following theorem.

Theorem 1. If $k \geq 1$, $p \geq 1$, and f has a (k, p) differential in E, it has a $(k, p)''$ differential almost everywhere in E.

We recommend that the reader see [3 and 4] for many related results on differentials. The point we will make is that Theorem 1 relies on a lemma that involves the differentiation basis B_2 and, just as Córdoba has shown in the above case, B_2 can be replaced by B_ϕ for this case as well. The consequence is that Theorem 1 can now be stated with $(k, p)''$ differential replaced by $(k, p)\phi$

differential. However, the result of Córdoba is not required and one can follow more established lines of argument.

Definition 1. The function f has a (k, p) differential at x if there is a polynomial $P(t)$ of degree k such that

$$\left(\frac{1}{|Q|}\int_Q |f(x+t) - P(t)|^p \, dt\right)^{1/p} = o(h^k) \text{ as } h \to 0, \tag{1}$$

where Q is a cube in $\mathbf{R}^n$ containing the origin with side length h.

Here and in the following, f is assumed to be in the appropriate Lebesgue class, in this case L^p_{loc}.

Let R be a rectangle containing the origin and have side length s for sides parallel to the axes corresponding to coordinates $x_1, \ldots, x_{j_1}$; side length, t for sides parallel to the axes corresponding to coordinates $x_{j_1+1}, \ldots, x_{j_2}$; and side length $\phi(s, t)$ for sides parallel to the axes corresponding to the coordinates $s_{j_2+1}, \ldots, x_n$. Let ω be the largest of these side lengths.

Definition 2. The function f has a $(k, p)_\phi$ differential at x if there is a polynomial $P(t)$ of degree k such that

$$\left(\frac{1}{|R|}\int_R |f(x+t) - P(t)|^p \, dt\right)^{1/p} = o(w^k) \text{ as } w \to 0, \tag{2}$$

where R is a rectangle described above.

If $\phi(s, t) \equiv s$, then a $(k, p)_\phi$ differential becomes a $(k, p)''$ differential. For what follows, we allow cubes and rectangles like those just described to be translated. The resulting rectangles form the differentiation basis B_ϕ. The lemma from which Theorem 1 can be obtained with $(k, p)_\phi$ replacing $(k, p)''$ is now given. The use of this lemma can be found in [4].

Lemma. There is a positive constant A depending only on the dimension n, j_1, j_2, and $\alpha > 0$ having the following property. Let $g(x) = g(x_1, \ldots, x_n)$ be defined on a cube Q_0, nonnegative and *integrable. Let U be the set of points in Q_0 that are contained in a cube Q of side length h such that*

$$\int_Q g(\xi)d\xi > h^{n+\alpha}. \tag{3}$$

Let V be the set of points in Q_0 that are contained in a rectangle $R \in B_\phi$ having longest side w such that

$$\int_R g(\xi)d\xi > w^\alpha |R|. \tag{4}$$

Then

$$|V| \leq A|U|. \tag{5}$$

Before we go to the proof, let $r = \phi(s, t)$. A rectangle can have three possible configurations, in the sense that we have that r is between the values of s and t, $r \geq \max(s, t)$ or $r \leq \min(s, t)$. We would like to consider only one situation among these; that is the case $r < t < s$. If we require that $\phi(s, t)$ be continuous and strictly monotone, then the case $t < r < s$ can be described as $\Psi_1(r, s) < r < s$ where Ψ_1 inverts the equation $r = \phi(s, t)$ by solving for t. Similarly, $t < s < r$ can be reduced to $t < \Psi_2(t, r) < r$ where Ψ_2 inverts the equation $r = \phi(s, t)$ by solving for s. In both of these cases, the resulting Ψ_i is also continuous and strictly increasing. Hence, we may assume $\phi(s, t) < t < s$ without loss of generality.

The proof of (5) follows very closely that of [3]. We will be short in describing the proof up to the point where it differs from that of [3].

As in [3], we may assume that there exists a sequence of cubes contained in Q_0 that are nonoverlapping, $\{Q_k\}$, with edge lengths h_1, h_2, ..., such that

$$1 < h_k^{-(n+\alpha)} \int_{Q_k} g(x)dx \leq 2^{n+\alpha}, \quad k = 1, 2, \ldots, \tag{6}$$

and $g(x) = 0$ almost everywhere on the complement of $\cup Q_k$. Again, as in [3], we have $\cup Q_k \subset U$ and we write $Q_k = Q_k' \times Q_k''$, where Q_k' is the j_1-dimensional cube that has sides parallel to those axes corresponding to the edge lengths s in B_ϕ. Let J_k'' be the cube obtained by expanding Q_k'' about its center by a factor γ, where γ is the smallest odd integer larger than 3 for which

$$(\gamma - 1)^\alpha \geq 2^{n+1+2\alpha} 3^n.$$

Let $\hat{Q}_k$ be Q_k expanded about its center by a factor of 3. Then V_2 is defined to be $V \cap \bigcup_k \hat{Q}_k$ and $V_1 = V - V_2$. One has

$$|V_2| \leq 3^n |U| \tag{7}$$

and with $V_1^{x''} = \{x' : (x', x'') \in V_1\}$,

$$|V_1| = \int_{Q_0''} |V_1^{x''}| \, dx''. \tag{8}$$

For most of the following, let x'' be fixed. Then $x' \in V_1^{x''}$ if and only if $(x', x'') \notin \cup \hat{Q}_k$ and there exists an $R = R' \times R''$ with longest edge length s and R' is a j_1-dimensional cube with edge length s such that

$$\int_R g(\xi) d\xi > s^{j_1+\alpha} |R''|, \text{ where } |R''| = t^{j_2} (\phi(s, t))^{n-j_1-j_2}. \tag{9}$$

By the elementary Vitali covering theorem, there exist a fixed constant $B > 0$ and a finite collection of rectangles R satisfying (q) for which the R''s are disjoint and

$$B \Sigma |R'| \geq |V_1^{x''}|. \tag{10}$$

Again following [3], we obtain

$$s^{j_1+\alpha} |R''| \leq 2 \Sigma_1 \int_{R \cap Q_k} g(\xi) d\xi, \tag{11}$$

where Σ_1 is the sum over those k for which $x'' \in J_k''$.

Let

$$g_k(\xi'') = \begin{cases} \int_{Q_k'} g(\xi', \xi'') d\xi' & \text{if } \xi'' \in Q_k'', \\ 0, & \text{otherwise,} \end{cases}$$

and for a fixed s let

$$g_k^{\star}(x'') = \sup\left\{\frac{1}{|R''|}\int_{R''} g_k(\xi'')\,d\xi'' : R'' \text{ with } x'' \in R''\right\}.$$

For a fixed k there are only 2^{j_1} R's for which $Q_k \cap R \neq \phi$. Let S_k be the set of values of s corresponding to the edge length of R' for which $Q_k \cap R \neq \phi$, where, if an s is repeated it is counted with its multiplicity in S_k. Note that S_k has a cardinality less than or equal to 2^{j_1}. Now we have

$$\begin{aligned} s^{j_1+\alpha} &\le 2 \sum_{Q_k \cap R \neq \phi} {}_1 \frac{1}{|R''|}\int_{R''\cap Q_k''} g_k(\xi'')\,d\xi'' \\ &= 2 \sum_{Q_k \cap R \neq \phi} \sum_{s \in S_k} \frac{1}{|R''|}\int_{R''\cap Q_k''} g_k(\xi'')\,d\xi'' \\ &\le 2 \sum_{Q_k \cap R \neq \phi} \sum_{s \in S_k} g_k^{\star(s)}(x''). \end{aligned}$$

Hence

$$|R'| = s^{j_1} \le 2 \sum_{Q_k \cap R \neq \phi} \sum_{s \in S_k} \left(g_k^{\star(s)}(x'')\right)^{j_1/j_1+\alpha}$$

and by (10),

$$|V_1^{x''}| \le 2B \sum_k \sum_{s \in S_k} \left(g_k^{\star(s)}(x'')\right)^{j_1/j_1+\alpha}. \tag{12}$$

Now for s fixed, the maximal function

$$f^{\star(s)}(x, y) = \sup_{0<t}\left\{t^{-j}\,\phi(s, t)^{-n+j_1+j_2}\iint_{|u|<t,\ |v|<\phi(s,t)} |f(x+u, y+v)|\,du\,dv\right\} \tag{13}$$

where $u \in \mathbf{R}^j$ and $v \in \mathbf{R}^{n-j_1-j_2}$, satisfies a weak-type inequality, as can be seen from [5, II, pp. 309-310]. The important fact is that the weak-type inequality does not depend on s or ϕ. Then by the results [2], it follows that if Q is a cube of edge length h, in $n - j_1$ dimensions, then

$$\int_Q [f^{\star(s)}(x, y)]^{\delta}\,dx\,dy \le Ch^{(n-j_1)(1-\delta)/\delta}\int_Q f(x, y)\,dx\,dy, \tag{14}$$

where C depends only on δ and $(n - j_1)$.

From (8), (11), (13), and (14), it follows that

$$|V_1| \leq 2B \sum_k \sum_{s \in S_k} \int_{J_k''} \left(g^{*(s)}(x'')\right)^{j_1/j_1+\alpha} dx''$$

$$\leq CB \sum_k \sum_{s \in S_k} h_k^{(n-j_1)(1-j_1/j_1+\alpha)} \left(\int_{Q_k''} g_k(x'')\,dx''\right)^{j_1/j_1+\alpha}$$

$$\leq C \sum_k h_k^{(n-j_1)(1-j_1/j_1+\alpha)} h_k^{(n+\alpha)(j_1/j_1+\alpha)}$$

$$= C \sum_k h_k^n \leq C|U|.$$

Combining this with (7), (5) follows.

REFERENCES

1. Córdoba, Antonio. Maximal functions, covering lemmas and Fourier multipliers. *Proc. Symp. in Pure Math.* 35, part 1 (1979):29-50.

2. Phillips, K. Maximal theorems of Hardy-Littlewood. *A. M. Monthly*, vol. 76 (June-July 1967).

3. Weiss, M. Strong differentials in L^p. *Studia Math.* 27 (1966): 49-72.

4. Weiss, M. Total and partial differentiation in L^p. *Studia Math.* 25 (1964):103-109.

5. Welland, G. V. Strong differentiation in L^p. *Rev. Acad. Ci. Madrid* 63 (1969):379-390.

6. Zygmund, A. *Trigonometric Series*, vols. I, II. New York: Cambridge Univ. Press, 1959.

PART SEVEN

PARTIAL DIFFERENTIAL EQUATIONS

ON THE LACK OF L^∞-ESTIMATES FOR SOLUTIONS OF ELLIPTIC SYSTEMS OR EQUATIONS WITH COMPLEX COEFFICIENTS

M. Cwikel
Technion I.I.T., Haifa

E.B. Fabes, C.E. Kenig
University of Minnesota

INTRODUCTION

In this paper we show the lack of uniform L^∞-estimates for solutions to the Dirichlet problem corresponding to strongly coercive second-order elliptic operators defined on a smooth domain $\Omega \subset R^n$. Precisely, our operators are of the form

$$Lu(x) = \operatorname{div}(A(x)\nabla u(x)),$$

where $A(x) = (a_{ij}(x))$ and strong coercivity means there exists $\lambda > 0$ such that for all $\varphi \in C_0^1(\Omega)$,

$$\int \operatorname{Re} \sum_{i,j=1}^{n} a_{ij}(x) D_{x_i}\varphi(x) D_{x_j}\overline{\varphi}(x)\,dx \geq \lambda \int |\nabla\varphi(x)|^2 dx$$

(see [1, p. 143]). In addition, we assume the coefficients to be bounded. We can then speak about the unique solution, u, of the Dirichlet problem

$$Lu = 0, \quad u|_{\partial\Omega} = g \tag{1}$$

in the class $H^1(\Omega) = \{u \in L^2(\Omega) : D_{x_i}u \in L^2(\Omega),\ i = 1, \ldots, n\}$ provided, say, $g \in H^1(\Omega)$. The statement $Lu = 0$ in (1) is in the weak sense; that is, for each $\varphi \in C_0^1(\Omega)$,

$$\sum_{i,j} \int_\Omega a_{ij}(x) D_{x_i} u(x) D_{x_j}\overline{\varphi}(x)\,dx = 0,$$

and the condition $u|_{\partial\Omega} = g$ means

$$u - g \in H_0^1(\Omega) = \text{closure of } C_0^1(\Omega) \text{ in } H^1(\Omega).$$

It is well known that when the coefficients of L are real, then the maximum principle holds for solutions of (1); that is, when $g \in H^1(\Omega) \cap C(\overline{\Omega})$,

$$\|u\|_{L^\infty(\Omega)} \leq \operatorname*{Max}_{\partial\Omega}|g|.$$

This fundamental result is crucial in constructing solutions to the Dirichlet problem for arbitrary $g \in C(\partial\Omega)$. By a solution, u, in this case we mean $u \in H^1_{loc}(\Omega) \cap C(\overline{\Omega})$, $Lu = 0$ in the weak sense, and $u|_{\partial\Omega} = g$ in the classical pointwise sense (see [6]). In fact, an L^∞-estimate similar to that given by the maximum principle is actually necessary if we have unique solvability for arbitrary $g \in C(\partial\Omega)$; for if this is the case, the closed graph theorem implies the existence of a constant, C_L, such that for all $g \in C(\partial\Omega)$ the solution u of $Lu = 0$, $u|_{\partial\Omega} = g$, satisfies

$$\|u\|_{L^\infty(\Omega)} \leq C_L \operatorname*{Max}_{\partial\Omega}|g|. \tag{2}$$

Conversely, the validity of (2) and the existence of solutions to the Dirichlet problem in the class $H^1(\Omega) \cap C(\overline{\Omega})$ when, say, g is smooth implies existence in the class $H^1_{loc}(\Omega) \cap C(\overline{\Omega})$ for arbitrary continuous boundary data.

When the coefficients of L are continuous in $\overline{\Omega}$, then indeed there exists a (unique) solution of the Dirichlet problem in $H^1(\Omega) \cap C(\overline{\Omega})$ when g is smooth (see [7, Chapter 6]). If, then, in the case of continuous coefficients, one wants to solve the Dirichlet problem for any $g \in C(\partial\Omega)$, we must be able to prove an L^∞-estimate as above. If the coefficients belong to $C^2(\overline{\Omega})$, (2) indeed holds with C_L depending only on the C^2-norm of the coefficients and the coercivity constant λ. However, it is the purpose of this paper to show that an estimate of the type,

$$\|u\|_{L^\infty(\Omega)} \leq C \operatorname*{Max}_{\partial\Omega}|g|, \tag{3}$$

cannot hold uniformly over all operators $L = \operatorname{div}(A\nabla)$ satisfying:

(a) $\sup_{|x-y|<s} |A(x) - A(y)| \le \omega(s)$, where $\omega(s)$ is a given continuous increasing function tending to 0 with s, and

(b) $\operatorname{Re} \int A(\nabla\varphi) \cdot \nabla\bar{\varphi}\, dx \ge \lambda \int |\nabla\varphi|^2 dx$ for all $\varphi \in C_0^\infty(\Omega)$, where λ is a fixed positive number.

The failure of the above L^∞-estimate, (3), will be shown in the case of R^2 and the proof rests upon the following.

Theorem 1. Fix a continuous increasing function $\eta(s)$, $s \ge 0$, *with* $\eta(0) = 0$. *There exists a smooth domain,* Ω, *contained in* $R_+^2 = R \times (0, \infty)$ *and a positive number* ℓ *satisfying*

(i) $\partial\Omega \cap R \supset [-2\ell, 2\ell]$ *and* $\Omega \supset (-2\ell, 2\ell) \times (0, 4\ell)$,

(ii) if $\int_0^1 \eta^2(s) \frac{ds}{s} < \infty$ *and* $L = \operatorname{div}(A(x, t)\nabla)$ *is a strongly coercive elliptic operator with bounded, uniformly continuous coefficients in* $R_+^2 = R \times (0, \infty)$ *such that*

$$\sup_{\substack{(x,t)\\ 0\le t\le s}} |I - A(x, t)| \le \eta(s),$$

then the solution, u, *of the Dirichlet problem,* $Lu = 0$ *in* Ω, $u|_{\partial\Omega} = g$, *for* $g \in C_0^\infty(-\ell, \ell)$, *satisfies the estimate*

$$\sup_{0<t\le\ell} \int_{-\ell}^{\ell} |u(x, t)|^2 dx \le c \int |g(x)|^2 dx,$$

where C *is independent of* u *and* g *(see Lemma 2),*

(iii) if, however, $\int_0^1 \eta^2(s) \frac{ds}{s} = \infty$, *then there exists an elliptic operator,* $L = \operatorname{div}(A(x, t)\nabla)$, *with real coefficients for which*

$$\sup_{\substack{(x,t)\\ 0\le t\le s}} |I - A(x, t)| = \eta(s)$$

and such that the L^p*-estimate*

$$\sup_{0<t\leq \ell}\int_{-\ell}^{\ell}|u(x,\ t)|^p dx \leq C_p|g(x)|^p\,dx \tag{4}$$

fails for any finite $p \geq 1$ *with* C_p *independent of* u *and* g *(see Lemma* 3 *and the discussion preceding it).*

The negative part of the above result, that is, the lack of an L^p-estimate, is really the consequence of the construction of an elliptic operator L, with real continuous coefficients whose L-harmonic measure (see [2]) is not absolutely continuous with respect to Lebesgue measure on any subinterval of the real line. The relation of this result to the L^∞-estimate, (3), is via interpolation of analytic families of operators. If, in fact, the L^∞-estimate is assumed with C depending only on the coercivity constant λ and the modulus of continuity, $w(s)$, of the coefficients, then the above theorem and an interpolation argument imply the validity of the L^p-estimate, (4), for some finite p and for the operator, L, just mentioned. This, in turn, implies the absolute continuity of the L-harmonic measure with respect to Lebesgue measure on the interval $(a,\ b)$, a fact that contradicts the known property of L.

Finally, we would like to point out the similarity of our theorem stated above to some of the results of Stein and Zygmund in [9] and of L. Carleson in [4]. In the former work the authors prove the following.

Theorem 2. Let $\eta(s)$ *be a function defined in* (0, 1] *monotonically decreasing to* 0 *with* s *and such that*

$$\int_0^1 \eta^2(s)\ \frac{ds}{s} < \infty.$$

If for each x *belonging to a set* $E \subset R$ *we have*

$$|F(x+s) + F(x-s) - 2F(x)| = 0\{s\eta(s)\} \qquad (s \to 0),$$

not necessarily uniformly in x, *then* F *is differentiable almost everywhere in* E.

However, if $\eta(s)$ *is a function monotonically decreasing in* s *satisfying the condition* $\frac{\eta(2s)}{\eta(s)} \to 1$ *for* $s \to 0$ *and*

$$\int_0^1 \eta^2(s)\, \frac{ds}{s} = \infty,$$

then there is a continuous $F(x)$ *satisfying for all* x *the estimate*

$$|F(x+s) + F(x-s) - 2F(x)| \le s\eta(s)$$

but F *is differentiable in a set of measure zero only.*

In his paper [4], Carleson proves the following.

Theorem 3. Let $\phi(x)$ *be a continuous, stictly increasing function on* R *such that*

$$\frac{\phi(x+s) - \phi(x)}{\phi(x) - \phi(x-s)} = 1 + O(\eta(\sqrt{s})) \qquad (s > 0)$$

uniformly on compact sets where

1° $\eta(s)$ *is continuous for* $s \ge 0$, $\eta(0) = 0$

2° $\eta(s)$ *and* $\eta(s)s^{-1/2} \searrow 0$ *for* $s > 0$.

If

$$\int_0^1 \eta^2(s)\, \frac{ds}{s} < \infty,$$

then ϕ *is absolutely continuous and* $\phi' \in L^2_{\mathrm{loc}}$.

Moreover, given any $\eta(s)$ *satisfying* 1° *and* 2°, *if*

$$\int_0^1 \eta^2(s)\, \frac{ds}{s} = \infty,$$

then there exists a continuous, strictly increasing $\phi(x)$ *for which*

$$\frac{\phi(x+s) - \phi(x)}{\phi(x) - \phi(x-s)} = 1 + O(\eta(\sqrt{s})),$$

uniformly on compact sets, and ϕ *is purely singular.*

All the theorems stated above are not merely similar in form. Carleson himself points out the close relation between his result and that of Stein and Zygmund. At the same time, our construction of an elliptic operator with real continuous coefficients whose "harmonic measure" is singular with respect to Lebesgue measure is based directly on the results of Carleson (see [2]).

STATEMENT AND PROOF OF MAIN RESULT

Let $A(t, s) = \left(a_{ij}(t, s)\right)$ denote an $n \times n$ matrix of complex-valued entries defined on $\overline{R^2_+}$. For such matrix-valued functions, $A(t, s)$, we set $|A(t, s)| = \max|a_{ij}(t, s)|$ and for each $\delta > 0$ we let

$$w(\delta, A) = \sup|A(t, s) - A(t', s')|$$
$$|(t, s) - (t', s')| < \delta$$
$$(t, s), (t', s') \in \overline{R^2_+}.$$

The main goal is to prove the following.

Theorem 4. There exists a smooth $\Omega \subset R^2_+$ *so that given* $0 < \varepsilon < \frac{1}{2}$, *we can find a family of symmetric matrices,* $A(t, s; y)$, *depending on the parameter* $y \in R$, *with the following properties:*

(i) $A(t, s; y)|_{\partial\Omega} = I$, *the identity matrix, and*

$$\sup_{(t,s) \in R^2_+} |A(t, s; y) - I| \le \varepsilon \; \Big(\textit{in particular,}$$

$$\sup_{(t, s) \in \overline{R^2_+}} |\operatorname{Im} A(t, s; y)| \le \varepsilon, \qquad \sup_{(t, s) \in \overline{R^2_+}} |A(t, s; y)| \le 2,$$

$$\operatorname{Re} \sum_{i, j=1}^{n} a_{ij}(t, s; y)\xi_i\xi_j \ge \frac{1}{2}|\xi|^2, \textit{ and}$$

$$\left|\sum_{i, j=1}^{2} \int_\Omega a_{ij}(t, s; y) D_i u(t, s) D_j \bar{u}(t, s)\, dt\, ds\right| \ge \frac{1}{4}\int_\Omega |\nabla u(t, s)|^2\, dt\, ds$$

for all $u \in C_0^\infty(\Omega)$. *As usual,* $D_1 u = \dfrac{\partial u}{\partial t}$ *and* $D_2 u = \dfrac{\partial u}{\partial s}\Big)$,

(ii) *there exists an increasing continuous function,* $\phi(\delta)$, *with* $\phi(0) = 0$, *such that for all* y,

$$w(\delta;\, A(\cdot,\ \cdot;\, y)) \le \phi(\delta),$$

(iii) the estimate

$$\|u(\cdot,\ \cdot;\, y)\|_{L^\infty(\Omega)} \le C\|g\|_{L^\infty(\partial\Omega)}$$

with C independent of y and $g \in C_0^\infty(R^2)$ does not *hold for the family of functions, $u(t,\, s;\, y)$, where for each y, $u(t,\, s;\, y)$ is the solution of the Dirichlet problem*

$$\operatorname{div}\big(A(t,\, s;\, y)\nabla u(t,\, s;\, y)\big) = 0 \text{ in } \Omega$$

$$u(t,\, s;\, y)\big|_{\partial\Omega} = g.$$

As mentioned in the Introduction, the proof of the above theorem is accomplished by a contradiction argument. The existence of a constant, C, making the inequality in (iii) above true for all y, is combined with an L^2-inequality stated in Lemma 2 below to imply, via interpolation, an L^p-estimate, in the form of Lemma 2, with $p < \infty$ for a specific operator, L, with real coefficients. This L^p-estimate then implies that a Poisson kernel exists for L that at least belongs to $L^1(\partial\Omega)$. However, for this particular L, it had been previously shown in [2] that such a kernel could not exist. This aspect of our proof is given below in Lemmas 2 and 3 and the discussion in between.

Consider now real-valued functions, $b(t,\, s)$, defined on the square $T = (0,\, \ell) \times (0,\, \ell)$ that are continuous on $\overline{T}$ and that *vanish* on ∂T. For such functions, b, we define

$$w_*(\delta,\, b) = \sup_{\substack{s,t \in [0,\ell] \\ 0 \le s \le \delta}} |b(t,\, s)|.$$

Let $p > 1$ and assume

$$\int_0^\ell \frac{W_*(\delta,\, b)^p}{\delta}\, d\delta < \infty.$$

We consider the strips

$$S = \{z : 0 \le \operatorname{Re} z \le 1\} \quad \text{and} \quad S_\tau = \{z : -\tau \le \operatorname{Re} z \le 1 + \tau\}$$

in the complex plane. Pick $q > p$, $0 < \tau < 1$, and for $z \in S_\tau$ set

$$b(t, s; z) = |b(t, s)|^{p(1-z)+\frac{p}{q}z}(\operatorname{sgn} b(t, s))\frac{1 + \tilde{\theta}}{1 + z},$$

where $\frac{1}{p} = (1 - \tilde{\theta}) + \frac{\tilde{\theta}}{q}$. The following lemma is easily verified.

Lemma 1. Let $b(t, s; \tilde{\theta})$ *be as above. There exists* $\tau_0 > 0$ *such that*

(i) $b(t, s; \tilde{\theta}) = b(t, s)$,

(ii) $$\sup_{z \in S_{\tau_0}} \|b(\cdot, \cdot; z)\|_{L^\infty(T)} \leq 2 \max\left\{\|b\|_{L^\infty(T)}^{pq/2}, \|b\|_{L^\infty(T)}^{p(1+\tau_0)-p\tau_0/q}\right\},$$

(iii) for each $z \in S_{\tau_0}$, $b(s, t; z)$ *is continuous on* $\overline{T}$, *vanishes on* ∂T, *and for all* $z \in S$,

$$w(\delta; b(\cdot, \cdot; z)) \leq C \max(w(\delta; b); w(\delta; b)^{p/q}),$$

(iv) for each $z = iy$, $-\infty < y < \infty$,

$$w_*(\delta, b(\cdot, \cdot; iy)) \leq 2w_*(\delta, b)^p,$$

and hence

$$\int_0^\ell w_*(\delta, b(\cdot, \cdot; iy))\frac{d\delta}{\delta} \leq 2\int_0^\ell w_*(\delta, b)^p\frac{d\delta}{\delta},$$

(v) uniformly for $(t, s) \in \overline{T}$ *we have*

$$\lim_{\zeta \to z} b(t, s; \zeta) = b(t, s; z)$$

for each $z \in S_{\tau_0}$ *and*

$$\lim_{\zeta \to z} \frac{b(t, s; \zeta) - b(t, s; z)}{\zeta - z} = \frac{d}{dz} b(t, s; z)$$

for each z *in the interior of* S_{τ_0}.

Lemma 2. For any fixed number ℓ, $0 < \ell < 1$, *let* Ω *denote a smooth bounded domain in* R_+^2 *such that*

$$\partial\Omega \cap R \supset [-2\ell, 2\ell] \quad \text{and} \quad \Omega \supset (-2\ell, 2\ell) \times (0, 4\ell).$$

Given a function $\eta(s)$ *with*

$$\int_0^1 \eta^2(s)\,\frac{ds}{s} < \infty,$$

there exists $\varepsilon_0 > 0$, *depending only on* η, *so that for all matrices* $A(t, s)$ *satisfying*

(a) $\omega_*(\delta; A - I) \leq \eta(\delta)$,

(b) $|A - I| \leq \varepsilon_0$,

(c) $A = I$ *on* $\overline{R}^2_+ \setminus \left[-\frac{3}{2}\ell, \frac{3}{2}\ell\right] \times [0, 3\ell]$,

we have

$$\left(\int_{-\ell}^{\ell} |u(t, s)|^2 dt\right)^{1/2} \leq C\left(\int |g|^2\right)^{1/2} \text{ for all } g \in C_0^\infty(-\ell, \ell),$$

where u *solves the Dirichlet problem*

$$\operatorname{div}(A(t, s)\nabla u(t, s)) = 0 \text{ in } \Omega$$

$$u|_{\partial\Omega} = g$$

and C *depends only on* η.

Proof. Set $P_s(t) = \frac{1}{\pi}\,\frac{s}{t^2 + s^2}$ and $G(t, s, r) = \log \frac{\sqrt{t^2 + (s + r)^2}}{\sqrt{t^2 + (s - r)^2}}$.

Given $g \in C_0^\infty(-\ell, \ell)$, extend the solution, u, of the Dirichlet problem described in Theorem 4 to be zero in $\overline{R}^2_+ \setminus \overline{\Omega}$. We now can write

$$\begin{aligned} u(t, s) &= \int_R P_s(t - r) g(r)\,dr \\ &\quad + \int_0^\infty \int_R \nabla_{y,r}(G(t - y, s, r)) \cdot (I - A(y, r))\nabla u(y, r)\,dy\,dr. \end{aligned}$$

Set $I(g)(t, s) = P_s * g(t)$. From the easy part of the Littlewood-Paley estimates in [8, p. 82],

$$\|I(g)(\cdot, s)\|_{L^2(R)} + \|s^{1/2}\nabla I(g)\|_{L^2(R^2_+)} \leq C\|g\|_{L^2(R)}.$$

Now set

$$V(f)(t, s) = \int_0^\infty \int_R DG(t - y, s, r) f(y, r)\,dy\,dr,$$

where D denotes a partial derivative of G with respect to t or r.

First Estimate. Assume $f(y, r) = 0$ for $r \geq 1$. There exists a constant C independent of f and η such that for any $0 < \delta < 1$,

$$\|Vf(\cdot, s)\|_{L^2(R)} \leq C\left[\left(\int_0^\delta \eta^2(s)\, \frac{ds}{s}\right)^{1/2} \|r^{1/2}\eta(r)^{-1}f(t, r)\|_{L^2(R_+^2)} + (\log 1/\delta)^{1/2}\|r^{1/2}f(t, r)\|_{L^2(R_+^2)}\right].$$

Proof. For each s and r,

$$\|DG(\cdot, s, r) * f(\cdot, r)\|_{L^2(R)} \leq C\|f(\cdot, r)\|_{L^2(R)},$$

where C is an absolute constant independent of s, r, and f. Hence, by Minkowski's inequality,

$$\begin{aligned}\|Vf(\cdot, s)\|_{L^2(R)} &\leq C\int_0^1 \|f(\cdot, r)\|_{L^2(R)}\, dr \\ &\leq C\left(\int_0^\delta \eta^2(r)\, \frac{dr}{r}\right)^{1/2} \|r^{1/2}\eta(r)^{-1}f(t, r)\|_{L^2(R_+^2)} \\ &\quad + C(\log 1/\delta)^{1/2}\|r^{1/2}f(t, r)\|_{L^2(R_+^2)}.\end{aligned}$$

Second Estimate. Again assume $f(y, r) = 0$ for $r \geq 1$. There exists a constant C independent of f and η such that for any $0 < \delta < 1$,

$$\|s^{1/2}DV(f)(t, s)\|_{L^2(R_+^2)} \leq C\left[\left(\int_0^\delta \eta^2(r)\, \frac{dr}{r}\right)^{1/2} \|r^{1/2}\eta(r)^{-1}f\|_{L^2(R_+^2)} + (\log 1/\delta)^{1/2}\|r^{1/2}f\|_{L^2(R_+^2)}\right].$$

Here D denotes the partial derivative with respect either to t or to s.

Proof.

$$\begin{aligned}s^{1/2}DVf(t, s) &= \int_0^\infty\int_R D^2G(t - y, s, r)r^{1/2}f(y, r)\,dy\,dr \\ &+ \int_0^\infty\int_R D^2G(t - y, s, r)(s^{1/2} - r^{1/2})f(y, r)\,dy\,dr \equiv Af + Bf.\end{aligned}$$

(D^2 denotes a general second-order partial.) Since $D^2 \log(t^2 + s^2)$

is a Calderón-Zygmund kernel in R^2, we have

$$\|Af\|_{L^2(R^2_+)} \le C\|r^{1/2}f\|_{L^2(R^2_+)}.$$

We write

$$Bf(t, s) = \int_0^{s/2}\int_0^{\infty} D^2G(t - y, s, r)(s^{1/2} - r^{1/2})f(y, r)\,dy\,dr$$

$$+ \int_{s/2}^{\infty}\int_0^{\infty} D^2G(t - y, s, r)(s^{1/2} - r^{1/2})f(y, r)\,dy\,dr$$

$$= B_1f + B_2f.$$

Since

$$\int |D^2G(y, s, r)|\,dy \le \frac{C}{|s - r|},$$

a combination of Minkowski's inequality and Young's theorem for convolution operators gives

$$\|B_2f(\cdot, s)\|_{L^2(R)} \le C\int_{s/2}^{\infty} \frac{|s^{1/2} - r^{1/2}|}{|s - r|r^{1/2}}\left(r^{1/2}\|f(\cdot, r)\|_{L^2(R)}\right)dr.$$

An application of Hardy's theorem [5, p. 227] yields

$$\|B_2f\|_{L^2(R^2_+)} \le C\|r^{1/2}f\|_{L^2(R^2_+)}.$$

Finally,

$$B_1f(t, s) = s^{1/2}\int_0^{s/2}\int_R (D^2G(t - y, s, r) - D^2G(t - y, s, 0))f(y, r)\,dy\,dr$$

$$+ s^{1/2}\int_R D^2G(t - y, s, 0)\left(\int_0^{s/2} f(y, r)\,dr\right)dy$$

$$- \int_0^{s/2}\int_R D^2G(t - y, s, r)r^{1/2}f(y, r)\,dy\,dr$$

$$= B_{11}f + B_{12}f + B_{13}f.$$

Using the mean-value theorem in the r-variable, we have

$$\|B_{11}f(\cdot, s)\|_{L^2(R)} \le C\,\frac{1}{s^{3/2}}\int_0^{s/2} \|f(\cdot, r)\|_{L^2(R)}\,r\,dr.$$

Also

$$\|B_{13}f(\cdot,\ s)\|_{L^2(R)} \leq \frac{c}{s}\int_0^{s/2} \|f(\cdot,\ r)\|_{L^2(R)} r^{1/2}\, dr.$$

If we once again use Hardy's theorem, we conclude that

$$\|B_{11}f\|_{L^2(R_+^2)} + \|B_{13}f\|_{L^2(R_+^2)} \leq C\|r^{1/2}f\|_{L^2(R_+^2)}.$$

In considering $B_{12}f$, observe first that

$$s^{1/2}D^2G(t,\ s,\ 0) = \frac{\psi(t/s)}{s^{3/2}},\ \text{where } \mathfrak{F}(\psi)(\zeta),$$

the Fourier transform of ψ, is bounded by $c|\xi|e^{-c|\xi|}$. Hence

$$\mathfrak{F}(B_{12}f(\cdot,\ s))(\xi) = \frac{\mathfrak{F}(\psi)(\xi s)}{s^{1/2}}\int_0^{s/2} \mathfrak{F}(f(\cdot,\ r))(\xi)dr$$

and so

$$|\mathfrak{F}(B_{12}f(\cdot,\ s))(\xi)| \leq \left|\frac{\mathfrak{F}(\psi)(\xi s)}{s^{1/2}}\right|\left[\left(\int_0^\delta \eta^2(r)\ \frac{dr}{r}\right)^{1/2}\right.$$

$$\cdot \left(\int_0^\delta |\mathfrak{F}(\cdot,\ r)\ (\xi)|^2 \frac{r}{\eta(r)^2}\ dr\right)^{1/2}$$

$$\left. + (\log 1/\delta)^{1/2}\left(\int_\delta^1 |\mathfrak{F}(f(\cdot,\ r))(\xi)|^2 r\ dr\right)^{1/2}\right].$$

Since $\int_0^\infty \left|\frac{\mathfrak{F}(\psi)(\xi s)}{s}\right|^2 ds \leq C$, independent of ξ, it readily follows that

$$\|B_{12}f\|_{L^2(R_+^2)} \leq C\left[\left(\int_0^\delta \eta^2(r)\ \frac{dr}{r}\right)^{1/2} \|r^{1/2}\eta(r)^{-1}f\|_{L^2(R_+^2)}\right.$$

$$\left. + (\log 1/\delta)^{1/2}\|r^{1/2}f\|_{L^2(R_+^2)}\right].$$

Our representation of the solution, u, together with the first and second estimates and the hypotheses concerning the matrix A, yields the inequality

$$\|u(\cdot, s)\|_{L^2(R)} + \|r^{1/2}Du\|_{L^2(R_+^2)} \leq C\|g\|_{L^2(R)}$$
$$+ C\left[\left(\int_0^\delta \eta^2(r)\, \frac{dr}{r}\right)^{1/2} \left\|\frac{(I - A)(Du)r^{1/2}}{\eta(r)}\right\|_{L^2(R_+^2)}\right.$$
$$\left. + (\log 1/\delta)^{1/2} \|(I - A)(Du)r^{1/2}\|_{L^2(R_+^2)}\right]$$
$$\leq C\|g\|_{L^2(R)} + C\left[\left(\int_0^\delta \eta^2(r)\, \frac{dr}{r}\right)^{1/2} + (\log 1/\delta)^{1/2}\varepsilon_0\right] \|r^{1/2}Du\|_{L^2(R_+^2)}.$$

It is clear that if we choose first $\delta < 1$ so that

$$C\left(\int_0^\delta \eta^2(r)\, \frac{dr}{r}\right)^{1/2} \leq 1/4$$

and then ε_0 so that

$$C(\log 1/\delta)^{1/2}\varepsilon_0 \leq 1/4,$$

the conclusion of Lemma 2 follows.

Because of the results in [4 and 2], given $\theta(s)$ continuous, increasing with $\theta(0) = 0$, $s^{-1/4}\theta(s)$ decreasing, and

$$\int_0^1 \theta^2(s)\, \frac{ds}{s} = \infty,$$

we can find a real symmetric matrix, $T = I + C$, in the upper half-plane (t, s), $s > 0$, such that the entries of C are $O(\theta(s))$ uniformly on compact sets in t, and such that the operator $\tilde{L} = \operatorname{div}(T\nabla)$ has a harmonic measure $w^p_{\tilde{L},\mathbf{R}^2_+}$, strictly singular with respect to Lebesgue measure on $\mathbf{R}$. Here, by the harmonic measure, $w^p_{\tilde{L},\mathbf{R}^2_+}$, we mean the finite Borel measure that gives the representation

$$u(P) = \int_{\mathbf{R}} g(Q)\, dw^p_{\tilde{L},\mathbf{R}^2_+}(Q)$$

of the solution of the Dirichlet problem for $\tilde{L}$ in $\mathbf{R}^2_+$, with boundary data $g \in C_0(\mathbf{R})$. (The existence of $w^p_{\tilde{L},\mathbf{R}^2_+}$ is guaranteed by the maximum principle, since T is real and symmetric.) Now fix one such θ as above, but with the additional property that

$$\int_0^1 \theta(s)^p\, \frac{ds}{s} < \infty$$

for some $p > 2$, which we also fix from now on. Set

$$\eta(s) = \theta(s)^p.$$

Given $\varepsilon_1 > 0$, we can find an interval, $[-\ell, \ell]$, such that in the box $[-2\ell, 2\ell] \times [0, 4\ell]$ we have the estimate

$$|I - T(t, s)| \leq \varepsilon_1.$$

Now let $\varphi \in C_0^\infty(R^2)$ satisfy $0 \leq \varphi \leq 1$ and $\varphi \equiv 1$ in $[-\ell, \ell] \times [0, 2\ell]$, $\varphi \equiv 0$ outside $[-3/2\ell, 3/2\ell] \times [0, 3\ell]$. The matrix

$$A(t, s) = (1 - \varphi(t, s))I + \varphi(t, s)T(t, s)$$

has the properties:

$$|I - A(t, s)| \leq \varepsilon_1, \text{ for all } (t, s) \in \overline{R}^2_+,$$

$A(t, 0) = I$ for all t, $A(t, s) = I$ in $\overline{R}^2_+ \backslash [-3/2\ell, 3/2\ell] \times [0, 3\ell]$, and $A(t, s) = T(t, s)$ on $[-\ell, \ell] \times [0, 2\ell]$.

We choose a smooth domain, $\Omega \subset R^2_+$, such that $\partial\Omega \cap R \supset [-2\ell, 2\ell]$ and $\Omega \supset (-2\ell, 2\ell) \times (0, 4\ell)$ (see Lemma 2). We set

$$\tilde{L} = \operatorname{div}(T\nabla) \quad \text{and} \quad L = \operatorname{div}(A\nabla).$$

By Theorem 1.4 in [3], $w^p_{\tilde{L},\Omega}$ is singular with respect to Lebesgue measure on $[-1/2\ell, 1/2\ell]$. By the same theorem, there exists a closed set $|F| \subset [-1/2\ell, 1/2\ell]$ with $|F| = 0$, but $w^p_{L,\Omega}(F) > 0$. An easy consequence of this fact, and Harnack's principle, is the following.

Lemma 3. The a priori estimate

$$\left(\int_{-\ell}^{\ell} |u(t, s_0)|^r \, dt\right)^{1/r} \leq C\left(\int |g(t)|^r \, dt\right)^{1/r}$$

cannot hold *for any fixed* $0 < s_0 < 2\ell$, $1 \leq r < \infty$, *with* C *independent of* $g \in C_0(-\ell, \ell)$ *and* u, *the solution of the Dirichlet problem*

$$Lu = 0 \text{ in } \Omega, \; u|_{\partial\Omega} = g.$$

Set $B(t, s) = I - A(t, s)$ and $b_{ij}(t, s)$ equal to a general entry in the matrix B. As in Lemma 1, define

$$b_{ij}(t, s; z) = |b_{ij}(t, s)|^{p(1-z)+\frac{p}{q}z} (\operatorname{sgn} b(t, s)) \frac{1 + \tilde{\theta}}{1 + z},$$

where $q > p$, $1/p = 1 - \tilde{\theta} + \tilde{\theta}/q$, $z \in S_{\tau_0} = \{z : -\tau_0 \leq \operatorname{Re}(z) \leq 1 + \tau_0\}$, and the positive number τ_0 has been chosen so that the conclusions of Lemma 1 are valid. Finally, we set

$$B(t, s; z) = (b_{ij}(t, s; z))$$

and

$$A(t, s; z) = I - B(t, s; z).$$

The main properties of the matrix $A(t, s; z)$ are stated in the following lemma, whose proof is immediate.

Lemma 4. Set $Q_\ell = [-\ell, \ell] \times [0, 2\ell]$. *Given* $\varepsilon_1 > 0$, *there exist* ℓ *so that*

(i) $A(t, s; \tilde{\theta}) = A(t, s)$,

(ii) $\sup_{z \in S_{\tau_0}} |I - A(t, s; z)| \leq \varepsilon$,

(iii) *for each* $z \in S_{\tau_0}$, $A(t, s; z)$ *is continuous in* $\overline{R}^2_+$, *equals* I *on* R, *and in* $R^2_+ \backslash Q_{3/2\ell}$ *also*

$$\sup_{z \in S = \{a : 0 \leq \operatorname{Re} z \leq 1\}} |w(\delta, A(\cdot, \cdot; z))| \leqq \phi(\delta)$$

where $\phi(\delta) \to 0$ *as* $\delta \to 0+$,

(iv) $\sup_{y \in (-\infty, \infty)} w_*(\delta; I - A(\cdot, \cdot; iy)) \leq \eta(\delta)$, *where* $\eta(s) = \theta(s)^p$ *and* $\theta(s)$ *was given following Lemma 2 and satisfies*

$$\int_0^1 \frac{\theta(s)^p}{s}\, ds < \infty;$$

in particular, $\displaystyle\int_0^1 \eta(s)\, \frac{ds}{s} < \infty$,

(v) *for each* (t, s), $A(t, s; z)$ *is analytic in* S_{τ_0}, *and for*

$$z \in S = \{z : 0 \leq \operatorname{Re} z \leq 1\},$$

$$\lim_{\zeta \to z} A(t, s; \zeta) = A(t, s; z)$$

and

$$\lim_{\zeta \to z} \frac{A(t, s; \zeta) - A(t, s; z)}{\zeta - z} = \frac{d}{dz} A(t, s; z)$$

with both limits occurring uniformly for $(t, s) \in \overline{R}^2_+$.

The family of matrices mentioned in Theorem 4 is the family $A(t, s; 1 + iy)$. The domain Ω mentioned in the same theorem is the domain Ω described just prior to Lemma 3. The properties of Ω that we need to recall are: $\Omega \subset R^2_+$, $\partial\Omega \cap R \supset [-2\ell, 2\ell]$, and $\Omega \supset (-2\ell, 2\ell) \times (0, 4\ell)$. The number ℓ is chosen so that the conclusions of Lemma 4 hold for a given positive number $\varepsilon_1 < \varepsilon_0$, ε_0 was determined by Lemma 2, and the function $\eta \equiv \theta^p$, and θ was described following Lemma 2.

The family of matrices $A(t, s; 1 + iy)$ satisfies conditions (i) and (ii) of the theorem by construction. By a contradiction argument, we will show that the estimate in part (iii) of the theorem cannot hold in a uniform manner over the family of associated elliptic operators.

Lemma 5. *For* $z \in S = \{z : 0 \leq \mathrm{Re}\, z \leq 1\}$ *and* $g \in C_0^\infty(-\ell, \ell)$, *set* $L_z = \mathrm{div}(A(t, s, z)\nabla)$ *and* u_z *equal to the solution of the Dirichlet problem*

$$L_z u_z = 0 \text{ in } \Omega, \ u_z|_{\partial\Omega} = g.$$

There exists a constant C *depending only on* $\theta(s)$ *(see Lemma 4(iv)) such that for all* $0 < s < \ell$,

$$\sup_{y \in R} \int_{-\ell}^{\ell} |u_{iy}(t, s)|^2 dt \leq C \int_{-\ell}^{\ell} |g(t)|^2 dt.$$

Proof. The proof is an immediate consequence of Lemma 2.

Lemma 6. *Let* $g \in C_0^\infty(-\ell, \ell)$ *and for* $z \in S_{\tau_0}$ *let* u_z *denote the solution to the Dirichlet problem*

$$L_z u_z \equiv \operatorname{div}(A(t, s; z)\nabla_{t,s} u_z(t, s)) = 0 \text{ in } \Omega,$$

where $\{A(t, s; z)\}$ is the family of matrices described in Lemma 4. If J is an interval $\subset (-\ell, \ell)$ and $0 < s_0 < \ell$, then

$$\int_J u_z(t, s_0)\,dt$$

is a well-defined bounded analytic function for $z \in \mathring{S}$, the interior of S, that is continuous in S.

Proof. We may assume that g is the restriction to $\partial\Omega$ of a function, which we again denote by g, belonging to $C_0^\infty(R^2)$. Let $v_z \in H_0^1(\Omega)$ be the solution of the problem

$$L_z v_z = \operatorname{div}(A(t, s; z)\nabla g(t, s)).$$

Then $u_z(t, s) = g(t, s) - v_z(t, s)$.

Since $v_z \in H_0^1(\Omega)$, the trace theorem [8, pp. 192-193] shows that

$$\int_J v_z(t, s_0)\,dt$$

is well defined. All we have to show is that it is analytic and bounded in $\mathring{S}$ and continuous in S. All this follows from the Lax-Milgram theorem [1].

Proof of Theorem. Consider the family of matrices $\{A(t, s, z);\ z \in S_{\tau_0}\}$ described in Lemma 4 and the domain Ω described just prior to Lemma 3. For each $g \in C_0^\infty(-\ell, \ell)$ and $z \in S_{\tau_0}$, let u_z denote the solution to the Dirichlet problem

$$L_z u_z = \operatorname{div}(A(z)\nabla u_z) = 0 \text{ in } \Omega$$

$$u_z|_{\partial\Omega} = g.$$

Assume

$$\sup_{y \in R} \|u(\cdot, \cdot, 1 + iy)\|_{L^\infty(\Omega)} \leq C\|g\|_{L^\infty(\partial\Omega)} \tag{5}$$

with C independent of g.

Recall that $p > 2$ and $\theta(s)$ have been chosen so that

$$\int_0^1 \frac{\theta^p(s)}{s}\, ds < \infty,$$

while $\int_0^1 \frac{\theta^2(s)}{s}\, ds = \infty$. We also have fixed $q > p$ and have chosen $\tilde{\theta}$ so that

$$\frac{1}{p} = (1 - \tilde{\theta}) + \frac{\tilde{\theta}}{q}.$$

With $\tilde{\theta}$ fixed, choose $r > 2$ satisfying $\frac{1}{r} = \frac{1 - \tilde{\theta}}{2}$. The assumed estimate (5) and the known estimate

$$\sup_y \|u(\cdot,\ s_0;\ iy)\|_{L^2(-\ell,\ \ell)} \leq C\|g\|_{L^2(-\ell,\ \ell)}$$

with C independent of g, given by Lemma 5, imply, via interpolation of analytic families of operators [10], the estimate

$$\|u(\cdot,\ s_0;\ \tilde{\theta})\|_{L^r(-\ell,\ \ell)} \leq C\|g\|_{L^r(-\ell,\ \ell)}$$

with C independent of g. This contradicts the results of Lemma 3 where $L \equiv L_{\tilde{\theta}}$. This contradiction is due to the assumed validity of the L^∞-estimate (5). This concludes the proof of the theorem.

Remark 1. In the above proof of the main theorem, if we replace the assumed estimate

$$\sup_y \|u(\cdot,\ \cdot;\ 1 + iy)\|_{L^\infty(\Omega)} \leq C\|g\|_{L^\infty(\partial\Omega)} \tag{6}$$

by the weaker inequality

$$\sup_y \|u(\cdot,\ s_0;\ 1 + iy)\|_{\mathrm{BMO}[-\ell,\ \ell]} \leq C\|g\|_{L^\infty(\partial\Omega)} \tag{7}$$

with C independent of g, we would still arrive at the same contradiction. (BMO$[-\ell,\ \ell]$ = the space of functions of bounded mean oscillation on $[-\ell,\ \ell]$.) Thus the inequality (7) cannot hold with C depending only on a fixed parameter of coercivity and a fixed modulus of continuity of the coefficients.

Remark 2. Interior estimates of the form

$$\|u\|_{L^p(\Omega')} \leq C\|g\|_{L^\infty(\partial\Omega)}$$

for some $1 < p < \infty$ and for each $\Omega' \subset\subset \Omega$ also cannot hold with C depending only on Ω', Ω, a fixed parameter of coercivity and a fixed modulus of continuity. This is because for continuous coefficients, solutions of the Dirichlet problem with data g satisfy the estimate

$$\|u\|_{L^\infty(\Omega')} \leq C\|u\|_{L^p(\Omega'')},$$

where $\Omega' \subset\subset \Omega'' \subset\subset \Omega$, and C depends only on Ω', Ω'', Ω, the modulus of continuity of the coefficients and the parameter of coercivity. If we now assume the validity of (8), we would obtain the estimate

$$\|u\|_{L^\infty(\Omega')} \leq C\|g\|_{L^\infty(\partial\Omega)}.$$

In particular for the domain Ω of our theorem, we would have the two estimates

$$\sup_y \|u(\cdot,\, s_0;\, 1 + iy)\|_{L^\infty(-\ell,\, \ell)} \leq C\|g\|_{L^\infty(-\ell,\, \ell)}$$

and

$$\sup_y \|u(\cdot,\, s_0;\, iy)\|_{L^2(-\ell,\, \ell)} \leq C\|g\|_{L^2(-\ell,\, \ell)}$$

$(g \in C_0^\infty(-\ell, \ell))$. Interpolation of analytic families of operators would again give

$$\|u(\cdot,\, s_0,\, \tilde{\theta})\|_{L^r(-\ell,\, \ell)} \leq C_{s_0,\ell,r}\|g\|_{L^r(-\ell,\, \ell)}$$

for some r, $1 < r < \infty$, and $C_{s_0,\ell,r}$ independent of g. This last estimate contradicts Lemma 3.

Remark 3. The L^2 bounds in Lemma 2 also hold for any L^p, $1 < p < \infty$. This can be done by the same technique of proof, using vector-valued L^p-estimates for singular integrals.

REFERENCES

1. Agmon, S. *Lectures on Elliptic Boundary Value Problems*. New York: Van Nostrand, 1965.

2. Caffarelli, L.; Fabes, E.; and Kenig, C. Completely singular elliptic-harmonic measures. To appear in the *Ind. Univ. Math. J.*, Vol. 30, No. 6 (1981):917-924.

3. Caffarelli, L.; Fabes, E.; Mortola, S.; and Salsa, S. Boundary behaviour of nonnegative solutions of elliptic operators in divergence form. *Ind. Univ. Math. J.*, Vol. 30, No. 4 (1981):621-639

4. Carleson, L. On mappings, conformal at the boundary. *J. D'Analyse* 19 (1967):1-13.

5. Hardy, G. H.; Littlewood, J. E.; and Polya, G. *Inequalities*. 2d ed. New York: Cambridge Univ. Press, 1964.

6. Littman, W.; Stampacchia, G.; and Weinberger, H. Regular points for elliptic equations with discontinuous coefficients. *Ann. Sc. Norm. Sup Pisa, Cl. Sci.*, Ser III, 17 (1963):45-79.

7. Morrey, C. B. *Multiple Integrals in the Calculus of Variations*. New York: Springer-Verlag, 1966.

8. Stein, E. *Singular Integrals and Differentiability Properties of Functions*. Princeton, N.J.: Princeton Univ. Press, 1970.

9. Stein, E., and Zygmund, A. On the differentiability of functions. *Studia Math.* 23 (1964):247-283.

10. Zygmund, A. *Trigonometric Series*. 2d ed. New York: Cambridge Univ. Press, 1968.

BOUNDARY BEHAVIOR OF SOLUTIONS TO DEGENERATE ELLIPTIC EQUATIONS

E.B. Fabes, C.E. Kenig
University of Minnesota

D. Jerison
University of Chicago

This is the third and last in a series of articles examining solutions to degenerate elliptic equations in divergence form,

$$Lu(x) = -\sum_{i,j=1}^{n} \frac{\partial}{\partial x_i}\left(a_{ij}(x)\frac{\partial}{\partial x_j}u(x)\right) = 0,$$

where the coefficients $a_{ij}(x)$ are real, measurable, symmetric, and satisfy

$$c^{-1}m(x)|\xi|^2 \leq \sum_{i,j=1}^{n} a_{ij}(x)\xi_i\xi_j \leq cm(x)|\xi|^2$$

for suitable $m(x)$. We will be concerned mainly with Fatou-type theorems, but we will also review briefly the aspects of potential theory treated in the previous two articles [6 and 5].

Denote $B(x, r) = \{y \in \mathbf{R}^n : |x - y| < r\}$ and $m(E) = \int_E m(x)dx$. The difference between the degenerate elliptic case (where m may vanish or become infinite) and the uniformly elliptic case (where m is constant) is that the function $\theta(x, r) = r^2/m(B(x, r))$ plays the role of r^{2-n}.

The property underlying most of the arguments in this subject is a scale-invariant version of Harnack's inequality. The

---

All authors were supported in part by the NSF. The second author is an NSF postdoctoral fellow. Univ. of Minnesota Math. Rpt. 81-106.

technique was introduced by L. Carleson [2] (see also J. Moser [11]) and extended by many authors (see [8, 9, 4, 13, 1, and 10]). In the process, there has been a progression to more general contexts. The study of harmonic functions in a Lipschitz domain is reduced by a change of variable to a special case of the study of solutions to a uniformly elliptic equation in a smooth domain. Similarly, the study of harmonic functions in the quasiconformal image of a ball is a special case of the study of solutions to a degenerate elliptic equation in the ball. The present version is more general still. It is only a slight exaggeration to say that it is based on estimates (1) and (2) alone.

We will not need to repeat most of the arguments of preceding work here, since they are the same in our case. Instead we will present an outline and refer the reader to [10] for details. Only one lemma (Lemma 3) requires a different proof.

NOTATIONS

We will assume that the weight $m(x)$ satisfies either Muckenhoupt's condition

$$(A_2) \qquad \sup_B\left(\frac{1}{|B|}\int_B m(x)dx\right)\left(\frac{1}{|B|}\int_B m(x)^{-1}dx\right) \leq c < \infty,$$

where the supremum is taken over the balls B and $|B| = \int_B dx$, or that $m(x) = |f'(x)|^{1-2/n}$, where f is a global quasiconformal mapping of $\mathbf{R}^n$ and $|f'(x)|$ denotes the absolute value of the Jacobian determinant of f. All functions $m(x)$ of the form $|x|^\alpha$, $\alpha > -n$, satisfy at least one of these conditions. For more details on the nature of the conditions, see [3, 6, and 7]. In particular, it is well known that $m(x)dx$ and dx are mutually absolutely continuous and

$$m(B(x, 2r)) \leq cm(B(x, r)) \qquad (\textit{doubling condition}). \qquad (1)$$

Ω will always denote an open, connected, bounded region in $\mathbf{R}^n$. Denote by $H(\Omega)$ [resp. $H_0(\Omega)$] the completion of $\mathcal{C}^\infty(\bar{\Omega})$ [resp. $\mathcal{C}_0^\infty(\Omega)$] in the norm

$$\|\varphi\|_\Omega = \left(\int_\Omega |\varphi|^2 m + \int_\Omega |\nabla\varphi|^2 m\right)^{1/2}.$$

An element u of $H(\Omega)$ need not be a distribution, but

$$\nabla u = \left(\frac{\partial u}{\partial x_1}, \ldots, \frac{\partial u}{\partial x_n}\right)$$

is still a well-defined n-tuple of elements of $L^2(\Omega, m)$ in an appropriate limiting sense [6, Property 1]. The Dirichlet form $D : H(\Omega) \times H(\Omega) \to \mathbf{R}$ is defined by

$$D(u, v) = \int a_{ij}(x)\frac{\partial u}{\partial x_i}(x)\frac{\partial u}{\partial x_j}(x)\,dx.$$

We will say that $Lu = 0$ in Ω if $u \in H(\Omega')$ and $D(u, v) = 0$ for every $v \in H_0(\Omega')$ and every $\Omega' \subset\subset \Omega$.

ESTIMATES

The first step is the inequality

$$\|\varphi\|_\Omega^2 \leq CD(\varphi, \varphi) \text{ for } \varphi \in \mathcal{C}_0^\infty(\Omega).$$

[See the stronger inequality (2) below.] It follows that $D(u, v)$ is an inner product for the Hilbert space $H_0(\Omega)$ and:

Theorem 1. *There exists a unique, bounded linear mapping $G : H_0(\Omega)^* \to H_0(\Omega)$ such that if $T \in H_0(\Omega)^*$ and $u \in G(T)$, then $D(u, v) = T(v)$ for every $v \in H_0(\Omega)$. [$H_0(\Omega)^*$ denotes the dual space to $H_0(\Omega)$.]*

It is easy to verify that the space $H_0(\Omega)^*$ is identified with the space of distributions $T = f_0 - \operatorname{div} \vec{f}$, where $\vec{f} = (f_1, \ldots, f_n)$ and $f_i/m \in L^2(\Omega, m)$, $i = 0, 1, \ldots, n$. The action of T on $H_0(\Omega)$

is given by

$$T(v) = \int_\Omega f_0(x)v(x)\,dx + \sum_{i=1}^{n} \int_\Omega f_i(x)\frac{\partial v}{\partial x_i}(x)\,dx.$$

Some of the main results of [6] are:

Theorem 2. If $Lu = 0$ in Ω, then u has a representative that is Hölder continuous in Ω, that is, there exists $\alpha > 0$ such that for any $\Omega' \subset\subset \Omega$, there exists c such that $|u(x) - u(x')| \le c|x - x'|^\alpha$ for x, x' in Ω'.

We will always take the continuous representative of u.

Harnack's Inequality. If $Lu = 0$ in $B(x, 2r)$ and $u \ge 0$ in $B(x, 2r)$, then $\max_{B(x,r)} u \le c \min_{B(x,r)} u$. ($c$ is independent of x and $r > 0$.)

Strong Maximum Principle. If $Lu = 0$ in Ω and u attains its maximum or minimum in Ω, then u is constant.

The main point of the proof is a kind of fractional integration estimate. There exist $k > 1$, $c < \infty$, such that

$$\left(\frac{1}{|B|}\int_B |\varphi|^{2k} m\right)^{1/2k} \le cr\left(\frac{1}{|B|}\int_B |\nabla\varphi|^2 m\right)^{1/2}, \quad \varphi \in C_0^\infty(B),\ B = B(x, r). \tag{2}$$

It is crucial that c and k are independent of r and x as well as φ. One can then follow Moser's proof of the theorem of de Georgi and Nash to obtain the scale-invariant version of Harnack's inequality above (with suitable interpretations of max and min) and Hölder continuity.

Let $\varphi \in C^\infty(\mathbf{R}^n)$ and $T = L\varphi = -\text{div}\,\vec{f}$, where $\vec{f} = (f_1, \ldots, f_n)$,

$$f_j(x) = \sum_i a_{ij}(x)\frac{\partial \varphi}{\partial x_i}(x).$$

Then T belongs to $H_0(\Omega)^*$ with the action

$$T(v) = \int a_{ij}\,\frac{\partial \varphi}{\partial x_i}\,\frac{\partial v}{\partial x_j}.$$

Let $u = \varphi - G(T)$. Then $Lu = 0$ in Ω and u is the (variational) solution to the Dirichlet problem: $Lu = 0$ in Ω, $u = \varphi$ on $\partial\Omega$. The weak maximum principle of Stampacchia [5, 1.18] implies that $\sup_{\Omega}|u| \leq \max_{\partial\Omega}|\varphi|$. It then follows from the methods of the previous theorems that:

Theorem 3. Suppose that $h \in \mathcal{C}(\partial\Omega)$, $\varphi_j \in \mathcal{C}^\infty(\mathbf{R}^n)$, $\varphi_j \to h$ *uniformly on* $\partial\Omega$; $Lu_j = 0$ *in* Ω, $u_j = \varphi_j$ *on* $\partial\Omega$ *in the same given above. Then there exists a unique Hölder continuous function* u *on* Ω *such that for any* $\Omega' \subset\subset \Omega$, $u_j \to u$ *uniformly on* Ω'. *Moreover,* $Lu = 0$ *in* Ω *and* $\sup_{\Omega}|u| \leq \max_{\partial\Omega}|h|$.

We will say that a point $y \in \partial\Omega$ is *regular for* Ω if $\lim_{x \to y} u(x) = h(y)$ for every $h \in \mathcal{C}(\partial\Omega)$.

THE GREEN FUNCTION

A complete description of regular points is based on interior estimates for the Green function. Denote $\Sigma = \{x : |x| < R\}$ for some large value of R. Here are some facts about the Green function proved in [5 and 6].

There is a function $g : \overline{\Sigma} \times \overline{\Sigma} \to [0, \infty]$ such that $g(x, y)$ is Hölder continuous in $\overline{\Sigma} \times \overline{\Sigma}\backslash\Delta$ $(\Delta = \{(y, y) : y \in \overline{\Sigma}\})$; $g(x, y) = g(y, x)$, $g(x, y) = 0$ for $y \in \partial\Sigma$, $g(x, \cdot) \in H(\Sigma\backslash\overline{B}(x, r))$, and $Lg(x, \cdot) = 0$ in $\Sigma\backslash\overline{B}(x, r)$ for any $r > 0$. The class of functions ψm such that $\psi \in L^\infty(\Sigma, dx)$ is dense in $H_0(\Sigma)^*$. Furthermore, $G(\psi m)$ is continuous. Let μ be a finite measure supported in Σ and $u(x) = \int_\Sigma g(x, y)\,d\mu(y)$. Then

$$\int_\Omega G(\psi m)\,d\mu = \int_\Omega \psi m u \text{ for all } \psi \in L^\infty(\Sigma, dx). \tag{3}$$

Recall the notation $\theta(x, r) = r^2/m(B(x, r))$.

*Theorem 4. Let $x \in \Sigma$ and let $d = \text{dist}(x, \partial\Sigma)$. If $|x - y| = r < \frac{1}{2}d$, then**

$$g(x, y) \simeq \int_r^d \theta(x, s)\frac{ds}{s}.$$

($\simeq$ means that the ratio of the two numbers is bounded above and below by positive constants that depend only on the (A_2) or quasi-conformal dilatation constant of m.)

The proof of Theorem 4 requires the notion of capacity

$$\text{cap}_\Sigma(K) = \inf\{D(\varphi, \varphi) : \varphi \geq 1 \text{ on } K, \varphi \in \mathcal{C}_0^\infty(\Sigma)\}.$$

(For many equivalent definitions, see [5].) Capacity also enters in the Wiener test, of course.

Theorem 5 (Wiener test [5, 5.1]). *Let $\Omega \subset\subset \Sigma$, $y \in \partial\Omega$ is regular for Ω if and only if either*

(a) $\displaystyle\int_0^R \theta(y, s)\frac{ds}{s} < \infty$ *or*

(b) $\displaystyle\int_0^R \text{cap}_\Sigma(\bar{B}(y, s) \cap \Sigma\backslash\Omega)\theta(y, s)\frac{ds}{s} = \infty.$

Conditions (a) and (b) are mutually exclusive. Condition (a) is equivalent to $\text{cap}_\Sigma(\{y\}) > 0$ and also to $\limsup_{x \to y} g(x, y) < \infty$. In fact, $\lim_{x \to y} g(x, y) = 1/\text{cap}_\Sigma(\{y\})$. Possibility (a) does not arise in the uniformly elliptic case. In keeping with the analogy of $\theta(y, s)$ with S^{2-n}, the "dimension" at points of positive capacity is less than two.

---

*This is very slightly different from the statement in [5], but the proof is the same.

HARMONIC MEASURE

For each $x \in \Omega$, *harmonic measure* (at x) is the measure ω^x on $\partial\Omega$ satisfying $u(x) = \int_{\partial\Omega} h\, d\omega^x$, where u and h are as in Theorem 3.

We call a domain Ω *nontangentially accessible* (abbreviated NTA) provided the following three properties hold.

(i) There exist constants $A > 1$ and $r_0 > 0$ such that for every r, $0 < r < r_0$ and every $y \in \partial\Omega$, there exists $e \in \Omega$ such that $|y - e| < Ar$ and $B(e, r/A) \subset \Omega$.

(ii) Property (i) for the complement of Ω.

(iii) For every $c > 0$, there exists N such that if $0 < \varepsilon < r_0$, $x, y \in \Omega$, $\operatorname{dist}(x, \partial\Omega) < \varepsilon$, $\operatorname{dist}(y, \partial\Omega) > \varepsilon$, and $|x - y| < c\varepsilon$, then there is a sequence of N points, $x = x_1, x_2, \ldots, x_N = y$, such that $|x_j - x_{j+1}| < \varepsilon/A$ and $B(x_j, 2\varepsilon/A) \subset \Omega$.

From now on Ω will denote an NTA domain.

It is easy to see from their proofs that the properties of the Green function for Σ, including Theorem 4, are valid without change for any NTA domain Ω in place of Σ. A consequence of the Wiener test (which can also be checked more directly) is that all boundary points of an NTA domain are regular. Thus, for every $h \in \mathcal{C}(\partial\Omega)$,

$$u(x) = \int_{\partial\Omega} h\, d\omega^{x_0} \text{ is continuous in } \overline{\Omega} \text{ and } u|_{\partial\Omega} = h. \tag{4}$$

Fix $x_0 \in \Omega$ and $\omega = \omega^x$. Harnack's inequality implies that harmonic measure ω^x differs from ω by a bounded factor. Hence, the Lebesgue classes $L^p(\partial\Omega, d\omega^x)$, $1 \le p \le \infty$, are equivalent for each value of x. The ***kernel function*** is defined for almost every (ω) $y \in \partial\Omega$ as the Radon Nikodym derivative $K(x, y) = \frac{d\omega^x}{d\omega}(y)$. A non-tangential approach region at $y \in \partial\Omega$ is

$$\Gamma_\alpha(y) = \{x : |x - y| < (1 + \alpha)\operatorname{dist}(x, \partial\Omega)\}.$$

The fact that Ω is an NTA domain implies that $\Gamma_\alpha(y)$ is nontrivial for sufficiently large α. The nontangential maximal function is given by $N(u)(y) = \sup\{|u(x)| : x \in \Gamma_\alpha(y)\}$. The particular value of $\alpha > 0$ turns out not to be important, so for simplicity we will not mention it further.

Here is a Fatou-type theorem.

Theorem 6. Let Ω be an NTA domain. If $f \in L^1(\partial\Omega, d\omega)$ and

$$u(x) = \int_{\partial\Omega} f\, d\omega^x,$$

then $Lu = 0$ in Ω, $\lim\limits_{\substack{x \to y \\ x \in \Gamma_\alpha(y)}} u(x) = f(y)$ for almost every (ω) $y \in \partial\Omega$. Moreover,

$$\omega(\{y \in \partial\Omega : N(u)(y) > \lambda\}) \leq c\lambda^{-1} \int_{\partial\Omega} |f|\, d\omega.$$

In addition, if $f \in L^p(\partial\Omega, d\omega)$ for $1 < p \leq \infty$, then

$$\|N(u)\|_{L^p(\partial\Omega,\, d\omega)} \leq c_p \|f\|_{L^p(\partial\Omega,\, d\omega)}.$$

Denote

$$Mf(y) = \sup\left\{\frac{1}{\omega(\Delta)} \int_\Delta |f|\, d\omega : \Delta = B(y, r) \cap \partial\Omega,\ r > 0\right\}.$$

The proof of Theorem 6 is a well-known argument based on (4) and

$$N(u)(y) \leq cMf(y) \text{ for } f \in L^1(\partial\Omega, d\omega), \tag{5}$$

and

$$\omega(\{y \in \partial\Omega : Mf(y) > \lambda\}) \leq c\lambda^{-1} \int_{\partial\Omega} |f|\, d\omega. \tag{6}$$

The estimate (6) follows from:

Lemma 1. For any $y \in \partial\Omega$, $r > 0$, $\omega(2\Delta) \leq c\omega(\Delta)$, where $\Delta = B(y, r) \cap \partial\Omega$ and $2\Delta = B(y, 2r) \cap \partial\Omega$.

The inequality (5) follows from:

Lemma 2. There exists $\alpha > 0$ such that if $y \in \partial\Omega$, $r > 0$ and $e \in \Omega$ are as in (i) *above, then*

$$K(e, y') \le c/\omega(\Delta)(\max(|y' - y|/r)^{\alpha}, 1),$$

where $\nabla = B(y, r) \cap \partial\Omega$.

Caffarelli et al. [1] observed that the simplest way to prove these two lemmas is by means of a comparison of harmonic measure with the Green function first used by Dahlberg [4] in the context of harmonic functions in Lipschitz domains. The comparison is:

Lemma 3. Let $y \in \partial\Omega$, $e \in \Omega$ *be as in* (i) *above, and let* $\nabla = B(y, r) \cap \partial\Omega$. *If* $x \in \Omega\backslash B(y, 4Ar)$, *then*

$$g(e, x) \simeq \omega^{x}(\Delta)\theta(e, r).$$

($\simeq$ *means that the ratio of the two sides is bounded above and below by positive constants independent of* y, e, *and* r.)

Lemma 1 follows immediately from Lemma 3 and Harnack's inequality. Lemma 2 depends on Lemma 3, (7), Remark 1 below, and the following consequence of Lemma 3 (proved by Dahlberg [4] for harmonic functions in Lipschitz domains [10, 4.10]).

Boundary Harnack Inequality. There are positive constants c_1 *and* c_2 *such that if* $y \in \partial\Omega$, u *and* v *are positive solutions to* $Lu = Lv = 0$ *in* Ω *that vanish continuously on* $B(y, 2r) \cap \partial\Omega$, *and* $u(e) = v(e)$ *with* e *as in* (i), *then* $c_1 < u(x)/v(x) < c_2$ *for* $x \in B(y, r) \cap \Omega$.

Proof of Lemma 3. An easy consequence of Hölder continuity at the boundary ([10, 4.1, 4.2]) is that for $z \in \partial B(e, r/2A)$, $\omega^{z}(\Delta) > c_0 > 0$. Furthermore, a special case of Theorem 4 is

$$g(e, z) \simeq \theta(e, r) \text{ for } z \in \partial B(e, r/2A). \tag{7}$$

The maximum principle in the domain $\Omega\backslash\overline{B}(e, r/2A)$ implies $g(e, x) \le c\omega^{x}(\Delta)\theta(e, r)$ for all $x \in \Omega\backslash B(y, 4Ar)$.

The opposite inequality requires three remarks. Let $B_1 = B(y, r)$, $B_2 = B(y, 2r)$, $B_3 = B(y, 2Ar)$, $B_4 = B(y, 4Ar)$.

Remark 1. Let u satisfy $Lu = 0$ in $\Omega \cap B_4$. Suppose that u vanishes continuously on $B_4 \cap \partial\Omega$; then $u(x) \le cu(e)$ for all $x \in B_3 \cap \Omega$.

Remark 1 originates with Carleson [2]. For a proof, see [10, 4.4]. Next we have the well-known inequality [6, 2.4.2].

Remark 2. Let u be as in Remark 1; then

$$\left(\int_{B_2 \cap \Omega} |\nabla u|^2 m\right)^{1/2} \le c\left(r^{-2}\int_{B_3 \cap \Omega} |u|^2 m\right)^{1/2}.$$

Finally, let $T = -\text{div}\, \vec{f}$ belong to $H_0(\Omega)^*$. Suppose that the support of T is disjoint from x. Then $G(T)$ is continuous at x. A limiting argument based on (3) with μ equal to the unit mass at x and the fact that $g(x, \cdot) \in H(\Omega\backslash\bar{B}(x, s))$ yields:

Remark 3. $G(T)(x) = T(g(x, \cdot)) = \int_\Omega f_j(z)\frac{\partial}{\partial z_j} g(x, z)dz.$

Let $\varphi \in \mathfrak{C}_0^\infty(B_2)$ so that $\varphi \ge 0$, $\varphi = 1$ on B_1, and $|\nabla\varphi| \le 2r^{-1}$. Let $x \in \Omega\backslash B_4$. Remark 3 implies

$$\omega^x(\Delta) \le \int \varphi\, d\omega^x = \varphi(x) - G(L\varphi)(x) = -G(L\varphi)(x)$$

$$= -\int a_{ij}(z)\frac{\partial\varphi}{\partial z_i}(z)\frac{\partial}{\partial z_j} g(x, z)dx.$$

By Schwarz's inequality, Remark 2, Remark 1, and (1),

$$\omega^x(\Delta) \le c\left(\int_\Omega |\nabla\varphi|^2 m\right)^{1/2}\left(\int_{B_2\cap\Omega} |\nabla_z g(x, z)|^2 m(z)dz\right)^{1/2}$$

$$\le c\left(\frac{m(B_2)}{r^2}\right)^{1/2}\left(r^{-2}\int_{B_3\cap\Omega} |g(x, z)|^2 m(z)dz\right)^{1/2}$$

$$\le c\left(\frac{m(B_2)}{r^2}\right)^{1/2}\left(\frac{m(B_3)}{r^2}\right)^{1/2} g(x, e) < c\theta(e, r)^{-1} g(x, e).$$

THE KERNEL FUNCTION

A consequence of the results of the previous section is that the kernel function $K(x, y)$ is continuous as a function of $g \in \partial\Omega$ (see [9, 1, and 10]). In addition:

Corollary 1. There is a one-to-one correspondence between positive solutions to $Lu = 0$ in Ω and positive measures on $\partial\Omega$ given by

$$u(x) = \int_{\partial\Omega} K(x, y)\,d\mu(y).$$

Also,

$$\lim_{\substack{x \to y \\ x \in \Gamma_\alpha(y)}} u(x) = h(y) \text{ a.e. } (\omega)\ y \in \partial\Omega,$$

where $\mu = h\omega + \nu$, $h \in L^1(\partial\Omega, d\omega)$, and ν is singular with respect to ω.

Following Jerison and Kenig [10], who treated the uniformly elliptic case, we can show that $K(x, y)$ is actually Hölder continuous as a function of y and deduce Hardy space estimates.

Theorem 7 [10, 7.1]. $g(x, z)/g(x_o, z)$ is Hölder continuous for $z \in \overline{\Omega}\backslash\{x, x_o\}$ and $K(x, y) = \lim_{z \to y} g(x, z)/g(x_o, z)$ for every $y \in \partial\Omega$. Furthermore, there is $\alpha > 0$ and $c < \infty$ such that

$$|K(x, y') - K(x, y)| \leq c\left(\frac{|y - y'|}{|x - y|}\right)^\alpha K(x, y)$$

for all $x \in \Omega$, $y, y' \in \partial\Omega$.

The proof is based on the boundary Harnack inequality above. An iteration (analogous to Moser's proof that the Harnack inequality implies Hölder continuity) yields:

Theorem 7'. Under the hypotheses of the statement of the boundary Harnack inequality above, $u(x)/v(x)$ is Hölder continuous in $B(y, r) \cap \overline{\Omega}$.

Theorem 7 is proved in the same way.

A function $a \in L^\infty(\partial\Omega, d\omega)$ is a *p-atom* if $\int a\, d\omega = 0$, a is supported in $\Delta = B(y, r) \cap \partial\Omega$ for some $y \in \partial\Omega$, $r > 0$, and $\sup|a| \leq 1/\omega(\Delta)^{1/p}$. The constant function $a \equiv 1$ is also assumed to be a p-atom. Let $p \leq 1$. The atomic Hardy space H^p_{at} is the class of f such that

$$\|f\|_{H^p_{at}} = \inf\{(\Sigma|\lambda_j|^p)^{1/p} : f = \Sigma\lambda_j a_j,\ a_j(x) \text{ are } p\text{-atoms}\}$$

is finite. A more formal definition requires us to make sense of the sum $\Sigma\lambda_j a_j$ in the dual space to a Hölder class. We will spare the reader the technical details. The estimate in Theorem 7 implies:

Corollary 2 [10, 8.13, 8.14]. *There exists* $p_0 < 1$ *such that if* $p_0 < p \leq 1$ *and* $f \in H^p_{at}$, *then* $u(x) = \int_{\partial\Omega} f\, d\omega^x$ *is a solution to* $Lu = 0$ *in* Ω *and*

$$\|N(u)\|_{L^p(\partial\Omega,\, d\omega)} \simeq \|f\|_{H^p_{at}}.$$

Conversely, if $N(u) \in L^1(\partial\Omega, d\omega)$, *then there is a unique* $f \in H^1_{at}$ *such that* $u(x) = \int_{\partial\Omega} f\, d\omega^x$.

The converse for $p < 1$ is open. To prove it one need only show that $L^1(\partial\Omega, d\omega)$ is dense in the space of boundary values of u such that $N(u) \in L^p(\partial\Omega, d\omega)$.

In closing, we pose the question of whether some of these theorems apply to solutions of nonlinear elliptic equations. In particular, it would be interesting to know if the components of a quasiregular mapping of the ball have such a theory (see [12]).

REFERENCES

1. Caffarelli, L.; Fabes, E.; Mortola, S.; and Salsa, S. Boundary behavior of nonnegative solutions of elliptic operators in divergence form. *Ind. Univ. Math. J.* 30 (1981):621-640.

2. Carleson, L. On the existence of boundary values for harmonic functions in several variables. *Ark. Mat.* 4 (1962):393-399.

3. Coifman, R., and Fefferman, C. Weighted norm inequalities for maximal functions and singular integrals. *Studia Math.* 51 (1974):241-250.

4. Dahlberg, B. E. J. Estimates of harmonic measure. *Arch. for Rat. Mech. and Anal.* 65 (1977):275-288.

5. Fabes, E.; Jerison, D.; and Kenig, C. The Weiner test for degenerate elliptic equations. To appear in *Ann. Inst. Fourier.*

6. Fabes, E.; Kenig, C.; and Serapioni, R. The local regularity of solutions of degenerate elliptic equations. *Communications in P.D.E.* 7 (1982):77-116.

7. Gehring, F. The L^p-integrability of the partial derivatives of a quasiconformal mapping. *Acta Math.* 130 (1973):266-277.

8. Hunt, R., and Wheeden, R. On the boundary values of harmonic functions. *I.A.M.S.* 132 (1968):307-322.

9. Hunt, R., and Wheeden, R. Positive harmonic functions on Lipschitz domains. *I.A.M.S.* 147 (1970):507-527.

10. Jerison, D., and Kenig, G. Boundary behavior of harmonic functions in nontangentially accessible domains. To appear in *Advances in Math.*

11. Moser, J. On Harnack's theorem for elliptic differential equations. *Comm. Pure Appl. Math.* 14 (1961):577-591.

12. Väisälä, J. A survey of quasiregular maps in $\mathbf{R}^n$. *Proc. Int. Cong. Math. Helsinki* (1978):685-691.

13. Wu, J. M. G. Comparisons of kernel functions, boundary Harnack principle and relative Faton theorem on Lipschitz domains. *Ann. Inst. Fourier*, Grenoble, 28.4 (1978):147-167.

SUBELLIPTIC EIGENVALUE PROBLEMS

C. Fefferman
Princeton University

D.H. Phong
Columbia University

This article studies subellipticity and asymptotic eigenvalue distribution of second-order partial differential operators with non-negative principal symbols. We obtain definitive results for self-adjoint operators with real coefficients. Deep theorems on subellipticity and related topics have been obtained by Hörmander [6], Kohn [7], and Rothschild and Stein [10] for sums of squares of vector fields; and by Olenik and Radkevitch [9] for general second-order equations; while subelliptic eigenvalue asymptotics were studied by Menikoff and Sjöstrand [8]. The interest of our work is that it gives sharp necessary and sufficient conditions for operators not assumed to be written as sums of squares. It would be interesting to understand also non-self-adjoint equations and equations with complex coefficients.

To fix the notation, let L be a second-order self-adjoint differential operator on a compact manifold M with smooth measure μ. Assume that in local coordinates

$$L = -\sum_{ij} a^{ij}(x)\frac{\partial^2}{\partial x_i \partial x_j} + \sum_j b_j(x)\frac{\partial}{\partial x_j} + c(x),$$

NSF Grant No. MCS80-03072.

NSF Grant No. MCS78-27119.

where a^{ij}, b_j, c are real and $(a^{ij}(x))$ is positive semidefinite. We want to understand the following:

(A) When does L satisfy the subelliptic estimates

$$\mathrm{Re}\langle Lu, u\rangle + C\|u\|^2 \geq c\|u\|^2_{(\varepsilon)} \quad \text{for } u \in C^\infty(M); \tag{1}$$

$$\|Lu\| + C\|u\| \geq c\|u\|_{(2\varepsilon)} \quad \text{for } u \in C^\infty(M)? \tag{2}$$

(B) What is the asymptotic behavior as $\lambda \to \infty$ of $N(\lambda, L)$, the number of eigenvalues of L which are $< \lambda$?

These questions are answered in terms of a family of "non-Euclidean balls," which we now define.

A tangent vector $X = \sum_j \gamma_j \frac{\partial}{\partial x_j}$ at $x \in M$ is said to be *subunit for L* if $(\gamma_j\gamma_k) \leq (a^{jk}(x))$ as matrices. One easily checks that this notion is independent of the particular coordinate chart. For $x \in M$ and $\rho > 0$, the "ball" $B_L(x, \rho)$ consists of all the points $y \in M$ that can be joined to x by a Lipschitz path $\gamma : [0, \rho] \to M$ for which $\frac{d}{dt}\gamma(t)$ is a subunit vector for L at $\gamma(t)$ for almost every t.

By $B_E(x, \rho)$ we denote an ordinary Euclidean ball of radius ρ about x. Note that if $-L$ is the Laplacian for a metric ds^2 on M, then $B_L(x, \rho)$ agrees with the usual ball of radius ρ in ds^2. See also Rothschild and Stein [10], in which suitable non-Euclidean balls play an important role.

Theorem 1. For L as above, the estimates (1) *and* (2) *are equivalent, and they both hold if and only if*

$$B_E(x, \rho) \subseteq B_L(x, C\rho^\varepsilon) \text{ for } x \in M,\ \rho > 0.$$

Theorem 2. Assume L satisfies (1) *and* (2). *Then for large λ, $N(\lambda, L)$ is bounded above and below by constant multiples of*

$$\tilde{N}(\lambda, L) = \int_M \frac{d\mu(x)}{\mu(B_L(x, \lambda^{-1/2}))}.$$

The rest of this paper sketches the proofs of Theorems 1 and 2. We make essential use of results and techniques in [2] and [5], with which we assume the reader is familiar. The new tools needed in our proofs are the following geometric lemmas.

Lemma 1. $B_E(x, \lambda) \subseteq B_L(x, C\lambda^{\varepsilon})$ *if and only if*

$$B_E(x, \lambda) \subseteq B_{L-\lambda^{2N}\Delta}(x, C\lambda^{\varepsilon}),$$

provided $N \geq \varepsilon^{-50}$. *Moreover, these inclusions imply that*

$$\mu(B_L(x, \lambda)) \sim \mu(B_{L-\lambda^{2N}\Delta}(x, \lambda)).$$

Lemma 2. *In* R^{n+1}, *set*

$$L_\lambda = -\left(\frac{\partial}{\partial t}\right)^2 - \sum_{ij} a^{ij}(t, x)\frac{\partial^2}{\partial x_i \partial x_j} - \lambda^{2N}\Delta_x;$$

in R^n, *set*

$$\overline{L}_\lambda = -\sum_{ij} \overline{a^{ij}}(x)\frac{\partial^2}{\partial x_i \partial x_j} - \lambda^{2N}\Delta_x,$$

where

$$\overline{a^{ij}}(x) = \lambda^{-1}\int_{|t|\leq\lambda} a^{ij}(t, x)\,dt.$$

Then the non-Euclidean balls B_{L_λ}, $B_{\overline{L}_\lambda}$ *about the origin are related by* $B_{L_\lambda}(c\lambda) \subseteq \{|t| \leq \lambda\} \times B_{\overline{L}_\lambda}(\lambda) \subseteq B_{L_\lambda}(C\lambda)$.

Lemma 1 and the terms $\lambda^{2N}\Delta_x$ in Lemma 2 are technicalities; while Lemma 2 contains new geometric information, which is essential both here and in our paper [4].

To prove Lemma 2, we need the following result, which forms an L^∞-analogue of the spectral decomposition theorem in [5].

For

$$L = -\sum_{ij} a^{ij}(x)\frac{\partial^2}{\partial x_i \partial x_j} + \cdots$$

defined on the unit cube Q^0, and for $u \in L^\infty(Q^0)$, define

$$L_K(u) = \inf_{v \in C^1(Q^0)}\left\{K\|u - v\|^2 + \left\| \sum_{ij} a^{ij}(x)\frac{\partial v}{\partial x_i}\frac{\partial v}{\partial x_j}\right\| + K^{-2N}\|\nabla u\|^2\right\},$$

where the norms are taken in $L^\infty(Q^0)$.

Lemma 3. $c_N(L + L')_K(u) \leq L_K(u) + L'_K(u)$.

We first sketch how Lemmas 1 and 2 yield Theorems 1 and 2 by using [2 and 5]. Next we prove Lemma 1 and reduce Lemma 2 to Lemma 3. Finally, we sketch the proof of Lemma 3.

Proof of Theorem 1. It is easy to show that (1) and (2) are equivalent. In fact, (2) asserts that $(L + CI)^2 > \Lambda^{4\varepsilon}$ as self-adjoint operators on $L^2(M)$. One knows from operator theory that $A^2 \geq B^2$ implies $A \geq B$ for positive operators A, B, so $L + CI \geq \Lambda^{2\varepsilon}$, which is (1). On the other hand, assume (1), and note that $\mathrm{Re}\langle Lu, \Lambda^{2\varepsilon}u\rangle \leq c_1\|u\|^2_{(2\varepsilon)} + 10c_1^{-1}\|Lu\|^2$ for a small c_1 to be picked later. However, $\mathrm{Re}\langle Lu, \Lambda^{2\varepsilon}u\rangle = \mathrm{Re}\langle L\Lambda^\varepsilon u, \Lambda^\varepsilon u\rangle + \mathrm{Re}\langle \Lambda^\varepsilon[\Lambda^\varepsilon, L]u, u\rangle$. One computes that $\Lambda^\varepsilon[\Lambda^\varepsilon, L] = T_1 + T_2$ with $T_1 \in S^{2\varepsilon+1}$ skew-adjoint and $T_2 \in S^{2\varepsilon}$. Therefore,

$$|\mathrm{Re}\langle \Lambda^\varepsilon[\Lambda^\varepsilon, L]u, u\rangle| = |\mathrm{Re}\langle T_2 u, u\rangle| \leq c_1\|u\|^2_{(2\varepsilon)} + 10c_1^{-1}\|u\|^2.$$

Moreover, our assumption (1) implies

$$\mathrm{Re}\langle L\Lambda^\varepsilon u, \Lambda^\varepsilon u\rangle \geq c\|u\|^2_{(2\varepsilon)} - C\|u\|^2_{(\varepsilon)} \geq \frac{1}{2}c\|u\|^2_{(2\varepsilon)} - C'\|u\|^2.$$

Putting these estimates together, we obtain

$$\frac{c}{2}\|u\|^2_{(2\varepsilon)} - C'\|u\|^2 - c_1\|u\|^2_{(2\varepsilon)} - 10c_1^{-1}\|u\|^2 \leq c_1\|u\|^2_{(2\varepsilon)} + 10c_1^{-1}\|Lu\|^2,$$

which implies (2) if we pick $c_1 < \frac{c}{10}$. We are indebted to J. J. Kohn and E. M. Stein for stimulating conversations on estimate (2).

The hard part of Theorem 1 is to show that (1) is equivalent to $B_E(x, \lambda) \leq B_L(x, C\lambda^\varepsilon)$. To prove this, we formulate a stronger result, which can be proved by induction on the dimension.

Given a second-order operator L defined on R^n, and large constants K, S, we say that $L \geq K$ *microlocally in* $|x| < 1$, $|\xi| \sim S$

if $\theta(x, D)^*(L - KI)\theta(x, D) \geq -CI$ for some symbol $\theta \in S^0$ satisfying $\theta(x, \xi) = 1$ for $|x| \leq 1$, $|\xi| \sim S$.

One checks easily that (1) amounts to saying that in local coordinates $L \geq (\text{const})K$ microlocally in $|x| \leq 1$, $|\xi| \sim S$, with $K = S^{2\varepsilon}$.

Therefore, Theorem 1 will follow at once from Lemma 1 and the

Main Estimate. Assume $K \geq S^{\delta}$ *and* $N \geq N_{\min}(\delta)$. *Then*

(i) $L \geq (\text{const})K$ *microlocally in* $|x| \leq 1$, $|\xi| \sim S$ *iff*

(ii) $B_E(x, (\text{const})S^{-1}) \subseteq B_{L - K^{-N}\Delta}(x, K^{-1/2})$.

So out task is to prove the main estimate. In one variable, the result is trivial; in n dimensions we proceed by induction on n as follows.

We first reduce matters to the case

$$L = -\left(\frac{\partial}{\partial t}\right)^2 + \tilde{L}\left(t, y, \frac{\partial}{\partial y}\right).$$

This is done by localizing. We make a Calderón-Zygmund decomposition of $\{|x| \leq 1\}$ into cubes $\{Q_\nu\}$ of diameters δ_ν, stopping at Q_ν when $\max_{ij} \max_{x \in Q_\nu^*} |a^{ij}(x)| \geq C\delta_\nu^2$. (Recall

$$L = -\sum_{ij} a^{ij}(x)\frac{\partial^2}{\partial x_i \partial x_j} + \cdots .)$$

We may assume all the $\delta_\nu \geq S^{-1}$, since otherwise (i) and (ii) are both false. In each Q_ν, make a change of coordinates from $x \in Q_\nu^*$ to $(t, y) \in$ unit cube, so that L goes over to

$$L_\nu = -\left(\frac{\partial}{\partial t}\right)^2 + \tilde{L}_\nu\left(t, y, \frac{\partial}{\partial y}\right).$$

Now the sumbolic calculus of pseudodifferential operators shows that $L \geq (\text{const})K$ microlocally for $|\xi| \sim S$ if and only if each $L_\nu \geq (\text{const})K$ microlocally for $|(\tau, \eta)| \sim S\delta_\nu$. Moreover, under the coordinate change $x \to (t, y)$, $B_E(x, S^{-1})$ goes over to $B_E((t, y), (S\delta_\nu)^{-1})$, while $B_{L + \text{junk}}(x, K^{-1/2})$ goes over to $B_{L + \text{junk}}((t, y), K^{-1/2})$. Thus, the main estimate holds for L if and only if it holds for L_ν. So it is enough to look at

$$L = -\left(\frac{\partial}{\partial t}\right)^2 + L(t, y, \partial y).$$

Also, we may assume $K \leq S^2$, since otherwise (i) and (ii) are again obviously false. Now, however, Lemma 2 shows that in place of $B_{L - K^{-N}\Delta}((\bar{t}, \bar{y}), K^{-1/2})$, we may take

$$\{|t - \bar{t}| \leq K^{-1/2}\} \times B_{\bar{L}_{\bar{t}} - K^{-N}\Delta}(\bar{y}, K^{-1/2})$$

in (ii). Here

$$\bar{L}_{\bar{t}}(y, \partial y) = K^{1/2} \int_{|t - \bar{t}| \leq K^{-1/2}} \tilde{L}(t, y, \partial_y)\, dt.$$

Also, the results in Fefferman and Phong [5] show that (i) is equivalent to saying that $\bar{L}_{\bar{t}} \geq (\text{const})K$ microlocally in $|y| \leq 1$, $|\eta| \sim S$ for each $\bar{t}$. Therefore, our main estimate holds for L, provided it holds for $\bar{L}_{\bar{t}}$ for every $\bar{t}$. Since $\bar{L}_{\bar{t}}$ is an operator in $n - 1$ variables, the induction step is complete. So we know both the main estimate and Theorem 1.

Proof of Theorem 2. Using symbolic calculus and the minimax formula for eigenvalues, one can easily localize the problem by assuming L to be elliptic outside $\{|x| \leq 1\}$. Another application of symbolic calculus localizes the problem still further to the operators L_ν in the proof of Theorem 1. So we may assume

$$L = -\left(\frac{\partial}{\partial t}\right)^2 + \tilde{L}(t, y, \partial y)$$

with L elliptic outside $\{|t|, |y| \leq 1\}$. Since L is subelliptic, one checks that $N(\lambda, L)$ is comparable to $N(\lambda; L - \lambda^{-2N}\Delta)$ for large N depending on the ε in (1). So we may also assume that

$$\tilde{L}(t, y, \partial y) = -\sum_{ij} a^{ij}(t, y)\frac{\partial^2}{\partial y_j \partial y_k} + \cdots,$$

with $(a^{ij}(t, y)) \geq \lambda^{-2N}(\delta_{ij})$. (This uses Lemma 1.) Now define

$$\overline{L}_{\overline{t}}(y, \partial_y) = \lambda^{+1/2}\int_{|t-\overline{t}|\leq\lambda^{+1/2}} \tilde{L}(t, y, \partial_y)dt.$$

Lemma 2 shows that $B_L((t, y), \lambda^{1/2})$ is comparable to $\{|t - \overline{t}| \leq \lambda^{-1/2}\} \times B_{\overline{L}_{\overline{t}}}(y, \lambda^{-1/2})$, while the results of Fefferman and Phong [2] show that $N(\lambda, L)$ is comparable to

$$\lambda^{+1/2}\int_{|\overline{t}|\leq 1} N(\lambda, \overline{L}_{\overline{t}})d\overline{t}$$

for large λ. Therefore, the formula

$$N(\lambda, L) \sim \tilde{N}(\lambda, L) = \int_M \frac{d\mu(x)}{\mu(B_L(x, \lambda^{-1/2}))}$$

is easily deduced from the corresponding formula for the $\overline{L}_{\overline{t}}$. Since $\overline{L}_{\overline{t}}$ is an operator in $n - 1$ variables, we can now proceed by induction on the dimension.

The above argument shows that N and $\tilde{N}$ are comparable in the sense that $c\tilde{N}(c\lambda, L) \leq N(\lambda, L) \leq C\tilde{N}(C\lambda, L)$. A final application of Lemma 2 gives $\tilde{N}(C\lambda, L) \leq C'\tilde{N}(\lambda, L)$ and $\tilde{N}(c\lambda, L) \geq c'N(\lambda, L)$, so that $c''\tilde{N}(\lambda, L) \leq N(\lambda, L) \leq C''\tilde{N}(\lambda, L)$ for large λ.

Proof of Lemma 1. Assuming $B_E(x, \lambda^{N/2}) \subseteq B_{L-\lambda^{2N}\Delta}(x, \lambda)$ for $x \in M$, $\lambda > 0$, we shall prove that

$$(+) \qquad B_{L-\lambda^{2N}\Delta}(x, \lambda) \subseteq B_L(x, C\lambda).$$

This easily implies Lemma 1. In fact, using $C\lambda^{\varepsilon}$ in place of λ and εN in place of N in (+), we obtain $B_{L-\lambda^{2N}\Delta}(x, C\lambda^{\varepsilon}) \subseteq B_L(x, C'\lambda^{\varepsilon})$, so that $B_E(\ldots) \subseteq B_{L-\lambda^{2N}\Delta}(\ldots)$ implies $B_E(\ldots) \subseteq B_L(\ldots)$. The converse implication is trivial, since $B_L(x, C\lambda^{\varepsilon}) \subseteq B_{L-\lambda^{2N}\Delta}(x, C\lambda^{\varepsilon})$. Inclusion (+) also yields the part of Lemma 1 on $\mu(B_L(x, \lambda))$, since $\mu(B_L(x, \lambda)) \leq \mu(B_{L-\lambda^{2N}\Delta}(x, \lambda)) \leq \mu(B_L(x, C\lambda))$. This implies $\mu(B_{L-c\lambda^{2N}\Delta}(x, c\lambda)) \leq \mu(B_L(x, \lambda)) \leq \mu(B_{L-\lambda^{2N}\Delta}(x, C\lambda))$.

An application of Lemma 2 shows that the terms on the extreme left and right are comparable, so that indeed (+) implies all of Lemma 1.

We proceed to prove (+).

Given a point $y \in B_{L-\lambda^{2N}\Delta}(x, \lambda)$, we may join x to y by a geodesic $\gamma : [0, \lambda] \to M$ in the metric

$$ds^2 = \sum_{ij} g_{ij}(x)\,dx_i\,dx_j,$$

where

$$(g_{ij})^{-1} = (a^{ij} + \lambda^{2N}\delta_{ij}) > 0,\ L = -\sum_{ij} a^{ij}(x)\frac{\partial^2}{\partial x_i \partial x_j} + \cdots .$$

Thus $\gamma(0) = x$, $\gamma(\lambda) = y$, and $\frac{d}{dt}\gamma(t)$ is a subunit vector for $L - \lambda^{2N}\Delta$.

We shall perturb $\gamma(t)$ to a broken path $\gamma_{\#}(t)$ that starts at x and ends very near y, so that $\frac{c d\gamma_{\#}(t)}{dt}$ is a subunit vector for L.

We construct $\gamma_{\#}$ using the following elementary remarks.

(a) PERTURBATION OF TANGENT VECTORS

Say that $X = \sum_j \gamma_j \frac{\partial}{\partial x_j}$ is a subunit vector for $L - \lambda^{2N}\Delta$ at x^0, and suppose $|x' - x^0| < c_1\lambda^N$. Then there is a tangent vector $Y = \sum_j \gamma_j' \frac{\partial}{\partial x_j}$ at x' satisfying

(i) $|\gamma_j - \gamma_j'| \le C\lambda^N$,

(ii) cY is a subunit vector for L at x',

(iii) $\gamma_i' = \sum_j a^{ij}(x')\xi_j'$ with $|\xi'| \le C\lambda^{-N}$.

For our hypothesis on X is that

$$\left(\sum_j \gamma_1\eta_1\right)^2 \le \sum_{ij}(a^{ij}(x^0) + \lambda^{2N}\delta_{ij})\eta_i\eta_j$$

for $\eta \in R^n$. Applying Lemma 4.1 in [5] to the function

$$f(t) = |\eta|^{-2}\sum_{ij}(a^{ij}(x' + t(x^0 - x')) + \lambda^{2N}\delta_{ij})\eta_i\eta_j \ge 0$$

yields

$$\sum_{ij}(a^{ij}(x^0)+\lambda^{2N}\delta_{ij})\eta_i\eta_j \le C\sum_{ij}(a^{ij}(x')+\lambda^{2N}\delta_{ij})\eta_i\eta_j + C|x'-x^0|^2|\eta|^2,$$

so that for $|x^0 - x'| \le c_1\lambda^N$, we have

$$c\Big(\sum_j \gamma_j\eta_j\Big)^2 \le \sum_{ij}(a^{ij}(x')+\lambda^{2N}\delta_{ij})\eta_i\eta_j.$$

Thus $c\tilde{X}$ is a subunit vector for $L - \lambda^{2N}\Delta$ at x', where $\tilde{X} = \sum_j \gamma_j \frac{\partial}{\partial x_j}$ at x'. Next we rotate the coordinate axes so that $a^{ij}(x')$ is diagonalized: $(a_{ij}(x')) = (\lambda_i\delta_{ij})$, $\lambda_i \ge 0$. Note that conclusions (i), (ii), and (iii) are unaffected by the rotation. Now we know that $c\tilde{X}$ is a subunit for $L - \lambda^{2N}\Delta$, that is,

$$\sum_i \frac{\gamma_i^2}{\lambda_i + \lambda^{2N}} \le C.$$

Define $Y = \sum_i \gamma_i' \frac{\partial}{\partial x_i}$ at x', where $\gamma_i' = \gamma_i$ if $\lambda_i \ge \lambda^{2N}$, $\gamma_i' = 0$ if $\lambda_i < \lambda^{2N}$. Setting $\xi_i' = \gamma^i/\lambda_i$ for $\lambda_i \ge \lambda^{2N}$, $\xi_i' = 0$ for $\lambda_i < \lambda^{2N}$, we see easily that (i), (ii), and (iii) hold.

(b) ESTIMATES FOR SECOND DERIVATIVES OF HAMILTONIAN PATHS

Say $H = \frac{1}{2}\sum_{ij} a^{ij}(x)\xi_i\xi_j$ and initially $|\xi| \le C\lambda^{-N}$. Hamiltonian's equations

$$\dot{x}^i = \sum_j a^{ij}(x)\xi_j, \qquad \dot{\xi}_1 = -\sum_{ij}\frac{\partial a^{ij}(x)}{\partial x_1}\xi_i\xi_j,$$

$$\ddot{x}^i = \sum_j a^{ij}(x)\dot{\xi}_j + \sum_{j\ell}\frac{\partial a^{ij}(x)}{\partial x_\ell}\dot{x}_\ell\xi_j$$

imply $|\dot{x}| \le C|\xi|$, $|\dot{\xi}| \le C|\xi|^2$, $|\ddot{x}| \le C|\xi|^2$. So if we flow for time $\Delta t \le c_1\lambda^N$, then $|\dot{\xi}| \le C'\lambda^{-N}$, $|\ddot{x}| \le C''\lambda^{-2N}$ along the path.

Now we can construct the broken path $\gamma_\#(t)$ mentioned above. Suppose we have constructed $\gamma_\#(t)$ for $0 \le t \le \tau_\kappa$, and that $\gamma_\#(0) = \gamma(0)$, $|\gamma(t) - \gamma_\#(t)| \le C_+\lambda^N\tau_\kappa$ for $0 \le t \le \tau_\kappa$. (The large constant C_+ will be picked later. Note that the assertion is vacuous for $\tau_0 = 0$.) Apply (a) above with $x^0 = \gamma(\tau_\kappa)$, $x' = \gamma_\#(\tau_\kappa)$, and

$X = \frac{d}{dt}\gamma(t)\Big|_{\tau_\kappa}$. Since λ is small, we have

$$|x^0 - x'| \le C_+\lambda^N\tau_\kappa \le C_+\lambda^{N+1},$$

so the hypotheses of (a) are satisfied. With Y, ξ' as in (a), we now define $\gamma_\#(t)$ for $\tau_\kappa \le t \le \tau_{\kappa+1}$ as the projection onto the x-coordinate of the Hamiltonian curve for

$$H = \frac{1}{2}\sum_{ij} a^{ij}(x)\xi_i\xi_j$$

starting at (x', ξ') for $t = \tau_\kappa$. Since $\frac{d}{dt}\gamma_\#\Big|_{\tau_\kappa} = Y$, and since H is conserved along the path, it follows from (a)(ii) above that $c\frac{d}{dt}\gamma_\#$ is a subunit vector for L throughout $\tau_\kappa \le t \le \tau_{\kappa+1}$. The estimates of (b) show that $\left|\frac{d^2}{dt^2}\gamma_\#\right| \le C\lambda^{-2N}$ for $\tau_\kappa \le t \le \tau_{\kappa+1}$, provided $\tau_{\kappa+1} - \tau_\kappa < c_1\lambda^N$. Similarly, $\left|\frac{d^2}{dt^2}\gamma\right| \le C\lambda^{-2N}$, while $\left|\frac{d\gamma}{dt}(\tau_\kappa) - \frac{d\gamma_\#}{dt}(\tau_\kappa)\right| = |\overline{X} - \overline{Y}| \le C\lambda^N$ by (a)(i). So if $\tau_{\kappa+1} - \tau_\kappa = \lambda^{3N}$, then we have for $\tau_\kappa \le t \le \tau_{\kappa+1}$ that

$$\left|\frac{d\gamma}{dt} - \frac{d\gamma_\#}{dt}\right| \le C\lambda^{+N} + C\lambda^{-2N}|t - \tau_\kappa| \le C'\lambda^N.$$

Our assumption $|\gamma(\tau_\kappa) - \gamma_\#(\tau_\kappa)| \le C_+\lambda^N\tau_\kappa$ now implies for $\tau_\kappa \le t \le \tau_{\kappa+1}$ that

$$|\gamma(t) - \gamma_\#(t)| \le C_+\lambda^N\tau_\kappa + C'\lambda^N(t - \tau_\kappa) \le C_+\lambda^N\tau_{\kappa+1},$$

provided we pick $C_+ \ge C'$. Our construction of $\gamma_\#(t)$ is complete, with $\tau_\kappa = \kappa \cdot \lambda^{3N}$.

The proof of (+) is now easy. Given $y \in B_{L-\lambda^{2N}\Delta}(x, \lambda)$ and γ as above, we constructed a broken curve $\gamma_\# : [0, \lambda] \to M$ so that $c\frac{d}{dt}\gamma_\#$ is subunit for L, while $\gamma_\#(0) = x$, $|\gamma_\#(\lambda) - y| \le C_+\lambda^{N+1}$. Thus $y_\# = \gamma_\#(\lambda) \in B_L(x, C\lambda)$, while $y \in B_E(y_\#, \lambda^N) \subseteq B_{L-(\lambda^2)^{2N}\Delta}(y^\#, \lambda^2)$. Repeating the process yields a sequence of paths $\gamma_\#^\mu : [t_\mu, t_{\mu+1}] \to M$ with tangent vectors subunit for L, so that $\gamma_\#^0$ joins $x = \gamma_\#^0(t_0 = 0)$ to $y_\#^1 = \gamma_\#(t_1)$, $\gamma_\#^\mu$ joins $y_\#^\mu$ to $y_\#^{\mu+1}$, $y_\#^\mu \to y$, and $\sum_\mu t_\mu \le C'\lambda$.

Combining the $\gamma_{\#}^{\mu}$ into a single Lipschitz path $\gamma_{\#\#}$, we see that $y \in B_L(x, C'\lambda)$ as required.

Proof of Lemma 2. Since $(a^{ij}(t, x)) \leq C(\overline{a^{ij}}(x) + \lambda^{2N}\delta_{ij})$ for $|t| \leq \lambda$, we see at once that $B_{L_\lambda}(c\lambda) \subseteq B_{L_\lambda^*}(\lambda) \subseteq \{|t| \leq \lambda\} \times B_{\overline{L}_\lambda}(\lambda)$, where $L_\lambda^* = (\partial/\partial t)^2 + \overline{L}_\lambda(y, \lambda y)$. So our problem is to show that $\{|t| \leq \lambda\} \times B_{\overline{L}_\lambda}(\lambda) \subseteq B_{L_\lambda}(C\lambda)$. To see this, let $u(t, y)$ be the distance from (t, y) to the origin induced by L_λ. Note that $u_\lambda \in \text{Lip}(1)$, since $(a^{ij} + \lambda^{2N}\delta_{ij}) > 0$. Since $\partial/\partial t$ is a subunit vector for L_λ, we have $t^{-2}|u_\lambda(t, x) - u_\lambda(0, x)|^2 \leq 1$, while also

$$\sum_{ij}(a^{ij}(t, y) + \lambda^{2N}\delta_{ij})\frac{\partial u_\lambda(t, y)}{\partial y_i}\frac{\partial u_\lambda(t, y)}{\partial y_j} \leq 1,$$

where u_λ is differentiable. By mollifying $u_\lambda(t, \cdot)$ slightly, we obtain $v = v_{\lambda, t} \in C^\infty$, so that

$$t^{-2}\|u_\lambda(0, \cdot) - v\|^2 + \left\|\sum_{ij} a^{ij}(t, \cdot)\frac{\partial v}{\partial y_i}\frac{\partial v}{\partial y_j}\right\| + \lambda^{2N}\|\nabla_y v\|^2 \leq 10,$$

the norms being taken in L^∞. That is, with

$$L(t) = -\sum_{ij} a^{ij}(t, y)\frac{\partial^2}{\partial y_i \partial y_j},$$

$(L(t))_{\lambda^{-2}}(u_\lambda(0, \cdot)) \leq 10$ for $|t| \leq \lambda$. Taking $t_\ell = \ell\lambda/N^2$, $\ell = 0, \ldots, N^2$, and applying Lemma 3 repeatedly, we obtain

$$\left(\sum_\ell L(t_\ell)\right)_{\lambda^{-2}}(u\ \ (0, \cdot)) \leq C_N,$$

which implies easily $(\overline{L}_\lambda)_{\lambda^{-2}}(u(0, \cdot)) \leq C_N'$. Thus, for a suitable C^1 function v we have

$$\lambda^{-2}|u_\lambda(0, y) - v(y)|^2 + \sum_{ij} a^{ij}(y)\frac{\partial v(y)}{\partial y_i}\frac{\partial v(y)}{\partial y_j} + \lambda^{2N}|\nabla v(y)|^2 \leq C$$

for all $y \in M$. In particular, $|v(0)| \leq C\lambda$. Moreover, let $\gamma : [0, \lambda] \to M$ be a Lipschitz path with $d\gamma/dt$ subunit for $\overline{L}_\lambda$. Our estimates on v and the definition of subunit vectors together yield

$$|v(\gamma(\lambda)) - v(\gamma(0))| \leq \int_0^\lambda |\langle dv, \dot{\gamma}(t)\rangle|dt,$$

while

$$\langle dv, \dot\gamma(t)\rangle^2 \le \sum_{ij} a^{ij}(y)\frac{\partial v}{\partial y_i}\frac{\partial v}{\partial y_j} + \lambda^{2N}|\nabla v(y)|^2 \text{ [with } y = \gamma(t)] \le C.$$

Therefore, $|v(\gamma(\lambda)) - v(\gamma(0))| \le C\lambda$, and it follows that $|v(y)| \le C\lambda$ for $y \in B_{\bar L_\lambda}(\lambda)$. The estimate defining v now shows that also $u_\lambda(0, y) \le C'\lambda$ for $y \in B_{\bar L_\lambda}(\lambda)$. Recalling that $|u_\lambda(t, y) - u_\lambda(0, y)| \le |t|$, we obtain $u_\lambda(t, y) \le C''\lambda$ for $(t, y) \in \{|t| \le \lambda\} \times B_{\bar L_\lambda}(\lambda)$. By definition of u_λ, this means that $\{|t| \le \lambda\} \times B_{\bar L_\lambda}(\lambda) \subseteq B_{L_\lambda}(C\lambda)$, as needed.

Sketch of the Proof of Lemma 3. We use induction on the dimension n. The result being trivial in $n = 0$, we assume Lemma 3 is known in R^{n-1} and deduce successively the following seven lemmas.

Lemma 3.1. Let $A = \sum_j a^j(x)\frac{\partial}{\partial x_j}$ *be a first-order operator in* R^{n-1}, *and let* $u, v, w : I \to C^1(R^{n-1})$, *where* I *is an interval of length* $\sim K^{-1/2}$. *Set* $u^0 = \frac{1}{|I|}\int_I u(t)dt$. *Then there exists* $u^+ \in C^1(R^{n-1})$ *for which*

$$K\|u^0 - u^+\|^2 + \|Au^+\|^2 + K^{-2N}\|\nabla_x u^+\|^2 \le C \sup_I\Big\{K\|u(t) - v(t)\|^2 + \left\|\frac{\partial v}{\partial t}\right\|^2 + K\|u(t) - w(t)\|^2 + \left\|\left(\frac{\partial}{\partial t} - A\right)w(t)\right\|^2 + K^{-2N}\|\nabla_x w\|^2\Big\},$$

the norms being taken in L^∞ (unit cube *in* R^{n-1}).

Lemma 3.2. Set $I_\ell = \{|t - \ell K^{-1/2}| \le 10K^{-1/2}\}$, I^+ = *union of consecutive* I_ℓ's,

$$L(t) = -\sum_{i,j=1}^{n-1} a^{ij}(t, x)\frac{\partial^2}{\partial x_i \partial x_j},$$

where a^{ij} *is smooth and* $(a^{ij}) \ge K^{-2N}(\delta_{ij})$. *Then given* C^1-*functions* $u(t, x)$ *on* $I^+ \times$ unit cube, u_ℓ *and* v_ℓ *on* $I_\ell \times$ unit cube, *we can find* $v \in C^1(I^+ \times$ cube) *so that*

$$K\|u - v\|^2 + \left\|\frac{\partial v}{\partial t}\right\|^2 + \left\|\sum_{ij} a^{ij}(t, x)\frac{\partial v}{\partial x_i}\frac{\partial v}{\partial x_j}\right\|$$
$$\leq C \max_\ell \left\{K\|u - v_\ell\|^2 + \left\|\frac{\partial v_\ell}{\partial t}\right\|^2 + K\|u - u_\ell\|^2 + \left\|\sum_{ij} a^{ij}(t, x)\frac{\partial u_\ell}{\partial x_i}\frac{\partial u_\ell}{\partial x_j}\right\|\right\}.$$

Here the norms on the left are taken in $L^\infty(I^+ \times \text{cube})$, *while those on the right are taken in* $L^\infty(I_\ell \times \text{cube})$.

Lemma 3.3. Set $A(t) = \sum_{j=1}^{n-1} a^j(t, x)\frac{\partial}{\partial x_j}$ *with* a^j *smooth, and let* I *be an interval of length* $\sim K^{-1/2}$. *Given* C^1*-functions* u, v, w *on* $I \times$ unit cube, *we can find* $u^\#$ *so that*

$$K\|u - u^\#\|^2 + \left\|\frac{\partial u^\#}{\partial t}\right\|^2 + \|A(t)u^\#\|^2 + K^{-2N}\|\nabla_x u^\#\|^2$$
$$\leq C\left\{K\|u - v\|^2 + \left\|\frac{\partial v}{\partial t}\right\|^2 + K\|u - w\|^2 + \left\|\left(\frac{\partial}{\partial t} + A(t)\right)w\right\|^2 + K^{-2N}\|\nabla_x w\|^2\right\},$$

the norms being taken in $L^\infty(I \times \text{cube})$.

Lemma 3.4. Let the unit cube in R^n *be partitioned into subcubes* $\{Q_\nu\}$ *of diameters* δ_ν, *where* $Q_\nu^* \cap Q_\mu^* \neq \emptyset$ *implies* $\delta_\nu \sim \delta_\mu$.

Let

$$L = -\sum_{ij} a^{ij}(x)\frac{\partial^2}{\partial x_i \partial x_j}$$

for smooth functions a^{ij} *satisfying* $K^{-2N}(\delta_{ij}) \leq (a^{ij}(x)) \leq C\delta_\nu^2(\delta_{ij})$ *for* $x \in Q_\nu$. *Then given* $u \in L^\infty$ *and* C^1*-functions* v_μ *defined on* Q_ν^*, *we have*

$$L_K(u) \leq C \max_\mu \left\{K\|u - v_\mu\|_{L^\infty(Q_\mu^*)} + \left\|\sum_{ij} a^{ij}(x)\frac{\partial v_\mu}{\partial x_i}\frac{\partial v_\mu}{\partial x_j}\right\|_{L^\infty(Q_\mu^*)}\right\}.$$

Lemma 3.5. Assume $L = -\sum_{ij} a^{ij}(x)\frac{\partial^2}{\partial x_i \partial x_j} + \cdots$ *with* $\max_{ij}|a^{ij}| \geq C$, *and set*

$$L' = -\left(\frac{\partial}{\partial x_1}\right)^2 - K^{-2N}\Delta.$$

Then

$$c(L + L')_K(u) \leq L_K(u) + L'_K(u).$$

Lemma 3.6. Set

$$L' = -\left(\frac{\partial}{\partial x_1}\right)^2 - K^{-2N}\Delta, \quad L = -\sum_{ij} a^{ij}(x)\frac{\partial^2}{\partial x_i \partial x_j},$$

where $(a^{ij}(x))$ *is smooth and positive semidefinite. Then*

$$c(L + L')_K(u) \leq L_K(u) + L'_K(u).$$

Lemma 3.7. Let

$$L = -\sum_{ij} a^{ij}(x)\frac{\partial^2}{\partial x_i \partial x_j}, \quad L' = -\sum_{ij} a'^{ij}(x)\frac{\partial^2}{\partial x_i \partial x_j},$$

and assume $\max_{ij}|a^{ij}| \geq C$. *Then*

$$c(L + L')_K(u) \leq L_K(u) + L'_K(u).$$

For the most part, these lemmas are proved in strict analogy with Lemmas 1, 2, 3, 5, 7, 8, and 9 in the proof of the spectral decomposition theorem in [5]. Instead of L^2-norms, one uses L^∞-norms; and for

$$L = -\sum_{ij} a^{ij}(x)\frac{\partial^2}{\partial x_i \partial x_j},$$

we work with $\left\| \sum_{ij} a^{ij}(x)\frac{\partial v}{\partial x_i}\frac{\partial v}{\partial x_j}\right\|_{L^\infty}$ in place of $\mathrm{Re}\langle Lv, v\rangle$.

In some ways, the present lemmas are easier than their L^2-analogues in [5]. For instance, where [5] uses the sharp Gårding inequality of [1], we use here merely the trivial estimate

$$\left\| \sum_{ij} b^{ij}(x)\frac{\partial v}{\partial x_i}\frac{\partial v}{\partial x_j}\right\|_{L^\infty} \leq \left\| \sum_{ij} a^{ij}(x)\frac{\partial v}{\partial x_i}\frac{\partial v}{\partial x_j}\right\|_{L^\infty}$$

for $(b^{ij}) \leq (a^{ij})$. Also, the Fourier integral operators of [5] are now replaced by simple changes of variable $y = \Phi(x)$.

It is now routine for a patient reader to reconstruct the proofs of Lemmas 3.2 through 3.7, given Lemma 3.1. Moreover,

Lemma 3.7 implies Lemma 3 in R^n, just as Lemma 6.9 implies the spectral decomposition theorem in [5]. Consequently, the proof of Lemma 3 is reduced to Lemma 3.1. Since the L^2-analogue of 3.1 in [5] was proved using the spectral theorem, we have to make a new argument to show (3.1). Since space is limited, we omit all details of Lemmas 3.2-3.7 and Lemma 3.7 → Lemma 3 in R^n. We close our article with the proof of Lemma 3.1. First suppose $A = \partial/\partial x_1$, and let Ω denote the right-hand side of the estimate in Lemma 3.1.

Since $K\|u - v\|^2 + \left\|\frac{\partial v}{\partial t}\right\|^2 \leq \Omega$, we have

$$K|u(t + h, x_1, x') - u(t, x_1, x')|^2 \leq C\Omega$$

for $|h| \leq K^{-1/2}$. Also $K\|u - w\|^2 + \left\|\left(\frac{\partial}{\partial t} - A\right)w\right\|^2 \leq \Omega$, so that

$$K|u(t + h, x_1 + h, x') - u(t, x_1, x')|^2 \leq C\Omega$$

for $|h| \leq K^{-1/2}$; and $K\|u - w\|^2 + K^{-2N}\|\nabla_x w\|^2 \leq \Omega$ so that

$$K|u(t, x_1, x') - u(t, x_1, y')|^2 \leq C\Omega$$

for $|x' - y'| \leq K^{-N}$. Now take

$$u^+(x_1, x') = \frac{1}{|I|}\int_I \int_{R^{n-1}} \psi(y_1, y')u(t, x_1 - y_1, x' - y')\,dt\,dy_1\,dy',$$

where $\psi(y_1, y')$ is a suitable approximate identity supported in $|y_1| \leq K^{-1/2}$, $|y'| \leq K^{-N}$. One checks easily using the previous estimates that

$$K\|u^0 - u^+\|^2 + \|Au^+\|^2 + K^{-2N}\|\nabla_x u^+\|^2 \leq C\Omega,$$

as needed.

Next we pass to the general case. With $A = \sum_j a^j(x)\frac{\partial}{\partial x_j}$, we make a Calderón-Zygmund decomposition of the unit cube in R^{n-1}, stopping at Q_ν with diameter δ_ν if

(i) $\max_{x \in Q_\nu^*} \max_j |a^j(x)| \geq 10\delta_\nu$

or

(ii) $\delta_\nu \leq CK^{-N}$.

In each Q_ν^* we can find a function u_ν^+ so that

$$K|u^0 - u^+|^2 + |Au^+|^2 + K^{-2N}|\nabla_x u_\nu^+|^2 \leq C\Omega \text{ on } Q_\nu^*.$$

For Q_ν arising from (i), this reduces by a change of variable to a slight variant of the case $A = \partial/\partial x_1$, in which the terms $K^{-2N}\|\nabla_x w\|^2$ and $K^{-2N}\|\nabla_x u^+\|^2$ are changed to $K^{-2N}\delta_\nu^{-2}\|\nabla_x w\|^2$ and $K^{-2N}\delta^{-2}\|\nabla_x u^+\|^2$— there is no trouble in adapting our discussion of $A = \partial/\partial x_1$ to this variant. If Q_ν arises from (ii), then we just take $u_\nu^+ = w$.

Now take a partition of unity $1 = \sum_\nu \phi_\nu$ with $\phi_\nu \in C_0^\infty(Q_\nu^*)$, $\|\nabla\phi_\nu\| \leq C\delta_\nu^{-1}$, and set $u^+ = \sum_\nu \phi_\nu u_\nu^+$. Clearly

$$K\|u^0 - u^+\|^2 \leq C\Omega.$$

We also have $Au^+ = \sum_\nu (Au_\nu^+)\phi_\nu + \sum_\nu (u_\nu^+ - u^0)(A\phi_\nu)$, while $|a^j(x)| \leq C\delta_\nu$ and $|\nabla\phi_\nu| \leq C\delta_\nu^{-1}$ in supp $\phi_\nu \subseteq Q_\nu^*$. Therefore,

$$\|Au^+\|^2 \leq C \max\{\|Au^+\|^2 + \|u_\nu^+ - u^0\|^2\} \leq C\Omega.$$

Similarly, $\nabla_x u^+ = \sum_\nu (\nabla_x u_\nu^+)\phi_\nu + \sum_\nu (u_\nu^+ - u^0)(\nabla\phi_\nu)$, while $|\nabla\phi_\nu| \leq \delta_\nu^{-1} \leq K^N$. So

$$K^{-2N}\|\nabla u^+\|^2 \leq C \max\{K^{-2N}\|\nabla_x u_\nu^+\|^2 + \|u_\nu^+ - u^0\|^2\} \leq C\Omega.$$

Thus

$$K\|u^+ - u^0\|^2 + \|Au^+\|^2 + K^{-2N}\|\nabla_x u^+\|^2 \leq C\Omega,$$

proving Lemma 3.1.

REFERENCES

1. Fefferman, C., and Phong, D. H. On positivity of pseudodifferential operators. *Proc. Nat. Acad. Sci.* 75 (1978):4673-4674.

2. Fefferman, C., and Phong, D. H. On the asymptotic eigenvalue distribution of a pseudodifferential operator. *Proc. Nat. Acad. Sci.* 77 (1980):5622-5625.

3. Fefferman, C., and Phong, D. H. On the lowest eigenvalue of a pseudodifferential operator. *Proc. Nat. Acad. Sci.* 76 (1979): 6055-6056.

4. Fefferman, C., and Phong, D. H. Symplectic geometry and positivity of pseudodifferential operators. *Proc. Nat. Acad. Sci.* 79 (1982):710-713.

5. Fefferman, C., and Phong, D. H. The uncertainty principle and sharp Gårding inequalities. *Comm. Pure Appl. Math.* 34 (1981): 285-331.

6. Hörmander, L. Hypoelliptic second-order differential equations. *Acta Math.* 119 (1967):147-171.

7. Kohn, J. J. Pseudodifferential operators and hypoellipticity. *Proc. Symp. Pure Math. AMS* 23 (1973):61-69.

8. Menikoff, A., and Sjöstrand, J. Eigenvalues for hypoelliptic operators and related methods. *Proc. Int. Cong. Math. Helsinki* (1978):797-801.

9. Olenik, O., and Radkevitch, E. Second-order equations with non-negative characteristic form.

10. Rothschild, L. and Stein, E. M. Hypoelliptic differential operators and nilpotent groups. *Acta Math.* 137 (1976):247-320.

SUMMABILITY AND BOUNDS FOR MULTIPLIERS OF ELLIPTIC OPERATORS ON R^n

David Gurarie
Oregon State University

Mark A. Kon
Boston University

In this paper we consider two related subjects: the properties of kernels of resolvents of elliptic operators on $\mathbf{R}^n$, and summability theory for these operators. For operators with variable (possibly singular) coefficients, we study summability of eigenfunction expansions using analytic summator functions. This involves consideration of resolvent kernels and yields functional analytic information (for example, closedness, semigroup generation) about the operators themselves. More comprehensive summability results are obtained for constant coefficient operators.

Classical summability theory originated in Fourier analysis [16] and has been studied systematically for Sturm-Liouville expansions [8 and 15] and more recently for eigenfunction expansions of pseudodifferential operators on compact manifolds [13 and 6]. The techniques used for these results cannot be adapted directly to the noncompact situation, or to operators with singular (nonsmooth) coefficients. Here we employ methods that are typical of functional as well as harmonic analysis. We consider the class $[RB]$ of integral operators on $\mathbf{R}^n$ whose kernels $K(x, y)$ are bounded by L^1 radial decreasing functions $h(|x - y|)$, and prove that suitable families

---

Partially supported by NSF Grant No. MCS-800 3407.

of these form approximate identities, in L^p and pointwise. For an elliptic operator A with coefficients in certain L^p classes, the resolvent and analytic multipliers are in [*RB*], which implies L^p-closedness, essential self-adjointness (if A is symmetric), and other regularity properties for A. In particular, the analytic semigroup $\{e^{-tA^s}\}_{\mathrm{Re}\, t>0}$ is continuous (in L_p and on the Lebesgue set) at $t = 0$; this yields convergence at the boundary for solutions of certain Cauchy and Dirichlet problems.

For brevity, most proofs have been sketched or omitted; details of these will appear elsewhere.

1. PRELIMINARIES: CONSTANT COEFFICIENT OPERATORS

Let $S_{1,0}^{m,N}$ denote the classical symbol class of functions $\phi(x, \xi)$ on $\mathbf{R}^n \times \mathbf{R}^n \backslash 0$ with N derivatives in ξ, such that

$$\|\phi\|_{S,\alpha} = \|(1 + |\xi|)^{-m+|\alpha|} \partial_\xi^\alpha \phi(x, \xi)\|_\infty < \infty$$

for $|\alpha| \leq N$, where $\alpha = (\alpha_1, \ldots, \alpha_n)$ is a multiindex, and ∂^α the corresponding partial derivative. Then

$$\phi(x, D)f(x) = \int \phi(x, \xi) e^{-ix\cdot\xi} \hat{f}(\xi)\, d\xi, \tag{1}$$

where $D^\alpha = i^{-|\alpha|}\partial^\alpha$; henceforth in Fourier integrals $d\xi = (2\pi)^{-n/2} d\lambda$, where λ is Lebesgue measure. For L^p-estimates a more useful form of (1) is

$$\phi(x, D)f(x) = \int K(x, y) f(y)\, dy,$$

where $K(x, y) = \int \phi(x, \xi) e^{i\xi\cdot(x-y)}\, d\xi$ is a distribution kernel.

The following (known) radial bounds on $K(x, y)$ are central to our estimates:

If $\phi \in S_{1,0}^{-m,N}$ and $N > n - m$, then

$$|K(x, y)| \leq h(|x - y|) = c_{s,t} \|\phi\|_S \begin{cases} |x|^{-s} ; & x \leq 1 \\ |x|^{-t} ; & |x| > 1 \end{cases}, \tag{2}$$

where $n - m < s,\ t \le N;\ s,\ t > 0$; and

$$\|\phi\|_s = \sum_{|\alpha| \le N} \|\phi\|_{s,\alpha}.$$

Consequently, for $1 \le p \le \infty$, and $\frac{n}{N} < p < \frac{n}{n - m}$ (if $m \le n$), and $\frac{n}{N} < p \le \infty$ (if $m > n$), h is in L^p and $\|h\|_p \le c_p \|\phi\|_s$.

These bounds follow easily from the fact that symbol class bounds interpolate correctly for fractional derivatives, that is,

$$|\Delta_\xi^{s/2} \phi(x, \xi)| \le c_s (1 + |\xi|)^{-m-s} \qquad (0 \le s \le N). \tag{3}$$

Similar calculations show, in fact, that for $\phi \in S^{m,N}_{\rho,0}$ $(0 \le \rho \le 1)$, the exponentials s and corresponding L^r classes vary within the limits $\frac{n - m}{\rho} < s \le N$; $\frac{n}{N} < r < \frac{n\rho}{n - m}$. An essential part of (2) is the uniformity for $\|\phi\|_s$ bounded. This, together with estimates for symbols singular near 0, will yield general summability for constant coefficient pseudodifferential operators.

Theorem 1. Let $a \in S_1^m$ $(m > 0)$ be a positive, constant coefficient elliptic symbol, and $A = a(D)$ the corresponding operator. Then:

(a) for any $\phi \in S_1^{-m'}$ $(m' > 0)$, $\phi(A)$ is convolution with the kernel $K(x) = (\widehat{\phi \circ a})(-x)$ in [RB].

(b) (summability) if $f \in L^p$, $1 \le p \le \infty$, then

$$\phi(\varepsilon A) f(x) \xrightarrow[\varepsilon \to 0]{} \phi(0) f(x)$$

in L^p (for $p < \infty$) and on the Lebesgue set of f.

(c) (resolvent summability) $(\zeta - A)^{-1}$ exists in L^p for $\zeta \notin \mathbf{R}^+$, and has kernel in [RB]. We have

$$\zeta(\zeta - A)^{-1} f(x) \xrightarrow[\zeta \to \infty]{} f(x)$$

in L^p $(p < \infty)$ and on the Lebesgue set of f, uniformly in each sector

$$\Omega_\theta = \{\zeta \in A \,|\, |\arg \zeta| \ge \theta\},\ \theta > 0.$$

Parts (a) and (b) generalize to

$$\phi(\varepsilon A_1, \ldots, \varepsilon A_k)f(x) \xrightarrow[\varepsilon \to 0]{} \phi(0)f(x)$$

for $\phi : \mathbf{R}^k \to \mathbf{C}$ and A_i real elliptic for each i. If $A_j = \frac{1}{i}\frac{\partial}{\partial x_j}$, this is a standard result on summation of the Fourier transform, and (b) easily reduces to this case if A is homogeneous.

In the general case, $\phi \circ a \in S_1^{-mm'}$ by the iterated chain rule, and its symbol class seminorms are estimated by those of ϕ, a, and a^{-1}. Thus, $\phi \circ a(\xi) \in [RB]$. One now considers the "approximation" $a_\varepsilon(\xi) = \varepsilon a(\varepsilon^{-1/m}\xi)$ of $a(\xi)$. In computing $\phi \circ a_\varepsilon$, the standard technique of "cutting off" a neighborhood of 0 takes care of a possible divergence at the origin in the uniform bound on $a_\varepsilon(\xi)$. There are no problems away from the origin, and by the chain and Leibniz rules,

$$D^\alpha[\chi\phi \circ a_\varepsilon] \leq d_\alpha(1 + |\xi|)^{-mm'-|\alpha|},$$

where $\chi(\xi) \in C^\infty$ is 1 for $|\xi| \geq 2$, and 0 for $|\xi| \leq 1$, and d_α depends only on the seminorms of a, a^{-1}, ϕ, and χ. Thus,

$$\phi \circ a_\varepsilon(x) \leq h(|x|) \in L^1.$$

Similarly,

$$\left(\frac{\zeta}{\zeta - a(|\zeta|^{1/m}\xi)}\right)^{\wedge}(x) \leq g(|x|) \in L^1 \tag{4}$$

with g independent of ζ for $|\zeta| \geq \rho_0 > 0$, $|\arg \zeta| \geq \theta_0 > 0$. The convolution kernels for $\phi(\varepsilon A)$ and $\zeta(\zeta - A)^{-1}$ (in the above region of the ζ-plane) are thus uniformly bounded by L^1-dilations of radial decreasing functions, while their integrals over $\mathbf{R}^n$ converge to 1. The proof is completed by the following, which extends a well-known result (see [12, Chapter 1]).

Proposition 1. If a family $\{K_\varepsilon\} \in [RB]$ *is bounded by* L^1*-dilations of a radially decreasing function* $\{\varepsilon^{-n}h(\varepsilon^{-1}|x - y|)\}$ *and* $\int K_\varepsilon(x, y)dy \to c$ *uniformly in* x*, then for all* $f \in L^p$*,* $K_\varepsilon(f) \to cf$ *in* L^p $(1 \leq p < \infty)$ *and on the Lebesgue set of* $f (1 \leq p \leq \infty)$.

The symbol classes of a and ϕ may clearly be relaxed to be $a \in S_1^{m,n+1}$, $\phi \in S_1^{-m',n+1}$.

If $\phi \in S_1^0$, $\phi(\varepsilon A)$ is uniformly bounded in L^p $(1 < p < \infty)$, by Mihlin's theorem [5]. It is not hard to show that $\phi(\varepsilon A)f \xrightarrow[\varepsilon \to 0]{} f$ in L^p if f is in the dense subset $\{f(x) \,|\, \hat{f}(\xi) \in C_K^\infty\}$, so that zero-class summability holds for all $f \in L^p$. However, summability on the Lebesgue set fails in this case: in $\mathbf{R}^1$, let $a(\xi) = \xi$, $\phi(\xi) = |\xi|^i$, and $f(x) = |x|^{-i}/\ell n|x|$. Then $\check{\phi}(x)$ is proportional to the principal value distribution $|x|^{-i-1}$, and $\check{\phi}(x) * f(x)\Big|_{x=0} = \infty$, even though f is continuous at $x = 0$. Failure on the Lebesgue set does not of course exclude the possibility of convergence a.e.

2. RESOLVENTS AND SUMMABILITY FOR PERTURBED OPERATORS

If $A = A_0 + B$ is a perturbed operator, the resolvent perturbation series

$$(\zeta - A)^{-1} = (\zeta - A_0)^{-1} \sum_{k=0}^{\infty} [B(\zeta - A_0)^{-1}]^k \tag{5}$$

is useful when the terms can be appropriately bounded. Let A_0 be a constant coefficient, positive and homogeneous differential operator of order m, and

$$B = \sum_{|\alpha| < m} b_\alpha(x) D^\alpha.$$

b_α may be defined on quotient spaces $V_\alpha = \mathbf{R}^n/U_\alpha$, with U_α a subspace of $\mathbf{R}^n$. We let $n_\alpha = \dim V_\alpha = n - \dim U_\alpha$, and require $b_\alpha \in L^{r_\alpha} + L^\infty$, for $n_\alpha/r_\alpha + |\alpha| < m$, $|\alpha| < m$. $U_\alpha \neq \{0\}$ is of interest, for example, in Schrödinger operators for Coulomb systems,

$$-\Delta + \sum_{i \neq j} \frac{c_{ij}}{|x_i - x_j|} + \sum_i \frac{c_i}{|x_i|},$$

with $x_i \in \mathbf{R}^3$ $(i = 1, \ldots, N)$, Δ the Laplacian on $\mathbf{R}^{3N}$.

Theorem 2. Let $A = A_0 + B$ be as above, and let

$$d = \max\{n_\alpha/r_\alpha + |\alpha| \,\big|\, |\alpha| < m\} < m.$$

There exists a constant $c > 0$ depending on A such that the resolvent set of A contains $\Omega = \{\zeta = \rho e^{i\theta} \,\big|\, c\rho^{d/m-1} < |\sin\theta/2|^{n+2}\}$ and $(\zeta - A)^{-1} \in [RB]$ for $\zeta \in \Omega$. Moreover, the kernel $\zeta L_\zeta(x, y)$ of $\zeta(\zeta - A)^{-1}$ is bounded as follows:

$$|\zeta L_\zeta(x, y)| \le \frac{c}{|\sin\theta/2|^{n+2}}\left(1 - \frac{c\rho^{d|m-1}}{|\sin(\theta/2)|^{n+2}}\right)^{-1} \times \rho^{n/m} h_{s,t}(\rho^{1/m}|x - y|),$$

where

$$h_{s,t}(x) = \begin{cases} |x|^{-s}; & |x| \le 1 \\ |x|^{-t}; & |x| > 1 \end{cases}.$$

The proof requires the notion of (p, q)*-mixed convolution* with respect to a decomposition

$$\mathbf{R}^n = U \oplus V = \{(x', x'') \,|\, x' \in U,\ x'' \in V\},$$

with $\dim V = n'$, $\dim U = n''$. Define

$$f *_{(p,q)} g = \left(\int_U \left(\int_V |f(y' + y'')g(x' + x'' - y' - y'')|^p \, dy'\right)^{q/p} dy''\right)^{1/q}.$$

After a rather lengthy calculation, one obtains:

Proposition 2. Let $\mathbf{R}^n = U_i \oplus V_i$ $(1 \le i \le k)$ be a sequence of partitions; let $\dim V_i = n_i$, $\dim U_i = n_i'$. If sequences of reals $\{s_i\}_1^k$, $\{t_i\}_1^k$; $\{p_i\}_1^k$, $\{q_i\}_1^k$ satisfy $t_i > n$;

$$\max\{s_i, \min(s_1, \ldots, s_{i-1})\} \le \frac{n_i}{p_i} + \frac{n_i'}{p_i'},$$

then the estimate

$$(h_{s_0,t_0} * h_{s_1,t_1}) * \cdots * h_{s_k,t_k} \le c_1^k h_{s',t'}$$

holds, where $s' = \min\{s_i\}$, $t' = \min\{t_i\}$ and c_1 is a constant. Above, the ith $*$ denotes a (p_i, q_i) mixed convolution.

The theorem is proved by bounding each term of (5) in [*RB*]. Note that

$$B(\zeta - A_0)^{-1} = \sum_{|\alpha|< m} b_\alpha(x)D^\alpha(\zeta - A_0)^{-1}$$

contains terms $\phi\alpha(D)$, whose symbol $\phi_\alpha(\xi) = \xi^\alpha(\zeta - a(\xi))^{-1}$, where $a(\xi)$ is the symbol of A_0. Let $K_\alpha(x)$ be the corresponding convolution kernels. Then

$$L_k = (\zeta - A_0)^{-1}[B(\zeta - A_0)]_k = \sum_{\alpha_1, \ldots, \alpha_k} L_{\alpha_1, \ldots, \alpha_k}.$$

Each term in the last sum has kernel

$$L_{\alpha_1, \ldots, \alpha_k}(x, y) = K_0 \star b_{\alpha_1}K_{\alpha_1} \star \cdots \star b_{\alpha_k}K_{\alpha_k}$$

obtained from multiplying out the left-hand side. Above, operations go from right to left and K_k has argument $x - y$. We can assume that $b_\alpha \in L^{r_\alpha}$, as the argument is the same for $r_\alpha \leq r \leq \infty$.

Application of a multiple Hölder inequality, two basic interpolation inequalities for multiplication and convolution of functions in L^r classes, and appropriate radial bounds yield (we let $b_i = b_{\alpha_i}$, and so on)

$$|L_{\alpha_1, \ldots, \alpha_k}(x, y)| \leq \prod_i \|b_i\|_{r_{\alpha_i}} \frac{c_2}{|\sin \theta/2|^{n+2}}|\zeta|^{n/m-1}\left(\frac{c_2}{|\sin \theta/2|^{n+2}}\zeta^{d/m-1}\right)^k$$

$$\times \left(h_{s_0, t} \star_{(p_1, 1)} h_{s_1, t} \star_{(p_2, 1)} \cdots \star_{(p_k, 1)} h_{s_k, t}\right)(|\zeta|^{1/m}|x - y|)$$

$$\equiv \prod_i \|b_i\|_{r_{\alpha_i}} H_{\zeta, k}(|x - y|),$$

where $d = \max\left(\frac{n_\alpha}{r_\alpha} + |\alpha|\right)$, $s_j > n - m - |\alpha|$, and $t > n$. According to Proposition 2, the multiple convolution is bounded by $c_1^k h_{s, t}(|\zeta|^{1/m}|x - y|)$, where $s = \min s_j \leq s_0$; hence

$$|L_k(x, y)| \leq \left(\sum_\alpha \|b_\alpha\|_{r_\alpha}\right)^k H_{\zeta, k}.$$

The geometric series $\sum L_k(x, y)$ converges (and hence is in [*RB*]) if

$$\frac{c_2}{|\sin \theta/2|^{n+2}} |\zeta|^{d/m-1} < 1;$$

the kernel is then estimated by summing the series.

Note that if A_0 is not positive and instead has range in a sector $D = \{\zeta | \theta_1 \leq \arg \zeta \leq \theta_2\}$, the theorem still holds with the replacement of $\sin \theta/2$ by a slightly more complicated function; the same will hold for later statements.

3. APPLICATIONS OF THEOREM 2

The radial bounds of Theorem 2 provide information on the smoothing properties of $(\zeta - A)^{-1}$, and additional global bounds on the kernel at infinity, giving uniform estimates of L^p, Sobolev, and Hölder bounds. These estimates extend to analytic functions $\phi(A)$ via Cauchy integration, yielding summability results and other interesting information about A and $\phi(A)$.

a. The Operator A

It follows immediately from the theorem that the L^p spectrum $(1 \leq p \leq \infty)$ of A is contained in

$$\Omega' = \left\{ \zeta = \rho e^{i\theta} \middle| \rho^{d/m-1} \geq \frac{|\sin \theta/2|^{n+2}}{c} \right\}$$

which is a "U-shaped" domain about $\mathbf{R}^+$. Consequently, A is closeable in all L^p. Furthermore, the norm of the resolvent has maximum decrease in all directions $\theta \neq 0$. If A_0 has range in D as above, Ω' encloses D in a similar way.

If A is formally self-adjoint, then A admits a self-adjoint extension in L^2; indeed A is clearly bounded from below. Such results have previously been obtained when $A = -\Delta + V(x)$ in the theory of Schrödinger operators. An argument similar to Theorem 2 yields a priori estimates for the pair (B, A_0) in certain L^p spaces.

Precisely, for $1 \leq p \leq \min\{r_\alpha\}$, $\|Bf\|_p \leq \varepsilon\|A_0 f\|_p + c_\varepsilon\|f\|_p$. If, in particular, $\min_\alpha\{r_\alpha\} \geq 2$ and A is formally self-adjoint, the Kato-Rellich theorem [7] applies to show that A is essentially self-adjoint on $\mathscr{L}_m^2$ or any essential domain of A_0 (c.f. [11, Chapter 10]).

b. Resolvent Summability

This has been studied in the context of regular ([14]) and singular [3 and 10] Sturm-Liouville expansions, with

$$A = \frac{-d^2}{dx^2} + q(x).$$

For eigenfunction expansions on $\mathbf{R}^n$, we have

Theorem 3. If $A = A_0 + B$ is as in Theorem 2 and $f \in L^p$, then $\zeta(\zeta - A)^{-1}f(x) \xrightarrow[\zeta\to\infty]{} f(x)$ uniformly for $|\arg\zeta| \geq \theta_0 > 0$, in L^p-norm $(1 \leq p < \infty)$ and on the Lebesgue set of f $(1 \leq p \leq \infty)$.

By Theorem 2 and Proposition 1, what needs to be proved is that $\int \zeta L_\zeta(x, y)dy \xrightarrow[\zeta\to\infty]{} 1$ uniformly for $|\arg\zeta| \geq \theta_0$; this can be done using decompositions like those in Theorem 2.

c. Summability for Analytic Multipliers

If $\phi_\varepsilon(\zeta)$ is analytic in a neighborhood of Ω', $\phi_\varepsilon(A)$ can be defined by

$$\frac{1}{2\pi i}\int_\Gamma \phi_\varepsilon(\zeta)(\zeta - A)^{-1}d\zeta,$$

where Γ is a contour containing Ω'. Such calculi for unbounded operators are treated in [4, Chapters 5 and 15]; one such calculus involves multipliers

$$\phi_\varepsilon(\zeta) = \int_0^\infty e^{-\zeta x}da(x),$$

where $a(x)$ is a function of strong bounded variation on $\mathbf{R}^+$. In order that $\phi_\varepsilon(A)$ be in [*RB*], it suffices that

$$\Gamma \subset \{\zeta \big| |\arg \zeta| \geq \theta_0 > 0\},$$

and

$$\int_\Gamma \left|\frac{\phi_\varepsilon(\zeta)}{\zeta}\right| |d\zeta| < \infty. \tag{6}$$

Writing down the obvious L^1 radial bound by integrating the resolvent bounds shows summability in L^p and pointwise for analytic dilations $\phi_\varepsilon(A) = \phi(\varepsilon A)$, via techniques of Theorem 3.

Let A be as before and $\phi_t(A) = e^{-tA^s}$ for $s \in \mathbf{R}$. A^s will then be well defined if Γ does not enclose the origin (for example, when A has spectrum in the right half-plane). Cauchy integration of $e^{-t\zeta^s}$ gives

Theorem 4. *A^s generates an analytic semigroup $\{e^{-tA^s}\}_{\mathrm{Re}\, t > 0}$ in all L^p spaces for which A^s is defined as above. Furthermore, we have L^p $(1 \leq p < \infty)$ and Lebesgue convergence $e^{-tA^s}f(x) \xrightarrow[t\to 0]{} f(x)$ uniformly in each sector $\{t \big| |\arg t| \leq \theta_0 < \pi/2\}$.*

For $s = 1$, this is a statement of Abel summability and is also equivalent to convergence at the boundary $u(x, t) \xrightarrow[t\to 0]{} f(x)$ in the generalized heat equation $u_t(x, t) = Au(x, t)$, with initial value $u(x, 0) = f(x)$.

Remark. The radial bounds in Theorem 2 can be sharpened to yield optimal (in the sense of being the best generic) radial bounds for $\zeta L_\zeta(x, y)$. These in turn integrate to give bounds on the kernel of $\phi(A)$, and via Hölder's inequality provide information about the (p, q) type of $\phi(A)$.

d. Dirichlet Problems on the Half-Space

Let $A = A_0 + \sum_{|\alpha| < 2m} b_\alpha(x) D^\alpha$ be an elliptic operator as before, this time with sufficiently smooth $b_\alpha \in \mathscr{L}_s^{r_\alpha}$ $(s \geq m)$, $\dfrac{n_\alpha}{r_\alpha} + |\alpha| < 2m$

($n_\alpha = \dim V_\alpha$), where $\mathscr{L}_s^{r_\alpha} = (1 - \Delta)^{-s/2} L^{r_\alpha}$ is an L^{r_α}-Sobolev space. Consider the generalized Dirichlet problem for the elliptic operator $\frac{\partial^{2m}}{\partial t^{2m}} + (-1)^m A$

$$\left(\frac{\partial^{2m}}{\partial t^{2m}} - (-1)^m A\right) u(x, t) = 0; \quad \left(\frac{\partial}{\partial t}\right)^j u\Big|_{t=0} = f_j(x); \qquad (7)$$
$$(j = 0, \ldots, m - 1),$$

with $f_j \in \mathscr{L}^p_{m-1-j}$. Let $\omega = e^{2\pi i/2m}$ or $e^{\pi i/2m}$, depending on whether m is odd or even. By considering the general decaying solution

$$u = \sum_{\mathrm{Re}\,\omega^j > 0} e^{-t\omega^j A^{1/2m}} g_j,$$

we can formally solve for g_j at $t = 0$ and obtain

$$u(x, t) = \sum_{\mathrm{Re}\,\omega^j > 0} e^{-t\omega^j A^{1/2m}} \sum_{0 \le k \le m-1} \omega_{jk} A^{-k/2m} f_k, \qquad (8)$$

where $\{\omega_{jk}\}$ is the inverse of the Wandermond matrix

$$W(\{\omega^j \mid \mathrm{Re}\,\omega^j > 0\}).$$

To make (8) rigorous we need well-defined fractional power semigroups e^{-tA^s} ($s \in \mathbf{R}$, $t \ge 0$). Theorem 4 supplies them, provided A has spectrum in the right half-plane. Henceforth, we assume that A is formally self-adjoint and positive, or constant coefficient and positive. In the first case, the L^2 and $\mathscr{L}_s^2$ ($s \ge 0$) spectrum of A is in $\mathbf{R}^+$, and in the second the same is true in $\mathscr{L}_s^p$ ($1 \le p < \infty$).

Theorem 5. In either of these cases, (7) *has a unique solution* (8) *that is regular* ($u(x, t) \in \mathscr{L}_s^p(\mathbf{R}_+^{n+1})$ *for all* s), *and approaches its boundary values, that is,* $\partial_t^k u(x, t) \xrightarrow[t \to 0]{} f_k(x)$ ($0 \le k \le 2\ell$) *in* $\mathscr{L}^p_{2\ell - k}$ *and on the Lebesgue set of* f_k.

Above, $p = 2$ in the first case, and $1 < p < \infty$ in the second.

Nontangential convergence can be proved using standard Hardy-Littlewood maximal function techniques.

REFERENCES

1. Calderón, A. P., and Zygmund, A. On singular integrals. *Acta Math.* 88 (1952):289-309.

2. Carmona, R. Regularity properties of Schrödinger and Dirichlet semigroups. *J. Func. Anal.* 35 (1980):215-229.

3. Diamond, H.; Kon, M. A.; and Raphael, L. Stable summation methods for a class of singular Sturm-Liouville expansions. *Proc. AMS* 81 (1981):279-286.

4. Hille, E., and Phillips, R. S. *Functional Analysis and Semigroups.* AMS Colloq. Publ. Vol. 31, 1957.

5. Hörmander, L. Estimates for translation invariant operators in L^p-spaces. *Acta Math.* 104 (1960):93-139.

6. Hörmander, L. On the Riesz means of spectral functions and eigenfunction expansions for elliptic differential operators. Recent Advances in the Basic Sciences, Yeshiva Univ. Conf., Nov. 1966, pp. 155-202.

7. Kato, T. *Perturbation Theory for Linear Operators.* New York: Springer-Verlag, 1976.

8. Levitan, B. M., and Sargsjan, I. S. *Introduction to Spectral Theory.* Providence, R.I.: A.M.S., 1975.

9. Nagel, A., and Stein, E. M. *Lectures on Pseudo-Differential Operators.* Princeton, N.J.: Princeton Univ. Press, 1979.

10. Raphael, L. A. A generalized Stieltjes transform summability method and summing Sturm-Liouville expansions. Preprint.

11. Reed, M., and Simon, B. *Methods of Modern Mathematical Physics*, vol. 2. New York: Academic Press, 1975.

12. Stein, E., and Weiss, G. *Introduction to Fourier Analysis on Euclidean Spaces.* Princeton, N.J.: Princeton Univ. Press, 1971.

13. Taylor, M. Fourier integral operators and harmonic analysis on compact manifolds. *AMS Proc. of Symposia in Pure Math.* 32 (1979):115-136.

14. Tikhonov, A. N. Stable methods for the summation of Fourier series. *Soviet Math. Dokl.* 5 (1964):641-644.

15. Titschmarsh, E. C. *Eigenfunction Expansions*. 2d ed. New York: Cambridge Univ. Press, 1959.

16. Zygmund, A. *Trigonometric Series*. 2d ed. New York: Cambridge Univ. Press, 1959.

ON THE DIRICHLET PROBLEM FOR HIGHER-ORDER EQUATIONS

Lars Inge Hedberg
University of Stockholm

The Dirichlet problem for higher-order elliptic partial differential equations is usually treated only for regions with smooth boundaries. The purpose of this expository paper is to discuss the problem for more general regions, and to present some new results, mostly without proofs. For simplicity, we treat the equation $\Delta^m u = 0$, $m \geq 2$, although the results carry over to much more general equations. Here Δ^m means the iterated Laplacian; $\Delta^2 u = \Delta(\Delta u)$, and so on.

The classical Dirichlet problem can be formulated in the following way.

Problem 1. Let $G \subset \mathbf{R}^d$ be a bounded region, and suppose that ∂G has a tangent hyperplane at each of its points. Let $g_0, g_1, \ldots, g_{m-1}$ be given continuous functions on ∂G. Find a function $u \in C^{m-1}(\bar{G}) \cap C^{2m}(G)$ such that $\Delta^m u = 0$ in G, and $\partial^k u/\partial n^k = g_k$ on ∂G for $0 \leq k \leq m - 1$. (Here $\partial^k/\partial n^k$ denotes the derivative in the exterior normal direction.)

To discuss the problem for more general regions, it is convenient to reformulate it.

Supported by the Swedish Natural Science Research Council.

Problem 2. Let $G \subset \mathbf{R}^d$ be an arbitrary bounded region. Let $g \in C^{m-1}(\overline{G})$ be given. Find $u \in C^{m-1}(\overline{G}) \cap C^{2m}(G)$ such that $\Delta^m u = 0$ in G, and $D^\alpha(u - g)|_{\partial G} = 0$ for all multiindices α, $0 \leq |\alpha| \leq m - 1$.

In this generality the problem is far from being solved. The highly developed theory of the Laplace equation breaks down completely because of the failure of the maximum principle. The following theorem is typical for what is known. See, for example, S. Agmon [3 and 2, Sec. 9], L. Nirenberg [20], and C. Miranda [19, Sec. 52].

Theorem 1. Let $G \subset \mathbf{R}^d$ be a bounded region with C^∞ boundary, and let $g \in C^\infty(\overline{G})$. Then Problem 1 *(and* 2*) has a unique solution u, and $u \in C^\infty(\overline{G})$.*

General criteria for the solvability of Problem 2 are unknown. To the author's knowledge, the only result in this direction is a sufficient condition of Wiener-type valid for $m = 2$ and $d \leq 7$ dimensions, due to V. G. Maz'ja [16 and 17].

Because of the difficulty of the classical Dirichlet problem, it is natural that various generalized formulations were considered long before Theorem 1 was proved.

Let $G \subset \mathbf{R}^d$ be an arbitrary open set, let $H^m(G)$ denote the closure of $C^m(G)$ in the norm $\|\cdot\|_m$ defined by

$$\|f\|_m^2 = \sum_{0 \leq |\alpha| \leq m} \int_G |D^\alpha f|^2 \, dx,$$

and denote by $H_0^m(G)$ the closure in $H^m(G)$ of $C_0^\infty(G)$, the test functions with support in G.

Equivalently, $H^m(G)$ consists of those functions in $L^2(G)$ whose distribution derivatives $D^\alpha f$ belong to $L^2(G)$ for all multiindices α with $|\alpha| \leq m$. That a function f belongs to $H_0^m(G)$ is interpreted as meaning that f and $D^\alpha f$, $|\alpha| \leq m - 1$, have boundary values zero, in a generalized sense. Thus, the following "generalized Dirichlet problem" is natural.

Problem 3. Let $G \subset \mathbf{R}^d$ be open and bounded, and let $g \in H^m(G)$ be given. Find $u \in H^m(G)$ such that $\Delta^m u = 0$ in G, and $u - g \in H_0^m(G)$.

Here $\Delta^m u = 0$ is interpreted in the sense of distributions, but by the well-known lemma of Weyl and Schwartz, this implies that $u \in C^\infty(G)$, and thus $\Delta^m u = 0$ in the ordinary sense.

Using projections in Hilbert space, one easily proves the following theorem.

Theorem 2. Problem 3 *always has a unique solution.*

This result is, of course, very satisfactory through its simplicity and generality, but on the other hand, it is difficult to understand in which way the boundary values are really taken. What does it mean that $u - g \in H_0^m(G)$ if, for example, G is a d-dimensional ball with a $(d - k)$-dimensional manifold removed from its interior?

Generalizing earlier results of K. Friedrichs [9], S. L. Sobolev studied this problem in the case when G is bounded by a finite union of manifolds of arbitrary codimension. His fundamental paper [23] appeared in 1937, but it was translated into English only in 1963. Sobolev's results also appear in his monograph [22] from 1950, which was also translated in 1963.

Let $f \in H^m(G)$, and let M be a compact piece of a $(d - k)$-dimensional smooth manifold in G. Here $1 \leq k \leq d$. Let f_h, $h > 0$, be the averaged function, obtained by taking the convolution with a smooth approximate identity with support in a ball with radius h. Then Sobolev proved an imbedding theorem of which the following is a special case.

Theorem 3. Suppose $k < 2m$. Then $\lim_{h \to 0} f_h(x) = \tilde{f}(x)$ exists a.e. on M in the $(d - k)$-dimensional sense, and $\lim_{h \to 0} \|f_h - \tilde{f}\|_{L^2(M)} = 0$. In particular, $f_h(x)$ converges pointwise, and $\tilde{f}(x)$ is continuous if $d < 2m$. The condition $k < 2m$ is sharp.

We shall denote the restriction of $\tilde{f}$ to M by $f|_M$. It is called the trace of f on M.

Now suppose that ∂G consists of a finite union of smooth manifolds, not forming cusps. We write

$$\partial G = \bigcup_{k=1}^{d} M_k,$$

where each M_k is a finite union of $(d - k)$-manifolds. Then Sobolev proved, by letting the above M approach the boundary, that $f|_{M_k}$ is also well defined provided $k < 2m$.

Since $D^\alpha f \in H^{m-|\alpha|}(G)$ if $f \in H^m(G)$, it follows that $D^\alpha f|_{M_k}$ is well defined, provided $k < 2(m - |\alpha|)$, that is, $|\alpha| < m - \frac{k}{2}$. Thus, the Dirichlet problem can be formulated in the following way.

Problem 4. Let $G \subset \mathbf{R}^d$ be open and bounded, and let

$$\partial G = \bigcup_{k=1}^{d} M_k$$

satisfy the above assumptions. Let $g \in H^m(G)$. Find $u \in H^m(G)$ such that $\Delta^m u = 0$ in G and $D^\alpha(u - g)|_{M_k} = 0$ for $1 \leq k < 2(m - |\alpha|)$, and $0 \leq |\alpha| \leq m - 1$.

It is clear from Theorem 2 that solutions always exist. In fact, it is easily seen that the above traces vanish if $u - g \in H_0^m(G)$. What is not evident is that this solution is the only one, but this was proved by Sobolev.

Theorem 4. Problem 4 always has a unique solution. In fact, $f \in H_0^m(G)$ if and only if $f \in H^m(G)$ and $D^\alpha f|_{M_k} = 0$ for $1 \leq k < 2(m - |\alpha|)$, and $0 \leq |\alpha| \leq m - 1$.

Example. Consider the biharmonic equation $\Delta^2 u = 0$ in $G \subset \mathbf{R}^3$. Then $G = M_1 \cup M_2 \cup M_3$, where M_1 is the two-dimensional part of the boundary. Let, for example, G be a ball with a one-dimensional interval (M_2) and a finite number of points (M_3) removed from its interior. Let $u \in H^2(G)$ be a solution. By Theorem 3, and the remarks following it, $u|_{M_k}$ is defined for $k = 1, 2, 3$, and $\nabla u|_{M_1}$ is defined. Theorem 4 implies that $u = 0$ in G if $u|_{M_k} = 0$, $k = 1, 2, 3$, and $\nabla u|_{M_1} = \dot{0}$.

One difficulty in extending Sobolev's theorem to more general regions lies in the definition of the trace. This problem can be overcome through the idea of corrected, or precisely defined, functions, which appeared, at least for H^1, in the 1950s in the work of Deny [6, Chapter 4] and Deny and Lions [8]. From now on, we shall consider the more general Sobolev spaces $H^{m,p}(G)$, $1 < p < \infty$, ($H^{m,2} = H^m$) defined as the closure of $C^m(G)$ in the norm $\|\cdot\|_{m,p}$ defined by

$$\|f\|_{m,p}^p = \sum_{0 \le |\alpha| \le m} \int_G |D^\alpha f|^p \, dx.$$

The extension of the theory of corrected functions to this context is due to V. G. Maz'ja and V. P. Havin [18].

Let $f \in H^{m,p}(\mathbf{R}^d)$. By the well-known imbedding theorem of Sobolev, f is continuous for $mp > d$ (after correction on a set of measure zero), but f can in general not be so corrected if $mp \le d$. The natural way of measuring the deviation from continuity is by means of an (m, p)-capacity, which we shall now define.

If K is compact, we set

$$C_{m,p}(K) = \inf\{\|\varphi\|_{m,p}^p;\ \varphi \in C_0^\infty,\ \varphi \ge 1 \text{ on } K\}.$$

$C_{m,p}$ is then extended to arbitrary sets as an outer capacity in the usual way; that is, for open G,

$$C_{m,p}(G) = \sup\{C_{m,p}(K);\ K \text{ compact},\ K \subset G\}$$

and for arbitrary E, $C_{m,p}(E) = \inf\{C_{m,p}(G);\ G \text{ open},\ G \supset E\}$.

There is an equivalent approach that is sometimes more convenient. By a theorem of A. P. Calderón [5], a function f belongs to $H^{m,p}(\mathbf{R}^d)$ for $1 < p < \infty$ if and only if $f = G_m * g$ for some $g \in L^p$, where G_m is the so-called Bessel kernel, defined as the inverse Fourier transform of $\hat{G}_m(\xi) = (1 + |\xi|^2)^{-m/2}$. There is also equivalence of norms, that is, $A^{-1}\|f\|_{m,p} \leq \|g\|_{0,p} \leq A\|f\|_{m,p}$. One can now define (m, p)-capacity at once for arbitrary sets by setting

$$C_{m,p}(E) = \inf\{\|g\|_{0,p}^p;\ g \in L^p,\ g \geq 0,\ G_m * g \geq 1 \text{ on } E\}.$$

The definition makes sense, since $G_m * g(x)$ is defined $(\leq +\infty)$ at all points, and one can show that this definition is equivalent to the previous one.

There are precise comparison theorems between (m, p)-capacities and Hausdorff measure (see [18]). Here we only quote the following result.

Theorem 5. *Let M_k be a $(d - k)$-dimensional smooth manifold. Then $C_{m,p}(M_k) = 0$ if and only if $k \geq mp$.*

Remark. It follows from Theorems 3 and 5 that if $f \in H^m(\mathbf{R}^d)$ and M_k is a $(d - k)$-manifold, then $D^\alpha f|_{M_k}$ is defined if $C_{m-|\alpha|,2}(M_k) > 0$.

Let

$$f(x) = G_m * g(x) = \int_{\mathbf{R}^d} G_m(x - y)g(y)dy,\ g \in L^p,$$

so that $f(x)$ is well defined wherever $G_m * |g|(x) < \infty$. Let E_λ denote the set where either $f(x)$ is undefined or $|f(x)| > \lambda$. Since $|f(x)| \leq G_m * |g|(x)$, it follows from the second definition of capacity that

$$C_{m,p}(E_\lambda) \leq \frac{1}{\lambda^p} \int_{\mathbf{R}^d} |g|^p dx.$$

In particular, $f(x)$ is well defined outside a set of zero (m, p)-capacity, that is, (m, p)-quasieverywhere [(m, p) - q.e.]. Let Mf denote the Hardy-Littlewood maximal function of f, that is,

$$Mf(x) = \sup_{B \ni x} \frac{1}{|B|} \int_B |f(y)|\,dy,$$

where the supremum is taken over all balls B containing x. It is not hard to prove that Mf satisfies a similar inequality (D. R. Adams [1]), that is, there is a constant A such that

$$C_{m,p}(\{x;\ Mf(x) > \lambda\}) \le \frac{A}{\lambda^p}\|f\|_{m,p}^p.$$

Let $\{\chi_h\}_{h>0}$ be a C^∞ approximate identity; that is, $\chi_h = h^{-d}\chi_1\left(\frac{x}{h}\right)$, $\chi_1 \ge 0$, $\int \chi_1 dx = 1$, and χ_1 has support in the unit ball. It follows in a well-known way from the above inequality that

$$\lim_{h\to 0} f * \chi_h(x) = \tilde{f}(x)$$

exists (m, p)-q.e. Moreover, $\tilde{f}(x) = f(x)$ a.e., and for any $\varepsilon > 0$, there is an open set G such that $C_{m,p}(G) < \varepsilon$, and $\tilde{f}|_{G^c}$ is continuous on G^c (the complement of G).

Such a function $\tilde{f}$ is called an (m, p)-quasicontinuous representative of f. By an extension of a theorem of Wallin [25] (see also Deny [7]), such representatives are unique up to sets of zero (m, p)-capacity. In particular, the potential $\int G_m(x - y)g(y)dy$ is such a representative.

If a quasicontinuous representative of f is chosen, f is said to be corrected, or precisely defined.

We now define the trace $f|_E$ of an element f in $H^{m,p}(\mathbf{R}^d)$ simply as the restriction to E of any (m, p)-quasicontinuous representative of f. Similarly, $D^\alpha f|_E$ is defined by taking the restriction to E of an $(m - |\alpha|, p)$-quasicontinuous representative of $D^\alpha f$. Here $D^\alpha f$ is the derivative in the sense of distributions.

In a manuscript from 1968, which appeared in print only in 1977 (see B. W. Schulze and G. Wildenhain [21, IX.5.1]), B. Fuglede proposed the following formulation of the Dirichlet problem.

Problem 5. Let $G \subset \mathbf{R}^d$ be open and bounded. Let $g \in H^m(\mathbf{R}^d)$. Find $u \in H^m(\mathbf{R}^d)$ such that $\Delta^m u = 0$ in G and $D^\alpha(u - g)|_{\partial G} = 0$, $0 \le |\alpha| \le m - 1$.

It is again easy to see that a solution exists. In fact, by Theorem 2 there is a solution u such that $u - g \in H_0^m(G)$, and it is easy to prove that this implies that $D^\alpha(u - g)|_{\partial G} = 0$, $0 \leq |\alpha| \leq m - 1$. The problem is again to prove that this solution is the only one, that is, that pathologic null solutions do not arise.

A bounded open G has this uniqueness property if and only if ∂G has a property that by analogy with harmonic analysis Fuglede called the $2m$-spectral synthesis property. More generally, we shall say that a closed set F admits (m, p)-spectral synthesis if every f in $H^{m,p}(\mathbf{R}^d)$ that satisfies $D^\alpha f|_F = 0$ for $0 \leq |\alpha| \leq m - 1$ belongs to $H_0^{m,p}(F^c)$. Thus, the Dirichlet problem in G has a unique solution if and only if ∂G admits $(m, 2)$-spectral synthesis.

The following theorem was proved in [13] after a partial result had been proved in [14].

Theorem 6. Problem 5 *always has a unique solution. In fact, every closed set* $F \subset \mathbf{R}^d$ *admits* (m, p)*-spectral synthesis, provided only* $p > 2 - \frac{1}{d}$.

Remark. Whether the result is true for all $p > 1$ remains an open problem. (Added in proof: The result is true. See the remark following Theorem 7.)

For the proof, more references, and other applications, we refer to [13 and 14]. See also the survey [12]. Here we only give a couple of remarks concerning the proof.

The proof depends on the idea of a thin set. A set $E \subset \mathbf{R}^d$ is called (k, p)-thin at x if $kp \leq d$, and

$$\sum_{n=1}^{\infty} [2^{n(d-kp)} C_{k,p}(E \cap B_n(x))]^{\frac{1}{p-1}} < \infty.$$

[$B_n(x)$ is the ball $\{y;\ |y - x| \leq 2^{-n}\}$.] For $k = 1$, $p = 2$, this agrees with the classical definition.

The importance of this definition is the following generalized Kellogg property.

Theorem 7. *If E' is the subset of E where E is (k, p)-thin, then $C_{k,p}(E') = 0$, provided $p > 2 - \frac{k}{d}$.*

Remark. If the restriction on p could be removed, the extension of Theorem 6 to $p > 1$ would follow. (Added in proof: T. Wolff has done this. See [26].)

Let $f \in H^{m,p}(\mathbf{R}^d)$, and suppose that $D^\alpha f|_F = 0$ for all α, $0 \leq |\alpha| \leq m - 1$. Assume that $p > 2 - \frac{1}{d}$, so that the Kellogg property holds for $C_{k,p}$, $k = 1, \ldots, m$. Set

$$\left\{\frac{1}{|B_n(x)|} \int_{B_n(x)} |f|^p \, dy\right\}^{1/p} = [f]_n(x),$$

and

$$\sum_{|\alpha|=m} |D^\alpha f| = |\nabla^m f|.$$

If $x \in F$, one can estimate $[f]_n(x)$ in terms of $C_{k,p}(F \cap B_n(x))$, $1 \leq k \leq m$, and using the Kellogg property, one can prove the following lemma.

Lemma 1. *Under the above assumptions, there are disjoint sets $E_0, E_1, \ldots, E_m$ such that*

$$F = \bigcup_{k=0}^{m} E_k,$$

and

(a) $$\sum_{n=1}^{\infty} \left\{\frac{2^{-nm}[\nabla^m f]_n(x)}{[f]_n(x)}\right\}^{p'} = \infty \text{ for all } x \in E_0, \quad \left(p' = \frac{p}{p-1}\right);$$

(b) $$\liminf_{n\to\infty} \frac{2^{-n(m-k)}[G_k * |\nabla^m f|]_n(x)}{[f]_n(x)} > 0 \text{ for all } x \in E_k,$$

and $C_{k,p}(E_k) = 0$, for $k = 1, 2, \ldots, m - 1$;

(c) $C_{m,p}(E_m) = 0$.

Suppose for simplicity that the sets E_k are compact, so that one can treat them separately. Using the divergent series in (a), one can construct a C^∞-function χ_0 so that $0 \le \chi_0 \le 1$, $\chi_0 = 0$ on a neighborhood of E_0, and $\|f - f\chi_0\|_{m,q} < \frac{\varepsilon}{2}$, for some arbitrarily given $\varepsilon > 0$.

Then one can use the fact that $C_{1,p}(E_1) = 0$ to construct a suitable function χ_1 so that $0 \le \chi_1 \le 1$, $\chi_1 = 0$ on a neighborhood of E_1, $\int |\nabla\chi_1|^p dx$ is small, and use (b) to show that

$$\|f\chi_0 - f\chi_0\chi_1\|_{m,p} < \frac{\varepsilon}{2^2}.$$

One proceeds step by step, until one has found $\chi_0, \ldots, \chi_m$ so that $\chi = \chi_0\chi_1\cdots\chi_m = 0$ on a neighborhood of F, and $\|f - f\chi\|_{m,p} < \varepsilon$. It follows that $f \in H_0^{m,p}(F^c)$.

We shall illustrate the result by showing that it contains the uniqueness theorem of Sobolev (Theorem 4).

Let G be bounded as before by a finite union of manifolds not forming cusps,

$$\partial G = \bigcup_{k=1}^{d} M_k, \ \dim M_k = d - k.$$

Let $f \in H^m(G)$.

In Theorem 6, f is required to belong to $H^m(\mathbf{R}^d)$, but this difference is only apparent. In fact, by a well-known extension theorem (see Stein [24]), a function f in $H^m(G)$ can be extended across M_1 if it is assumed that M_1 satisfies a local Lipschitz condition.

That f can be extended to $H^m(\mathbf{R}^d)$ follows from the following lemma, which is probably well known.

Lemma 2. Let $f \in H^{m,p}(G\backslash F)$, *where* F *is a closed set with zero* $(d - 1)$*-dimensional Hausdorff measure. Then* $f \in H^{m,p}(G)$.

Proof. We can assume without loss of generality that $G = \mathbf{R}^d$. By assumption, almost every line parallel to the x_i-axis is disjoint from F. In fact, the perpendicular projection of F on the hyperplane $x_i = 0$ has $(d - 1)$-dimensional measure zero.

Let $f \in H^{m,p}(\mathbf{R}^d \setminus F)$. Then f and $D^\alpha f$, $|\alpha| \leq m$, are defined almost everywhere in $\mathbf{R}^d$, so $D^\alpha f \in L^p(\mathbf{R}^d)$, $0 \leq |\alpha| \leq m$. It is enough to prove that

$$\int_{\mathbf{R}^d} \varphi D^\alpha f \, dx = (-1)^{|\alpha|} \int_{\mathbf{R}^d} f D^\alpha \varphi \, dx \text{ for all } \varphi \in C_0^\infty.$$

First let $|\alpha| = 1$. By Fubini's theorem,

$$\int_{\mathbf{R}^d} \varphi \frac{\partial f}{\partial x_i} \, dx = \int_{\mathbf{R}^{d-1}} dx' \int_{-\infty}^{\infty} \varphi \frac{\partial f}{\partial x_i} \, dx_i$$

$(dx' = dx_1 \dots dx_{i-1} dx_{i+1} \dots dx_d)$. For almost every $a' \in \mathbf{R}^{d-1}$, the line $x' = a'$ lies in $\mathbf{R}^d \setminus F$. Thus, by a well-known property of Sobolev spaces, f is absolutely continuous on almost all such lines. By integration by parts,

$$\int_{\mathbf{R}^{d-1}} dx' \int_{-\infty}^{\infty} \varphi \frac{\partial f}{\partial x_i} \, dx_i = -\int_{\mathbf{R}^{d-1}} dx' \int_{-\infty}^{\infty} f \frac{\partial \varphi}{\partial x_i} \, dx_i = -\int_{\mathbf{R}^d} f \frac{\partial \varphi}{\partial x_i} \, dx.$$

The lemma follows by induction.

Now let $f \in H^m(G)$, where

$$\partial G = \bigcup_{k=1}^{d} M_k,$$

and suppose that $D^\alpha f|_{M_k} = 0$, in the sense of Sobolev, for $1 \leq k < 2(m - |\alpha|)$, $0 \leq |\alpha| \leq m - 1$. It is easily seen that $D^\alpha f|_{M_k} = 0$ in the sense of corrected functions. But by Theorem 5, $C_{m-|\alpha|,2}(M_k) = 0$ for $k \geq 2(m - |\alpha|)$. It follows that $D^\alpha f|_{\partial G} = 0$, $0 \leq |\alpha| \leq m - 1$, in the sense of corrected functions, and thus Theorem 6 applies.

The formulation in Problem 5 of the Dirichlet problem is less general than the generalized Dirichlet problem (Problem 3) in the sense that the boundary values are supposed to be given through a function g that is defined in all of $\mathbf{R}^d$, or at least across ∂G. One can remove this restriction by formulating a "fine Dirichlet problem."

Let $f \in H^{m,p}(\mathbf{R}^d)$, and suppose that f is corrected. Then one can prove (Fuglede [11]) that for (m, p)-quasievery point x, the set

$\{y;\ |f(y) - f(x)| \geq \varepsilon\}$ is (m, p)-thin at x for all $\varepsilon > 0$. One says that (m, p)-fine $\lim_{y \to x} f(y) = f(x)$, or one introduces a fine topology and says that f is (m, p)-finely continuous at x.

The fine Dirichlet problem can be formulated as follows.

Problem 6. Let $G \subset \mathbf{R}^d$ be open and bounded, and let $g \in H^m(G)$. Find $u \in H^m(G)$ so that $\Delta^m u = 0$ in G, and

$$(m - |\alpha|, 2)\text{-fine} \lim_{G \ni x \to x_0} D^\alpha(u(x) - g(x)) = 0$$

for

$$(m - |\alpha|, 2)\text{-q.e. } x_0 \in \partial G,\ 0 \leq |\alpha| \leq m - 1.$$

It is clear from the above that a solution exists. Using Theorem 6, T. Kolsrud [15] has proved the following theorem.

Theorem 8. Problem 6 *always has a unique solution. In fact, a function f in $H^{m,p}(G)$ belongs to $H_0^{m,p}(G)$ if and only if*

$$(m - |\alpha|, p)\text{-fine} \lim_{G \ni x \to x_0} D^\alpha f(x) = 0$$

for $(m - |\alpha|, 2)$-q.e. $x_0 \in \partial G$, $0 \leq |\alpha| \leq m - 1$, *provided* $p > 2 - \frac{1}{d}$. (Added in proof: The theorem is true for $p > 1$. See [26].)

This characterization of $H_0^1(G)$ is due to Deny and Lions [8, Theorem 5.1], who deduced it from a uniqueness theorem for harmonic functions of Brelot [4]. See also Fuglede [10, Theorem 9.1]. Here the procedure is the opposite one; the uniqueness theorem is proved using a characterization of $H_0^m(G)$.

REFERENCES

1. Adams, D. R. Maximal operators and capacity. *Proc. Amer. Math. Soc.* 34 (1972):152-156.

2. Agmon, S. *Lectures on Elliptic Boundary Value Problems.* New York: Van Nostrand, 1965.

3. Agmon, S. Maximum theorems for solutions of higher order elliptic equations. *Bull. Amer. Math. Soc.* 66 (1960):77-80.

4. Brelot, M. Sur l'allure des fonctions harmoniques et sous-harmoniques à la frontière. *Math. Nachr.* 4 (1950-51):298-307.

5. Calderón, A. P. Lebesgue spaces of differentiable functions and distributions. *Proc. Symp. Pure Math.* 4 (1961):33-49.

6. Deny, J. Les potentiels d'énergie finie. *Acta Math.* 82 (1950): 107-183.

7. Deny, J. Théorie de la capacité dans les espaces fonctionnels. *Séminaire de Théorie du Potentiel* 9 (1964-65), No. 1.

8. Deny, J., and Lions, J.-L. Les espaces du type de Beppo Levi. *Ann. Inst. Fourier* (Grenoble) 5 (1953-54):305-370.

9. Friedrichs, K. Die randwert- und eigenwertprobleme aus der theorie der elastischen platten. *Math. Ann.* 98 (1928):205-247.

10. Fuglede, B. Finely harmonic functions. *Lecture Notes in Mathematics* 289. New York: Springer-Verlag, 1972.

11. Fuglede, B. Quasi topology and fine topology. *Séminaire de Théorie du Potentiel* 10 (1965-66), No. 12.

12. Hedberg, L. I. Spectral synthesis and stability in Sobolev spaces, Euclidean harmonic analysis. Proc. Univ. of Maryland, 1979. *Lecture Notes in Mathematics* 779. New York: Springer-Verlag, 1980, pp. 73-103.

13. Hedberg, L. I. Spectral synthesis in Sobolev spaces, and uniqueness of solutions of the Dirichlet problem. *Acta Math.* 147 (1981):237-264.

14. Hedberg, L. I. Two approximation problems in function spaces. *Ark. Mat.* 16 (1978):51-81.

15. Kolsrud, T. A uniqueness theorem for higher order elliptic partial differential equations. To appear in *Math. Scand.*

16. Maz'ja, V. G. Behavior of solutions to the Dirichlet problem for the biharmonic operator at the boundary point. Technische Hochschule Karl-Marx-Stadt, Sektion Mathematik, *Wissenschaftliche Informationen* 10 (1979):1-16.

17. Maz'ja, V. G. On the behavior near the boundary of solutions of the Dirichlet problem for the biharmonic operator. *Dokl. Akad. Nauk SSSR* 235:6 (1977):1263-1266.

18. Maz'ja, V. G., and Havin, V. P. Non-linear potential theory. *Uspehi Mat. Nauk* 27:6 (1972):67-138.

19. Miranda, C. Partial differential equations of elliptic type. 2d ed. *Ergebnisse der Math.*, vol. 2. New York: Springer-Verlag, 1970.

20. Nirenberg, L. Remarks on strongly elliptic partial differential equations. *Comm. Pure Appl. Math.* 8 (1955):648-674.

21. Schulze, B.-W., and Wildenhain, G. *Methoden der Potentialtheorie für Elliptische Differentialgleichungen beliebiger Ordnung*. Berlin: Akademie-Verlag, 1977.

22. Sobolev, S. L. Applications of functional analysis in mathematical physics. *Izd. LGU*, Leningrad, 1950. (English translation, *Amer. Math. Soc.*, Providence, R.I., 1963).

23. Sobolev, S. L. On a boundary value problem for polyharmonic equations. *Mat. Sb.* (N.S.) 2 (44) (1937):467-499. [English translation, *Amer. Math. Soc. Translations* (2) 33 (1963):1-40.]

24. Stein, E. M. *Singular Integrals and Differentiability Properties of Functions*. Princeton, N.J.: Princeton Univ. Press, 1970.

25. Wallin, H. Continuous functions and potential theory. *Ark. Mat.* 5 (1963):55-84.

26. Hedberg, L. I., and Wolff, T. H. Thin sets in nonlinear potential theory. To appear.

L^p-ESTIMATES FOR A SINGULAR HYPERBOLIC EQUATION

Jeff E. Lewis
University of Illinois at Chicago

Cesare Parenti
University of Bologna

Let A be a self-adjoint operator on a Hilbert space H with domain $D(A)$ and compact resolvent. We study the singular hyperbolic equation

$$tu'(t) - (iA + bI)u(t) = f(t) \text{ on } R^+, \tag{1}$$

with b real.

As a simple model for (1), consider the scalar equation

$$tu'(t) - (ic + b)u(t) = f(t) \in C_0^\infty(R^+). \tag{2}$$

Equation (2) has two solutions given by

$$\begin{aligned} F_{b+ic}f(t) &= \int_0^t (t/s)^{b+ic} f(s)\,ds/s, \\ B_{b+ic}f(t) &= -\int_t^\infty (t/s)^{b+ic} f(s)\,ds/s. \end{aligned} \tag{3}$$

If supp $f \subset (T_1, T_2)$, then $F_{b+ic}f(t) = 0$ on $(0, T_1)$ and $B_{b+ic}f(t) = 0$ on (T_2, ∞); moreover,

$$F_{b+ic}f(t) - B_{b+ic}f(t) = t^{b+ic}\tilde{f}(-(b + ic)),$$

where $\tilde{f}(z) = \int_0^\infty t^{z-1}f(t)\,dt$ is the Mellin transform of f.

This work was completed while the second author was visiting the University of Illinois at Chicago. Both authors have been partially supported by the Italian C.N.R., gruppo G.N.A.F.A.

By Hardy's inequality,

if $1/p + a + b < 0$, then

$$\|t^a F_{b+ic} f;\ L^p(0,\ T)\| \leq C\|t^a f;\ L^p(0,\ T)\|, \tag{5}$$

if $1/p + a + b > 0$, then

$$\|t^a B_{b+ic} f;\ L^p(T,\ \infty)\| \leq C\|t^a f;\ L^p(T,\ \infty)\|. \tag{6}$$

The constant C in (5) or (6) may be chosen as $C = |1/p + a + b|^{-1}$.

For $I = R^+$, $(0,\ T)$, or $(T,\ \infty)$, and $t^a f(t) \in L^p(I;\ H)$, a distribution $u \in \mathscr{D}'(I;\ H)$ is an $L^{p,\,a}$ solution of

$$Pu = tu' - (iA + bI)u = f \text{ on } I \tag{7}$$

if

$$\begin{gathered} t^a u \in L^p(I;\ H), \\ \int_0^\infty (u,\ P^\star v)\,dt = \int_0^\infty (f,\ v)\,dt, \text{ for all } v \in C_0^\infty(I;\ D(A)). \end{gathered} \tag{8}$$

The formal adjoint of P is

$$P^\star v = -\{tv' - (iA - (1 + b)I)v\}. \tag{9}$$

We now give our main result.

Theorem. *Let* $t^a f \in L^p(R^+;\ H)$. *Then:*

(a) *If* $1/p + a + b < 0$, *then for any* T, $0 < T \leq \infty$, *there is a unique* $L^{p,\,a}$ *solution of* $Pu = f$ *on* $(0,\ T)$; *the solution* $u = Ff$ *satisfies*

$$\|t^a u;\ L^p(0,\ T;\ H)\| \leq C\|t^a f;\ L^p(0,\ T;\ H)\|. \tag{10}$$

(b) *If* $1/p + a + b > 0$, *then for any* T, $0 \leq T < \infty$, *there is a unique* $L^{p,\,a}$ *solution of* $Pu = f$ *on* $(T,\ \infty)$; *the solution* $u = Bf$ *satisfies*

$$\|t^a u;\ L^p(T,\ \infty;\ H)\| \leq C\|t^a f;\ L^p(T,\ \infty;\ H)\|. \tag{11}$$

(c) *The constant* C *in* (10) *or* (11) *may be taken as*

$$C = |1/p + a + b|^{-1}. \tag{12}$$

Proof. Since A has compact resolvent, denote the eigenvalues of A as $\{c_k\}$ with a corresponding complete orthonormal basis of eigenfunctions $\{y_k\}$. If

$$f(t) = \sum_{\text{finite}} f_k(t)y_k, \; f_k \in C_0^\infty(R^+),$$

define

$$Ff(t) = \Sigma F_{b+ic_k} f_k y_k, \tag{13}$$

$$Bf(t) = \Sigma B_{b+ic_k} f_k y_k. \tag{14}$$

Obviously, $PFf = f$ and $PBf = f$ on R^+. Since

$$t^{a-ic}F_{b+ic}f(t) = \int_0^t (t/s)^{a+b}(s^{a-ic}f(s))ds/s,$$

we have that

$$|t^a Ff(t)|_H^2 = \Sigma|t^a F_{b+ic_k} f_k|^2 = \Sigma|F_{a+b}(s^{a-ic}f_k)|^2.$$

If $1/p + a + b < 0$, using the bound $C = |1/p + a + b|^{-1}$ for F_{a+b} on $L^p(0, T)$ and the Littlewood-Paley lemma of Zygmund [2, vol. 2, Lemma 2.10], we obtain that $\|t^a Ff;\, L^p(0, T;\, H)\|$ is dominated by C times

$$\|(\Sigma|t^{a-ic_k}f_k|^2)^{1/2};\, L^p(0, T)\| = \|t^a f;\, L^p(0, T;\, H)\|. \tag{15}$$

A similar argument yields (11) for Bf.

To prove the uniqueness, suppose that $1/p + a + b < 0$ and that u is an $L^{p,a}$ solution of $Pu = 0$ on $(0, T)$. Fix $c = c_k$ and for any $g \in C_0^\infty(0, T)$ such that

$$\tilde{g}(-(ic - (1 + b))) = 0, \tag{16}$$

define $v_k = (B_{ic-(1+b)}g(t))y_k$. By (16) and (4), $v_k \in C_0^\infty(0, T)$ and $P^* v_k = -gy_k$. Since $Pu = 0$ on $(0, T)$,

$$\int_0^T (u, y_k)g(t)dt = 0. \tag{17}$$

Since $1/p + a + b < 0$, the linear functional

$$g \to \tilde{g}(1 + b - ic)$$

is not continuous on $L^{q,-a}(0, T)$, $1/p + 1/q = 1$; hence functions

$g(t)$ satisfying (16) are dense in $L^{q,-a}(0, T)$ and (17) holds for all $g \in C_0^\infty(0, T)$. Hence $(u, y_k) = 0$ for every k.

Uniqueness in the case $1/p + a + b > 0$ is handled in a similar manner.

Remarks.

1. When $p = 2$ the equation (1) on $(0, T)$ and many generalizations were studied by Alinhac and Baouendi [1] by the method of energy inequalities.

2. Under the change of variables $s = \ln(1/t)$, equation (1) becomes

$$-(d/ds)u - (iA + bI)u = f, \tag{18}$$

and inequality (10) becomes for $1/p + a + b < 0$

$$\|e^{-as}u;\ L^p(S, \infty;\ H)\| \leq C\|e^{-as}f;\ L^p(S, \infty;\ H)\|$$

for solutions of (18).

REFERENCES

1. Alinhac, S., and Baouendi, M. S. Uniqueness for the characteristic Cauchy problem and strong unique continuation for higher order partial differential inequalities. *Amer. J. of Math.* 102 (1980):179-217.

2. Zygmund, A. *Trigonometric Series.* New York: Cambridge Univ. Press, 1959.

MIXED NORM ESTIMATES FOR THE KLEIN-GORDON EQUATION

Bernard Marshall
McGill University

Let $u(x, t)$ be the solution of the Klein-Gordon equation

$$\begin{cases} u_{tt} - \Delta_x u + u = 0 \\ u(x, 0) = 0 \qquad u_t(x, 0) = f(x) \end{cases}$$

where $x \in \mathbf{R}^n$, $t > 0$. For what values of q and r does there exist α such that

$$\|(1 + t)^{-\alpha} u\|_{q,r} \equiv \left(\int_0^\infty ((1 + t)^{-\alpha} \|u(\cdot, t)\|_q)^r dt\right)^{1/r} \le C\|f\|_2, \tag{1}$$

and what is the smallest possible value of α? Thus $\alpha - \frac{1}{r}$ is a measure of the decay of the solution u.

Theorem 1.

(i) *There can exist an estimate of the form* (1) *only if* $\left(\frac{1}{q}, \frac{1}{r}\right)$ *is in the region* R *determined by*

$$0 \le \frac{1}{r} \le 1, \ \frac{1}{2} - \frac{3}{2n} \le \frac{1}{q} \le \frac{1}{2}, \ \frac{n}{q} + \frac{1}{r} \ge \frac{n - 2}{2}.$$

(ii) *For* $\left(\frac{1}{q}, \frac{1}{r}\right)$ *in* R, *if* (1) *holds, then* $\alpha \ge \alpha_0\left(\frac{1}{q}, \frac{1}{r}\right)$, *where* α_0 *is defined by*

$$\alpha_0\left(\frac{1}{q}, \frac{1}{r}\right) = \max\left\{0, \frac{n}{2q} - \frac{n}{4} + \frac{1}{r}, \frac{1}{r} - \frac{1}{2}\right\}.$$

This research was supported by NSERC Grant u0074.

To describe the corresponding positive result, let R_0 be the region in the $\left(\frac{1}{q}, \frac{1}{r}\right)$ plane defined by

$$0 \leq \frac{1}{r} \leq 1, \ \frac{1}{2} - \frac{3}{2n} < \frac{1}{q} \leq \frac{1}{2}, \ \frac{n}{q} + \frac{1}{r} > \frac{n-2}{2},$$

together with the points satisfying

$$\frac{n}{q} + \frac{1}{r} = \frac{n-2}{2}, \ \frac{1}{r} \leq \frac{1}{2} - \frac{1}{n+1}. \tag{2}$$

Note that the closure of R_0 is R. These regions in Figure 1 have vertices P_0, P_1, P_4, P_7, P_5.

The coordinates of the points in Figure 1 are as follows:

$$P_0 = \left(\frac{1}{2}, 0\right) \qquad P_1 = \left(\frac{1}{2} - \frac{1}{n}, 0\right) \qquad P_2 = \left(\frac{1}{2} - \frac{1}{n}, \frac{1}{2}\right)$$

$$P_3 = \left(\frac{1}{2} - \frac{1}{n-1}, \frac{1}{2}\right) \qquad P_4 = \left(\frac{1}{2} - \frac{3}{2n}, \frac{1}{2}\right) \qquad P_5 = \left(\frac{1}{2}, 1\right)$$

$$P_6 = \left(\frac{1}{2} - \frac{1}{n}, 1\right) \qquad P_7 = \left(\frac{1}{2} - \frac{3}{2n}, 1\right)$$

Theorem 2. For $\left(\frac{1}{q}, \frac{1}{r}\right)$ *in* R_0, *if* $\alpha > \alpha_0\left(\frac{1}{q}, \frac{1}{r}\right)$, *then*

$$\|(1+t)^{-\alpha}u\|_{q,r} \leq C\|f\|_2.$$

For $\alpha = \alpha_0$, the operator $Tf = u$ may or may not be bounded. This can be seen quickly in the case

$$\|t^{-1/2}u\|_{2,2} = \int_{\mathbf{R}^n}\int_0^\infty |\hat{f}(\xi)|^2(|\xi|^2+1)^{-1}\int_0^\infty \sin^2(t\sqrt{1+|\xi|^2})\,t^{-1}dt\,d\xi = \infty,$$

where we used the Plancherel theorem and the fact that

$$\widehat{T_t f}(\xi) = (1+|\xi|^2)^{-1/2}\sin(t\sqrt{1+|\xi|^2})\hat{f}(\xi).$$

Of particular importance in applications is the quadrilateral in R where $\alpha_0 = 0$. Define

$$Q = \left\{\left(\frac{1}{q}, \frac{1}{r}\right) \in R_0 : 0 \leq \frac{1}{r} \leq \frac{1}{2}, \ \frac{1}{2} - \frac{1}{n} - \frac{1}{nr} \leq \frac{1}{q} < \frac{1}{2} - \frac{1}{nr}\right\}.$$

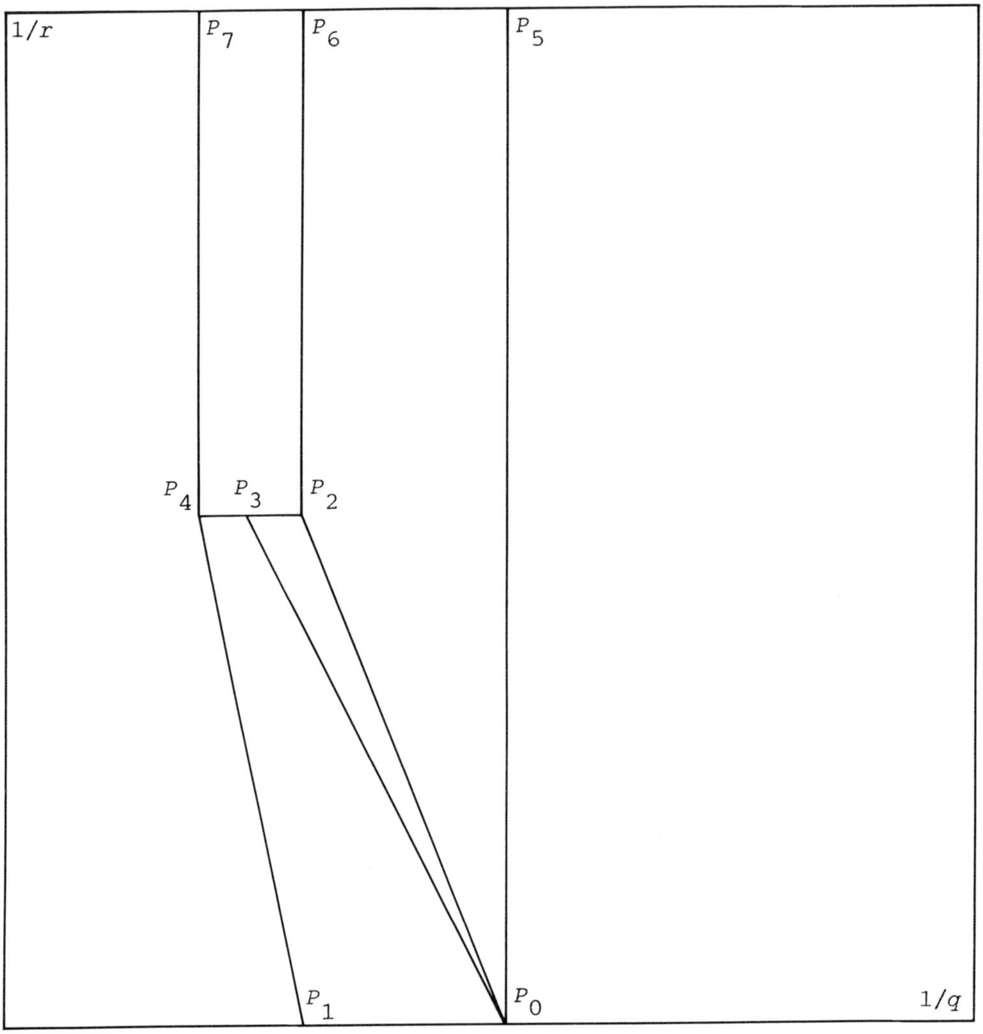

Figure 1

The vertices of Q are P_0, P_1, P_2, P_4. The closure of Q is the region where $\alpha_0 = 0$.

Theorem 3. *If* $\left(\frac{1}{q}, \frac{1}{r}\right) \in Q$, *then* $\|u\|_{q,r} \leq C\|f\|_2$. *If* $\left(\frac{1}{q}, \frac{1}{r}\right) \notin \bar{Q}$, *then* $Tf = u$ *is not bounded from* L^2 *to* $L^r(L^q)$.

These estimates are a consequence of estimates involving derivatives of u. Let $\mathscr{T}$ be a triangle

$$\mathscr{T} = \left\{\left(\frac{1}{q}, \frac{1}{r}\right) : 0 \leq r \leq \frac{1}{2}, \frac{1}{2} - \frac{2}{(n-1)r} < \frac{1}{q} < \frac{1}{2} - \frac{2}{nr}\right\}.$$

The vertices of $\mathscr{T}$ are P_0, P_2, P_3.

Theorem 4. *If* $\left(\frac{1}{q}, \frac{1}{r}\right) \in \mathscr{T}$ *and* $\gamma < \frac{1}{2} + \frac{1}{q} - \frac{1}{r}$, *then*

$$\|(I - \Delta)^{\gamma/2} u\|_{q,r} \leq C\|f\|_2. \tag{3}$$

When $q = \infty$, the L^∞-norm is replaced by the BMO norm. Also, by Sobolev's theorem, if $\left(\frac{1}{q}, \frac{1}{r}\right) \in Q - \mathscr{T}$ and $0 \leq \gamma < 1 - \frac{n}{2} + \frac{n}{q} + \frac{1}{r}$, then (3) holds.

In the case $r = \infty$, the estimate (1) takes the form

$$\|u(\cdot, t)\|_q \leq Ct^\alpha \|f\|_p.$$

The estimates of this type have been determined completely in [3].

In the case of $q = r$, $\alpha = 0$, Strichartz [8] has shown that $(I - \Delta)^{1/4} T$ is bounded from $L^2(\mathbf{R}^n)$ to $L^q(\mathbf{R}^{n+1})$ if

$$2 + \frac{4}{n} \leq q \leq 2 + \frac{4}{n-1}.$$

This result is stronger than Theorem 4, which states that $(I - \Delta)^\sigma T$ is bounded if $\sigma < \frac{1}{4}$ and $2 + \frac{4}{n} < q < 2 + \frac{4}{n-1}$. Thus, in Theorem 4, for $q \leq r$ we can have $\mathscr{T}$ closed and $\gamma = \frac{1}{2} + \frac{1}{q} - \frac{1}{r}$. In [8], Strichartz used duality to take advantage of a device of Tomas [9] for handling the restrictions of Fourier transforms. We follow the same approach for Theorem 4 but use a different analytic family of operators, $(I - \Delta)^z T$, to control the different behavior in the x- and t-variables. The estimates needed, then, are those in [3].

The stimulus for studying these estimates came from the scattering theory of various nonlinear variants of the Klein-Gordon equation. In particular, the estimates of Theorem 3 have been used in Strauss [7] in the study of nonlinearities of the form $(V \star u^2)u$.

1. NEGATIVE RESULTS (THEOREM 1)

Proof of Theorem 1. Part (i)a. There are three estimates to be proved:

$$\frac{1}{q} \le \frac{1}{2}; \quad \frac{1}{2} - \frac{3}{2n} \le \frac{1}{q}; \quad \frac{n}{q} + \frac{1}{r} \ge \frac{n-2}{2}.$$

Fix $\ell > 1$ and let $k = (\ell^2 - 1)^{1/2}$. Let $f \in L^2$ be such that $f = 1$ in $|x| < 1$ and let $g(x) = \cos(kx_1)$ and $f_s(x) = s^{n/2} f(sx)$ for s small. If $u(x, t)$ denotes the solution with $u(x, 0) = 0$, $u_t(x, 0) = g(x) f_s(x)$, then $u_t(x, 0) = \cos(kx_1)$ for $|x| < s^{-1}$ and so in the domain of dependence,

$$T_t(g \cdot f_s)(x) = \ell^{-1} \sin \ell t \cos kx_1 \text{ for } A = \{x : |x| < s^{-1} - t\}.$$

Consider $\frac{\pi}{2} \le \ell t \le \frac{3\pi}{4}$ and $s^{-1} > 2t$. Then

$$\begin{aligned} \|t^{-\alpha}u\|_{q,r}^r &\ge s^{nr/2} \int_{\pi/4\ell}^{3\pi/4\ell} \left|\frac{\sin \ell t}{\ell t^\alpha}\right|^r \left(\int_A |\cos kx_1|^q\, dx\right)^{r/q} dx \\ &\ge C s^{nr/2} \left(\int_{|x| < (2s)^{-1}} |\cos kx_1|^q\, dx\right)^{r/q} \\ &\ge C s^{nr\left(\frac{1}{2} - \frac{1}{q}\right)} \to \infty \text{ as } s \to 0 \end{aligned}$$

whenever $q < 2$. Since $\|gf_s\|_2 \le \|f_s\|_2 = \|f\|_2$, this shows that $t^{-\alpha}T$ is unbounded for $\frac{1}{q} > \frac{1}{2}$.

Part (i)b. Let $f(x) = \sin \frac{1}{\sigma}(|x| - 1)$ for $1 \leq |x| \leq M$ and $f(x) = 0$ otherwise. Then $\|f\|_2 \leq CM^{n/2}$. We calculate $T_t f = K_t * f$ by using the explicit formulas for K_t in [1, pp. 693-694]. Suppose that $1 < t < 1 + s$ and $|x| \leq C\sigma$. In odd dimensions, the main term of $K_t * f(x)$ is

$$\left| t^{-\frac{n-1}{2}} \left(\frac{\partial}{\partial t}\right)^{\frac{n-1}{2}} X_t * f(x) \right| \geq Ct^{\frac{n-1}{2}} \sigma^{-\frac{n-3}{2}},$$

where X_t is the characteristic function of $\{|x| \leq t\}$. Similarly, in all dimensions

$$|T_t f(x)| \geq Ct^{\frac{n-1}{2}} \sigma^{-\frac{n-3}{2}} \qquad |x| \leq C\sigma$$

for $1 < t < 1 + \sigma$ and σ sufficiently small. Therefore,

$$\begin{aligned} \|t^{-\alpha} Tf\|_{q,r} &\geq \left(\int_1^M \left(t^{-\alpha} Ct^{\frac{n-1}{2}} \sigma^{\frac{n}{q} - \frac{n-3}{2}} \right)^r dt \right)^{1/r} \\ &\geq C\sigma^{\frac{n}{q} - \frac{n-3}{2}} \left(1 + M^{-\alpha + \frac{n-1}{2} + \frac{1}{r}} \right). \end{aligned}$$

By letting $\sigma \to 0$ or $M \to \infty$, this shows that $t^{-\alpha} T$ is unbounded if either

$$-\alpha + \frac{n-1}{2} + \frac{1}{r} - \frac{n}{2} > 0 \qquad \text{or} \qquad \frac{n}{q} - \frac{n-3}{2} < 0,$$

that is, $\alpha < \frac{1}{r} - \frac{1}{2}$ or $\frac{1}{q} < \frac{n-3}{2n}$.

Part (i)c. From Lemma 10, p. 434 of [3], there exist f, a, b, c_t such that

$$|T_t f_s(x)| \geq c_t s^{-1/2} \text{ for } t - b/s < |x| < t - a/s, \ s \geq 1,$$

where $f_s(x) = s^{n/2} f(sx)$. If we consider only $\frac{1}{2} \leq t \leq 1$, then c_t can be replaced by an absolute constant. Consider s fixed for the moment and let g be the characteristic function of the region $A = \{x : 1 - b_1/s \leq |x| \leq 1 - a_1/s\}$ and let $v(x, t) = f_s(x)$ if $t \in A_2 = \{t : 1 - b_2/s < t < 1 - a_2/s\}$ and $v(x, t) = 0$ otherwise. Choose the constants a_1, a_2, b_1, b_2 so that if $x \in A$ and $t \in A_2$, then $t - b/s < |x| < t - a/s$ and so we can use the estimate from [3]. $\langle\langle\, ,\, \rangle\rangle$ will denote the inner product over $\mathbf{R}^{n+1}$ and $\langle\, ,\, \rangle$ will be the inner product over $\mathbf{R}^n$. Since $T_t g = u(\cdot, t)$ is self-adjoint,

$$\langle\langle t^{-\alpha}Tg,\ v\rangle\rangle = \int_{A_2} t^{-\alpha}\langle T_t g,\ f_s\rangle\, dt$$

$$= \int_{A_2}\int_A t^{-\alpha}(T_t f_s)\, dx\, dt \geq Cs^{-2}s^{-1/2}.$$

Note above that if f is chosen to be real-valued, then $T_t f_s$ is always positive or always negative and so $T_t f_s$ could be replaced by $|T_t f_s|$. But because

$$\|v\|_{q',r'} = Cs^{\frac{n}{2}-\frac{n}{q'}-\frac{1}{r'}} \text{ and } \|g\|_2 = Cs^{-1/2},$$

we have

$$\|t^{-\alpha}T\| \geq Cs^{-\frac{n}{q}-\frac{1}{r}+\frac{n-2}{2}}.$$

Letting $s \to \infty$ shows that $t^{-\alpha}T$ is not bounded for $\dfrac{n}{q} + \dfrac{1}{r} < \dfrac{n-2}{2}$.

Part (ii)a. ($\alpha < 0$). From [3, p. 434] there exist f, a, b, c such that the measure of the set of x that satisfy

$$|T_t f(x)| \geq ct^{-n/2} \quad \text{and} \quad at < |x| < bt$$

is at least ct^n for $t \geq 1$.

Suppose g is the characteristic function of $A = \{x : s + as < |x| < s + bs\}$ and $v(x, t) = f(x)$ for $t \in A_2 = \{t : s + d < t < s + e\}$ and $v(x, t) = 0$ otherwise. Then

$$|\langle\langle t^{-\alpha}Tg,\ v\rangle\rangle| \geq \int_{A_2}\int_A |T_t f|\, t^{-\alpha}\, dt \geq Cs^{-\alpha+\frac{n}{2}}$$

and $\|v\|_{q',r'} = C$, $\|g\|_2 = Cs^{n/2}$. Therefore,

$$\|t^{-\alpha}Tg\| \geq Cs^{-\alpha}\|g\|_2 \text{ for all } s \geq 1.$$

Hence $t^{-\alpha}T$ is not bounded if $\alpha < 0$.

Part (ii)b. $\left(\alpha < \dfrac{n}{2q} - \dfrac{n}{4} + \dfrac{1}{r}\right)$. This estimate follows an argument of A. W. Knapp in [8]. Let Q be the rectangle centered at

$(0, 1) \in \mathbf{R}^n \times \mathbf{R}$ whose sides are parallel to the axes. The first n sides have length $\sqrt{\varepsilon}$ and the length parallel to the t-axis is ε. Let v be the inverse Fourier transform of the characteristic function of Q. Let f be the function on $\mathbf{R}^n$ whose Fourier transform is the characteristic function of the cube centered at the origin with sides of length $\sqrt{\varepsilon}$ parallel to the axes.

Then by the Plancherel theorem on $\mathbf{R}^{n+1}$,

$$\langle\langle Tf, v\rangle\rangle = \langle\langle \widehat{Tf}, \hat{v}\rangle\rangle = \int \chi_Q d\mu \geq C\varepsilon^{n/2},$$

where $d\mu$ is the measure on $S = \{(\xi, \tau) : |\xi|^2 + 1 = \tau^2\}$ given by $d\mu = (1 + |\xi|^2)^{-1} d\xi_1 \ldots d\xi_n$. But f and v are essentially equal to

$$\prod_{j=1}^{n} \frac{\sin(\sqrt{\varepsilon}\, x_j)}{x_j} \quad \text{and} \quad \frac{\sin \varepsilon t}{t} \prod_{j=1}^{n} \frac{\sin(\sqrt{\varepsilon}\, x_j)}{x_j}$$

and so $\|f\|_2 \leq C\varepsilon^{n/4}$ and $\|t^{\alpha}v\|_{q', r'} \leq C\varepsilon^{\frac{n}{2} - \frac{1}{q} + \frac{1}{r} - \alpha}$. Therefore,

$$\|t^{-\alpha}t\| \geq \langle\langle Tf, v\rangle\rangle / \|f\|_2 \|t^{\alpha}v\|_{q', r'} \geq C\varepsilon^{\frac{n}{4} - \frac{n}{2} + \frac{1}{q} - \frac{1}{r} + \alpha}.$$

By letting $\varepsilon \to 0$, we see that $t^{-\alpha}T$ is unbounded for

$$\frac{n}{4} - \frac{n}{2q} - \frac{1}{r} + \alpha < 0.$$

This completes the proof of Theorem 1.

2. THE POSITIVE RESULTS

The proofs of Theorems 2, 3, and 4 will be given together. If estimates can be obtained at the vertices P_0 to P_7, then the rest of the results follow by interpolation. Since the multipliers for T_t and $\sqrt{I - \Delta}\, T_t$, namely, $(1 + |\xi|^2)^{-1/2} \sin(t\sqrt{1 + |\xi|^2})$ and $\sin(t\sqrt{1 + |\xi|^2})$, are bounded, then T and $\sqrt{I - \Delta}\, T$ map L^2 to $L^{\infty}(L^2)$. Also, by Sobolev's theorem, T maps L^2 to $L^{\infty}(L^q)$, where $\frac{1}{q} = \frac{1}{2} - \frac{1}{n}$. This takes care of P_0 and P_1.

The estimates for P_5, P_6, P_7 follow from those of P_0, P_2, P_4, respectively, by Hölder's inequality, since

$$\|(1 + t)^{-\alpha - a}u\|_{q,1} \le C\|(1 + t)^{-\alpha}u\|_{q,r_1}$$

so long as $(1 + t)^{-a} \in L^{r_1'}(dt)$.

Observe now that Theorem 4 implies Theorem 3 by Sobolev's theorem. The result of Theorem 3 is stronger than that of Theorem 2 in Q because the former claims boundedness of $\alpha = \alpha_0 = 0$, whereas the latter is for $\alpha > \alpha_0$. Theorem 3 gives the estimates near P_2 and P_4, which complete the proof of Theorem 2. The estimates (2) on the boundary of Q follow by Sobolev's theorem from Strichartz' results [8].

Therefore, it remains only to prove Theorem 4 for

$$\left\{0 < \frac{1}{r} \le \frac{1}{2}, \ \frac{1}{2} - \frac{2}{(n - 1)r} < \frac{1}{q} < \frac{1}{2} - \frac{2}{nr}\right\}.$$

If $T_t f = u(\cdot, t) = K_t * f$ and $D_* = (I - \Delta)^{1/2}$, then it was shown in [3, Lemma 1, p. 421] that for any $0 \le a \le \frac{1}{2}$, if $\nu = a + (n - 1)/2$, then

$$\|D_*^{-\nu}K_t\|_{\text{BMO}} \le Ct^{-\nu} \quad \text{for } t \ge 1.$$

For small t,

$$\|D_*^{-\nu}K_t\|_{\text{BMO}} \le Ct^{-\frac{n-1}{2}} \quad \text{for } t \le 1.$$

Since $D_*^{-\nu}T_t$ is a bounded operator from H^1 to L^∞ with norm $\le Ct^{-\frac{n-1}{2}}(1 + |t|)^{-a}$, and D_*T_t is bounded from L^2 to L^2 with norm ≤ 1, then by interpolation, $D_*^{-b}T_t$ is bounded from $L^{q'}$ to L^q with norm $\le Ct^{\varepsilon-1}(1 + |t|)^{-\pi}$. The parameters are given by

$$\frac{1}{q'} = \frac{1}{2} + \frac{1 - \varepsilon}{n - 1}, \ b = \left(\frac{n + 1 + 2a}{n - 1}\right)(1 - \varepsilon) - 1, \ \pi = \frac{1 - \varepsilon}{n - 1}\,2a. \tag{4}$$

Let $k(t)$ be the norm of $D_*^{-b}T_t$ as an operator from $L^{q'}$ to L^q. Since $|k(t)| \le Ct^{\varepsilon-1}(1 + |t|)^{-\pi}$, then $k(t) \in L^R(dt)$ if

$$(1 - \varepsilon)R < 1 \quad \text{and} \quad R(1 + \pi - \varepsilon) > 1. \tag{5}$$

Let $d\nu$ be the measure on $|\xi|^2 + 1 = \tau^2$ determined by

$$d\nu = (1 + |\xi|^2)^{-(b+1)/2}\, d\xi_1 \ldots d\xi_n .$$

If $\widehat{d\nu}$ denotes the Fourier transform of $d\nu$ in both space and time variables, then $\widehat{d\nu}(x, t) = D_\star^{-b} K_t(x)$. The next restriction theorem follows an argument of P. Tomas [9].

Lemma. *Under the conditions* (4), (5), $0 \le a \le 1/2$, *and* $R = r/2$, *for all* $F \in L^{r'}(L^{q'})$ *we have*

$$\int |\hat{F}|^2 d\nu \le C\|F\|^2_{q', r'}.$$

Proof. To avoid confusion, ^ and * will denote the Fourier transform and convolution in the n space variables while ~ and # will denote the same operators in $(n + 1)$ variables.

First we show that the hypotheses imply that $\|\widetilde{d\nu} \# F\|_{q,r} \le C\|F\|_{q',r'}$ where the convolution is in $(n + 1)$ variables. Let $F_s(x) = F(x, s)$. Then

$$\|\widetilde{d\nu} \# F\|_{q,r} = \left\| \int D_\star^{-b} K_{t-s} * F_s\, ds \right\|_{q,r}$$
$$\le \left(\int \left(\int k(t - s) \|F_s\|_{q'}\, ds \right)^r dt \right)^{1/r} .$$

If $h(s) = \|F_s\|_{q'}$ then the last term is $\|k * h\|_r$, which is dominated by $\|k\|_R \|h\|_{r'}$ since

$$\frac{1}{r} = \frac{1}{R} + \frac{1}{r'} - 1 \quad \text{or} \quad R = \frac{r}{2}.$$

This shows that

$$\|\widetilde{d\nu} \# F\|_{q,r} \le C\|h\|_{r'} = C\|F\|_{q',r'}.$$

The proof of the lemma is completed with an application of the Plancherel theorem,

$$\int |\tilde{F}|^2 d\nu = \int \bar{\tilde{F}}\, \widetilde{F \# \widetilde{d\nu}} \le C\|F\|_{q',r'} \|F \# \widetilde{d\nu}\|_{q,r} \le C\|F\|^2_{q',r'}.$$

By duality, the lemma is equivalent to

$$\|\widetilde{F d\nu}\|_{q,r} \le C\left(\int |F|^2 d\nu \right)^{1/2}. \tag{6}$$

If $u(x, t)$ is the solution of the Klein-Gordon equation with initial velocity $f(x)$ and $\xi_\star = (1 + |\xi|^2)^{1/2}$, then $\hat{u}(\xi, t) = \xi_\star^{-1} \sin(t\xi_\star)\hat{f}(\xi)$. Define $F(\xi, \tau) = \hat{f}(\xi)$. Since the Fourier transform with respect to t of $\xi_\star^{-1-b} \sin(t\xi_\star)\hat{f}(\xi)$ is $\hat{f}d\nu = Fd\nu$, then $D_\star^{-b}u = \widetilde{Fd\nu}$. By the Plancherel theorem,

$$\int |F|^2 d\nu = \int |\hat{f}(\xi)|^2 (1 + |\xi|^2)^{-(b+1)/2} d\xi = \|D_\star^{-(b+1)/2} f\|_2^2.$$

Therefore, (6) becomes $\|D_\star^{-b}u\|_{q,r} \leq C\|D_\star^{-(b+1)/2}f\|_2$. Equivalently,

$$\|D_\star^{\gamma}u\|_{q,r} \leq C\|f\|_2,$$

where $\gamma = (1 - b)/2$.

Now to complete the proof of Theorem 4, we sort out the values of the various parameters. Use (4) to write b, ε, π in terms of n, q, a. Then

$$\gamma = \frac{1}{2}(1 - b) = 1 - \frac{1}{2}(n + 1 + 2a)\left(\frac{1}{2} - \frac{1}{q}\right). \tag{7}$$

The inequalities in (5) become

$$\frac{n - 1}{2}\left(\frac{1}{2} - \frac{1}{q}\right) < \frac{1}{r} < \frac{1}{2}(n - 1 + 2a)\left(\frac{1}{2} - \frac{1}{q}\right)$$

or

$$\frac{1}{2} - \frac{2}{(n - 1)r} < \frac{1}{q} < \frac{1}{2} - \frac{2}{r(n - 1 + 2a)}. \tag{8}$$

If $\frac{1}{q}$ equals the right-hand endpoint of (8), then $\gamma = \frac{1}{2} + \frac{1}{q} - \frac{1}{r}$. Thus (7) and (8) are equivalent to

$$\gamma < \frac{1}{2} + \frac{1}{q} - \frac{1}{r}, \quad \frac{1}{2} - \frac{2}{(n - 1)r} < \frac{1}{q} < \frac{1}{2} - \frac{2}{nr}.$$

Also $R = r/2 \geq 1$ implies $\frac{1}{r} < \frac{1}{2}$.

In the case $q = \infty$, we have shown that

$$\|D_\star^{\gamma}u\|_{\mathrm{BMO},r} \leq C\|f\|_2$$

for $\gamma < 1 - \frac{1}{2}(n + 1 + 2a)\left(\frac{1}{2} - \frac{1}{q}\right)$. But because $\|D_\star^{\gamma}u\|_{\infty,r} \leq C\|D_\star^{\gamma+\varepsilon}u\|_{\mathrm{BMO},r}$ for any $\varepsilon > 0$, we can replace BMO by L^∞ in all estimates above. This completes the proof of Theorem 2.

REFERENCES

1. Courant, R., and Hilbert, D. *Methods of Mathematical Physics*, vol. 2. New York: Interscience, 1962.

2. Littman, W. The wave operator and L^p norms. *J. Math. Mech.* 12 (1963):55-68.

3. Marshall, B.; Strauss, W.; and Wainger, S. $L^p - L^q$ estimates for the Klein-Gordon equation. *J. Math. Pures et Appl.* 59 (1980):417-440.

4. Peral, J. C. L^p estimates for the wave equation. *J. Func. Anal.* 36 (1980):114-145.

5. Segal, I. E. Space time decay for solutions of wave equations. *Adv. in Math.* 22 (1976):305-311.

6. Strauss, W. A. Everywhere defined wave operators. In *Nonlinear Evolution Equations*, M. G. Crandall, ed. New York: Academic Press, 1978, pp. 85-102.

7. Strauss, W. A. Nonlinear scattering theory at low energy, I. *J. Func. Anal.* 41 (1981):110-133. II. 43 (1981):281-293.

8. Strichartz, R. S. Restrictions of Fourier transforms to quadratic surfaces . . . . *Duke Math. J.* 44 (1977):705-714.

9. Tomas, P. A restriction theorem for the Fourier transform. *Bull. Amer. Math. Soc.* 81 (1975):477-478.

PART EIGHT

OTHER TOPICS RELATED TO HARMONIC ANALYSIS

SCATTERING AND INVERSE SCATTERING RELATED TO EVOLUTION EQUATIONS

R. Beals
R.R. Coifman
Yale University

It is now well known that many important nonlinear evolution equations are closely related to spectral problems for ordinary differential equations [1, 7, 8]. In principle, the former problems can be linearized and solved by solving associated scattering and inverse scattering problems for the ordinary differential equations. These latter problems are well understood for the second-order (Sturm-Liouville) case [3] and for certain first-order systems when the Gelfand-Levitan-Marcenko theory is applicable [1]. We summarize here a different approach to scattering and inverse scattering that applies to general first-order systems and contributes to the study of higher-order problems. We also discuss the question of the exact scope of the inverse scattering method for evolution equations and obtain a complete solution to a model case.

1. FIRST-ORDER SYSTEMS

Consider a system of the form

$$Du - Qu = f, \qquad D = d/dx, \tag{1}$$

where $Q = \left(q_{ij}(x)\right)$, $u = (u_1, \ldots, u_n)^t$, $f = (f_1, \ldots, f_n)^t$. We assume that the matrix-valued function Q satisfies

$$q_{ij} \in L^1(\mathbf{R}), \qquad q_{ii} \equiv 0, \tag{2}$$

and denote the set of such Q again by $L^1(\mathbf{R})$. Consider an associated spectral problem

$$Du - Qu = zJu, \qquad J = \text{diag}(\lambda_1, \ldots, \lambda_n), \tag{3}$$

where the λ_j are distinct complex constants. We seek a fundamental matrix solution $\Psi(x, z)$,

$$D\Psi = Q\Psi + zJ\Psi, \qquad \det \Psi \neq 0. \tag{4}$$

If $Q \equiv 0$ then (4) has the standard solution exp xzJ. If Q has compact support then (4) has a unique solution Ψ such that

$$\Psi(x, z) = e^{xzJ} \text{ if } x << 0. \tag{5}$$

Then for each $z \in C$ there is a matrix $S(z)$ such that

$$\Psi(x, z) = e^{xzJ} S(z) \text{ if } x >> 0. \tag{6}$$

Associated evolution equations may be described as follows. Let μ be a constant diagonal matrix and assume Q is a (matrix-valued) test function. Then there is an asymptotic expansion

$$\Psi(x, z)\mu\Psi(x, z)^{-1} \sim \sum_{k=0}^{\infty} \mu_{k-1}(x)z^{-k}, \quad z \to \infty, \tag{7}$$

where the coefficients depend on Q, $\mu_j = \mu_j(Q)$. *Formally*, if Q depends on a parameter and evolves according to

$$\frac{\partial Q}{\partial t} = [J, \mu_j(Q)], \tag{8}$$

then the matrix S of (6) evolves according to

$$\frac{\partial}{\partial t} S(z, t) = z^j[\mu, S(z, t)]. \tag{9}$$

Thus the initial value problem for the (generally) nonlinear problem (8) may be solved exactly if we can recover Q from the asymptotic data S. This formal procedure as described so far is well known, but there are several obstructions. First, S as defined will make sense in general only for compactly supported Q, and compact support is generally incompatible with (8). If $J + J^* = 0$, then $S(\xi)$ can be defined for all *real* ξ and all $Q \in L^1(\mathbf{R})$, but the map $Q \mapsto S$ is not injective, while for general J the map $Q \mapsto S(z)$ may not extend to $L^1(\mathbf{R})$ for any z.

A well-known transformation of the problem seeks Ψ of the form

$$\Psi(x, z) = \mathcal{M}(x, z)e^{xzJ}, \qquad \mathcal{M}(\cdot, z) \text{ bounded.} \tag{10}$$

Thus

$$D\mathcal{M} = Q\mathcal{M} + z[J, \mathcal{M}], \qquad \mathcal{M}(\cdot, z) \text{ bounded,} \tag{11}$$

and we normalize by requiring

$$\mathcal{M}(x, z) \to I \text{ as } x \to -\infty. \tag{12}$$

Let Σ be the following union of lines in $\mathbf{C}$:

$$\Sigma = \{z \in \mathbf{C} : \operatorname{Re}(z\lambda_j - z\lambda_k) = 0, \text{ some } j \neq k\}. \tag{13}$$

Theorem 1. Given $Q \in L^1(\mathbf{R})$, *the problem* (12), (13) *has a unique solution for every* $z \in \mathbf{C}\setminus\Sigma$ *except for a countable set. For* $x \in \mathbf{R}$, *the function* $\mathcal{M}(x, \cdot)$ *is meromorphic on* $\mathbf{C}\setminus\Sigma$ *with poles at* $\{z_i\}$. *Moreover,*

$$\lim_{z \to \infty} \mathcal{M}(x, z) = I. \tag{14}$$

Definition. $Q \in L^1(\mathbf{R})$ is *generic* is the associated $\mathcal{M}$ has only simple poles, different columns have different poles, and for each sector Ω of $\mathbf{C}\setminus\Sigma$, $\mathcal{M}$ has a continuous extension to $\overline{\Omega}\setminus\{z_1, \ldots, z_n\}$.

Proposition. The set of generic Q *is a dense open set in* $L^1(\mathbf{R})$.

If Q is generic, $\mathcal{M}(x, \cdot)$ may be considered as a matrix-valued distribution on $\mathbf{C}$. The distribution derivative $\partial\mathcal{M}/\partial\bar{z}$, which has support in $\Sigma \cup \{z_1, \ldots, z_N\}$, also satisfies (11). Formally, this implies that

$$D\left(\mathcal{M}^{-1}\frac{\partial \mathcal{M}}{\partial \bar{z}}\right) = \left[J, \mathcal{M}^{-1}\frac{\partial \mathcal{M}}{\partial \bar{z}}\right], \tag{15}$$

so that $\mathcal{M}^{-1}\partial\mathcal{M}/\partial\bar{z}$ must have the form

$$\mathcal{M}^{-1}\frac{\partial \mathcal{M}}{\partial \bar{z}} = W_x(z) \equiv e^{xzJ}W(z)e^{-xzJ}. \tag{16}$$

We interpret W as a density on Σ, and an isolated set of residues $W(z_j)$, and call it the *spectral transform* of Q.

Theorem 2. The map $Q \to W$ is injective.

In fact, if Q_1 and Q_2 yield W, then one may justify the following formal calculation and apply Liouville's theorem:

$$\frac{\partial}{\partial \bar{z}}(\mathcal{M}_1\mathcal{M}_2^{-1}) = \frac{\partial \mathcal{M}_1}{\partial \bar{z}}\mathcal{M}_2^{-1} - \mathcal{M}_1\mathcal{M}_2^{-1}\frac{\partial \mathcal{M}_2}{\partial \bar{z}}\mathcal{M}_2^{-1} \tag{17}$$

$$= \mathcal{M}_1\{W_x - W_x\}\mathcal{M}_2^{-1} \equiv 0.$$

To determine Q from $\mathcal{M}$, we use (14) and (16) to see that

$$\mathcal{M}(x, z) = I + \frac{1}{2\pi i}\int_{\mathbf{C}}(\zeta - z)^{-1}\frac{\partial \mathcal{M}}{\partial \bar{z}}(x, \zeta)\,d\zeta \wedge d\bar{\zeta} \tag{18}$$

so

$$\mathcal{M}(x, z) = I + \frac{1}{2\pi i}\int_{\Sigma}(\zeta - z)^{-1}\mathcal{M}(x, \zeta)W_x(\zeta)\,d\zeta \tag{19}$$

$$+ \Sigma\frac{1}{z_j - z}\mathcal{M}(x, \zeta_j)W_x(\zeta_j).$$

Then

$$Q(x) = \left[J, \frac{1}{2\pi i}\int_{\Sigma}\mathcal{M}(x, \zeta)W_x(\zeta)\,d\zeta + \Sigma\mathcal{M}(x, \zeta_j)W_x(\zeta_j)\right]. \tag{20}$$

Not all matrix-valued functions W on Σ can arise as spectral transforms, even those which are test functions with small norm: there are necessary and sufficient algebraic conditions. These conditions involve the shape of the matrix W on the various rays of Σ (relative to an appropriate ordering of the basis of eigenvectors of J) as well as the possibility of solving a certain finite linear system of equations. Here again there exists a class

S of generic scattering data for which the inverse problem is always solvable and which is dense in the class of all scattering transforms (see [2]). When $W \equiv 0$ on Σ, one sees from (19) and (20) that as functions of x, $\mathcal{M}$ and Q are rational functions of exponentials. Special cases give the familiar, soliton, multi-soliton, and breather potentials [6].

In the special case $n = 2$ and $J + J^* = 0$, the restriction of the spectral transform W to $\Sigma = R$ is essentially the "reduced scattering data" of [1]. In all cases it plays the same role in evolution as the ill-defined function S. Thus the evolution (8) corresponds to

$$\frac{\partial W}{\partial t} = z^j[\mu, W], \tag{21}$$

which has the solution

$$W(z, t) = e^{tz^j\mu} W(z, 0)e^{-tz^j\mu}. \tag{22}$$

It follows that a necessary condition for stability of (8) in any reasonable Banach space is the boundedness of the exponentials occurring in (22), for $z \in \Sigma$, $z \to \infty$, when W satisfies the algebraic constraints mentioned above. [These constraints are preserved by (22).]

2. HIGHER-ORDER EQUATIONS

Consider an eigenvalue problem

$$\left(D^n + \sum_{j=0}^{n-2} q_j(x)D^j\right)u = z^n u. \tag{23}$$

This can at least formally be reduced in various ways to a first-order system of the type above, with $J = \operatorname{diag}(\alpha, \alpha^2, \ldots, \alpha^{n-1}, 1)$, $\alpha = \exp \frac{1}{n} 2\pi i$.

Through such a reduction or by a more direct argument, one is led to construct eigenfunctions

$$\Psi_j(x, z) = \mathcal{M}_j(x, z)\exp(\alpha^j xz) \tag{24}$$

via an integral equation for $\mathcal{M} = (\mathcal{M}_1, \mathcal{M}_2, \ldots, \mathcal{M}_n)$:

$$\mathcal{M}(x, z) = (1, 1, \ldots, 1) + \frac{1}{2\pi i}\int_\Sigma (\zeta - z)^{-1}\mathcal{M}(x, \zeta)W_x(\zeta)d\zeta \tag{25}$$

$$+ \text{ (sum over singularities)}$$

where $W_x(z)$ has the form (16) and W is subject to the symmetry requirement

$$W_{j+1,k+1}(z) = W_{j,k}(\alpha z) \tag{26}$$

where we take $W_{n+1,k} = W_{1,k}$, $W_{j,n+1} = W_{j,1}$. For W a test function of small norm and, for simplicity, no singularities, the solution of (25) is unique and gives eigenfunctions for (23), where, of course, the potentials q_j depend on W. The direct problem in this case is more delicate than in the case of a first-order system because of the difficulties in constructing normalized eigenfunctions as the eigenvalue parameter tends to zero. It appears, however, that the picture is essentially the same as in the case of a first-order system.

3. THE SCOPE OF THE INVERSE SCATTERING METHOD

The results described here are joint work with B. Dahlberg. A considerable amount is known about inverse scattering in more than one dimension, but there is as yet no systematic way to associate tractable evolution equations. The possibility of doing so in one dimension may be due to the particularly simple way in which eigenfunctions may be generated. For example, for any small test function w, the solutions of the integral equations

$$\Psi(x, z) = e^{ixz}\left\{1 + \int_{\mathbf{R}} \frac{e^{ix\zeta}\Psi(x, \zeta)}{z + \zeta} w(\zeta)d\zeta\right\}$$

are eigenfunctions of

$$(D^2 + z^2)\Psi = q\Psi, \qquad D = 1/i\ d/dx, \tag{28}$$

where q depends on w. This is a version of (25) when $n = 2$, and the connection between evolutions of w and evolutions of q can be established directly from (27).

As a general model of this in the scalar case we can ask for a function or measure Φ and a function Q of two variables such that for all small test functions w, the solutions of

$$\Psi(x, z) = e^{ixz}\left\{1 + \int \Phi(x, z, \zeta)\Psi(x, \zeta)w(\zeta)d\zeta\right\} \tag{29}$$

are the solutions of the pseudodifferential equation

$$Q(D, z)\Psi(x, \zeta) = q(x)\Psi(x, \zeta) \tag{30}$$

with q depending on w. Under some regularity assumptions, the problem reduces to the case when $\Phi(x, z, \cdot)$ is a density of the form $e^{ix\zeta}p(z, \zeta)$ and the pair Q, p satisfy the functional equation

$$[Q(\xi + \zeta + z, z) - Q(\xi, \zeta)]p(z, \zeta) \equiv r(\xi, \zeta) \tag{31}$$

for some function r.

The functional equation (31) has three types of solution, which may be normalized as follows. The first is a trivial family in which $Q(\xi, z)$ is a function of $\xi - z$ and one may, therefore, take $p \equiv 1$. This leads, for example, to the Cole-Hopf transformation and has Burger's equation as an associated evolution. The second is the Sturm-Liouville case

$$Q(\xi, z) = \xi^2 - z^2, \; p(z, \zeta) = (z + \zeta)^{-1}, \tag{32}$$

which has *KdV* as an associated evolution.

Finally, there is a remarkable one-parameter family of solutions giving pseudodifferential operators $Q_\delta(D, z)$, $\delta \neq 0$, where

$$Q_\delta(\xi, z) = \xi - z \coth 2\delta z + z \operatorname{csch} 2\delta z \cdot e^{-2\delta\xi}. \tag{33}$$

An associated evolution is

$$\frac{\partial u}{\partial t} + \delta^{-1} \frac{\partial u}{\partial x} + 2u \frac{\partial u}{\partial x} = 2(\cosh \delta D)u.$$

Despite the artificial appearance of this last family of pseudo-differential equations, it occurs in the physics literature in connection with waves in a finite depth fluid [4], and the associated eigenvalue problem (30), (33) has been treated in [5].

REFERENCES

1. Ablowitz, M. J.; Kaup, D. J.: Newell, A. C.; and Segur, H. The inverse scattering transform—Fourier analysis for nonlinear problems. *Studies Appl. Math.* 53 (1974):249-315.

2. Beals, R., and Coifman, R. R. Scattering and inverse scattering for first-order systems. To appear.

3. Deift, P., and Trubowitz, E. Inverse scattering on the line. *Comm. Pure Appl. Math.* 32 (1979):121-253.

4. Joseph, R. I. Solitary waves in a finite depth fluid. *J. Physics* A 10 (1977):L225-L227.

5. Kodama, Y.; Satsuma, J.; and Ablowitz, M. J. Nonlinear intermediate long-wave equation: Analysis and method of solution. *Phys. Rev. Letters* 46 (1981):687-690.

6. Lamb, G. L., Jr. *Elements of Soliton Theory.* New York: Wiley, 1980.

7. Newell, A. C. The inverse scattering transform. In Bullough and Caudrey, eds., *Solitons.* New York: Springer-Verlag, 1980, pp. 177-242.

8. Zakharov, V. E. The inverse scattering method. In Bullough and Caudrey, eds., *Solitons.* New York: Springer-Verlag, 1980, pp. 243-286.

NORM INEQUALITIES FOR HOLOMORPHIC FUNCTIONS OF SEVERAL COMPLEX VARIABLES

Jacob Burbea
University of Pittsburgh

1. INTRODUCTION

We consider the n-dimensional complex space $\mathbf{C}^n$ with the usual inner product

$$z \cdot \bar{\zeta} = z_1\bar{\zeta}_1 + \cdots + z_n\bar{\zeta}_n; \qquad |z|^2 = z \cdot \bar{z}.$$

The unit ball Δ_n of $\mathbf{C}^n$ is then the set of all $z \in \mathbf{C}^n$ with $|z| < 1$. The class of all holomorphic functions with domain Δ_n shall be denoted by $H(\Delta_n)$. It is a linear space over $\mathbf{C}$. The following standard notation is employed: if $\alpha = (\alpha_1, \ldots, \alpha_n) \in \mathbf{Z}_+^n$ (that is, an n-tuple of nonnegative integers), then $|\alpha| = \alpha_1 + \cdots + \alpha_n$ and $\alpha! = \alpha_1! \ldots \alpha_n!$. The notation $a_\alpha z^\alpha$ is an abbreviation for the monomial

$$a_{\alpha_1 \ldots \alpha_n} z_1^{\alpha_1} \ldots z_n^{\alpha_n}; \qquad a_{\alpha_1 \ldots \alpha_n} \in \mathbf{C},\ z \in \mathbf{C}^n.$$

Any $f \in H(\Delta_n)$ can be represented as

$$f(z) = \sum_{\alpha \geq 0} a_\alpha z^\alpha,\ z \in \Delta_n,$$

where the power series converges locally uniformly in Δ_n. We may therefore regard $H(\Delta_n)$ as the span of all z^α, $\alpha \geq 0$, in the uniform topology of Δ_n.

We shall introduce a "derivation-operator" $\partial_n : H(\Delta_n) \to H(\Delta_n)$ as follows: we require ∂_n to be linear on $H(\Delta_n)$, and we specify its action on z^α, $\alpha \geq 0$. If $\alpha = 0$, we let

$$\partial_n(z^0) = \partial_n(1) = 0. \tag{1}$$

On the other hand, if $\alpha \geq 1$, we let $\alpha_{k_1}, \ldots, \alpha_{k_m}$, $1 \leq k_1 \leq \cdots \leq k_m \leq n$ $(1 \leq m \leq n)$ be the nonzero (≥ 1) components of α. Thus $z^\alpha = z_{k_1}^{\alpha_{k_1}} \ldots z_{k_m}^{\alpha_{k_m}}$. We then define

$$\partial_n(z^\alpha) = \sqrt{\alpha_{k_1} \ldots \alpha_{k_m} (|\alpha|)_{n-m+1}}\, z_{k_1}^{\alpha_{k_1}-1} \ldots z_{k_m}^{\alpha_{k_m}-1}, \tag{2}$$

where, for a real $q > 0$,

$$(q)_0 = 1; \ (q)_m = q(q+1) \ldots (q+m-1), \ m \geq 1.$$

This derivation-operator ∂_n is the n-dimensional generalization of the usual derivative $\partial \equiv \partial_1$ on $H(\Delta_1)$. In fact, for $f \in H(\Delta_1)$, $\partial f = f'$.

By $d\sigma_n(z)$ we denote the Lebesgue measure of $\mathbf{C}^n \equiv \mathbf{R}^{2n}$ and by $ds_n(z)$ we denote the Euclidean measure on the boundary $\partial\Delta_n$ of Δ_n. For a real $q > 0$, we introduce the "(n, q)-Dirichlet space"

$$D_{n,q} = \left\{ f \in H(\Delta_n) : f(0) = 0, \ \|f\|_{D_{n,q}} < \infty \right\},$$

where

$$\|f\|^2_{D_{n,q}} = \frac{1}{\pi^n} \frac{2}{(n+1)q} \int_{\Delta_n} |\partial_n f(z)|^2 d\sigma_n(z). \tag{3}$$

It will be verified later that $D_{n,q}$ is a functional Hilbert space of holomorphic functions in Δ_n with the reproducing kernel

$$k_{n,q}(z, \overline{\zeta}) = -\frac{(n+1)q}{2} \log(1 - z \cdot \overline{\zeta}); \qquad z, \zeta \in \Delta_n. \tag{4}$$

Another related functional Hilbert space is the so-called (generalized) "(n, q)-Hardy space" $H_{n,q}$. This is the space of all holomorphic functions g in Δ_n normed by $\|g\|_{n,q}$. The norm $\|g\|_{n,q}$ for arbitrary $q > 0$ is represented by means of a suitable weighted ℓ_2-norm that will be described later (see also Burbea [1] for the case $n = 1$). When, however, $q > \frac{2n}{n+1}$, this norm can also be realized as the following weighted L_2-norm:

$$\|g\|^2_{n,\,q} = \frac{1}{\pi^n}\left(\frac{n+1}{2}q - n\right)_n \int_{\Delta_n} |g(z)|^2 (1 - |z|^2)^{\frac{n+1}{2}(q-2)}\, d\sigma_n(z). \tag{5}$$

On the other hand, the limiting case of $q = \dfrac{2n}{n+1}$ is given by

$$\|g\|^2_{n,\,\frac{2n}{n+1}} = \frac{(n-1)!}{2\pi n} \int_{\partial\Delta_n} |g(z)|^2 ds_n(z), \tag{6}$$

where in the last integral, g stands for the nontangential boundary values of the holomorphic function $g(z)$ in Δ_n. Thus, $H_{n,\,\frac{2n}{n+1}}$ is the usual "Hardy-Szegö space," $H_{n,\,2}$ is the customary "Bergman space," and $H_{n,\,q}$, $q > \dfrac{2n}{n+1}$, is the "Bergman-Selberg space" (see [3]). The reproducing kernel of $H_{n,\,q}$, $q > 0$, is, as we also shall see later,

$$K_{n,\,q}(z, \bar{\zeta}) = (1 - z \cdot \bar{\zeta})^{-\frac{(n+1)q}{2}}; \qquad z, \zeta \in \Delta_n,$$

and we note the relationship $K_{n,\,q}(z, \bar{\zeta}) = \exp[k_{n,\,q}(z, \bar{\zeta})]$, $z, \zeta \in \Delta_n$.

The main purpose of this paper is in the establishment of the following theorem.

Theorem. Let $q > 0$ *and let* $f \in D_{n,\,q}$. *Then*

$$\|\exp f\|^2_{n,\,q} \leq \exp\|f\|^2_{D_{n,\,q}}.$$

Equality holds if and only if f *is of the form* $f = k_{n,\,q}(\cdot, \bar{\zeta})$ *for some* $\zeta \in \Delta_n$.

The inequality and the sufficient condition in the equality statement of this theorem constitute a simple consequence of a more abstract theorem in the general theory of reproducing kernels (see Burbea [1] for details). The essence, therefore, of this theorem is in the necessity of the equality statement. In the special case that $n = 1$ and $q \geq 1$, a proof of the theorem was given by Saitoh [2]. His proof, however, is difficult and rather involved. In [1], Burbea has extended the result of Saitoh [2] to cover the case of $0 < q < 1$ with a proof that is substantially easier and shorter than that of [2]. The method of proof in Burbea [1] conceals in it the main ingredients for establishing the present theorem. To do so, however, it is found necessary to derive certain multinomial and combinatorial identities that are of some interest in their own right. Apart from that, the proof given in this paper is tailored after that of Burbea [1] for the case $n = 1$.

We should also remark that it is possible to extend the theorem to cover the case where Δ_n is replaced by one of the classical Cartan domains once some minor modifications in the present proof are made. However, we shall not pursue this point here.

2. WEIGHTED l_2-SPACES

We first note the following two elementary identities: For $q > 0$ and $\alpha, \beta \in \mathbf{Z}_+^n$, we have

$$\frac{1}{\pi^n}\int_{\Delta_n} z^\alpha \bar{z}^\beta (1 - |z|^2)^{q-1} d\sigma_n(z) = \frac{\alpha!}{(q)_{|\alpha|+n}}\delta_{\alpha\beta} \tag{8}$$

and

$$\frac{1}{2\pi^n}\int_{\partial\Delta_n} z^\alpha \bar{z}^\beta ds_n(z) = \frac{\alpha!}{(1)_{|\alpha|+n}}\delta_{\alpha\beta}, \tag{9}$$

where $\delta_{\alpha\beta}$ is the Kronecker delta. With the aid of (8) and the derivation operator of (1) and (2), the space $D_{n,q}$ can be realized as the weighted ℓ_2-space described below. Let $f \in D_{n,q}$ so that

$$f(z) = \sum_{\alpha \geq 1} a_\alpha z^\alpha, \quad z \in \Delta_n.$$

By (1)-(3) and (8),

$$\begin{aligned}
\|f\|^2_{D_{n,q}} &= \frac{1}{\pi^n}\frac{2}{(n+1)q}\sum_{\alpha\geq 1}|a_\alpha|^2 \alpha_{k_1}\cdots\alpha_{k_m}|\alpha|_{n-m+1}\int_{\Delta_n} z^{\alpha-1}\bar{z}^{\alpha-1}d\sigma_n(z) \\
&= \frac{2}{(n+1)q}\sum_{\alpha\geq 1}|a_\alpha|^2 \alpha_{k_1}\cdots\alpha_{k_m}|\alpha|_{n-m+1}\frac{(\alpha-1)!}{(1)_{n+|\alpha|-m}} \\
&= \frac{2}{(n+1)q}\sum_{\alpha\geq 1}|a_\alpha|^2 \alpha!\,\frac{(|\alpha|)_{n-m+1}}{(|\alpha|+n-m)!},
\end{aligned}$$

where here $\alpha_{k_1} \ldots \alpha_{k_m}$, $1 \le k_1 \le \cdots \le k_m \le n$ $(1 \le m \le n)$ are the nonzero components of $\alpha \ge 1$. It follows that

$$\|f\|^2_{D_{n,q}} = \frac{2}{(n+1)q} \sum_{\alpha \ge 1} |a_\alpha|^2 \frac{\alpha!}{(|\alpha|-1)!}. \tag{10}$$

From (10) we deduce that $D_{n,q}$ is a functional Hilbert space of holomorphic functions f in Δ_n with $f(0) = 0$. The reproducing kernel of this space is

$$\begin{aligned} k_{n,q}(z, \overline{\zeta}) &= \frac{(n+1)q}{2} \sum_{\alpha \ge 1} \frac{(|\alpha|-1)!}{\alpha!} z^\alpha \overline{\zeta}^\alpha \\ &= \frac{(n+1)q}{2} \sum_{\alpha \ge 1} \frac{1}{\alpha!} \left(\sum_{m=|\alpha|} \frac{1}{m} m!\, z^\alpha \overline{\zeta}^\alpha \right) \\ &= \frac{(n+1)q}{2} \sum_{m=1}^{\infty} \frac{1}{m} \left(\sum_{|\alpha|=m} \frac{m!}{\alpha!} z^\alpha \overline{\zeta}^\alpha \right) \\ &= \frac{(n+1)q}{2} \sum_{m=1}^{\infty} \frac{1}{m} (z \cdot \overline{\zeta})^m = -\frac{(n+1)q}{2} \log(1 - z \cdot \overline{\zeta}). \end{aligned}$$

This is in agreement with (4). In a similar fashion, the space $H_{n,q}$ can be realized as follows. Let $g \in H(\Delta_n)$ with

$$g(z) = \sum_{\alpha \ge 0} b_\alpha z^\alpha, \quad z \in \Delta_n.$$

For $q > 0$, we write

$$Q_n \equiv \frac{(n+1)q}{2} \tag{11}$$

and

$$\|g\|^2_{n,q} \equiv \sum_{\alpha > 0} |b_\alpha|^2 \frac{\alpha!}{(Q_n)_{|\alpha|}}. \tag{12}$$

The space $H_{n,q}$ can then be defined as

$$H_{n,q} \equiv \left\{ g \in H(\Delta_n) : \|g\|_{n,q} < \infty \right\},$$

and therefore $H_{n,q}$ is a functional Hilbert space of holomorphic functions in Δ_n whose reproducing kernel is

$$k_{n,q}(z, \overline{\zeta}) = \sum_{\alpha \ge 0} \frac{(Q_n)_{|\alpha|}}{\alpha!} z^\alpha \overline{\zeta}^\alpha = \sum_{m=0}^{\infty} \frac{(Q_n)m}{m!} \left(\sum_{|\alpha|=m} \frac{m!}{\alpha!} z^\alpha \overline{\zeta}^\alpha \right)$$

$$= \sum_{m=0}^{\infty} \frac{(Q_n)m}{m!}(z \cdot \overline{\zeta})^m = (1 - z \cdot \overline{\zeta})^{-Q_n},$$

which is in agreement with (7). By virtue of identities (8) and (9), it is easily verified that for $Q_n \geq n$, (11) and (12) reduce to (5) and (6).

3. SOME COMBINATORIAL IDENTITIES

For a real $q > 0$ and $m \in \mathbf{Z}_+$, we use the notation

$$\frac{(q)_m}{m!} = \frac{\Gamma(q + m)}{\Gamma(m + 1)\Gamma(q)} = \binom{q + m - 1}{m} = (-1)^m\binom{-q}{m}, \tag{13}$$

and thus, for $m = 0$,

$$\binom{q - 1}{0} = \binom{-q}{0} = 1.$$

Formula (13) may, of course, be extended to include the case where q is any real number. A basic combinatorial identity is the "Vandermonde convolution formula"

$$\binom{q}{m} = \sum_{k=0}^{m}\binom{q - p}{m - k}\binom{p}{k}; \quad q,\ p \in \mathbf{R},\ m \in \mathbf{Z}_+. \tag{14}$$

This formula is easily derived from the identity $(1 + x)^q = (1 + x)^{q-p}(1 + x)^p$. From (13) and (14) we deduce that

$$\binom{q}{m} = \binom{q - m + 1 + m - 1}{m} = (-1)^m\binom{-q + m - 1}{m}$$

$$= (-1)^m \sum_{k=0}^{m}\binom{-q + m - 1 + p}{m - k}\binom{-p}{k}$$

$$= (-1)^m \sum_{k=0}^{m}(-1)^{m-k}\binom{q - m + 1 - p + m - k - 1}{m - k}(-1)^k\binom{p + k - 1}{k},$$

and therefore,

$$\binom{q}{m} = \sum_{k=0}^{m}\binom{q - p - k}{m - k}\binom{p + k - 1}{k}; \qquad q,\ p \in \mathbf{R},\ m \in \mathbf{Z}_+. \tag{15}$$

Another consequence of (14) is the identity

$$\binom{q+m}{m} = \sum_{k=0}^{m} \binom{q+k-1}{k}; \quad q \in \mathbf{R},\ m \in \mathbf{Z}_{+}. \qquad (16)$$

This is obtained as follows. By (13) and (14), with $p = -q$,

$$\binom{q+m}{m} = (-1)^m \binom{-q-1}{m} = (-1)^m \sum_{k=0}^{m} \binom{-q-1+q}{m-k}\binom{-q}{k}$$

$$= (-1)^m \sum_{k=0}^{m} \binom{-1}{m-k}\binom{-q}{k}$$

$$= (-1)^m \sum_{k=0}^{m} (-1)^{m-k} \binom{1+m-k-1}{m-k} (-1)^k \binom{q+k-1}{k}$$

and (16) follows.

Identity (16) admits a multinomial extension described below. For $\alpha, \beta \in \mathbf{Z}_{+}^{n}$ with $\beta \leq \alpha$, we define

$$\binom{\alpha}{\beta} \equiv \binom{\alpha_1}{\beta_1}\binom{\alpha_2}{\beta_2} \cdots \binom{\alpha_n}{\beta_n} = \frac{\alpha!}{\beta!(\alpha-\beta)!} \qquad (17)$$

Then

$$\binom{q+|\alpha|}{|\alpha|} = \sum_{0 \leq \beta \leq \alpha} \binom{\alpha}{\beta}\binom{|\alpha|}{|\beta|}^{-1} \binom{q+|\beta|-1}{|\beta|}; \quad q \in \mathbf{R},\ \alpha \in \mathbf{Z}_{+}^{n}. \qquad (18)$$

For $n = 1$, (18) clearly reduces to (16), while for $n > 1$, (18) may be established as follows. From (13) and (17), we have

$$\sum_{0 \leq \beta \leq \alpha} \binom{\alpha}{\beta}\binom{|\alpha|}{|\beta|}^{-1} \binom{q+|\beta|-1}{|\beta|} = \frac{\alpha!}{|\alpha|!} \sum_{0 \leq \beta \leq \alpha} \frac{|\beta|!(|\alpha|-|\beta|)!}{\beta!(\alpha-\beta)!} \binom{q+|\beta|-1}{|\beta|}$$

$$= \frac{\alpha!}{|\alpha|!} \sum_{\beta_1=0}^{\alpha_1} \binom{q+\beta_1-1}{\beta_1} \sum_{\beta_2=0}^{\alpha_2} \binom{\alpha_1-\beta_1+\alpha_2-\beta_2}{\alpha_2-\beta_2}\binom{q+\beta_1+\beta_2-1}{\beta_2} \cdots$$

$$\sum_{\beta_n=0}^{\alpha_n} \binom{\alpha_1-\beta_1+\cdots+\alpha_n-\beta_n}{\alpha_n-\beta_n}\binom{q+\beta_1+\cdots+\beta_n-1}{\beta_n},$$

which by a successive application of (15) becomes

$$= \frac{\alpha!}{|\alpha|!} \sum_{\beta_1=0}^{\alpha_1} \binom{q+\beta_1-1}{\beta_1}\binom{q+\alpha_1+\alpha_2}{\alpha_2} \cdots \binom{q+\alpha_1+\alpha_2+\cdots+\alpha_n}{\alpha_n}$$

$$= \frac{\alpha_1!}{|\alpha|!} \frac{\Gamma(q+\alpha_1+\cdots+\alpha_n+1)}{\Gamma(q+\alpha_1+1)} \sum_{\beta_1=0}^{\alpha_1} \binom{q+\beta_1-1}{\beta_1}.$$

Finally, using identities (13) and (16), we see that (18) follows.

4. AN INEQUALITY

Let $\alpha \in \mathbf{Z}_+^n$, $\alpha \neq 0$, and let α_k, $1 \leq k \leq n$, be a nonzero component of α. We then write

$$\alpha - 1_k = (\alpha_1, \ldots, \alpha_k - 1, \ldots, \alpha_n);\ \alpha_k \geq 1,\ 1 \leq k \leq n.$$

Let

$$f(z) = \sum_{\alpha \geq 1} a_\alpha z^\alpha,\ z \in \Delta_n, \tag{19}$$

and let

$$g(z) = \exp[f(z)] = \sum_{\alpha \geq 0} b_\alpha z^\alpha \tag{20}$$

be its exponential transform. For an arbitrary k, $1 \leq k \leq n$, (20) shows that

$$\frac{\partial}{\partial z_k} g(z) = g(z) \frac{\partial}{\partial z_k} f(z)$$

and therefore, by (19) and (20),

$$\sum_{\alpha \geq 0} \alpha_k b_\alpha z^{\alpha - 1k} = \left(\sum_{\alpha \geq 0} b_\alpha z^\alpha\right)\left(\sum_{\alpha \geq 0} \alpha_k a_\alpha z^{\alpha - 1}k\right)$$

$$= \sum_{\substack{\alpha \geq 0 \\ \alpha_k \geq 1}} \left(\sum_{\beta \leq \alpha} \beta_k a_\beta b_{\alpha - \beta}\right) z^{\alpha - 1}k.$$

This shows that

$$b_\alpha = \frac{1}{\alpha_k} \sum_{\beta \leq \alpha} \beta_k a_\beta b_{\alpha - \beta};\ \alpha_k \geq 1\ (b_0 = 1). \tag{21}$$

In view of (10)-(12) and (19) and (20), the theorem is equivalent to the following sharp inequality.

Lemma. Let the notation of (19) *and* (20) *apply and let* $q > 0$. *Define*

$$Q_n \equiv \frac{(n + 1)q}{2}.$$

Then

$$\sum_{\alpha \geq 0} \frac{\alpha!}{(Q_n)_{|\alpha|}} |b_\alpha|^2 \leq \exp\left\{\frac{1}{Q_n} \sum_{\alpha \geq 1} \frac{\alpha!}{(|\alpha| - 1)!} |a_\alpha|^2\right\} \tag{22}$$

if the right-hand side is finite. Equality holds if and only if

$$a_\alpha = Q_n \frac{(|\alpha| - 1)!}{\alpha!} \bar{\zeta}^\alpha, \ \alpha \geq 1; \ b_\alpha = \frac{(Q_n)_{|\alpha|}}{\alpha!} \bar{\zeta}^\alpha, \quad \alpha \geq 0,$$

for some $\zeta \in \Delta_n$.

Proof. For $r = (r_1, \ldots, r_n) \in [0, 1)^n$, define

$$A(r) = \frac{1}{Q_n} \sum_{\alpha \geq 1} \frac{\alpha!}{(|\alpha| - 1)!} |a_\alpha|^2 r^\alpha, \ B(r) = \sum_{\alpha \geq 0} \frac{\alpha!}{(Q_n)_{|\alpha|}} |b_\alpha|^2 r^\alpha.$$

Let k, $1 \leq k \leq n$, be fixed. By (21) and the Cauchy-Schwarz inequality,

$$|b_\alpha|^2 = \frac{1}{\alpha_k^2} \left| \sum_{\substack{\beta \leq \alpha \\ \beta_k \geq 1}} \beta_k \left(\frac{\beta!(\alpha - \beta)!}{(|\beta| - 1)!\beta_k (Q_n)_{|\alpha - \beta|}} \right)^{1/2} \right.$$
$$\left. \times\ a_\beta b_{\alpha - \beta} \left(\frac{(|\beta| - 1)!\beta_k (Q_n)_{|\alpha - \beta|}}{\beta!(\alpha - \beta)!} \right)^{1/2} \right|$$
$$\leq \frac{1}{\alpha_k^2} \left(\sum_{\substack{\beta \leq \alpha \\ \beta_k \geq 1}} \frac{\beta_k \beta!}{(|\beta| - 1)!} |a_\beta|^2 \frac{(\alpha - \beta)!}{(Q_n)_{|\alpha - \beta|}} |b_{\alpha - \beta}|^2 \right)$$
$$\times \left(\sum_{\substack{\beta \leq \alpha \\ \beta_k \geq 1}} \frac{\beta_k (|\beta| - 1)!}{\beta!} \frac{(Q_n)_{|\alpha - \beta|}}{(\alpha - \beta)!} \right).$$

On the other hand, by changing to variables

$$\gamma = \alpha - 1_k \geq 0, \quad \delta = \gamma - \beta,$$

we obtain

$$\sum_{\substack{\beta \leq \alpha \\ \beta_k \geq 1}} \frac{\beta_k (|\beta| - 1)!}{\beta!} \frac{(Q_n)_{|\alpha - \beta|}}{(\alpha - \beta)!} = \sum_{0 \leq \delta \leq \gamma} \frac{(|\gamma| - |\delta|)!}{\delta!(\gamma - \delta)!} (Q_n)_{|\delta|},$$

which, by virtue of (13), (17), and (18), is equal to

$$\frac{|\gamma|!}{\gamma!} \sum_{0 \leq \delta \leq \gamma} \binom{\gamma}{\delta} \binom{|\gamma|}{|\delta|}^{-1} \binom{Q_n + |\delta| - 1}{|\delta|}$$
$$= \frac{|\gamma|!}{\gamma!} \binom{Q_n + |\gamma|}{|\gamma|} = \frac{1}{\gamma!} (Q_n + 1)_{|\gamma|}.$$

Therefore,

$$\sum_{\beta \le \alpha} \frac{\beta_k(|\beta| - 1)!}{\beta!} \frac{(Q_n)_{|\alpha-\beta|}}{(\alpha - \beta)!} = \frac{\alpha_k}{\alpha!} \frac{(Q_n)_{|\alpha|}}{Q_n} \tag{23}$$

and

$$\frac{\alpha_k \alpha!}{(Q_n)_{|\alpha|}} |b_\alpha|^2 \le \frac{1}{Q_n} \sum_{\substack{\beta \le \alpha \\ \beta_k \ge 1}} \frac{\beta_k \beta!}{(|\beta| - 1)!} |a_\beta|^2 \frac{(\alpha - \beta)!}{(Q_n)_{|\alpha-\beta|}} |b_{\alpha-\beta}|^2. \tag{24}$$

The last inequality shows that

$$\frac{\partial}{\partial r_k} B(r) \le B(r) \frac{\partial}{\partial r_k} A(r)$$

or that

$$\frac{\partial}{\partial r_k} \log B(r) \le \frac{\partial}{\partial r_k} A(r); \qquad 1 \le k \le n, \tag{25}$$

as formal power series. Note that if the right-hand side of (22) is finite, then $A(r)$ converges for $r \in [0, 1)^n$, and therefore, by (25), $B(r)$ also converges for $r \in [0, 1)^n$. Let $\rho \in [0, 1)$ and choose a path $r_k = r_k(t)$, where $r(t)$ is increasing in $t \in [0, 1]$ and $r_k(0) = 0$, $r_k(1) = \rho$ for $1 \le k \le n$. Since $A(0, \ldots, 0) = 0$ and $B(0, \ldots, 0) = 1$, it follows from (25) that

$$\begin{aligned} \log B(\rho, \ldots, \rho) &= \int_0^1 \frac{d}{dt} \log B[r(t)]dt \\ &\le \int_0^1 \frac{d}{dt} A[r(t)]dt = A(\rho, \ldots, \rho) \end{aligned}$$

and (22) follows by letting $\rho \to 1 - 0$. In view of (25), equality in (22) holds if and only if for any $r \in [0, 1)^n$ and any k, $1 \le k \le n$,

$$\frac{\partial}{\partial r_k} B(r) = B(r) \frac{\partial}{\partial r_k} A(r),$$

which is equivalent to having equality in (24) for every component $\alpha_k \ge 1$. This is obviously equivalent to an existence of $\lambda_\alpha \in \mathbf{C}$ so that

$$\beta_k a_\beta b_{\alpha-\beta} \left(\frac{\beta!(\alpha - \beta)!}{(|\beta| - 1)!\beta_k (Q_n)_{|\alpha-\beta|}} \right)^{1/2} = \lambda_\alpha \left(\frac{(|\beta| - 1)!\beta_k (Q_n)_{|\alpha-\beta|}}{\beta!(\alpha - \beta)!} \right)^{1/2}$$

for all $\beta \le \alpha$ and $\beta_k \ge 1$. Therefore,

$$a_\beta b_{\alpha-\beta} = \lambda_\alpha \frac{(|\beta| - 1)!(Q_n)_{|\alpha-\beta|}}{\beta!(\alpha - \beta)!}, \ 1 \le \beta \le \alpha. \tag{26}$$

Putting $\beta = \alpha$ in (26) results in

$$\lambda_\alpha = \frac{\alpha!}{(|\alpha| - 1)!} a_\alpha, \quad \alpha \geq 1. \tag{27}$$

Also, by (21), (24), and (26),

$$b_\alpha = \frac{\lambda_\alpha}{\alpha_k} \sum_{\substack{\beta \leq \alpha \\ \beta_k \geq 1}} \frac{\beta_k (|\beta| - 1)!}{\beta!} \frac{(Q_n)_{|\alpha - \beta|}}{(\alpha - \beta)!} = \lambda_\alpha \frac{(Q_n)_{|\alpha|}}{\alpha! Q_n},$$

and therefore,

$$\lambda_\alpha = \frac{\alpha! Q_n}{(Q_n)_{|\alpha|}} b_\alpha, \ \alpha \geq 1. \tag{28}$$

Relations (27) and (28) give

$$\frac{1}{(|\alpha| - 1)!} a_\alpha = \frac{Q_n}{(Q_n)_{|\alpha|}} b_\alpha, \quad \alpha \geq 1. \tag{29}$$

For $1 \leq k \leq n$, we denote by e_k the kth unit vector $(0, \ldots, 1_k, \ldots, 0) \in \mathbf{Z}_+^n$, and we define $\zeta = (\zeta_1, \ldots, \zeta_n) \in \mathbf{C}^n$ by

$$\zeta_k = Q_n^{-1} \bar{a}_{e_k}, \ 1 \leq k \leq n. \tag{30}$$

We shall show that the only solution of (29) subject to (21) is

$$a_\alpha = Q_n \frac{(|\alpha| - 1)!}{\alpha!} \bar{\zeta}^\alpha, \quad \alpha \geq 1. \tag{31}$$

and therefore, $b_\alpha = [(Q_n)_{|\alpha|}/\alpha!]\bar{\zeta}^\alpha$, $\alpha \geq 0$. Indeed, for $\alpha \geq 1$ with a component $\alpha_k \geq 1$, $1 \leq k \leq n$, we have from (21) and (29),

$$\frac{1}{(|\alpha| - 1)!} a_\alpha = \frac{Q_n}{(Q_n)_{|\alpha|}} \frac{1}{\alpha_k} \left[\alpha_k a_\alpha + \sum_{\beta < \alpha} \beta_k a_\beta b_{\alpha - \beta}\right],$$

and therefore,

$$\left[\frac{1}{(|\alpha| - 1)!} - \frac{Q_n}{(Q_n)_{|\alpha|}}\right] a_\alpha = \frac{1}{(Q_n)_{|\alpha|}} \frac{1}{\alpha_k} \sum_{\beta < \alpha} \beta_k \frac{(Q_n)_{|\alpha - \beta|}}{(|\alpha - \beta| - 1)!} a_\beta a_{\alpha - \beta}. \tag{32}$$

To prove (31), we use induction on the weight $|\alpha| \geq 1$. For $|\alpha| = 1$, (31) holds true by virtue of the definition (30). We assume that (31) is true for α, $1 \leq |\alpha| < m$. When $|\alpha| = m$, there is a component $\alpha_k \geq 1$, $1 \leq k \leq n$, of α so that (23) and (32) apply. By the induction assumption, (23), and (32),

$$\left[\frac{1}{(|\alpha| - 1)!} - \frac{Q_n}{(Q_n)_{|\alpha|}}\right]a_\alpha = \frac{1}{(Q_n)_{|\alpha|}} \frac{1}{\alpha_k} Q_n^2 \bar{\zeta}^\alpha \sum_{\beta<\alpha} \frac{\beta_k(|\beta| - 1)!}{\beta!} \frac{(Q_n)_{|\alpha-\beta|}}{(\alpha - \beta)!}$$

$$= \frac{1}{(Q_n)_{|\alpha|}} \frac{1}{\alpha_k} Q_n^2 \bar{\zeta}^\alpha \left[\frac{\alpha_k}{\alpha!} \frac{(Q_n)_{|\alpha|}}{Q_n} - \frac{\alpha_k}{\alpha!}(|\alpha| - 1)!\right]$$

$$= \left[\frac{1}{(|\alpha| - 1)!} - \frac{Q_n}{(Q_n)_{|\alpha|}}\right]Q_n \frac{(|\alpha| - 1)!}{\alpha!} \bar{\zeta}^\alpha.$$

Since the common factor in the square bracket is positive, (31) follows at once. Also, since the right-hand side of (22) is finite, $\zeta = (\zeta_1, \ldots, \zeta_n)$ of (30) must belong to Δ_n. This concludes the proof of the lemma.

The proof of the theorem is now complete.

REFERENCES

1. Burbea, J. A Dirichlet norm inequality and some inequalities for reproducing kernel spaces. *Proc. Amer. Math. Soc.* 83 (1981): 279-285.

2. Saitoh, S. Some inequalities for analytic functions with finite Dirichlet integral on the unit disc. *Math. Ann.* 246 (1979):69-77.

3. Selberg, A. *Automorphic Functions and Integral Operators.* Seminars on Analytic Functions, Institute for Advanced Study, Princeton, N.J., 1957, vol. 2, pp. 152-161.

ON THE RADON TRANSFORM AND SOME OF ITS GENERALIZATIONS

A.P. Calderón
University of Chicago

1. INTRODUCTION

Let $f(x)$ be a function on the n-dimensional Euclidean space $\mathbf{R}^n$ and let N denote a k-dimensional affine subspace of $\mathbf{R}^n$, $0 < k < n$. The k-plane transform of f is the function of N defined by

$$\mathscr{R}_k(f)(N) = \int_N f \, d\sigma,$$

where $d\sigma$ denotes the ordinary measure on N. For $k = n - 1$ this is the well-known Radon transform. Our goal is to obtain existence, integrability, regularity, and support properties of such transforms. Some of the results presented here have been obtained previously or independently by D. M. Oberlin and E. M. Stein [8], S. W. Drury [4 and 5], K. T. Smith and D. C. Solmon [9 and 10], and R. Strichartz [11].

2. STATEMENT OF RESULTS

Before we proceed, we shall describe the k-plane transforms in terms of the group $\mathscr{O}_n$ of proper rotations of $\mathbf{R}^n$ around the origin. Let us

The author was partially supported by a grant from the National Science Foundation.

fix a subspace M of $\mathbf{R}^n$ through the origin and of dimension $n - k$, and for $y \in M$, let N_y be the orthogonal complement of M through y. Thus, if $\omega \in \mathcal{O}_n$ and $N = \omega(N_0 + y) = \Phi(\omega, y)$,

$$\mathcal{R}_k(f)(N) = \int_{\omega N_y} f(x)\,d\sigma = \int_{N_y} f(\omega^{-1}x)\,d\sigma = F_k(\omega, y)$$

is a function on the Cartesian product $\mathcal{O}_n \times M$, which we assume to be equipped with the product of the measure on M and the normalized Haar measure $d\omega$ on $\mathcal{O}_n$. The set of subspaces N carries a measure dN that is invariant under the action of rigid motions of $\mathbf{R}^n$ and is defined by

$$\int F(N)\,dN = \int_{\mathcal{O}_n} d\omega \int_M F[\omega(N_0 + y)]\,dy.$$

This measure is clearly invariant under rotations. That it is also invariant under translations w follows readily by observing that if P denotes the orthogonal projection on M, then

$$\int F(N + w)\,dN = \int_{\mathcal{O}_n} d\omega \int_M F[\omega(N_0 + y + P\omega^{-1}w)\,dy$$
$$= \int_{\mathcal{O}_n} d\omega \int_M F[\omega(N_0 + y)]\,dy = \int F(N)\,dN.$$

The measure dN is uniquely determined by its invariance under rigid motions and the fact that the measure of the set of all N intersecting a ball of measure 1 is 1. Such an invariant measure we call normalized.

We shall also use the natural topology of the set of subspaces N, which is the strongest that makes the map $(\omega, y) \to N = \Phi(\omega, y)$ continuous.

Theorem 1 (D. C. Solomon [10]). *Let $f(x) \geq 0$ and $\alpha < n - k$. Then $F_k(\omega, y)$ is well defined almost everywhere and $F_k(\omega, y)(1 + |y|)^{-\alpha}$ is integrable over $\mathcal{O}_n \times M$ if and only if $f(x)(1 + |x|)^{-\alpha}$ is integrable over $\mathbf{R}^n$ and the ratio of the corresponding integrals is between two positive constants.*

Suppose now that $F_k(\omega, y)$ is well defined almost everywhere for $|y| \leq \varepsilon$, $\varepsilon > 0$, and integrable there; then $f(x)(1 + |x|)^{-n+k}$ is integrable. Conversely, if $f(x)(1 + |x|)^{-n+k}$ is integrable, then $F_k(\omega, y)$ is well defined almost everywhere and if $\varphi(t)$ is positive and decreasing in $(0, \infty)$,

$$\int_{\mathcal{O}_n} d\omega \int F_k(\omega, y)(1 + |y|)^{-n+k} \varphi(|y|) dy$$
$$\leq c \left[\int_{\mathbf{R}^n} f(x)(1 + |x|)^{-n+k} \, dx \right] \left[\int_0^\infty \varphi(t) \, \frac{dt}{t} \right].$$

Remark. The result implies in particular that $F_k(\omega, y)$ is well defined almost everywhere if $f \in L^p$, $p < n/k$. However, this is false if $p = n/k$, as the example

$$f(x) = (1 + |x|)^{-k} [\log(1 + |x|)]^{-1}$$

shows.

Theorem 2. Let $f(x) \in L^{p_1} \cap L^{p_2}$, $1 < p_1 \leq p_2 \leq 2$, $p_1 < n/k$. *Then* $F_k(\omega, y)$ *is a well-defined function of* y *for almost all* ω *and its Fourier transform with respect to* y *is a locally integrable function* $\hat{F}_k(\omega, z)$ *for almost all* ω. *Thus, the operator* L

$$[(L F_k)(\omega, y)]^\wedge = m(x) \hat{F}_k(\omega, z),$$

where $m(z)$ *is a positively homogeneous function of complex degree* $s + it$, $s \geq 0$, *is well defined for almost all* ω. *Now let* $p_1 \leq p \leq p_2$, $p > 1$, *and let* $m(z)$ *be infinitely differentiable in* $|z| > 0$ *and its degree of homogeneity be such that*

$$k - \frac{n}{p} < s < k - \frac{k}{p}.$$

Then

$$\int_{\mathcal{O}_n} \| (L F_k)(\omega, y) \|_r^{p'} d\omega \leq c \|f\|_p^{p'}$$

where the norm in the integral is the norm of $L^r(M)$, $p' = p/(p - 1)$ *and*

$$\frac{1}{r} = \frac{1}{p} \frac{n}{n - k} - \frac{k - s}{n - k}.$$

In the extreme case $p = 1$ and $s = 0$, we have the following substitute result. If f belongs to the Hardy space $H^1(\mathbf{R}^n)$, then $F_k(\omega, y)$ and $(L\, F_k)(\omega, y)$ are well defined for every ω and

$$\|(L\, F_k)(\omega, y)\|_{H^1(M)} \leq c\|f\|_{H^1(\mathbf{R}^n)}.$$

Now if $s < k - n/p_2$ [which requires that $p_2 > n/(k - s)$], then $(L\, F_k)(\omega, y)$ is bounded for almost all ω and

$$\|L\, F_k\|_\infty^u d\omega \leq c\left(\|f\|_{p_1} + \|f\|_{p_2}\right)^u, \quad u = n/(n - k + s).$$

Finally, if $s < k - n/p_2 < s + 1$, then $(L\, F_k)(\omega, y)$ satisfies a uniform Hölder condition of order $k - n/p_2 - s$ in y for almost all ω.

Remark. If $k - n/p_2 > 0$ and we specialize $m(x)$ to be monomial, we find that F_k has bounded derivatives of all orders h, $0 \leq h < k - n/p_2$, in y for almost all ω and derivatives of order h, $h < k - n/p_2 < h + 1$, satisfying a Hölder condition of order $k - n/p_2 - h$ in y. In general, F_k has derivatives with respect to y of orders less than $k - k/p_2$ in the corresponding classes described in the theorem. What happens for $p > 2$ remains an open problem. However, if $f(x)$ is the characteristic function of the unit ball, then

$$F_k(\omega, y) = c(1 - |y|^2)^{k/2}, \quad |y| \leq 1,$$

and if k is even, this function does not have functions as derivatives of order higher than $k/2$.

Theorem 3. Let C be a closed subset of $\mathbf{R}^n$ with the following property: for every x not in C, there is a subspace L_x through x of dimension $k + 1$, $(k < n)$, such that no $(k + 1)$-dimensional subspace L in a neighborhood of L_x intersects C. Let $f(x)(1 + |x|)^{-n+k}$ be integrable and suppose that

$$\mathcal{R}_k(f)(N) = 0$$

for almost all N not intersecting C. Then $f(x)$ vanishes almost everywhere outside C.

Remark. If C is convex and bounded, the theorem asserts that if $k < n - 1$ and $\mathscr{R}_k(fg)(N) = 0$ for almost every N not intersecting C, then f vanishes almost everywhere outside C. This is no longer true if $k = n - 1$. In fact, if $k = n - 1$, then for every α, $\alpha > n$, there exists a continuous function $f(x)$, $|f(x)| \leq (1 + x)^{-\alpha}$, such that $\mathscr{R}_k(f)(N) = 0$ for every N not intersecting a ball and the support of f is unbounded. For positive results in the case $k = n - 1$, see [6 and 7].

3. PROOF OF THEOREM 1

Let M and N_y be as in the beginning of Sec. 2, and let P denote the orthogonal projection on M. Then we have

$$\int_M F_k(\omega, y)(1 + |y|)^{-\alpha}\,dy = \int_M (1 + |y|)^{-\alpha}\int_{N_y} f(\omega^{-1}x)\,d\sigma\,dy$$
$$= \int_{\mathbf{R}^n} (1 + |Px|)^{-\alpha} f(\omega^{-1}x)\,dx$$
$$= \int_{\mathbf{R}^n} (1 + |P\omega x|)^{-\alpha}\,d\omega\,dx.$$

Integrating with respect to ω and interchanging the order of integration, we obtain

$$\int_{M \times \mathcal{O}_n} F_k(\omega, y)(1 + |y|)^{-\alpha}dy\,d\omega = \int_{\mathbf{R}^n} f(x)\int_{\mathcal{O}_n} (1 + |P\omega x|)^{-\alpha}\,d\omega\,dx.$$

But the inner integral in the last expression is nothing but the mean value of $(1 + |Px|)^{\alpha}$ over the sphere of radius $|x|$, which an elementary calculation shows to be between two positive constants times $(1 + |x|)^{-\alpha}$. To prove the second part of the theorem, we argue as above, replacing the function $(1 + |y|)^{-\alpha}$ by $\chi(|y|/t)$, where χ is the characteristic function of the interval [0, 1] and t is a positive real number, and obtain

$$\int_{\mathcal{O}_n} d\omega\int_{|y|\leq t} F_k(\omega, y)\,dy = \int_{\mathbf{R}^n} f(x)\int_{\mathcal{O}_n} \chi\left[\frac{|P\omega x|}{t}\right]d\omega\,dx.$$

Again, the inner integral on the right is the mean value of the function $\chi(|Px|)$ over the sphere of radius $|x|/t$, which is readily seen to be of the order of

$$\left(1 + \frac{|x|}{t}\right)^{-n+k}.$$

Thus, the integrability of

$$f(x)\left(1 + \frac{|x|}{t}\right)^{-n+k}$$

is equivalent to that of $F_k(\omega, y)$ in $|y| \leq t$. But the integrability of

$$f(x)\left(1 + \frac{|x|}{t_1}\right)^{-n+k}$$

is equivalent to that of

$$f(x)\left(1 + \frac{|x|}{t_2}\right)^{-n+k}, \quad t_1 > 0,\ t_2 > 0,$$

whence setting $t_1 = \varepsilon$, $t_2 = 1$, the first assertion follows. Finally, suppose that $f(x)(1 + |x|)^{-n+k}$ is integrable. Then if $\varphi(t)$ is positive and decreasing, we have

$$\int_{\mathcal{O}_n} d\omega \int_M F_k(\omega, y)(1 + |y|)^{-n+k}\varphi(|y|)\,dy$$

$$= -\int_{\mathcal{O}_{n\infty}} d\omega \int_0^\infty \left[\int_{|y|\leq t} F_k(\omega, y)\,dy\right] d(1 + t)^{-n+k}\varphi(t)$$

$$\leq -c\int_0^\infty \left[\int_{\mathbf{R}^n} f(x)\left(1 + \frac{|x|}{t}\right)^{-n+k} dx\right] d(1 + t)^{-n+k}\varphi(t)$$

$$= c(n - k)\int_{\mathbf{R}^n} f(x)\int_0^\infty (1 + t)^{-n+k}\left(1 + \frac{|x|}{t}\right)^{-n+k-1}\frac{|x|}{t^2}\varphi(t)\,dt.$$

But, as is readily verified,

$$(1 + t)^{-n+k}\left(1 + \frac{|x|}{t}\right)^{-n+k-1}\frac{|x|}{t} \leq c(1 + |x|)^{-n+k},$$

whence, substituting in the preceding integral, the desired result follows.

4. PROOF OF THEOREM 2

First let us consider the case when f is a function in $H^1(\mathbf{R}^n)$. An atom (see [1]) is a bounded function with support in a ball and integral equal to 0. The norm of the atom is the least upper bound of $|f|$ times the measure of the ball. Every sum of atoms with finite sum of norms is a function in H^1 with norm less than or equal to a constant (depending on the choice of the norm for H^1) times the sum of the norms of the atoms. Conversely, every function in H^1 is a sum of atoms with sum of norms less than or equal to a constant times the norm of the function. Now, if f is an atom, so is $F_k(\omega, y)$ as a function of y for every ω, and the norm of F as an atom is less than or equal to a constant times the norm of f as an atom. This is readily seen and is left to the reader to verify. From this there follows by addition that if f is in $H^1(\mathbf{R}^n)$, then $F_k(\omega, y)$ is in $H^1(M)$ for every ω and

$$\|F_k(\omega, y)\|_{H^1} \leq c\|f\|_{H^1}.$$

Assume now that the Fourier transform of f is infinitely differentiable, has compact support, and vanishes near the origin. For z a point of M, we have

$$\begin{aligned}\hat{F}_k(\omega, z) &= \int_M F_k(\omega, y)\exp 2\pi i(y \cdot z)dy \\ &= \int_M \exp 2\pi i(y \cdot z)\int_{N_y} f(\omega^{-1}x)d\sigma \\ &= \int_{\mathbf{R}^n} f(\omega^{-1}x)\exp 2\pi i(x \cdot z)dx \\ &= \int_{\mathbf{R}^n} f(x)\exp 2\pi i(x \cdot \omega z)dx = \hat{f}(\omega z).\end{aligned}$$

Thus, $\hat{F}_k(\omega, z) = \hat{f}(\omega z)$. For a complex number ξ, $\xi = s + it$, let D^ξ be the operator defined on functions F in M with Fourier transform vanishing near the origin by

$$(D^{\xi}F)\hat{} = |z|^{\xi}\hat{F}(z).$$

Then for $s = k/2$, we have

$$\|D^{\xi}F_k\|_2^2 = \||z|^{k/2}\hat{f}(\omega z)\|_2^2,$$

and integrating with respect to ω, we obtain

$$\int_{\mathcal{O}_n}\|D^{\xi}F_k\|_2^2 d\omega = \int_{\mathcal{O}_n} d\omega\int_M |z|^k|\hat{f}(\omega z)|^2 dz = \int_{\mathcal{O}_n} d\omega\int_{\omega^{-1}M}|x|^k|\hat{f}(x)|^2 d\sigma.$$

The last expression is the integral of $|\hat{f}|$ with respect to a certain measure μ in $\mathbf{R}^n$ that clearly is rotation invariant and homogeneous of degree n in the sense that

$$t^n\int g(tx)\,d\mu = \int g(x)\,d\mu,\ t > 0.$$

Now, every measure with these properties is a multiple of the ordinary measure dx. Consequently,

$$\int_{\mathcal{O}_n}\|D^{\xi}F_k\|_2^2 d\omega = c\int|\hat{f}|^2 dx = c\|f\|_2^2.$$

Now let $s = 0$. Then, since $|z|^{it}$ is a multiplier for $H(M)$ with norm less than or equal to $c(1 + |t|)^{\alpha}$ for some constants c and α (see [3, Theorem 4.6]), we find that

$$\|D^{\xi}F_k\|_1 \le c\|D^{\xi}F_k\|_{H^1(M)} \le c(1 + |t|)^{\alpha}\|F_k\|_{H^1(M)}$$

$$\le c(1 + |t|)^{\alpha}\|f\|_{H^1(\mathbf{R}^n)}.$$

Thus the operator taking f to the function $(1 + \xi)^{-\alpha}D^{\xi}F_k$ in $M \times \mathcal{O}_n$ has the property of being continuous with respect to the norms of $H^1(\mathbf{R}^n)$ and $\sup_{\omega}\|F(\omega, y)\|_1$ for $s = 0$, and the norms of $L^2(M \times \mathcal{O}_n)$ for $s = k/2$, uniformly in t. Thus, by interpolation (see, for example, [3, Theorem 3.4, and 2, Theorem 13.6]), there follows that

$$\int_{\mathcal{O}_n}\|D^{k/p'}F_k\|_p^{p'} d\omega \le c\|f\|_p^{p'},\ 1 < p < 2.$$

Now let p, r, and s be as in the first part of the theorem. Then

$$k/p' - s > 0 \quad \text{and} \quad 1/p > (k/p' - s)/(n - k),$$

and since $D^{-(k/p'-s)}$ is fractional integration of order $k/p' - s$, by Sobolev's theorem we have

$$\|D^s F_k\|_r = \|D^{-(k/p'-s)} D^{k/p'} F_k\|_r < c\|D^{k/p'} F_k\|_p,$$

and substituting above, we obtain

$$\int_{\mathcal{O}_n} \|D^s F_k\|_r^{p'} \, d\omega \leq c\|f\|_p^{p'}.$$

Now, since $m(z)|z|^{-s}$ is a multiplier for $L^r(M)$ (we have $r > p > 1$),

$$\|L\ F_k\|_r \leq c\|D^s F_k\|_r.$$

Combining this with the preceding inequality, we obtain the first inequality of the theorem for functions f whose Fourier transforms are infinitely differentiable, have compact support, and vanish near the origin. Such functions are dense in the intersection of any two L^p spaces, $p > 1$. To see this, we observe first that bounded functions with compact support have this property, and if $(\phi_\varepsilon)^\wedge = \hat{\phi}(\varepsilon z)$ where $\hat{\phi}$ is infinitely differentiable, has compact support, and equals 1 near the origin, and if g is bounded and has compact support, then, by the maximal theorem, the convolution $g * \phi_\varepsilon$ will be majorized by a function belonging to every L^p, $p > 1$, and will converge to g as $\varepsilon \to 0$ and to 0 as $\varepsilon \to \infty$ in the norm of every L^p, $p > 1$. Thus, $g * \phi_\varepsilon - g * \phi_\delta$ will converge to g in every such norm as $\varepsilon \to 0$ and $\delta \to \infty$, and its Fourier transform will be infinitely differentiable, have compact support, and will vanish near the origin. Thus, by passage to the limit, we obtain the first inequality of the theorem in the general case.

To prove our assertion in the case $s < k - n/p_2$, we need the following result, which we state without proof.

Lemma. Let F be a function on M belonging to $L^{q_1} \cap L^{q_2}$, $1 < q_1 < q_2 \leq \infty$, and let

$$\frac{s}{n-k} = \frac{v}{q_1} + \frac{1-v}{q_2}, \quad 0 < v < 1.$$

Then $D^{-s}F$ is bounded and

$$\|D^{-s}F\|_\infty \le c\|F\|_{q_1}^{v}\ \|F\|_{q_2}^{1-v}.$$

If, in addition,

$$\frac{1}{q_2} < \frac{s}{n-k} < \frac{1}{q_2} + \frac{1}{n-k}$$

then $D^{-s}F$ satisfies a Hölder condition of order $s - (n-k)/q_2$.

Returning to our proof, since $1 < p_1 \le p_2 \le 2$, $p_1 < n/k$, and $s \ge 0$, we must have $p_2 > p_1$, for if $p_2 = p_1$, then $p_2 < n/k$ and $k - n/p_2 < 0$. Now choose q_1, q_2, $p_1 \le q_1 < q_2 \le p_2$, and σ so that

$$k - \frac{k}{q_1} > \sigma,\ 1 + k - \frac{n}{q_1} > \sigma > k - \frac{n}{q_2} > s > k - \frac{n}{q_1}.$$

This is clearly possible. Then from the last inequalities above, we obtain

$$\frac{1}{r_2} = \frac{1}{q_2}\frac{n}{n-k} - \frac{k-\sigma}{n-k} < \frac{\sigma - s}{n-k} < \frac{1}{q_1}\frac{n}{n-k} - \frac{k-\sigma}{n-k} = \frac{1}{r_1} \tag{1}$$

from the third $0 < r_2 < \infty$, and from the second $r_1 < 1$. Now since

$$k - \frac{n}{q_i} < \sigma < k - \frac{k}{q_i},\ i = 1, 2,$$

and $f \in L^{p_1} \cap L^{p_2} \subset L^{q_1} \cap L^{q_2}$; from the first part of the theorem it follows that

$$\int \|D^{\sigma - s}L\ F_k(\omega, y)\|_{r_i}^{q_i'} \le c\|f\|_{q_i}^{q_i'}, \tag{2}$$

and from the lemma and (1),

$$\|L\ F_k(\omega, y)\|_\infty \le c\|D^{\sigma-s}L\ F_k(\omega, y)\|_{r_1}^{v}\ \|D^{\sigma-s}L\ F_k(\omega, y)\|_{r_2}^{1-v}, \tag{3}$$

where

$$\frac{\sigma - s}{n-k} = \frac{v}{r_1} + \frac{1-v}{r_2},\ 0 < v < 1,$$

which, as follows from (1), is equivalent to

$$\frac{n}{n-k+s}\frac{1}{q_1'}v + \frac{n}{n-k+s}\frac{1}{q_2'}(1-v) = 1.$$

Thus, setting

$$u = \frac{n}{n - k + s}, \quad w = \frac{n}{n - k + s}\frac{1}{q_1'}v, \quad 1 - w = \frac{n}{n - k + s}\frac{1}{q_2'}(1 - v),$$

applying Hölder's inequality to (3), we obtain

$$\int_{\mathcal{O}_n} \|LF_k(\omega, y)\|_\infty^\mu d\omega \le c\left[\int_{\mathcal{O}_n} \|D^{\sigma-s}LF_k(\omega, y)\|_{r_1}^{\frac{uv}{w}} d\omega\right]^w$$
$$\times\left[\int \|D^{\sigma-s}LF_k(\omega, y)\|_{r_2}^{\frac{u(1-v)}{1-w}} d\omega\right]^{1-w};$$

and from (2), since $\frac{uv}{w} = q_2'$, $\frac{u(1 - v)}{1 - w} = q_2'$, and $\|f\|_{q_i} \le \|f\|_{p_1} + \|f\|_{p_2}$, $i = 1, 2$, there follows that

$$\int_{\mathcal{O}_n} \|LF_k(\omega, y)\|_\infty^u d\omega \le c(\|f\|_{p_1} + \|f\|_{p_2})^u.$$

To prove the last assertion of the theorem we argue as above, now setting $q_2 = p_2$. If r_2 is as in (1), there will follow that

$$\frac{1}{r_2} + \frac{1}{n - k} > \frac{\sigma - s}{n - k},$$

and applying the lemma, one concludes that $LF_k(y, \omega)$, as a function of y, satisfies a Hölder condition of order

$$\sigma - s - \frac{n - k}{r_2} = k - \frac{n}{p_2} - s$$

whenever $\|D^{\sigma-s}LF_k(\omega, y)\|_{r_1}$ and $\|D^{\sigma-s}LF_k(\omega, y)\|_{r_2}$ are finite, that is, for almost all ω. This concludes the proof of Theorem 2.

5. PROOF OF THEOREM 3

First we shall prove a special case of our theorem, namely, when $k = n - 1$. In this case C must be empty and the theorem asserts that if $f(x)(1 + |x|)^{-1}$ is integrable and f has vanishing integral on almost all N, then $f = 0$ almost everywhere. If f were integrable, then, as was indicated at the beginning of the proof of Theorem 2, it would follow that $\hat{f} = 0$, which would imply the desired conclusion. In the general case we must proceed otherwise.

Assume first that the Fourier transform $\hat{f}$ of f coincides with an infinitely differntiable function in $|x| \neq 0$. Let ω, M, N_y, N_0 be as in Sec. 2, and $\varphi(y)$, $\psi(z)$, respectively, be functions in the Schwartz classes of M and its orthogonal complement through the origin N_0, with $\psi(0) = 1$. Set

$$\eta_\varepsilon(x) = \varphi(y)\psi(\varepsilon z), \; x = y + z.$$

Then we have

$$\int_{\mathbf{R}^n} f(\omega^{-1}x)\eta_\varepsilon(x)\,dx = \int_{M \times N_0} f[\omega^{-1}(y + z)]\varphi(y)\psi(\varepsilon z)\,dy\,dz.$$

Now, according to Theorem 1, $f[\omega^{-1}(y + z)]\varphi(y)$ is integrable in $M \times N_0 = \mathbf{R}^n$ for almost all ω. For such an ω, we have

$$\lim_{\varepsilon \to 0} \int_{M \times N_0} f[\omega^{-1}(y + z)]\varphi(y)\psi(\varepsilon z)\,dy\,dz$$

$$= \int_{M \times N_0} f[\omega^{-1}(y + z)]\varphi(y)\,dy\,dz = \int_M \varphi(y)\int_{N_y} f(\omega^{-1}x)\,d\sigma\,dy,$$

and according to our hypotheses, for almost all ω the inner integral in the last expression vanishes for almost all y. Thus

$$\lim_{\varepsilon \to 0} \int_{\mathbf{R}^n} f(\omega^{-1}x)\eta_\varepsilon(x)\,dx = 0$$

for almost all ω. Suppose now that the Fourier transform $\hat{\varphi}(y)$ of $\varphi(y)$ vanishes near the origin. Then $\hat{\eta}_\varepsilon(x) = \hat{\varphi}(y)\varepsilon^{-n}\hat{\psi}(\varepsilon^{-1}z)$, $x = y + z$, also vanishes near the origin, and since $\hat{f}(x)$ was assumed to coincide with an infinitely differentiable function in $|x| \neq 0$, if $\hat{\varphi}(y)$ and $\hat{\psi}(z)$ have compact support, we have

$$\lim_{\varepsilon \to 0} \int f(\omega^{-1}x)\eta_\varepsilon(x)\,dx = \lim_{\varepsilon \to 0} \int \hat{f}(\omega^{-1}x)\hat{\eta}_\varepsilon(-x)\,dx$$

$$= \lim_{\varepsilon \to 0} \int_{M \times N_0} \hat{f}[\omega^{-1}(y + z)]\hat{\varphi}(-y)\varepsilon^{-n}\hat{\psi}(-\varepsilon^{-1}z)\,dy\,dz$$

$$= \int_M \hat{f}(\omega^{-1}y)\hat{\varphi}(-y)\,dy.$$

Thus, for almost every ω, the last integral vanishes whenever $\hat{\varphi}(y)$ has compact support and vanishes near the origin. Since $\hat{f}$

coincides with a continuous function in $|x| \neq 0$, this clearly implies that $\hat{f}$ is a distribution supported at the origin and f is a polynomial. But $f(x)(1 + |x|)^{-1}$ is integrable and this is possible only if $f = 0$.

Now let us pass to the case when $f(x)(1 + |x|)^{-1}$ is merely assumed to be integrable. Let A denote a linear transformation of $\mathbf{R}^n$ and I the identity transformation. Let $\eta(A)$ be an infinitely differentiable function of A with compact support, that is, a C_0^∞-function of the entries a_{ij} of the matrix describing A in terms of a coordinate system $(x_1, \ldots, x_n)$ in $\mathbf{R}^n$, such that

$$\int \eta(A)\,dA = 1$$

where $dA = \prod(da_{ij})$. Let

$$f_\varepsilon(x) = \int f[(I + A)x]\,\varepsilon^{-n^2}\,\eta(\varepsilon^{-1}A)\,dA, \quad \varepsilon > 0.$$

If A is sufficiently small, that is, the a_{ij} are sufficiently small, then $I + A$ is invertible, $f[(I + A)x]$ is well defined, $f[(I + A)x](1 + |x|)^{-1}$ is integrable, and the integral of $f[(I + A)x]$ on subspaces N of dimension $n - 1$ vanishes for almost all N (we leave the proof of this to the reader). Thus, if ε is sufficiently small, the same is true for $f_\varepsilon(x)$. Now, as we shall see below, for sufficiently small ε, $\hat{f}_\varepsilon$ coincides with an infinitely differentiable function in $|x| \neq 0$, and this, as we have shown, implies that $f_\varepsilon = 0$. Now, from the definition of f_ε, it follows that for every $\varphi(x)$ in C_0^∞,

$$\int f(x)\varphi(x)\,dx = \lim_{\varepsilon \to 0} \int f_\varepsilon(x)\,\varphi(x)\,dx = 0.$$

Thus, we conclude that $f = 0$ almost everywhere.

To prove that $\hat{f}_\varepsilon$ coincides with an infinitely differentiable function in an open set $\mathcal{O}$ will suffice to show that for every compact set C contained in $\mathcal{O}$ and every monomial differential operator $(\partial/\partial x)^\alpha$, there is a constant c such that if φ is a testing function with support in C, then

$$\left| \left\langle \hat{f}_\varepsilon, \left(\frac{\partial}{\partial x}\right)^\alpha \varphi \right\rangle \right| \leq c\|\varphi\|_\infty.$$

Let us take a coordinate system $(x_1, \ldots, x_n)$ in $\mathbf{R}^n$ and let the linear transformation $\bar{x} = Ax$ be given by $\bar{x}_i = \Sigma a_{ij}x_j$. If B is a linear transformation and ψ a function on $\mathbf{R}^n$, let us write

$$\psi(Bx) = (\psi \circ B)(x).$$

Then if B_t denotes the transpose of B,

$$\hat{\psi} \circ B^{-1} = (\det B)(\psi \circ B_t)^\wedge.$$

Thus, denoting $\varepsilon^{-n^2}\eta(\varepsilon^{-1}A)$ simply by $\eta_\varepsilon(A)$, we have

$$\begin{aligned}\langle \hat{f}_\varepsilon, \varphi\rangle &= \int f[(I + A)x]\hat{\varphi}(x)\eta_\varepsilon(A)\,dA\,dx \\ &= \int f(x)\hat{\varphi}[(I + A)^{-1}x]\det(I + A)^{-1}\eta_\varepsilon(A)\,dA\,dx \\ &= \int \langle \hat{f}, \varphi \circ (I + A_t)\rangle \eta_\varepsilon(A)\,dA \\ &= \langle \hat{f}, \int [\varphi \circ (I + A_t)]\eta_\varepsilon(A)\,dA\rangle\end{aligned}$$

and more generally,

$$\left\langle \hat{f}_\varepsilon, \left(\frac{\partial}{\partial x}\right)^\alpha \varphi \right\rangle = \left\langle \hat{f}, \int \left[\left(\left(\frac{\partial}{\partial x}\right)^\alpha \varphi\right) \circ (I + A_t)\right]\eta_\varepsilon(A)\,dA \right\rangle.$$

Now, if φ has support in a compact set C, then the preceding integral is a function ψ of x with support in a compact set C_ε and

$$|\langle \hat{f}, \psi\rangle| \leq c \sup_{|\beta| \leq M} \left\| \left(\frac{\partial}{\partial x}\right)^\beta \psi \right\|_\infty$$

for some M.

Assume now that C is contained in $|x_1| > \delta > 0$. Then differentiating under the integral sign in

$$\left(\frac{\partial}{\partial x}\right)^\beta \psi = \left(\frac{\partial}{\partial x}\right)^\beta \int \left(\frac{\partial}{\partial x}\right)^\alpha \varphi \circ (I + A_t)(x)\eta_\varepsilon(A)\,dA,$$

using the identity

$$\left(\frac{\partial}{\partial x_j}\xi\right)[(I + A_t)x] = \frac{1}{x_1}\frac{\partial}{\partial a_{1j}}\xi[(I + A_t)x],$$

and integrating by parts with respect to a_{1j}, we find that

$$\left(\frac{\partial}{\partial x}\right)^{\beta}\psi = \int x_1^{-|\beta|}\varphi[(I + A_t)x](D\eta_\varepsilon)(A)\,dA,$$

where D is a differential operator with polynomial coefficients acting on the function η_ε of the variables a_{ij}. Thus, since for ε sufficiently small and A in the support of $\eta_\varepsilon(A)$, the support of $\varphi[(I + A_t)x]$ is also contained in $|x_1| > \delta > 0$, we find that for small ε,

$$\left\|\left(\frac{\partial}{\partial x}\right)^{\beta}\psi\right\|_\infty \le c\|\varphi\|_\infty,$$

where c depends on β, δ, and η_ε, but not on φ. Thus, for small ε,

$$\left|\left\langle \hat{f}_\varepsilon, \left(\frac{\partial}{\partial x}\right)^{\alpha}\varphi\right\rangle\right| = |\langle \hat{f}, \psi\rangle| \le \sup_{|\beta|\le M}\left\|\left(\frac{\partial}{\partial x}\right)^{\beta}\psi\right\|_\infty \le c\|\varphi\|_\infty$$

and we conclude that $\hat{f}_\varepsilon$ coincides with an infinitely differentiable function in $|x_1| > \delta > 0$. Since the same is true in $|x_j| > \delta > 0$, the validity of our theorem in the special case is established.

Now let us pass to the general case. Since $f(x)(1+|x|)^{-n+k}$ is integrable, the function $f(x)(1 + |x|)^{-1}$ is integrable on almost every subspace L of dimension $k + 1$. Let $x \notin C$, L_x a subspace of dimension $k + 1$ through x and not intersecting C, U a neighborhood of L_x such that $L \in U$ implies that L does not intersect C. Let dL denote the normalized invariant measure on the set of $(k + 1)$-dimensional subspaces of $\mathbf{R}^n$, dN the normalized invariant measure on the set of k-dimensional subspaces of $\mathbf{R}^n$, and for each L, d_LN the normalized invariant measure on the set of k-dimensional subspaces of L. Let $F(N)$ be a continuous function with compact support on the set of k-dimensional subspaces of $\mathbf{R}^n$ and

$$G(L) = \int_{\{N|N\subset L\}} F(N)\,d_LN.$$

Then $G(L)$ is continuous and

$$\int G(L)\,dL = \int dL \int_{\{N|N\subset L\}} F(N)\,d_LN = \int F(N)\,dN.$$

To see this, one merely has to observe that the iterated integral above gives a linear functional of F that is invariant under the action of rigid motions of $\mathbf{R}^n$ and therefore must coincide with a multiple of

$$\int F(N)\,dN.$$

That is actually coincides with this integral follows by setting

$$F(N) = \mathscr{R}_k(g)(N) = \int_N g\,d\sigma,$$

where g is now a continuous function in $\mathbf{R}^n$ with compact support and nonvanishing integral, in which case, as is readily verified,

$$\int F(N)\,dN = \int_{\mathbf{R}^n} g(x)\,dx$$

$$G(L) = \int_{\{N|N \cap L\}} F(N)\,d_L N = \int_L g\,d\sigma$$

$$\int G(L)\,dL = \int_{\mathbf{R}^n} g(x)\,dx.$$

Now, by passage to the limit through monotone sequences of functions, we find that if $F(N)$ is measurable, nonnegative, then

$$G(L) = \int_{\{N|N \cap L\}} F(N)\,d_L N$$

is well defined for almost all L and

$$\int G(L)\,dL = \int F(N)\,dN.$$

Now let $V = \{N|N \subset L,\ L \in U\}$ and $\chi_v(N)$ the characteristic function of this set. Let

$$F(N) = \chi_v(N)\left|\int_N f\,d\sigma\right| = \chi_v(N)\,|\,\mathscr{R}_k(f)(N)\,|.$$

Since if $N \in V$, then N does not intersect C, our assumption on $\mathscr{R}_k(f)(N)$ implies that $F(N)$ vanishes almost everywhere, and, consequently, so does

$$G(L) = \int_{\{N|N \subset L\}} F(N)\,d_L N.$$

Thus, for almost every L in U, $f(x)(1 + |x|)^{-1}$ is integrable and the integral of f vanishes on almost every subspace of L of codimension 1. But, as we have shown, this implies that f vanishes almost everywhere on L, and from this follows that f vanishes almost everywhere in a neighborhood of x. Thus our assertion is established.

REFERENCES

1. Calderón, A. P. An atomic decomposition of distributions in parabolic H^p spaces. *Adv. in Math.* 25 (1977):216-225.

2. Calderón, A. P. Intermediate spaces and interpolation, the complex method. *Studia Math.* 24 (1964):113-190.

3. Calderón, A. P., and Torchinsky, A. Parabolic maximal functions associated with a distribution, II. *Adv. in Math.* 214 (1977):101-171.

4. Drury, S. W. L^p estimates for the X-ray transform. To appear.

5. Drury, S. W. Some remarks on k-plane transforms. Preprint.

6. Helgason, S. The Radon transform on Euclidean spaces, compact two-point homogeneous spaces and Grassmann manifolds. *Acta Math.* 113 (1965):153-180.

7. Ludwig, D. The Radon transform on Euclidean spaces. *Comm. Pure Appl. Math.* 23 (1966):49-81.

8. Oberlin, D. M., and Stein, E. M. Mapping properties of the Radon transform. Preprint.

9. Smith, K. T., and Solmon, D. C. Lower dimensional integrability of L^2 functions. *J. Math. Anal. and Appl.* 51 (1975):539-549.

10. Solmon, D. C. A note on k-plane integral transforms. *J. Math. Anal. and Appl.* 71 (1979):351-358.

11. Strichartz, R. L^p estimates for Radon transforms in Euclidean and non-Euclidean spaces. Preprint.

AN ADDITION THEOREM FOR HEISENBERG HARMONICS

Charles F. Dunkl
University of Virginia

1. INTRODUCTION

The subelliptic Laplacian on the Heisenberg group has been studied by G. Folland, P. Greiner, and E. Stein [3, 6, 7]. Greiner [6] proposed the study of harmonic polynomials as a way of attacking the Dirichlet problem for the ball. This led him to a class of generalized ultraspherical polynomials and a situation analogous to ordinary spherical harmonics in $\mathbf{R}^N$. In this paper, after a brief description of the general situation (Greiner's paper discusses only H_1), an addition theorem is obtained that explicitly shows the effect of translation on harmonic polynomials and also leads to the harmonic polynomial expansion of an arbitrary translate of the fundamental solution. It is shown that this series converges (in an appropriate sense) in the largest ball avoiding the singularity. The disc polynomials (some spherical functions of the unitary group) are important in the analysis.

The Heisenberg group H_N is the space $\mathbf{C}^N \times \mathbf{R}$ furnished with the group operation $(z, t) \cdot (w, s) := (z + w, s + t + 2\,\mathrm{Im}\langle z, w\rangle)$, where $N = 1, 2, \ldots,$ and

$$\langle z, w\rangle := \sum_{j=1}^{N} z_j \bar{w}_j .$$

Two important functions on H_N are the norm $\|(z, t)\| := (t^2 + |z|^4)^{1/4}$ and $S(z, t) := t + i|z|^2$.

Two groups of automorphisms of H_N are the unitary group $U(N)$ (on $\mathbf{C}^N$) acting by $U(z, t) := (Uz, t)$ and the dilation group $(\{r > 0\})$ acting by $r: (z, t) \leftrightarrow (rz, r^2t)$.

If p is a polynomial function on H_N (in the variables z_j, $\bar{z}_j$, t, $1 \leq j \leq N$), say p is homogeneous of degree n if $p(rz, r^2t) = r^n p(z, t)$, $r > 0$.

The left-invariant tangent fields are spanned by

$$Z_j := \frac{\partial}{\partial z_j} + i\bar{z}_j \frac{\partial}{\partial t},\ \bar{Z}_j := \frac{\partial}{\partial \bar{z}_j} - iz_j \frac{\partial}{\partial t},\ 1 \leq j \leq N,\ T := \frac{\partial}{\partial t},$$

and the subelliptic Laplacian is

$$L := -\frac{1}{2}\sum_{j=1}^{N}(Z_j\bar{Z}_j + \bar{Z}_j Z_j)$$

(see Folland and Stein [3]). We will be concerned with the operator $L_\gamma := L + i\gamma T$, for $\gamma \in \mathbf{C}$, which arises for certain values of γ when applying $\square_b$ to forms [3]. Say a function on H_N (or some open subset) is L_γ-harmonic if it is twice differentiable and annihilated by L_γ. Note that L_γ applied to a homogeneous polynomial decreases the degree by 2. Also L_γ commutes with the action of $U(N)$. Greiner found the L_γ-harmonic homogeneous polynomials for $N = 1$; here we will extend his ideas to $N \geq 2$ [the main difference is due to $U(N)$ being non-Abelian for $N \geq 2$].

When the space of polynomials on $\mathbf{C}^N$ is split as a $U(N)$-module, the following irreducible components arise.

Definition 1. For $k, \ell \geq 0$ let $V_{k\ell} := \Big\{p(z) : p$ is a homogeneous polynomial on $\mathbf{C}^N$ of bidegree (k, ℓ), and

$$\sum_{j=1}^{N} \frac{\partial^2}{\partial z_j \partial \bar{z}_j} p = 0\Big\},$$

[bidegree (k, ℓ) means $p(cz) = c^k \bar{c}^\ell p(z)$ for $c \in \mathbf{C}$].

Definition 2. The generalized ultraspherical polynomial of degree n and index (α, β) (for $\alpha, \beta \in \mathbf{C}$) is

$$C_n^{(\alpha, \beta)}(\zeta) := \sum_{j=0}^{n} \frac{(\alpha)_j (\beta)_{n-j}}{j!(n-j)!} \bar{\zeta}^j \zeta^{n-j} \quad (\zeta \in \mathbf{C}),$$

where the shifted factorial $(x)_m$ is defined by

$$(x)_0 = 1, \ (x)_{m+1} = (x)_m (x+m), m = 0, 1, 2, \ldots .$$

We are using the notation of Gasper [4], who found an orthogonality relation. However, this relation uses all values of ζ on $|\zeta| = 1$, whereas in our context $\operatorname{Im} \zeta \geq 0$ always. Also the (finite) linear span of $\{C_n^{(\alpha, \beta)} : n \geq 0\}$ is not an algebra when $\alpha \neq \beta$, so this is not a family of polynomials in the usual sense.

Proposition 1. Let $p \in V_{k\ell}$; *then* $L_\gamma(p(z)C_n^{(\alpha+\ell, \beta+k)}(t + i|z|^2)) = 0$, *where* $\alpha := (N - \gamma)/2$, $\beta := (N + \gamma)/2$; *and every* L_γ*-harmonic polynomial is a linear combination of such terms (all values of* k, ℓ, n*).*

This expansion for a left translate of the basic L_γ-harmonic function $C_n^{(\alpha, \beta)}(t + i|z|^2)$ is exactly the addition theorem. Specifically, we want the expansion of $C_n^{(\alpha, \beta)}(S((-w, -s) \cdot (z, t)))$ [note that $S((-w, -s) \cdot (z, t)) = t - s + i(|w|^2 + |z|^2 - 2\langle z, w\rangle)$]. To expand functions in $\langle z, w\rangle$ (for fixed w) in terms of $\{V_{k\ell}\}$, the spherical functions for $U(N)/U(N-1)$, namely, the disc polynomials, are needed (see Sec. 4 for more details).

Definition 3. For $k, \ell \geq 0$, $\lambda > -1$, $\xi \in \mathbf{C}$,

$$R_{k\ell}^{(\lambda)}(\xi) := \frac{(\lambda+1)_{k+\ell}}{(\lambda+1)_k (\lambda+1)_\ell} \xi^k \bar{\xi}^\ell {}_2F_1(-k, -\ell; -\lambda - k - \ell; 1/\xi\bar{\xi}).$$

Also, for $z, w \in \mathbf{C}^N$, let

$$F_{k\ell}(z, w) := (|z||w|)^{k+\ell} R_{k\ell}^{(N-2)}(\langle z, w\rangle/(|z||w|)).$$

For fixed $w \neq 0$, the polynomial $z \mapsto F_{k\ell}(z, w)$ is in $V_{k\ell}$. Also, $|F_{k\ell}(z, w)| \leq (|z||w|)^{k+\ell}$ with equality at $z = w$ [see (14) and (16)].

Theorem 1 (Addition formula). For $\alpha + \beta = N$,

$$C_n^{(\alpha,\beta)}(S((-w, -s)\cdot(z, t))) = \sum_{k+\ell\leq n} (2|w||z|)^{k+\ell} R_{k\ell}^{(N-2)}(\langle z, w\rangle/|z||w|) \cdot \frac{i^{\ell-k}(\alpha)_\ell(\beta)_k(N-1)_k(N-1)_\ell}{k!\,\ell!\,(N-1)_{k+\ell}} \cdot \sum_{m=0}^{n-k-\ell} \frac{(N+k+\ell)_{n-k-\ell}}{(N+k+\ell)_m(N+k+\ell)_{n-k-\ell-m}} \cdot C_m^{(\alpha+\ell,\beta+k)}(t + i|z|^2)\, C_{n-k-\ell-m}^{(\alpha+\ell,\beta+k)}(-s + i|w|^2).$$

Folland and Stein [3] showed that a certain multiple of the function $\psi_\gamma(z, t) := (-t + i|z|^2)^{-\alpha}(-t - i|z|^2)^{-\beta}$ is the fundamental solution for L_γ (provided $\pm\gamma$ avoids the values N, $N + 2$, $N + 4$, ..., and $\gamma = \beta - \alpha$, $N = \alpha + \beta$). Thus $L_\gamma\psi = 0$ except at $(0, 0) \in H_N$.

Remark on Branch Points. Whenever a power function with a branch point is used, we will choose a branch cut to lie in some half-plane (Im ζ > or < 0) as appropriate, and the determination that is positive for all large positive arguments.

Theorem 2. The L_γ*-harmonic polynomial expansion of a translate of* ψ_γ *is*

$$\psi_\gamma((-w, -s)\cdot(z, t)) = \sum_{k,\ell\geq 0} (2|z||w|)^{k+\ell} R_{k\ell}^{(N-2)}(\langle z, w\rangle/|z||w|) \cdot \frac{i^{\ell-k}(\alpha)_\ell(\beta)_k(N-1)_k(N-1)_\ell}{k!\,\ell!\,(N-1)_{k+\ell}} \cdot \sum_{m=0}^{\infty} \frac{m!}{(N+k+\ell)_m} C_m^{(\alpha+\ell,\beta+k)}(t + i|z|^2)\, C_m^{(\alpha+\ell,\beta+k)}(s + i|w|^2) \cdot (s - i|w|^2)^{-\beta-k-m}(s + i|w|^2)^{-\alpha-\ell-m}.$$

For each k, ℓ the m-sum converges absolutely for $\|(z, t)\| < \|(w, s)\|$, and the (k, ℓ)-sum converges in the same region in the Abel or Cesaro $(C, 2N - 1)$ methods applied to the partial sums $\sum_{k+\ell \leq n}$ (see Theorem 5).

2. THE L_γ-HARMONIC POLYNOMIALS AND THE ADDITION THEOREM

Since the space of L_γ-harmonic polynomials is a $U(N)$-module, it can be decomposed into copies of $V_{k\ell}$, with $k, \ell \geq 0$, so that every L_γ-harmonic is a sum of terms like $p(z)g_1(z, t)$ where $p \in V_{k\ell}$ and g_1 is invariant under $U(N)$. Clearly $g_1(z, t) = g(|z|^2, t)$ for some function g.

Lemma 1. $L_\gamma = -\Delta - |z|^2\left(\frac{\partial}{\partial t}\right)^2 + i\left(\gamma + \sum_{j=1}^{N}\left(z_j\frac{\partial}{\partial z_j} - \bar{z}_j\frac{\partial}{\partial \bar{z}_j}\right)\right)\frac{\partial}{\partial t}$, *where*

$$\Delta := \sum_{j=1}^{N} \frac{\partial^2}{\partial z_j \partial \bar{z}_j}.$$

Further, if $p \in V_{k\ell}$ and g is twice differentiable, then

$$L_\gamma(p(z)g(|z|^2, t)) = p(z)\left(-s\left(\frac{\partial^2}{\partial s^2} + \frac{\partial^2}{\partial t^2}\right) - (k + \ell + N)\frac{\partial}{\partial s} + i(k - \ell + \gamma)\frac{\partial}{\partial t}\right)g(s, t),$$

where $s = |z|^2$.

Proof. The calculation uses the product formula for Δ, the fact that $\Delta p = 0$, Euler's formula

$$\sum_{j=1}^{N} z_j \frac{\partial}{\partial z_j} p(z) = kp(z),$$

and the similar $\bar{z}$-result.

To find polynomial solutions for g, we use Greiner's trick, and let $g(|z|^2, t) = g_2(t + i|z|^2)$.

Lemma 2. Let $p \in V_{k\ell}$; then $L_\gamma(p(z)g_2(t + i|z|^2)) = 0$ if and only if g_2 satisfies

$$\left((\zeta - \bar{\zeta})\frac{\partial^2}{\partial\zeta\partial\bar{\zeta}} - (\alpha + \ell)\frac{\partial}{\partial\zeta} + (\beta + k)\frac{\partial}{\partial\bar{\zeta}}\right)g_2(\zeta) = 0$$

$(\alpha + \beta = N,\ \beta - \alpha = \gamma)$. *The polynomial solutions for* g_2 *are arbitrary linear combinations of* $C_n^{(\alpha+\ell,\ \beta+k)}(\zeta)$, $n = 0, 1, 2, \ldots$.

Proof. The change of variable $\zeta = t + i|z|^2$ leads to this equation. Polynomial solutions can be split by degree of homogeneity, and the substitution

$$g_2(\zeta) = \sum_{j=0}^{n} a_j \bar{\zeta}^j \zeta^{n-j}$$

leads to the two-term recurrence

$$(j + 1)(\beta + k + n - j - 1)a_{j+1} - (n - j)(\alpha + \ell + j)a_j = 0,$$

whose unique solution is

$$a_j = c\,\frac{(\alpha + \ell)_j\,(\beta + k)_{n-j}}{j!\,(n - j)!},$$

some constant c.

Another solution for g_2 is given by $(c - \bar{\zeta})^{-\alpha-\ell}(c - \zeta)^{-\beta-k}$, some constant c. To obtain the addition formula, we first deduce the general form of the expansion and then find the coefficients by a Lie technique. Note that $C_n^{(\alpha,\ \beta)}(S((-w, -s)\cdot(z, t)))$ is homogeneous of degree $2n$, in the sense that the replacement of (z, w, t, s) by (rz, rw, r^2t, r^2s) multiplies the value by r^{2n} $(r > 0)$. Further, if this function is to be expanded as a function of z, for fixed w, with respect to $U(N)$, then the terms must be invariant under the subgroup fixing w, hence functions of $\langle z, w\rangle$, $|z|$, $|w|$ only (namely, the $F_{k\ell}$ functions; see Definition 3). Since L_γ is left-invariant, each term in the $U(N)$-expansion must be L_γ-harmonic (in z), so the multipliers of $F_{k\ell}(z, w)$ must be multiples of $C_m^{(\alpha+\ell,\ \beta+k)}(t + i|z|^2)$ $(0 \le m \le n - k - \ell)$.

On the other hand, for fixed (z, t), the left side of Theorem 1 is a left translate of the function

$$(s,\ w) \leftrightarrow C_n^{(\alpha,\ \beta)}(S((s,\ w)^{-1})) = C_n^{(\alpha,\ \beta)}(-s + i|w|^2)$$
$$= (-1)^n C_n^{(\beta,\ \alpha)}(s + i|w|^2),$$

which (by interchanging α and β, or γ and $-\gamma$) can similarly be expanded in a series of

$$F_{\ell k}(w,\ z) C_m^{(\beta + k,\ \alpha + \ell)}(s + i|w|^2) = F_{k\ell}(z,\ w)(-1)^m C_m^{(\alpha + \ell,\ \beta + k)}(-s + i|w|^2).$$

Thus

$$C_n^{(\alpha,\ \beta)}(S((-w,\ -s)\cdot(z,\ t))) = \sum_{k+\ell+m \le n} A_{k\ell m}^n F_{k\ell}(z,\ w) \cdot C_m^{(\alpha+\ell,\ \beta+k)}(t + i|z|^2) C_{n-k-\ell-m}^{(\alpha+\ell,\ \beta+k)}(-s + i|w|^2) \tag{1}$$

with the constants $A_{k\ell m}^n$ to be determined.

To do this, we will use two differential operators that commute with left translation by a fixed $(w,\ s) \in H_N$. Accordingly, when such an operator is applied to the right-hand side of identity (1), the same sum will be obtained regardless of whether the operator acts on the $(z,\ t)$ or the $(w,\ s)$ variables. Equating the two results will give recurrence relations for $A_{k\ell m}^n$.

The operators are T, defined by

$$Tf(z,\ t) := \frac{\partial}{\partial\theta} f((0,\ \theta)\cdot(z,\ t))\Big|_{\theta=0} \quad \text{(that is, } \partial/\partial t\text{)},$$

and R, defined by

$$Rf(z,\ t) := \frac{\partial}{\partial\theta} f((\theta w,\ 0)\cdot(z,\ t))\Big|_{\theta=0}$$

[for f differentiable, $(z,\ t) \in H_N$, θ real].

Equating the two results of applying T to (1), we obtain

$$\sum_{k+\ell+m \le n} A_{k\ell m}^n F_{k\ell}(z,\ w)(N + n - m - 1) C_m^{(\alpha+\ell,\ \beta+k)}(t + i|z|^2) \cdot C_{n-k-\ell-m-1}^{(\alpha+\ell,\ \beta+k)}(-s + i|w|^2) = \sum_{k+\ell+m \le n} A_{k\ell m}^n F_{k\ell}(z,\ w) \cdot (N + k + \ell + m - 1) C_{m-1}^{(\alpha+\ell,\ \beta+k)}(t + i|z|^2) C_{n-k-\ell-m}^{(\alpha+\ell,\ \beta+k)}(-s + i|w|^2)$$

[using $\partial/\partial t = \partial/\partial\zeta + \partial/\partial\bar{\zeta}$, when $\zeta = t + i|z|^2$ and the identity (9)]. This leads to the recurrence

$$A^n_{k\ell,\,m+1} = \frac{(N + n - m - 1)}{(N + k + \ell + m)} A^n_{k\ell m}$$

with solution

$$A^n_{k\ell m} = A^n_{k\ell 0} \frac{(N + k + \ell)_{n-k-\ell}}{(N + k + \ell)_m \, (N + k + \ell)_{n-k-\ell-m}} \tag{2}$$

(note $A^n_{k\ell,\,n-k-\ell} = A^n_{k\ell 0}$).

We use R to get relations among the $A^n_{k\ell 0}$ coefficients. We already know $A^n_{000} = 1$ [set $(w, s) = (0, 0)$]. Since

$$(-w, -s) \cdot \big((\theta w, 0) \cdot (z, t)\big) = \big((\theta - 1)w, -s\big) \cdot (z, t),$$

we apply $\partial/\partial\theta$ to the right-hand side of (1) with $(-w, -s)$ replaced by $\big((\theta - 1)w, -s\big)$ and obtain

$$\sum_{k+\ell+m\le n} A^n_{k\ell m} C_m^{(\alpha+\ell,\, \beta+k)}(t + i|z|^2) F_{k\ell}(z, w)\big(-(k + \ell) + R\big) C_{n-k-\ell-m}^{(\alpha+\ell,\, \beta+k)}(-s + i|w|^2)$$

(setting $\theta = 0$). It will suffice to use the $m = n - k - \ell$ terms. On the other hand, we have:

Lemma 3. $RF_{k\ell}(z, w) C_m^{(\alpha+\ell,\, \beta+k)}(t + i|z|^2)$

$$\begin{aligned} &= \big(2i(\beta + k)(N + k - 1)F_{k+1,\,\ell}\, C_{m-1}^{(\alpha+\ell,\, \beta+k+1)} \\ &\quad - 2i(\alpha + \ell)(N + \ell - 1)F_{k,\,\ell+1}\, C_{m-1}^{(\alpha+\ell+1,\, \beta+k)} \\ &\quad + (N + k + \ell + m - 1)|w|^2\big(k\, F_{k-1,\,\ell}\, C_m^{(\alpha+\ell,\, \beta+k-1)} \\ &\quad + \ell F_{k,\,\ell-1} C_m^{(\alpha+\ell-1,\, \beta+k)}\big)\big)/(N + k + \ell - 1), \end{aligned} \tag{3}$$

with arguments (z, w) and $S(z, t)$ for the F and C functions, respectively.

Proof. We use differential, recurrence, and contiguity relations for F- and C-functions. First we note that

$$RF(z) = \sum_{j=1}^{N} \left(w_j \frac{\partial}{\partial z_j} + \bar{w}_j \frac{\partial}{\partial \bar{z}_j} \right) F(z),$$

so

$$RF_{k\ell}(z, w) = k|w|^2 F_{k-1,\ell}(z, w) + \ell|w|^2 F_{k,\ell-1}(z, w),$$

where we use $R\langle z, w\rangle = R\langle w, z\rangle = |w|^2$, $R(|z|^2) = 2\mathrm{Re}\langle z, w\rangle$, and a direct computation with the definition of $F_{k\ell}$ (see Definition 3).

Let $\zeta := t + i|z|^2$; then

$$Rg(\zeta) = 2i\left(\langle z, w\rangle\frac{\partial}{\partial\zeta} - \langle w, z\rangle\frac{\partial}{\partial\bar{\zeta}}\right)g(\zeta).$$

Using these relations, the recurrence for $\langle z, w\rangle F_{k\ell}$ (18) and $\frac{\partial}{\partial\zeta} C_m^{(\alpha,\beta)}(\zeta) = \beta\, C_{m-1}^{(\alpha,\beta+1)}(\zeta)$ (8), we obtain the term $F_{k+1,\ell}$ with coefficient as in (3) and the term $F_{k,\ell-1}$ with coefficient

$$\ell|w|^2(C_m^{(\alpha+\ell,\beta+k)}(\zeta) + (2i|z|^2(\beta+k)/(N+k+\ell-1))C_{m-1}^{(\alpha+\ell,\beta+k+1)}(\zeta)),$$

which equals

$$\ell|w|^2((N+k+\ell+m-1)/(N+k+\ell-1))C_m^{(\alpha+\ell-1,\beta+k)}(\zeta)$$

by (10). A similar computation applies to $\langle w, z\rangle\frac{\partial}{\partial\bar{\zeta}} C_m^{(\alpha,\beta)}(\zeta)$, $F_{k,\ell+1}$, and $F_{k-1,\ell}$.

Theorem 3. The coefficient in (1),

$$A_{k\ell m}^n = \frac{2^{k+\ell} i^{\ell-k}(\alpha)_\ell(\beta)_k(N-1)_k(N-1)_\ell(N+k+\ell)_{n-k-\ell}}{(N-1)_{k+\ell}k!\ell!(N+k+\ell)_m(N+k+\ell)_{n-k-\ell-m}},$$

$$(k+\ell+m \le n).$$

Proof. We equate the coefficients of $F_{k\ell}(z, w)C_{n-k-\ell}^{(\alpha+\ell,\beta+k)}(t+i|z|^2)$ in the two expressions, which result from applying R to (1). This gives the recurrence

$$\frac{2i((\beta+k-1)(N+k-2)A_{k-1,\ell,n-k-\ell+1} - (\alpha+\ell-1)(N+\ell-2)A_{k,\ell-1,n-k-\ell+1})}{(N+k+\ell-2)}$$

$$= -(k+\ell)A_{k,\ell,n-k-\ell}$$

[left side from (z, t), right from (w, s)]. We make the substitution (arrived at by judicious guessing)

$$A_{k,\, \ell,\, n-k-\ell} = \frac{2^{k+\ell} i^{\ell-k} (\alpha)_\ell (\beta)_k (N-1)_k (N-1)_\ell}{(N-1)_{k+\ell} (k+\ell)!} C_{k\ell}$$

(for $k + \ell \leq n$) and obtain the equation $C_{k-1,\, \ell} + C_{k,\, \ell-1} = C_{k\ell}$. Since $A_{00n} = A_{000} = C_{00} = 1$, the solution is obviously $C_{k\ell} = \binom{k+\ell}{\ell}$. This, together with (2), proves the theorem.

3. THE FUNDAMENTAL SOLUTION OF L_γ

We will use a trick that exploits the addition formula to derive an expansion for an arbitrary translate of the L_γ-harmonic function $\psi_\gamma(z, t) = (-\bar{S}(z, t))^{-\alpha}(-S(z, t))^{-\beta}$.

Lemma 4. *For $m = 0, 1, 2, \ldots, r$ real,*

$$(r - \bar{\zeta})^{-\alpha-m}(r - \zeta)^{-\beta-m} C_m^{(\alpha,\, \beta)}(r - \bar{\zeta}) r^{m+N} m! = \sum_{n=0}^{\infty} r^{-n}(N + n)_m C_n^{(\alpha,\, \beta)}(\zeta), \tag{4}$$

with absolute convergence in $|z| < |r|$, and for fixed r the function is L_γ-harmonic ($\zeta = t + i|z|^2$).

Proof. We apply T^m to both sides of the generating function

$$(r - \bar{\zeta})^{-\alpha}(r - \zeta)^{-\beta} r^N = \sum_{n=0}^{\infty} r^{-n} C_n^{(\alpha,\, \beta)}(\zeta)$$

(in $|\zeta| < |r|$, r real), where $T = \partial/\partial\zeta + \partial/\partial\bar{\zeta}$. We have

$$TC_n^{(\alpha,\, \beta)}(\zeta) = (N + n - 1) C_{n-1}^{(\alpha,\, \beta)}(\zeta),$$

and inductively,

$$TC_m^{(\alpha,\, \beta)}(r - \bar{\zeta})(r - \bar{\zeta})^{-\beta-m}(r - \zeta)^{-\alpha-m} = (r - \bar{\zeta})^{-\alpha-m-1}(r - \zeta)^{-\beta-m-1}$$

$$\cdot \{((r - \bar{\zeta})(\beta + m) + (r - \zeta)(\alpha + m)) C_m^{(\alpha,\, \beta)}(r - \bar{\zeta})$$

$$- (r - \zeta)(r - \bar{\zeta})(N + m - 1) C_{m-1}^{(\alpha,\, \beta)}(r - \bar{\zeta})\}$$

$$= (m + 1)(r - \bar{\zeta})^{-\alpha-m-1}(r - \zeta)^{-\beta-m-1} C_{m+1}^{(\alpha,\, \beta)}(r - \bar{\zeta})$$

by (12).

If ζ is replaced by $r - \zeta_1$ in Lemma 4, then absolute convergence holds in $|\zeta_1 + r| < |r|$, which is possible only if $\operatorname{sgn} r = -\operatorname{sgn}(\operatorname{Re} \zeta_1) \neq 0$.

The argument will use three regions, all subsets of $H_N \times H_N$,

$$\Omega := \{((z, t), (w, s)) : \|(z, t)\| < \|(w, s)\|\}$$

$$\Omega' := \{((z, t), (w, s)) : \|(z, t)\|^2 + 2|z||w| < \|(w, s)\|^2\}$$

$$\Omega'' := \{((z, t), (w, s)) : \|(z, t)\|^2 + 2|z||w| < s/2\}$$

in decreasing order. We will use an auxiliary variable in Ω'' and extend the result to Ω' and Ω, in turn, by analytic continuation.

Let

$$B_{k\ell} := \frac{2^{k+\ell} i^{\ell-k} (\alpha)_\ell (\beta)_k (N-1)_k (N-1)_\ell}{k!\,\ell!\,(N-1)_{k+\ell}}.$$

Lemma 5. Suppose $((z, t), (w, s)) \in \Omega''$ *and* $\alpha > 0$, $\beta > 0$ *(that is,* $-N < \gamma < N$*). Then there exists* r *such that*

$$r - |r - s + i|w|^2| > s/2 \quad \text{and} \quad |S((w, r-s) \cdot (z, t))| < r;$$

and

$$\psi_\gamma((-w, -s) \cdot (z, t)) = \sum_{k,\ell \geq 0} F_{k\ell}(z, w) B_{k\ell}$$

$$\cdot \sum_{m=0}^{\infty} \frac{1}{(N+k+\ell)_m} C_m^{(\alpha+\ell,\, \beta+k)}(t + i|z|^2) r^{-(k+\ell+m+N)} \tag{5}$$

$$\cdot \sum_{j=0}^{\infty} r^{-j} (N+k+\ell+j)_m C_j^{(\alpha+\ell,\, \beta+k)}(r - s + i|w|^2),$$

with absolute convergence.

Proof. Fix a point in Ω''. Since $s > 0$, $\lim_{r\to\infty} (r - |r - s + i|w|^2|) = s$ $(r > 0)$, there exists an r with $r - |r - s + i|w|^2| > s/2$, and so $|S((-w, r-s) \cdot (z, t))| \leq |r - s + i|w|^2| + \|(z, t)\|^2 + 2|z||w| < r$ (by definition of Ω''). Using (4) with $m = 0$, we have

$$\psi_\gamma((-w, -s) \cdot (z, t)) = (r - \bar{S}((-w, r-s) \cdot (z, t)))^{-\alpha}$$

$$(r - S((-w, r-s) \cdot (z, t)))^{-\beta}$$

$$= \sum_{n=0}^{\infty} r^{-n-N} C_n^{(\alpha,\, \beta)}(S((-w, r-s) \cdot (z, t)))$$

(by the addition theorem)

$$= \sum_{n=0}^{\infty} r^{-n-N} \sum_{k+\ell+m \leq n} A^n_{k\ell m} F_{k\ell}(z, w) C_m^{(\alpha+\ell,\, \beta+k)}(t + i|z|^2)$$

$$\cdot\, C^{(\alpha+\ell,\, \beta+k)}_{n-k-\ell-m}(r - s + i|w|^2)$$

$$= \sum_{k,\,\ell \geq 0} F_{k\ell}(z, w) \sum_{m=0}^{\infty} C_m^{(\alpha+\ell,\, \beta+k)}(t + i|z|^2) r^{-(k+\ell+m+N)}$$

$$\cdot \sum_{j=0}^{\infty} r^{-j} A^{j+k+\ell+m}_{k\ell m} C_j^{(\alpha+\ell,\, \beta+k)}(r - s + i|w|^2),$$

which is the sum (5). The interchange of summation needs to be justified. The sum is dominated by

$$\sum_{k,\,\ell,\,m,\,j} (|z|\,|w|)^{k+\ell} |B_{k\ell}| \frac{(N + k + \ell)_{m+j}}{m!\,j!} r^{-(k+\ell+m+j+N)}$$

$$\|(z, t)\|^{2m} |r - s + i|w|^2|^j$$

[see (13)]. The sums over j, then m, are negative binomial series, giving the sum

$$\sum_{k,\,\ell} (|z|\,|w|)^{k+\ell} |B_{k\ell}| (r - |r - s + i|w|^2| - \|(z, t)\|^2)^{-N-k-\ell}$$

(the last term is positive by construction). Finally,

$$|B_{k\ell}| \leq K(k + \ell)^M 2^{k+\ell},$$

where K and M are constants, so the (k, ℓ) series converges absolutely.

Lemma 6. $\psi_\gamma((-w, -s) \cdot (z, t)) = \sum_{k,\,\ell \geq 0} F_{k\ell}(z, w) B_{k\ell} \sum_{m=0}^{\infty} \frac{m!}{(N + k + \ell)_m}$

$$\cdot\, C_m^{(\alpha+\ell,\, \beta+k)}(t + i|z|^2) C_m^{(\alpha+\ell,\, \beta+k)}(s + i|w|^2) \tag{6}$$

$$\cdot\, (s + i|w|^2)^{-\alpha-\ell-m}(s - i|w|^2)^{-\beta-k-m},$$

with absolute convergence in Ω'.

Proof. With the variables as in Lemma 5, sum (6) follows immediately from (5), because the sum over j is done by Lemma 4 with

$\zeta = r - s + i|w|^2$ (appropriate α, β). We claim that the right-hand side of (6) converges uniformly and absolutely on compact subsets of Ω' and γ in any compact subset of $\mathbf{C}$. Indeed, let M be an integer so that $|\alpha| + |\beta| \leq N + M$. The m-sum in (6) is dominated term-wise by the series

$$\sum_{m=0}^{\infty} \frac{((|\alpha| + |\beta| + k + \ell)_m)^2}{(N + k + \ell)_m m!} \|(z, t)\|^{2m} \|(w, s)\|^{-2(m+k+\ell)} W,$$

which in turn is dominated by the series

$$W\|(w, s)\|^{-2(k+\ell)} {}_2F_1(N + M + k + \ell, N + M + k + \ell; N + k + \ell; x)$$
$$= W\|(w, s)\|^{-2(k+\ell)} (1 - x)^{-N-2M-k-\ell} {}_2F_1(-M, -M; N + k + \ell; x)$$

(by Euler's transformation), where

$$W = |(s + i|w|^2)^{-\alpha} (s - i|w|^2)^{-\beta}|$$

and

$$x = (\|(z, t)\| / \|(w, s)\|)^?.$$

The last series terminates and has positive terms only, so

$${}_2F_1(-M, -M; N + k + \ell; x) < {}_2F_1(-M, -M; N + k + \ell; 1)$$
$$= \frac{(N + M + k + \ell)_M}{(N + k + \ell)_M},$$

which is bounded in $k + \ell$. Since $\log|B_{k\ell}| = O((k + \ell)\log 2)$ [see (5)], the (k, ℓ)-sum converges absolutely if

$$2|z||w| < \|(s, w)\|^2 - \|(t, z)\|^2.$$

Consider both sides of (6) as functions of the variables s, t, $|z|$, $|w|$, ξ, $\bar{\xi}$ (where $\xi := \langle z, w\rangle / |z||w|$), γ. Any compact subset of $\{t^2 + |z|^4 < s^2 + |w|^4, |\xi| \leq 1, \gamma \in \mathbf{C}, s, t, |z|, |w|$ real$\}$ can be fattened slightly to allow complex values and still preserve the uniform and absolute convergence in the sum as above. Thus, by analytic continuation, the identity (6) holds throughout Ω', $\gamma \in \mathbf{C}$.

Theorem 4. The expansion (6) *holds throughout* Ω *in the disc polynomial sense; in particular, for fixed* $|z|$, t, $|w|$, s,

$$\psi_\gamma((-s, -w) \cdot (t, z)) = \lim_{r \to 1^-} \sum_{k, \ell \geq 0} r^{k+\ell} F_{k\ell}(z, w) B_{k\ell} G_{k\ell}(t + i|z|^2, s + i|w|^2),$$

where $G_{k\ell}$ *denotes the value of the* m*-sum in* (6). *The Cesaro* $(C, 2N - 1)$ *sums in the* $(k + \ell)$*-index also converge.*

Proof. Let $\Psi(\xi; |z|, t, |w|, s) := (s - t + i|z|^2 + i|w|^2 - 2i|z||w|\bar{\xi})^{-\alpha}(s - t - i|z|^2 - i|w|^2 + 2i|z||w|\xi)^{-\beta} = \psi_\gamma((-w, -s) \cdot (z, t))$ (where $\xi = \langle z, w\rangle/|z||w|$). Then Ψ is a smooth function in ξ provided $(|z|, t) \neq (|w|, s)$, and as such, it has a disc polynomial expansion [see (19) and (20)]

$$\sum_{k, \ell \geq 0} h_{k\ell}^{(N-2)} \hat{\Psi}_{k\ell}(|z|, t, |w|, s) R_{k\ell}^{(N-2)}(\xi),$$

where

$$\hat{\Psi}_{k\ell} = \frac{(N - 1)}{\pi} \int_D \Psi(\xi; |z|, t, |w|, s) \cdot \overline{R_{k\ell}(\xi)}(1 - |\xi|^2)^{N-2} \, dm(\xi).$$

In Lemma 6 we showed that

$$\hat{\Psi}_{k\ell} = (2|z||w|)^{k+\ell} i^{\ell-k} \frac{(\alpha)_\ell (\beta)_k}{(N)_{k+\ell}} G_{k\ell}$$

holds in Ω'. But $\hat{\Psi}_{k\ell}$ and $G_{k\ell}$ are analytic in the variables $|z|$, t, $|w|$, s in the domain Ω (by an argument similar to that of Lemma 6) and agree in Ω', so by analytic continuation agree in Ω. The convergence follows from Theorem 5.

Greiner studied the L_γ-harmonics primarily because of their hoped-for applicability to the Dirichlet problem (which is solvable with $\gamma = 0$, for the ball $\|(z, t)\| \leq 1$, Gaveau [5]). He pointed out for a particular example $[(w, s) = (i, 0), N = 1]$ that the expansion (6) arranged by degree of homogeneity of $\|(z, t)\|$ (first sum over $k + \ell + 2m = n$ then n) converges only in the region Ω' [an argument based on power series in $\|(z, t)\|$]. However, the fact that the m-series, for each k, ℓ, converges absolutely in Ω still supports the conjecture that the functions $\{C_n^{(\alpha,\beta)}(\zeta) : n = 0, 1, 2, \ldots\}$ are a complete set on $|\zeta| = 1$, $\operatorname{Im}\zeta \geq 0$, for most values of α and β.

4. SOME RESULTS ON $C_n^{\alpha,\beta}$ AND THE DISC POLYNOMIALS

The generating function for the $C_n^{(\alpha,\beta)}$-functions is

$$(1 - r\bar{\zeta})^{-\alpha}(1 - r\zeta)^{-\beta} = \sum_{n=0}^{\infty} r^n C_n^{(\alpha,\beta)}(\zeta),$$

for $|r\zeta| < 1$. These hold (all can be directly verified from Definition 2):

$$\left((\zeta - \bar{\zeta})\frac{\partial^2}{\partial\zeta\partial\bar{\zeta}} - \alpha\frac{\partial}{\partial\zeta} + \beta\frac{\partial}{\partial\bar{\zeta}}\right)C_n^{(\alpha,\beta)}(\zeta) = 0; \tag{7}$$

$$\begin{aligned} &\text{(i)} \quad \frac{\partial}{\partial\zeta} C_n^{(\alpha,\beta)}(\zeta) = \beta C_{n-1}^{(\alpha,\beta+1)}(\zeta), \\ &\text{(ii)} \quad \frac{\partial}{\partial\bar{\zeta}} C_n^{(\alpha,\beta)}(\zeta) = \alpha C_{n-1}^{(\alpha+1,\beta)}(\zeta); \end{aligned} \tag{8}$$

$$\left(\frac{\partial}{\partial\zeta} + \frac{\partial}{\partial\bar{\zeta}}\right)C_n^{(\alpha,\beta)}(\zeta) = (\alpha + \beta + n - 1)C_{n-1}^{(\alpha,\beta)}(\zeta); \tag{9}$$

$$C_n^{(\alpha,\beta+1)}(\zeta) - (\zeta - \bar{\zeta})\frac{\alpha}{\alpha + \beta} C_{n-1}^{(\alpha+1,\beta+1)}(\zeta) = \frac{\alpha + \beta + n}{\alpha + \beta} C_n^{(\alpha,\beta)}(\zeta); \tag{10}$$

$$C_n^{(\alpha+1,\beta)}(\zeta) + (\zeta - \bar{\zeta})\frac{\beta}{\alpha + \beta} C_{n-1}^{(\alpha+1,\beta+1)}(\zeta) = \frac{\alpha + \beta + n}{\alpha + \beta} C_n^{(\alpha,\beta)}(\zeta); \tag{11}$$

$$\begin{aligned} &(\zeta(\beta + n) + \bar{\zeta}(\alpha + n))C_n^{(\alpha,\beta)}(\zeta) - (\alpha + \beta + n - 1)\zeta\bar{\zeta}C_{n-1}^{(\alpha,\beta)}(\zeta) \\ &= (n + 1)C_{n+1}^{(\alpha,\beta)}(\zeta); \end{aligned} \tag{12}$$

$$\begin{aligned} &\text{(i)} \quad C_n^{(\alpha,\beta)}(1) = \frac{(\alpha + \beta)_n}{n!}, \\ &\text{(ii)} \quad |C_n^{(\alpha,\beta)}(\zeta)| \leq \frac{(|\alpha| + |\beta|)_n}{n!}|\zeta|^n. \end{aligned} \tag{13}$$

The disc polynomials $R_{k\ell}^{\lambda}$ (see Definition 3) satisfy (for $\lambda > -1$):

$$R_{k\ell}^{(\lambda)}(1) = 1; \tag{14}$$

$$\bar{R}_{k\ell}^{(\lambda)}(\xi) = R_{\ell k}^{(\lambda)}(\xi); \tag{15}$$

$$|R_{k\ell}^{(\lambda)}(\xi)| \leq 1, \text{ for } |\xi| \leq 1 \text{ and } \lambda \geq 0; \tag{16}$$

$$R_{k\ell}(\xi) = \frac{\ell!\xi^{k-\ell}}{(\lambda+1)_\ell} P_\ell^{(\lambda, k-\ell)}(2|\xi|^2 - 1) \tag{17}$$

[the Jacobi polynomial, for $k \geq \ell$; use (15) for $\ell \geq k$];

$$\xi R_{k\ell}^{(\lambda)}(\xi) = \left((\lambda + k + 1)R_{k+1,\ell}^{(\lambda)}(\xi) + \ell R_{k,\ell-1}^{(\lambda)}(\xi)\right)/(\lambda + k + \ell + 1) \tag{18}$$

and a similar relation for $\bar{\xi} R_{k\ell}^{(\lambda)}(\xi)$ [use (15)], (Boyd [1]). They form an orthogonal basis for $L^2(D, \mu_\lambda)$, where $D := \{\xi \in \mathbf{C}: |\xi| \leq 1\}$ and $d\mu_\lambda(\xi) := ((\lambda+1)/\pi)(1 - |\xi|^2)^\lambda dm(\xi)$ (dm is Lebesgue measure in $\mathbf{R}^2$); μ_λ is a probability measure. The orthogonality relation is:

$$\int_D R_{k\ell}^{(\lambda)} \bar{R}_{mn}^{(\lambda)} d\mu_\lambda = \delta_{km}\delta_{\ell n}/h_{k\ell}^{(\lambda)}$$

$$\text{where } h_{k\ell}^{(\lambda)} := \frac{k + \ell + \lambda + 1}{\lambda + 1} \frac{(\lambda+1)_k (\lambda+1)_\ell}{k!\ell!}.$$

For $\lambda = N - 2$, $N = 2, 3, \ldots$, they are spherical functions for $U(N)/U(N-1)$ (Ikeda [8], see also Koornwinder [10]) and $h_{k\ell}^{(N-2)} = \dim V_{k\ell}$ (see Definition 1). Folland [2] used them to expand the Poisson-Szegö kernel for the ball in $\mathbf{C}^N$.

Each $f \in L^1(D, \mu_\lambda)$ has a formal series

$$\sum_{k,\ell \geq 0} h_{k\ell}^{(\lambda)} \hat{f}_{k\ell} R_{k\ell}^{(\lambda)}, \text{ where } \hat{f}_{k\ell} = \int_D f\bar{R}_{k\ell}^{(\lambda)} d\mu_\lambda. \tag{20}$$

Two summation processes are suggested by the interpretation of the unit sphere in $\mathbf{C}^N$ as the unit sphere in $\mathbf{R}^{2N}$ and the fact that $\sum_{k+\ell=n} V_{k\ell}$ is the space of spherical harmonics of degree n on $\mathbf{R}^{2N}$. The following is proved by using the corresponding result for ultraspherical series of index $\lambda + 1$ and the usual convolution methods; the Cesaro summability was established by Kogbetliantz [9, p. 169].

Theorem 5. For $f \in C(D)$, define the functions

$$P_r f(\xi) := \sum_{n=0}^{\infty} r^n \sum_{k+\ell=n} h_{k\ell}^{(\lambda)} \hat{f}_{k\ell} R_{k\ell}^{(\lambda)}(\xi), \quad 0 \le r < 1,$$

$$S_m^{(\delta)} f(\xi) := \sum_{n=0}^{m} \frac{(-m)_n}{(-\delta - m)_n} \sum_{k+\ell=n} h_{k\ell}^{(\lambda)} \hat{f}_{k\ell} R_{k\ell}^{(\lambda)}(\xi),$$

$m = 1, 2, \ldots, \delta \ge 2\lambda + 3$. *Then $P_r f \to f$ and $S_m^{(\delta)} f \to f$ uniformly on D as $r \to 1-$ and $m \to \infty$, respectively, and if $f \ge 0$, then so are $P_r f$ and $S_m^{(\delta)} f$.*

REFERENCES

1. Boyd, J. Orthogonal polynomials on the disc. M.A. Thesis, Univ. of Virvinia, 1972.

2. Folland, G. B. Spherical harmonic expansion of the Poisson-Szegö kernel for the ball. *Proc. Amer. Math. Soc.* 47 (1975): 401-408.

3. Folland, G. B., and Stein, E. M. Estimates for the $\bar{\partial}_b$ complex and analysis on the Heisenberg group. *Comm. Pure and Appl. Math.* 27 (1974):429-522.

4. Gasper, G. Orthogonality of certain functions with respect to complex valued weights. To appear.

5. Gaveau, B. Principe de moindre action, propagation de chaleur et estimées sous elliptiques sur certains groupes nilpotents. *Acta Math.* 139 (1977):95-153.

6. Greiner, P. C. Spherical harmonics on the Heisenberg group. *Canad. Math. Bull.* 23 (1980):383-396.

7. Greiner, P. C., and Stein, E. M. *Estimates for the $\bar{\partial}$-Neumann Problem. Mathematical Notes.* Princeton, N.J.: Princeton Univ. Press, 1977.

8. Ikeda, M. On spherical functions for the unitary group, I, II, III. *Mem. Fac. Engrg. Hiroshima Univ.* 3 (1967):17-75.

9. Kogbetliantz, E. Recherches sur la sommabilité de séries ultrasphériques par la methode des moyennes arithmétiques. *J.de Math* . (9)3 (1924):107-187.

10. Koornwinder, T. Two-variable analogues of the classical orthogonal polynomials. In *Theory and Application of Special Functions*, R, Askey, Ed. New York: Academic Press, 1975, pp. 435-495.

SUR LES DIFFEOMORPHISMES DU CERCLE DE NOMBRE DE ROTATION DE TYPE CONSTANT

M.R. Herman
Centre de Mathématiques de l'Ecole Polytechnique, Palaiseau

1. INTRODUCTION

Pour presque tout nombre α $\mathbf{T}^1 - (\mathbf{Q}/\mathbf{Z})$ (presque tout pour la mesure de Lebesgue $m = d\theta$), si f est un difféomorphisme du cercle $\mathbf{T}^1 = \mathbf{R}/\mathbf{Z}$ de classe $C^{2+\varepsilon}$, $\varepsilon > 0$, si $\rho(f) = \alpha$, et si f est $C^{2+\varepsilon}$-proche de la translation $R_\alpha : \mathbf{T}^1 \to \mathbf{T}^1$, $R_\alpha(x) = x + \alpha$, alors f est C^1-conjugué à R_α. Le difféomorphisme f laisse invariant une mesure de densité de classe C^0, strictement positive (voir [4, Appendice]).

De plus, si f est de classe C^3, il n'y a pas de restriction de proximité (voir [4, IX]).

J. Hawkins et K. Schmidt [3] ont montré que, si α n'est pas un nombre de type constant [i.e., $\operatorname{Inf}_{p/q}(q^2|\alpha - (p/q)|) = 0$, où $p/q \in \mathbf{Q}$, $(p, q) = 1$] et si α est un nombre irrationnel, alors il existe un difféomorphisme f de classe C^2 vérifiant $\rho(f) = \alpha$; f est aussi C^2-proche de R_α qu'on veut; et de plus f ne laisse pas de mesure σ-finie invariante absolument par rapport à la mesure de Lebesgue.

Y. Katznelson [7] a montré que, pour tout module de continuité w différent de Lipschitz (Lip_1) et tout nombre $\alpha \in \mathbf{T}^1 - (\mathbf{Q}/\mathbf{Z})$

(vérifiant l'hypothèse suivante: si $\alpha = [a_0, a_1, \ldots]$, où les a_i sont les quotients partiels du développement en fraction continue de α, alors il existe une infinité de $a = a$, avec $a_i > 15$), il existe un difféomorphisme f de classe C^1 du cercle, $Df > 0$, $Df \in \mathrm{Lip}(w)$, f est C^{1+w}-proche de R_α, $\rho(f) = R_\alpha$, et telle que f ne laisse pas invariant de mesure σ-finie absolument continue par rapport à la mesure de Lebesgue. (Pour le théorème de Denjoy en classe C^{1+w}, voir [4, X]).

Nous proposons de démontrer que si α est un nombre de type constant, alors il existe $\varepsilon > 0$, tel que si f est un difféomorphisme de classe C^2 du cercle vérifiant $\|f - R_\alpha\|_{C^2} < \varepsilon$ et $\rho(f) = \alpha$, alors f laisse invariant une mesure strictement positive de densité L^1 équivalente à la mesure de Lebesgue.

Ceci est équivalent à ce que f soit conjugué à R_α par un homéomorphisme du cercle h qui soit un homéomorphisme absolument continu: si $f = h^{-1} \circ R_\alpha \circ h$, alors

$$h(x) = \int_0^x d\mu(t),$$

où μ est l'unique mesure de probabilité de $\mathbf{T}^1$ invariante par f [si $\rho(f) = a \notin \mathbf{Q}/\mathbf{Z}$, μ est unique et sans atome, et on a $h_*\mu = m$].

La mesure μ est donc la dérivée au sens des distributions de la fonction à variation bornée h.

On peut "raisonnablement" conjecturer qu'en général pour la catégorie de Baire un difféomorphisme de classe C^2 de nombre de rotation α de type constant n'est pas C^1-conjugué à la rotation R_α (cf. [4, XIII 4.6]).

Ce travail repose de façon essentielle sur le lemme d'Yves Meyer [8], et je désire lui exprimer ma très grande reconnaissance d'avoir démontré ce lemme. Je tiens aussi à le remercier de m'avoir signalé le résultat de Peter Jones [6].

2. RAPPEL SUR LES ESPACES DE SOBOLEV

2.1

Soient r un entier, $r \geq 1$, et $p > 1$ ($p \in \mathbf{R}$). On considère l'espace de Sobolev $W^{r,p} = \{\varphi \in C^{r-1}(\mathbf{T}^1) \mid D\varphi \in L^p\}$, où $D\varphi$ est la dérivée de φ au sens des distributions et $L^p = L^p(\mathbf{T}^1, d\theta, \mathbf{R})$, et $d\theta$ est la mesure de Lebesgue (ou de Haar), de $\mathbf{T}^1 = \mathbf{R}/\mathbf{Z}$. On note aussi la mesure de Lebesgue par m ($m = d\theta$).

Si $\varphi \in L^p$, on note $\|\varphi\|_{L^p} = \left(\int_0^1 |\varphi(\theta)|^p d\theta\right)^{1/p}$.

2.2

Par Zygmund [11, VI 3.8, p. 242], si $\varphi \in W^{1,p}$,

(a) φ est absolument continue (φ est donc à variation bornée sur $\mathbf{T}^1$),

(b) La série de Fourier de

$$\varphi(x) = \sum_{k \in \mathbf{Z}} a_k e^{2\pi i k x}$$

vérifie $\Sigma |a_k| < +\infty$ (i.e., $\varphi \in$ classe A),

(c) $\varphi \in C^{1/p'}(\mathbf{T}^1)$ avec $(1/p) + (1/p') = 1$, en effet, on a

$$|\varphi(x + h) - \varphi(x)| \leq \int_x^{x+h} |D\varphi(t)| dt \leq \left(\int_0^1 |D\varphi(t)|^p dt\right)^{1/p} h^{1/p'}$$

et donc

$$|\varphi|_{1/p'} = \sup_{x \neq y} \frac{|\varphi(x) - \varphi(y)|}{|x - y|^{1/p'}} \leq \|D\varphi\|_{L^p}.$$

2.3

Par [10, V], on a (i) $\Leftrightarrow$ (ii):

(i) $\varphi \in L^p$ vérifie $\|\varphi(x + t) - \varphi(x)\|_{L^p} \leq t$, $t > 0$,

(ii) $\varphi \in W^{1/p}$ et $\|D\varphi\|_{L^p} \leq 1$.

2.4

On fixe $q > 1$. Soit

$$K_q = \left\{ \varphi \in L^q \,\middle|\, \int_0^1 \varphi(x)\,dx = 1, \quad \|D\varphi\|_{L^q} \leq 2, \ \varphi \geq 0 \right\}$$

($\varphi \geq 0$ veut dire Lebesgue presque partout). Pour la topologie faible induite par la dualité L^q, $L^{q'}$ $\left(\frac{1}{q} + \frac{1}{q'} = 1\right)$, K_q est faiblement compact convexe et faiblement métrisable. Noter que

$$\varphi \geq 0 \Leftrightarrow \forall\, \eta \in L^{q'},\ \eta \geq 0, \text{ on a } \int_0^1 \varphi\eta\, d\theta \geq 0.$$

2.5

A chaque $\varphi \in K_q$ on associe

$$h(x) = \int_0^x \varphi(t)\,dt.$$

L'application continue h de $\mathbf{R}$ dans $\mathbf{R}$ induit une application continue du cercle $\mathbf{T}^1$ dans lui-même, monotone non décroissante, homotope à l'identité et vérifiant $h(0) = 0$. En effet, on a $h(x + 1) = h(x) + 1$ puisque

$$\int_0^1 \varphi(t)\,dt = 1.$$

2.6

Lemme 1. Soit $(\varphi_i)_{i \in \mathbf{N}} \subset K_q$ *une suite convergeant faiblement vers* φ. *Soit* (h_i) *la suite d'applications continues associées à la suite* (φ_i). *Alors, si* $i \to +\infty$, $h_i \to h$ *pour la topologie de la convergence uniforme dans* $C^0(\mathbf{T}^1, \mathbf{T}^1)$.

Démonstration. Puisque la suite (φ_i) est faiblement convergente si $i \to +\infty$, $h_i \to h$ pour la topologie de la convergence simple, i.e.,

$$\forall\, x \in \mathbf{R}, \int_0^x \varphi_i(t)\,dt \to \int_0^x \varphi(t)\,dt.$$

Par 2.2(c), la suite $(h_i)_i$ est équicontinue et la résultat suit.

2.7

Lemme 2. Soient $(\varphi_i)_i \subset K_q$ *une suite faiblement convergente vers* φ *et* $(\psi_i)_i \subset L^\infty$ *une suite convergente dans* L^∞ *vers* ψ *pour la topologie de la norme* L^∞. *La suite* $(\psi_i \cdot \varphi_i) \in L^q$ *converge faiblement dans* L^q *vers* $\varphi\psi$.

Démonstration. Soit $\eta \in L^{q'}$. On doit montrer que, si $i \to +\infty$,

$$\int_0^1 (\psi_i \varphi_i \eta)(x)\,dx \to \int_0^1 (\psi\varphi\eta)(x)\,dx.$$

On a

$$\left| \int_0^1 \psi_i \varphi_i \eta \; dx - \int_0^1 \psi \varphi_i \eta \; dx \right| \leq \|\psi - \psi_i\|_{L^\infty} \|\varphi_i \eta\|_{L^1}$$

$$\leq \|\psi - \psi_i\|_{L^\infty} \|\varphi_i\|_{L^q} \|\eta\|_{L^{q'}} \leq \|\psi - \psi_i\|_{L^\infty} 2\|\eta\|_{L^{q'}},$$

et donc si $i \to +\infty$, $\|\psi - \psi_i\|_{L^\infty} \|\varphi_i \eta\|_{L^1} \to 0$.

Puisque $\psi\eta \in L^{q'}$ est une fonction fixée, et comme $\varphi_i \to \varphi$ faiblement dans L^q on a, si $i \to +\infty$,

$$\int_0^1 \psi\varphi_i\eta \; dx \to \int_0^1 \psi\varphi\eta \; dx. \quad \blacksquare$$

2.8

Soit la rotation $R_\alpha : \mathbf{T}^1 \to \mathbf{T}^1$, $R_\alpha(\theta) = \theta + \alpha$. L'application $\varphi \in L^q \to \varphi \circ R_\alpha \in L^q$ est une *isométrie* de L^q *faiblement continue* pour la topologie faible induite par la dualité entre L^q et $L^{q'}$.

3. RAPPELS SUR BMO

3.1

On désigne par BMO ou aussi BMO($\mathbf{T}^1$) l'espace des fonctions $\varphi \in L^1(\mathbf{T}^1)$ tel que

$$\sup_J \frac{1}{|J|} \int_J |\varphi(x) - \varphi_J| \, dx = \|\varphi\|_{\mathrm{BMO}} < +\infty,$$

où J est un intervalle de $\mathbf{T}^1$, $|J|$ sa longuer et

$$\varphi_J = \frac{1}{|J|} \int_J \varphi(x)\,dx.$$

Pour la norme $\|\varphi\|_* = \|\varphi\|_{L^1} + \|\varphi\|_{\text{BMO}}$, BMO est un espace de Banach. Comme $\|C\|_{\text{BMO}} = 0$, si C est une constante, $\|\ \|_{\text{BMO}}$ est seulement une seminorme.

Si $\varphi \in$ BMO vérifie $\int_0^1 \varphi(t)\,dt = 0$, on a $\|\varphi\|_{L^1} \leq \|\varphi\|_{\text{BMO}}$.

3.2

Le théorème suivant est de John et Nirenberg [5].

Théorème 1. Il existe des constantes $b > 0$, $B > 0$, telles que, si $\varphi \in$ BMO,

$$\kappa = \|\varphi\|_{\text{BMO}}$$

verifie $b' \leq \kappa^{-1} b$, alors $e^{b'|\varphi - \varphi_I|} \in L^1$ (où $I = [0, 1]$) et on a

$$\int_0^1 e^{b'|\varphi - \varphi_I|} dx \leq 1 + \frac{Bb'\kappa}{b - \kappa b'}.$$

3.3

Il suit que si $\varphi_I = 0$, $\kappa^{-1} b > b'$, on a

$$\int_0^1 e^{b'\varphi}\, dx \leq 1 + \frac{Bb'\kappa}{b - \kappa b'},$$

$$\int_0^1 e^{-b'\varphi}\, dx \leq 1 + \frac{Bb'\kappa}{b - \kappa b'}.$$

Il en résulte que si $q \geq 1$ est fixé, alors, pour tout $\varepsilon > 0$, il existe $\eta(q) > 0$ tel que, si φ vérifie $\|\varphi\|_{\text{BMO}} \leq \eta(q)$, $\int_0^1 \varphi(x)\,dx = 0$, on ait

$$\int_0^1 e^{q\varphi}\, dx \leq 1 + \varepsilon, \quad \int_0^1 e^{-q\varphi}\, dx \leq 1 + \varepsilon.$$

4. UNE LEMME REMARQUABLE DE REGULARITE (D'YVES MEYER)

4.1

Soit α un nombre de type constant, il existe $\gamma > 0$ tel que pour tout $p/q \in \mathbf{Q}$, $(p,q) = 1$, on ait $|\alpha = (p/q)| \geq \gamma q^{-2}$. La constante de Markoff de α est $\operatorname{Inf}_{p/q}(q^2|\alpha - (p/q)|) = \gamma$. Par le principe de Dirichlet, on a $\gamma < 1$.

Le lemme suivant est dû à Yves Meyer [8].

4.2

Lemme 3. Soit $p > 1$, alors il existe une constante C_p telle que, si α est un nombre de type constant, γ étant la constante de Markoff de α, pour tout $\varphi \in W^{1,p}$ vérifiant $\int_0^1 \varphi(x)\,dx = 0$, il existe un unique $\eta \in$ BMO,* $\int_0^1 \eta(x)\,dx = 0$, tel que l'on ait

$$\eta - \eta \circ R_\alpha = \varphi,$$

$$\|\eta\|_{\mathrm{BMO}} \leq C_p \gamma^{-1} \|D\varphi\|_{L^p}.$$

Remarques.

(1) La constante C_p ne dépend pas de α ni de γ, mais seulement de p.

(2) Si $p \geq 2$, il est évident que $\eta \in L^2$; ce qui est remarquable, c'est qu'on a beaucoup mieux.

Le lemme d'Yves Meyer résulte de [8] et du lemme suivant:

Lemme 4. Soit α un nombre de type constant, de constante de Markoff γ, et $p > 1$. Il existe une constante C_p^1 dépendant seulement de p, telle que, si $N \geq 1$, on ait

---

*On peut aussi montrer que η appartient à la classe $\exp(\lambda x^2)$: il existe $\lambda > 0$ tel que l'on ait $\int \exp(\lambda|\eta|^2)\,dx < +\infty$. Je remercie Yves Meyer de m'avoir aussi communiqué ce résultat.

$$\left(\sum_{N \leq |n| \leq 2N} \frac{1}{|n(1 - e^{2\pi i n\alpha})|^p}\right)^{1/p} \leq C_p^1 \gamma^{-1}.$$

Démonstration. Par la même démonstration [4, IX 6.7.1], on a

$$\sum_{0 < |n| \leq N} \frac{1}{|1 - e^{2\pi i n\alpha}|^p} \leq C_p'' \gamma^{-p} N^p,$$

ou C_p'' est une constante ne dépendant que de $p > 1$ (cette inégalité a été remarquée semble-t-il pour la première par Rüssmann [9] dans le cas $p = 2$). D'où:

$$\sum_{N \leq |n| \leq 2N} \frac{1}{|n(1 - e^{2\pi i n\alpha})|^p} \leq \frac{1}{N^p} \sum_{N \leq |n| \leq 2N} \frac{1}{|1 - e^{2\pi i n\alpha}|^p}$$

$$\leq \frac{1}{N^p} \sum_{0 < |n| \leq 2N} \frac{1}{|1 - e^{2\pi i n\alpha}|^p} \leq C_p'' \gamma^{-p} \frac{(2N)^p}{N^p} = 2^p C_p'' \gamma^{-p}.$$

5. DIFFEOMORPHISMES DE CLASSE SOBOLEV

5.1

Soient r un entier $r \geq 1$ et $D^r(\mathbf{T}^1) = \{f \in \mathrm{Diff}^r_+(\mathbf{R}) \,|\, f - \mathrm{Id} \in C^r(\mathbf{T}^1)\}$. $D^r(\mathbf{T}^1)$ est le groupe revêtement universel du groupe des difféomorphismes de classe C^r du cercle préservant l'orientation. Pour la C^r-topologie, c'est un groupe topologique.

5.2

Soient $r \geq 2$ et $p \geq 1$, on définit

$$D^{r,p}(\mathbf{T}^1) = \{f \in D^{r-1}(\mathbf{T}^1) \,|\, f - \mathrm{Id} \in W^{r,p}\}.$$

Comme $D^{r,p}(\mathbf{T}^1) \subset D^{r-1}(\mathbf{T}^1) \subset D^1(\mathbf{T}^1)$ et en utilisant la formule de Leibnitz et de Faa-di Bruno, on vérifie que $D^{r,p}(\mathbf{T}^1)$ est un groupe (c'est même un groupe topologique si on munit $W^{r,p}$ de sa topologie naturelle d'espace de Banach). On a $D^r(\mathbf{T}^1) \subset D^{r,p}(\mathbf{T}^1)$.

5.3

A chaque $f \in D^{r,p}(\mathbf{T}^1)$ est associé son nombre de rotation $\alpha = \rho(f) \in \mathbf{R}$ et l'application $f \to \rho(f)$ est continue (voir [4, II]).

5.4

Si $f \in D^{2,p}(\mathbf{T}^1)$ vérifie $\rho(f) = \alpha \notin \mathbf{Q}$, alors par le théorème de Denjoy, qui s'applique puisque $f \in D^1(\mathbf{T}^1)$ et que la fonction Df est à variation bornée (et même absolument continue par 2.2), alors f est C^0-conjugué par $h \in D^0(\mathbf{T}^1) = \{f \in \text{Homéo}_+(\mathbf{R}),\ f - \text{Id} \in C^0(\mathbf{T}^1)\}$ à la translation $R_\alpha : x \to x + \alpha$. Soit $f = h \circ R_\alpha \circ h^{-1}$. De plus, h est unique si on impose $h(0) = 0$ (si on a $h_1 \circ R_\alpha \circ h_1^{-1} = h \circ R_\alpha \circ h^{-1}$, alors $h_1^{-1} = R_\lambda \circ h^{-1}$, $\lambda \in \mathbf{R}$).

5.5

Si on n'impose pas de restriction sur $\alpha \notin \mathbf{Q}/\mathbf{Z}$, en général (pour la catégorie de Baire), même si $f \in D^\omega(\mathbf{T}^1)$, h est un homéomorphisme singulier pour la mesure de Lebesgue m: il existe $B \subset \mathbf{R}$ un borélien $m(B) = 0$ tel que $m(\mathbf{R} - h(B)) = 0$ (voir [4, XII]).

5.6

Soit $h \in D^0(\mathbf{T}^1)$ un homéomorphisme.

Définition. L'homeomorphisme h est appelé homéomorphisme absolument continu si les fonctions $h : [0, 1] \to \mathbf{R}$ et $h^{-1} : [0, 1] \to \mathbf{R}$ sont des fonctions absolument continues.

Puisque h et h^{-1} sont à variations bornées sur $[0, 1]$, il suffit de voir que si B est un borélien de mesure de Lebesgue nulle [i.e., $m(B) = 0$], alors $m(h(B)) = m(h^{-1}(B)) = 0$.

En utilisant ceci, on vérifie que les homéomorphismes absolument continus forment un groupe.

On a d'autre part que, si h est un homéomorphisme tel que $h|_{[0,1]}$ soit absolument continue, en général h^{-1} n'est pas absolument continue.

Exemple. Soient $K \subset [0, 1]$ un ensemble de Cantor de mesure de Lebesgue $m(K) > 0$ et φ une fonction de classe C^∞ sur $[0, 1]$ vérifiant:

$$\begin{cases} \varphi(x) > 0 & \text{si } x \notin K \\ \varphi(x) = 0 & \text{si } x \in K \end{cases}$$

$$\int_0^1 \varphi(x)\,dx = 1.$$

On pose

$$h(x) = \int_0^x \varphi(t)\,dt.$$

L'homéomorphisme $h: [0, 1] \to [0, 1]$ est une application de classe C^∞ mais h^{-1} n'est pas absolument continue.

5.7

Remarques.

(1) Soit $h \in D^0(\mathbf{T}^1)$ un homéomorphisme tel que l'application $h: [0, 1] \to \mathbf{R}$ soit absolument continue. Si $Dh(t) > 0$ pour presque tout $t \in [0, 1]$, alors $h^{-1}: [0, 1] \to \mathbf{R}$ est une fonction absolument continue. Il suit que h est un homéomorphisme absolument continu. L'hypothèse est vérifiée, par exemple, si $\log Df \in L^1$. (En effet, si h^{-1} n'était pas absolument continue, alors il existerait un ensemble de Cantor $K \subset [0, 1]$, $m(K) > 0$ tel que l'on ait $m(h(K)) = 0$ et donc $Dh(x) = 0$ pour presque tout $x \in K$.)

(2) Si $h \in D^0(\mathbf{T}^1)$ est un homéomorphisme, tel que l'application $h: [0, 1] \to \mathbf{R}$ soit singulière par rapport à m, il en est de même pour h^{-1} (une application à variation bornée est dite singulière si sa dérivée au sens de distributions

est une mesure singulière par rapport à la mesure de Lebesgue).

(3) Si f est un difféomorphisme de classe C^1 vérifiant $\rho(f) = \alpha \notin \mathbf{Q}/\mathbf{Z}$, alors f laisse invariant une unique mesure de probabilité μ. On décompose la mesure $\mu = \mu_{ab} + \mu_{\text{sing}}$ (la décomposition étant unique). Puisque f est de classe C^1, il suit que les mesures (≥ 0) μ_{ab} et μ_{sing} sont invariantes par f. De l'unicité de la mesure μ on a $\mu_{\text{sing}} = 0$ ou bien $\mu_{ab} = 0$.

Il résulte de $h \circ f = R_\alpha \circ h$, où $h(x) = \int_0^x d\mu(t)$, que si h n'est pas absolument continue alors h est singulière pour la mesure de Lebesgue.

Si h est absolument continue, en général, la mesure μ n'est pas équivalente à la mesure de Lebesgue (par exemple, un contre-exemple de Denjoy, de classe C^1, construit en [4, X-3] laissant invariant un ensemble de Cantor de mesure de Lebesgue strictement positive obtenu en perturbant une rotation R_α). Si f est un difféomorphisme de classe C^2, $\rho(f) = \alpha \notin \mathbf{Q}/\mathbf{Z}$, alors par le théorème de Denjoy, h est un homéomorphisme (voir [4, VI]). Par la m-ergodicité de f, si la fonction $h|_{[0,1]}$ est absolument continue, alors la mesure μ est équivalente à la mesure de Lebesgue (cf. [4, VII.1, XII.1]). Il en résulte que h est un homéomorphisme absolument continu.

6. THEOREME PRINCIPAL

On se propose de démontrer le théorème suivant:

6.1

Théorème 2. Soient $p > 1$ fixé, et α un nombre de type constant de Markoff γ. Il existe $\varepsilon_0 > 0$ (ne dépendant que de p et pas de α), tel que si $f \in D^{2,p}(\mathbf{T}^1)$ vérifie

(a) $\|D^2(f - R_\alpha)\|_{L^p} \leq \varepsilon_0\gamma$,

(b) $\rho(f) = \alpha$,

alors f est conjugué par un homéomorphisme $h \in D^0(\mathbf{T}^1)$ à R_α avec $h - \mathrm{Id} \in W^{1,q}$ et $h^{-1} - \mathrm{Id} \in W^{1,q}$, $q > 1$.

6.2

Remarques.

(1) Il suit que h est un homéomorphisme absolument continu.

(2) En fait, on a mieux; h appartient au groupe définit par Peter Jones [6]: le groupe des homéomorphismes f (absolument continus) de $\mathbf{T}^1$ qui préservent BMO (i.e., si $\varphi \in$ BMO, alors $\varphi \circ \mathbf{f} \in$ BMO). Ceci équivaut à $Df \in \bigcup_{p>1} A_p$.

(3) En utilisant 4.2 et [4] (en particulier le théorème fondamental de [4]), on peut montrer que si $f \in D^{r,p}(\mathbf{T}^1)$ ($r \in \mathbf{N}$, $r \geq 3$, et $p > 1$) vérifie $\rho(f) = \alpha$ est un nombre de type constant, alors $f = h \circ R_\alpha \circ h^{-1}$ où $h \in D^{r-1,p}(\mathbf{T}^1)$.

(4) Par une démonstration analogue à celle de 6.1 en utilisant 4.2 (et si $p = 2$, la remarque 2 de 4 suffit), on peut démontrer très simplement une version locale de la remarque précédente.

6.3

Démonstration.

① Soit $1 < p_1 < p$. On pose $p/p_1 = r > 1$, $r' = r/(r-1)$, et $q = (p - (1/r))r'$. On a $p > r$ et comme $q = (p-1)/(r-1)$, $q > 1$. Soit K_q l'ensmeble défini en 2.4.

② Par 2.5, à $\varphi \in K_q$ on associe $h(x) = \int_0^x \varphi(t)\,dt$. Nous allons construire une application $\Phi : K_q \to K_q$.

③ On se fixe f vérifiant $\|D^2 f\|_{L^p} < \varepsilon\gamma$, $\varepsilon > 0$ petit. Il en résulte que $\|Df - 1\|_{C^0} < 1/2$, si $\varepsilon < 1/2$ est assez petit et comme $\gamma < 1$ ceci ne dépend pas de γ. On a donc $Df > 0$. Soit

$$\lambda(f, h) = -\int_0^1 \log Df \circ h(t)\,dt;\ \lambda(f, h) \in \mathbf{R}.$$

L'application $(h, f) \to \lambda(f, h)$ est continue pour la C^0-topologie. La fonction $\log Df \circ h + \lambda(f, h)$ est d'intégrale nulle et $\log Df \circ h + \lambda(f, h) \in W^{1,p_1}$, pour cela on utilise les lemmes suivants:

Lemme 5. Si $f \in D^{2,p}(\mathbf{T}^1)$, alors $\log Df \in W^{1,p}$.

Démonstration. Par [1], on a

$$\log Df(x) - \log Df(0) = \int_0^x \frac{D^2 f}{Df}(t)\,dt.$$

(En effet, $\log Df$ est à variation bornée puisque la fonction log est Lipschitzienne sur $[\min Df, \max Df]$ et de plus, si A est un ensemble borélien de mesure de Lebesgue nulle, alors l'ensemble $\log Df(A)$ est mesure de Lebesgue nulle. Il suit que $\log Df$ est absolument continue.) Comme $1/Df \in L^\infty$ et que $D^2 f \in L^p$, on a $D \log Df \in W^{1,p}$.

Lemme 6. Soit $h(x) = \int_0^x \varphi(t)\,dt$, avec $\varphi \in L^1$, $\varphi \geq 0$, $\int_0^1 \varphi(t)\,dt = 1$. Soit $\eta \in L^1(\mathbf{T}^1)$, alors $Dh\eta \circ h \in L^1$, et on a

$$\int_0^1 \eta(t)dt = \int_0^1 \eta(h(t))Dh(t)dt.$$

Démonstration. [1]. ■

④ On a bien $\log Df \circ h + \lambda(f, h) \in W^{1,p_1}$ puisque

$$\int_0^1 |(D \log Df) \circ hDh|^{p_1} dx = \int |((D \log Df) \circ h)^{p_1} Dh^{\frac{1}{r}}|Dh^{p_1 - \frac{1}{r}} dx$$

$$\underset{\text{inégalité de Hölder}}{\leq} \left(\int |((D \log Df) \circ h)^{p_1 r} Dh| dx\right)^{\frac{1}{r}} \left(\int Dh^{\left(p_1 - \frac{1}{r}\right)r'} dx\right)^{\frac{1}{r'}}.$$

Or,

$$\int |(D \log Df) \circ h|^{p_1 r} Dh\, dx = \int |D \log Df|^{p_1 r} dx = \int |D \log Df|^p dx.$$

On obtient finalement

$$\|(D \log Df) \circ hDh\|_{L^{p_1}} \leq \|D \log Df\|_{L^p} \left(\int Dh^{\left(p_1 - \frac{1}{r}\right)r'} dx\right)^{\frac{1}{r' p_1}},$$

$$\boxed{\|(D \log Df \circ h)Dh\|_{L^{p_1}} \leq \|D \log Df\|_{L^p} \|Dh\|_{L^q}^{q/(r' p_1)}}.$$

⑤ Comme $\|Dh\|_{L^q} \leq 2$ et qu'on suppose p et p_1 fixés, par 4.2 il existe $\varphi_1 \in$ BMO vérifiant $\int_0^1 \varphi_1(x)dx = 0$ et

$$\log Df \circ h + \lambda(f, h) = \varphi_1 \circ R_\alpha - \varphi_1 \qquad (+)$$

et on a

$$\boxed{\|\varphi_1\|_{\text{BMO}} \leq C_1 \gamma^{-1} \|D \log Df\|_{L^p}}$$

où C_1 est une constante ne dépendant que de p et p_1. Si on remplace φ_1 par $\varphi_1 + c$ ($c \in \mathbf{R}$), alors $\varphi_1 + c$ vérifie encore (+).

⑥ Soit $0 < \eta < 1/2$, par 3.3 et ⑤, il existe $\varepsilon > 0$ tel que si $\|D \log Df\|_{L^p} \leq \varepsilon\gamma$, alors $e^{\varphi_1} \in L^q$, $e^{-\varphi_1} \in L^q$,

$$\int_0^1 e^{q\varphi_1} dx \leq 1 + \eta,$$

$$\int_0^1 e^{-\varphi_1} dx \leq 1 + \eta.$$

Par l'inégalité de Hölder, il en résulte

$$\int_0^1 e^{\varphi_1}\,dx \geq \left(\int_0^1 e^{-\varphi_1}\,dx\right)^{-1} \geq (1+\eta)^{-1};$$

soit

$$\left(\int_0^1 e^{\varphi_1}\,dx\right)^{-1} \leq 1+\eta.$$

⑦ On pose $\eta_0 = \mathrm{Inf}\left(\frac{1}{2}, \eta_1\right)$ où $\eta_1 > 0$ vérifie $(1+\eta_1)^{\frac{q+1}{q}} = 2$. On détermine $0 < \varepsilon_0 < \frac{1}{2}$ comme en ⑥ et associée à η_0. La constante ε_0 ne dépend que de p, p_1 et du théorème de John et Nirenberg.

⑧ Si f vérifie $\|D \log Df\|_{L^p} \leq \varepsilon_0\gamma$ et si $\varphi \in K_q$, on pose

$$\Phi(\varphi) = \left(\int_0^1 e^{\varphi_1}\,dx\right)^{-1} e^{\varphi_1}.$$

On a $\Phi(\varphi) \in K_q$ où encore $\Phi : K_q \to K_q$ puisque $\Phi(\varphi) \geq 0$,

$$\int_0^1 \Phi(\varphi)\,dx = 1$$

et

$$\|\Phi(\varphi)\|_{L^q} \leq (1+\eta)^{\frac{q+1}{q}} \leq 2.$$

⑨ Nous avons ainsi défini une application $\Phi : K_q \to K_q$ telle que, $\Phi(\varphi) = \psi$ vérifie:

$$\boxed{\psi \circ R_\alpha = e^{\lambda(f,\,h)} Df \circ h\psi}. \qquad (*)$$

⑩ Nous allons maintenant montrer que *l'application* Φ *est continue pour la topologie faible sur* K_q.

En effet, si la suite (φ_i) tend faiblement vers $\varphi \in K_q$, alors il faut montrer que la suite $(\Phi(\varphi_i))_i$ tend faiblement vers $\Phi(\varphi)$. Puisque $\Phi(\varphi_i) \in K_q$ et que K_q est faiblement compact métrisable, il suffit démontrer que toute valeur d'adhérence $\tilde{\psi}$ de la suite $(\Phi(\varphi_i))_i$ est égale à $\Phi(\varphi)$, ou ce qui revient au même si $\varphi_i \to \varphi$ et $\psi_i = \Phi(\varphi_i) \to \tilde{\psi}$, alors on a $\Phi(\varphi) = \tilde{\psi}$. Or par $(*)$, on a

$$\psi_i \circ R_\alpha = e^{\lambda(f,\,h_i)} Df \circ h_i\psi_i.$$

Par 2.6-2.8, si $i \to +\infty$, on a

$$\tilde{\psi} \circ R_\alpha = e^{\lambda(f,\, h)} Df \circ h\tilde{\psi}.$$

Or $\tilde{\psi} \geq 0$, $\int_0^1 \tilde{\psi}\, dx = 1$, et donc $\tilde{\psi} \neq 0$. Par l'ergodicité de la rotation R_α on obtient $\tilde{\psi} = \Phi(\varphi)$.

(11) Par le théorème de Schauder-Tychonoff [2, p. 456] l'application Φ a un point fixe φ. En fait, par la démonstration de (10), φ est > 0 et de plus φ est unique. Il en résulte: $\varphi = Dh$ avec

$$h(x) = \int_0^x Dh(t)\, dt;$$

h est un homéomorphisme puisque $Dh > 0$ (presque partout).

On a la relation

$$Dh \circ R_\alpha = e^{\lambda(f,\, h)} Df \circ hDh.$$

(12) Puisque $\int_0^1 Dh \circ R_\alpha\, dx = 1$ et que

$$\int_0^1 e^{\lambda(f,\, h)} Df \circ hDh\, dx = e^{\lambda(f,\, h)},$$

on a $e^{\lambda(f,\, h)} = 1$, et donc

$$Dh \circ R_\alpha = D(f \circ h)$$

d'où $h \circ R_\alpha = f \circ h + a$, avec $a \in \mathbf{R}$.

Puisque $\rho(f) = \alpha$ est irrationnel par [4, III.4], on a $a = 0$. On a finalement:

$$f = h \circ R_\alpha \circ h^{-1}.$$

(13) Puisque $Dh > 0$ (presque partout), h est un homéomorphisme absolument continu (voir 5.7). On a de plus

$$Dh \in L^q \quad \text{et} \quad Dh^{-1} \in L^q,$$

en effet

$$\int_0^1 (Dh^{-1})^q\, dx = \int_0^1 \left(\frac{1}{Dh} \circ h^{-1}\right)^q dx = \int_0^1 \left(\frac{1}{Dh}\right)^{q-1} dx$$

et il suffit d'appliquer (6) pour conclure.

Le théorème est ainsi démontré. ∎

7. UNE CONSEQUENCE

Soient $\alpha \in \mathbf{R}$, et $p > 1$, on note

$$F_\alpha^{2,p} = \{f \in D^{2,p}(\mathbf{T}^1) \,|\, \rho(f) = \alpha\}$$

et

$$F_\alpha^\infty = \{f \in D^\infty(\mathbf{T}^1) \,|\, \rho(f) = \alpha\}.$$

Par [4, III], $F_\alpha^{2,p}$ est un sous-ensemble fermé de $D^{2,p}(\mathbf{T}^1)$ pour la C^0-topologie. Par le même raisonnement qu'en [4, III.4], si $\alpha \notin \mathbf{Q}$, F_α^∞ est dense dans $F_\alpha^{2,p}$ pour la $W^{2,p}$-topologie.

Proposition. Si α *est un nombre de type constant, alors l'ensemble* $V = \{f \in F_\alpha^{2,p} \,|\, f$ *est conjugué par un homéomorphisme* $h \in D^0(\mathbf{T}^1)$ *qui est un homéomorphisme absolument continu*$\}$*, contient un ouvert dense dans* $F_\alpha^{2,p}$.

Démonstration. Puisque α est de type constant, par le théorème fondamental de [4], tout $f \in F_\alpha^\infty$ est C^∞-conjugué à R_α et la proposition suit du théorème 6 par conjugaison C^∞ et de la densité de F_α^∞ dans $F_\alpha^{2,p}$.

REFERENCES

1. Bourbaki, N. Integration. Chap. 5, Sec. 6, No. 5.

2. Dunford, N., and Schwartz, J. T. *Linear Operators*, part I. New York: John Wiley & Sons, 1957.

3. Hawkins, J., and Schmidt, K. On C^2-diffeomorphisms of the circle which are of type III_1. *Invent. Math.* 66 (1982):511-518.

4. Herman, M. Sur la conjugaison différentiable des difféomorphismes du cercle à des rotations. *Pub. I.H.E.S.* 49 (1979):5-233.

5. John, F., and Nirenberg, L. On functions of bounded mean oscillation. *Comm. Pure Appl. Math.* 14 (1961):415-426.

6. Jones, P. W. Homeomorphisms of the line which preserve BMO. A paraître. *Comm. Math. Helv.*

7. Katznelson, Y. Smooth mappings of the circle without sigma-finite invariant measures. *Symposia Math.* 22:363-369.

8. Meyer, Y. Sur un problème de Michael Herman. Preprint, Centre de Mathématiques de l'Ecole Polytechnique, 1980.

9. Russmann, H. On optimal estimates for the solutions of linear differential equations of the circle. *Celestial Mechanics* 14 (1976):33-37.

10. Stein, E. M. *Singular Integrals and Differentiability Properties of Functions.* Princeton, N.J.: Princeton Univ. Press, 1970.

11. Zygmund, A. *Trigonometric Series.* 2d ed. New York: Cambridge Univ. Press, 1968.

SUR UN PROBLÈME DE MICHAEL HERMAN

Yves Meyer
Ecole Polytechnique, Plateau de Palaiseau

Soient $\mathbf{T} = \mathbf{R}/\mathbf{Z}$ et $\alpha \notin \mathbf{Q}/\mathbf{Z}$ un nombre irrationnel.

Nous nous proposons d'étudier la régularité des solutions $f : \mathbf{T} \to \mathbf{C}$ de l'équation

$$f(x + \alpha) - f(x) = g(x) \tag{1}$$

en fonction de celle du second membre $g(x)$.

Désignons, pour $p \in (1, +\infty)$, par $L_1^p = L_1^p(\mathbf{T})$ l'espace de Banach des fonctions g appartenant à $L^p(\mathbf{T})$ et dont la dérivée g' (prise au sens des distributions) appartient aussi à L^p.

Théorème 1. Les quatre propriétés suivantes sont équivalentes:

(i) *il existe* $\delta > 0$ *tel que pour tout* $p \in \mathbf{Z}$ *et tout* $q \geq 1$ *on ait*

$$|\alpha - p/q| \geq \delta/q^2; \tag{2}$$

(ii) *il existe* $C > 0$ *tel que, pour toute fonction* $g \in L_1^2$ *d'intégrale nulle,* (1) *ait une solution* $f \in L^2$ *telle que* $\|f\|_2 \leq C\|g'\|_2$;

Laboratoire associé au C. N. R. S. n° 169

(iii) *il existe* $C > 0$ *tel que, pour toute fonction* $g \in L_1^2$ *d'intégrale nulle, toute solution* f *de* (1) *vérifie* $\|f\|_{\mathrm{BMO}} \le C\|g'\|_2$;

(iv) *pour tout* $p > 1$, *il existe une constante* C_p *telle que* $\|f\|_{\mathrm{BMO}} \le C_p \|g'\|_p$ *pour toute solution* f *de* (1) *telle que* $g \in L_1^p$.

Naturellement, (iv) ⇒ (iii) ⇒ (ii). Vérifions que (ii) ⇒ (i) ce qui est facile puis démontrons (i) ⇒ (iv). Pour vérifier que (ii) ⇒ (i), on teste (ii) sur $f(x) = e^{2\pi iqx}$ et l'on obtient $1 \le C2\pi q|e^{2\pi iq\alpha} - 1|$ ce qui est équivalent à (i). Montrons que (i) ⇒ (iv). Nous allons nous passer du théorème de Fefferman-Stein sur la dualité entre H^1 et BMO et de la caractérisation due à E. Stein de H^1 par la fonction de Littlewood-Paley. La preuve qui suit est plus "élémentaire." Elle repose sur le Theoreme 2 suivant (d'intérêt indépendant) dont voici les notations. Pour $f \in L^2(\mathbf{T})$ et tout $k \ge 0$, on pose $f_k(x) = \sum_{2^k \le |j| < 2.2^k} c_j e_j(x)$ où $c_j = \hat{f}(j)$ et $e_j(x) = e^{2\pi ijx}$. On a donc $f(x) = a + \sum_{k\ge 0} f_k(x)$.

Théorème 2. [2] *Avec les notations précédentes, il existe une constante* $C > 0$ *telle que, pour toute série de Fourier, on ait*

$$\|f\|_{\mathrm{BMO}} \le C \left\| \left(\sum_{k\ge 0} |f_k(x)|^2 \right)^{1/2} \right\|_\infty . \tag{3}$$

Naturellement, on peut alors se demander si la norme L^∞ du second membre de (3) peut être remplacée par la norme BMO. Il n'en est rien à cause du lemme suivant.

Lemme 1. Pour toute fonction continue $\varphi : \mathbf{T} \to \mathbf{C}$, *on a*

$$\|\varphi\|_\infty = \lim_{N\to+\infty} \|e_N \varphi\|_{\mathrm{BMO}} \quad (\text{où } e_N(x) = e^{2\pi iNx}).$$

C'est immédiat et laissé au lecteur.

Si φ est une somme trigonométrique finie, on teste (3) sur $f(x) = e_N(x)\varphi(x)$, $N = 2^k + k$. Alors, avec les notations du

Théorème 2, $f = f_k$ et il en résulte que dans le second membre de (3), la norme L^∞ ne peut être remplacée par la norme BMO.

La preuve du Théorème 2 découle presque immédiatement du lemme suivant.

Lemme 2. Il existe un constante $C > 0$ telle que pour tout entier $m \geq 0$ et tout intervalle $I \subset \mathbf{T}$ de longueur $\geq 2^{-m}$, *on ait, pour toute série* (2),

$$\left(\frac{1}{|I|}\int_I \Big|\sum_{k\geq m} f_k\Big|^2\,dx\right)^{1/2} \leq C\,\Big\|\Big(\sum_{k\geq m}|f_k(x)|^2\Big)^{1/2}\Big\|_\infty. \tag{4}$$

Pour démontrer le Lemme 2, on observe que si (4) est satisfaite pour un couple (m, I) et une constante C, il en est, a fortiori, de même si m est remplacé par $m_1 \geq m$. Cette remarque permet de supposer $2^{-m} \leq |I| < 2.2^{-m}$. De plus on peut évidemment se ramener au cas où

$$I = [-2^{-m}, 2^{-m}].$$

On définit le noyau de Fejer $K_N(x) = \frac{1}{N}|1 + \cdots + e_{N-1}(x)|^2 = \frac{1}{N}\frac{\sin^2 \pi Nx}{\sin^2 \pi x}$. On a, si $N = 2^{m-1}$ et $x \in I$, $K_N(x) \geq \frac{4}{\pi^2} N$. En d'autres termes, en appelant Π_I la fonction indicatrice de I, on a $\frac{1}{|I|}\Pi_I \leq CK_N$ lorsque $N = 2^{m-1}$ et $2^{-m} \leq |I| < 2.2^{-m}$, I étant centré en 0. On pose alors $A_N(x) = N^{-1/2}(1 + \cdots + e_{N-1}(x))$ et l'on a

$$\frac{1}{|I|}\int_I \Big|\sum_{k\geq m} f_k\Big|^2\,dx \leq C^2\Big\|\sum_{k\geq m} A_N f_k\Big\|_2^2 .$$

Or les fréquences du produit $A_N f_k$ appartiennent à $E_k = \{2^k - 2^{m-1} \leq |j| < 2.2^k + 2^{m-1}\}$ et ces ensembles E_k sont deux à deux disjoints lorsque $k \in 3\mathbf{N}$ (ou $k \in 3\mathbf{N} + 1$ ou $k \in 3\mathbf{N} + 2$). Il en résulte que

$$\| \sum_{3\ell \geq m} A_N f_{3\ell} \|_2^2 = \sum_{3\ell \geq m} \int_{\mathbf{T}} K_N(x) |f_{3\ell}(x)|^2 \, dx$$
$$\leq \sup_{x \in \mathbf{T}} \sum_{3\ell \geq m} |f_{3\ell}(x)|^2 \int_{\mathbf{T}} K_N(x) \, dx$$
$$= \| \sum_{3\ell \geq m} |f_{3\ell}(x)|^2 \|_\infty .$$

Puisque les inégalités analogues sont vraies si $k = 3\ell + 1$ (ou $k = 3\ell + 2$), le Lemme 2 est démontré. La preuve du Théorème 2 est alors calquée sur la démonstration directe du fait que $\sum_0^\infty a_n \cos 2^n x \in \mathrm{BMO}$ lorsque $a_n \in \ell^2$. On suppose que $2^{-m} \leq |I| < 2.2^{-m}$ et l'on écrit

$$f(x) = \sum_{k \geq 0} f_k(x) = S_m(x) + R_m(x) \quad \text{où } R_m(x) = \sum_{k \geq m+1} f_k(x).$$

Soit x_0 le centre de I. Montrons que

$$\frac{1}{|I|} \int_I |f(x) - S_m(x_0)|^2 \, dx \leq C \, \| (\sum_0^\infty |f_k(x)|^2)^{1/2} \|_\infty^2 , \tag{5}$$

où C est une constante absolue. On a, d'une part, $|S_m(x) - S_m(x_0)| \leq |x - x_0| \| S'_m \|_\infty$. Or $\| S'_m \|_\infty \leq \| f'_0 \|_\infty + \cdots + \| f'_m \|_\infty \leq 2 \| f_m \|_\infty + \cdots + 2^{m+1} \| f_m \|_\infty$ (grâce à la célèbre inégalité de S. Bernstein) et cette somme ne dépasse pas

$$2^{m+2} \| (\sum_0^\infty |f_k(x)|^2)^{1/2} \|_\infty .$$

Finalement, $|S_m(x) - S_m(x_0)| \leq 4 \| (\sum_0^\infty |f_k(x)|^2)^{1/2} \|_\infty$. En ce qui concerne $\frac{1}{|I|} \int_I |R_m(x)|^2 \, dx$, le Lemme 2 donne l'estimation cherchée ce qui achève de démontrer le Théorème 2.

Revenons maintenant au Théorème 1. On utilise le lemme suivant.

Lemme 3. Soit $(\mu_j)_{j \in \mathbf{Z}}$ une suite bornée de nombres complexes. Supposons qu'il existe une constante $C > 0$ et $p \in (1, 2)$ tels que, pour tout $k \geq 0$, on ait $\sum_{2^k \leq |j| < 2.2^k} |\mu_j|^p \leq C^p$.

Alors l'opérateur M défini par $M(e_j) = \mu_j e_j$ *se prolonge en un opérateur linéaire continu de* $L^p(\mathbf{T})$ *dans* BMO($\mathbf{T}$).

En effet, si $f \in L^p(\mathbf{T})$, on a $\sum_{k\geq 0} \| f_k\|_p^2 \leq C_p^2 \ \| f\|_p^2$ ([1]). Ecrivons $f_k(x) = \sum_{2^k\leq|j|<2.2^k} c_j e_j(x)$. Il vient, en appelant q l'exposant conjugué de p,

$$\Big(\sum_{2^k\leq|j|<2.2^k} |c_j|^q\Big)^{1/q} \leq \| f_k \|_p \quad \text{et donc}$$

$$\sum_{2^k\leq|j|<2.2^k} |c_j \| \mu_j| \leq C\| f_k \|_p .$$

On pose $g_k(x) = \sum_{2^k\leq|j|<2.2^k} c_j \mu_j e_j(x)$ et l'on a

$$\| g_k \|_\infty \leq \sum_{2^k\leq|j|<2.2^k} |c_j \| \mu_j| \leq C \| f_k\|_p$$

qui entraîne $$\sum_0^\infty \| g_k \|_\infty^2 \leq C_p'^{\,2} \| f \|_p^2$$

et, a fortiori, $$\Big\| \Big(\sum_0^\infty |g_k(x)|^2\Big)^{1/2} \Big\|_\infty \leq C_p' \| f \|_p .$$

Dans le cas qui nous intéresse $\mu_j = \dfrac{1}{j(e^{2\pi ij\alpha} - 1)}$. Lorsque $2^k \leq j < 2.2^k$ les distances mutuelles entre les points $e^{2\pi ij\alpha}$ dépassent $\delta 2^{-k}$ ($\delta > 0$) et il en résulte de façon très simple que $\sum_{2^k\leq j<2.2^k} |\mu_j|^p \leq C_p$ pour tout $p > 1$. Il en est de même pour $\mu_{-j} = -\bar{\mu}_j$.

RÉFÉRENCES

1. Zygmund, A. *Trigonometric Series*. 2d ed. New York: Cambridge Univ. Press, 1968.

2. Remarquons qu'avec les notations du Théorème 2, la condition

$$(*) \quad \frac{1}{|I|} \int_I \left(|f_k(x)|^2 + |f_{k+1}(x)|^2 + \cdots \right) dx \leq C, \text{ satisfaite}$$

pour tous les intervalles I de longueur 2^{-k}, est suffisante pour que $f \in \mathrm{BMO}$. Cette condition est nécessaire si les f_k sont définies de façon "raisonnable"; par exemple si $\hat{f}_k(j) = \hat{f}(j)\Psi\left(\frac{j}{2^k}\right)$ où Ψ est continûment dérivable et est nulle hors de [1/3, 3]. La condition (*) est la version dyadique de la caractérisation de BMO par les mesures de Carleson.

MINIMAL AND MAXIMAL METHODS OF INTERPOLATION OF BANACH SPACES

Svante Janson
Uppsala University

Most theorems on interpolation of linear operators can be stated in the following form, where X_0, Y_0, X_1, Y_1, X, Y are some normed (or quasinormed) linear spaces such that $X \subset X_0 + X_1$ and $Y \subset Y_0 + Y_1$. (All inclusions in this paper are assumed to be continuous.)

If T is a linear operator that maps X_0 into Y_0 and X_1 into Y_1 boundedly, then T maps X into Y boundedly.

For simplicity, we will only consider Banach spaces in this paper. $X = Y$ means only that the spaces are isomorphic, that is, the norms are equivalent. Also, we will only consider linear operators, although some results, for example, the Marcinkiewicz interpolation theorem [9], extend to suitable classes of nonlinear operators. (See also [7].) The first results of this type were the well-known theorems by Riesz (L^p spaces), Marcinkiewicz (L^p and weak L^p) and Salem-Zygmund (H^p and L^p); see [9, Chapter 12] for details and references.

About 1960 the interpolation theory was transformed by the works of Calderón, Gagliardo, Krein, Lions, Peetre, and others. The concept of interpolation methods was introduced, turning theorems of the above type into a definition.

Definitions. A Banach couple (X_0, X_1) is a pair of Banach spaces such that X_0 and X_1 both are continuously included in some Hausdorff vector space (thus $X_0 + X_1$ is defined). F is an interpolation method if, for any Banach couple (X_0, X_1), $F(X_0, X_1)$ is a Banach space such that $X_0 \cap X_1 \subset F(X_0, X_1) \subset X_0 + X_1$, and for any two Banach couples (X_0, X_1) and (Y_0, Y_1), every linear operator that maps X_0 boundedly into Y_0 and X_1 into Y_1 also maps $F(X_0, X_1)$ boundedly into $F(Y_0, Y_1)$.

The search for interpolation theorems of the above type is now broken down into two steps; first an interpolation method has to be constructed; second, the interpolation spaces $F(X_0, X_1)$ and $F(Y_0, Y_1)$ have to be computed.

There exist plenty of interpolation methods, but two suffice for most applications, viz. the complex method $(\ )_\theta$ and the real method $(\ )_{\theta q}$ $(0 < \theta < 1,\ 1 \leq q \leq \infty)$. See [2] for the definitions. (According to the definition above, these are families of interpolation methods, one for each choice of θ (θ and q).)

Aronszajn and Gagliardo [1] gave the following two constructions of interpolation methods. Given a Banach couple (A_0, A_1) and a Banach space A such that $A_0 \cap A_1 \subset A \subset A_0 + A_1$, we define for any Banach couple (X_0, X_1),

$$G(X_0, X_1) = \left\{ \sum_1^\infty T_n a_n : a_n \in A,\ T_n \in L(A_0, X_0) \cap L(A_1, X_1) \quad \text{and} \right. \\ \left. \sum \|a_n\|_A \max(\|T_n\|_{A_0,X_0}, \|T_n\|_{A_1,X_1}) < \infty \right\}$$

and

$$H(X_0, X_1) \\ = \{x \in X_0 + X_1 : Tx \in A \text{ for every } T \in L(X_0, A_0) \cap L(X_1, A_1)\}.$$

G and H are interpolation methods and $G(A_0, A_1) \supset A$, $H(A_0, A_1) \subset A$. Furthermore, G is minimal for this property; that is, if F is any interpolation method such that $F(A_0, A_1) \supset A$, then $F(X_0, X_1) \supset G(X_0, X_1)$ for all couples (X_0, X_1). Similarly, H is maximal.

The purpose of this paper is to show that the real and complex methods can be characterized in this way, and that this simplifies many proofs of properties of the methods. This is mainly a summary of [6], where many other results and proofs are given, including duality theorems and applications to other interpolation methods.

THE REAL METHOD

Let $\ell_\theta^p = \{\{a_n\}_{-\infty}^{\infty} : (\sum_{-\infty}^{\infty} |2^{-n\theta} a_n|^p)^{1/p} < \infty\}$. Thus $\ell_0^p = \ell^p$.

Theorem 1. *The real method* $(\ )_{\theta q}$ *coincides with the minimal method such that* $G(\ell_0^1, \ell_1^1) \supset \ell_\theta^q$ *, as well as with the maximal method such that* $H(\ell_0^\infty, \ell_1^\infty) \subset \ell_\theta^q$. *(In fact, equalities hold.)*

Sketch of proof. A linear operator $T : \ell_0^1 \to X_0$ and $\ell_1^1 \to X_1$ is given by $T\{a_n\} = \Sigma a_n x_n$ for some sequence $\{x_n\}$ with $\sup J(2^n, x_n) = \max(\|T\|_{\ell_0^1, X_0}, \|T\|_{\ell_1^1, X_1}) < \infty$. It follows that the construction by Aronszajn and Gagliardo yields the same result as the discrete J-method [2, Chapter 3.2], which is one version of the real method. Similarly, the maximal construction with ℓ_0^∞, ℓ_1^∞, ℓ_θ^∞ yields the discrete K-method.

Theorem 2. *Let F be an interpolation method,* $0 < \theta < 1$, $1 \leq q \leq \infty$.

(i) *If* $F(\ell_0^1, \ell_1^1) \supset \ell_\theta^q$, *then* $F(X_0, X_1) \supset (X_0, X_1)_{\theta q}$ *for all couples* (X_0, X_1).

(ii) *If* $F(\ell_0^\infty, \ell_1^\infty) \subset \ell_\theta^q$, *then* $F(X_0, X_1) \subset (X_0, X_1)_{\theta q}$ *for all couples* (X_0, X_1).

Theorem 3. $(\ )_{\theta q}$ *is the only interpolation method F such that* $F(\ell_0^1, \ell_1^1) = F(\ell_0^\infty, \ell_1^\infty) = \ell_\theta^q$.

One application of this characterization is the reiteration theorem [2]. For $0 < \theta_0, \theta_1, \theta < 1$ and $1 \le q_0, q_1, q \le \infty$, let $F(X_0, X_1) = \big((X_0, X_1)_{\theta_0 q_0}, (X_0, X_1)_{\theta_1 q_1}\big)_{\theta q}$. F is an interpolation method and, if $\theta_0 \ne \theta_1$, $F(\ell_0^1, \ell_1^1) = F(\ell_0^\infty, \ell_1^\infty) = (\ell_{\theta_0}^{q_0}, \ell_{\theta_1}^{q_1})_{\theta q} = \ell_\eta^q$, where $\eta = (1 - \theta)\theta_0 + \theta\theta_1$, whence $F = (\ )_{\eta q}$. Similarly, $F = (\ )_{\theta_0 q}$ if $\theta_0 = \theta_1$ and $\dfrac{1}{q} = \dfrac{1-\theta}{q_0} + \dfrac{\theta}{q_1}$. The same method proves mixed reiteration theorems combining the real method and another interpolation method, for example,

$$\big((X_0, X_1)_{\theta_0 q_0}, (X_0, X_1)_{\theta_1 q_1}\big)_\theta = (X_0, X_1)_{\eta q}$$
$$\eta = (1 - \theta)\theta_0 + \theta\theta_1,$$
$$\frac{1}{q} = \frac{1-\theta}{q_0} + \frac{\theta}{q_1} > 0.$$

In general, a mixed reiteration including the real method gives back the real method (cf. [3]).

We can also prove a similar reiteration theorem for interpolation of families of Banach spaces as described in [5], where the corresponding theorem for the complex method is proved. In brief, the definition uses a domain D in the complex plane with nice boundary Γ and a family $\{B(\gamma) \mid \gamma \in \Gamma\}$ of Banach spaces satisfying a measurability condition. The interpolation spaces $B(z)$, $z \in D$, are defined to be the values assumed at z by a certain class of analytic functions having boundary values in $B(\gamma)$ with bounded norms; see [5] for details. Let dP_z be the harmonic measure on Γ with respect to $z \in D$. We will use the fact that if the norms on the spaces $B(\gamma)$ are changed within factors $k(\gamma)$ (retaining the measurability condition) and $\int_\Gamma \log k(\gamma)\, dP_z < \infty$, then the interpolation spaces remain the same.

Theorem 4. Let $0 < \theta(\gamma) < 1$ *and* $1 \le q(\gamma) \le \infty$ *be measurable functions on* Γ *such that* $\int_\Gamma \log \theta(\gamma)\, dP_z(\gamma) > -\infty$, $\int_\Gamma \log\big(1 - \theta(\gamma)\big)\, dP_z(\gamma) > -\infty$ *and* $\int_\Gamma \dfrac{1}{q(\gamma)}\, dP_z(\gamma) > 0$ *for some (and hence all)* $z \in D$. *Let*

X_0, X_1 be a Banach couple and let $B(\gamma) = (X_0, X_1)_{\theta(\gamma),q(\gamma)}$ with the K-method norm. Then the interpolation spaces are $B(z) = (X_0, X_1)_{\theta(z),q(z)}$, where $\theta(z)$ and $1/q(z)$ are the harmonic extensions of $\theta(\gamma)$ and $1/q(\gamma)$ to D.

Proof. It is easy to see that $\{(X_0, X_1)_{\theta(\gamma),q(\gamma)}\}$ satisfy the measurability condition of [5]. (This is the reason for choosing the K-method norm.) Fix $z_0 \in D$ and let $F(X_0, X_1) = B(z_0)$. F is an interpolation method, and we test the couples $(X_0, X_1) = (\ell_0^1, \ell_1^1)$ and $(\ell_0^\infty, \ell_1^\infty)$. In both cases $B(\gamma) = (X_0, X_1)_{\theta(\gamma),q(\gamma)} = \ell_{\theta(\gamma)}^{q(\gamma)}$ and the norms are equivalent within a factor $3/\theta(\gamma)\big(1 - \theta(\gamma)\big)$. Since $\int \log\big(3/\theta(\gamma)(1 - \theta(\gamma))\big)\, dP_z < \infty$, we may use the $\ell_{\theta(\gamma)}^{q(\gamma)}$ norm on $B(\gamma)$.

Let $\alpha(z) = \theta(z) + i\tilde{\theta}(z)$ and $\beta(z) = 1/q(z) + i\big(1/q(z)\big)^{\sim}$ be analytic in D and real at z_0.

If $\{a_n\}_{-N}^{N}$ is a finite sequence and $\|\{a_n\}\|_{\ell_{\theta(z_0)}^{q(z_0)}} = 1$, define $F(z) = \{2^{n\alpha(z)}|2^{-n\theta(z_0)}a_n|^{q(z_0)\beta(z)}\text{ sign } a_n\}$. F is analytic in D, $F(z_0) = \{a_n\}$, and $\|F(\gamma)\|_{\ell_{\theta(\gamma)}^{q(\gamma)}} = 1$. Consequently, $\|\{a_n\}\|_{B(z_0)} \leq 1$ and thus $\ell_{\theta(z_0)}^{q(z_0)} \subset B(z_0)$.

Conversely, assume that $F(z) = \{f_n(z)\}$ is analytic and that $\|F(\gamma)\|_{B(\gamma)} \leq 1$. If $\{b_n\}_{-N}^{N}$ is any finite sequence such that $\|\{b_n\}\|_{-\theta(z_0)}^{q(z_0)'} = 1$, define

$$G(z) = \{g_n(z)\} = \{2^{-n\alpha(z)}|2^{n\theta(z_0)}b_n|^{q(z_0)'(1-\beta(z))}\text{ sign } b_n\}.$$

$\Sigma f_n(z)g_n(z)$ is analytic and

$$|\Sigma f_n(\gamma)g_n(\gamma)| \leq \|F(\gamma)\|_{\ell_{\theta(\gamma)}^{q(\gamma)}}\|G(\gamma)\|_{\ell_{-\theta(\gamma)}^{q(\gamma)'}} = \|F(\gamma)\|_{B(\gamma)} \leq 1,\ \gamma \in \Gamma.$$

By the maximum principle, $|\Sigma f_n(z_0)g_n(z_0)| \leq 1$. Since $\{g_n(z_0)\} = \{b_n\}$ is arbitrary, this implies that $\|F(z_0)\|_{\ell_{\theta(z_0)}^{q(z_0)}} \leq 1$. Consequently, $B(z_0) \subset \ell_{\theta(z_0)}^{q(z_0)}$.

We have proved that $F(\ell_0^1, \ell_1^1) = F(\ell_0^\infty, \ell_1^\infty) = \ell_{\theta(z_0)}^{q(z_0)}$. By Theorem 3, $F(X_0, X_1) = (X_0, X_1)_{q(z_0),\theta(z_0)}$ for any X_0, X_1.

Remark. If $q(\gamma) = \infty$ a.e. on Γ, it follows similarly that $B(z)$ equals the closure of $X_0 \cap X_1$ in $(X_0, X_1)_{\theta(z)\infty}$ (except when $\theta(\gamma)$ is constant a.e.).

Corollary. Let $0 \leq 1/p(\gamma)$, $1/q(\gamma) \leq 1$ be measurable functions on Γ and let $1/p(z)$, $1/q(z)$ be their harmonic extensions to D. Let $B(\gamma)$ be the Lorentz space $L^{p(\gamma),q(\gamma)}$. If $\int_\Gamma |\log(p(\gamma) - 1)|\, dP_z < \infty$ and $q(z) < \infty$, then $B(z) = L^{p(z),q(z)}$, $z \in D$.

Proof. Take $X_0 = L^\infty$, $X_1 = L^1$ and $\theta(\gamma) = 1/p(\gamma)$ in Theorem 4.

THE COMPLEX METHOD

Let $\hat{L}_\theta^1 = \{\{a_n\} : \{2^{-n\theta} a_n\} \in \hat{L}^1\} = \{\{a_n\} : \sum_{-\infty}^{\infty} 2^{-n\theta} a_n e^{int} \in L^1(0, 2\pi)\}$ and define $\hat{M}_\theta$, $\hat{L}_\theta^\infty$ similarly.

Theorem 5. The complex method $(\ )_\theta$ *coincides with the minimal interpolation method* G *such that* $G(\hat{L}_0^1, \hat{L}_1^1) \supset \hat{L}_\theta^1$. $(\ )^\theta$ *coincides with the minimal method* G *such that* $G(\hat{L}_0^1, \hat{L}_1^1) \supset \hat{M}_\theta$, $0 < \theta < 1$.

Let ${}^\theta(\ )$ *be the maximal method* H *such that* $H(\hat{L}_0^\infty, \hat{L}_1^\infty) = \hat{L}_\theta^\infty$. *Then* ${}^\theta(X_0, X_1) \supset (X_0, X_1)^\theta$, *but we do not know whether strict inclusion may occur. In any case,* $(X_0, X_1)_\theta$ *is a closed subspace of* ${}^\theta(X_0, X_1)$; *viz. the closure of* $X_0 \cap X_1$.

This yields:

Theorem 6. If F *is an interpolation method such that* $F(\hat{L}_0^1, \hat{L}_1^1) \supset \hat{L}_\theta^1$, $\hat{F}(L_0^\infty, \hat{L}_1^\infty) \subset \hat{L}_\theta^\infty$, *and* $X_0 \cap X_1$ *is dense in* $F(X_0, X_1)$, *then* $F(X_0, X_1) = (X_0, X_1)_\theta$.

One application of this is that $(\tilde{X}_0, \tilde{X}_1)_\theta = (X_0, X_1)_\theta$ for any Banach couple, where $\tilde{X}_0$ and $\tilde{X}_1$ are the Gagliardo completions [2] of

X_0 and X_1 in $X_0 + X_1$. This leads (see [6 and 8]) to the following abstract version of the Stein interpolation theorem (cf. [4 and 5]). Let $\bar{S}$ be the strip $\{z : 0 \leq \text{Re } z \leq 1\}$.

Theorem 7. Let T_z, $z \in \bar{S}$, be a family of linear operators from $X_0 \cap X_1$ into $Y_0 + Y_1$ such that for every $x \in X_0 \cap X_1$ and $u \in (Y_0 + Y_1)^*$, $\langle u, T_z x \rangle$ is continuous on $\bar{S}$ and analytic in the interior and $\sup(\| T_{it} \|_{X_0, Y_0}, \| T_{1+it} \|_{X_1, Y_1}) < \infty$. Assume further that Y_0 (or Y_1) is separable or that $t \to \langle u, T_{it} x \rangle$ is continuous for every $x \in X_0 \cap X_1$ and $u \in Y_0^*$ (or the corresponding condition for Y_1).

Then there is a unique extension of T_θ mapping $(X_0, X_1)_\theta$ continuously into $(Y_0, Y_1)_\theta$.

Note that the extra condition involves only one of the spaces Y_0, Y_1. It is automatically satisfied if $Y_0 \supset Y_1$ (or $Y_0 \subset Y_1$). (It is not known whether this condition is needed at all.)

[Note added in proof: Some condition on Y_0 or Y_1 is necessary in Theorem 7. See Cwikel, M., and Janson, S., Interpolation of analytic families of operators (to appear).]

REFERENCES

1. Aronszajn, N., and Gagliardo, E. Interpolation spaces and interpolation methods. *Ann. Math. Pure Appl.* 68 (1965): 51-118.

2. Bergh, J., and Löfström, J. *Interpolation Spaces*. New York: Springer, 1976.

3. Brudnyi, Yu, A., and Krugljak, N. Ja. Real interpolation functors, *Dokl. Akad. Nauk SSSR* 256 (1981): 14-17. (Russian.)

4. Calderón, A. P., and Torchinsky, A. Parabolic maximal functions associated with a distribution, II. *Adv. Math.* 24 (1977): 101-171.

5. Coifman, R.; Cwikel, M.; Rochberg, R.; Sagher, Y.; and Weiss, G. A theory of complex interpolation for families of Banach spaces. *Adv. Math.* 43 (1982): 203-229.

6. Janson, S. Minimal and maximal methods of interpolation. *J. Functional Anal.* 44 (1981): 50-73.

7. Janson, S. On the interpolation of sublinear operators. *Studia Math.* 75 (1982): 51-53.

8. Janson, S., and Jones, P. Interpolation of H^p-spaces: the complex method. To appear.

9. Zygmund, A. *Trigonometric Series*. 2d ed. New York: Cambridge Univ. Press, 1959.

A THEOREM OF KHRUSHCHEV AND PELLER ON RESTRICTIONS OF ANALYTIC FUNCTIONS HAVING FINITE DIRICHLET INTEGRAL TO CLOSED SUBSETS OF THE UNIT CIRCUMFERENCE

Paul Koosis
University of California, Los Angeles

The following exposition is based on some recent work of Khrushchev and Peller [1, Sec. 3.6].

A well-known theorem of Rudin says that every continuous function defined on a closed subset E of the unit circumference coincides thereon with a function analytic in the unit disc and continuous up to its boundary if and only if E has Lebesgue measure zero. S. Khrushchev and V. Peller have asked what additional requirements must be imposed on E in order that some of the above-mentioned analytic extensions of any continuous function defined on it have *finite Dirichlet integral over the unit disc*. Their answer, namely, that this is the case *if and only if* E *has logarithmic capacity zero,* is a direct consequence of their theorem, given in Sec. 1 below, together with a known result of H. Wallin discussed in Sec. 2.

Our discussion makes use of only the bare rudiments of potential theory; everything needed can be found in Chapter 3 of [3]. We will use the following *notation*:

$\mathbf{T}$ is the interval $[-\pi, \pi]$ with the endpoints identified.

This work was partially supported by NSF Grant NO. MC580-02955.

If E is a closed subset of $\mathbf{T}$, $\mathscr{C}(E)$ is the Banach space of continuous functions on E, equipped with the usual sup norm $\| \ \|_E$.

If $f \in L_1(\mathbf{T})$, we write $f(\theta) \sim \sum_{-\infty}^{\infty} \hat{f}(n) e^{in\theta}$ to indicate that the expression on the right is the Fourier series of $f(\theta)$.

The Dirichlet norm, $\| f \|_{\mathscr{D}}$, of an $f \in L_1(\mathbf{T})$ is given by

$$\| f \|_{\mathscr{D}}^2 = \pi \sum_{-\infty}^{\infty} |n| \, |\hat{f}(n)|^2 .$$

$\mathscr{D}$ denotes the set of f with $\| f \|_{\mathscr{D}} < \infty$.

$\mathscr{D}_A$ denotes the set of f in $\mathscr{D}$ with $\hat{f}(n) = 0$ for all $n < 0$; these are the "analytic" f in $\mathscr{D}$.

$\mathscr{M}(E)$ denotes the set of complex-valued finite Borel measures carried on the closed set E; the total variation norm for such measures is denoted by $\| \ \|$.

We write $\hat{\mu}(n) = \frac{1}{2\pi} \int_{-\pi}^{\pi} e^{-in\theta} \, d\mu(\theta)$ for $\mu \in \mathscr{M}(\mathbf{T})$.

If $\mu \in \mathscr{M}(\mathbf{T})$, the *energy*, $\mathscr{E}(\mu)$, of μ is defined by

$$\mathscr{E}(\mu) = \pi \sum_{n \neq 0} \frac{|\hat{\mu}(n)|^2}{|n|} .$$

If $\mu \in \mathscr{M}(\mathbf{T})$, its logarithmic potential is

$$U_\mu(\theta) = \frac{1}{\pi} \int_{-\pi}^{\pi} \log \left\{ \operatorname{cosec} \left| \frac{\theta - t}{2} \right| \right\} d\mu(t) .$$

The logarithmic potential of a measure always belongs to $L_1(\mathbf{T})$, and we always have

$$\| U_\mu \|_{\mathscr{D}}^2 = \mathscr{E}(\mu) = \frac{1}{2} \int_{-\pi}^{\pi} U_\mu(\theta) \, \overline{d\mu(\theta)} .$$

1.

Theorem of Khrushchev and Peller. *If $E \subseteq \mathbf{T}$ is closed and of logarithmic capacity zero, the restrictions to E of the functions in $\mathscr{D}_A \cap \mathscr{C}(\mathbf{T})$ fill out $\mathscr{C}(E)$.*

Khrushchev and Peller prove this result by using an ingenious duality argument, which will be given shortly. Two lemmas are needed.

Lemma 1. *Let* $\mu \in \mathscr{M}(\mathbf{T})$ *and* $\mathscr{E}(\mu) < \infty$. *If a closed* $F \subseteq \mathbf{T}$ *has logarithmic capacity zero,* $\mu(F) = 0$.

Lemma 2. *Let* $\mu \in \mathscr{M}(\mathbf{T})$. *If* $\sum_1^\infty \frac{|\hat{\mu}(n)|^2}{n} < \infty$, *then* $\sum_{-\infty}^{-1} \frac{|\hat{\mu}(n)|^2}{|n|} < \infty$.

Assuming, for the time being, the two lemmas, let us prove the theorem. The idea of Khrushchev and Peller is to use the product space $\mathscr{D}_A \times \mathscr{C}(\mathbf{T})$ with its natural norm $\|(f, g)\|_\times = \|f\|_{\mathscr{D}} + \|g\|_{\mathbf{T}}$ and to examine the special subspace $\mathscr{L}$ of it consisting of the pairs (f, f) with $f \in \mathscr{D}_A \cap \mathscr{C}(\mathbf{T})$. Making every such pair (f, f) correspond to the restriction of f to E gives us an obvious continuous map of $\mathscr{L}$ into $\mathscr{C}(E)$; we are to prove that this map is *onto*. Let Σ be the unit sphere in $\mathscr{L}$; it is *more* than enough to show that the restriction map just described takes Σ onto a *dense subset* of the unit sphere in $\mathscr{C}(E)$. Indeed, then the map's image of Σ will contain any $\varphi \in \mathscr{C}(E)$ with $\|\varphi\|_E < 1$, as is shown by a well-known approximation argument.

Reasoning by contradiction, let us assume that the restrictions of f to E with $(f, f) \in \Sigma$ are *not* dense in the unit sphere of $\mathscr{C}(E)$. The Hahn-Banach theorem then gives us a $\nu \in \mathscr{M}(E)$, $\|\nu\| = 1$, with $\left|\int_{\mathbf{T}} \bar{f} d\nu\right| <$ some $\lambda < 1$ for all $(f, f) \in \Sigma$. To ν corresponds a *functional* $(0, \nu)$ on $\mathscr{L}$, and since Σ is the *unit sphere* of $\mathscr{L}$, the inequality just written means that the *norm* $\|(0, \nu)\|_{\mathscr{L}}^*$ *of this functional on* $\mathscr{L}$ *is* $< \lambda$.

Let us describe the dual $\mathscr{L}^*$ of $\mathscr{L}$. The dual of $\mathscr{C}(\mathbf{T})$ is of course $\mathscr{M}(\mathbf{T})$. The dual $\mathscr{D}_A^*$ of $\mathscr{D}_A$ consists of all distributions $S(\theta) \sim \sum_1^\infty \hat{S}(n) e^{in\theta}$ with finite norm $\|S\|_{\mathscr{D}}^* = \sqrt{\frac{1}{\pi} \sum_1^\infty \frac{|\hat{S}(n)|^2}{n}}$; action on $\mathscr{D}_A$ is through the inner product $\langle S, f \rangle = \sum_1^\infty \overline{\hat{S}(n)} \hat{f}(n)$. The dual of $\mathscr{D}_A \times \mathscr{C}(\mathbf{T})$ is thus $\mathscr{D}_A^* \times \mathscr{M}(\mathbf{T})$, and, finally, the dual of $\mathscr{L}$ is the quotient space $(\mathscr{D}_A^* \times \mathscr{M}(\mathbf{T}))/\mathscr{L}^\perp$, where $\mathscr{L}^\perp$ is the subspace of $\mathcal{D}_A^* \times \mathcal{M}(\mathbf{T})$ consisting of the (S, μ) with $\langle S, f\rangle + \int_{\mathbf{T}} \overline{f d\mu} = 0$ for all $(f, f) \in \mathcal{L}$.

Returning to our $\nu \in \mathcal{M}(E)$, let us compute $\|(0, \nu)\|_{\mathcal{L}}^*$. It is the infimum of $\|(0, \nu) - (S, \mu)\|_{\times}^*$ with (S, μ) ranging over $\mathcal{L}^{\perp}$. So, since $\|(0, \nu)\|_{\mathcal{L}}^* < \lambda$, there is an $(S, \mu) \in \mathcal{L}^{\perp}$ with $\sup\{\|S\|_{\mathcal{D}}^*, \|\nu - \mu\|\} < \lambda$. *This, however, is impossible.* The pairs $(e^{in\theta}, e^{in\theta})$ with $n = 0, 1, 2, \ldots$ all belong to $\mathcal{L}$, so, from $(S, \mu) \in \mathcal{L}^{\perp}$ we get $2\pi\overline{\hat{\mu}(n)} + \overline{\hat{S}(n)} = 0$ for such n. Since $S \in \mathcal{D}_A^*$, this makes $\sum_1^{\infty} \frac{|\hat{\mu}(n)|^2}{n} < \infty$. From Lemma 2, we now obtain $\mathcal{E}(\mu) < \infty$. However, E has logarithmic capacity zero by hypothesis. So Lemma 1 implies that $\mu(F) = 0$ for any closed subset F of E. Since $\nu \in \mathcal{M}(E)$, we see from this that $\|\nu - \mu\| \geq \|\nu\| = 1$, contradicting the inequality $\sup\{\|S\|_{\mathcal{D}}^*, \|\nu - \mu\|\} < \lambda$.

The theorem is thus established; we have in fact proved *more*. If $f \in \mathcal{D}_A \cap \mathcal{C}(\mathbf{T})$, let φ be the restriction of f to E. Clearly $\|\varphi\|_E \leq \|f\|_{\mathbf{T}}$. Our argument has shown that for *given* $\varphi \in \mathcal{C}(E)$, we can *find* such f with $\|f\|_{\mathcal{D}} + \|f\|_{\mathbf{T}}$ *arbitrarily close to* $\|\varphi\|_E$; this means that the *Dirichlet norms* $\|f\|_{\mathcal{D}}$ *of the* $f \in \mathcal{D}_A \cap \mathcal{C}(\mathbf{T})$ *equal to* φ *on* E *can be taken arbitrarily small.*

We return now to the lemmas .

The *first* one is routine and follows almost from the definition of capacity. For F closed and of capacity zero, take any open neighborhood $\mathcal{O}$ of F on $\mathbf{T}$ and any $\varepsilon > 0$. It is enough to construct a $\varphi \in \mathcal{D} \cap \mathcal{C}(\mathbf{T})$ with $0 \leq \varphi \leq 1$, $\varphi \equiv 1$ on F and $\varphi < \varepsilon$ outside $\mathcal{O}$, such that $|\hat{\varphi}(0)|^2 + \|\varphi\|_{\mathcal{D}}^2 < \varepsilon^2$, for then $\left|\int_{\mathbf{T}} \bar{\varphi} d\mu\right|$, which differs but little from $|\mu(F)|$, will be $\leq 2\varepsilon\sqrt{\pi^2|\hat{\mu}(0)|^2 + \mathcal{E}(\mu)}$ by Schwarz's inequality.

Let $2\delta = \operatorname{dis}(\sim\mathcal{O}, F)$. Since F has logarithmic capacity zero, we can find a neighborhood $\mathcal{N}$ of F with $\operatorname{dis}(\sim\mathcal{N}, F) < \delta$ such that the conductor potential of $\bar{\mathcal{N}}$ is a U_σ with a positive $\sigma \in \mathcal{M}(\bar{\mathcal{N}})$ having $\|\sigma\|$ *as small as we please*. We need only put $\varphi(\theta) = U_\sigma(\theta)$, for $U_\sigma(\theta) \equiv 1$ on $\bar{\mathcal{N}}$ and $0 \leq U_\sigma \leq 1$ everywhere on $\mathbf{T}$. ($\mathcal{N}$ can be taken as a finite union of intervals and then has no irregular points.) Also, for $\theta \notin \mathcal{O}$, $U_\sigma(\theta) \leq \frac{1}{\pi}\|\sigma\| \log \operatorname{cosec} \frac{2}{\delta}$, and this is

$< \varepsilon$ if $\|\sigma\|$ is small. Finally, $|\hat{\varphi}(0)| = |\hat{\sigma}(0)|\log 4$ and $\|\varphi\|^2_{\mathscr{D}} = \frac{1}{2}\int_{\mathbf{T}} U_\sigma d\sigma \leq \frac{1}{2}\|\sigma\|$, showing that $|\hat{\varphi}(0)|^2 + \|\varphi\|^2_{\mathscr{D}} < \varepsilon$ if $\|\sigma\|$ is small enough. Lemma 1 is thus proved.

There remains Lemma 2. Khrushchev and Peller remark that it can be obtained by duality from a known result in the theory of Gaussian processes (see [2, Chapter 4, Sec. 4]). They also have a quite complicated proof of their own; it uses BMO and the theory of Hankel and Toeplitz operators. This argument is a good deal longer than the one used to prove the theorem.

While walking on the UCLA campus, S. Pichorides and I came upon the following very easy proof (my real motivation for preparing this exposition!); it is based on the idea used to prove Hardy's theorem about the Fourier coefficients of H_1 functions.

It is enough to establish the inequality

$$\left|\sum_1^\infty \frac{|\hat{\mu}(n)|^2}{n} - \sum_{-\infty}^{-1} \frac{|\hat{\mu}(n)|^2}{|n|}\right| \leq \frac{\|\mu\|^2}{4\pi}$$

for *absolutely continuous measures* μ. Let $d\mu(\theta) = f(\theta)\,d\theta$ with $f \in L_1(\mathbf{T})$; put $f^*(\theta) = \overline{f(-\theta)}$, and let $G = f * f^*$, the convolution of f and f^*. We have $\hat{G}(n) = 2\pi|\hat{f}(n)|^2 = 2\pi|\hat{\mu}(n)|^2$. Take $G(\theta)$, multiply it by

$$\frac{\pi - |\theta|}{2} \operatorname{sgn}\theta \sim \sum_1^\infty \frac{\sin n\theta}{n},$$

and integrate from $-\pi$ to π!. We get $2\pi^2 i \sum_1^\infty \frac{1}{n}\left(|\hat{\mu}(n)|^2 - |\hat{\mu}(-n)|^2\right)$. This must, however, have absolute value $\leq \frac{\pi}{2}\|G\|_1 \leq \frac{\pi}{2}\|\mu\|^2$. That's all there is to it.

2.

To complete this discussion, it is necessary to show that if *E has logarithmic capacity* > 0, *then the restrictions of functions in* $\mathscr{D}_A \cap \mathscr{C}(\mathbf{T})$ *to E do not fill out* $\mathscr{C}(E)$. Wallin proved, however, in

1963 that *in these circumstances not even the restrictions of functions in* $\mathcal{D} \cap \mathscr{C}(\mathbf{T})$ *to* E *can exhaust* $\mathscr{C}(E)$. Because Wallin's proof [4] is rather long, I give here a different, easier one, based on duality.

Use the norm $\|f\|_{\mathcal{D}} + \|f\|_{\mathbf{T}}$ for functions $f \in \mathcal{D} \cap \mathscr{C}(\mathbf{T})$. For this norm, restriction of f to E is a *continuous* mapping from $\mathcal{D} \cap \mathscr{C}(\mathbf{T})$ into $\mathscr{C}(E)$. It is claimed that this mapping cannot be *onto*.

Suppose that it *were* onto. Then, by the Baire category theorem, for some N, the restrictions to E of the f with $\|f\|_{\mathcal{D}} + \|f\|_{\mathbf{T}} \leq N$ would be *dense* in some nonempty open *sphere* of $\mathscr{C}(E)$; by taking N larger, we may suppose that sphere to have its center at 0. This means that there is a constant K such that, if $\varphi \in \mathscr{C}(E)$ and $\|\varphi\|_E \leq 1$, we have $f \in \mathcal{D} \cap \mathscr{C}(\mathbf{T})$ with $\|f\|_{\mathcal{D}} + \|f\|_{\mathbf{T}} \leq K$, whose restrictions to E are *arbitrarily close* to φ in norm $\|\ \|_E$.

We proceed to dualize this last statement. Take any $\nu \in \mathscr{M}(E)$; we can choose a $\varphi \in \mathscr{C}(E)$ with $\|\varphi\|_E = 1$ and $\int_{\mathbf{T}} \bar{\varphi} d\nu$ as close as we like to $\|\nu\|$; by approximating φ with restrictions of f to E in the manner just described, we see that $\|\nu\|$ is a limit of some expressions of the form $2\pi \sum_{-\infty}^{\infty} \overline{\hat{f}(n)}\hat{\nu}(n)$ with $f \in \mathcal{D} \cap \mathscr{C}(\mathbf{T})$ and $\|f\|_{\mathcal{D}} + \|f\|_{\mathbf{T}} \leq K$. Suppose now that $\mathscr{E}(\nu) < \infty$. Then Schwarz's inequality shows that

$$\pi\left|\sum_{-\infty}^{\infty} \overline{\hat{f}(n)}\hat{\nu}(n)\right| \leq \sqrt{\left(|\hat{f}(0)|^2 + \|f\|_{\mathcal{D}}^2\right)\left(\pi^2|\hat{\nu}(0)|^2 + \mathscr{E}(\nu)\right)},$$

yielding

$$\|\nu\| \leq 2K\left(\pi^2|\hat{\nu}(0)|^2 + \mathscr{E}(\nu)\right)^{1/2} \qquad (*)$$

for $\nu \in \mathscr{M}(E)$ with $\mathscr{E}(\nu) < \infty$.

We will therefore be done if we show that a relation like (*) *cannot hold*; to do this we use the fact that E has nonzero logarithmic capacity. By definition, there *is* some positive $\mu \in \mathscr{M}(E)$ with $\mathscr{E}(\mu) < \infty$; there is no loss of generality in supposing $\|\mu\| = 1$. Assume, for the moment, that we can construct μ-measurable functions φ_n on $\mathbf{T}$ with $|\varphi_n(\theta)| \equiv 1$ such that the measures ν_n

given by $d\nu_n(\theta) = \varphi_n(\theta)\, d\mu(\theta)$ tend to zero weak* in $\mathscr{M}(E)$. Then (*) cannot hold for the ν_n; indeed, $\|\nu_n\| = \|\mu\| = 1$, but as we now show, $\mathscr{E}(\nu_n) \xrightarrow[n]{} 0$.

We have

$$\mathscr{E}(\mu) = \frac{1}{2}\int_{-\pi}^{\pi}\int_{-\pi}^{\pi} \log\left\{\operatorname{cosec}\left|\frac{s-t}{2}\right|\right\} d\mu(s)\, d\mu(t).$$

Write $K(s, t) = \frac{1}{2}\log\left\{\operatorname{cosec}\left|\frac{s-t}{2}\right|\right\}$; because μ is positive and $\mathscr{E}(\mu) < \infty$, we can, for any given $\varepsilon > 0$, get an M such that $\iint_{\Delta_M} K(s, t)\, d\mu(s)\, d\mu(t) < \varepsilon$, where Δ_M is the closed subset of $\mathbb{T}^2$ on which $K(s, t) \geq M$. Put $K_M(s, t) = \min[K(s, t), M]$; then $K_M(s, t)$ is continuous on $\mathbb{T}^2$, so, since the product measures $d\nu_n(s)\overline{d\nu_n(t)}$, converge weak* to zero on $\mathbb{T}^2$, $\iint_{\mathbb{T}^2} K_M(s, t)\, d\nu_n(s)\, \overline{d\nu_n(t)} \xrightarrow[n]{} 0$. Here, the difference $K(s, t) - K_M(s, t)$ is ≥ 0 on $\mathbb{T}^2$ and different from zero only on Δ_M, where it is, however, $\leq K(s, t)$. So

$$\left|\iint_{\mathbb{T}^2}(K(s, t) - K_M(s, t))d\nu_n(s)\overline{d\nu_n(t)}\right| \leq \iint_{\Delta_M} K(s, t)d\mu(s)d\mu(t) < \varepsilon,$$

since $|d\nu_n(t)| = d\mu(t)$. This shows that $\mathscr{E}(\nu_n) = \iint_{\mathbb{T}^2} K(s, t)\, d\nu_n(s)\, \overline{d\nu_n(t)}$ is $< \varepsilon$ for all sufficiently large n, invalidating (*) and thereby establishing our result.

It remains to construct the φ_n. Take any n, and look at the intervals $J_k = \left(-\pi + \frac{2\pi}{n}(k-1),\ -\pi + \frac{2\pi}{n}k\right]$, $k = 1, 2, \ldots, n$. If $\mu(J_k) = 0$, put $\varphi_n \equiv 1$ on J_k. If $\mu(J_k) > 0$, there must be some point $m_k \in J_k$ for which $\mu(J_k^-) = \mu(J_k^+) = \frac{1}{2}\mu(J_k)$, where J_k^- consists of the $t \in J_k$ that are $\leq m_k$ and J_k^+ of those that are $> m_k$. That's because the positive measure μ can contain no point mass, since $\mathscr{E}(\mu) = \frac{1}{2}\int_{\mathbb{T}} U_\mu\, d\mu < \infty$. For such a J_k, take $\varphi_n(t) \equiv 1$ on J_k^- and $\varphi_n(t) \equiv -1$ on J_k^+. The function φ_n is now defined on all the J_k, and it is clear that $\varphi_n(\theta)\, d\mu(\theta) \xrightarrow[n]{} 0$ weak*. We are done.

3. A CONCLUDING REMARK

It seems like a good idea to try and give an analogous characterization of the closed sets $E \subseteq \mathbf{T}$ having α-capacity equal to zero, replacing the Dirichlet norm for our analytic functions by one corresponding to the Riesz potentials of order α. In this situation, the reasoning of Sec. 2 goes through as is; one need only make the necessary changes in the formulas. Lemma 1 of Sec. 1 also holds; however, *the proof of Lemma 2 breaks down, because* $\sum_1^\infty \frac{\sin n\theta}{n^{1-\alpha}}$ *is no longer the Fourier series of a bounded function if* $\alpha > 0$. It is therefore not clear that an analogue of the Khrushchev-Peller theorem, referring to *analytic functions*, will hold.

The converse of the result in Sec. 2 *is*, however, true in the present situation, namely, if E has α-capacity zero, the restrictions to E of *all* continuous f having $\sum |n|^{1-\alpha}|\hat{f}(n)|^2 < \infty$ (*not* just the *analytic* ones) do fill out $\mathscr{C}(E)$. This result is also due to Wallin [4]; it can be obtained by running the duality argument of Sec. 2 backwards. Such reasoning resembles that followed in proving the theorem of Sec. 1; it is, however, simpler, using only an analogue of Lemma 1.

It should be mentioned, finally, that by using their own methods, Khrushchev and Peller obtain [1, Sec. 3.6] a result analogous to the one in Sec. 1 about restrictions to E of continuous functions f analytic in the unit disc with

$$D_p(f) = \sqrt[p]{\int_{-\pi}^{\pi}\int_{-\pi}^{\pi} \frac{|f(e^{i(x+t)}) - f(e^{ix})|^p}{|t|^2}\, dx\, dt} < \infty.$$

They find that these restrictions fill out $\mathscr{C}(E)$ provided that E has capacity zero with respect to the norm D_p. For $p = 2$, this reduces to the theorem of Sec. 1.

REFERENCES

1. Khrushchev, S. V., and Peller, V. V. Hankel operators, best approximations and stationary Gaussian processes, II. LOMI preprint E-5-81, Leningrad Department, Steklov Mathematical Institute, Leningrad, 1981. In Russian, this material is now published in *Uspekhi Matem. Nauk* 37 (1982): 53-124.

2. Ibragimov, I. A., and Rosanov, Y. A. *Gaussian Random Processes.* New York: Springer, 1978. English translation of *Gaussovskie sluchaĭnye protesessy,* Nauka, Moscow, 1970.

3. Kahane, J. P., and Salem, R. *Ensembles parfaits et séries trigonométriques.* Paris: Hermann, 1963.

4. Wallin, H. Continuous functions and potential theory. *Arkiv för Matem.* 5 (1963): 55-84.

SEMIEMBEDDINGS IN LINEAR TOPOLOGICAL SPACES

Heinrich P. Lotz
N.T. Peck
Horacio Porta
University of Illinois

INTRODUCTION

The starting point of this paper is an observation about metrics on the space of measurable functions. If L_0 denotes the space of (equivalence classes of) Lebesgue measurable functions on [0, 1], two metrics on L_0 that give the topology of convergence in measure are $d_\sigma(f, g) = \sigma(f - g)$ and $d_\delta(f, g) = \delta(f - g)$, where $\sigma(f) = \inf\{\varepsilon + m(|f| > \varepsilon); \varepsilon > 0\}$, and $\delta(f) = \int_0^1 |f(t)|(1 + |f(t)|)^{-1}\, dt$ (here and in the following, $m(E)$ denotes the measure of a set E). The observation is that σ and δ are badly nonhomogeneous for small scalars. In other words, the quotients $\sigma(rf)/r$ and $\delta(rf)/r$ considered as functions of r are far from being constant. In fact, it is easy to see that for $f(t) = 1/t$ $(0 < t \leq 1)$ and $r > 0$ near 0, they behave, respectively, like $r^{-1/2}$ and $-\ell n\, r$. In particular, they are unbounded.

It turns out that we can identify rather simply the functions f for which these quotients *are* bounded (see below, Proposition 3): $\sigma(rf)/r$ (resp., $\delta(rf)/r$) is bounded if and only if $f \in L_\infty$ (resp., $f \in L_1$), and in that case, $\|f\|_\infty = \lim_{r\to 0^+} \sigma(rf)/r$ (resp., $\|f\|_1 = \lim_{r\to 0^+} \delta(rf)/r$).

Quite naturally, this raises the question: which Banach spaces of measurable functions can be recovered from suitable metrics on L_0 that induce the convergence in measure? We consider this question below (see Sec. 2).

At this point, two lines of generalization suggest themselves:

(i) identify the Banach spaces that are isomorphic to the Banach spaces of measurable functions recovered from L_0;

(ii) consider the corresponding situation with L_0 replaced by an arbitrary linear metric space.

Concerning (ii), we show in Theorem 1 that a normed space E can be obtained by this procedure from a metric linear space L if and only if E can be semiembedded in L.

In Sec. 3 we consider semiembeddings of Banach lattices in L_p, $0 \leq p \leq 1$, and in Sec. 4 we make some remarks about semiembeddings of spaces of continuous functions in L_0.

1. NORMED SPACES ASSOCIATED TO METRIC LINEAR SPACES

Let L be a real vector space and ρ a *pseudonorm* on L. This simply means that ρ is a function from L into the nonnegative reals such that

$$\rho(x) = 0 \quad \text{if and only if} \quad x = 0; \tag{1}$$

$$\rho(x + y) \leq \rho(x) + \rho(y); \tag{2}$$

$$\rho(-x) = \rho(x); \tag{3}$$

$$\rho(tx) \text{ is a continuous function of } t \text{ for each fixed } x \in L. \tag{4}$$

We will say that L provided with ρ is a *metric linear space*. This is justified by the fact that $d(x, y) = \rho(x - y)$ is an invariant metric on L making the operations continuous.

A characteristic property of metric linear spaces is that balls of different sizes also have different shapes. To compare them, we have to magnify the balls to the same scale with a procedure that can be described as follows. First, for each $r > 0$, define

$$\rho_r(x) = \rho(rx)/r. \tag{5}$$

We have:

Proposition 1. Let ρ *be a pseudonorm on the linear space* L*. Then:*

(i) *for each* $r > 0$, ρ_r *is a pseudonorm on* L *defining the same metric topology as* ρ*;*

(ii) *for each* $x \in L$*, the limit* $\|x\|_\rho = \lim_{r\to 0^+} \rho_r(x)$ *exists in* $[0, +\infty]$*, and* $\rho_r(x) \leq \|x\|_\rho$ *for all* $r > 0$*;*

(iii) *the set* L_ρ *of all* $x \in L$ *with* $\|x\|_\rho < +\infty$ *is a linear subspace of* L *and* $\|\ \|_\rho$ *is a norm on* L*;*

(iv) L_ρ *is the largest subspace supporting a norm* $\|\ \|$ satisfying $\|x\| \geq \rho(x)$, *and* $\|\ \|_\rho$ *is the smallest such norm;*

(v) *let* $B_r = \{x \in L;\ \rho(x) \leq r\}$*; then* $(1/r)B_r$ *is the unit ball of* ρ_r *and the unit ball of* $\|\ \|_\rho$ *is the intersection* $\bigcap_{r>0} (1/r)B_r$.

Proof. We leave the proof to the reader except for the technical point in (ii), which can be obtained as a consequence of Theorem 7.11.1 in [2] or directly in the following way:

Let $a = \liminf_{r\to 0^+} g(r)$, where $g(r) = \rho_r(x)$, and $x \in L$ is fixed. Let $\varepsilon > 0$, and pick $r_n > 0$, $r_n \to 0$ such that $g(r_n) \leq a + \varepsilon$. Since $g(kr) \leq g(r)$ for k a positive integer (use (2)), we conclude that also $g(kr_n) \leq a + \varepsilon$, for all k, $n = 1, 2, \ldots$. Finally, since $\{kr_n\}$ is dense in $[0, +\infty]$, it follows from continuity that $g(r) \leq a + \varepsilon$ for all $r > 0$, and so $\sup_{r>0} g(r) \leq a$. Thus, $\limsup_{r\to 0^+} g(r) \leq \sup_{r>0} g(r) \leq \liminf_{r\to 0^+} g(r)$, and (ii) of Proposition 1 follows. Observe that $\|x\|_\rho$ coincides also with $\sup_{r>0} \rho_r(x)$, although $\rho_r(x)$ is not a decreasing function of r in general.

Corollary 1. All closed balls of L_ρ *are closed in* L *(for the* ρ*-topology).* This is clear from (v) of Proposition 1.

Proposition 2. If L *is complete under* ρ*, then* $(L_\rho, \|\ \|_\rho)$ *is a Banach space.*

Proof. From $\rho = \rho_1 \leq \|\ \|_\rho$ (see (ii) of Proposition 1) it follows that a $\|\ \|_\rho$-Cauchy sequence $\{x_n\}$ is also ρ-Cauchy, and therefore ρ-convergent to, say, $\tilde{x}$. Now let $\varepsilon > 0$ and pick N such that $\|x_n - x_m\|_\rho \leq \varepsilon$ for $n, m \geq N$. Assume that $n \geq N$ is fixed. Since the ball $B = \{x \in L_\rho; \|x_n - x\|_\rho \leq \varepsilon\}$ is closed for the ρ-convergence (Corollary 1) and $x_m \in B$ when $m \geq N$, it follows that $\tilde{x} \in B$, or $\|x_n - \tilde{x}\|_\rho \leq \varepsilon$, which proves that $\|x_n - x\|_\rho \to 0$ as $n \to \infty$.

Suppose now that L is a vector space provided with a metric linear topology. For each pseudonorm ρ defining the topology of L, a normed space $(L_\rho, \|\ \|_\rho)$ is determined by ρ. We will say that the normed space $(E, \|\ \|)$ is a *normed space associated to the metric linear space* L if $(E, \|\ \|)$ is isometric to some $(L_\rho, \|\ \|_\rho)$. Recall that a continuous linear operator $T : E \to L$ from a normed space E into a topological linear space L is a *semiembedding* if T is one-to-one and the image under T of the unit ball of E (equivalently, any closed ball of E) is a closed subset of L.

The main result of this section is the following:

Theorem 1. Let L *be a metric linear space and* B *a normed space. The following conditions are equivalent:*

(i) B *is a normed space associated to* L*;*

(ii) *there is a semiembedding from* B *into* L.

Proof. Corollary (1) shows that (i) implies (ii). Assume now that $T : B \to L$ is a semiembedding of B (whose norm we denote by $\|\ \|$) into L, and let ρ be any pseudonorm defining the topology of L. Set, for $x \in L$:

$$\rho'(x) = \inf_{b \in B} \{\|b\| + \rho(x - Tb)^{1/2}\}.$$

Routine calculations show that ρ' is a pseudonorm, and taking $b = 0$, we also get $\rho'(x) \leq \rho(x)^{1/2}$, so that ρ-convergent sequences are also ρ'-convergent. Assume next that $\rho'(x_n) \to 0$. Then there exist elements $b_n \in B$ such that also $\|b_n\| + \rho(x_n - Tb_n)^{1/2} \to 0$ and, in particular, $b_n \to 0$ and $\rho(x_n - Tb_n) \to 0$. But then also $\rho(Tb_n) \to 0$ by continuity of T and therefore $\rho(x_n) \leq \rho(x_n - Tb_n) + \rho(Tb_n)$ implies that $\rho(x_n) \to 0$. Hence ρ' defines the same topology as ρ.

Assume now that $0 < \|x\|_{\rho'} < +\infty$. We claim that $x = Tb$ for some b and that $\|b\| = \|x\|_{\rho'}$. In fact, let $\varepsilon > 0$. Picking $r = 1/n$ in (ii) of Proposition 1, we conclude that for n large, there are elements $b_n \in B$ such that

$$n[\|b_n\| + \rho(x/n - Tb_n)^{1/2}] \leq \|x\|_{\rho'} + \varepsilon .$$

It follows that $\|nb_n\| \leq \|x\|_{\rho'} + \varepsilon$ and also $n\rho(x/n - Tb_n)^{1/2} \leq \|x\|_{\rho'} + \varepsilon$. Hence,

$$\rho(x - nTb_n)^{1/2} = \rho\left(n\left[\frac{x}{n} - Tb_n\right]\right)^{1/2} \leq n^{1/2}\rho\left(\frac{x}{n} - Tb_n\right)^{1/2}$$
$$= n^{-1/2}\left[n\rho\left(\frac{x}{n} - Tb_n\right)^{1/2}\right] \leq n^{-1/2}[\|x\|_{\rho'} + \varepsilon] \to 0.$$

This shows that x is the limit (for the pseudonorm ρ) of the sequence $\{T(nb_n)\}$, which is the image under T of the bounded sequence $\{nb_n\}$. Then, the image under T of a closed ball being closed, we conclude that x is of the form $x = Tb$ with $\|b\| \leq \|x\|_{\rho'} + \varepsilon$. But T being one-to-one, the element b can not depend on ε, and so $\|b\| \leq \|x\|_{\rho'}$.

The reverse inequality follows from

$$\rho'(rx)/r = \rho'(rTb)/r \leq [\|rb\| + \rho(rTb - Trb)^{1/2}]/r$$
$$= \|rb\|/r = \|b\| .$$

and the theorem follows.

2. BANACH SPACES OF MEASURABLE FUNCTIONS

Throughout this section we consider the metric linear space L_0 of (classes of) measurable real functions defined on [0, 1] under the convergence in measure. Since L_0 is complete, all normed spaces associated to L_0 are also complete, that is, Banach spaces.

We begin by proving our claim in the Introduction about the spaces L_1 and L_∞ (the pseudonorms δ and σ were defined in the Introduction).

Proposition 3. The Banach spaces L_1 and L_∞ are associated to L_0 (by means of δ and σ, respectively).

Proof. Consider the case of σ. We want to prove that

$$\operatorname{ess}\sup_{0\le t\le 1} |f(t)| = \lim_{r\to 0^+} \sigma(rf)/r \tag{6}$$

for each measurable function f defined on [0, 1]. Set $M = \operatorname{ess}\sup |f(t)|$, $0 \le t \le 1$ (which is either finite or $+\infty$), and $N = \| f \|_\sigma$ (and again, N could be $+\infty$).

By definition, $\sigma(rf) = \inf\{\varepsilon + m(|f| > \varepsilon/r);\ \varepsilon > 0\}$ so that picking $\varepsilon = rM$ when $M < +\infty$, we conclude that $\sigma(rf) \le rM$. Thus $N \le M$.

Assume now that $N < M$, and pick $r_n \to 0$, $a > 0$ and $\varepsilon_n > 0$ so that $N + a < M$ and

$$[\varepsilon_n + m(|f| > \varepsilon_n/r_n)]/r_n \le N + a$$

for each $n = 1, 2, \ldots$. In particular, $m(|f| > \varepsilon_n/r_n) \le r_n(N + a)$ and $\varepsilon_n \le r_n(N + a)$. But then

$$m(|f| > N + a) \le m(|f| > \varepsilon_n/r_n) \le r_n(N + a) \to 0$$

whence $m(|f| > N + a) = 0$ and so $M \le N + a$, a contradiction. Thus $\| \ \|_\sigma = \| \ \|_\infty$. For the case of δ, we observe that Beppo Levi's theorem implies that

$$\int_0^1 |f(t)|\, dt = \lim_{r\to 0^+} \int_0^1 |f(t)|(1 + r|f(t)|)^{-1} dt$$

and then the equality $\| \|_\delta = \| \|_1$ follows easily.

The next theorem gives a list of classical Banach spaces that can be recovered from L_0 and the convergence in measure by this procedure.

Theorem 2. The following Banach spaces are associated to L_0:

(i) L_1;

(ii) L_∞;

(iii) *all closed subspaces of* L_p, $1 < p < \infty$;

(iv) *all reflexive subspaces of* L_1;

(v) H_1;

(vi) Lip(α), $0 < \alpha < 1$.

Proof. In all cases we will simply prove that the natural inclusion $B \subset L_0$, where B is one of the above, is a semiembedding, so that the theorem will follow from Theorem 1. Typically, then, we consider a sequence $\{f_n\}$ in the unit ball U of B converging in measure to $f \in L_0$. The desired conclusion is that $f \in U$.

Proofs of (*i*) and (ii) are given above.

Proof of (iii): Suppose, taking a subsequence if necessary, that f_n converges weakly in L_p to g. For each $\varepsilon > 0$, Egoroff's theorem implies that $f = g$ on a set E_ε with $m([0, 1] - E_\varepsilon) < \varepsilon$. But then $f = g$ almost everywhere, and $f \in U$ as claimed.

Proof of (iv): The argument just used applies verbatim.

Proof of (v): Here we consider, for simplicity, the interval $[0, 2\pi]$ rather than $[0, 1]$. Using (i), we conclude that $f \in L_1$ and that $\| f \|_1 \le 1$. Egoroff's theorem again and the Cauchy-Schwarz inequality imply that

$$\lim_n \int_0^{2\pi} |f_n(t) - f(t)|^{1/2} dt = 0$$

so that, by passing to a subsequence, we may assume that

$$\int_0^{2\pi} |f_n(t) - f(t)|^{1/2}\,dt \le 2^{-n}. \tag{7}$$

Observing that $\|f_n\|_1 \le 1$ implies that the functions $f_n(z)$ are uniformly bounded on compact sets of the unit disc $|z| \le 1$, we also assume (normal families) that there is an analytic function g defined on $|z| < 1$ such that the sequence $\{f_n(z)\}$ converges uniformly on compact subsets of the unit disc to g.

We shall now show that $g \in H_{1/2}$ and that $f = g$ a.e. on the circle $|z| = 1$. This, in conjunction with Theorem 2.11 in [1], will imply that $f \in H_1$ as desired.

A special case of the Hardy-Littlewood maximal theorem (see 1.9 in [1], for example) is that there is a constant K such that for all $h \in H_{1/2}$,

$$\int_0^{2\pi} |Mh(t)|^{1/2}\,dt \le K \int_0^{2\pi} |h(t)|^{1/2}\,dt, \tag{8}$$

where Mh is the maximal function $Mh(t) = \sup_{0<r<1} |h(re^{it})|$. Combining (7) and (8), we get

$$\int_0^{2\pi} |M(g - f_n)(t)|^{1/2}\,dt$$

$$\le \int_0^{2\pi} \sup_{0<r<1} \left\{ \sum_{k=n}^{\infty} |f_k(re^{it}) - f_{k+1}(re^{it})|^{1/2} \right\} dt$$

$$\le K \int_0^{2\pi} \sum_{k=n}^{\infty} |f_k(t) - f_{k+1}(t)|^{1/2}\,dt \le K2 \sum_{k=n}^{\infty} 2^{-k} = 2^{-n+2}K.$$

In particular,

$$\int_0^{2\pi} |g(re^{it}) - f_n(re^{it})|^{1/2}\,dt \le 2^{-n+2}K$$

for all $r < 1$, and therefore $g = (g - f_n) + f_n$ is in $H_{1/2}$. Taking now limits for $r \to 1$ in the last inequality, we get $g = f$ a.e., and the proof of (v) is complete.

Proof of (vi): Recall that $f \in \mathrm{Lip}(\alpha)$ if and only if $|f(x) - f(y)| \leq k|x - y|^{\alpha}$ for some k and all $x, y \in [0, 1]$, with the norm in $\mathrm{Lip}(\alpha)$ defined by $\| f \| = \| f \|_{\infty} + k_0$, where $k_0 = \inf\{k\}$ and $\| f \|_{\infty}$ is the supnorm. In particular, $|f(x) - f(y)| \leq \| f \| \cdot \| x - y \|^{\alpha}$.

Taking a subsequence if necessary, we assume now that $f_n(x) \to f(x)$ for all $x \in E$ where $[0, 1] - E$ has measure zero. Then

$$|f(x) - f(y)| \leq |x - y|^{\alpha} \tag{9}$$

holds for all $x, y \in E$. Since E is dense, and the above condition implies uniform continuity of f on E, we can find a (unique) continuous function $F : [0, 1] \to \mathbf{R}$ satisfying $F|_E = f$. Next we observe that $\{F, f_1, f_2, \ldots\}$ is an equicontinuous family such that $f_n \to F$ on a dense set ($=E$). Thus f_n converges uniformly on $[0, 1]$ to F, and so F also satisfies a Lipschitz condition. Since $F = f$ a.e., we conclude that $f \in U$ as desired.

3. BANACH LATTICES

A Banach lattice E is called σ-1-*boundedly order complete* if for every increasing sequence $\{x_n\}$ of positive elements in E with $\| x_n \| \leq 1$, the supremum $x = \sup_n x_n$ exists and $\| x \| \leq 1$. A positive element u in a Banach lattice E is a *weak order unit* if $u \wedge x = 0$ implies $x = 0$. A positive element x' in the dual E^* of E is a *strictly positive* linear form on E if $x'(|x|) = 0$ for some x implies $x = 0$; also, an x' in E^* is *order continuous* if for every decreasing net $\{x_a\}$ in E with $\inf_a x_a = 0$ we have $\lim_a x'(x_a) = 0$.

Proposition 4. Let E be a Banach lattice. Then there exist a finite measure space (Ω, Σ, μ) and a semiembedding T from E in $L_1(\mu)$ that is a lattice homomorphism onto an ideal if and only if the following conditions hold:

(i) *E is σ-1-boundedly order complete;*

(ii) *E has a weak order unit;*

(iii) *there exist strictly positive order continuous linear forms on E.*

Proof. We begin with the necessity implication: let T be a lattice isomorphism from E onto an ideal of $L_1(\mu)$. Then $x \longmapsto \int Tx\, d\mu$ is a strictly positive order continuous linear form on E and (iii) holds. Since every maximal family of positive pairwise disjoint elements in $L_1(\mu)$ is at most countable, this is also true in E and therefore E has a weak order unit, which means that (ii) holds. Suppose now that $\{x_n\}$ is a sequence of positive elements in E with $x_n \leq x_{n+1}$ and $\|x_n\| \leq 1$ for each n. Then $\{Tx_n\}$ is a Cauchy sequence in $L_1(\mu)$, and T being a semiembedding, $f = \lim Tx_n = Tx$ for some $x \in E$ with $\|x\| \leq 1$. But since T is an order continuous lattice isomorphism, $x = \sup x_n$, that is, E is σ-1-boundedly order complete.

Sufficiency: Let $w' \in E^*$ be a strictly positive order continuous linear form on E. Then we may consider E to be an ideal in the AL - space (E, w'), since w' is strictly positive and order continuous (see [6, p. 113 and IV.9.3]). Let u be a weak order unit in E; u is also an order unit in (E, w'). Hence, by Kakutani's theorem [6, II.8.5], we may identify (E, w') with a space $L_1(\mu)$, μ finite. Hence we have a lattice isomorphism T from E onto an ideal of $L_1(\mu)$ with $Tu = 1$. We shall show that this T is a semiembedding. Let $\{x_n\}$ be a sequence in the unit ball of E such that $Tx_n = f_n$ converges in measure to $f \in L_0(\mu)$. Then $T(|x_n| \wedge ku) = |f_n| \wedge k$ converges to $|f| \wedge k$ for every $k > 0$. Hence there exist $z_k \in [0, ku]$ with $Tz_k = |f| \wedge k$. Passing to a subsequence we may assume that $|f_n| \wedge k$ converges μ-almost everywhere to $|f| \wedge k$, that is,

$$\bigvee_{m=1}^{i} \bigwedge_{n=m} (|f_n| \wedge k) = |f| \wedge k,$$

which implies that

$$\bigvee_{m=1} \bigwedge_{n=m} (|x_n| \wedge ku) = z_k.$$

Since E is σ-1-boundedly complete, $\| z_k \| \leq 1$. Moreover, since $\{z_k\}$ is an increasing sequence, $\bar{z} = \sup z_k$ exists, $\| \bar{z} \| \leq 1$, and $T\bar{z} = |f|$. Hence there exists $z \in E$ with $|z| \leq |\bar{z}|$ and such that $Tz = f$. Clearly $\| z \| \leq 1$, which completes the proof.

Corollary 2. Let E be a separable Banach lattice. Then there exists a semiembedding in L_1 *if and only if E is weakly sequentially complete.*

Proof. If there exists a semiembedding from E into L_1, then no closed subspace of E is isomorphic to c_0 by Corollary 12 in [4]. Hence E is weakly sequentially complete [3, 2.4]. Now suppose that E is a separable weakly sequentially complete Banach lattice. Then E has a weak order unit and also E^* has a weak order unit w' (see [6, II.6.2, II.6.5, and II.6.6]). It follows easily from (2.1) and (2.4) in [3] that w' is strictly positive and order continuous and that E is σ-1-boundedly complete. Hence there exists a semiembedding from E into a space $L_1(\mu)$ by Proposition 4 above. Since we may assume that $L_1(\mu)$ is separable, $L_1(\mu)$ is lattice isomorphic to a closed sublattice of L_1, which completes the proof.

Observe that in the proof of Proposition 4, the sequence f_n was assumed to converge *in measure* (rather than in L_1). This means that the proof applies to all L_p, $0 \leq p \leq 1$. More precisely:

Corollary 3. Let E be a Banach lattice that satisfies (i), (ii), *and* (iii) *of Proposition* 4. *Then:*

(i) *for each p,* $0 \leq p \leq 1$, *there exists a measure space with measure* μ *and a lattice isomorphism from E onto an ideal of* $L_p(\mu)$ *that is a semiembedding;*

(ii) *for each* $0 \leq p < 1$, *there is a semiembedding from E into* L_p.

Corollary 4. If E is a separable Banach lattice that does not contain c_0 *as a sublattice, then there is a semiembedding of E in* L_p *for each* $0 \le p \le 1$.

4. SPACES OF CONTINUOUS FUNCTIONS

We now take up the semiembeddings of spaces $C(X)$ (continuous real functions on a compact space X) in the space $L_0 = L_0[0, 1]$. Recall that X is *scattered* if it contains no (nonempty) perfect sets. The main result of this section is the following.

Theorem 3. Let X be a scattered space. If $C(X)$ *has infinite dimension, then there are no semiembeddings of* $C(X)$ *in* L_0.

We indicate some results that are needed for the proof of Theorem 3:

If $T : c_0 \to L_0$ is linear and continuous for the norm in c_0 (and the convergence in measure in L_0), then T is also continuous for the weak topology on any ball in c_0 (and again the convergence in measure in L_0). (10)

All continuous linear maps from c_0 into L_0 are compact. (11)

With (10) and (11), we can prove the following proposition of independent interest (cf. Prop. 4 in [4]):

Proposition 5. Let E be a Banach space and $U : E \to L_0$ *a semiembedding. Then any continuous linear map* $S : c_0 \to E$ *has an extension* $\tilde{S} : \ell^\infty \to E$ *with the same norm.*

Proof of Proposition 5. Write $T = US : c_0 \to L_0$. For $x'' \in \ell^\infty$, let $\{x_n\}$ be a sequence in c_0 converging weak* to x'' and with $\|x_n\| \le \|x''\|$. By (11), there is a subsequence $\{y_n\}$ such that $Ty_n = USy_n$ is convergent in L_0, and, U being a semiembedding, there is a $w \in E$ such that $Uw = \lim Ty_n$. We claim that w depends only on x''. In fact, if

y'_n is a subsequence of a different x'_n also converging weak* to x'', then $y_n - y'_n \to 0$ weakly in c_0, and by (10), $T(y_n - y'_n) \to 0$ so that $Uw = \lim Ty_n$ is well defined. But U is one-to-one, and therefore w is also well defined, as claimed. Observe further that using again the fact that U is a semiembedding, we have $\| w \| \leq \sup \| Sx_n \| \leq \| S \| \| x'' \|$, and therefore $\tilde{S}x'' = w$ is an extension of S satisfying $\| \tilde{S} \| = \| S \|$.

Proof of Theorem 3. Suppose that $U : C(X) \to L_0$ is a semiembedding. Using the remark on page 238 of [4], we see that there is a subspace $V \subset C(X)$ isomorphic to c_0 and such that V is complemented in $C(X)$. According to Proposition 5, the injection $S : V \to C(X)$ has an extension $S : V^{**} \to C(X)$; but then the composition of S with the projection of $C(X)$ onto V gives a projection of V^{**} onto V. Since there are no such projections, we obtain Theorem 3.

For the sake of completeness, we give a proof of (10). Suppose that $\{x_n\}$ is a sequence in c_0 converging to 0 weakly. If $\| x_n \| \to 0$ also, then there is nothing to prove. Assume then (passing to a subsequence) that $0 < a \leq \| x_n \| \leq b$. The usual argument allows us to pass to a second subsequence (also denoted by x_n) equivalent to the standard basis in c_0. This means that for any sequence of coefficients $u_n \to 0$, the series $\sum_{n=1}^{\infty} u_n x_n$ is subseries convergent in c_0. Let $f_n = Tx_n$. Then $\sum_{n=1}^{\infty} u_n f_n$ is also subseries convergent in L_0. Using the Hilfsatz on page 29 of [5], we conclude that $\sum_{n=1}^{\infty} u_n^2 f_n^2 < +\infty$ almost everywhere if $u_n \to 0$. If a subsequence of the f_n converges to 0 in measure, the result follows. Suppose then that there are numbers $r, s > 0$ such that $m(f_n^2 \geq r) \geq s$ for all n. Pick now $F_1, F_2, \ldots$ such that:

$$0 \leq F_n \leq f_n^2; \tag{12}$$

$$F_n = r \quad \text{on a set} \quad E_n \quad \text{of measure} \quad s; \tag{13}$$

$$F_n = 0 \quad \text{off} \quad E_n. \tag{14}$$

Then the functions $g_N = \sum_{n=1}^{N} u_n^2 F_n$ satisfy:

$$\int g_N = rs \sum_{n=1}^{N} u_n^2; \tag{15}$$

$$0 \le g_1 \le g_2 \le \cdots \quad \le \sum_{n=1}^{\infty} u_n^2 F_n \le \sum_{n=1}^{\infty} u_n^2 f_n^2. \tag{16}$$

However, if g is a measurable function satisfying $\int g = B$ and $0 \le g \le A$, then for $0 \le C \le B$, the set $\{g \ge C\}$ has measure at least $(B - C)/(A - C)$. This applied to g_N with the values $A = r \sum_{1 \le n \le N} u_n^2$, $B = rs \sum_{1 \le n \le N} u_n^2$, and $C = B/2$ yields

$$m\left(g_N \ge \frac{1}{2} rs \sum_{1}^{N} u_n^2\right) \ge \frac{s}{2 - s},$$

and therefore, using (16),

$$m\left(\sum_{n=1}^{\infty} u_n^2 f_n^2 \ge \frac{1}{2} rs \sum_{n=1}^{N} u_n^2\right) \ge \frac{s}{2 - s}.$$

Choosing now $u_n = n^{-1/2}$, we get that $\sum_{n=1}^{\infty} n^{-1} f_n^2$ cannot be finite almost everywhere, a contradiction.

Finally, (11) follows from (10), since every bounded sequence in c_0 has a subsequence that is weak Cauchy.

REFERENCES

1. Duren, P. *The Theory of H^p-spaces*. New York: Academic Press, 1970.

2. Hille, E., and Phillips, R. *Functional Analysis and Semigroups*. Providence, R.I.: A.M.S. Colloquium Publications, 1957.

3. Lotz, H. Minimal and reflexive Banach lattices. *Math. Ann.* 209 (1974): 117-126.

4. Lotz, H. P.; Peck, N. T.; and Porta, H. Semi-embeddings of Banach spaces. *Proc. Edinburgh Math. Soc.* 22 (1979): 233-240.

5. Orlicz, W. Über die Divergenz von allgemeinen Orthogonalreihen. *Studia Math.* 4 (1932): 27-32.

6. Schaefer, H. H. *Banach Lattices and Positive Operators.* New York: Springer, 1974.

DIFFUSION ON THE LOOPS

Paul Malliavin
Institut Henri Poincaré, Paris

Given a Riemannian compact simply connected manifold M, the study of the space of closed curves $L(M)$ with a base point had been introduced by Marston Morse [10]. On some smoothness condition $L(M)$ becomes a Hilbert Riemann manifold [3]. Therefore, it has a natural Laplace operator and a diffusion process can be constructed on $L(M)$.

Starting at $t = 0$ from the zero loop, this diffusion makes possible to construct for each $t \in \mathbf{R}^+$ a natural measure γ_t on $L(M)$. Given a vector fiber bundle F of rank r on M, its holonomy defines a natural map h of $L(M)$ in the orthogonal group of dimension r; it is natural to ask what is the asymptotic, for t large, of the image measure $h_\star \gamma_t$.

A construction of γ_t had been given by B. Gaveau and E. Mazet [4] using the structure of infinite dimensional manifold of $L(M)$. In our approach we shall not use functional analysis; a path on $L(M)$ will be a function of two variables with values in M. This means that we will use more the resources of stochastic calculus for two-parameter processes than the theory of infinite dimensional manifolds.

Our starting point will be the loop space on $\mathbf{R}^n$, that is, the study of some random Fourier series of Wiener's type. We shall relate the *loop space on* $\mathbf{R}^n$ to the *Brownian sheet*.

The development on $L(M)$ in the *smooth* case can be reduced, by the moving frame, to the solution of a *nonlinear* hyperbolic system. We will have to solve this nonlinear system when the driving force is a *distribution* belonging to a Sobolev space of order -1. The existence of this solution will be proved using the double stochastic integral of Cairolli and Walsh [2] and an iterative procedure.

In the case of one-dimensional time processes, the Itô's integral makes possible to solve nonlinear systems for one-dimensional time processes.

1. TIME EVOLUTION OF WIENER-FOURIER SERIES

The Brownian Loop

It has been defined by Wiener by the Fourier series

$$u_\omega(\tau) = 2^{1/2} \sum_{k\geq 1} \frac{\sin k\pi\tau}{k} G_k, \qquad (1)$$

where G_k are independent normal variables. We denote by ω the random. The probability space Ω corresponds to $\mathbf{R}^{\mathbf{N}}$ with the product measure of the normal law on each component.

The covariance is defined by $\varphi(\tau, \tau') = E\big(u_\omega(\tau)u_\omega(\tau')\big)$.

Lemma 1. $\varphi(\tau, \tau') = \tau'(1 - \tau)$ *for* $0 < \tau' < \tau < 1$.

Regularization of u_ω

It will be convenient to use the convolution by a smooth periodic function as mollifier of u_ω. In fact, we shall use the Poisson kernel and denote, for $0 \leq \rho \leq 1$,

$$u_{\omega,\rho}(\tau) = 2^{1/2} \sum_{k=1}^{+\infty} \frac{\rho^k}{k} G_k \sin k\pi\tau.$$

We remark that 1, $\{2^{1/2} \sin k\pi\tau\}$ form an orthonormal basis of $L^2([0, 1])$. Define, for $f \in L^2$,

$$\hat{f}(0) = \int f(\tau)\, d\tau, \quad \hat{f}(k) = 2^{1/2} \int f(\tau) \sin k\pi\tau\, d\tau.$$

For $f \in L^2$ and $\rho < 1$, we have then

$$I(\omega, f, \rho) = \int_0^{2\pi} f(\tau) u'_{\omega,\rho}(\tau)\, d\tau = \sum_{k\geq 1} \hat{f}(k) G_k \rho^k.$$

Then $I(\omega, f, \rho)$ is a Gaussian variable with variance

$$\sigma_\rho(f) = E\big((I(f, \rho))^2\big) = \sum_{k=1}^{+\infty} (\hat{f}_1(k))^2 \rho^{2k},$$

then, when $\rho \to 1$, we have

$$\lim \sigma_\rho(f) = \int_0^1 |f(\tau) - \hat{f}(0)|^2\, d\tau.$$

White Noise

Given an abstract real Hilbert Space $\mathcal{H}$, we shall call the white noise $W_{\mathcal{H}}$ the canonical cylindrical measure on $\mathcal{H}$. For every finite set $h_i \in \mathcal{H}$, $i \in [1, n]$, the map of $\mathcal{H}$ into $\mathbf{R}^n$ defined by

$$h \to (h|h_i) \qquad i \in [1, n]$$

sent $W_{\mathcal{H}}$ into a Gaussian variable on $\mathbf{R}^n$ having as covariance matrix $(h_i|h_j)$.

Proposition 1. Denote by $\mathcal{H}$ the Hilbert space that is the orthogonal in $L^2([0, 1])$ to the constants. Let $P_{\mathcal{H}}$ be the orthogonal projection of L^2 on $\mathcal{H}$. We define the stochastic integral

$$\int f\, du_\omega = (h|W_{\mathcal{H}}) \quad \textit{where} \quad h = P_{\mathcal{H}} f.$$

Then

$$\lim_{r\to 1} \int f\, du_{r,\omega} = \int f\ du_{\omega}.$$

Conditioning White Noise

We can consider the Hilbert space $\mathcal{H}_1 = L^2([0, 1])$. Then the white noise $W_{\mathcal{H}}$ considered can be interpreted as the white noise $W_{\mathcal{H}_1}$, *under the conditioning* $W_{\mathcal{H}}(\mathbf{1}) = 0$.

More generally, given an abstract Hilbert space $\mathcal{H}_2$ and the canonical white noise defined on $\mathcal{H}_2$, if we give a finite system of linear form $h_{\alpha_1} \ldots h_{\alpha_s}$ on $\mathcal{H}_2$, we can interpret the white noise $W_{\mathcal{H}_2}$ under the conditioning

$$(W_{\mathcal{H}_2} | h_{\alpha_\ell}) = \beta_\ell, \quad 1 \le \ell \le s,$$

where β_ℓ are given constants. We denote by $W_{\mathcal{H}_2}^{h_\alpha}$ this conditioned white noise. Given h_i, $1 \le i \le n$, we consider the map

$$\mathcal{H}_2 \to \mathbf{R}^{s+n} \qquad \text{defined by}$$

$$h \to (h|h_\alpha),\ (h|h_i).$$

The image of $W_{\mathcal{H}_2}$ under this map is a Gaussian measure; we shall consider its conditioned law on the affine linear variety $\xi_j = \beta_j |1 \le j \le s|$; we will get a Gaussian measure.

On the other hand, we could introduce the orthogonal $\mathcal{H}_3$ to the linear subspace K generated by the h_α. Then

$$W_{\mathcal{H}_2}^{h_\alpha} = W_{\mathcal{H}_3}. \tag{2}$$

In this form the conditional white noise does not depend on the values of the conditioning. Denote by W_K the white noise associated to the finite dimensional subspace K (it can be identified to the normal Gaussian variable on K); then we have

$$W_{\mathscr{H}_2} = W_{\mathscr{H}_3} + W_K \tag{3}$$

this sum being a sum of independent variables.

Markov Property for the Brownian Loop

Proposition 2. Let γ *be a subinterval of* $[0, 1]$, γ^c *its complement in* $[0, 1]$. *Consider the conditioning:*

$u_\omega(\tau)$ *is given at the two boundary points of* γ; *we shall denote this conditioning by*

$$u_\omega|_{\partial\gamma}.$$

Then the two processes $u_\omega|_\gamma$ *and* $u_\omega|_{\gamma^c}$ *are independent under the conditioning* $u_\omega|_{\partial\gamma}$.

Proof. The differential du_ω, in the sense of distribution, is the white noise W on $L^2([0, 1])$, under the conditioning

$$W(\mathbf{1}) = 0.$$

As $\mathbf{1}_\gamma + \mathbf{1}_{\gamma^c} = \mathbf{1}$, this conditioning can be written as

$$W(\mathbf{1}_\gamma) = -W(\mathbf{1}_{\gamma^c}).$$

Let P_1, P_2 be the orthogonal projections on $L^2([0, 1])$ defined by $P_2 f = \mathbf{1}_\gamma f$, $P_3 f = \mathbf{1}_{\gamma^c} f$. Let $\mathscr{H}_i$ be the range of P_i, W_i the white noise associated to $\mathscr{H}_i$; then the decomposition $L^2 = \mathscr{H}^1 \overset{\perp}{\oplus} \mathscr{H}^2$ implies:

$$W = W_2 + W_3 \quad \text{(sum of independent variables)}.$$

Let $\mathscr{H}$ be the orthogonal of $\mathbf{1}$ in $L^2([0, 1])$. Define a map of $\mathscr{H}_2 \oplus \mathscr{H}_3 \to \mathbf{R}^2$ by $(h_2, h_3) \to ((h_2|\mathbf{1}), (h_3|\mathbf{1}))$.

Note that $(h_2|\mathbf{1}) = (h_2|P_2\mathbf{1}) = (h_2|\mathbf{1}_\gamma)$.

Denote by K the two-dimensional subspace of $L^2([0, 1])$ generated by $\mathbf{1}_\gamma$, $\mathbf{1}_{\gamma^c}$:

$$K^{\perp} \cap \mathcal{H}^2 = \{h_2 \in \mathcal{H}_2;\ (\mathbf{1}_\gamma | h) = 0\}.$$

As $(K^{\perp} \cap \mathcal{H}_2)$ and $(K^{\perp} \cap \mathcal{H}_3)$ are orthogonal, the associated white noise are independent; we shall denote them by W_4 and W_5. The conditioning by $\mathbf{1}_\gamma$, $\mathbf{1}_{\gamma^c}$ is stronger than the conditioning by $\mathbf{1}_\gamma + \mathbf{1}_{\gamma^c}$, and therefore

$$W^{\mathbf{1}=0} = \int_{\mathbf{R}} (W_4^{\mathbf{1}_\gamma} + W_5^{\mathbf{1}_{\gamma^c}})\, d\sigma,$$

where $d\sigma$ is the law of $W^{\mathbf{1}=0}(\mathbf{1}_{\gamma^c})$.

We have then shown the proposition at the level of stochastic differential. As, by the continuity a.s. of $u_\omega(.)$ we have, if $\gamma = [\alpha, \beta]$, $u_\omega(\alpha) + \int_\alpha^\tau du_\omega = u_\omega(\tau)$, for $\tau \in [\alpha, \beta]$. We have therefore proved the proposition.

Conditional Laws

Proposition 3. Let $u_\omega(\tau)$ *be the Brownian loop. Let* $0 < \xi_1 < \xi_3 < \xi_2 < 1$, $\xi_3 = \alpha\xi_1 + \beta\xi_2$, *where* $\alpha + \beta = 1$. *The variable* $u_\omega(\tau_3)$, *conditioned by* $u_\omega(\tau_1)$ *and* $u_\omega(\tau_2)$, *can be written*

$$u_\omega(\tau_3) = \alpha u_\omega(\tau_1) + \beta u_\omega(\tau_2) + \gamma G,$$

where G *is a normal variable, independent of the conditioning, and where*

$$\gamma = (\xi_2 - \xi_1)^{1/2}.$$

Proof. The covariance computed in Lemma 1.

The Brownian Loop and the Brownian Path

Proposition 4. Let $b_{\omega_1}(\tau)$ *be the Brownian path starting at zero. Then there exists a mapping* $\omega \to \omega_1$ *preserving the probability such that*

$$b_{\omega_1}(\tau) - \tau b_{\omega_1}(1) = u_\omega(\tau).$$

Proof. We can use the Wiener's series for the Brownian motion:

$$b_{\omega_1}(\tau) = \tau G_0 + 2^{1/2} \sum_{k>1} G_k \frac{\sin k\pi\tau}{k}$$

(where the G_k are independent normal variables).

Another way to see the same fact will be to use the differential db_{ω_1}, which is the unconditioned white noise on [0, 1].

Stochastic Differential Equation Driven by the Brownian Loop

Let $u^1_\omega, \ldots, u^n_\omega$ n independent samples of the Brownian loop, $n^2 + n$ smooth functions $a^i_j(x)$, $c^i(x)$ on $\mathbf{R}^n$ been given, we want to solve the S.D.E. system defined by

$$dx^i_\omega(\tau) = a^i_j(x_\omega(\tau))\, du^j_\omega(\tau) + c^i(x_\omega(\tau))\, d\tau.$$

Using Proposition 4, this system can be written

$$dx^i_{\omega_1}(\tau) = a^i_j(x_{\omega_1}(\tau))\, db^j_{\omega_1}(\tau) + [c^i(x_{\omega_1}(\tau)) - a^i_j(x_{\omega_1}(\tau) b^j_{\omega_1}(1)]\, d\tau.$$

This system can be solved by iteration of the corresponding integral equation (see J. M. Bismut [1]).

Time Evolution of the Brownian Loop

Our objective is to consider the parameter $\tau \in T$ as "a geometric variable" and to introduce a "time parameter," taking its values in R^+. The increasing in time will correspond to the augmentation of the variance of the Brownian loop. At time $t = 0$, we will start with the function identically to zero. This point of view had been used in [6]. Define the function

$$q_\xi(t, \tau) = \sum_{k\geq 1} \frac{1}{k} G_k(t) \sin k\pi\tau, \tag{4}$$

where $G_k(t)$ are independent samples of the Brownian motion on $\mathbf{R}$, starting at time $t = 0$ from zero. We shall denote by Ξ the infinite product of the probability space of the Brownian on **R** starting from zero. Then $\xi \in \Xi$ denotes the generic point of Ξ.

Theorem 1. Denote $\mathscr{C}^\beta$ the class of β-Hölderian *function of* τ *defined on* [0, 1], *then a.s. in* ξ *the mapping*

$$t \to q_\xi(t, .)$$

is a continuous map from R^+ *to* $\mathscr{C}^\beta$ *for* $\beta < \frac{1}{2}$.

Proof. Cf [7], Proposition 5.1, p. 246].

Definition. We shall call the map

$$(t, \tau) \to q_\xi(t, \tau)$$

the *loop-sheet*.

The Brownian Sheet

We define the Brownian sheet by

$$p_{\xi_1}(t, \tau) = \tau G_o(t) + 2^{1/2} \sum_{k\geq 1} \frac{\sin k\pi\tau}{k\pi} G_k(t).$$

Then the random parameter $\xi_1 \in \Xi \times \Omega$, where the last factor of this product of measure spaces is the probability space of the Brownian on **R**. We have by Proposition 4,

$$q_\xi(t, \tau) = p_{\xi_1}(t, \tau) - \tau p_{\xi_1}(t, 1).$$

The $\mathbf{R}^n$-Valued Sheets

We shall define the *Brownian sheet* on $\mathbf{R}^n$ taking $p^1_{\xi_1}, \ldots, p^n_{\xi_1}$ n independent components of the p_{ξ_1}. In the same way, the loop-sheet on $\mathbf{R}^n$ is defined by taking n independent components $q^1_\xi, \ldots, q^n_\xi$.

2. RIEMANNIAN DEVELOPMENTS

Let M be a compact Riemannian manifold of dimension n. Let $m_0 \in M$ be fixed. We shall consider $P_{m_0}(M)$ the path space on M starting from m_0 that is the continuous maps u of $[0, 1] \to M$, such that $u(0) = m_0$. We shall denote by

$$L_{m_0}(M) = \{u \in P_{m_0}(M) ;\ u(1) = m_0\}$$

the loop space with base point m_0.

The development problem is to take some sample path of the $p_\xi(t, \tau)$ in the Euclidian tangent space $T_{m_0}(M)$ and to *develop* it on the manifold, solving some stochastic system, obtaining then a map $\tilde{p}_\xi(t, \tau)$ with values in M.

In [8], such development is realized using the variation of the stochastic differential system defining the parallelism transport. The development that we shall construct will be different; its machinery is based on the natural Riemannian metric on the space of path (or curves) of Sobolev class 1.

The Riemann Hilbert Structure on $P^1(M)$

We shall denote

$$P^1_{m_0}(M) = \left\{ u \in P_{m_0}(M) ;\ \int_0^1 \left\| \frac{du}{d\tau} \right\|^2 d\tau < +\infty \right\}$$

$$L^1_{m_0}(M) = L_{m_0} \cap P^1_{m_0} .$$

Finally, we denote P^∞ the space of smooth path.

Let φ_t be a smooth curve on $P^\infty_{m_0}$. Then φ_t can be seen as a function of two parameters $\varphi(t, \tau)$. We want to compute its length on L^1

$$\int_{t_0}^{t_1} \| d\varphi_t \|_{L^1} .$$

We take by definition

$$\| d\varphi_t \|^2_{L^1} = \int_0^1 \| \frac{\partial \varphi}{\partial t} \|^2 d\tau + \int_0^1 \| \nabla \left(\frac{\partial \varphi}{\partial t}\right) \|^2 d\tau, \tag{5}$$

where ∇ denotes the covariant derivative along the curve $\tau \to \varphi_t(\tau)$.

Then it can be proved [3] that this metric defines a Riemann-Hilbert structure on $P^1_m(M)$.

The Normal Charts on $P^1(M)$.

Given $\varphi \in P_{m_0}(M)$, let $\Gamma_\varphi = \varphi_\star(T(M))$ be the pull back bundle of the tangent bundle $T(M)$.

Then, given a continuous section s of Γ_φ, with $s(0) = 0$, we can define

$$\exp_\varphi(s) = \psi \in P_{m_0}(M) \quad \text{by the formula}$$

$$\psi(\tau) = \exp_{\varphi(\tau)}(s(\tau)),$$

where $\exp_m$ denotes the normal chart relative to M with center m.

Development of $\Gamma\varphi$

If $\varphi \in L^\infty$, then the Riemannian connection makes possible to identify a section of Γ_φ with a function *defined* on [0, 1] with values in $P_{m_0}(M)$.

The space of sections $\mathcal{H}_1$ are the sections for which the following norm is finite:

$$\| s \|^2_{\mathcal{H}} = \int_0^1 \| s(\tau) \|^2 d\tau + \int_0^1 \| \nabla s \|^2 d\tau,$$

where ∇ is the covariant derivative along the smooth path φ. Denote

$$0_{\varphi,\varepsilon} = \{\varphi; \|\varphi\|_{\mathcal{H}_1} < \varepsilon\}.$$

Then

$$s \to \exp_\varphi(s)$$

defines, for ε small enough, a normal chart of the Riemann-Hilbert structure on $P^1_{m_0}(M)$ (cf. [3]).

The Development on M

Let p be a C^∞-map of $(R^+)^2$ into $T_{m_0}(M)$, equal to zero on the boundary. We shall define the development $\tilde{p}$ on M by

$$\tilde{p}(0, \tau) = \tilde{p}(t, 0) = m_0$$

$$\frac{\partial \tilde{p}}{\partial t} = \frac{\partial p}{\partial t}, \tag{6}$$

equation (6) being understood under the development of Γ_φ of the first member, which is a section of Γ_φ (where $\varphi(\tau) = \tilde{p}(t, \tau)$), to $T_{m_0}(M)$-valued function. We shall write equation (6) more explicitly using a lifted curve to the frame bundle.

The Development on $0(M)$

Denote by $0(M)$ the $0(n)$-principal bundle of orthonormal frames on M. Denote by $\mathbf{p}$ the projection of $0(M)$ over M. Recall that a frame at $m \in M$ is a Euclidian isomorphism r of R^n into $T_m(M)$. Then the orthogonal group $0(n)$ operates on the right by

$$r \to r \circ g, \quad g \in 0(n), \quad r \in 0(M).$$

We can consider the canonic horizontal vector fields on $0(M)$, $A_1, \ldots, A_n$ defined by the Riemannian connection.

Given $r_1, r_2 \in p^{-1}(m)$, we have two bijection φ_1, φ_2 of $0(n)$ onto $p^{-1}(m)$ defined by $\varphi_i(g) = r_i \circ g$. As $r_2 = r_1 \circ \gamma$, we deduce that $\varphi_2(g) = \varphi_1(\gamma g)$, which means that $p^{-1}(m)$ can be identified to $0(n)$, in a canonic way, up to left translation. We can therefore identify vertical tangent vectors with the Lie algebra of *left invariant* vector field on $0(n)$. Denote by $\underline{0}(n)$ this Lie algebra. We shall denote by π the differential form with values in $\mathbf{R}^n \times \underline{0}(n)$ defining the parallelism on $0(M)$.

The component of π will be denoted by π^i, π^i_j. We shall also denote by $\dot{\pi}$ the component on $\mathbf{R}^n$, $\ddot{\pi}$ the component on $\underline{O}(n)$.

The horizontal lift of $\tau \to \tilde{p}(t, \tau)$ is defined by the equation

$$\begin{cases} \left< \frac{\partial r}{\partial \tau}, \dot{\pi} \right> = r^{-1}\left(\frac{\partial \tilde{p}}{\partial \tau}\right) \\ \left< \frac{\partial r}{\partial \tau}, \ddot{\pi} \right> = 0. \end{cases} \tag{7}$$

Then the parallel transport along $\tau \to \tilde{p}(t, \tau)$ is defined by

$$r(t, \tau) \circ r_0^{-1}.$$

Therefore, equation (6) can be written

$$\left< \frac{\partial r}{\partial t}, \pi \right> = r_0\left(\frac{\partial p}{\partial t}\right). \tag{8}$$

The two equations (7) and (8) will define the development on $O(M)$. We shall write these equations using the pull back form by r of the parallelism form:

$$(r)^{\star}\pi = \Theta.$$

Then writing the component of Θ on $\mathbf{R}^n$ and $\underline{O}(n)$ and using (7) and (8), we obtain the following Darboux equation:

$$\begin{cases} \Theta = (\dot{a} + \ddot{a})\, dt + (\dot{c} + \ddot{c})\, d\tau \\ \ddot{c} = 0 \\ \dot{a} = r_0\left(\frac{\partial p}{\partial t}\right) \end{cases} \tag{9}$$

with the boundary condition

$$\begin{cases} \dot{c}(0, \tau) = \ddot{c}(0, \tau) = 0 \\ \dot{a}(t, 0) = \ddot{a}(t, 0). \end{cases} \tag{10}$$

The Structure Equation on $O(M)$

The structure equation can be written $R^j_{i,k,\ell}$ denoting the curvature tensor,

$$< d\pi, z_1 \wedge z_2 > + [\ddot{\pi}(z_1)\dot{\pi}(z_2) - \ddot{\pi}(z_2)\dot{\pi}(z_1)] = 0$$

$$< d\ddot{\pi}, z_1 \wedge z_2 > +[\ddot{\pi}(z_1)\ddot{\pi}(z_2) - \ddot{\pi}(z_2)\ddot{\pi}(z_1)] + R_{k,\ell}\dot{\pi}^k(z_1)\dot{\pi}^\ell(z_2) = 0.$$

3. THE SMOOTH DEVELOPMENT

The system (9) gives *two equations* when we have *three components* to determine: $\dot{c}$, $\dot{a}$, $\ddot{a}$. To complete our computation, we shall use the structure equations

$$r^\star d\pi = d(r^\star \pi) = d\Theta,$$

which give, taking in account the fact that $\ddot{c} = 0$,

$$\left(\frac{\partial \dot{a}}{\partial \tau} - \frac{\partial \dot{c}}{\partial t}\right) d\tau \wedge dt = -\ddot{a}\dot{c}\ d\tau \wedge dt$$

$$\left(\frac{\partial \ddot{a}}{\partial \tau}\right) d\tau \wedge dt = (R_{k,\ell}\dot{a}^k\dot{c}^\ell)\ d\tau \wedge dt.$$

We obtain finally the following system:

$$\dot{a} = P_t \tag{11}$$

$$\frac{\partial \dot{c}}{\partial t} = -\ddot{a}\dot{c} + P_{t\tau}, \quad c(0, \tau) = 0 \tag{12}$$

$$\frac{\partial \ddot{a}}{\partial \tau} = R_{k,\ell}\dot{a}^k\dot{c}^\ell, \quad a(t, 0) = 0 \tag{13}$$

4. CONSTRUCTION OF THE DEVELOPMENT

We shall introduce the matrix $X(t)$ solution of the ordinary differential matrix equation

$$\begin{cases} \dfrac{dX}{dt} = -\ddot{a}X \\ X(0, \tau) = \text{Identity}. \end{cases} \tag{14}$$

Then $X(t, \tau) \in O(n)$ and we can write the solution of the linear equation (12) on the integral form

$$c(t, \tau) = X(t, \tau) \int_0^t \tilde{X}(t' \tau) p_{t,\tau}(t', \tau)\, dt', \tag{15}$$

where $\tilde{X}$ denotes the adjoint of X.

Introduce a matrix $B(t, \tau) \in O(n)$ by the following P.D.E.:

$$\frac{\partial B}{\partial t \partial \tau} = R_{k,\ell} p_t^k(t, \tau)\left[X_{\underline{j}}^k(t, \tau) \int_0^t X_{\underline{j}}^{\underline{s}}(t', \tau) p_{t,\tau}^{\underline{s}}(t', \tau)\, dt'\right],$$

$$B(0, \tau) = B(t, 0) = 0.$$

(The underlined indices correspond to a summation not included in Einstein convention.) In this formula we consider that the second member of the equation is known, that is finally, *B is defined by the following integral*:

$$B(t_0, \tau_0) \tag{16}$$

$$= \int_0^{t_0}\int_0^{\tau_0} R_{k,\ell}[r(t, \tau)] p_t \, dt\, d\tau \left[\int_0^t X_{\underline{j}}^{\ell}(t, \tau) X_{\underline{j}}^{\underline{s}}(t', \tau) p_{t,\tau}^{\underline{s}}(t', \tau)\, dt'\right].$$

Our purpose is to express everything in terms of $p_{t,\tau} = u$. For instance, we have

$$p_t^k = \int_0^\tau u^k(t, \tau)\, dt.$$

On $(R^+)^2$ we introduce the order relation

$$(t', \tau') < (t, \tau) \Longleftrightarrow t' \leq t \quad \text{and} \quad \tau' \leq \tau.$$

We denote by $T_1 = (t, \tau)$ a point of $(R^+)^2$.

We define for this order relation

$$T_1 \vee T_2 = \sup(T_1, T_2)$$

$$Q_T = \{(t', \tau');\ (t', \tau') < T\}$$

Define on $(R^+)^2 \times (R^+)^2$ the following function h by

$$h(T_1, T_2) = 0 \quad \text{if} \quad T_1 < T_2 \quad \text{or} \quad T_2 < T_1 \text{ or } \quad t_1 < t_2$$

$$h(T_1, T_2) = 1 \quad \text{otherwise.}$$

Lemma 2. *We have*

$$B(t, \tau) = \iiiint_{Q_T^2} R_{k,\ell}(r(T_1 \vee T_2))X_{\underline{j}}^{\ell}(T_1 \vee T_2)X_{\underline{j}}^{\underline{s}}(T_2)h(T_1, T_2)u^k(dT_1)u^{\underline{s}}(dT_2), \tag{17}$$

where $$u^s(dT) = p_{t,\tau}^s(t, \tau)\, dt\, d\tau.$$

Proof. We can fix $T_1 \vee T_2 = T_0$ and integrate in the remaining variables. We will then have to compute a double integral. The introduction of h means that we have only to integrate on the set

$$T_1 = (t_0, \tau) \quad \text{and} \quad T_2 = (t, \tau_0), \qquad t < t_0,\ \tau < \tau_0. \tag{18}$$

Denote $\varphi : (T_1, T_2) \to T_1 \vee T_2$, the relation (18) means that, for the set on which we integrate, we have

$$\varphi(T_1, T_2) = (t_2, \tau_1).$$

This reduces to conditioning by φ to the usual Fubini:

$$\iiint_{Q_T^2} = \iint_{(t_2,\tau_1)<T} dt_1\, d\tau_2 R_{k,\ell}(r(t_1, \tau_2))X_{\underline{j}}^{k}(t_1, \tau_2)$$
$$\left[\int_0^{t_1} X_{\overline{j}}^{\underline{s}}(t_2, \tau_2)p_{t,\tau}^{\overline{s}}(t_2, \tau_2)\, dt_2\right]\left[\int_0^{\tau_2} p_{t,\tau}^{k}(t_1, \tau_1)\, d\tau_1\right].$$

Performing the last integration, we recognize (16).

<u>Computation of X</u>

We use the matrix differential equation

$$\frac{\partial X}{\partial t} = \frac{\partial B}{\partial t} X, \qquad X(0, \tau) = \text{Identity}. \tag{19}$$

<u>Computation of r</u>

We will have two different ways to compute $r(t, \tau)$: either to choose a differential equation in τ or in t. For the development of the loop-sheet, the second point of view will be better. This means that r will be defined by the solution of the differential equation

$$\begin{cases} \langle \frac{\partial r}{\partial t}, \dot{\pi} \rangle = p_t \\ \langle \frac{\partial r}{\partial t}, \ddot{\pi} \rangle = \frac{\partial B}{\partial t} \\ r(0, \tau) = r_0 \end{cases} \quad (20)$$

The equations (17), (19), and (20) form an integro-differential system of the development of smooth sheet.

Geometric Signification of X

Suppose that we are developing a smooth loop-sheet that is

$$p(t, 0) = p(t, 1) = m_0 \quad \text{for all } t; \text{ then solve}$$

$$\frac{dZ}{dt} = Z\ddot{a}, \; Z(t, 0) = I.$$

We have then that $r(t, 1) = r_0 \circ Z(t, 1)$, that is, Z is the *stochastic holonomy*. We have obviously that Z is the transposed of X.

Integration of the System

We shall use an iteration method of perturbative nature. The dominant part of the system will be (17).

We shall take

$$B_1(t, \tau) = R_{k,\ell}(r_0)p^k(t, \tau)p^\ell(t, \tau).$$

Suppose that we have defined $B_{n-1}(t, \tau)$; define X_n by the matrix differential equation

$$\frac{\partial X_n}{\partial t} = \frac{\partial B_{n-1}}{\partial t} X_n.$$

Using $\frac{\partial B_{n-1}}{\partial t}$ instead of $\frac{\partial B}{\partial t}$ in (20), we construct in the same way r_n. Then B_{n+1} is defined by

$$B_{n+1}(t, \tau) = \iiiint_{Q_T^2} R_{k,\ell}(r_n(T_1 \vee T_2))X^{\ell}_{\underline{j},n}(T_1 \vee T_2)X^{s}_{\underline{j}}(T_2)\mathbf{h}(T_1, T_2)$$
$$\cdot\; p^{k}_{t,\tau}(T_1)p^{s}_{t,\tau}(T_2)\; dT_1\, dT_2 .$$

Then if $p_{t,\tau}$ is a continuous function of (t, τ), and if $R_{k,\ell}$ is smooth, then $B_n(t, \tau)$ will converge to the solution of the development.

5. THE REGULARIZATION PROCEDURE AND THE STOCHASTIC SHEET

Denote by W the white noise constructed on the Hilbert space $L^2(R^2)$. Then it is well known that the Brownian sheet on R^n can be written

$$p^k(t, \tau) = W^k([0, t] \times [0, \tau]), \qquad k \in [1, n],$$

where the W^k are independent white noise.

Then equation (17) can be written formally

$$\tilde{B}(t, \tau)$$
$$= \iiiint_{Q_T^2} R_{k,\ell}(r(T_1 \vee T_2))X^{\ell}_{\underline{j}}(T_1 \vee T_2)X^{s}_{\underline{j}}(T_2)h(T_1, T_2)W^k(dT_1)W^{s}(dT_2),$$

which is the form of a Cairolli-Walsh double stochastic integral [2]. We have to use a regularization procedure as in [7] to derive from the ordinary integral equation the stochastic integral equation. We shall denote by $v(\lambda)$ a smooth function with support $[-1, +1]$ and satisfying $\int v(\lambda)\, d\lambda = 1$, $\int v(\lambda)\lambda\, d\lambda = 1$.

We denote $\qquad v_\varepsilon(\tau) = \frac{1}{\varepsilon}v\left(\frac{t}{\varepsilon}\right), \quad v_\eta(\tau) = \frac{1}{\eta}v\left(\frac{\tau}{\eta}\right)$

$$p_{\varepsilon,\eta}(t, \tau) = \iint v_\varepsilon(t + t')v_\eta(\tau + \tau')W(dT').$$

Then $p_{\varepsilon,\eta}$ is a C^∞-function of (t, τ) and we can write the corresponding system defining the development on M. We shall first fix ε and let $\eta \to 0$. We shall after let $\eta \to 0$ and then get the equation of the stochastic development. To obtain the stochastic equation in one variable, we shall use a transfer principle to compute a limit of O.D.E. as the solution of an S.D.E. [7]. To compute the contracted term, we will have to introduce others S.D.E. of the form of (19) and (20). The contracted expression of B will involve those solutions depending themselves on B. The iterative scheme indicated in "Integration of the System" will prove the existence of the Brownian sheet. The Brownian loop can be obtained using the point of view of the section "Stochastic Differential Equation Driven by the Brownian Loop." To finish, we shall compare the development discussed here with the construction of [8]. In this last paper, the smooth equation was

$$\begin{cases} \langle \dot{\pi}, \dfrac{\partial r}{\partial \tau} \rangle = \dfrac{\partial p}{\partial \tau} \\ \langle \ddot{\pi}, \dfrac{\partial r}{\partial \tau} \rangle = 0 \end{cases}$$

and the use of a structure equation was not necessary for the definition of the sheet; it was appearing only for the computation of the stochastic differential of the sheet map.

REFERENCES

1. Bismut, J. M. Mécanique aléatoire. Lecture Note, 866, Springer-Verlag.

2. Cairolli, R., and Walsh, J. B. Stochastic integrals in the plane. *Acta Math.* 134 (1975): 111-183.

3. Flaschel, P., and Klingenberg, W. Riemannsche Hilbert mannigfaltigkeiten. Lecture Note, 282, Springer-Verlag.

4. Gaveau, B., and Mazet, E. Diffusions sur des variétés de chemins. *Comptes Rendus Acad. Sciences Paris* 289 (1979): 643-645.

5. Malliavin, P. *Géométrie Différentielle Stochastique*. Presses de l'Université de Montréal, 1978.

6. Malliavin, P. Cacul des variation stochastiques subordonies. *Comptes Rendus Acad. des Sciences* 295 (1982).

7. Malliavin, P. Stochastic calculus of variation. *Proc. of the International Conf. on S.D.E., Kyoto,* 1976, Kinokuniya et John Wiley, 1978.

8. Malliavin, P. Stochastic Jacobi field. *Proc. International Conf. on P.D.E. on Manifold,* Salt Lake City, 1978.

9. Michel, D. Formule de Stockes stochastique. *Bull. Sciences Math.* 103 (1979): 193-240.

10. Morse, M. Calculus of variation in the Large. *Colloquium Amer. Math. Soc.* (1932).

WEIGHTED ANALOGUES OF NIKOLSKII-TYPE INEQUALITIES AND THEIR APPLICATIONS

H.N. Mhaskar
California State University, Los Angeles

1. INTRODUCTION

Nikolskii inequalities relate different L^p-metrics of a function in a specified class. Thus, for example, in the treatise of Timan [15, pp. 229-231] one finds the following inequality concerning trigonometric polynomials.

Theorem 1. *Let* $0 < p \leq q \leq \infty$. *Then, for every trigonometric polynomial* T_n *of order at most n, we have*

$$\| T_n \|_{L^q[0,2\Pi]} \leq 2n^{\frac{1}{p}-\frac{1}{q}} \| T_N \|_{L^p[0,2\Pi]}, \quad n \geq 1.$$

This theorem was proved by S. M. Nikolskii for $p \geq 1$. It was also extended for trigonometric polynomials in several variables. Nikolskii has also extended these results to the case of entire functions of finite exponential type [12].

In this paper, we introduce similar type of inequalities for algebraic polynomials and for suitably weighted norms. Of particular interest is the case when the weight is supported on the whole real line. The methods used to prove the inequalities above use a direct estimation of the norms in question. For a general weight function, these methods are not quite suitable to apply. Instead, we shall make use of some estimates on the Christoffel functions and de la Vallee Poussin means associated with polynomials orthogonal with respect to the weight function in question.

Our inequalities play an important role in investigations concerning the convergence of orthogonal polynomial expansions, characterization of entire functions of finite exponential type in terms of the constructive properties of their restrictions to the real line [9], and so on. In this paper, we shall discuss an application to Fourier transforms.

It is well known that the Fourier transform of a function f in $L^p(R)$, $p > 2$, need not exist as a function. We shall give some conditions on the constructive properties of f, under which the Fourier transform $\hat{f}$ of f will be a function in $L^r(R)$ for certain values of r that we shall specify. We shall also establish a relationship between the smoothness of f and that of $\hat{f}$. This smoothness will be measured by a certain modification of Besov space norms, which we shall introduce. As an application of our inequalities, we shall also prove embedding theorems involving these modified Besov spaces.

2. MAIN RESULTS

Let w be a weight function such that $wx^n \in L^1(R) \cap L^\infty(R)$, $n = 0, 1, 2, \ldots.$ Let $\{p_k\}_{k=1}^{\infty}$ be the sequence of orthonormalized polynomials with respect to w^2;

$$p_k(x) = \gamma_k x^k + \cdots, \quad \gamma_k > 0 \tag{1a}$$

$$\int_R p_k(x) p_j(x) w^2(x)\, dx = \delta_{jk} \tag{1b}$$

$$\Gamma_n = \sup_{1 \le m \le n} \frac{\gamma_{m-1}}{\gamma_m} \tag{1c}$$

$$M_n = \| w^2 \sum_{k=0}^{2n} p_k^2 \|_\infty \tag{1d}$$

$$N_n = \frac{\Gamma_{2n}}{2n} M_n . \tag{1e}$$

Here and elsewhere in the sequel, $\| . \|_p$ will denote the norm of the $L^p(R)$ space, unless otherwise specified. We shall denote the class of all polynomials of degree at most n by π_n. Also, by $A \ll B$, we mean that $A \leq cB$ for some positive constant c depending on w and the metrics involved but on nothing else.

The most general form of our analogue of the Nikolskii-type inequalities is given in the following theorem.

Theorem 2. Let $P \in \pi_n$.

(a) *If* $1 \leq p \leq r \leq 2$,

$$\| wP \|_r \ll M_n^{\frac{1}{p}-\frac{1}{r}} N_n^{\frac{1}{r}-\frac{1}{2}} \| wP \|_p . \tag{2}$$

(b) *If* $1 \leq p \leq 2 \leq r$,

$$\| wP \|_r \ll M_n^{\frac{1}{p}-\frac{1}{r}} \| wP \|_p . \tag{3}$$

(c) *If* $2 \leq p \leq r \leq \infty$,

$$\| wP \|_r \ll M_n^{\frac{1}{p}-\frac{1}{r}} N_n^{\frac{1}{2}-\frac{1}{p}} \| wP \|_p . \tag{4}$$

The most interesting case occurs when the sequence $\{N_n\}$ is bounded. Weights that satisfy this condition have already risen to importance in the investigations concerning weighted polynomial approximation, convergence of orthogonal polynomial expansion, zeros of orthogonal polynomials, and so on, recently carried out by Freud [5]. Examples of such weight functions are $w_\alpha(x) = \exp(-|x|^\alpha)$, $\alpha \geq 2$. For such weights, we have the following.

Corollary 1. If N_n *is a bounded sequence and* $P \in \pi_n$, $1 \leq p \leq r \leq \infty$,

$$\| wP \|_r \ll M_n^{\frac{1}{p} - \frac{1}{r}} \| wP \|_p \ll \left(\frac{2n}{\Gamma_{2n}}\right)^{\frac{1}{p} - \frac{1}{r}} \| wP \|_p . \tag{5}$$

In the remaining theorems of this paper, we shall restrict our attention to a special class of weight functions, also including all the weights $\exp(-|x|^{\alpha})$, $\alpha \geq 2$, for which the theory of weighted polynomial approximation has been more fruitfully investigated. Thus, we do not yet know Bernstein's inequality and converse theorems in weighted polynomial approximation for weights satisfying only the hypothesis of Corollary 1; but these are known for the more special class of weights that we now introduce.

We say that a weight function $w_Q(x) = \exp(-Q(x))$ is *superregular* if:

(i) Q is even, convex, twice continuously differentiable on $(0, \infty)$ and $Q'(x) \to \infty$ as $x \to \infty$.

(ii) Q'' is increasing.

(iii)
$$t \frac{Q''(t)}{Q'(t)} \leq K, \quad \text{for all} \quad t \geq a > 0, \tag{6}$$

where K and a are constants.

(iv) Let

$$q_n Q'(q_n) = n. \tag{*}$$

Then there exist constants c and d such that

$$q_n < cq_n \leq q_{2n} \leq dq_n < 2q_n \tag{7}$$

Corollary 2. Let w_Q be superregular. Let $P \in \pi_n$, $1 \leq p \leq r \leq \infty$, and q_n be as in (*). Then

$$\| w_Q P \|_r \ll \left(\frac{n}{q_n}\right)^{\frac{1}{p} - \frac{1}{r}} \| w_Q P \|_p . \tag{8}$$

For superregular weights, we also have the following inequality.

Theorem 3. *Let* w_Q *be superregular,* $P \in \pi_n$, $1 \leq p \leq r \leq \infty$, *and* q_n *be as in* (*). *Then*

$$\| w_Q P \|_p \ll q_n^{\frac{1}{p}-\frac{1}{r}} \| w_Q P \|_r \ll \left(\frac{n}{q_n}\right)^{\frac{1}{p}-\frac{1}{r}} \| w_Q P \|_r. \tag{9}$$

We now turn our attention to the definition of Besov spaces modified to incorporate the weight function (Definition 1).

If w_Q is superregular and $w_Q f \in L^p(R)$, we define the modulus of smoothness of f as follows (equation (10)):

$$\omega_2(L^p, w_Q; f, \delta)$$

$$= \sup_{|t|\leq\delta} \| \Delta_t^2(w_Q f) \|_p + \delta \sup_{|t|\leq\delta} \| Q'_\delta \Delta_t(w_Q f) \|_p + \delta^2 \| Q'^2_\delta w_Q f \|_p,$$

where

$$\Delta_t f(x) = f(x+t) - f(x),$$

$$\Delta_t^2 f = \Delta_t(\Delta_t f), \quad Q'_\delta(x) = \min\left\{\frac{1}{\delta}, \left(1 + Q'^2(x)\right)^{\frac{1}{2}}\right\}$$

$$\Omega_2(f, \delta) \equiv \Omega_2(L^p, w_Q; f, \delta)$$

$$= \inf\{\omega_2(L^p, w_Q; f - a - bx, \delta) : a, b \in R\}. \tag{10}$$

Let

$$\varepsilon_n(L^p, w_Q, f) = \inf\{ \| w_Q(f - P) \|_p : P \in \pi_n \}. \tag{11}$$

In [7], we proved direct and converse theorems for weighted polynomial approximation where Ω_2 plays the same role that the ordinary second-order modulus of continuity plays in direct and converse theorems for trigonometric approximation. In particular, we proved the following equivalence.

Proposition 1. *Let* w_Q *be superregular,* $1 \leq p \leq \infty$, $w_Q f \in L^p(R)$, $0 < \alpha \leq 1$. *The following are equivalent:*

(a) *f is r-times differentiable,* $w_Q f^{(r)} \in L^p(R)$, *and*

$$\sup_{0<\delta\leq 1}\left\{\delta^{-\alpha}\Omega_2(L^p, w_Q; f^{(r)}, \delta)\right\} < \infty.$$

(b) $$\sup_{n\geq 1}\left\{\left(\frac{n}{q_n}\right)^{r+\alpha} \varepsilon_n (L^p, w_Q, f)\right\} < \infty.$$

The Besov spaces consist of functions satisfying smoothness conditions that are somewhat more refined than the Lipschitz-type conditions above.

Definition 1. A function f is said to be in the Besov space $B(w_Q, p, \rho, \theta)$ if $w_Q f \in L^p(R)$, f is r-times differentiable where $\rho = r + \alpha$, $(0 < \alpha \leq 1)$, $w_Q f^{(r)} \in L^p(R)$, and

$$\| f \|_{b,Q,p,\rho,\theta} = \left\{\int_0^1 \left(\frac{\Omega_2(L^p, w_Q, f^{(r)}, \delta)}{\delta^\alpha}\right)^\theta \frac{d\delta}{\delta}\right\}^{\frac{1}{\theta}}, \quad 1 \leq \theta < \infty \tag{12}$$

or

$$\| f \|_{b,Q,p,\rho,\infty} = \| f \|_{h,Q,p,\rho}$$
$$= \sup_{0<\delta<1}\left\{\delta^{-\alpha}\Omega_2(L^p, w_Q; f^{(r)}, \delta)\right\} < \infty. \tag{13}$$

Put

$$\| f \|_{B,Q,p,\rho,\theta} = \| w_Q f \|_p + \| f \|_{b,Q,p,\rho,\theta}, \quad 1 \leq \theta \leq \infty \tag{14}$$

$$\| f \|_{H,Q,p,\rho} = \| f \|_{B,Q,p,\rho,\infty}. \tag{15}$$

A discussion of the nonweighted Besov spaces, especially the embedding theorems can be found in [12]. See also [13, Chapter 5]. The embeddings for the weighted analogues are expressed in our Theorem 4 below.

Theorem 4.

(a) *Suppose* w_{Q_1}, w_{Q_2} *are superregular weights and* $Q_1(x) \leq Q_2(x)$ *if* $|x| > K$ *for some* K. *Then*

$$\|f\|_{B,Q_2,p,\rho,\theta} \ll \|f\|_{B,Q_1,p,\rho,\theta},$$
$$1 \le p \le \infty,\ 1 \le \theta \le \infty,\ \rho > 0. \tag{16}$$

(b) *If* $1 \le \theta_1 \le \theta_2 \le \infty$ and $\varepsilon > 0$, *and* w_Q *is superregular, then*

$$\|f\|_{B,Q,p,\rho,\infty} \ll \|f\|_{B,Q,p,\rho,\theta_2} \ll \|f\|_{B,Q,p,\rho,\theta_1} \ll \|f\|_{B,Q,p,\rho+\varepsilon,\infty}. \tag{17}$$

(c) (*Embeddings of different metrics*): *If* $1 \le p,\ q,\ \theta \le \infty$, *then*

$$\|f\|_{B,Q,q,\rho-\left|\frac{1}{p}-\frac{1}{q}\right|,\theta} \ll \|f\|_{B,Q,p,\rho,\theta}, \tag{18}$$

provided $\rho > \left|\frac{1}{p} - \frac{1}{q}\right|$ *and* w_Q *is superregular.*

We are now in a position to state our result about the Fourier transform.

Let, for $f \in L^p(R)$, $f^*(x) = \exp\left(\frac{x^2}{2}\right)f(x)$.

Theorem 5. Suppose $f \in L^2(R)$ *and* g *is its Fourier transform (or inverse Fourier transform) defined as a principal value. We have*

$$\|g^*\|_{B,\frac{x^2}{2},2,\rho,\theta} \ll \|f^*\|_{B,\frac{x^2}{2},2,\rho,\theta} \ll \|g^*\|_{B,\frac{x^2}{2},2,\rho,\theta}. \tag{19}$$

Remark. With the help of Theorem 4, we can now find the Besov spaces where g^* will lie, given that f^* is in a certain Besov space. Thus, for example, suppose $f \in L^p(R)$, $p > 2$ and satisfies the modified Lipschitz condition

$$\Omega_2\left(L^p, \exp\left(-\frac{x^2}{2}\right), f^*, \delta\right) = O(\delta^\alpha)$$

for some α where $\frac{1}{2} - \frac{1}{p} < \alpha \le 1 - \frac{1}{p}$. Then the Fourier transform g of f will exist as a function in $L^r(R)$ for all r with $\frac{p}{1+\alpha p} < r < \frac{p}{p - \alpha p - 1}$. Further, for all such r, g will satisfy the modified Lipschitz condition

$$\Omega_2(L^r, \exp(-\frac{x^2}{2}), g^*, \delta) = O(\delta^\rho)$$

where $\rho = \alpha + \frac{1}{p} - \frac{1}{r}$ if $r \leq 2$ and $\alpha + \frac{1}{p} + \frac{1}{r} - 1$ if $r \geq 2$.

Continuing the same chain of thought, we also note the following corollary.

Corollary 3. Let $1 < p < \infty$; $\frac{1}{p} + \frac{1}{p'} = 1$, $\alpha = \max\left(\frac{1}{p}, \frac{1}{p'}\right)$; $\beta = \min\left(\frac{1}{p}, \frac{1}{p'}\right)$. *Suppose* $f \in L^p(R)$ and f^* *satisfies*

$$\Omega_2\left(L^p, \exp\left(-\frac{x^2}{2}\right); f^*, \delta\right) = O(\delta^\alpha).$$

Then the Fourier transform g *of* f *exists as a function in* $L^r(R)$ *for all* r *with* $1 < r < \infty$, *and, in particular,* g^* *will satisfy in* $L^{p'}$ *the condition*

$$\Omega_2\left(L^{p'}, \exp\left(-\frac{x^2}{2}\right), g^*, \delta\right) = O(\delta^\beta).$$

When $1 \leq p \leq 2$ and thus we know the existence of the Fourier transform in $L^{p'}(R)$, where $\frac{1}{p} + \frac{1}{p'} = 1$, a similar relationship between the smoothness of f in $L^p(R)$ and the smoothness of g in $L^{p'}(R)$ was given by *Ky* [8].

Our results neither include his theorem nor are included in it. However, when the hypotheses of both our theorem and Ky's theorem are satisfied, our estimates are sharper than those of Ky.

3. PROOFS

Let $wf \in L^p R$, $p \geq 1$. We then have the orthogonal polynomial expansion, $f \sim \sum c_k(f) p_k(w^2, x)$, where

$$c_k(f) = \int_R w^2(x) f(x) p_k(w^2, x)\, dx. \tag{20}$$

Denote

$$s_n(w^2, f) = \sum_{k=0}^{n-1} c_k(f) p_k(w^2) \in \pi_{n-1} \tag{21a}$$

$$v_n(w^2, f) = \frac{1}{n} \sum_{k=n+1}^{2n} s_k(f) \in \pi_{2n-1}. \tag{21b}$$

For $g \in L^p(R)$, let

$$T_n(g) = wv_n(w^2, w^{-1}g). \tag{22}$$

Proposition 2.

(a) *If* $P \in \pi_n$, $v_n(P) = P$.

(b) *The operator* T_n *is self-adjoint in the sense that if* $g \in L^p(R)$, $h \in L^{p'}(R)$, $\frac{1}{p} + \frac{1}{p'} = 1$, *then*

$$\int_R T_n(g)h\, dx = \int_R gT_n(h)\, dx. \tag{23}$$

In particular,

$$\| T_n \|_{L^p(R)\to L^r(R)} = \| T_n \|_{L^{r'}(R)\to L^{p'}(R)}. \tag{24}$$

(c) $\| T_n \|_{L^p(R)\to L^p(R)} \ll N_n^{\left|\frac{1}{p}-\frac{1}{2}\right|}$, $p \geq 1$, $n \geq 1$.

(Here, $\ll$ *is, in fact, independent of* p.)

Proof. Part (a) is obvious. The first part of (b) follows easily by substituting from (22), (21), (20) in the expression on the left of (23) and interchanging the order of integration, which is permissible, since $T_n(g) \in L^p(R)$ and $h \in L^{p'}(R)$. The equation (24) follows from (23) and the formula

$$\| f \|_p = \sup\{\int_R fg\, dx : \| g \|_{p'} = 1\}. \tag{25}$$

Part (c) was proved by Freud in [4], although it was not stated there in this form. We reproduce a sketch of the proof. Let $g \in L^\infty(R)$, $f = w^{-1}g$, $x \in R$. Put $f = f_1 + f_2$, where

$$f_1(t) = \begin{cases} f(t) & \text{if } |x - t| \leq \dfrac{\Gamma_{2n}}{2n} \\ 0 & \text{otherwise.} \end{cases}$$

Then

$$s_k(f) = s_k(f_1) + s_k(f_2) \tag{26a}$$

$$v_k(f) = v_k(f_1) + v_k(f_2). \tag{26b}$$

From (21),

$$w(x)\,|s_k(f_1, x)| \leq M_n^{\frac{1}{2}} \| wf_1 \|_2 \ll N_n^{\frac{1}{2}} \| wf \|_\infty, \quad 1 \leq k \leq 2n.$$

Hence,

$$|w(x)v_n(f_1, x)| \ll N_n^{\frac{1}{2}} \| wf \|_\infty \tag{27}$$

Let $f_2{}^* = \dfrac{f_2(t)}{x - t}$. We have [4]

$$w(x)s_k(f_2, x) = w(x)\,\frac{\gamma_{k-1}}{\gamma_k}\left[c_{k-1}(f_2^*)p_k(x) - c_k(f_2^*)p_{k-1}(x)\right].$$

Hence,

$$\sum_{k=n}^{2n} w(x)\,|s_k(f_2, x)| \ll \Gamma_{2n} M_n^{\frac{1}{2}} \| wf_2^* \|_2 \ll N_n^{\frac{1}{2}} n \| wf \|_\infty .$$

Thus,

$$|v_n(f_2, x)|\,w(x) \ll N_n^{\frac{1}{2}} \| wf \|_\infty . \tag{28}$$

From (26), (27), (28),

$$\| T_n(g) \|_\infty = \| wv_n(w^{-1}g) \|_\infty = \| wv_n(f) \|_\infty \ll N_n^{\frac{1}{2}} \| g \|_\infty. \tag{29}$$

Clearly,

$$\| T_n(g) \|_2 = \| wv_n(f) \|_2 \ll \| wf \|_2 = \| g \|_2. \tag{30}$$

Part (c) now follows for $p \geq 2$ by the Riesz-Thorin interpolation theorem and for $1 \leq p \leq$ by part (b).

Proof of Theorem 2. It is convenient to prove parts (c), (a), and (b) of the theorem in that order.

(c) Let $2 \le p \le r \le \infty$. Observe that if $P \in \pi_n$, we have

$$P(x) = \sum_{k=0}^{n} c_k(P)p_k(x), \quad \text{where} \quad p_k(x) \equiv p_k(w^2, x).$$

So

$$w(x)\,|P(x)| \le \left(\sum_{k=0}^{n} c_k(P)^2\right)^{\frac{1}{2}} \cdot \left(\sum_{k=0}^{n} w^2(x)p_k^2(x)\right)^{\frac{1}{2}}; \; x \in R,$$

that is,

$$\| wP \|_\infty \le \| w^2 \sum_{k=0}^{n} p_k^2 \|_\infty^{\frac{1}{2}} \| wP \|_2 .$$

Hence, for $g \in L^2(R)$,

$$\| T_n(g) \|_\infty \le M_n^{\frac{1}{2}} \| T_n(g) \|_2 \ll M_n^{\frac{1}{2}} \| g \|_2 . \tag{31}$$

Also, for $g \in L^\infty(R)$, we have by Proposition 2(c),

$$\| T_n(g) \|_\infty \ll N_n^{\frac{1}{2}} \| g \|_\infty . \tag{32}$$

Then by the Riesz-Thorin interpolation theorem, for all $g \in L^p(R)$,

$$\| T_n(g) \|_\infty \ll N_n^{\frac{1}{2} - \frac{1}{p}} M_n^{\frac{1}{p}} \| g \|_p . \tag{33}$$

But, by Proposition 2(c), for all $g \in L^p(R)$, we also have

$$\| T_n(g) \|_p \ll N_n^{\frac{1}{2} - \frac{1}{p}} \| g \|_p . \tag{34}$$

Again, by the Riesz-Thorin interpolation theorem, we get, for all $g \in L^p(R)$,

$$\| T_n(g) \|_r \ll N_n^{\frac{1}{2} - \frac{1}{p}} M_n^{\frac{1}{p} - \frac{1}{r}} \| g \|_p . \tag{35}$$

In view of Proposition 2(a), this completes the proof of part (c).

(a) Let $1 \le p \le r \le 2$. Let

$$\frac{1}{p} + \frac{1}{p'} = \frac{1}{r} + \frac{1}{r'} = 1, \quad \text{so that} \quad 2 \le r' \le p' \le \infty.$$

Hence, by part (c),

$$\| T_n(g) \|_{p'} \ll N_n^{\frac{1}{2} - \frac{1}{r'}} M_n^{\frac{1}{r'} - \frac{1}{p'}} \| g \|_{r'},$$

for all $g \in L^{r'}(R)$; that is,

$$\| T_n \|_{L^{r'}(R) \to L^{p'}(R)} = \| T_n \|_{L^p(R) \to L^r(R)} \ll N_n^{\frac{1}{r} - \frac{1}{2}} M_n^{\frac{1}{p} - \frac{1}{r}}.$$

(The first equation above uses Proposition 2(b).) Thus, for all $g \in L^p(R)$,

$$\| T_n(g) \|_r \ll N_n^{\frac{1}{r} - \frac{1}{2}} M_n^{\frac{1}{p} - \frac{1}{r}} \| g \|_p. \tag{36}$$

In view of Proposition 2(a), this proves part (a) of the theorem.

(b) If $1 \le p \le 2 \le r \le \infty$ and $P \in \pi_n$, we have by parts (c) and (a) of this theorem,

$$\| wP \|_r \ll M_n^{\frac{1}{2} - \frac{1}{r}} \| wP \|_2 \ll M_n^{\frac{1}{2} - \frac{1}{r}} M_n^{\frac{1}{p} - \frac{1}{2}} \| wP \|_p$$

$$= M_n^{\frac{1}{p} - \frac{1}{r}} \| wP \|_p.$$

This completes the proof.

Corollary 1 is obvious. Corollary 2 becomes obvious if we note the inequality

$$M_n \ll \frac{n}{q_n} \tag{37}$$

proved by Freud for the weights in question [3].

Proof of Theorem 3. If w_Q is superregular, then the sequence N_n is bounded. Hence, for all $n \geq 1$,

$$\| T_n \|_{L^p(R) \to L^p(R)} \leq K, \tag{38}$$

where K is a constant depending on Q alone.

If Q is even, positive, differentiable and $Q''(t) > 0$ for all $t > 0$ and $Q'(t) \to \infty$ as $t \to \infty$; in particular, when w_Q is superregular, we have for every $p \in \pi_n$,

$$\int_R P^2(x) w_Q^2(x)\, dx \ll \int_{-cq_{2n}}^{cq_{2n}} P^2(w) w_Q^2(x)\, dx, \tag{39}$$

where c is a constant depending on Q alone [4, p. 157]. From (39),

$$\| w_Q P \|_2 \ll q_{2n}^{\frac{1}{2}} \| w_Q P \|_\infty. \tag{40}$$

Since w_Q is superregular, equations (7), (38), and (40) imply that

$$\| T_n(g) \|_2 \ll q_n^{\frac{1}{2}} \| g \|_\infty , \quad g \in L^\infty(R). \tag{41}$$

The remainder of the proof is similar to that of Theorem 2:

$$\| T_n(g) \|_2 \ll \| g \|_2 , \quad g \in L^2(R). \tag{42}$$

If $2 \leq r \leq \infty$, (41) and (42) give, for all $g \in L^r(R)$,

$$\| T_n(g) \|_2 \ll q_n^{\frac{1}{2} - \frac{1}{r}} \| g \|_r \tag{43}$$

$$\| T_n(g) \|_r \ll \| g \|_r \tag{44}$$

If $2 \leq p \leq r \leq \infty$, (43) and (44) give for all $g \in L^r(R)$,

$$\| T_n(g) \|_p \ll q_n^{\frac{1}{p} - \frac{1}{r}} \| g \|_r . \tag{45}$$

By the duality relation in Proposition 2(b), (45) holds even when $1 \leq p \leq r \leq 2$. Then by Proposition 2(a), for all $P \in \pi_n$,

$$\| wP \|_p \ll q_n^{\frac{1}{p} - \frac{1}{r}} \| wP \|_r$$

when p and r are on the same side of 2. If $1 \leq p \leq 2 \leq r \leq \infty$, we get

$$\| wP \|_p \ll q_n^{\frac{1}{p} - \frac{1}{2}} \| wP \|_2 \ll q_n^{\frac{1}{p} - \frac{1}{r}} \| wP \|_r .$$

This proves the first inequality in (9) for all p, r, $1 \leq p \leq r \leq \infty$. Finally, observe that since

$$\frac{xQ''(x)}{Q'(x)} \leq K,$$

$$\frac{q_n^2 Q''(q_n)}{q_n Q'(q_n)} = \frac{q_n^2 Q''(q_n)}{n} \leq K.$$

Since Q'' is increasing,

$$q_n \ll \frac{n}{q_n Q''(q_n)} \ll \frac{n}{q_n} .$$

This completes the proof of Theorem 3.

To prove Theorem 4, we need a lemma, which we proved in [10], and which gives some equivalent formulas for the Besov space norms. Since these norms are refinements of usual Lipschitz-space norms, the lemma also embodies a refinement of Proposition 1.

Lemma 1. *Let w_Q be superregular and $w_Q f \in L^p(R)$. Define*

$$\|f\|_{b,1,Q,p,\rho,\theta} = \left\{\int_0^1 \left(\frac{\Omega_2(f^{(r)}, \delta)}{\delta^\alpha}\right)^\theta \frac{d\delta}{\delta}\right\}^{\frac{1}{\theta}}$$

$$\|f\|_{b,2,Q,p,\rho,\theta} = \left\{\sum_{n=0}^{\infty} \left[\left(\frac{2^n}{q_{2^n}}\right)^\alpha \Omega_2\left(f^{(r)}, \frac{q_{2^n}}{2^n}\right)\right]^\theta\right\}^{\frac{1}{\theta}}$$

$$\|f\|_{b,3,Q,p,\rho,\theta} = \left\{\sum_{n=0}^{\infty} \left[\left(\frac{2^n}{q_{2^n}}\right)^\rho \varepsilon_{2^n}(L^p, w_Q, f)\right]^\theta\right\}^{\frac{1}{\theta}}$$

$$\|f\|_{b,4,Q,p,\rho,\theta} = \inf\left\{\sum_{n=0}^{\infty} \left[\left(\frac{2^n}{q_{2^n}}\right)^\rho \|w_Q P_{2^n}\|_p\right]^\theta\right\}^{\frac{1}{\theta}},$$

where the inf is taken over all sequences of polynomials $\{P_{2^n}\}$ such that

$$P_{2^n} \in \pi_{2^n} \quad \text{and} \quad \left\|w_Q\left(\sum_{n=0}^{N} P_{2^n} - f\right)\right\| \to 0 \quad \text{as} \quad N \to \infty$$

$$\|f\|_{B,k,Q,p,\rho,\theta} = \|w_Q f\|_p + \|f\|_{b,k,Q,p,\rho,\theta}, \quad k = 1, 2, 3, 4.$$

In all of these definitions, $1 \leq \theta \leq \infty$ and when $\theta = \infty$, the usual convention of taking sup is followed.

Then, for $j, k = 1, 2, 3, 4$,

$$\|f\|_{B,k,Q,p,\rho,\theta} \sim \|f\|_{B,j,Q,p,\rho,\theta}.$$

Here and in the sequel, $A \sim B$ will mean $A \ll B$ and $B \ll A$.

Proof of Theorem 4.

(a) We notice the following equivalence proved in [6]:

$$\Omega_2(L^p, w_Q, f, \theta) \sim \inf\{\|w_Q f_1\|_p + \delta^2\|w_Q f_2''\|_p\}, \tag{46}$$

where the inf is over all f_1, f_2 such that $f = f_1 + f_2$, $w_Q f_1 \in L^p(R)$ and f_2 is twice differentiable with

$w_Q f_2'' \in L^p(R)$. Part (a) now follows from the version $\| f \|_{B,1,Q,p,\rho,\theta}$ of the Besov space norms.

(b) This part is clear from the version $\| f \|_{B,3,Q,p,\rho,\theta}$ of the Besov space norms if we observe the inequality

$$\left\{ \sum_{k=0}^{\infty} |a_k|^{\theta_2} \right\}^{\frac{1}{\theta_2}} \leq \left\{ \sum_{k=0}^{\infty} |a_k|^{\theta_1} \right\}^{\frac{1}{\theta_1}} \tag{47}$$

for $1 \leq \theta_1 \leq \theta_2 \leq \infty$ and for all sequences $\{a_k\}$.

(c) Note that in view of Corollary 2 and Theorem 3, if w_Q is superregular and $P \in \pi_n$, then

$$\| w_Q P \|_p << \left(\frac{n}{q_n} \right)^{|\frac{1}{p} - \frac{1}{q}|} \| w_Q P \|_q , \quad 1 \leq p, q \leq \infty.$$

Now let $f \in B(w_Q, p, \rho, \theta)$ and $\rho > |\frac{1}{p} - \frac{1}{q}|$.

Choose a sequence $P_{2^k} \in \pi_{2^k}$ such that

$$\lim_{n\to\infty} \| w_Q(f - \sum_{k=0}^{n} P_{2^k}) \|_p = 0 \tag{48}$$

$$\left\{ \sum_{k=0}^{\infty} \left[\left(\frac{2^k}{q_{2^k}} \right)^{\rho} \| w_Q P_{2^k} \|_p \right]^{\theta} \right\}^{\frac{1}{\theta}} \leq 2 \| f \|_{b,4,Q,p,\rho,\theta}. \tag{49}$$

We have
$$\left\{ \sum_{k=0}^{\infty} \left[\left(\frac{2^k}{q_{2^k}} \right)^{\rho - |\frac{1}{p} - \frac{1}{q}|} \| w_Q P_{2^k} \|_q \right]^{\theta} \right\}^{\frac{1}{\theta}}$$

$$<< \left\{ \sum_{k=0}^{\infty} \left[\left(\frac{2^k}{q_{2^k}} \right)^{\rho} \| w_Q P_{2^k} \|_p \right]^{\theta} \right\}^{\frac{1}{\theta}} \leq 2 \| f \|_{b,4,Q,p,\rho,\theta}. \tag{50}$$

Since $\rho - |\frac{1}{p} - \frac{1}{q}| > 0$, we also conclude that

$\lim_{n\to\infty} \| w_Q(f - \sum_{k=0}^{n} P_{2^k})\|_q$ is 0; in particular, $w_Q f \in L^q(R)$.

Further, if $\rho - |\frac{1}{p} - \frac{1}{q}| = \beta$, then

$$\| w_Q f \|_q \le \sum_{k=0}^{\infty} \| w_Q P_{2^k} \|_q$$

$$\le \left\{ \sum_{k=0}^{\infty} \left[\left(\frac{2^k}{q_{2^k}} \right)^{\beta} \| w_Q P_{2^k} \|_q \right]^{\theta} \right\}^{\frac{1}{\theta}} \left\{ \sum_{k=0}^{\infty} \left(\frac{q_{2^k}}{2^k} \right)^{\beta\theta'} \right\}^{\frac{1}{\theta'}}$$

$$\left(\frac{1}{\theta} + \frac{1}{\theta'} = 1 \right)$$

$$\ll \| f \|_{B,4,Q,p,\rho,\theta}. \tag{51}$$

Part (c) follows from (50) and (51). This completes the proof of Theorem 4.

If h_k is orthonormal Hermite polynomial of degree k, and $\psi_k(x) = \exp\left(-\frac{x^2}{2}\right) h_k(x)$, it is well known that

$$\hat{\psi}_k = \lambda_k \psi_k , \tag{52}$$

where $\hat{\psi}_k$ is the Fourier transform of ψ_k and $|\lambda_k| = 1$ [16]. Further if $f \in L^2(R)$, by the completeness of the Hermite polynomial system, we have

$$f(x) = \sum_{k=0}^{\infty} c_k \psi_k(x), \quad \text{where} \quad c_k = \int_R f(x) \psi_k(x)\, dx. \tag{53}$$

The series in (53) is convergent in $L^2(R)$. Then by Plancherel's theorem, if $\hat{f} = g$ is the Fourier transform of f, we get, from (52),

$$g(x) = \sum_{k=0}^{\infty} \lambda_k c_k \psi_k(x) \quad \text{and} \quad \lambda_k c_k = \int_R g \psi_k \, dx. \tag{54}$$

From (53), (52), and (54),

$$\varepsilon_n\left(L^2, \exp\left(-\frac{x^2}{2}\right), g^*\right) = \varepsilon_n\left(L^2, \exp\left(-\frac{x^2}{2}\right), f^*\right), \tag{55}$$

where $f^*(x) = \exp\left(\frac{x^2}{2}\right) f(x)$ and $g^*(x) = \exp\left(\frac{x^2}{2}\right) g(x)$. Theorem 5 now follows from the version $\| f \|_{B,3,\exp(-\frac{x^2}{2}),2,\rho,\theta}$ of the Besov space norms.

REFERENCES

1. Freud, G. Extension of the Dirichlet-Jordan criterion to a general class of orthogonal polynomial expansions. *Acta Math. Acad. Sci. Hung.* 25 (1974): 109-122.

2. Freud, G. Markov-Bernstein type inequalities in $L_p(-\infty, \infty)$. In *Approximation Theory II* (Lorentz, Chui, Shumaker, eds.). New York, Academic Press, 1976, pp. 369-377.

3. Freud, G. On Markov-Bernstein type inequalities and their applications. *J. of Approx. Theo.* 19 (1977): 22-37.

4. Freud, G. On polynomial approximation with respect to general weights. In *Lecture Notes* 399 (Garnir, Unni, Williamson, eds.). New York: Springer, 1974, pp. 149-179.

5. Freud, G. On the extension of the Fejer-Lebesgue theorem to orthogonal polynomial series. *Iliev Festschrift* (1975), Sofia, 257-265.

6. Freud, G., and Mhaskar, H. N. *K*-functionals and moduli of continuity in weighted polynomial approximation. To appear in *Arkiv för Matematik.*

7. Freud, G., and Mhaskar, H. N. Weighted polynomial approximation in rearrangement invariant Banach function spaces on the whole real line. To appear in *Ind. J. of Math.*

8. Ky, N. X. Fourier-transzformaltak modositott folytonossagi modulusarol. *Matem. Lapok* 23 (1972): 99-102.

9. Mhaskar, H. N. Weighted polynomial approximation of entire functions I, II. To appear in *J. of Approx. Theo.*

10. Mhaskar, H. N. Weighted polynomial approximation on the whole real line and related topics. Ph.D. dissertation, The Ohio State Univ. 1980.

11. Nevai, G. P. Orthogonal Polynomials; Memoirs of the American Mathematical Society, 213, 1979.

12. Nikolskii, S. M. *Approximation of Functions of Several Variables and Imbedding Theorems.* New York: Springer, 1975.

13. Stein, E. M. *Singular Integrals and Differentiability Properties of Functions.* Princeton, N.J.: Princeton Univ. Press, 1970.

14. Szegö, G. *Orthogonal Polynomials*. American Mathematical Society Colloquium Publications, 23, 4th ed., 1975.

15. Timan, A. F. *Theory of Approximation of Functions of a Real Variable*. New York: Pergamon Press and Macmillan Company, 1963.

16. Wiener, N. *The Fourier Integral and Certain of Its Applications*. New York: Cambridge Univ. Press, 1933.

17. Zygmund, A. *Trigonometric Series*, vol. 1. New York: Cambridge Univ. Press, 1959.

AN APPLICATION OF THE APPROXIMATION FUNCTIONAL IN INTERPOLATION THEORY

Yoram Sagher
University of Illinois at Chicago Circle

Real interpolation theory has been dominated by the J and K functionals, while the approximation functional, defined

$$E(t, a; A_0, A_1) = \inf\{|a - a_0|_{A_1} / |a_0|_{A_0} \leqq t\}, \tag{1}$$

has taken a secondary role. This is understandable, since J is a quasinorm on $A_0 \cap A_1$, K is a quasinorm on $A_0 + A_1$, while E lacks any such interpretation. Moreover, the statements of some important theorems, most notably the reiteration theorem, in terms of the E functional alone, are quite cumbersome.

We shall see, however, that the approximation functional is useful in the incorporation of W (see [1]) in the $L(p, q)$ interpolation scale (see [4]), as well as in the incorporation of the analogues of W in other interpolation scales. W consist of all functions on a σ-finite measurable space (X, μ), satisfying

$$f^{**}(t) - f^{*}(t) \leqq C, \tag{2}$$

where f^* is the nonincreasing rearrangement of f, and $f^{**}(t) = \frac{1}{t}\int_0^t f^*(u)\,du$. Using the K functional and

$$\int_0^t f^*(u)\,du = K(t, f; L_1, L_\infty), \tag{3}$$

the following theorem is proved in [4]:

Theorem 1. If

$$T : A_0 \to L(p_0, \infty)$$
$$T : A_1 \to W$$

is quasilinear: $|T(a_0 + a_1)|(x) \leqq k(|Ta_0|(x) + |Ta_1|(x))$ *a.e., then, for* $0 < \theta < 1$, $0 < q \leqq \infty$, $\frac{1}{p} = \frac{1-\theta}{p_0}$, *we have*

$$T : (A_0, A_1)_{\theta,q;K} \to L(p, q).$$

An annoying feature of the proof in [4] is that since (3) involves L_1, one must deal separately with the cases $p_0 < 1$, $p_0 = 1$, $1 < p_0$. Using the approximation functional, E, we will get the theorem for all three cases at once. There is another reason to use the E functional. The condition replacing (2) in the general theorem is

$$\frac{1}{t} K(t, b; B_0, B_1) - K'(t, b; B_0, B_1) \leqq C. \tag{4}$$

The derivative, contrary to the first impression, presents no great difficulty. As is true almost everywhere in interpolation, the formulas containing the continuous parameter t are in fact just shorthand notation for the series definitions where all problems of measurability and differentiability vanish. The aspect of (4) that presents a practical problem is the subtraction of K', for it dictates a precise determination of K.

This is where the approximation functional proves useful. In some cases it can be determined precisely, almost by observation. We recall two results presented in [3]:

$$E(t, f; L_0, L_\infty) = f^*(t) \tag{5}$$

$$E(t, f; \Lambda_\omega, L_p) = \left(\int_{\{\omega(x)>t\}} |f(x)|^p\, dx\right)^{1/p}. \tag{6}$$

Motivated by these considerations, we present an approach to the inclusion of W and its analogues in interpolation theory, based on the E functional. Since we cannot hope to make a self-contained presentation, we refer the reader to [3] for the definitions and theorems used here. Our notation, except for some minor details, is consistent with that of [3].

Definition. A_j, B_j are quasinormed Abelian groups. $T: A_0 + A_1 \to B_0 + B_1$ is E-quasilinear iff there exist $0 < c, d$ so that for all $a_j \in A_j$ and all $0 < t_0, t_1$, we have

$$E(t_0 + t_1, T(a_0 + a_1); B_0, B_1) \leqq c[E(dt_0, Ta_0; B_0, B_1) + E(dt_1, Ta_1; B_0, B_1)]. \tag{7}$$

Theorem 2. If $T: A_0 + A_1 \to Lp_0 + Lp_1$, $0 \leqq p_0 < p_1 \leqq \infty$, satisfies

$$|T(a_0 + a_1)|(x) \leqq k(|Ta_0|(x) + |Ta_1|(x)) \quad \text{a.e.}, \tag{8}$$

then T is E-quasilinear.

Proof. The following facts are easily verified from the definitions:

$$E(t_0 + t_1, f_0 + f_1; Lp_0, Lp_1) \leqq C_1E(t_0/C_0, f_0; Lp_0, Lp_1) + C_1E(t_1/C_0, f_1; Lp_0, Lp_1), \tag{9}$$

C_0, C_1 depending on p_0 and p_1.

$$E(t, kf; Lp_0, Lp_1) = kE(t/k, f; Lp_0, Lp_1); \; 0 < k \tag{10}$$

if $0 < p_0$. If $p_0 = 0$, we have:

$$E(t, kf; Lp_0, Lp_1) = kE(t, f; Lp_0, Lp_1). \tag{11}$$

$$|f| \leqq |g| \text{ a.e. implies } E(t, f; Lp_0, Lp_1) \leqq E(t, g; Lp_0, Lp_1). \tag{12}$$

Using (9)-(12), the theorem follows easily:

$$\begin{aligned} E(t_0 + t_1, T(a_0 + a_1); Lp_0, Lp_1) &= E(t_0 + t_1, |T(a_0 + a_1)|; Lp_0, Lp_1) \\ &\leqq E(t_0 + t_1, k|Ta_0| + k|Ta_1|; Lp_0, Lp_1) \\ &\leqq C_1kE(t_0/kC_0, |Ta_0|; Lp_0, Lp_1) + C_1kE(t_1/kC_0, |Ta_1|; Lp_0, Lp_1) \\ &= CE(t_0d, Ta_0; Lp_0, Lp_1) + CE(t_1d, Ta_1; Lp_0, Lp_1). \end{aligned}$$

Theorem 3. $T : A_0 + A_2 \to \Lambda\omega + Lp$ *satisfies*

$$|T(a_0 + a_1)|(x) \leqq k(|Ta_0|(x) + |Ta_1|(x)) \quad a.e.$$

Then T is E-quasilinear.

Proof. Given a function g, denote by χ_g the characteristic function of its support. Recall:

$$|g|_{\Lambda\omega} = |\omega(x)\chi_g(x)|_{L_\infty}. \tag{13}$$

One verifies that (9)-(12) hold when Lp_0 is replaced by $\Lambda\omega$, and Lp_1 by Lp, and the proof concludes as that of Theorem 2.

Theorem 4. (A_0, A_1), (B_0, B_1) *are quasinormed Abelian couples.* $T : A_0 + A_1 \to B_0 + B_1$ *is E-quasilinear,* $1 \leqq p \leqq \infty$, *and*

$$\left(\frac{1}{t}\int_0^t E^p(s, Ta; B_0, B_1)\, ds\right)^{1/p} - E(t, Ta; B_0, B_1) \leqq C_1|a|_{A_1} \tag{14}$$

$$E(t, Ta; B_0, B_1) \leqq C_0 t^{-\beta}|a|_{A_0}. \tag{15}$$

Then:

$$|Ta|_{\beta(1-\theta),q;E} \leqq C(\beta, \theta, q, C_0, C_1, c, d)|a|_{\theta,q;K} \tag{16}$$

where $0 < \theta < 1$ *and* $0 < q \leqq \infty$.

Proof. Since $\frac{1}{t}\int_0^t E(s, Ta; B_0, B_1)\, ds - E(t, Ta; B_0, B_1) \leqq \left(\frac{1}{t}\int_0^t E^p(s, Ta; B_0, B_1)\right)^{1/p} - E(t, Ta; B_0, B_1)$, we can assume in (14) $p = 1$. (The statement with $1 < p$ will be useful later.) Denote $g(t) = \frac{1}{t}\int_0^t E(s, Ta; B_0, B_1)\, ds$. Clearly, $0 \leqq -g'(t) \leqq \frac{C_1}{t}|a|_{A_1}$.

For $t < s$, we have

$$0 \leqq g(t) - g(s) = -\int_t^s g'(u)\, du \leqq C_1|a|_{A_1} \log s/t,$$

and so:

$$g(t) \leqq C_1|a|_{A_1} \log s/t + g(s) \leqq C_1|a|_{A_1} \log s/t + C_1|a|_{A_1}$$

$$+ E(s, Ta; B_0, B_1) \leqq C_0 s^{-\beta}|a|_{A_0} + C_1|a|_{A_1}(1 + \log s/t).$$

Therefore,

$$\int_0^s t^\alpha E(t, Ta; B_0, B_1)\frac{dt}{t} \leqq \int_0^s t^\alpha g(t)\frac{dt}{t} \leqq C \int_0^s t^\alpha (s^{-\beta}|a|_{A_0}$$

$$+ (1 + \log s/t)|a|_{A_1})\frac{dt}{t}$$

$$\leqq C(C_0, C_1, \alpha, \beta)[s^{\alpha-\beta}|a|_{A_0} + s^\alpha |a|_{A_1}].$$

For $s < t$, use (15):

$$\int_s^\infty t^\alpha E(t, Ta; B_0, B_1)\frac{dt}{t} \leqq C_0|a|_{A_0} \int_s^\infty t^{-(\beta-\alpha)}\frac{dt}{t}$$

$$= C(\alpha, \beta, C_0)|a|_{A_0} s^{\alpha-\beta}.$$

Taking: $s = \left(\dfrac{|a|_{A_0}}{|a|_{A_1}}\right)^{1/\beta}$, we have

$$|Ta|_{\alpha,1;E} \leqq C(\alpha, \beta, C_0, C_1)|a|_{A_0}^{\alpha/\beta}|a|_{A_1}^{1-\alpha/\beta}. \tag{17}$$

$a \to |Ta|_{\alpha,1;E}$ is a semi-quasinorm on $A_0 \cap A_1$, and by the reiteration theorem, using (17) with different values of α,

$$T : (A_0, A_1)_{\theta,q;k} \to (B_{\alpha_0,1;E}, B_{\alpha_1,1;E})_{\lambda,q;k} \tag{18}$$

with $\theta = (1 - \lambda)(1 - \alpha_0/\beta) + \lambda(1 - \alpha_1/\beta)$.

Further,

$$(B_{\alpha_0,1;E}, B_{\alpha_1,1;E})_{\lambda,q;K} = B_{\alpha,q;E} \tag{19}$$

with $\alpha = (1 - \lambda)\alpha_0 + \lambda\alpha_1$, so that $\alpha = \beta(1 - \theta)$, and we have

$$T : (A_0, A_1)_{\theta,q;K} \to (B_0, B_1)_{\beta(1-\theta),q;E}. \tag{20}$$

The theorem is proved.

The first corollary is Theorem 1, which follows simply from $E(t, f; L_0, L_\infty) = f^*(t)$.

We next consider interpolation of weighted Lp spaces, with $1 \leqq p \leqq \infty$. We want first to characterize $(\Lambda\omega, Lp)_{\alpha,q;E}$. For this we shall need a theorem of J. E. Gilbert [2]:

Theorem 5. *X is a Banach function space. $\omega(x) > 0$ a.e. Denote by G the class of piecewise continuous functions $\sigma : R^+ \to R^+$, so that both $\sigma(t) \in L_\infty(R^+)$ and $t\sigma(t) \in L_\infty(R^+)$. Denote also $|f|_{X_\omega} = |\omega f|_X$ and $X_\omega = \{|f/|f|_{X_\omega} < \infty\}$. Then, for $0 < \eta < 1$, $0 < q \leqq \infty$ and any $\sigma \in G$:*

$$|f|_{(X,X_\omega)_{\eta,q;K}} \sim \left(\int_0^\infty t^{(1-\eta)q} |\omega(x)\sigma(t\omega(x))f(x)|_X^q \frac{dt}{t}\right)^{1/q}. \tag{21}$$

Theorem 6. *Let $0 < \alpha < \infty$, $0 < \eta < 1$, $0 < q \leqq \infty$, and $\sigma \in G$. Then*

$$|f|_{(\Lambda_\omega, Lp)_{\eta\alpha,q;E}} \sim \left(\int_0^\infty t^{(1-\eta)q}\left(\int|\omega^\alpha(x)\sigma(t\omega^\alpha(x))f(x)|^p\, dx\right)^{q/p}\frac{dt}{t}\right)^{1/q}. \tag{22}$$

Proof. From (6) we have for all γ:

$(\Lambda_\omega, Lp)_{\gamma,p;E} = Lp,\omega^\gamma$. Take any $\alpha_0 < \eta\alpha < \alpha$, and let $0 < \lambda < 1$ be defined by $\eta\alpha = (1 - \lambda)\alpha_0 + \lambda\alpha$. Using reiteration we have

$$\begin{aligned}(\Lambda_\omega, Lp)_{\eta\alpha,q;E} &= \left((\Lambda_\omega, Lp)_{\alpha_0,p;E}, (\Lambda_\omega, Lp)_{\alpha,p;E}\right)_{\lambda,q;K} \\ &= (Lp,\omega^{\alpha_0}, Lp,\omega^\alpha)_{\lambda,q;K}.\end{aligned}$$

From the case $p = q$ of Theorem 5, we have

$$Lp,\omega^{\alpha_0} = (Lp, Lp,\omega^\alpha)_{\frac{\alpha_0}{\alpha},p;K}$$

Using this, and reiterating again, we get

$$(\Lambda_\omega,\ Lp)_{\eta\alpha,q;E} = (Lp,\ Lp,\omega^\alpha)_{\eta,q;K}. \tag{23}$$

Equation (22) is the corresponding norm equivalence.

Theorem 7. *If* $T : A_0 + A_1 \to \Lambda_\omega + Lp$ *is E-quasilinear.* $1 \leqq p \leqq \infty$, $0 < \theta < 1$, $0 < q \leqq \infty$, $0 < \beta$, *and if*

$$\Big(\frac{1}{t}\int_{\omega(x)\leqq t} \omega(x)\,|Ta|^p(x)\,dx\Big)^{1/p} \leqq c_1|a|_{A_1} \tag{24}$$

$$\Big(\int_{t<\omega(x)} |Ta|^p(x)\,dx\Big)^{1/p} \leqq c_0 t^{-\beta}|a|_{A_0}, \tag{25}$$

then

$$\Big(\int_0^\infty t^{\theta q}\Big(\int|\omega^\beta(x)\sigma(t\omega^\beta(x))(Ta)(x)|^p\,dx\Big)^{q/p}\frac{dt}{t}\Big)^{1/q} \leqq c|a|_{(A_0,A_1)_{\theta,q;K}} \tag{26}$$

Proof. We want to apply Theorem 4:

$$\Big(\frac{1}{t}\int_0^t E^p(s,\ Ta;\ \Lambda_\omega,\ Lp)\,ds\Big)^{1/p} - E(t,\ Ta;\ \Lambda_\omega,\ Lp)$$
$$= \Big(\frac{1}{t}\int_0^t\Big(\int_{s<\omega(x)} |Ta|^p(x)\,dx\Big)ds\Big)^{1/p} - \Big(\int_{t<\omega(x)} |Ta|^p(x)\,dx\Big)^{1/p}$$
$$= \Big(\frac{1}{t}\int|Ta|^p(x)\Big(\int_0^{\omega(x)\wedge t} ds\Big)dx\Big)^{1/p} - \Big(\int_{t<\omega(x)} |Ta|^p(x)\,dx\Big)^{1/p}$$
$$= \Big(\frac{1}{t}\int_{\omega(x)\leqq t} |Ta|^p(x)\omega(x)\,dx + \int_{t<\omega(x)} |Ta|^p(x)\,dx\Big)^{1/p}$$
$$- \Big(\int_{t<\omega(x)} |Ta|^p(x)\,dx\Big)^{1/p} \leqq \Big(\frac{1}{t}\int_{\omega(x)\leqq t} |Ta|^p(x)\omega(x)\,dx\Big)^{1/p}.$$

Condition (24) therefore implies (14).

Condition (25) is the same as (15) for our interpolation couple, so that we have

$$T : (A_0,\ A_1)_{\theta,q;K} \to (\Lambda_\omega,\ Lp)_{\beta(1-\theta),q;E}.$$

(26) is the corresponding norm inequality, by Theorem 6.

Finally, let us compare Theorem 7 with Gilbert's theorem. (24) is clearly a weaker hypothesis than

$$T : A_1 \to Lp. \tag{27}$$

Since $(\Lambda_\omega, Lp)_{\beta,p;E} \subset (\Lambda_\omega, Lp)_{\beta,\infty;E}$, condition (25), which is $T : A_0 \to (\Lambda_\omega, Lp)_{\beta,\infty;E}$, is weaker than

$$T : A_0 \to Lp,\omega^\beta. \tag{28}$$

Conditions (27) and (28) yield

$$T : (A_0, A_1)_{\theta,q;K} \to (Lp, Lp,\omega^\beta)_{1-\theta,q;K} = (\Lambda_\omega, Lp)_{\beta(1-\theta),q;E}. \tag{29}$$

The equality of the interpolation spaces above is (23).

We therefore have, from weaker hypotheses, the conclusion of Gilbert's theorem. The situation is analogous to the relation between the Marcinkiewicz and the M. Riesz interpolation theorems.

REFERENCES

1. Bennett, C.; DeVore, R. A.; and Sharpley, R. Weak L^∞ and BMO. *Ann. Math.* 113 (1981): 601-611.

2. Gilbert, J. E. Interpolation between weighted *Lp* spaces. *Arkiv för Mat.* 10 (1972): 234-249.

3. Peetre, J., and Sparr, G. Interpolation of normed Abelian groups. *Annali di Matematica pura ed Applicata* (*IV*) 92 (1972): 217-262.

4. Sagher, Y. A new interpolation theorem. *Proc. of Conf. in Harmonic Analysis, Minneapolis,* 1981. Springer-Verlag lecture notes in *Mathematics* 908: 189-198.

AN ANALOGUE OF THE FEJER-RIESZ THEOREM FOR THE DIRICHLET SPACE

Allen L. Shields
University of Michigan

L. Fejér and F. Riesz proved that if $f \in H^2$, then

$$\int_0^1 |f(x)|^2 \, dx \leq \pi \|f\|_2^2 . \tag{1}$$

(See [5] or [4, Theorem 3.13].) Here H^2 denotes the Hardy space of square-summable power series. We shall prove an analogue for the Dirichlet space, D. This is the space of analytic functions $f(z) = \Sigma a_n z^n$ in the open unit disc Δ, with $f(0) = 0$, such that

$$\|f\|^2 = \sum_1^\infty n|a_n|^2 = \frac{1}{\pi} \iint_\Delta |f'(re^{i\theta})|^2 \, r dr \, d\theta < \infty. \tag{2}$$

We prove the following result, which answers a question of A. J. Nagel.

Theorem 1. If $f \in D$, then

$$\int_0^1 |f(x)|^2 \frac{x^2}{1-x} \left(\log \frac{1}{1-x}\right)^{-2} dx \leq c\|f\|^2. \tag{3}$$

Here c is a constant independent of f.

This research was supported in part by the National Science Foundation.

We note that the exponent -2 on the logarithm cannot be replaced by $-\alpha$ for any $\alpha < 2$, since for $\beta < 1/2$, the function $(\log(1 - z))^{\beta}$ is in D. On the other hand, the result is trivially true if -2 is replaced by $-\alpha$ for any $\alpha > 2$. This is a consequence of the following growth estimate for functions in the Dirichlet space. Applying the Cauchy (-Buniakowsky-Schwarz) inequality to the power series for f, evaluated at x, one has:

$$|f(x)|^2 = |\sum_1^{\infty} (n^{1/2}a_n)(x^n/n^{1/2})|^2 \\ \leq \|f\|^2 \log \frac{1}{1 - x^2}, \quad 0 \leq x < 1. \tag{4}$$

For any fixed x, this result cannot be improved (that is, there is a nonzero $f \in D$ for which equality is obtained). However, this result does not imply (3). (Similarly, in the Fejér-Riesz theorem, if $f \in H^2$, then $|f(x)|^2 \leq 1/(1 - x^2)$, which is not integrable.) Theorem 1 is a consequence of the following result.

Theorem 2. Let μ be a positive finite Borel measure on the half-open interval $0 \leq x < 1$. Then there is a constant c_1 such that

$$\int |f(x)|^2 \, d\mu(x) \leq c_1 \|f\|^2 \qquad (f \in D), \tag{5}$$

if and only if there is a constant c_2 such that

$$\mu([x, 1)) \leq c_2[-\log(1 - x)]^{-1}, \qquad (x < 1). \tag{6}$$

Note. If μ is a finite positive Borel measure on the open unit disc Δ such that $\int |f|^2 \, d\mu < \infty$ for all $f \in D$, then there is a constant c such that $\int |f|^2 \, d\mu \leq c \, \|f\|^2$; this follows from the closed graph theorem. These measures have been characterized in a very nice paper by David Stegenga [10, Theorem 2.3]; the characterization involves comparing the logarithmic capacity of a finite union of open intervals on $\partial\Delta$ with the μ-measure of a certain associated set in Δ. In an important paper by Nagel, Rudin, and Shapiro [8, Theorem 4.5], a simpler sufficient condition is given (in a more general context). When their result is specialized to our situation, it says that (6) implies (5) (which is the difficult half of Theorem 2). Since our proof is simpler than either of the above, it seems worth presenting it; we use the theory of quadratic forms. Afterward we point out the connection of Theorem 1 to the theory of multipliers.

Warning: The same symbol c will be used to denote various constants, not necessarily the same at each occurrence.

We require two lemmas; μ will denote a fixed positive finite Borel measure on $[0, 1)$.

Lemma 1. If μ *satisfies* (6), *then*

$$\int_0^1 x^n \, d\mu(x) \leq c[\log n]^{-1} \qquad (n \geq 2).$$

Proof. Let $\psi(x) = \mu\big([x, 1)\big)$. Integrating by parts, we have

$$\int_0^1 x^n \, d\mu(x) = n \int_0^1 x^{n-1} \psi(x)\, dx = n\left(\int_0^{t_n} + \int_{t_n}^1\right) x^{n-1} \psi(x)\, dx,$$

where $t_n = 1 - n^{-1}$. Then

$$n \int_{t_n}^1 x^{n-1} \psi(x)\, dx \leq n\psi(t_n)(1 - t_n) \leq c[\log n]^{-1}.$$

Next, one verifies that the function $\varphi(x) = (1 - x)[-x^{-1} \log(1-x)]$ is decreasing on $[0, 1)$. Hence

$$\begin{aligned} n \int_0^{t_n} x^{n-1} \psi(x)\, dx &\leq cn \int_0^{t_n} x^{n-2}(1 - x)[\varphi(x)]^{-1}\, dx \\ &\leq cn[\varphi(t_n)]^{-1} \int_0^1 x^{n-2}(1 - x)\, dx \\ &= cn[\log n]^{-1}(n - 1)^{-1}. \end{aligned}$$

This completes the proof.

Lemma 2. There is a constant c such that

$$\sum_{n=1}^{\infty} \frac{1}{n[\log(n+1)]^{1/2}\log(n+m)} \leq c/[\log(m+1)]^{1/2},$$

$m = 1, 2, \ldots.$

Proof. Let m be given. For simplicity, the summand on the left side of (9) will be omitted. Then

$$\sum_{n=1}^{m} \leq \frac{1}{\log(m+1)} \sum_{1}^{m} \frac{1}{n[\log(n+1)]^{1/2}} \leq c/[\log(m+1)]^{1/2},$$

as one sees by comparing the sum to an integral. Also,

$$\sum_{n=m+1}^{\infty} \leq \sum_{m+1}^{\infty} \frac{1}{n[\log(n+1)]^{3/2}} \leq c/[\log(m+1)]^{1/2},$$

again by comparison with an integral. This completes the proof of the lemma.

We also require the following sufficient condition for a matrix operator to be bounded.

Proposition 1 (Schur Test). Let $a_{nm} \geq 0$ $(n, m \geq 1)$ *be given. If there is a function* $p(n) > 0$ $(n \geq 1)$ *such that*

(i) $\sum_n a_{nm} p(n) \leq c_1 p(m)$ *(for all* m*),*

(ii) $\sum_m a_{nm} p(m) \leq c_2 p(n)$ *(for all* n*),*

then

$$\left|\sum_{n,m} a_{nm} x_n \bar{y}_m\right| \leq (c_1 c_2)^{1/2} \left(\sum |x_n|^2\right)^{1/2} \left(\sum |y_m|^2\right)^{1/2}$$

for all square-summable sequences $\{x_n\}$ *and* $\{y_m\}$.

For a proof, see, for example, [2, p. 126] (where the designation Schur test is used). See [7, Problems 37, 38, pp. 22, 23] for an application of the test to the Hilbert matrix $a_{nm} = (n + m + 1)^{-1}$ $(n, m \geq 0)$. This is the matrix arising from (1); in this case the test gives the exact bound, π, and thus furnishes another proof of the Fejér-Riesz theorem (the usual proof uses Cauchy's integral theorem). Schur himself considered the case $p(n) \equiv 1$ (see [9, I, p. 6]). This case can be regarded as an early example of the interpolation of linear operators, since now the conditions (i) and (ii) state that the matrix $A = (a_{nm})$ is a bounded operator on the spaces ℓ^1 and ℓ^∞, respectively, with bounds c_1 and c_2. Schur's conclusion, then, is that A is bounded on ℓ^2 with bound $(c_1c_2)^{1/2}$. I do not know when the general function $p(n)$ was first introduced, but Carleman [3, Chapter 3, p. 112] establishes the test in this form (for integral transforms with real and symmetric kernel). There is a still more general form with two functions $p(n)$ and $q(m)$, where $q(m)$ replaces $p(m)$ on the right side of (i) and on the left side of (ii), and the other sides are unchanged. See [1, Theorem 6.3, especially (6.8), p. 240], where the condition (suitably modified) is shown to imply boundedness from L^p to L^p. In this form the test is necessary as well as sufficient (see E. Gagliardo [6]).

Proof of Theorem 2. We first show that (6) implies (5). So let μ be given and assume that (6) is satisfied. For $f(z) = \sum a_n z^n \in D$, we have, by Lemma 1,

$$\int |f|^2 \, d\mu \leq \sum_{n,m=1}^{\infty} |a_n a_m| \int x^{n+m} \, d\mu(x) \leq c \sum \alpha_{nm} b_n b_m ,$$

where $b_n = n^{1/2} a_n$, and $\alpha_{nm} = (nm)^{-1/2}[\log(n + m)]^{-1}$. We apply the Schur test with $p(n) = [n \log(n + 1)]^{-1/2}$, $n = 1, 2, \ldots$. By Lemma 2 we have $\sum \alpha_{nm} p(n) \leq cp(m)$, $m = 1, 2, \ldots$; since $\alpha_{nm} = \alpha_{mn}$, we have the same inequality when n and m are interchanged. This completes the proof of (5).

For the converse, assume that μ satisfies condition (5). For $w \in \Delta$, let $k_w(z) = -\log(1 - \bar{w}z)$. This function is the "reproducing kernel" for the space D; that is, in terms of the inner product induced by (2), we have

$$f(w) = (f, k_w) \qquad (f \in D).$$

Hence, $\| k_w \|^2 = (k_w, k_w) = k_w(w) = -\log(1 - |w|^2)$. Now choose $t \in [0, 1)$, and let $I = [t, 1)$. Since $\| k_t \|^4 \le |k_t(x)|^2$ for $x \in I$, we have, by (5),

$$\mu([t, 1)) \| k_t \|^4 \le \int_I |k_t(x)|^2 \, d\mu(x) \le \int_0^1 |k_t|^2 \, d\mu \le c \| k_t \|^2 .$$

Hence $\mu([t, 1)) \le c[-\log(1 - t^2)]^{-1}$ and (6) follows from this. (There is a slight problem, since we have t^2 on the right side, unlike (6). And $\log(1 - t)/\log(1 - t^2) \to \infty$ as $t \to 0^+$. However, there is no problem away from $t = 0$, and (6) is automatically satisfied near 0 for any finite measure μ. This completes the proof of the theorem.

Next we discuss Theorem 1 from the viewpoint of multipliers. A function φ, analytic in Δ, is said to multiply D into H^2 if $\varphi D \subset H^2$. These multipliers (and many others) are classified by Stegenga in [10] (see Theorems 1.1(b) and 2.3). Namely, he showed that φ is such a multiplier if and only if the measure $d\mu(z) = |\varphi'(z)|^2 (1 - |z|) \, dx\,dy$ satisfies the condition: $\int |f|^2 \, d\mu < \infty$ for all f in D. And as pointed out earlier, in the same paper he gave a characterization of all such measures. We now show that Theorem 1 is equivalent to a multiplier result. The factor z in the function φ below is inserted to cancel the pole at the origin; this could be omitted, since $f(0) = 0$ for $f \in D$.

Proposition 2. The following two statements are equivalent:

(a) *The function* $\phi(z) = z(1 - z)^{-1/2}[-\log(1 - z)]^{-1}$ *is a multiplier from* D *to* H^2.

(b) *Inequality* (3) *is valid.*

Proof. In view of the Fejér-Riesz inequality (1), we see that the first condition above implies the second.

For the reverse implication, we use the fact that an analytic function g in the unit disc is in H^2 if and only if

$\iint |g'|^2(1 - r)\,r\,dr\,d\theta < \infty$ (this may be seen, for example, by expressing the integral in terms of Taylor coefficients). Thus, an analytic function ψ is a multiplier from D to H^2 if and only if $g = \psi f$ satisfies the above condition. Since the function f is in D, its derivative is square-integrable with respect to area measure (see (2)). Thus a sufficient condition for $|\psi f'|^2(1 - r)$ to have a finite area integral is that $(1 - |z|)|\psi(z)|^2$ be bounded. Actually, the stronger condition

$$|\psi(z)|^2 \leq \frac{c}{(1 - |z|^2)(-\log(1 - |z|^2))}, \qquad |z| < 1, \tag{7}$$

is necessary for ψ to multiply D into H^2. This follows from Theorem 1 of [10] if one specializes to the particular spaces we are concerned with. (The logarithmic factor comes from inequality (4) for D.) And of course the function ϕ of the present proposition satisfies an even stronger condition. Thus a general function ψ satisfying (7) will multiply D into H^2 if and only if

$$\iint |f\psi'|^2(1 - r)\,r\,dr\,d\theta < \infty \qquad (f \in D). \tag{8}$$

To prove this for ϕ, we need the following result.

Lemma 3. *If* $[\log(1 - z)]^2 = \sum_2^\infty a_n z^n$, *then* $a_2 > a_3 > \cdots \to 0$.

Proof. We have

$$a_n = \sum_1^{n-1} \frac{1}{k(n - k)} = \frac{1}{n}\sum_1^{n-1}\left(\frac{1}{k} + \frac{1}{n - k}\right) = \frac{2}{n}\sum_1^{n-1}\frac{1}{k}.$$

Clearly, $a_n \to 0$. Thus it remains to show that

$$(n + 1)\sum_1^{n-1}\frac{1}{k} > n\sum_1^{n}\frac{1}{k} = 1 + n\sum_1^{n-1}\frac{1}{k},$$

and this is obvious.

Corollary. *If* $g(z) = (1 - z)[\log(1 - z)]^2$, *then* $|g(z)| \geq g(|z|)$ *for* $|z| < 1$.

Proof. With the notations of the lemma, we have

$$g(z) = a_2 z^2 - \sum_3^\infty (a_{n-1} - a_n) z^n.$$

Hence, $|g(z)| \geq a_2|z|^2 - \Sigma(a_{n-1} - a_n)|z|^n = g(|z|)$.

To complete the proof of the proposition, we must show that (8) holds with ψ replaced by ϕ. Let $f = \Sigma\, b_n z^n$ be in D and let $h = \Sigma\, |b_n| z^n$. Then h is in D and has the same norm as f. Also, $|f(z)| \leq h(|z|)$. We use the function g of the above corollary. Also,

$$\phi'(z) = (1 - z)^{-3/2}(-\log(1 - z))^{-1}[z/2 - z(-\log(1 - z))^{-1} + 1 - z].$$

By the above corollary, and since the function in square brackets is bounded, we have (with $r = |z|$): $|\phi'(z)|^2 \leq cg(r)^{-1}|1 - z|^2$. Thus

$$\iint |f\phi'|^2(1 - r) \leq c \int_0^1 h(r)^2 g(r)^{-1}(1 - r) \int_0^{2\pi} |1 - z|^{-2} \delta\theta\; rdr$$
$$\leq c \int_0^1 h(r)^2 g(r)^{-1}\, dr \leq c\, \| f \|^2$$

by (3). This completes the proof.

Finally, we would like to say something about the extent to which Theorem 1 is a "best possible" result. Theorem 2 really gives the whole story and we shall just make a few comments. First, if $\psi(x)$ is a positive measurable function on the interval $0 \leq x < 1$ such that $\int |f\psi|^2\, dx < \infty$ for all f in D, then

$$\varlimsup_{x\uparrow 1}\left(\frac{\psi(x)}{\varphi(x)}\right) < \infty, \tag{9}$$

where φ is the function from the preceding proposition. This is an easy consequence of condition (6), applied to the measure $d\mu = \psi^2\, dx$.

For the Fejér-Riesz theorem, one has a better result: if ψ is a positive increasing function on the half-open interval [0, 1), and if $\int |f\psi|^2 < \infty$ for all $f \in H^2$, then ψ is bounded. This follows, since $\psi^2\, dx$ is a Carleson measure (see [4, Theorem 9.3, p. 157]).

It turns out that the analogue of this result is false for the Dirichlet space.

Proposition 3. There is an increasing function $\psi(x)$, $0 \leq x < 1$, *such that*

(i) $\psi(x) \geq \phi(x)$, $0 \leq x < 1$, *where* ϕ *is the function defined in Theorem 2;*

(ii) $\overline{\lim}\ \psi(x)/\phi(x) = \infty, \quad x \to 1;$

(iii) $\int_0^1 |(\psi f)(x)|^2\, dx < \infty, \quad f \in D.$

Proof. We shall choose a sequence of numbers $\{\varepsilon_n\}$ such that $\varepsilon_1 < 1$ and $n\varepsilon_n < \varepsilon_{n-1}$ $(n > 1)$. Hence $\varepsilon_n < 1/n!$ There will be an additional condition to be specified later. Let $x_n = 1 - \varepsilon_n$, $x_n' = 1 - \varepsilon_n/n$. Then $0 < x_1 \leq x_1' < x_2 < x_2' < \cdots \to 1$.
We define $\psi(x)$ as follows:

$$\psi(x) = \phi(x_n'), \qquad x_n \leq x \leq x_n',$$

$$\psi(x) = \phi(x), \qquad x_n' \leq x < x_{n+1},$$

$n = 1, 2, \ldots$, and $\psi(x) = \phi(x)$ for $0 \leq x \leq x_1$. Then clearly ψ is increasing and $\psi(x) \geq \phi(x)$ for all x. (If desired, ψ could be modified slightly to be continuous, without affecting the proof.) Also,

$$\frac{\psi(x_n)}{\phi(x_n)} = \frac{\phi(x_n')}{\phi(x_n)} = n^{1/2}\,\frac{x_n'}{x_n}\,\frac{\log(1/\varepsilon_n)}{\log(n/\varepsilon_n)} \to \infty,$$

since the factor multiplying $n^{1/2}$ tends to 1.

Thus, to complete the proof it only remains to establish (iii). In view of inequality (3), to prove (iii) we only need to consider those intervals where ψ is larger than ϕ. Let $f \in D$ with $\|f\| \leq 1$. Then for $x \in [x_n, x_n']$, we have, by (4),

$$|f(x)|^2 \leq -\log(1 - x_n'^2) < -\log(1 - x_n').$$

Hence

$$\sum_{n=2}^{\infty} \int_{x_n}^{x'_n} |(\psi f)(x)|^2 dx < \sum_{n>2} [\phi(x'_n)]^2 \left(\log \frac{1}{1 - x'_n}\right)(x'_n - x_n)$$

$$< \sum_{n>2} \frac{n}{\varepsilon_n}\left(\log \frac{n}{\varepsilon_n}\right)^{-1} \varepsilon_n < \Sigma\, n\left(\log \frac{1}{\varepsilon_n}\right)^{-1}.$$

This series will converge if ε_n approaches 0 fast enough, for example, if $\varepsilon_n = \exp(-n^3)$. This completes the proof.

REFERENCES

1. Aronszajn, N.; Mulla, F.; and Szeptycki, P. On spaces of potentials connected with L^p spaces. *Ann. Inst. Fourier (Grenoble)*, 13, Fasc. 2 (1963): 211-306.

2. Brown, Arlen; Halmos, P. R.; and Shields, A. L. Cesàro operators. *Acta Scient. Math. (Szeged)* 26 (1965): 125-137.

3. Carleman, T. *Sur les Équations Intégrales Singulières a Noyau Rèel et Symétrique*. Uppsala: Almqvist and Wiksells, 1923.

4. Duren, Peter L. *Theory of H^p Spaces*. New York: Academic Press, 1970.

5. Fejér, L., and Riesz, F. Über einige funktionentheoretische Ungleichungen. *Math. Zeitschrift* 11 (1921): 305-314.

6. Gagliardo, Emilio. On integral transformations with positive kernel. *Proc. Amer. Math. Soc.* 16 (1965): 429-434.

7. Halmos, P. R. *A Hilbert Space Problem Book*. New York: Van Nostrand, 1967.

8. Nagel, Alexander; Rudin, Walter; and Shapiro, Joel H. Tangential boundary behaviour of functions in Dirichlet-type spaces. To appear in *Ann. of Math.*

9. Schur, I. Bemerkungen zur theorie der beschränkten Bilinearformen mit unendlichvielen Veränderlichen. *J. reine angew. Math.* 140 (1911): 1-28.

10. Stegenga, D. A. Multipliers of the Dirichlet space. *Ill. J. Math.* 24 (1980): 113-139.

11. Taylor, G. D. Multipliers on D_α. *Trans. Amer. Math. Soc.* 123 (1966): 229-240.

POTENTIAL THEORY AND DIFFUSION ON RIEMANNIAN MANIFOLDS

N.T. Varopoulos
Université Paris

INTRODUCTION

Let M be a complete connected Riemannian manifold and let us denote by Δ = divgrad its invarient Laplacian. I shall follow [1] in recalling the definitions and the basic properties of the Green's function and the harmonic measure on M.

Let $p \in M$ be fixed and let V_p be the family of functions that are defined and subharmonic on $M \setminus p$ and identically zero outside some compact subset of M and such that for all $v \in V_p$ we can find some $E = E_p(m)$, $m \in \Omega$, where Ω is an open *Nhd* of p, that satisfies

$$\frac{1}{2}\Delta E = -\delta_p$$

in the distribution sense in Ω (that is, E is an elementary solution of Δ in Ω) and such that

$$\overline{\lim_{m\to p}}\,\big(v(m) - E(m)\big) < +\infty.$$

V_p is clearly a Perron family of functions (cf. [1, Chapter 9], for a definition), and therefore the function

$$g(p, .) = \sup\{v;\ v \in V_p\},$$

when not identically $+\infty$, is harmonic on $M \setminus p$. When $g \neq +\infty$, we say that Green's function exists on M (at p); g is then the Green's function. Here are some of the basic properties of the Green's function (cf. [1, Chapter 9] for the proofs).

If g exists at $p \in M$, it also exists at every other point $p_1 \in M$. When g exists, it satisfies

$$\frac{1}{2}\Delta g(p, .) = -\delta p$$

in the distribution sense. In fact, g is then the minimal positive solution of the above equation, in the sense that if $0 \leq G \in L^1_{loc}(M)$ is any other (distributional) solution of $\frac{1}{2}\Delta G = -\delta_p$, we then have $g(p, m) \leq G(m)$ $(m \in M \setminus p)$. Indeed, by the local behavior of G (near p) and the definition of V, we see that

$$(1 + \varepsilon)G(m) - u(m) \geq 0, \quad \varepsilon > 0,$$

when $u \in V$ and $m \neq p$ lies in some *Nhd* of p. The above inequality also holds in some *Nhd* of ∞ of M and therefore everywhere. Letting $\varepsilon \to 0$, we get that

$$G(m) \geq u(m).$$

Now let K be a compact subset of M with nonempty interior ($K^\circ \neq \emptyset$). Let also U_K be the Perron family of functions that are defined and subharmonic on $M \setminus K$ and such that each $u \in U_K$ satisfies

$$u \leq 1 \quad \text{on} \quad M \setminus K; \ \overline{\lim_{m\to\infty}}\, u(m) \leq 0.$$

$u_K = \sup\{u;\ u \in U_K\}$ is then a harmonic function on $M \setminus K$, and when $u_K \neq 1$, we call it the harmonic measure of K (observe that $0 \leq u_K \leq 1$ and $0 \neq u_K$ by $K^\circ \neq \emptyset$) and say that harmonic measure exists (for K). It is then a well-known fact (cf. [1, Chapter 9]) that if harmonic measure exists for K, it exists for every other compact subset K_1 ($K_1^\circ \neq \emptyset$) and that on any Riemannian manifold harmonic measure exists if and only if the Green's function exists.

Now let $\{z(t);\ 0 \leq t \leq e\}$ be the canonical diffusion on M with $\frac{1}{2}\Delta$ as an infinitesimal generator, and let

$$p_t(x, y)\, dV = \mathbf{P}_x[z(t) \in dV,\ t < e]$$

be the heat diffusion Kernel (dV denotes the Riemannian volume element). The above diffusion, which is sometimes called the Brownian motion of M, has been constructed by Ito (cf. [8, Chapter 4]). e is a random time called explosion time and we have $z(t) \xrightarrow[t \to e]{} \infty_M$, and if $f(x) = \mathbf{P}_x[e = +\infty] = 1$ for some $x \in M$, then $f(y) = \mathbf{P}_y[e = +\infty] = 1$ for all $y \in M$. This is because $f(x)$ is a harmonic function on M (cf. [8, 4.4]). We then say that there is no explosion. The nonexplosion of Brownian motion on M is equivalent to

$$\int_M p_t(x, y)\, dV(y) = 1,\ \forall\, x \in M,\quad t \geq 0, \tag{1}$$

and in fact, everything that will be said about Brownian motion here has an interpretation in terms of the heat diffusion semigroup and the language of P.D.E. and semigroups. (1) has been proved under the condition $\mathrm{Ric}(M) \geq -K$ (for some $K \geq 0$) by Yau [11].

We have:

Theorem 1. Let $C(r)\,(r \geq 0)$ be a smooth (C^5, say) positive nondecreasing function such that

$$\int_1^\infty \frac{dr}{C^{1/2}(r)} = +\infty. \tag{2}$$

Let also M be a complete connected Riemannian manifold that satisfies

$$\lambda(x) \geq -C(r),\quad r = d(x, m),\quad x \in M, \tag{3}$$

where m is some fixed point on M, d denotes Riemannian distance, and λ denotes the lowest eigenvalue of Ric(M). *Under the above conditions, Brownian motion does not explode.*

We shall also see that if we replace the Ricci curvature by the sectional curvature, then the condition (2) becomes optimal.

Explosion or not on M, let us define for a fixed $x \in M$

$$0 \leq R_\lambda(x, y) = \int_0^\infty \bar{e}^{\lambda t} p_t(x, y)\, dt \in L^1_{\text{loc}}\big(M; dV(y)\big)$$

the resolvant operator. By the general theory (cf. [8, Sec. 1.5, and 13, IX 4]) we have

$$\frac{1}{2}\Delta_y R_\lambda(x, y) = \lambda R_\lambda(x, y) - \delta_x. \tag{4}$$

It follows that if we make the assumption that

$$G(x, y) = \int_0^\infty p_t(x, y)\, dt \in L^1_{\text{loc}}\big(M; dV(y)\big)$$

($x \in M$ fixed), then (let $\lambda \to 0$ in (4)) G satisfies $\frac{1}{2}\Delta G = -\delta$. The conclusion is that the Green's function g exists on M and that $g(x, y) \leq (G\ x, y)$.

Conversely, if the Green's function g exists on M, then $G(x, y) \in L^1_{\text{loc}}\big(M; dV(y)\big)$ (for $\forall\ x \in M$ fixed) and we have, in fact, $G(x, y) = g(x, y)$. To see this, we can argue as follows.

Let $x \in \Omega \subset M$ be a relatively compact subset of M with smooth boundary and let

$$P_t^\Omega(x, y)\, dV(y) = \mathbf{P}_x[z(t) \in dV(y);\ z(s) \in \Omega \quad 0 \leq s \leq t]$$

be the relative heat kernel. By the general theory of boundary value problems in P.D.E.'s, we know then that (the relative Green's function)

$$G_\Omega(x, y) = \int_0^\infty p_t^\Omega(x, y)\, dt \in L^1\big(\Omega; dV(y)\big)$$

satisfies

$$\frac{1}{2}\Delta_y G_\Omega(x, y) = -\delta_x$$

in Ω. It follows from the definition of g that

$$G_\Omega(x, y) \leq g(x, y).$$

It is clear, on the other hand, that for $\Omega_1 \subset \Omega_2 \subset \cdots \subset \Omega_n \subset \cdots M = U\Omega_n$ relatively compact subsets as above, we have

$$p_t^{\Omega_n}(x, y) \xrightarrow[n\to\infty]{} p_t(x, y)$$

and that the p^{Ω_n}'s are an increasing sequence. Our assertion follows.

Let $\Omega \subset M$ be a relatively compact subset with smooth boundary and let

$$\widetilde{u}_\Omega(x) = \mathbf{P}_x[\exists\ t \geq 0 \quad z(t) \in \bar{\Omega}].$$

It is a standard fact (cf. [8, Sec. 3]) that $\widetilde{u}_\Omega(x)$ is a harmonic function on $M \setminus \bar{\Omega}$ that extends continuously to $\partial\Omega$ with $\widetilde{u}_\Omega|_{\partial\Omega} \equiv 1$.

We say that the Brownian motion is *recurrent* if $\tilde{u}_\Omega(x) = 1\ \forall\ x$ and $\forall\ \Omega$ as above. We say that the Brownian motion is *transient* if $\widetilde{u}_\Omega(x) < 1\ \forall\ \Omega$ and $\forall\ x \notin \bar{\Omega}$.

From what has been said, it follows that if for some x and Ω as above with $x \notin \bar{\Omega}$ we have $\widetilde{u}_\Omega(x) < 1$, then harmonic measure exists on M. Conversely, if for some $x \notin \bar{\Omega}$ as above we have $\widetilde{u}_\Omega(x) = 1$, then harmonic measure does not exist. Brownian motion is therefore transient (resp. recurrent) if and only if harmonic measure exits (resp. does not exist) on M. To see that last point, we can argue as follows.

Let $\Omega = \Omega_0 \subset \Omega_1 \subset \cdots \subset M = U\Omega_j$ be an increasing family of relatively compact open subsets with smooth boundary and let $u_j \in C(\bar{\Omega}_j \setminus \Omega_0)$ be the solution of the Dirichlet problem with $u_j|_{\partial\Omega_0} = 1 u_j|_{\partial\Omega_j} = 0$. We have then (cf. [8, Sec. 3])

$$u_j(m) = \mathbf{P}_m[z(t) \quad \text{hits} \quad \partial\Omega_0 \quad \text{before it hits} \quad \partial\Omega_0].$$

Now Brownian motion cannot stay in a compact set; if it did, it would do so with probability one (by the zero one law) and this is absurd. It follows, therefore, that $u_j(x) \xrightarrow[j\to\infty]{} \widetilde{u}_\Omega$ (on $M \setminus \Omega$)

and so by our hypothesis, $u_j(x) \xrightarrow[j\to\infty]{} 1$ and therefore we have $u_k(x) = 1$ (by definition of u_k). This proves our assertion.

Now let M be a complete Riemannian manifold as before, and let

$$B_t(m) = \{x \in M;\ d(x, m) \leq t\}$$

be the Riemannian ball ($t \geq 0$; $m \in M$). We shall denote by $V_m(t) = \mathrm{Vol}\big(B_t(m)\big)$ the Riemannian volume of that ball. We have then:

Theorem 2. Let M be as above. Then a necessary condition for the existence of Green's function on M is that we have

$$\int_1^\infty \frac{t\,dt}{V_m(t)} < +\infty \tag{5}$$

for all $m \in M$. Conversely, if we suppose in addition that $\mathrm{Ric}(M) \geq 0$ *and if* (5) *is verified for some $m \in M$, the Green's function exists on M.*

The first half of the above theorem (that is, the necessity of the condition) for the two-dimensional case (that is, Riemann surfaces) is a classical theorem of Ahlfors [2].

By the following example (which is a simplified version of an example given to me by R. Greene), we see that (5) is not in general sufficient to ensure the existence of Green's function.

Example. Let $M = \mathbf{C}$ be the complex plane and let $ds^2 = \varphi(y)|dz|^2$ ($z = x + iy \in \mathbf{C}$), where $0 \leq \varphi \in C^\infty(M)$, $\varphi(y) = 1/y$ ($y \geq 1/10$), and $\varphi(y) = 1$ ($y < 0$). Green's function clearly does not exist (since the conformal structural is that of $\mathbf{C}$). An easy computation in hyperbolic geometry, on the other hand, shows that (5) is very generously verified ($V(t)$ increases exponentially!).

If we assume that (5) holds on M and that M has nonnegative *sectional* curvature everywhere, then we can even obtain a nice estimate on the Green's function $g(m, x)$. Indeed, I shall show elsewhere that under these conditions for all $m \in M$, we can find $C = C(m) > 0$ such that:

$$C^{-1} \leq g(x, m) \left\{ \int_{d(x,m)}^{\infty} \frac{tdt}{V_m(t)} \right\}^{-1} \leq C.$$

PROOF OF THEOREM 1

Let M be a smooth complete connected Riemannian manifold. Theorem 1 is a consequence of the following two propositions.

Proposition 1. Let M be as above and let us suppose that there exists $U > 0$ a locally Lipschitz function on M and two contents $A, B > 0$ such that $U(x) \to \infty$ $(x \to \infty)$ and such that

$$\Delta U \leq AU + B \tag{6}$$

in the distribution sense; that is, we suppose that

$$\int_M \Delta U \cdot \varphi \, dV \leq \int_M (AU + B)\varphi \, dV$$

for all $0 \leq \varphi \in C_0^\infty(M)$. Under the above conditions, there is no explosion on M; that is, we have $\mathbf{P}_m[e = +\infty] = 1$ $(m \in M)$.

Proposition 2. Let $0 < C(r) \in C^5([0, +\infty))$ be a nondecreasing function that satisfies condition (2), and let M be a complete Riemannian manifold that satisfies

$$\mathrm{Ric}_x(X, X) \geq -C(r)|X|^2 \tag{7}$$

$\forall\, x \in M$, $X \in T_xM$, where $r = d(x, m)$ denotes the distance from some fixed point m. Then a function U exists on M that satisfies the conditions of Proposition 1.

Proof of Proposition 1. The proof is based on two distinct technical components. The first regularizes $U(x)$ locally, that is, on compact subsets, and the second deals with the singularity at infinity.

Lemma 1. Let U, A, B be as in Proposition 1 and let $K \subset M$ be a compact subset of M. There exists then $0 \leq \varphi_n \in C_0^\infty(M)$ such that

$$\Delta \varphi_n \leq A\varphi_n + B + 1, \quad on\ K \quad (n \geq 1)$$

$$\varphi_n \xrightarrow[n\to\infty]{} U \quad uniformly\ on\ K.$$

This lemma is essentially contained in [6], and its proof is based on the following idea of P. Malliavin.

Let us first modify U and consider $U_K \in C_0(M)$, which is Lipschitz and coincides with U on some *Nhd* of K. Let $\{P_t;\ t \geq 0\}$ denote the heat diffusion semigroup on M and let us set

$$\varphi_\varepsilon = (P_\varepsilon U_K)\Psi, \ 0 < \varepsilon < 1,$$

with $\Psi \in C_0^\infty(M)$, $0 \leq \Psi \leq 1$, and $\Psi \equiv 1$ on some *Nhd* of K. If we then set $\varphi_n = \varphi_{1/n}$, we have our sequence (at least for n large enough). The reason for this is that the heat diffusion semigroup commutes with the Laplacian, that is, $(\Delta P_t - P_t\Delta)\varphi = 0$, $\forall\ \varphi \in C_0^\infty(M)$, and that both Δ and P_t are self-adjoint.

If the above proof frightens you, you can look at [10, Lemma 2.1], where I prove a weaker lemma with more conventional methods (the Friedrich mollifier), which is also sufficient for our purpose.

To cope with the singularity at ∞, it is very advantageous to use probability theory (rather than try to cope directly with the formulation (1)).

Let $\{z(t);\ 0 < t < e\}$ be the Brownian motion on M and let

$$T_a = \inf\{t/d(m, z(t)) \geq a\}$$

(for some fixed point $m \in M$) be the exit time from the ball $B_t(m)$.

Lemma 2. Let U be as in Proposition 1. *We then have*

$$\mathbf{E}_m U[z(t \wedge T_a)] \leq \left(U(m) + 2 + \frac{B+1}{A}\right)e^{At} = K(m)e^{At},$$

where m is an arbitrary fixed point of M.

Proof. Let m and $a > 0$ be fixed and let $\varphi_n \in C_0^\infty(M)\ (n \geq 1)$ be a sequence that satisfies the conditions of Lemma 1 with $K =$

$\{x \in M/d(x, m) \leq 2a\}$. Dynkin's formula (cf. [8, Sec. 2, n.6]) applies then, since $z(t)$ has the strong Markov property (the strong Markov property is already contained in Ito's construction (cf. [8, Chapter 4]) and we get

$$\mathbf{E}_m \varphi_n\big(z(t \wedge T_a)\big) - \varphi_n(m) = \mathbf{E}_m \int_0^{t\wedge T_a} \Delta \varphi_n\big(z(s)\big)\, ds$$

$$\leq \mathbf{E}_m \int_0^{t\wedge T_a} \big(A\varphi_n\big(z(s)\big) + B + 1\big)\, ds$$

$$\leq \mathbf{E}_m \int_0^{t} \big[A\varphi_n\big(z(s \wedge T_a)\big) + B + 1\big]\, ds.$$

This implies that the expectation

$$\Psi_n(t) = \mathbf{E}_m \varphi_n[z(t \wedge T_a)], \quad t \geq 0$$

satisfies the differential inequality.

$$\Psi_n(t) \leq U(m) + 1 + \int_0^t \big(A\Psi_n(s) + B + 1\big)\, ds$$

for n large enough.

Let θ be the unique solution of the equation:

$$\theta(t) = U(m) + 2 + \int_0^t \big(A\theta(s) + B + 1\big)\, ds, \quad t \geq 0.$$

It follows that $\Psi_n - \theta \leq 0$ for $t \geq 0$ and n large enough.

If we let $n \to \infty$, we obtain that $\mathbf{E}_m U[z(t \wedge T_a)] \leq \theta(t)$ $(t \geq 0)$. If we substitute the explicit value of $\theta(t)$, we obtain our lemma.

We are now in a position to give the:

End of the Proof of Proposition 1. By Lemma 2, it follows that for all $t \geq 0$ and $a \geq 0$ (and m fixed), we have

$$\inf\{U(z);\ z \in M;\ d(m, z) = a\} \cdot \mathbf{P}_m[T_a \leq t] \leq K(m)e^{At}.$$

By our hypothesis on U, it follows therefore that

$$\mathbf{P}_m[T_a \leq t] \xrightarrow[a\to\infty]{} 0.$$

On the other hand, we clearly have $[e < t] = \bigcap_{a\geq 0} [T_a < t]$, which is a monotone intersection. Our proposition follows.

Now let $F(r)$ $(r \geq 0)$ be a smooth function on $[0, +\infty)$ (say, C^{10}) such that $F(0) = 0$ $F'(0) = 1$. The quadratic differential

$$ds^2 = dr^2 + F(r)^2\, d^2Ⓗ$$

on $\mathbf{R}^n \setminus 0$, where $r^2 = \sum_{r=1}^n x_r^2$ and $d^2Ⓗ$ is the standard Euclidean metric on the unit (Euclidean) sphere, determines then a smooth metric on $\mathbf{R}^n \setminus 0$ that extends to a continuous (but not necessarily smooth) metric on $\mathbf{R}^n$. The Jacobi equations give us then

$$F''(r) = -K(r)F(r),$$

where $K(r)$ is the curvature (the *radial* curvature) at the point $x \neq 0$ with $r = d(0, x)$ for any section that contains the ray $0x$. (We see, therefore, that the metric will certainly not be smooth at 0 unless $F''(r) \xrightarrow[r \to 0]{} 0$.) We have also $\mathrm{Ric}\left(\frac{\partial}{\partial r}, \frac{\partial}{\partial r}\right) = (n - 1)K(r)$. The above Riemannian space is called an (n-dimensional) model.

For $u(r)$ a smooth radial function on $\mathbf{R}^n \setminus 0$, the Laplacian with respect to the above metric takes the form

$$\Delta u = \frac{d^2u}{dr^2} + (n - 1)\frac{F'(r)}{F(r)}\frac{du}{dr}, \quad r > 0.$$

We shall base the proof of Proposition 2 on the following classical theorem (cf. [7, Sec. 2]).

Comparison Theorem. Let M be a smooth manifold and M_0 a (smooth at 0) model of the same dimension, and let us suppose that $\lambda(x)$ the lowest eigenvalue of Ric(M) *at the point $x \in M$ satisfies*

$$\lambda(x) \geq \mathrm{Ric}_0\left(\frac{\partial}{\partial r}, \frac{\partial}{\partial r}\right) = (n - 1)K(r), \tag{8}$$

where $r = d(m, x)$ with m some fixed point on M, and Ric_0 *is the Ricci curvature of M_0. Let us furthermore assume that $u(r)$ $(r > 0)$ is a nondecreasing smooth radial function on $M_0 \setminus 0$ such that $\Delta_0 u \leq Au + B$ on $M_0 \setminus 0$ for two positive constants A and B. Under the above conditions, the function $U(x) = u(d(m, x))$, $x \in M \setminus m$, satisfies $\Delta U \leq AU + B$ in the distribution sense on $M \setminus m$.*

In fact, for $M \ni x \neq m$ and $x \notin C\, m$ = the cut locus of m, we have the pointwise estimate $\Delta\, U(x) \leq \Delta\, u(d(m, x))$ (cf. [7, Sec. 2]). To obtain the distributional inequality on the whole of $M \setminus m$, we can use the general principle formulated by Yau at the very end of [12] (and which is essentially due to Cheeger and Gromoll [4]) and which says just this: namely, that we can extend an inequality, like the one we have, across the cut locus of m provided that we interpret that inequality in the distribution sense. We are now in a position to give the:

Proof of Proposition 2. Let $0 < \varphi(t) \in C^{10}([0, +\infty))$ be fixed and let

$$\Phi(t) = \frac{1}{t} + \varphi(t),\ t > 0;\ F(r) = A \exp\left[\int_1^r \Phi(t)\, dt\right],\ r > 0,$$

where we can extend the definition of F to $F(0) = 0$ and also choose A such that $F'(0) = \lim_{r \to 0} F'(r) = 1$ so as to be able to define a model M_0 by $ds^2 = dr^2 + F^2(r)\, d^2Ⓗ$. The radial curvature of M_0 is then given by

$$-K(r) = \frac{F''}{F} = \Phi' + \Phi^2 = \varphi' + \frac{2\varphi}{t} + \varphi^2.$$

If we set $\varphi(t) = \sqrt{C(t)}$, where $C(t)$ is as in the prosposition for $t \geq t_0 > 0$ and $\varphi(t) = \alpha t$ for $t < t_0/2$, then for t_0 small enough and $\alpha > 0$ large enough, we can complete the definition of $\varphi(f)$ in $[t_0/2, t_0]$ so that M_0 becomes smooth at 0 and satisfies (cf. [7], pp. 58-61)

$$\mathrm{Ric}_0\left(\frac{\partial}{\partial r}, \frac{\partial}{\partial r}\right) = (n-1)K(r) \leq -C(r) \tag{9}$$

for that model.

If we set

$$u(r) = \exp\left(\int_1^r \frac{dt}{\Phi(t)}\right),\ r > 0,$$

we have for the above choice $\varphi = \sqrt{C(r)}$

$$\Delta u = u'' + (n - 1)\Phi u' = \left[-\frac{\Phi'(r)}{\Phi^2(r)} + \frac{1}{\Phi^2(r)} + n - 1\right], \quad u \leq Au \quad (10)$$

for some fixed $A > 0$ (by conditions on $C(t)$). We also have $u(r) \xrightarrow[r\to\infty]{} \infty$, again by the conditions on $C(t)$.

Now let M satisfy condition (7) of Proposition 2. By (9) it then satisfies condition (8). The comparison theorem and (10) tell us, therefore, that $U(x) = u(d(m, x))$ satisfies (6) on $M \setminus m$. We can therefore smooth U on some small *Nhd* of m on M and obtain condition (6) everywhere on M. This completes the proof of the proposition.

To show that Theorem 1 is optimal, we shall prove the following.

Theorem 3. Let M be a complete connected Riemannian manifold and let $m \in M$ be a fixed point of M such that the sectional curvature $K(V)$ of M satisfies

$$K(V) \leq -C(r), \quad r \geq 0,$$

for every two-dimensional section $V \subset T_x(M)$ with $d(m, x) = r \geq 0$, where now $C(r)$ is assumed to be a positive C^1-function on $r \geq 0$ that satisfies

$$\int_1^\infty \frac{dr}{\sqrt{C(r)}} < +\infty$$

and

$$C'(r) \leq aC^{3/2}(r), \quad r \geq 0 \quad (11)$$

for some positive a. Under the above conditions, the explosion time e satisfies

$$\mathbf{P}[e < +\infty] = 1. \quad (12)$$

Proof. Observe first that by renormalizing the metric, we can suppose that a in (11) is as small as we like. Let then

$$\varphi(t) = \lambda t, \; t \in [0, t_0]; \; \varphi(t) = u\sqrt{C(t)}, \; t \in [t_0, +\infty],$$

where t_0 is the first point if anywhere $y = \lambda t$ intersects $y = \mu\sqrt{C(t)}$ and where $\lambda, \mu > 0$. By choosing λ and μ appropriately and then a appropriately small and smoothing the definition of $\varphi(t)$ out at the point t_0, we see that $\varphi(t)$ then verifies

$$\varphi'(t) + \frac{2\varphi(t)}{t} + \varphi^2(t) \leq C(t)$$

$$\left(\frac{1}{t^2} - \varphi'(t) + 1\right) + \left(\frac{1}{t} + \varphi(t)\right)^{-2} + n - 1 \geq \frac{1}{2} > 0.$$

A model M_0 can be constructed then as in the proof of Proposition 2 by $F(r) = A \exp[\int_1^r (\frac{1}{t} + \varphi(t))\, dt]$. With the use of that model and the Hessian comparison theorem of [7, p. 19] (this is where the full strength of the sectional curvature is used), we can then construct, just as before, $u(x) \in C^2(M) \cap L^\infty(M)$ that satisfies

$$\Delta u(x) \geq \alpha u(x) \geq 0, \ \forall\ x \in M; \ \lim_{m\to\infty} u(m) = \| u \|_\infty \qquad (13)$$

for some fixed positive $\alpha > 0$.

To terminate the proof, we use the following.

Lemma 3. Let M complete connected manifold be such that $u \in C^2 \cap L^\infty(M)$ exists that satisfies (13). *Then* (12) *holds.*

Proof of the Lemma. Fix $m \in M$ and set

$$T = \inf\{t/d(z(t), m) \geq r\}$$

$$X(t) = \exp(-\alpha(T \wedge t))\, u(z(t \wedge T))$$

($r \geq 0$ and $z(t)$ is the Brownian motion on M). The process $X(t)$ is a submartingale (compute dX by ITO calculus), by letting $r \to \infty$ in the inequality

$$u(m) \leq \mathbf{E}_m[\exp(-\alpha T) u(z(T))];$$

we then obtain

$$u(m) \leq \| u \|_\infty \mathbf{E}_m(\exp(-\alpha e)) \leq \| U \|_\infty \mathbf{P}_m[e < +\infty].$$

The lemma follows (observe that $v(m) = \mathbf{P}_m[e < +\infty]$ is a harmonic function on M).

PROOF OF THEOREM 2

Let M be an arbitrary complete Riemannian manifold, let $m \in M$ be a fixed point on M, and let us suppose that

$$L(R) = \int_1^R \frac{t\,dt}{\text{Vol } B_t(m)} \xrightarrow[R\to\infty]{} +\infty.$$

I shall proceed to show that harmonic measure does not exist on M.

Let

$$F(r) = F_R(r) = 1 - \frac{L(r)}{L(R)}, \quad (r > 0),$$

and

$$\Phi(x) = \Phi_R(x) = F[d(m, x)] \quad (x \in M \quad x \neq m).$$

Clearly, then, Φ is a locally Lipschitz function on $M \setminus m$ and C^1-smooth on the complement of $C(m)$ the cut locus of m. For $x \neq m$, $x \notin C(m)$, we have

$$d\Phi(x) = -\frac{1}{L(R)} \frac{r}{\text{Vol } B_r(m)} dr, \quad r = d(x, m).$$

We have, therefore, if we denote by $|\ |$ the Riemannian norm on $T^\star M$,

$$\begin{aligned}
I_R &= \int_M [x \notin C(m);\ \tfrac{1}{2} \leq d(x, m) \leq R] |d\Phi|^2 \, d\text{Vol}(x) \\
&\leq \frac{1}{L^2(R)} \int_{1/2 \leq d(x,m) \leq R} \left[\frac{r}{\text{Vol } B_r(m)}\right]^2 d\text{Vol}(x) \\
&= \frac{1}{L^2(R)} \int_{1/2}^R r^2 \frac{1}{(\text{Vol } B_r(m))^2} \frac{d}{dr}(\text{Vol } B_r(m))\, dr \\
&= -\frac{}{L^2(R)} \left[\frac{r^2}{\text{Vol } B_r(m)}\right]_{1/2}^R + 2\,\frac{L(R) - L(1/2)}{L^2(R)} \\
&\leq \frac{1}{4L^2(R) \text{Vol } B_{1/2}(M)} + 2\,\frac{L(R) - L(1/2)}{L^2(R)}.
\end{aligned}$$

The last step is obtained by integration by parts. The upshot, by our hypothesis, is that

$$I_R \xrightarrow[R\to\infty]{} 0. \tag{14}$$

Using the fact that Φ is locally Lipschitz and that $C(m)$ is of measure zero on M, we see that we can find $\varphi_\varepsilon \in C_0^\infty(M \setminus m)$ such that $\varphi_\varepsilon \xrightarrow[\varepsilon\to 0]{} \Phi$ uniformly on the compact of $M \setminus m$ and such that

$$\int_K |d\varphi_\varepsilon|^2 \, d \text{ Vol} \xrightarrow[\varepsilon\to 0]{} \int_{K\setminus C(m)} |d\Phi|^2 \, d \text{ Vol}$$

for every compact subset $K \subset M \setminus m$. This is done by the classical Friedrich mollifier (cf. [10, Lemma 2.1]). This together with (14) shows that for every $\varepsilon > 0$, we can find some $R > 1$ and some $\varphi \in C_0^\infty(M)$ such that

$$\begin{aligned} &1 < \varphi(x) < 1 + \varepsilon; \quad x \in M,\ d(m, x) = 1 \\ &0 < \varphi(x) < \varepsilon; \quad x \in M,\ d(m, x) = R \\ &\int_{1\le d(x,m)\le R} |d\varphi|^2 \, dV \le \varepsilon . \end{aligned}$$

If we denote then by u_R $(R > 0)$ the solution of the obvious Dirichlet problem in $C_R = \{x \in M;\ 1 \le d(x, m) \le R\}$, it follows from the above and the minimizing energy property of u_R (for example, ch. [3]) that

$$\int_{C_{R_j}} |du_{R_j}|^2 \, d\text{Vol} \xrightarrow[j\to\infty]{} 0$$

for some sequence $R_j \xrightarrow[j\to\infty]{} \infty$. The conclusion is of course that harmonic measure does not exist on M. This proves the first half of this theorem. The second and (by far) harder half has been proved in [10] and I shall not repeat myself here. I shall finish up with a few remarks, allowing myself to become sketchy and incomplete in my presentation.

Proposition 3. Let $a_j > 0$ $(j = 1, 2, \ldots)$. *Then we have*

$$\left(\frac{1}{a_1} + \frac{1}{a_2} + \cdots + \frac{1}{a_n}\right) \ge \frac{1}{100}\left(\sum_{j=1}^{n} \frac{j}{a_1 + \cdots + a_j}\right), \quad n \ge 1.$$

The above inequality is clear when $a_1 \leq a_2 \leq \cdots$. The proof consists in reducing the general case to the special nondecreasing one. Indeed, if for some $1 \leq j < n$ we have $a_j > a_{j+1}$, without changing any other a_r we can replace a_j by $a_j^\star$ and a_{j+1} by $a_{j+1}^\star$ so that

$$a_j > a_j^\star > a_{j+1}^\star > a_{j+1}, \quad \frac{1}{a_j} + \frac{1}{a_{j+1}} = \frac{1}{a_j^\star} + \frac{1}{a_{j+1}^\star}.$$

This operation does not change the left-hand side of the above inequality while it increases the right-hand side. This completes the proof.

The same inequality holds, therefore, for a locally integrable $f \geq 0$:

$$\int_a^b \frac{dt}{f(t)} \geq \frac{1}{100} \int_a^b \frac{t\,dt}{\int_0^t f(s)\,ds} \tag{15}$$

(at least this is certainly so if f is Riemann integrable).

Let us now return to our manifold M and let us denote by $\sigma_m(t)$ for some fixed $m \in M$ the $(n - 1)$-dimensional Hausdorff measure (induced by the Riemannian metric (cf. [5, 3.2.46]) of the "sphere of radius t around m") $\{x \in M;\ d(x, m) = t\}$. We then have

$$\sigma(t) = \frac{d}{dt} \operatorname{Vol} B_t(m), \qquad \text{a.a. } t \geq 0,$$

and we can even use this as a definition for $\sigma(t)$. It follows from (15) that

$$\Lambda(R) = \int_1^R \frac{dt}{\sigma(t)} \geq \frac{1}{100} \int_1^R \frac{t}{\operatorname{Vol} B_t(m)}\,dt. \tag{16}$$

We can then repeat the proof of Theorem 2 with $F(r) = 1 - \frac{\Lambda(r)}{\Lambda(R)}$ (this is technically more difficult, since $F(r)$ is no longer locally Lipschitz) and deduce that when

$$\int^\infty \frac{dt}{\sigma(t)} = +\infty,$$

there is no harmonic measure on *M*. This in view of (16) is a sharper result and it is in this form that L. Ahlfors proved it for $n = 2$. (His proof is different; it uses the conformal structure of a Riemann surface.)

REFERENCES

1. Ahlfors, L. V. Conformal invariants. In *Topics in Geometric Function Theory*. New York: McGraw Hill.

2. Ahlfors, L. V. Sur le type d'une surface de Riemann. *C.R.A.S.* (1935).

3. Brelot, M. Elements de la theorie classique du potentiel. Cours de la Sorbonne, 1959.

4. Cheeger, J., and Gromoll, D. The splitting theorem for manifolds of nonnegative Ricci curvature. *J. Diff. Geometry* 6 (1971): 119-128.

5. Federer, H. *Geometric Measure Theory*. New York: Springer-Verlag, 1969.

6. Greene, R. E., and Wu, H. C^{∞} approximation of convex subharmonic and plurisabharmonic functions. *Ann. Scient. Ec. Norm. Sup.*, 4ème série t. 12 (1979): 47-84.

7. Greene, R. E., and Wu, H. Function theory on manifolds which possess a pole. *Lecture Notes in Math.*, 699.

8. Ito, K. *Lectures on Stochastic Processes*. Tata Institute of Fundamental Research, Bombay.

9. McKean, H. P., Jr. *Stochastic Integrals*. New York: Academic Press, 1969.

10. Varopoulos, N. T. The Poisson kernel on positively curved manifolds. To appear in *J. Func. Anal.*

11. Yau, S. T. On the heat kernel of a complete Riemannian manifold. *J. Math. Pure et Appl.* 57 (1978): 191-201.

12. Yau, S. T. Some function-theoretic properties of complete Riemannian manifolds and their applications to geometry. *Ind. Univ. Math. J.* 25 (1976): 659-670.

13. Yosida, K. *Functional Analysis*. New York: Springer-Verlag, 1978.